Basin Modeling: New Horizons in Research and Applications

Edited by
Kenneth E. Peters
David J. Curry
Marek Kacewicz

AAPG Hedberg Series, No. 4

Published by
The American Association of Petroleum Geologists
Tulsa, Oklahoma, U.S.A.

ISBN13: 978-0-89181-903-5

AAPG Editor: Stephen E. Laubach
AAPG Geoscience Director: James B. Blankenship

COVER: Present-day geometry of the Alaskan North Slope petroleum system modeling study (Schenk et al., pg. 317) with north-south cut-planes showing the slope from the Barrow Arch toward the foothills of the Brooks Range and west-east cut-planes along the Barrow Arch pointing to the Prudhoe Bay structure.

This publication is available from:

The AAPG Bookstore
P.O. Box 979
Tulsa, OK U.S.A. 74101-0979
Phone: 1-918-584-2555 or 1-800-364-AAPG (U.S.A. only)
E-mail: bookstore@aapg.org
www.aapg.org

Canadian Society of Petroleum Geologists
600, 640 – 8th Avenue S.W.
Calgary, Alberta T2P 1G7
Canada
Phone: 1-403-264-5610
Fax: 1-403-264-5898
E-mail: reception@cspg.org
www.cspg.org

Geological Society Publishing House
Unit 7, Brassmill Enterprise Centre
Brassmill Lane, Bath BA13JN
United Kingdom
Phone: +44-1225-445046
Fax: +44-1225-442836
E-mail: sales@geolsoc.org.uk
www.geolsoc.org.uk

Affiliated East-West Press Private Ltd.
G-1/16 Ansari Road, Darya Gaaj
New Delhi 110-002
India
Phone: +91-11-23279113
Fax: +91-11-23260538
E-mail: affiliate@vsnl.com

The American Association of Petroleum Geologists

The American Association of Petroleum Geologists Books Refereeing Procedures

The Association makes every effort to ensure that the scientific and production quality of its books matches that of its journals. Since 1937, all book proposals have been refereed by specialist reviewers as well as by the Association's Publications Committee. If the referees identify weaknesses in the proposal, these must be addressed before the proposal is accepted.

Once the book is accepted, the Association Book Editors ensure that the volume editors follow strict guidelines on refereeing and quality control. We insist that individual papers can only be accepted after satisfactory review by two independent referees. The questions on the review forms are similar to those for the AAPG Bulletin. The referees' forms and comments must be available to the Association's Book Editors on request.

Although many of the books result from meetings, the editors are expected to commission papers that were not presented at the meeting to ensure that the book provides a balanced coverage of the subject. Being accepted for presentation at the meeting does not guarantee inclusion in the book.

More information about submitting a proposal and producing a book for The American Association of Petroleum Geologists can be found on its web site: www.aapg.org.

Sponsors

The conference conveners and AAPG gratefully acknowledge the generous support of the following corporate sponsors:

 MAERSK OIL

StatoilHydro

About the Editors

Kenneth E. Peters is science advisor for Schlumberger Information Solutions (SIS) where he uses geochemistry and numerical modeling to study petroleum systems. He has more than 32 years of experience working for Chevron, Mobil, ExxonMobil, USGS, and Schlumberger and has taught petroleum geochemistry and basin modeling at Chevron; Mobil; ExxonMobil; Oil & Gas Consultants International; University of California Berkeley; and Stanford University. Ken is principal author of *The Biomarker Guide* (2005, Cambridge U. Press) and consulting professor in the Geological and Environmental Sciences Department at Stanford University where he leads the Basin and Petroleum System Modeling Industrial Affiliates Program. He was chair of the AAPG Research Committee (2007–2010), AAPG Distinguished Lecturer for 2009 and 2010, and editor for the 2009 AAPG compact disk *Getting Started in Basin and Petroleum System Modeling*. He is an associate editor for *AAPG Bulletin* and *Organic Geochemistry*. In 2009, he received the Schlumberger Henri Doll Prize for Innovation and the Alfred E. Treibs Award presented on behalf of the Organic Geochemistry Division of the Geochemical Society to scientists who have had a major impact on the field of organic geochemistry through long-standing contributions. Ken has B.S. and M.S. degrees in geology from University of California, Santa Barbara and a Ph.D. in geochemistry from University of California, Los Angeles.

David J. Curry is a specialist in integrated petroleum systems analysis (geochemistry and basin modeling) and is currently with HRT America Inc., in Houston, Texas. With over 25 years of industry experience, he has also worked in research and exploration positions in petroleum geochemistry and basin modeling for Devon Energy, ExxonMobil Exploration Company, ExxonMobil Upstream Research Company as well as other organizations. Among his research interests and areas of expertise are depositional controls on source rock occurrence and organic matter type; petroleum generation and expulsion processes (including maturation and kinetics); petroleum quality and alteration processes; basin modeling and the characterization and behavior of polar components in petroleum systems. David has a B.S. in chemistry from the Virginia Military Institute; a master's from Rice University, and a Ph.D. from the University of Texas at Austin, both with specialization in organic geochemistry.

Marek Kacewicz is research consultant and basin modeler at Chevron Energy Technology Company in Houston, Texas. His primary responsibilities include technology applications and research integrating petroleum systems modeling, seismic inversion, velocity modeling, pressure prediction, geomechanics, and structural modeling. Prior to Chevron, Marek worked as a research geologist at ARCO Exploration and Production Research Center in Plano (Texas, USA); as a basin modeler at Unocal Exploration & Exploitation Technology in Sugar Land (Texas, USA); Alexander von Humboldt Fellow at the Freie Universitaet Berlin (Berlin, Germany); and research assistant at the University of Warsaw (Warsaw, Poland). Marek has over 20 years of experience in petroleum systems modeling, exploration, and research. His experience includes both conventional and unconventional resources and covers a wide range of sedimentary basins worldwide. Some of Marek's professional honors include receiving the 1986 International Association for Mathematical Geology Vistelius Research Award, being selected for the Alexander von Humboldt Fellowship (Germany); and receiving the 2005 AAPG Gabriel Dengo Memorial award. He has an M.S. degree in numerical mathematics/computer science and a Ph.D. in earth sciences, both from the University of Warsaw (Poland).

Table of Contents

Peters, K. E., D. J. Curry, and M. Kacewicz, 2012, An overview of basin and petroleum system modeling: Definitions and Concepts, *in* K. E. Peters, D. J. Curry, and M. Kacewicz, eds., Basin Modeling: New Horizons in Research and Applications: AAPG Hedberg Series, no. 4, p. 1–16.

An Overview of Basin and Petroleum System Modeling: Definitions and Concepts

Kenneth E. Peters
Schlumberger; Stanford University; Mill Valley, California, U.S.A.

Marek Kacewicz
Chevron Energy Technology Company, Houston, Texas, U.S.A.

David J. Curry
HRT America Inc., Houston, Texas, U.S.A.

ABSTRACT

This special volume contains a selection of articles presented at the AAPG Hedberg Research Conference on Basin and Petroleum System Modeling (BPSM) held in Napa, California, on May 3–8, 2009. These articles provide an overview of the current state of the science. The main purpose of the conference was to facilitate the exchange of ideas and enhance cooperation among the leading experts from industry, academia, and government to improve modeling concepts, tools, and workflows required to optimize exploration and production. Kacewicz et al. (2010) summarize the conference proceedings and findings. This conference was a direct result of the previous heavily oversubscribed and highly successful Hedberg Conference on Basin Modeling held in the Netherlands (May 6–9, 2006), at which the Dutch organizers suggested the need for another meeting on the same topic within 2 to 3 yr, preferably in North America.

The substantial target audience for this volume consists of (1) geochemists and BPSM modelers in industry, government, and academia; (2) geoscientists with interest in region-specific BPSM studies; and (3) specialists in different geoscience disciplines that are significant to the development of BPSM input parameters and simulation algorithms. Basin and petroleum system modeling is one of the most rapidly growing disciplines within the geosciences because of the high demand by industry for continuous improvement in predictions of petroleum generation, migration, accumulation, and alteration. Virtually all major oil companies now have substantial geochemistry and BPSM groups.

DOI:10.1306/13311426H4139

INTRODUCTION

Basin and petroleum system modeling (BPSM) is a rapidly growing research and application tool that spans many disciplines within the geosciences (Hantschel and Kauerauf, 2009). The principal aim of BPSM is to reduce the risk associated with exploration, production, and development of petroleum. The purpose of this chapter is to introduce the articles from those sessions included in this volume, describe some of the basic concepts of BPSM, define key terms, and provide an overview of some future directions for this multidisciplinary research.

Although many workers use the term "basin modeling" to include petroleum system modeling, basin modeling and petroleum system modeling are different. In basin modeling, mathematical equations are used to reconstruct the deposition, compaction, and erosion of rock layers through time and space and the concomitant evolution of the thermal history of the basin. Although basin modeling deals with rocks, petroleum system modeling deals with hydrocarbon fluids. Hydrocarbons are composed mainly of hydrogen and carbon, especially thermogenic oil and gas, but also including biogenic methane. The term "petroleum system" is commonly misused, although it is rigorously defined. A petroleum system consists of all the elements and processes responsible for accumulations and all of the genetically related petroleum that originates from one pod of active source rock (Magoon and Dow, 1994). Of course, the petroleum industry is mainly interested in the hydrocarbon fluids, but to model petroleum systems, one must first model the rocks and their geologic histories.

Basin and petroleum system modeling uses forward deterministic computations to simulate the thermal histories of the rocks and the related generation, migration, and accumulation of petroleum, that is, processes are modeled from past to present using inferred starting conditions (Figure 1). These computations require a conceptual model of basin history that has been subdivided into an uninterrupted sequence of events in space and time, including the deposition and erosion of rock units. Model simulations are performed on discretized numerical representations of the available geologic and geochemical data, that is, spatial data are subdivided into grid cells having constant property distributions within each cell. For example, a grid cell might measure 1 km (0.6 mi) on each side and 400 m (1312 ft) in thickness. Numerical values are required for all input parameters in Figure 1. Input data include gridded surfaces of buried rock units from seismic and well-log interpretations, ages of units, present and past rock unit thicknesses, structural restoration through time, lithology and physical properties of units, porosity, permeability, and various boundary conditions, such as present and past water depths, basal heat flow, and sediment-water interface temperatures. The type and amount of organic matter in the source rocks and the kinetics for the conversion of source rock organic matter to petroleum are also required. Model output, such as predicted temperature, pressure, or porosity, can be compared with independent calibration parameters, thus allowing the conceptual model to be adjusted to improve the match between the calibration data and simulation output.

During the last decade, the value and advantages of BPSM have become widely recognized throughout the geoscience community because BPSM can:

1. Organize and archive input data, for example, subsurface structure from seismic surveys, lithology, and present-day temperature from well logs, paleowater depth, and formation age from micropaleontology, heat flow interpreted from tectonic history, and organic richness and thermal maturity from geochemical measurements
2. Help to quantify the essential risk elements (source, reservoir, seal, and overburden) and processes of the petroleum system (trap formation and generation-migration-accumulation; Magoon and Dow, 1994) and focus further work
3. Quantify key petroleum system elements and processes that control petroleum generation, migration, and accumulation within basins, which helps to focus work on the parameters that most affect simulation results
4. Convert raw seismic, geochemical, and geologic data to interpretations that can be tested to assess the range of possible model outcomes
5. Provide a consistent approach to rank or risk prospects and to allocate appropriate levels of professional staff or resources

DISCUSSION

This book consists of six sections similar to those in the conference. The articles in each section are briefly described.

Geochemistry: Challenges in the Future

Clement N. Uguna et al. conducted high-temperature, high-pressure experiments in supercritical water (as high as 420°C) to achieve high thermal maturities (vitrinite reflectance [R_o], >2% R_o) for samples of Upper Jurassic Kimmeridge Clay source rock. Pressures between 390 and 500 bar, the highest used, retard both hydrocarbon generation and maturation after the well-documented promotional effects of water on maturation reach a maximum at 200 to 300 bar. The results provide further evidence that the first-order

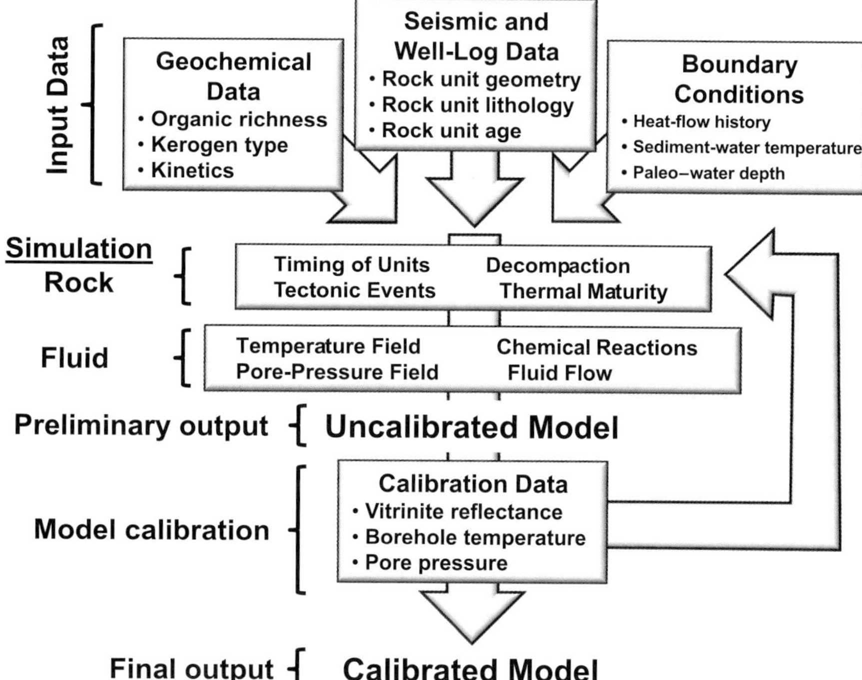

FIGURE 1. Process workflow for basin and petroleum system modeling (from Peters, 2009).

kinetic models used in petroleum system modeling need to incorporate the effects of pressure.

Frank Haeseler et al. present the results of a compositional model called BioClass 0D that predicts total hydrocarbon losses and the residual composition of biodegraded oil. This model was applied to three nonbiodegraded oils representative of the source rocks that contain the three main types of organic matter: types I, II, and III.

Frederic Schneider et al. show that more care should be exercised in the determination of kinetic parameters for source rocks to account for phenomena that occur in the diagenetic zone (<0.6% R_o). Petroleum system models that use kinetic parameters determined on samples at the end of the diagenetic zone (~0.6% R_o) underestimate the amounts of hydrocarbons produced.

Seismic/Pressure/Gravity Magnetics Revisited

Heinz-Juergen Brink et al. used the common reflection surface (CRS) stack technique for reprocessing seismic data to better image the Moho and support an alternate view of the geodynamic origin of the North German Basin. The data allow a revised interpretation of the structural setting and evolution of different salt features in the basin and yield new insights into petroleum systems in the basin.

Frederic Monnier et al. used local grid refinement (LGR) to better define key areas within a larger regional model from northern Kuwait, where they compared it with other refinement methods. The study illustrates the linkage of LGR and compositional three-dimensional (3-D) Darcy flow modeling to predict the distribution of hydrocarbon composition and properties in local reservoir rock areas, where accumulations were predicted. The LGR approach bridges the gap between conventional basin modeling and reservoir modeling and was used to predict API gravities, gas/oil ratios, and the oil-water contact in new prospects.

Structural and Tectonic Processes

Richard Gibson outlines a simple method to realistically describe variations in subsurface salt geometry through time and shows how to implement these variations into basin models without requiring full 3-D structural restoration as an input. This approach was used to produce a basin-scale numerical simulation model of the northern Gulf of Mexico that restores back through time with minimal geometric artifacts, despite complex structural geometries.

Carolyn Lampe et al. studied the Brooks Range Foothills in Alaska using a new approach to integrate three classic disciplines of modeling: structural balancing, fault property analysis, and BPSM. The combination of these disciplines improves understanding, assessment, and prediction of the extent, timing, and generative potential of petroleum systems in tectonically complex areas.

Susanne Nelskamp et al. combined structural modeling with petroleum system modeling in a two-dimensional section of the Dutch onshore region. The model results demonstrate the effects of tectonic inversion, erosion, and subsidence on the geohistory of the thermal field and petroleum generation.

New-Generation Methods/ Unconventional Approaches

Jan Derks et al. describe a local grid refinement method that combines regional-scale petroleum system models and local reservoir- or prospect-scale models. The method was applied to an oil field in Kuwait, where heavy oil zones occur at the original oil-water contact and also in stratigraphic and structural positions above it. The study demonstrates that the heavy oil distribution in those layers can be explained by the petroleum charge history.

Hanneke Verweij et al. modeled the geohistory of the Terschelling Basin and the southern part of the Dutch Central Graben by incorporating a tectonic heat flow boundary condition and detailed knowledge of Tertiary climate changes. The 3-D simulation shows a marked difference in generated hydrocarbon volumes and a shift in the timing of Tertiary petroleum generation compared with simulations using a default surface temperature boundary condition based on paleolatitude reconstruction. The results improved understanding of the thermal maturity and timing of hydrocarbon generation from the Jurassic and Carboniferous source rocks.

Sylvie Wolf et al. developed a new time-stepping method to model oil generation in sedimentary basins based on separating pressure from oil saturation. The method is capable of yielding accurate solutions compared with those computed by the fully implicit scheme, but with an effective reduction of computation time by as much as a factor of five.

Uncertainty in Basin Modeling

Paul Hicks et al. present a basin modeling workflow to identify key input parameters and to quantify uncertainties in these input parameters to evaluate the model results in light of a business question. They demonstrate this workflow using a hypothetical case in which uncertainties in key input parameters that control hydrocarbon generation, volumes, and timing are identified, quantified, and propagated through a basin model.

Sylvie Pegaz-Fiornet et al. focus on the Darcy and invasion percolation approaches to quantify secondary and tertiary petroleum migration and compare the advantages and limitations of each method. The Darcy method accounts for buoyancy, capillary pressure, pressure gradients, and transient physics. The calculation can be time consuming, but it provides a good description of caprock leakage. Invasion percolation is less comprehensive than the Darcy method, but the calculation is faster, and it is especially useful to simulate secondary migration.

Noelle Schoellkopf describes the role of the basin modeler in new ventures exploration risk assessment. Although we can rank model input parameters, such as fetch area, depth, source thickness, total organic carbon (TOC), hydrogen index (HI), temperature gradient or heat flow, and the impact on fluid phase and volumes, we should be aware of pitfalls. Underlying geologic assumptions may account for the greatest uncertainty in new basin areas, where data are sparse and models are poorly calibrated. Modeling tools must be flexible enough to allow multiple working hypotheses within the project timeframe. These hypotheses are best evaluated by an integrated project team that includes a basin modeler.

Andre Vayssaire examines the premise that kilometer-scale vertical migration through fine-grained sediments is the principal mode of transport for hydrocarbons from source rock to reservoir in the Malvinas Basin of Argentina. He compares simulation results based on Darcy flow with those of invasion percolation methods. Based on physics, Darcy flow is more appropriate to simulate kilometer-scale hydrocarbon migration across fine-grained sediments. However, the relative permeability curves need to be upscaled to mimic a migration process that likely occurs along thin stringers. Under these conditions, viscous forces can be ignored, and the invasion percolation method seems to be more appropriate.

Case Studies and Workflows

Friedemann Baur et al. show that gas/oil ratios and API gravities of petroleum accumulations in the Jeanne d'Arc Basin, offshore eastern Canada, can be accurately reproduced by 3-D numerical models that use multicompositional phase-reproducing reaction kinetics measured on samples from the Egret Formation source rock. The chapter reconstructs the detailed filling history of the Terra Nova oil field and deduces the locations of the kitchen areas based on mass-balance calculations and migration pathway analysis.

John Guthrie et al. used one-dimensional and multisurface thermal modeling to help resolve the effects of hydrocarbon charge timing, charge rate, timing of trap formation, and reservoir temperature history on the quality of oil in reservoirs of the Roncador and Frade fields, offshore Campos Basin, Brazil. An interactive biodegradation tool in the modeling software was used to predict API gravity, and the results are constrained by the geology and the geochemical composition of the present-day fluids in the reservoirs.

Oliver Schenk et al. present a detailed 3-D BPSM study of the North Slope of Alaska (275,000 km^2 [106,178 mi^2]) designed to reduce exploration risk and assess remaining petroleum potential. The model is based on more than 48,000 km (29,826 mi) of seismic data and formation top, geochemical, and calibration data from more than 400 wells. The study includes reconstruction of complex basin paleogeometry, including diachronous deposition of overburden rock and several localized erosion events.

CONCLUSIONS

It is appropriate to end this introduction with a brief discussion of several key topics where further research might significantly improve the use of BPSM. A more comprehensive list of targeted research topics is available in the summary of the conference proceedings (Kacewicz et al., 2010).

Basin and petroleum system modeling can be used to accurately predict the extent of the pod of active source rock and the thermal maturity and timing of petroleum generation from that source rock. It is currently less useful to predict volumes of trapped petroleum, their detailed compositions, or the effects of postgeneration processes, such as biodegradation. However, improved prediction of these parameters could have a major impact on domestic and world economies. Basin and petroleum system modeling will continue to expand because of the potential for high-impact solutions to these problems with respect to exploration, development, and assessment.

Modeled predictions of the volumes and compositions of trapped petroleum are limited by current understanding of migration efficiency out of the source rock and from the source rock to the trap. This gap in knowledge will become even more evident with increased exploration for unconventional resources, such as shale gas and shale oil. Further research is needed to better understand the heterogeneity, geochemistry, and mechanical properties of source rocks and how these influence the expulsion and retention of hydrocarbons. This research will require the integration of many disciplines, including seismic lithofacies analysis, sedimentary petrology, geochemistry, petrophysics, rock mechanics, and BPSM. Readers will note that one deficiency in this volume is a lack of articles on rock properties and lithofacies distributions. More studies are needed to improve the link between seismic lithofacies analysis and model input. Lithofacies distributions are rarely known, so better integration of BPSM and basin filling software is needed. Although the properties of clastic rocks have been studied for some time, carbonate source rocks and migration through carbonate carrier rocks deserve much more attention.

Rates of petroleum generation are controlled by kinetic parameters for the organic matter in different source rocks, which in many studies are poorly known because of the lack of adequate rock samples for kinetic measurements. Further study is needed to investigate the potential of asphaltenes isolated from oil samples to infer source rock kinetics. In addition, most models are calibrated using R_O and corrected bottomhole temperatures. In the future, more compositional kinetic models will be calibrated using measured reservoir pressures and the compositions of reservoir fluids. This will require further development of algorithms to address secondary processes that affect petroleum composition, including biodegradation, thermochemical sulfate reduction, and secondary cracking in the source and reservoir rocks.

Simulation of fluid migration through faults requires more research. Most simulators treat fault behavior as input. For example, faults can be designated by the user as open or closed during various time intervals. Deterministic models are needed that predict evolving fault behavior with respect to migrating fluids. This goal can only be achieved by parallel development of viable 3-D rock movement simulators. Current simulators are designed to model vertical rock movement, including burial, uplift, and insertion or removal of salt layers. However, an urgent need for 3-D simulators that can routinely reconstruct nonvertical rock movement, such as thrust faulting, exists.

Despite recent advances in computing speed, personal computers remain inadequate for complex petroleum system models, especially those designed to evaluate risk based on several parameters having large uncertainties. Computations involving large models will continue to be impractical without the speed offered by parallel processing. Nevertheless, continued advances in computing power will allow more models to be conveniently run using personal computers. Expanded options for uncertainty analysis are needed in most simulators, such as uncertainty-related subsurface geometry or the ages of sedimentary units.

GLOSSARY

The purpose of the following glossary is to assist readers of this volume who may have various levels of expertise in basin and petroleum system modeling. The glossary draws upon definitions in Peters and Cassa (1994), Magoon (2004), and Peters et al. (2005). As much as possible, the glossary includes terms directly relevant to basin and petroleum system modeling and excludes terms that are familiar to most geologists.

Activation energy

The energy required for chemical transformations, commonly expressed in kilocalories per mole or kilojoules per mole. Activation energy distributions are used in maturation modeling to calculate the degree of transformation of kerogen in the source rock to oil and gas (Baur et al., Section VI).

Active source rock

See Source rock.

American Petroleum Institute (API) gravity

A convenient scale of the American Petroleum Institute (API) that is inversely related to the density of liquid petroleum: API gravity = (141.5°/[specific gravity at 16°C] − 131.5°). A higher API indicates lighter oil. Fresh water has a gravity of 10° API. Heavy oils have less than 25° API, medium oils have 25 to 35° API, light oils have 35 to 45° API, and condensates have more than 45° API.

Asphaltenes

A complex mixture of heavy organic compounds in crude oils that precipitate under natural conditions due to the admixture of light hydrocarbons to the reservoir or in the laboratory by addition of excess n-pentane, n-hexane, or n-heptane. After precipitation of asphaltenes, the remaining oil consists of saturates, aromatics, and nitrogen, sulfur, and/or oxygen (NSO) compounds.

Backstripping

A key step in basin modeling, where successively older rock layers are removed to decompact underlying layers, thus allowing reconstruction of the burial history in a basin. The method requires decompaction of the remaining sequence of layers after each stage of backstripping to account for the amount of compaction at each step of burial.

Biodegradation

The microbial alteration of organic matter. Petroleum can undergo aerobic or anaerobic (e.g., sulfate-reducing or methanogenic) biodegradation during migration or while in the reservoir, or at surface seeps, typically at temperatures approximately less than 80°C (Haeseler et al., Section I; Guthrie et al., Section VI).

Bulk kinetics

A kinetic model where the generation of oil and gas from kerogen in the source rock is simulated using a single set of activation energies and frequency factors.

Burial history chart

A diagram that shows the depth of burial corrected for compaction of each rock unit (subsidence curve) versus the timing of the essential elements in a petroleum system, including the time interval and critical moment for petroleum generation. The calculated temperature history and the timing of petroleum generation based on calculated vitrinite reflectance or other maturity parameters are commonly overlain on the burial history diagram.

Calibration

A process where calculated model results are fit to measured data by adjusting key input parameters. For example, calculated present-day temperature and thermal maturity (e.g., vitrinite reflectance) at a given location within a sedimentary basin might be fit to measured temperatures and vitrinite reflectance values from a nearby well by adjusting paleoheat flow (Schenk et al., Section VI).

Capillary entry pressure

The pressure difference across the interface that separates two immiscible fluids, which is related to the force needed to move gas or a droplet of oil through a water wet pore. Capillary pressure increases for smaller pore diameters. The minimum capillary pressure measured in seal rock, such as shale, defines the maximum height of an oil or gas column that can be trapped beneath the seal, that is, the trap capacity.

Carrier bed

A permeable rock that conducts migrating petroleum.

Catagenesis

Thermal alteration of organic matter in the range of 50 to 150°C, typically requiring millions of years; equivalent to approximately 0.6 to 2.0% vitrinite reflectance.

Charge

The volume of petroleum expelled from the source rock that is available for entrapment.

Chromatography

Separation of mixtures of compounds based on physicochemical properties. Liquid chromatography and gas chromatography are used to separate petroleum compounds by retention time based on their tendency to partition between a mobile and stationary phase during movement through a chromatographic column.

Coal

A rock that contains greater than 50 wt. % organic matter. Most coals in North America and Europe originate mainly from higher plants and consist of type III (gas-prone) kerogen dominated by vitrinite group macerals. Thermal maturation transforms peat to lignite, bituminous coal, and then anthracite coal.

Common reflectance surface stack method

Common reflectance surface (CRS) sums more seismic traces and improves signal-to-noise ratio of subsurface images compared with classical common midpoint (CMP) stack methods of seismic analysis (Brink et al., Section II).

Condensate

Light oil with API gravity greater than 45°. The term was originally used for petroleum that is gaseous under reservoir conditions but liquid at surface temperatures and pressures.

Cracking

A thermal process where large molecules break apart into smaller molecules during burial maturation or in a refinery. Kerogen in source rocks and oil in reservoirs are cracked to generate lighter petroleum products at high temperatures.

Critical moment

The time that best depicts the generation-migration-accumulation of hydrocarbons in a petroleum system (Magoon and Dow, 1994). The geographic and stratigraphic extents of the system are best evaluated using a map and cross section drawn at the critical moment.

Crude oil

A natural mixture of hydrocarbons and other compounds that have not been refined. Produced crude oil commonly contains solution gas and closely resembles the original oil from the reservoir rock. Hydrocarbons are the most abundant compounds in crude oils, but they also contain nitrogen, sulfur, and/or oxygen (NSO) compounds and variable concentrations of trace elements, such as vanadium and nickel.

Darcy flow

A formula for single-phase fluid flow through porous media (Pegaz-Fiornet et al., Vayssaire, Section V).

Deasphalting

The process where asphaltenes precipitate from crude oil. Laboratory or refinery deasphalting is achieved by adding light hydrocarbons, such as pentane or hexane, to the oil. A similar process can occur in nature when methane and other light hydrocarbons that escape from deep reservoirs enter a shallower oil reservoir.

Decompaction

A burial history diagram requires that the present-day thickness of rock units be corrected to their original thickness before burial using a specific porosity-depth function for each lithologic unit. This involves sequential backstripping of each rock unit and restoration of the thickness of the immediately underlying unit before burial from the surface to basement until the entire sedimentary succession is restored to its original thickness.

Diagenesis

Chemical, physical, and biological changes that affect sediments during and after deposition and lithification but before oil generation or other significant changes caused by heat.

Discretization

The process of converting continuous features into discrete components. For example, various physical properties of each sedimentary unit in a basin model covering a large area, such as thermal conductivity or total organic carbon (TOC), can be divided into discrete values for each model cell, might measure 1 km wide × 1 km long × 400 m thick (0.6 mi × 0.6 mi × 1312 ft) (Wolf et al., Section IV).

Drilling mud

A mixture of clay, water, and chemicals that is pumped into and out of the borehole during drilling. Circulation of drilling mud cools the drill bit, flushes rock cuttings produced by the drill bit to the surface, and maintains pressure in the well bore.

Dry gas

Microbial or highly mature gas that is dominated by methane (>95% by volume) with little or no natural gas liquids. Microbial gas is depleted in ^{13}C compared with a highly mature gas.

Easy vitrinite reflectance (Easy% R_o)

An Arrhenius first-order kinetic model that uses a distribution of activation energies to calculate equivalent vitrinite reflectance (% R_o) for given time-temperature conditions (Sweeney and Burnham, 1990).

Effective source rock

See Source rock.

Essential elements

The source, reservoir, seal, and overburden rocks in a petroleum system (Magoon and Dow, 1994). The essential elements and the processes of generation-migration-accumulation and trap formation control the distribution of petroleum in the subsurface.

Events chart

A chart (also called timing-risk chart) that shows the timing of essential elements and processes for a petroleum system as well as the critical moment and the length of time required to preserve the accumulated petroleum until present day.

Expulsion

The process of primary migration where petroleum escapes from the source rock because of increased pressure and temperature. Expulsion generally involves short distances (meters to tens of meters).

First-order kinetics

See Kinetics.

Flowpath modeling

Simulation of instantaneous fluid flow through porous and permeable carrier beds as driven by buoyancy with essentially no resistance.

Frequency factor

A parameter in the Arrhenius equation that reflects reaction probability and is commonly expressed in geochemical kinetic modeling as s^{-1} or $m.y.^{-1}$. The frequency factor is also known as the A factor or, more rigorously, as the preexponential factor.

Gas hydrate

Various crystalline phases of water that contain gas (mainly methane) in arctic and deep-water settings.

Gas-prone

Organic matter that generates mainly hydrocarbon gases instead of oil during thermal maturation. Type III kerogen is gas prone.

Gas-to-oil ratio

The amount of hydrocarbon gas relative to oil in a reservoir, commonly measured in cubic feet per barrel.

Gas wetness

Expressed in various ways, but generally represents the amount of methane (C_1) relative to the total hydrocarbon gases ($C_1 + C_{2+}$) in a sample. For example, C_1/(C_1 to C_5) ratios greater than and less than 98% are dry and wet gases, respectively.

Generation-migration-accumulation

A petroleum system process that includes the generation, expulsion, and movement of petroleum from the pod of active source rock to the petroleum seep, show, or accumulation (Magoon and Dow, 1994).

Geographic extent

The area of occurrence of a petroleum system as defined by a line that encloses the pod of active source rock and all discovered shows, seeps, and petroleum accumulations that originated from that pod; the geographic extent is mapped at the critical moment (Magoon and Dow, 1994).

Geohistory diagram

See Burial history chart.

Geothermal gradient

The increase in temperature with depth in the Earth, commonly in degrees Celsius per kilometer or degrees Fahrenheit per 100 feet. Gradients are sensitive to basal heat flow, lithology, circulating groundwater, and the cooling effect of drilling fluids. Worldwide average geothermal gradients are from 24 to 41°C/km (1.3–2.2°F/100 ft), with extremes outside this range.

Gravity segregation

A process where heavier and lighter petroleum components accumulate near the bottom and top of the reservoir, respectively.

Heavy oil

Crude oil with less than 25° API.

Humic coal

Gas-prone (type III) coal that consists mainly of higher plant detritus, including vitrinite and inertinite group macerals, with little or no liptinite macerals.

Hydrogen index

A Rock-Eval pyrolysis measure of oil-generative potential defined as 100(S2/TOC) and measured in milligram hydrocarbons per gram total organic carbon (TOC). The HI is used on Van Krevelen–type plots of HI versus oxygen index (OI) to describe organic matter type, and a general relationship exists between HI and atomic hydrogen/carbon (H/C).

Hydrous pyrolysis

A laboratory technique where potential source rocks are heated without air, under pressure, and with water to artificially increase the level of thermal maturity.

Immature

Organic matter where conditions were too cool or too short in duration for thermal generation of petroleum, for example, vitrinite reflectance less than 0.6%.

Inactive source rock

See Source rock.

Inertinite

A maceral group composed of inert hydrogen-poor organic matter with little or no petroleum-generative potential. Type IV kerogen is dominated by inertinite.

Invasion percolation

Petroleum migration based on the distribution of capillary pressures in sediments and their evolution through time. Invasion percolation (IP) assumes instantaneous movement of petroleum under the effects of buoyancy and capillary pressure (Pegaz-Fiornet et al., Vayssaire, Section V).

Kerogen types I, II, IIS, III, IV

See Type.

Kinetics

Kinetic parameters that describe chemical reaction rates include the activation energy and frequency factor. Most petroleum system modeling assumes that oil and gas generation can be described by a series of parallel first-order kinetics, where the rate of each reaction depends on the concentration of only one reactant (a unimolecular reaction; Uguna et al., Schneider et al., Section I). See also Multicomponent kinetics.

Kitchen

See Pod of active source rock.

Level of certainty

A measure of the degree of confidence that petroleum originated from a specific pod of active source rock; three levels include known (!), hypothetical (.), and speculative (?), depending on the level of geochemical, geophysical, and geologic evidence (Magoon and Dow, 1994).

Light hydrocarbons

Gases and volatile liquids at standard temperature and pressure, which range from methane to octane, including normal, iso-, and cyclic alkanes, and aromatic compounds.

Light oil

Crude oil with 35 to 45° API.

Liptinite

A maceral group composed of oil-prone hydrogen-rich kerogen that fluoresces under ultraviolet light. Both structured (e.g., resinite, sporinite, or cutinite) and unstructured or amorphous liptinites, sometimes called amorphinite, can occur.

Maceral

Microscopically recognizable organic particles in kerogen. The three main maceral groups include liptinite, vitrinite, and inertinite.

Mantle

Part of the Earth composed mainly of solid silicate rock that extends from the base of the crust (Moho) to the core-mantle boundary at approximately 2900 km (~1802 mi) in depth.

Mature

Organic matter that is in the oil window; thermal maturity equivalent to the range of 0.6 to 1.4% vitrinite reflectance. Organic matter can also be mature with respect to the gas window (~0.9–2.0% vitrinite reflectance).

Maturity

See Thermal maturity.

McKenzie model

A mathematic expression for symmetrical tectonic rifting that assumes (1) an initial stretching phase with constant thinning of the crust and upper mantle caused by upwelling of the underlying asthenosphere (lower mantle) followed by (2) a cooling phase with near or full restoration of the thickness of the lithosphere (crust and upper mantle). The total thinning of the crust is described by a stretching factor, β.

Metagenesis

Thermal destruction of organic molecules by cracking to gas in the range approximately 150 to 200°C, which occurs after catagenesis, but before greenschist metamorphism (>200°C); thermal maturity equivalent to the range of 2.0 to 4.0% vitrinite reflectance.

Microbial gas

Also called marsh gas or biogenic gas; typically greater than 99% methane produced by methanogenic microbes in shallow sediments at temperatures less than 80°C. Microbial methane is generally depleted in ^{13}C compared with thermogenic gas.

Microscale sealed vessel pyrolysis (MSSV)

A laboratory heating procedure in the absence of air used to investigate the compositional variability of petroleum during increasing thermal maturation.

Migration

Movement of petroleum from source rock toward a reservoir or seep. Primary migration is expulsion of petroleum from fine-grained source rock, whereas secondary migration moves petroleum through a coarse-grained carrier bed or fault to a reservoir or seep. Tertiary migration occurs when petroleum moves from one trap to another or to a seep (Pegaz-Fiornet et al., Section V).

Mixing and upscaling

Single fluid or mineral properties can be used to derive bulk properties of mixtures, such as oil-wet limestone, by mathematically mixing the components in the same proportions thought to exist in nature. Arithmetic, harmonic, and geometric averaging are three methods used to mix components. Porosity is always arithmetically mixed, whereas grains (e.g., thermal conductivity) can be mixed geometrically for homogeneous sorting or arithmetically or harmonically for layered strata. Upscaling transforms microscale measurements, such as laboratory capillary entry pressure, to macroscale values that can be used in full-scale models. For example, capillary entry pressures for clastic and carbonate rocks can be estimated by dividing the laboratory measurements by an upscaling factor of 2.56 (Hantschel and Kauerauf, 2009).

Mohorovicic discontinuity

A zone that separates the Earth's crust from the underlying mantle (Brink et al., Section II). The Moho typically occurs approximately 35 km (~19 mi) below the continents and 5 to 10 km (3–5 mi) below the floor of the ocean.

Monte Carlo simulation

A set of model runs that requires the distribution of uncertainty for one or more variables, such as heat flow or thickness of a rock layer (see Uncertainty). Random values within the uncertainty distribution for each variable are used to derive the complete set of simulation results. Output parameters can be visualized and interpreted using statistical tools, such as histograms. Confidence intervals related to risking can be derived from these histograms, for example, "based on 1000 simulations, a 95% probability that this source rock generated at least 100 million barrels of petroleum exists."

Multicomponent kinetics

Kinetic models in which the generation of different subcomponents of the kerogen are individually simulated (i.e., each subcomponent has a different set of frequency factors and distribution of activation energies). These subcomponents can be defined by molecular weight range (e.g., C_7–C_{15}, C_{16}–C_{25}, C_{26}–C_{35}, C_{36}–C_{45}, C_{46}–C_{55}, C_{56}–C_{80}), compound type, or a combination thereof (Baur et al., Section VI).

Natural gas

Gaseous petroleum that can consist of C_1–C_5 hydrocarbons, CO_2, N_2, H_2, H_2S, Ar, and He. When natural gas occurs with oil, it is called associated gas.

Nitrogen, sulfur, and/or oxygen (NSO) compounds (resins)

The NSO compounds (resins) are pentane-soluble compounds in petroleum that contain various elements in addition to hydrogen and carbon, for example, nitrogen, sulfur, and/or oxygen (NSO). Compounds in this fraction are sometimes called heterocompounds or nonhydrocarbons. Other fractions include saturates, aromatics, and asphaltenes.

Nonassociated gas

Natural gas that is not associated with crude oil in the reservoir.

Nonhydrocarbon gases

Mainly carbon dioxide (CO_2), nitrogen (N_2), and hydrogen sulfide (H_2S), but also including helium (He), argon (Ar), and hydrogen (H_2).

Oil deadline

The depth where oil no longer exists as a liquid phase in petroleum reservoirs, generally corresponding to a gas-to-oil ratio (GOR) greater than 5000 SCF/barrel or more than 150°C (typically in the range 165–185°C).

Oil-oil correlation

A comparison of chemical compositions to describe the genetic relationships among crude oils based on source-related geochemical data such as biomarkers, isotopes, and metal distributions, although the source rock may not be defined.

Oil prone

Organic matter that generates significant quantities of oil during catagenesis. Oil-prone organic matter is typically also more gas prone than gas-prone organic matter.

Oil–source rock correlation

A comparison of chemical compositions to describe the genetic relationships among crude oils and source rock extracts based on source-related geochemical data such as biomarkers, isotopes, and metal distributions.

Oil window

The maturity range where oil is generated from oil-prone organic matter (~0.6–1.4% vitrinite reflectance), that is, within the catagenesis zone (~0.5–2.0% vitrinite reflectance).

Organic facies (organofacies)

A mappable rock unit that contains a distinctive assemblage of organic matter without regard to the mineralogy (Jones, 1987).

Organic matter

Biogenic carbonaceous materials. Organic matter preserved in rocks including kerogen, bitumen, pyrobitumen, oil, and gas.

Overburden rock

Sedimentary or other rock that compresses and contributes to the thermal maturation of the underlying source rock.

Overmature

See Postmature.

Oxygen index (OI)

A Rock-Eval pyrolysis measure of the amount of oxygen in organic matter defined as 100(S3/TOC) and measured in milligrams carbon dioxide per gram total organic carbon (TOC). The OI is used on Van Krevelen-type plots of hydrogen index (HI) versus OI to describe organic matter type, and a general relationship exists between OI and atomic oxygen/carbon (O/C).

Palynomorphs

Organic-walled acid-resistant microfossils that provide information on age, paleoenvironment, and thermal maturity (e.g., thermal alteration index [TAI]).

Permeability

The capacity of a rock layer to transmit water or other fluids, such as oil. The standard unit for permeability is the Darcy (d) or, more commonly, the millidarcy (md). Relative permeability is a dimensionless ratio that reflects the capability of oil, water, or gas to move through a formation compared with that of a single-phase fluid, commonly water. If a single fluid moves through rock, its relative permeability is 1.0. Two or more fluids generally inhibit flow through rock compared with that of a single phase of each component (Vayssaire, Section V).

Petroleum

A mixture of organic compounds composed mainly of hydrogen and carbon and found in the gaseous, liquid, or solid state in the Earth; includes hydrocarbon gases, bitumen, migrated oil, pyrobitumen, and their refined products, but not kerogen. In European usage, the term is sometimes restricted to refined products only.

Petroleum system

The essential elements (source, reservoir, seal, and overburden rock) and processes (trap formation, generation-migration-accumulation) and all genetically related petroleum that originated from one pod of active source rock and occurs in shows, seeps, or accumulations (Magoon and Dow, 1994); also called hydrocarbon system.

Petroleum system name

Systematic nomenclature (Magoon and Dow, 1994) that names (1) the formation containing the pod of active source rock followed by a hyphen, and (2) the reservoir rock that contains the largest volume of genetically related petroleum, followed by (3) a symbol for the level of certainty in the petroleum system, for example, Bazhenov-Neocomian(!)

Petroleum system processes

Trap formation and generation-migration-accumulation. Biodegradation and thermal destruction are omitted as processes because they occur after a petroleum system forms.

PhaseKinetics

A method based on pyrolysis experiments to determine compositional kinetic models that allows prediction of petroleum-phase properties and behavior during thermal modeling (di Primio and Horsfield, 2006).

Phytoclast

An identifiable particle or maceral in kerogen, for example, phytoclasts of vitrinite are used to measure vitrinite reflectance.

Pod of active source rock

A contiguous volume of fine-grained organic-rich rock that generated and expelled petroleum at the critical moment. A pod of thermally mature source rock may be active, inactive, or spent.

Porosity

The volume percent of a rock that consists of open pore space.

Postmature (for oil)

High maturity at which no further oil generation occurs (i.e., >1.3% vitrinite reflectance).

Potential source rock

See Source rock.

Primary migration

See Expulsion or Migration.

Production index

A Rock-Eval pyrolysis measure of thermal maturity or contamination measured as S1/(S1 + S2).

Pseudo–Van Krevelen diagram

See Van Krevelen diagram.

Pyrolysis

Breakdown of organic matter by heating in the absence of oxygen; Rock-Eval instrumentation uses programmed-temperature pyrolysis because the temperature is programmed to increase at a selected rate during analysis (25°C/min). Hydrous pyrolysis uses a constant temperature for each experiment (e.g., 330°C).

Ray tracing

A migration approach that assumes petroleum migrates instantaneously under the influence of buoyancy through a fault network or porous carrier bed.

Relative permeability

See Permeability.

Reservoir

A porous and permeable sedimentary rock formation that contains oil and/or natural gas enclosed or surrounded by layers of less permeable rock. An oil pool consists of a reservoir or group of reservoirs. However, the term is misleading because petroleum does not exist in pools, but in pores between rock grains.

Reservoir characterization

Integrating and interpreting geologic, geophysical, petrophysical, fluid and performance data to describe a reservoir.

Reservoir rock

Porous and permeable rock, such as sandstone, vuggy carbonate, or fractured shale, that permits migration and accumulation of petroleum provided that an overlying or updip trap is present.

R_o

See Vitrinite reflectance.

Rock-Eval

A commercially available pyrolysis system used as a rapid screening tool to evaluate the quantity, quality, and thermal maturity of rock samples.

S1, S2, S3

Rock-Eval pyrolysis parameters where S1 = volatile organic compounds (mg hydrocarbons/g rock); S2 = organic compounds generated by cracking of the kerogen (mg hydrocarbons/g rock); and S3 = organic carbon dioxide generated from the kerogen up to 390°C (mg CO_2/g rock).

Seal rock (cap rock)

Fine-grained rock that is relatively impermeable to migrating petroleum in the subsurface, such as shale or anhydrite, thus slowing or preventing leakage of petroleum to the surface.

Secondary migration

See Migration.

Solid bitumen

Solid bitumen includes pyrobitumen and bitumen formed by nonthermal processes, such as deasphalting or biodegradation (Derks et al., Section IV).

Source rock

Rock that contains sufficient organic matter of the proper composition to generate and expel petroleum during diagenesis (microbial methane) or catagenesis. Source rock can be effective (is generating or has generated and expelled), potential (sufficient quantity and quality of organic matter, but has not yet expelled), active (is generating and expelling petroleum at the critical moment), inactive (stopped generating because of uplift and/or reduced thermal stress, although petroleum potential remains), or spent (postmature for oil but can still generate methane and wet gas).

Spent source rock

See Source rock.

Standard temperature and pressure (STP)

Defined as room temperature (15.6°C or 60°F) and one atmosphere pressure (14.7 psi).

Suboxic

Refers to water column or sediments with molecular oxygen contents of 0 to 0.2 mL/L of interstitial water.

Subsidence curve

See Burial history chart.

Tectonic inversion

A process where extensional normal faults are reactivated by compression (positive inversion), or where reverse faults are reactivated by extension (negative inversion). Positive inversion results in uplift of formerly low-lying areas (Nelskampe et al., Section III).

Tertiary migration

See Migration.

Thermal maturity

The extent of heat-driven reactions that alter the composition of organic matter (e.g., conversion of sedimentary organic matter to petroleum or cracking of oil to gas). Different geochemical scales, such as vitrinite reflectance, pyrolysis T_{max}, and biomarker maturity ratios can be used to indicate the level of thermal maturity of organic matter.

Total organic carbon (TOC)

Total organic carbon ([TOC] wt. %) in a rock sample as determined by various combustion methods.

Transformation ratio

Ranges from 0 to 1.0 and is the ratio of the generated petroleum to the original petroleum potential of a sample before maturation (Derks et al., Section IV).

Trap

A subsurface arrangement of relatively impermeable rock that allows accumulation of petroleum in underlying or downdip, relatively porous, reservoir rock. Traps can be structural (e.g., domes, anticlines), stratigraphic (pinch-outs, permeability or diagenetic changes), or combinations of both.

Type I kerogen

Highly oil-prone organic matter having Rock-Eval pyrolysis hydrogen indices greater than 600 mg hydrocarbon/g TOC when thermally immature; contains algal and bacterial input dominated by amorphous liptinite macerals; common in both marine and lacustrine settings; type I pathway on Van Krevelen diagram.

Type II kerogen

Oil-prone organic matter having Rock-Eval pyrolysis hydrogen indices in the range of 300 to 600 mg hydrocarbon/g TOC when thermally immature; contains algal and bacterial organic matter dominated by liptinite macerals, such as exinite and sporinite; common but not restricted to marine settings; Type II pathway on Van Krevelen diagram.

Type IIS kerogen

Hydrogen indices in the range of type II kerogen, but sulfur-rich type IIS kerogens contain uncommonly high organic sulfur (8–14 wt. %, atomic $S/C \geq 0.04$) and begin to generate oil at a lower thermal maturity than typical type II kerogens with less than 6 wt. % sulfur (Orr, 1986).

Type III kerogen

Gas-prone organic matter having Rock-Eval pyrolysis hydrogen indices in the range of 50 to 200 mg hydrocarbon/g TOC when thermally immature; generally contains higher plant organic matter dominated by vitrinite macerals, common but not restricted to paralic marine settings, type III pathway on Van Krevelen diagram.

Type IV kerogen

Inert organic matter having Rock-Eval pyrolysis hydrogen indices below 50 mg hydrocarbon/g TOC in thermally immature rocks; dominated by organic matter that was recycled or extensively oxidized during deposition. Sometimes the type IV pathway is not shown on Van Krevelen diagrams, but it lies below the type III pathway.

Uncertainty

The possible distribution of values for one number, such as the thermal conductivity of shale. Uncertainty distributions must be specified for Monte Carlo simulations, for example, normal (Gaussian), logarithmic normal, uniform, triangular, or exponential (Hicks et al., Schoellkopf et al., Section V).

Uniform stretching model

See McKenzie model.

Upscaling

See Mixing and upscaling.

Van Krevelen diagram

A plot of atomic oxygen/carbon (O/C) versus atomic hydrogen/carbon (H/C) originally used to classify coals and predict compositional evolution during thermal maturation, but adapted to classify kerogen types I, II, III, and IV. More common is the pseudo–Van Krevelen diagram, where Rock-Eval pyrolysis oxygen index ([OI] mg HC/g TOC) is plotted versus hydrogen index ([HI] mg HC/g TOC).

Viscosity

Fluid viscosity is a measure of the resistance of a fluid to flow. The viscosity of liquid oil depends on its composition, temperature, and pressure.

Vitrinite

A group of gas-prone macerals derived from land plant tissues. Phytoclasts of vitrinite are used for vitrinite reflectance (R_o) determination of thermal maturity.

Vitrinite reflectance (R_o)

A thermal maturation parameter based on microscopy of organic matter in fine-grained rocks. Each R_o value represents the percent of incident light (546 nm) reflected from a single vitrinite phytoclast in a polished slide of kerogen, coal, or whole rock under an oil-immersion microscope objective as measured by a photometer. Average R_o is typically measured on at least 20 randomly oriented vitrinite phytoclasts. Some microscopes have rotating stages that allow measurements of anisotropy. Thus, R_m, R_{max}, R_{min}, and R_r indicate mean, maximum, minimum, and random vitrinite reflectance, respectively.

Wet gas

Natural gas that contains ethane, propane, and heavier hydrocarbons and less than 98% methane/total hydrocarbons. The wet gas zone occurs during catagenesis below the bottom of the oil window and above the top of the gas window (1.4–2.0% vitrinite reflectance).

ACKNOWLEDGEMENTS

The AAPG Hedberg Research Conference on Basin and Petroleum System Modeling received record-breaking financial support from industry sponsors despite a difficult economic climate. We thank the following sponsors for their generous support: Anadarko, BHP Billiton, British Petroleum, Chevron, ConocoPhillips, Devon, ENI, Maersk Oil, Marathon, Schlumberger, and StatoilHydro. Allegra Hosford Scheirer, Les Magoon, and Ron Nelson provided useful reviews of the draft manuscript.

APPENDIX

Section I: Geochemistry: Challenges in the Future

1. Retardation of hydrocarbon generation and maturation by water pressure in geologic basins: An experimental investigation by C. N. Uguna, A. D. Carr, W. Meredith, C. E. Snape, I. C. Scotchman, and R. C. Davis
2. First stoichiometric model of oil biodegradation in natural petroleum systems. Part II: Application of the BioClass 0D approach to oils from various sources by F. Haeseler, F. Behar, and D. Garnier
3. Model of low-maturity generation of hydrocarbons applied to the Carupano Basin, offshore Venezuela by F. J. Schneider, J. Noya, and C. Magnier

Section II: Seismic/Pressure/Gravity Magnetics Revisited

4. Moho, basin dynamics, salt stock family development, and hydrocarbon system examples of the North German Basin revisited by applying seismic common reflectance surface processing by H.-J. Brink, D. Gajeswki, M. Baykulov, and M.-K. Yoon
5. Prediction of fluid compositional heterogeneities in fields using local grid refinement: Example from the Jurassic of northern Kuwait by F. Monnier, J.-M. Laigle, S. Pegaz-Fiornet, P.-Y. Chenet, F. Lorant, and A. Al-Khamiss

Section III: Structural and Tectonic Processes

6. A methodology to incorporate dynamic salt evolution in three-dimensional basin models: Application to regional modeling of the Gulf of Mexico by R. G. Gibson
7. Modeling^3: Integrating structural modeling, fault property analysis, and petroleum systems modeling—An example from the Brooks Range foothills of the Alaska North Slope by C. Lampe, K. J. Bird, T. E. Moore, R. A. Ratliff, and B. Freeman
8. Structural evolution, temperature, and maturity of sedimentary basins in the Netherlands: Results of combined structural and thermal two-dimensional modeling by S. Nelskamp, J. D. Van Wees, and R. Littke

Section IV: New-Generation Methods/ Unconventional Approaches

9. Three-dimensional basin and petroleum system model of the Cretaceous Burgan formation, Kuwait: Model-in-model, high-resolution charge modeling by J. F. Derks, M. Al-Saeed, T. Fuchs, M. Al-Hadjeri, M. Al-Quattan, O. Swientek, and A. Kauerauf
10. Reconstruction of basal heat flow, surface temperature, source rock maturity, and hydrocarbon generation in salt-dominated Dutch basins by H. Verweij, M. S. C. Echternach, N. Witmans, and R. Abdul Fattah
11. A new efficient scheme to model hydrocarbon migration at basin scale: A pressure-saturation splitting by S. Wolf, I. Faille, S. Pegaz-Fiornet, F. Willien, and B. Carpentier

Section V: Uncertainty in Basin Modeling

12. Identifying and quantifying significant uncertainties in basin modeling by P. J. Hicks Jr., J. D. Shosa, C. M. Fraticelli, M. J. Hardy, and M. B. Townsley
13. Comparison between the different approaches of secondary and tertiary hydrocarbon migration modeling in basin simulators by S. Pegaz-Fiornet, B. Carpentier, A. Michel, and S. Wolf
14. Quantitative assessment of hydrocarbon charge risk in new ventures: Are we fooling ourselves? by N. B. Schoellkopf
15. Simulation of petroleum migration in fine-grained rock by upscaling relative permeability curves: The Malvinas Basin, offshore Argentina by A. Vayssaire

Section VI: Case Studies and Workflows

16. Prediction of reservoir fluid composition using basin and petroleum system modeling: A study from the Jeanne d'Arc Basin, eastern Canada by F. Baur, R. di Primio, H. Wielens, and R. Littke

17. Integrating geochemistry, charge rate and timing, trap timing, and reservoir temperature history to model fluid properties in the Frade and Roncador fields, Campos Basin, offshore Brazil by J. Guthrie, C. Nino, and H. Hassan
18. Petroleum system modeling of northern Alaska by O. Schenk, K. J. Bird, L. B. Magoon, and K. E. Peters

REFERENCES CITED

di Primio, R., and B. Horsfield, 2006, From petroleum-type organofacies to hydrocarbon phase prediction: AAPG Bulletin, v. 90, p. 1031–1058, doi:10.1306/02140605129.

Hantschel, T., and A. I. Kauerauf, 2009, Fundamentals of basin and petroleum systems modeling: Berlin, Germany, Springer, 476 p.

Jones, R. W., 1987, Organic facies, *in* J. Brooks and D. Welte, eds., Advances in petroleum geochemistry: New York, Academic Press, p. 1–90.

Kacewicz, M., K. E. Peters, and D. Curry, 2010, The 2009 Napa Hedberg conference on basin and petroleum system modeling: AAPG Bulletin, v. 94, p. 773–789, doi:10.1306/10270909128.

Magoon, L. B., 2004, Petroleum system: Nature's distribution system for oil and gas: Encyclopedia of Energy, Elsevier, v. 4, p. 823–836.

Magoon, L. B., and W. G. Dow, 1994, The petroleum system: From source to trap: AAPG Memoir 60, 655 p.

Orr, W. L., 1986, Kerogen/asphaltene/sulfur relationships in sulfur-rich Monterey oils: Organic Geochemistry, v. 10, p. 499–516, doi:10.1016/0146-6380(86) 90049-5.

Peters, K. E., ed., 2009, Basin and petroleum system modeling: AAPG Getting Started Series 16, AAPG/Datapages.

Peters, K. E., and M. R. Cassa, 1994, Applied source rock geochemistry, *in* L. B. Magoon and W. G. Dow, eds., The petroleum system: From source to trap: AAPG Memoir 60, p. 93–120.

Peters, K. E., C. C. Walters, and J. M. Moldowan, 2005, The biomarker guide: Cambridge, Cambridge University Press, 1155 p.

Sweeney, J. J., and A. K. Burnham, 1990, Evaluation of a simple model of vitrinite reflectance based on chemical kinetics: AAPG Bulletin, v. 74, p. 1559–1570.

Venue for the AAPG Hedberg Research Conference on Basin and Petroleum System Modeling was the elegant Meritage Resort and Spa in Napa, gateway to the scenic California Wine Country and more than 300 outstanding wineries.

The Napa Valley is world famous for vineyards and wineries. Today, Napa Valley features more than 300 wineries and grows many different grape varieties including Cabernet Sauvignon, Chardonnay, Zinfandel, and Merlot. *Courtesy of Meritage Resort and Spa.*

SECTION 1

Geochemistry–Challenges in the Future

1

Ugana, C. N., C. E. Snape, W. Meredith, A. D. Carr, I. C. Scotchman, and R. C. Davis, 2012, Retardation of hydrocarbon generation and maturation by water pressure in geologic basins: An experimental investigation, *in* K. E. Peters, D. J. Curry, and M. Kacewicz, eds., Basin Modeling: New Horizons in Research and Applications: AAPG Hedberg Series, no. 4, p. 19–37.

Retardation of Hydrocarbon Generation and Maturation by Water Pressure in Geologic Basins: An Experimental Investigation

Clement N. Ugana, Colin E. Snape, and Will Meredith

Department of Chemical and Environmental Engineering, Faculty of Engineering, University of Nottingham, Nottingham, United Kingdom

Andrew D. Carr[1]

Advanced Geochemical Systems Ltd., Leicestershire, United Kingdom

Iain C. Scotchman

Statoil (UK) Ltd., London, United Kingdom

Robert C. Davis

Woodside Energy (USA) Inc., Houston, Texas, U.S.A.

ABSTRACT

Temperature-time–based first-order kinetic models are currently used to predict hydrocarbon generation and maturation in basin modeling. Physical chemical theory, however, indicates that water pressure should exert significant control on the extent of these hydrocarbon generation and maturation reactions. We previously heated type II Kimmeridge Clay source rock in the range of 310 to 350°C at a water pressure of 500 bar to show that pressure retarded hydrocarbon generation. This study extended a previous study on hydrocarbon generation from the Kimmeridge Clay that investigated the effects of temperature in the range of 350 to 420°C at water pressures as much as 500 bar and for periods of 6, 12, and 24 hr. Although hydrocarbon generation reactions at temperatures of 420°C are controlled mostly by the high temperature, pressure is found to have a significant effect on the phase and the amounts of hydrocarbons generated.

In addition to hydrocarbon yields, this study also includes the effect of temperature, time, and pressure on maturation. Water pressure of 390 bar or higher retards the vitrinite reflectance by an average of *ca.* 0.3% Ro compared with the values obtained under low pressure hydrous conditions across the temperature range investigated. Temperature,

[1]*Present address:* British Geological Survey, Keyworth, Nottingham, United Kingdom.

DOI:10.1306/13311427H43461

pressure, and time all control the vitrinite reflectance. Therefore, models to predict hydrocarbon generation and maturation in geological basins must include pressure in the kinetic models used to predict the extent of these reactions.

INTRODUCTION

The growth of petroleum system modeling is caused by its ability to predict the timing of maturation and hydrocarbon generation in geologic basins and also advances in computer technology that allow for petroleum system models to be constructed and run in reasonable amounts of time. Although initially predicted using temperature-time calibrations, both maturation and hydrocarbon generation are now predicted using Arrhenius first-order parallel kinetic models derived mainly from laboratory pyrolysis studies. From the initial pioneering work on coals (Van Krevelen et al., 1951; Pitt, 1962), the subsequent 20 yr have resulted in sophisticated kinetic models that predict hydrocarbon composition and gas-oil ratios (di Primio and Horsfield, 2006) and hydrocarbon expulsion (Stainforth, 2009). These parallel kinetic models are inappropriate because they contain forced compensation effects produced by heat and mass transfer effects (Braun and Burnham, 1987; Burnham, 1993; Barth et al., 1996; Schenk and Dieckmann, 2004; Stainforth, 2009).

Arrhenius kinetics also ignores the fact that transition state and thermodynamic theories both show that water pressure reduces the reaction rates of endothermic volume expansion reactions, such as hydrocarbon generation and maturation (Laidler, 1987). The Arrhenius equation arises from empirical observations that ignore mechanistic considerations involved in the reaction. An increase in water pressure increases the activation energy caused by increased activation enthalpy because of the additional energy required to displace the pressurized water in the kerogen pore systems (Carr et al., 2009). Energy that is used to displace pressurized water is no longer available to be used to overcome the activation energy barrier because that energy (thermal) is leaving the kerogen and being transferred to the water column as potential energy. In addition, an increase in pressure reduces the entropy, thereby reducing the preexponential factor (Carr et al., 2009). The combined effect of increasing the enthalpy and reducing the entropy increases the Gibbs free energy (ΔG), meaning that the reactants instead of the products are favored, that is, less reaction occurs. Indeed, experiments conducted at temperatures between 310 and 350°C under 500-bar water pressure always produced lower yields than those obtained under nonhydrous or low-pressure hydrous conditions (Carr et al., 2009). The results at 450 to 500 bar water pressure are entirely consistent with the predictions derived from transition state and thermodynamic theories.

Several workers using laboratory experimental studies (e.g., Price and Wenger, 1992; Dalla Torre et al., 1997) and data from geologic basins (McTavish, 1978, 1998; Hao et al., 1995; Huijun et al., 2004) found evidence that pressure retards either hydrocarbon generation or maturation. Other workers, however, did not find any laboratory evidence to support the retardation of either hydrocarbon generation or maturation (see Zou and Peng, 2001; Carr et al., 2009, for details).

Carr et al. (2009) suggested that the different conclusions reached by studies involving laboratory experiments could be explained by the different experimental methods used, with the vapor pressure and gold bag studies generally finding no evidence for pressure retardation. This lack of evidence for retardation may be caused by the combined effects of the compressibility of the vapor phase (thereby allowing volume expansion against the compressible vapor) and the malleability of the gold bag, which allows the volatiles generated by the maturation reactions to expand the gold bag, so that the internal pressure remains relatively constant against the externally applied pressure. In contrast, high pressure water pyrolysis experiments conducted in closed autoclave vessels in which the pressurized water is in direct contact with the sample being pyrolyzed (typical of the conditions in geologic basins) show significant retardation effects (Price and Wenger, 1992; Landais et al., 1994; Michels et al., 1995a, b; Carr et al., 2009).

Previously, Carr et al. (2009) showed that water pressure retarded hydrocarbon generation when Kimmeridge Clay samples were pyrolyzed in the temperature range of 310 to 350°C and at water pressures as much as 500 bar. However, as identified first by Lewan (1993, 1997) and found in our previous study (Carr et al., 2009), water first promotes bitumen-plus-oil and gas generation, and under laboratory conditions at 350°C, the maximum conversion to these products was observed at the steam pressure of approximately 160 bar before pressure retardation effects were observed at 500 bar. In this study, temperature in the range between 350 and 420°C, pressures of 45 to 500 bar, and periods of 6, 12, and 24 hr were used to evaluate the effects of temperature, pressure, and time on bitumen, bitumen-plus-oil, and gas yields as well as vitrinite reflectance (VR R_o). The specific aim of this study was to evaluate pressure retardation effects in water at higher levels of maturity than has been previously investigated. This

Table 1. Geochemical data for the Kimmeridge Clay sample.

*TOC** (%)	*S1 (mg HC/g rock)*	*S2 (mg HC/g rock)*	*S3 (mg HC/g rock)*	*HI (mg HC/g TOC)*	*OI (mg CO_2/g TOC)*	*PI S1/ (S1 + S2)*	T_{max} (°C)	% R_o
25.5	3.17	174.47	2.67	684	10	0.02	416	0.36

**TOC = total organic carbon; R_o = vitrinite reflectance; HC = hydrocarbon; HI = hydrogen index; OI = oxygen index; PI = production index.*

has necessitated using supercritical water at temperatures up to 420°C because only moderate maturities can be achieved in liquid water at 350°C in reasonable experimental periods. This study is the first in which high maturities were achieved under high-pressure hydrous conditions.

EXPERIMENT

Sample

The source rock investigated in this study is from the fresh samples of organic-rich Kimmeridge Clay Formation from the coastal outcrop at Kimmeridge Bay, Dorset, United Kingdom. The source rock is immature and oil prone with a hydrogen index (HI) of 684 mg HC/g TOC, T_{max} of 416°C, and 0.36% R_o (Table 1). Whole rock samples instead of kerogens were used in an attempt to reproduce the mineral-kerogen interactions that occur in source rock kitchens (Eglinton et al., 1986).

Pyrolysis Experiments

Pyrolysis experiments were conducted on 4-g samples of Kimmeridge Clay source rock (180- to 425-μm particle size) at temperatures in the range of 350 to 420°C (temperature accuracy ±1°C) and durations of 6 to 24 hr under nonhydrous (no water added), low-pressure hydrous, and high–water pressure hydrous pyrolysis conditions. The pyrolysis equipment (Figure 1) used a Parr 4740 series Hastalloy (22 mL cylindrical) pressure vessel (rated to 648 bar at 350°C) connected to a pressure gauge rated to 690 bar. Previously, Carr et al. (2009) reported the results from a 22-mL Parr 4740 stainless steel vessel. The change in the vessel was required to achieve temperatures of 420°C while using a water pressure of as much as 500 bar.

Heat was applied by means of a fluidized sand bath controlled by a temperature gauge. Temperature was monitored by means of an additional K-type thermocouple connected externally to a computer that recorded the temperature every 10 s. Hydrous experiments were conducted by the addition of 5 to 20 mL of water to the vessel, whereas the nonhydrous runs were conducted without water. The unextracted rock sample to be pyrolyzed was first weighed and transferred to the vessel, after which the volume of water needed for the experiment was added before the pressure vessel was assembled for the hydrous runs. For all experiments, the reaction vessel was flushed with nitrogen gas to replace air in the reactor head space, after which 2 bar pressure of nitrogen was pumped into the vessel to produce an inert atmosphere during the pyrolysis runs. The sand bath was connected to a compressed air source and was preheated to the required temperature and left to equilibrate. After equilibration, the pressure vessel was then lowered onto the sand bath and the experiment was left to run with constant air flow through the sand bath. The amount of water used for the experiments was varied to control the pressure at all temperatures, except for the high pressure (500 bar) runs at 350°C, where the added water was insufficient to generate the required pressure and additional water was pumped into the vessel to increase the pressure.

The high-pressure (500 bar) runs at 350°C were performed in a similar fashion to the low pressure hydrous experiments, with the vessel initially filled with 20 mL water. After lowering the pressure vessel onto the sand bath, the vessel was connected to the high-pressure line and allowed to attain its maximum vapor pressure (in about 20 min) before adding more water to increase the pressure. This is to ensure that excess water was not added to the vessel that could result in excessive pressure, thereby creating a safety issue. To apply a high liquid water pressure to the system with the aid of a compressed air-driven liquid pump, the emergency pressure release valve B was first closed (Figure 1), and valve A opened until a pressure slightly higher than the vapor pressure of the experiment was displayed on the external pressure gauge (to avoid losing any content of the vessel when the reactor valve C was opened). A high liquid water pressure was then applied to the system by first opening valve C and immediately gradually opening valve A to add more distilled water to the reaction vessel. When the required pressure was attained, valve C was closed to isolate the reactor from the high-pressure line, and valve A was also closed to prevent more water going to the

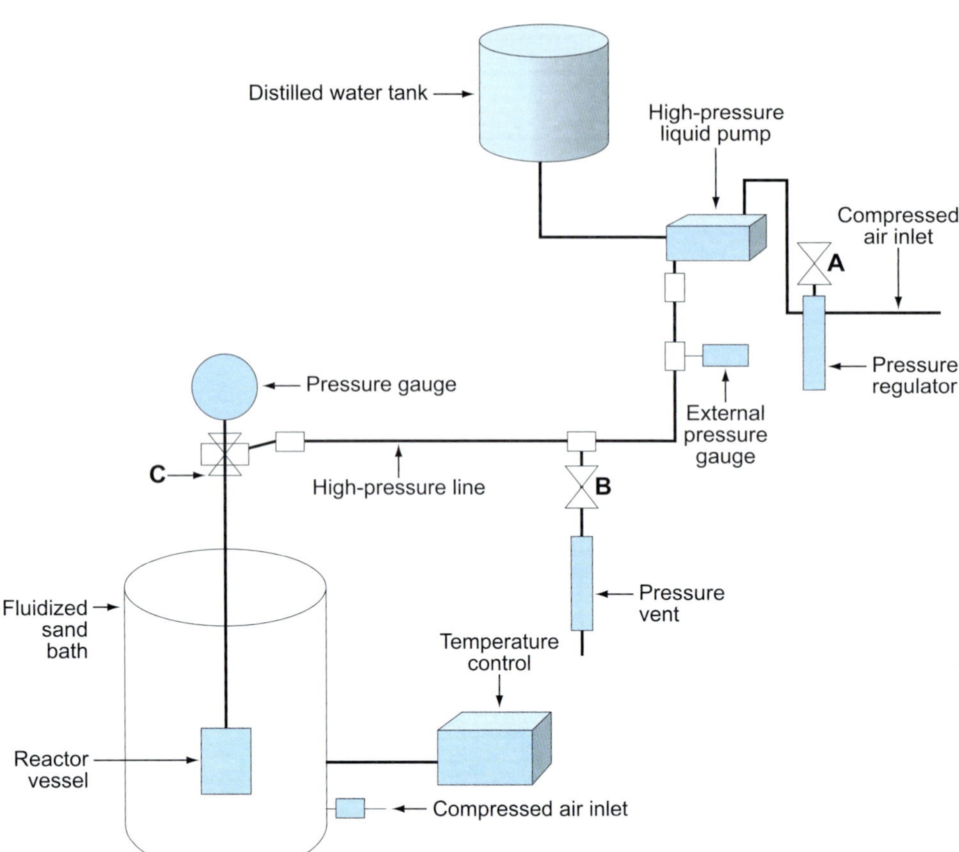

FIGURE 1. Schematic diagram of pyrolysis equipment.

pressure line. Then valve B was opened to vent the excess pressure in the line. The experiment was then allowed to run (leaving valve C tightly closed to avoid losing generated products) for the required time, after which the sand bath was switched off and left to cool to ambient temperatures before product recovery.

To attain a high pressure (e.g., 500 bar) at 350°C for 24 hr, the vessel was initially pressurized to about 440 bar, and the final pressure obtained was caused by thermal expansion of the additional water and generated gas. The volume of water giving a pressure of around 500 bar at 350°C was measured to be about 25 mL at ambient temperature from a blank run. The total internal volume of the empty pressure vessel and its associated pipe work (pressure gauge connected) was estimated to be about 34 mL by pressurizing with nitrogen gas to 2 bar and measuring the volume of gas released. The water in the hydrous experiments did not expand fully as expected considering the high temperatures used. This is because the head space of the reactor (gauge block), which is not immersed in the sand bath, is at a temperature lower than each of the experimental temperatures, and the extent of convection is not sufficient to overcome the reduced expansion that results from any temperature gradient that might exist above the reactor. Note that an overpressure between 40 and 60 bar was generated because of gas generation for the 380 and 420°C high-pressure (390 bar and above) experiments.

To test whether any gas in the system was lost during pressurizing the vessel, a control experiment was conducted at 350°C for 50 min at 500 bar pressure using a light (35° API) crude oil, which is more thermally stable than the source rock being investigated. The short period used for the control experiment was to ensure that the oil did not thermally crack. The volume of gas collected after the control experiment was found to be equal to the volume of nitrogen pumped into the system at 2 bar before the experiment started. This indicated that gas was not lost during pressurization, pressure buildup in the vessel, or during gas sampling from the reactor after the experiments.

Gas Analysis

The gaseous product was collected from the pressure vessel with the aid of a gas-tight syringe and transferred to a gas bag. Generated gases were analyzed immediately on a Carlo Erba HRGC 5300 gas chromatograph fitted with a FID detector operating at 200°C. Ten microliters of gas sample was injected at 100°C, with separation performed on a Varian Plot fused silica 25 m × 0.32 mm column, with helium as the carrier gas,

Table 2. Bitumen and hydrocarbon gas yields (milligrams per gram) total organic carbon from pyrolysis at 350°C for 6 and 12 hr.

Experiment	*Pressure (bar)*	*Time (hr)*	CH_4	C_2H_4	C_2H_6	C_3H_6	C_3H_8	C_4H_{10}	*Total* C_1-C_4	*Bitumen Plus Oil*
5 mL	45	6	3.31	0.18	3.20	0.59	2.61	1.27	11.16	975
10 mL	180	6	3.36	0.16	3.22	0.49	2.58	1.20	11.01	998
20 mL	210	6	2.84	0.05	2.50	0.16	1.85	0.71	8.11	997
485 bar	485	6	1.63	0.01	1.41	0.09	1.31	0.76	5.21	944
5 mL	45	12	6.22	0.13	6.70	0.66	5.13	2.04	20.88	881
10 mL	180	12	7.51	0.13	7.55	0.68	5.53	2.35	23.75	851
20 mL	210	12	6.82	0.04	6.67	0.32	5.16	2.63	21.64	818
470 bar	470	12	5.64	0.01	4.98	0.12	3.82	2.06	16.63	854

and an oven temperature program of 70 (2 min) to 90°C (3 min) at 40°C min^{-1}, then to 140°C (3min) at 40°C min^{-1}, and finally to 180°C (49 min) at 40°C min^{-1}. Individual gas yields were determined quantitatively in relation to an external gas standard.

Recovery of Generated Bitumen Plus Oil

After gas analysis, the pressure vessel was dismantled and the water in the vessel was decanted. For experiments where expelled oil was generated, the floating oil was separated from the water with the aid of a separating funnel, after which the vessel was transferred to a vacuum oven where the reacted source rock was vacuum dried at 45°C for 3 to 4 hr. The dried rock sample was then transferred to a pre-extracted cellulose thimble and Soxhlet extracted using 150 mL dichloromethane (DCM)-methanol mixture (93:7 vol/vol) for 48 hr. After extraction, the DCM-methanol mixture was gently evaporated (almost to dryness) by rotary evaporation and transferred to a pre-weighed glass vial with a Pasteur pipette. The vial was then left to air-dry to remove any DCM and/or methanol left in the sample. The generated product stuck to the side of the vessel (recovered by rinsing with DCM), and any expelled oil was combined with the Soxhlet extract and termed the "generated bitumen" or "bitumen plus oil."

Vitrinite Reflectance Measurement

Vitrinite reflectance measurements were obtained from extracted reacted rock residues and unreacted rock sample mounted in epoxy resin. Before reflectance measurements, samples were ground and polished using diamond solutions to produce a scratch-free polish. Measurements were made on a Leitz Ortholux microscope fitted with an MPV control and photometer head. The samples were examined in white light using oil-immersion objectives, whereas reflectance was measured using a green filter with a peak transmission of 547 nm and a measuring aperture of 16 μm^2. The results were expressed as arithmetic means (% R_o).

RESULTS

The pyrolysis of Kimmeridge Clay Formation source rock at 350 to 420°C between 6 and 24 hr was used to investigate the effects of temperature, pressure, and time on both hydrocarbon generation and maturation.

Effect of Water, Pressure, and Time on Product Yield at 350°C

Experiments conducted at 350°C for 6 hr were used to investigate kerogen conversion to bitumen, whereas the 12- and 24-hr experiments were used to investigate bitumen cracking to oil. Table 2 and Figure 2 show the amounts of gas and bitumen generated in 6 and 12 hr. Bitumen yield increased from 975 mg/g at 45 bar to a maximum of 998 mg/g and 997 mg/g at 180 and 210 bar, respectively, for the 6-hr run, and the data are consistent with those of Carr et al., (2009), who also reported maximum bitumen generation from an identical Kimmeridge Clay sample at 320°C for 24 hr under low-pressure hydrous conditions. At 485 bar for 6 hr, the bitumen yield decreased by about 5% when compared with the yields obtained at 180 and 210 bar. The total C_1-C_4 gas yield was maximum and identical at 45 and 180 bar, but decreased by about 27% at 210 bar. At 485 bar, the total C_1-C_4 gas yield further decreased by about 50% when compared with the 180-bar experiment. The higher yields of gas and bitumen obtained under low pressure are caused by water promoting hydrocarbon generation. The increased bitumen and hydrocarbon gas yields obtained during low-pressure hydrous pyrolysis have been widely reported (Lewan, 1997; Price and

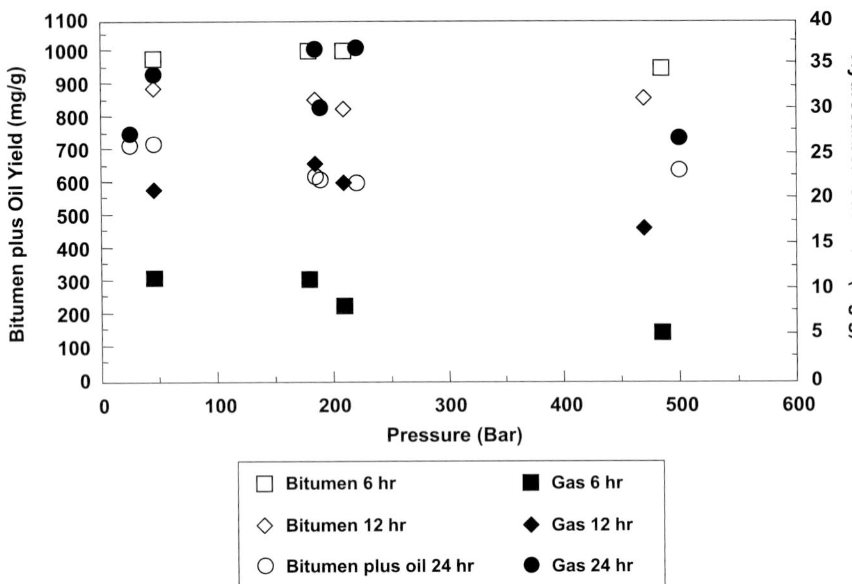

FIGURE 2. Bitumen, bitumen-plus-oil, and hydrocarbon gas yields (mg/g total organic carbon) obtained by pyrolysis at 350°C.

Wenger, 1992). The lower yields obtained at high pressure are caused by the water retarding hydrocarbon generation. Increasing the run time to 12 hr resulted in a decrease in the bitumen yield and an increase in the gas yield under all experimental conditions in relation to the 6-hr data. The reduced bitumen and increased gas yield obtained for the 12-hr experiment are the results of the bitumen starting to crack to oil. The total C_1-C_4 gas yield was maximum and about the same under 45-, 180-, and 210-bar conditions but decreased with increase in pressure to 470 bar. The bitumen yields initially decreased from 45 bar to a minimum at 210 bar and then increased with increase in pressure to 470 bar. The increase in bitumen and decrease in gas yield observed with increase in pressure for 12 hr are caused by water pressure retarding the cracking of bitumen to oil and gas. The liquid product from the 12-hr experiments was also classified as bitumen because the source rock had not started expelling oil at this stage, and the nongaseous product was also solid as the 6-hr product.

At 350°C for 24 hr, the experiment is well into the oil-generating stage, and the liquid product (termed bitumen plus oil) was a combination of oil expelled during the experiment and unexpelled oil and bitumen extracted from the source rock after the experiments. The expelled oil generated from the experiments was not treated separately for two reasons. First, the small particle size (180–425 μm) of the sample made it difficult to separate the expelled oil adsorbed on the rock because any attempt to separate the oil from the rock by washing with solvent also dissolves the generated bitumen. Second, because of the small amount of sample used for the experiments (4 g), the expelled oil generated was also small, which would produce an increased experimental error if treated separately.

The bitumen-plus-oil and gas yields obtained from the 24-hr experiments at 350°C are presented in Figure 2 and Table 3. The bitumen-plus-oil yields at 24 hr are significantly lower, and the total C_1-C_4 gas yields significantly higher when compared with the 350°C 6-hr data. This reduction in bitumen-plus-oil yield at 350°C for 24 hr compared with 350°C for 6 hr is caused by cracking of the generated bitumen to lighter oil and gas. The nonhydrous and 45-bar products at 350°C for 24 hr is solid bitumen because of the lack of expelled oil during the experiment, unlike the product from hydrous (185 and 220 bar) and high (500 bar)–water pressure experiments, which is a viscous liquid comprising of expelled oil, extracted unexpelled oil, and bitumen. The bitumen-plus-oil yield decreased as the conditions changed from nonhydrous to low-pressure hydrous (185 and 220 bar) conditions, but at 500-bar pressure, a reversal in the trend was observed with the bitumen-plus-oil yield increasing slightly.

The total C_1-C_4 gas yields at 24 hr increased from nonhydrous to a maximum under low-pressure hydrous (185 and 220 bar) conditions, before decreasing when the water pressure was raised to 500 bar. The unsaturated ethene and propene yields decrease from the value obtained under nonhydrous pyrolysis to become either zero (below the detection limit) at 500 bar in the case of ethene or 0.11 mg/g in the case of propene. Methane, ethane, propane, and butane yields reach maxima at either 185 or 220 bar of pressure when compared with the 500-bar experiment (Table 3).

Effect of Water, Pressure, and Time on Product Yield at 380°C

At 380°C for 6 and 12 hr, the experiment is well into the oil-generating stage, with the bitumen already

Table 3. Bitumen-plus-oil and hydrocarbon gas yields (milligrams per gram) total organic carbon from pyrolysis at 350°C for 24 hr.

Experiment	*Pressure (bar)*	*Time (hr)*	CH_4	C_2H_4	C_2H_6	C_3H_6	C_3H_8	C_4H_{10}	*Total* C_1-C_4	*Bitumen Plus Oil*
Nonhydrous	22	24	7.19	0.10	8.46	0.67	7.47	3.36	27.25	709
5 mL	45	24	9.38	0.06	10.68	0.71	8.89	4.17	33.89	714
10 mL	185	24	9.89	0.04	10.83	0.78	9.44	5.59	36.57	614
20 mL	220	24	12.62	0.01	11.46	0.26	8.51	3.72	36.58	597
500 bar	500	24	8.82	0.00	8.23	0.11	6.35	3.12	26.63	632

cracking to oil and gas, whereas for the 24-hr experiment, virtually all the bitumen is cracked to oil and gas. Liquid products obtained at 380°C for 6 and 12 hr will be referred to as "bitumen plus oil" because the product is highly viscous and is a mixture of oil and bitumen yet to be cracked to oil, whereas the 24-hr product will be referred to as "oil" because the product is less viscous when compared with the 6- and 12-hr product. The cracking of the bitumen at 6 and 12 hr resulted to lower bitumen-plus-oil yield and higher gas yield when compared with the bitumen yield obtained at 350°C for 6 hr at the end of bitumen generation. The bitumen-plus-oil and C_1-C_4 gas yields obtained at 380°C for 6 and 12 hr are shown in Figure 3 and Table 4. The bitumen-plus-oil yield decreased while the gas yield increased with increase in residence time because of cracking of the generated bitumen. The bitumen-plus-oil and gas yields from the 6- and 12-hr data show the same trend, both the bitumen-plus-oil yield and the C_1-C_4 gas yield increased from low supercritical water pressure (60 bar) to high supercritical water pressure (390 bar).

The oil and gas yield obtained at 380°C for 24 hr are shown in Figure 3 and Table 5. Increasing the experimental time to 24 hr resulted in the oil yield further decreasing and the gas yield increasing when compared with the 6- and 12-hr data and is a result of the bitumen cracking to oil. The C_1-C_4 gas yield increased for the 24-hr experiment going from nonhydrous to 60- and 140-bar hydrous conditions. At 235 bar supercritical water pressure, the C_1-C_4 gas yield decreased, and a further reduction in gas yield was observed at the 450-bar supercritical water pressure. The gas yield obtained at 380°C follows the same trend as the 350°C data, with the yield increasing initially at low pressures (60 and 140 bar) in the presence of water before decreasing with increase in pressure. The oil yield is similar under nonhydrous and 45-bar hydrous conditions but increased slightly for the 140-bar experiment. Increasing the supercritical water pressures of the experiments to 235 and 450 bar by increasing the volumes of water added to the experiments to 10 and 20 mL, respectively, caused the oil yield to increase significantly, with the increase most significant at 450 bar.

The individual gas compositions in the 24-hr experiments show a maximum under either 45- or 140-bar hydrous conditions. The results at 6 and 12 hr are

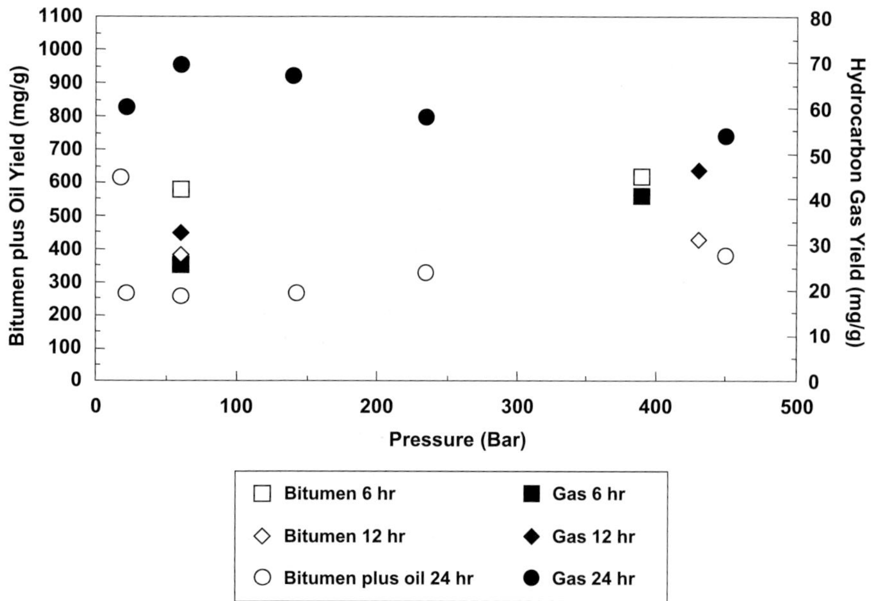

FIGURE 3. Bitumen-plus-oil and hydrocarbon gas yields (mg/g total organic carbon) obtained by pyrolysis at 380°C.

Table 4. Bitumen-plus-oil and hydrocarbon gas yields (milligrams per gram) total organic carbon from pyrolysis at 380°C for 6 and 12 hr.

Experiment	*Pressure (bar)*	*Time (hr)*	CH_4	C_2H_4	C_2H_6	C_3H_6	C_3H_8	C_4H_{10}	*Total* C_1-C_4	*Bitumen Plus Oil*
5 mL	60	6	7.25	0.09	8.11	0.80	6.54	2.97	25.76	580
20 mL	390	6	14.21	0.03	12.82	0.45	9.42	3.83	40.76	615
5 mL	60	12	9.20	0.06	10.41	0.66	8.63	3.92	32.88	385
20 mL	430	12	17.09	0.00	14.84	0.30	10.50	3.69	46.42	427

generally lower than the 24-hr experiment values regardless of whether low-hydrous (60 bar) or high–supercritical water pressure was used. The exceptions appear to be methane, ethane, and propene, all of which seem to be the same at 12 hr as at 24 hr. Comparisons between the 6- and 12-hr values show increases for the saturated C_1 to C_3 gases, whereas butane seems to remain relatively constant.

Effect of Water, Pressure, and Time on Product Yield at 420°C

As for the 380°C product, the liquid product of the 420°C pyrolysis at 6 and 12 hr is also called "bitumen plus oil" and that obtained for 24 hr is called "oil." The bitumen-plus-oil and C_1-C_4 gas yields for 6 and 12 hr from the 420°C pyrolysis are presented in Figure 4 and Table 6, whereas the oil and C_1-C_4 gas yields for the 24-hr pyrolysis are shown in Figure 4 and Table 7. Although no quantitative measure of the viscosity of the oil generated existed, the oil produced by the experiments at 420°C appears to be less viscous than the oil generated at 380°C, and the 380°C oil also appears less viscous than that generated at 350°C for identical periods. The bitumen-plus-oil yield decreased whereas the gas yield increased with the increase in experimental period from 6 to 12 hr. Increase in residence time to 24 hr also resulted in a decrease in oil yield and increase in gas yield as observed for the 380°C runs. The bitumen-plus-oil and the total C_1-C_4 gas yields for the 6- and 12-hr runs were lower for the 60-bar experiments than those obtained from the supercritical water pressure (430–435 bar) runs, a characteristic that was also observed in the 380°C experiments. The 24-hr data show that the gas yield increased with increase in water content and pressure to a maximum at 450 bar, a trend that is opposite to that observed at 380°C. However, the oil yield followed the same trend as at 380°C, increasing from nonhydrous (25 bar) and low pressures (60 and 150 bar) hydrous conditions to a maximum at 450 bar.

The individual gas compositions show increases in all the saturated components as the time of the experiment was increased from 6 to 12 hr and then to 24 hr. Ethene, although present in the 60-bar hydrous gases, is always absent from the 430- to 450-bar supercritical water pressure runs. Propene shows a slight reduction with increasing time for 6 to 24 hr. Methane shows the largest increase as the experiment was extended from 12 to 24 hr, with the amount of increase decreasing as the carbon number of the saturated gas component increased.

Vitrinite Reflectance

The VR values obtained from the source rock residues for experiments conducted at temperatures in the range of 310 to 420°C are shown in Figures 5 and 6 and in Table 8. The 310 and 320°C samples were taken from the experiments described in Carr et al. (2009). The mean VR values are based on only 1 to 7 counts because the Kimmeridge Clay source rock was deposited under anaerobic conditions with very little terrigenous kerogen. Such small numbers of primary particles of vitrinite, as

Table 5. Oil and hydrocarbon gas yields (milligrams per gram) total organic carbon from pyrolysis at 380°C for 24 hr.

Experiment	*Pressure (bar)*	*Time (hr)*	CH_4	C_2H_4	C_2H_6	C_3H_6	C_3H_8	C_4H_{10}	*Total* C_1-C_4	*Bitumen Plus Oil*
Nonhydrous	22	24	17.35	0.17	19.24	1.00	15.58	6.84	60.18	264
5 mL	60	24	20.82	0.02	23.15	0.87	18.24	6.68	69.78	255
7 mL	140	24	20.21	0.02	21.44	0.59	17.39	7.50	67.15	276
10 mL	235	24	19.34	0.04	18.61	0.77	13.91	5.51	58.18	329
20 mL	450	24	16.88	0.00	16.10	0.27	12.74	8.13	54.12	381

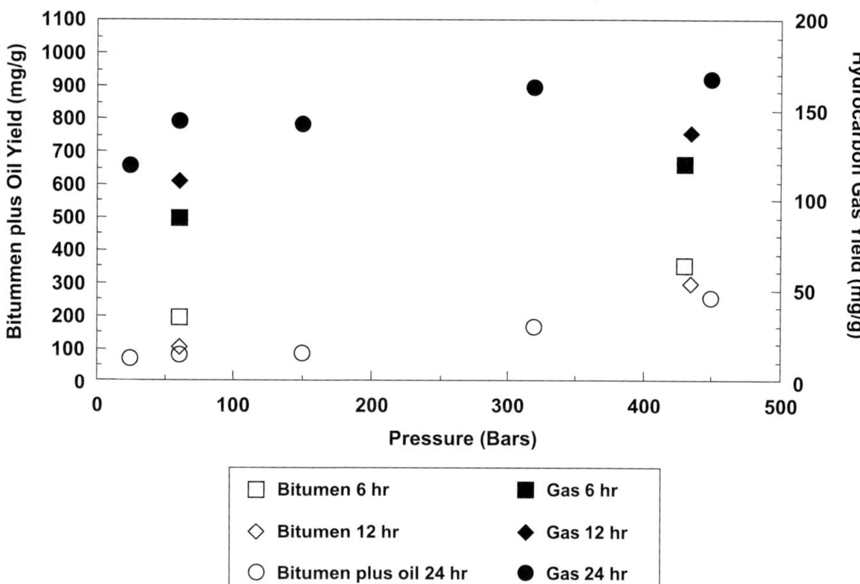

FIGURE 4. Bitumen-plus-oil and hydrocarbon gas yields (mg/g total organic carbon) obtained by pyrolysis at 420°C.

opposed to other types of kerogen, are not unusual for the Kimmeridge Clay.

The VR data for the 24-hr experiment at 350°C show that the mean reflectance value increased significantly by 0.6% R_o as the experimental conditions changed from nonhydrous to 185-bar hydrous conditions. At 220-bar hydrous conditions, a slight decrease in VR was observed relative to the 185-bar VR. Increasing the water pressure to 500 bar resulted in VR of 0.66% R_o lower than the maximum value obtained under 185-bar hydrous conditions.

At 380°C for 6 hr, the VR decreased from 1.23% R_o at low pressure (60 bar) to 1.07% R_o at 390 bar. However, the 12-hr result shows that VR increases from 0.99% R_o at low pressure (60 bar) to 1.19% R_o at 430 bar. Comparison of the 6- and 12-hr VR values indicates that the source rock was at a similar level of maturity between 6 and 12 hr. However, extending the run time to 24 hr at 380°C showed that the VR increased compared with the 6- and 12-hr values for identical heating conditions. The VR values increased with increases in water content and pressure to a maximum of 1.61% R_o at 235 bar. At 450 bar, the measured VR was lower (1.48% R_o), with the trend at 380°C for 24 hr mirroring that observed at 350°C for 24 hr.

At 420°C, the VR increased with increasing time, and a much higher VR was also achieved compared with the 380°C values. The 420°C values for 6 and 12 hr show lower VR values at high pressure (430 bar) than at low pressure (60 bar). The 24-hr experiments also show the same trend as the 350 and 380°C data. The VR increased from 1.98% R_o under nonhydrous conditions to a maximum of 2.09% and 2.18% R_o in the 60- and 320-bar experiments, whereas a lower value of 1.71% R_o was recorded at 450 bar.

Experimental Mass Balances

Mass balances were performed for the different temperature experiments at low and high pressures. Tables 9–11 list the mass balances obtained at 350°C for 6 hr, 350°C for 24 hr, and 380 to 420°C for 24 hr, respectively. As previously mentioned in the experiment section, the initial rock contains 25% carbon, and using 4 g of sample for the experiments equates to 1 g or 1000 mg

Table 6. Bitumen-plus-oil and hydrocarbon gas yields (milligrams per gram) total organic carbon from pyrolysis at 420°C for 6 and 12 hr.

Experiment	*Pressure (bar)*	*Time (hr)*	*CH_4*	*C_2H_4*	*C_2H_6*	*C_3H_6*	*C_3H_8*	*C_4H_{10}*	*Total C_1-C_4*	*Bitumen Plus Oil*
5 mL	60	6	27.53	0.05	28.69	0.97	22.63	8.09	87.96	186
15 mL	430	6	43.69	0.01	36.86	0.55	28.84	10.82	120.77	351
5 mL	60	12	35.78	0.02	35.12	0.77	28.24	9.84	109.77	103
15 mL	435	12	50.35	0.00	39.90	0.39	33.33	13.51	137.48	296

Table 7. Oil and hydrocarbon gas yields (milligrams per gram) total organic carbon from pyrolysis at 420°C for 24 hr.

Experiment	*Pressure (bar)*	*Time (hr)*	CH_4	C_2H_4	C_2H_6	C_3H_6	C_3H_8	C_4H_{10}	*Total* C_1-C_4	*Bitumen Plus Oil*
Nonhydrous	25	24	36.33	0.20	35.48	1.34	31.21	14.12	118.68	84
5 mL	60	24	45.83	0.00	44.62	0.55	38.05	15.17	144.22	78
7 mL	150	24	47.53	0.02	42.57	0.32	37.11	14.52	142.07	83
10 mL	320	24	57.35	0.05	48.31	0.48	39.83	15.80	161.82	167
15 mL	450	24	61.08	0.00	47.72	0.31	41.33	16.96	167.40	252

of starting carbon. At 350°C for 6 hr (bitumen-generation stage), where the products are mainly gas and bitumen, almost all of the starting carbon was recovered after the experiments for both the 45- and 485-bar experiments.

Increasing the residence time of the experiment to 24 hr at 350°C (bitumen cracking to oil stage) where the products are gas, bitumen, and oil, 80% of the starting carbon was recovered for both the low- and high-pressure experiments. This is caused by the loss of the lighter liquid products during sample drying and Soxhlet extraction of the recovered rock after the pyrolysis experiments. The observed increase in the carbon content of residual rock from 192 mg (low pressure) and 180 mg (high pressure) at 350°C for 6 hr to 277 mg (low pressure) and 265 mg (high pressure) at 350°C for 24 hr is caused by the onset of coke or pyrobitumen formation as the bitumen is cracked to oil.

At 380°C for 24 hr, only about 70% of the starting carbon was recovered for the 235- and 450-bar experiments, 10% less when compared with the amount recovered at 350°C for 24 hr. This is caused by more loss of lighter liquid product at 380°C as the liquid product obtained is less viscous than the 350°C for 24 hr products. The carbon content of the residual rock also increased further at 380°C for 24 hr, indicating that more pyrobitumen or coke formed with increased cracking of the generated bitumen and oil. The recovered carbon for the 420°C experiments for 24 hr is 79 and 70% for the 320- and 450-bar experiments, respectively. The lower amount of carbon recovered when compared with the starting carbon is also a result of the loss of lighter liquid products as observed for the 350 and 380°C experiments for 24 hr. The carbon content of the residual rock also increased further with temperature when compared with the 380°C experiment and is also caused by more pyrobitumen or coke formed with increased cracking of the generated bitumen and oil at 420°C. The fact that similar amounts of total carbon were recovered for both the low- and high-pressure experiments at all three temperatures, apart from the

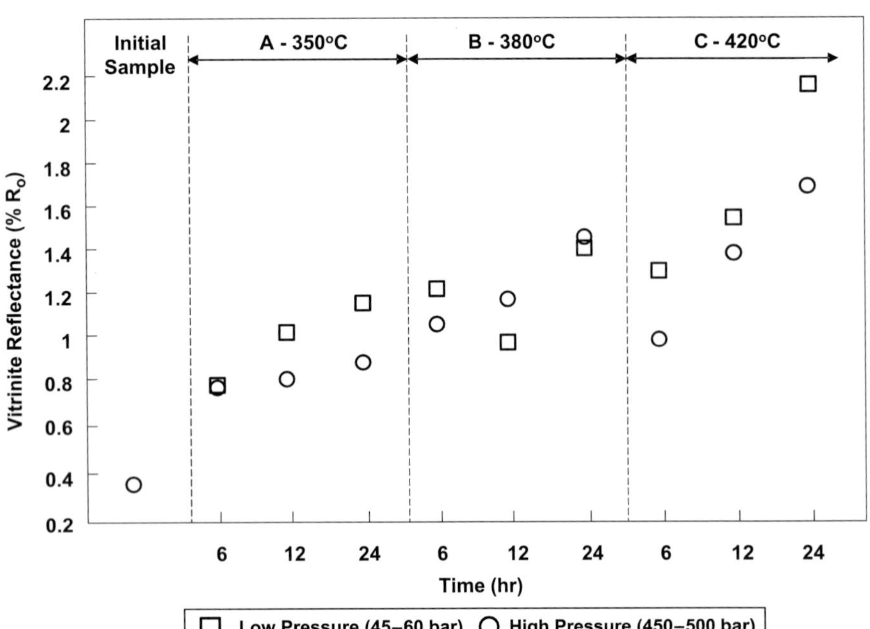

FIGURE 5. Vitrinite reflectance (R_o) results for the 6-, 12-, and 24-hr experiments at low (45–60 bar)– and high (450–500 bar)–water or supercritical water pressures.

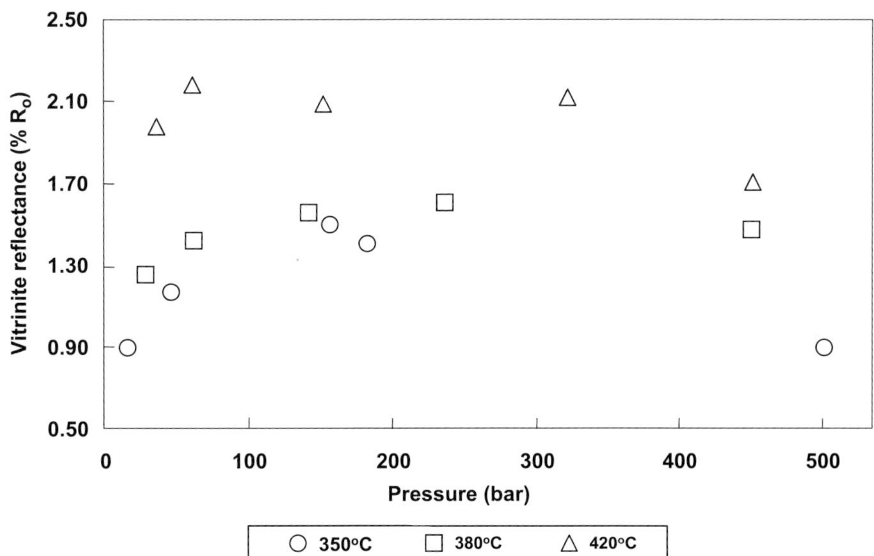

FIGURE 6. Vitrinite reflectance (R_o) results for the 24-hr experiments undertaken at 350, 380, and 420°C in the pressure range of 15 (nonhydrous) to 450 and 500 (water or supercritical water pressure) bar.

380 and 420°C runs where slightly more carbon was recovered at low pressure, indicates that product recovery was good.

Note that the carbon content of the residual rock was always higher for the low-pressure compared with the high-pressure experiments for 24 hr at 350, 380, and 420°C, with the effect most significant at 380 and 420°C. This is because of a high water pressure retarding the cracking of the generated bitumen or oil to gas and insoluble pyrobitumen or coke. The high pyrobitumen or coke content of the residual rock from the low-pressure runs is consistent with bitumen-plus-oil or oil yield being lower, whereas the low pyrobitumen or coke content of high-pressure residual rock is also consistent with bitumen-plus-oil or oil yield being higher.

DISCUSSION

Carr et al. (2009) showed that the pyrolysis of Kimmeridge Clay samples under 500 bar of water pressure at 310 and 320°C for 7 and 24 hr resulted in retardation of hydrocarbon (bitumen and gas) generation. At 350°C and 500 bar of water pressure for 24 hr, retardation of hydrocarbon generation still occurred, although the increase in temperature initiated some cracking of the bitumen to oil and gas. In the experimental results reported in this study, the results from the 350°C experiments follow the same general trends as those reported by Carr et al. (2009), although the different Kimmeridge Clay sample and the different water-rock ratio in the pressure vessel (with different internal volume) used in this study resulted in values slightly different to those reported in the previous study. To understand the effects of temperature, pressure, and time on hydrocarbon generation, the temperature and pressure effects will be discussed separately from time and pressure effects.

Hydrocarbon Generation: Temperature and Pressure Effects

In this section, the results mainly from the 24-hr experiments will be discussed. At 350°C, the decrease in bitumen-plus-oil yield and the increase in gas yield going from 22-bar (nonhydrous) to 185- and 220-bar hydrous conditions are the results of water promoting the cracking of bitumen to oil and gas. Within source rocks in geologic basins and laboratory pyrolysis experiments, the conversion of kerogen into thermogenic hydrocarbons can be described by three somewhat overlapping stages (Erdmann and Horsfield, 2006). Thus, at 350°C for 6 hr, the kerogen is converted into bitumen and gas, with the bitumen itself converting into both oil and gas and the minor conversion of oil into gas after 12 hr (stages 1, 2, and 3, respectively, of Erdmann and Horsfield, 2006). The observed increase in bitumen-plus-oil yield and decrease in gas yield observed at 500 bar for 24 hr were caused by water pressure retarding the cracking of bitumen to oil and gas (stage 3 of Erdmann and Horsfield, 2006). Under 185 and 220 bar hydrous conditions, water promoted bitumen cracking to oil by donating hydrogen to the bitumen, a trend similar to that observed by Carr et al. (2009) during pyrolysis of the Kimmeridge Clay at 350°C using 155- and 180-bar hydrous conditions. At 500-bar water pressure, bitumen cracking is retarded because of the lack of sufficient vapor space in the vessel to accommodate the volume expansion required by cracking of bitumen to oil and gas. This contrasts to the 185-and 220-bar hydrous experiments

Table 8. Vitrinite reflectance values for the matured source rock after pyrolysis at 350, 380, and 420°C.

Experiment	*Pressure (bar)*	*Time (hr)*	*Temperature (°C)*	*Mean VR (% R_o)**	*Other Values (% R_o)*
Nonhydrous	15	7	310	0.56 (1)	
10 mL	90	7	310	0.58 (1)	
500 bar	500	7	310	0.43 (2)	
Nonhydrous	18	24	320	0.58 (2)	0.13 (2) B, 0.31 L
10 mL	115	24	320	0.72 (2)	0.54 B
20 mL	125	24	320	0.89 (3)	
5 mL	45	6	350	0.77 (2)	
500 bar	500	6	350	0.80 (3)	0.43 B, 1.43 (7) I
5 mL	45	12	350	0.82 (2)	
20 mL	210	12	350	1.00 (1)	
470 bar	470	12	350	1.04 (3)	0.68 (2) B, 1.28 (2) I
Nonhydrous	22	24	350	0.90 (2)	1.29 I
5 mL	45	24	350	1.17 (3)	
10 mL	185	24	350	1.50 (2)	
20 mL	220	24	350	1.41 (1)	
500 bar	500	24	350	0.90 (8)	0.48 (15) B
5 mL	60	6	380	1.23 (2)	
20 mL	390	6	380	1.07 (3)	
5 mL	60	12	380	0.99 (2)	
20 mL	430	12	380	1.19 (2)	
Nonhydrous	22	24	380	1.25 (6)	
5 mL	60	24	380	1.42 (3)	1.73, 1.84, 1.77 I
7 mL	140	24	380	1.56 (2)	
10 mL	235	24	380	1.61 (2)	
16 mL	305	24	380	1.36 (2)	2.13 I
20 mL	450	24	380	1.48 (2)	1.86 (2) I
5 mL	60	6	420	1.34 (2)	
15 mL	430	6	420	1.02 (2)	
5 mL	60	12	420	1.56 (3)	
15 mL	435	12	420	1.40 (1)	
Nonhydrous	25	24	420	1.98 (6)	
5 mL	60	24	420	2.18 (7)	
7 mL	150	24	420	2.09 (2)	
10 mL	320	24	420	2.12 (2)	
15 mL	450	24	420	1.71 (3)	

**Mean random reflectance value, % Ro (no. of readings); B = bitumen; I = inertinite.*

Table 9. Mass balance (product and residue in milligrams of carbon) for pyrolysis at 350°C for 6 hr.

Experiment	*Pressure (bar)*	*Total C_1-C_4*	*Bitumen*	*Residual Rock*	*Total Recovered*	*(%) Recovery*
5 mL	45	9	780	192	981	98
485 bar	485	4	755	180	939	94

Table 10. Mass balance (product and residue in milligrams of carbon) for pyrolysis at 350°C for 24 hr.

Experiment	*Pressure (bar)*	*Total C_1-C_4*	*Bitumen Plus Oil*	*Residual Rock*	*Total Recovered*	*(%) Recovery*
10 mL	185	29	491	277	797	80
500 bar	500	21	506	265	792	79

where enough compressible vapor space still exists in the system to accommodate the generated gases and give the higher gas yields obtained under low-pressure hydrous conditions. The retardation effect of pressure on gas generation from bitumen cracking mirrors was observed at 500 bar for 24 hr at 320°C (Carr et al., 2009). This result is consistent with previous studies (Brooks et al., 1971; Hesp and Rigby, 1973) in which water and inert gas pressure retarded the cracking of oil to gas.

The results for the 380°C experiments at 24 hr follow the same trend as the 350°C experiments. At 380°C, the conversion of bitumen into oil and gas (stage 2) and the conversion of oil into gas (stage 3) become increasingly important. Water initially promoted bitumen cracking, in that the yield decreases as the conditions change from nonhydrous (22 bar) to low-pressure hydrous conditions (<235 bar). The maximum effect of water was observed between 60 and 140 bar, producing more gas and less oil, before cracking was retarded by the increase in supercritical water pressure to 235 and 450 bar. At 450 bar, the total hydrocarbon C_1-C_4 gas yield (54.12 mg/g) was about 20% lower than the gas yield (69.78 and 67.15 mg/g) obtained at low pressures (60 and 140 bar, respectively), whereas the oil yield remained about 50% higher than that from the low-pressure (<235 bar) experiments.

The retardation effect of pressure on oil cracking at 420°C is more significant than at 380°C, when the oil remaining at high pressures (450 bar) is compared with that obtained at low pressures for both sets of temperatures. This effect can be calculated by subtracting the bitumen plus oil yield at 380°C, 140 bar and 420°C, 150 bar (where maximum promotional effect of water was observed) from the values at 450 bar. At 420°C, the difference is 169 mg/g, whereas at 380°C, the difference is 105 mg/g. In addition, subtracting the bitumen-plus-oil yield (597 mg/g) at 220 bar and 350°C (where maximum promotional effect of water was observed) from the 500 bar value (632 mg/g) gives a difference of only 35 mg/g, showing that the retardation effect of 450 to 500 bar of water or supercritical water pressure increases as the temperature increases. This occurs because the products obtained at higher temperatures (380 and 420°C) contain mainly oil (which is more thermally stable and more difficult to crack), whereas the 350°C products contain more bitumen than oil.

The increase in total C_1-C_4 gas yields at 420°C with increasing water pressure is in contrast to the 350 and 380°C data, which show that water pressure retarded bitumen cracking to oil and gas. At 420°C, the increase in gas and bitumen-plus-oil yield with increase in water pressure to 450 bar resulted from the promotional effect of water on bitumen cracking to oil and gas, and not pressure promoting gas generation. The 450-bar pressure attained at 420°C was insufficient to prevent gas generation from bitumen cracking to oil, with the effect that chemical effect of water dominated the retardation of pressure because of the higher temperature and lower pressure conditions used. Pressures higher than 450 bar would need to be used before the retardation effect of pressure on bitumen cracking to oil and gas can be observed, which will be our future work. This indicates that at 450 bar, the retardation effect of water pressure on gas generation from bitumen cracking becomes less significant as the temperature increases from 350 to 420°C. However, 450 bar of pressure does significantly retard the cracking of the oil formed from bitumen cracking as the temperature increases from 350 to 420°C, resulting in more oil than in the low-pressure hydrous runs.

Table 11. Mass balance (product and residue in milligrams of carbon) for pyrolysis at 380 and 420°C for 24 hr.

Experiment	*Temperature (°C)*	*Pressure (bar)*	*Total C_1-C_4*	*Oil*	*Residual Rock*	*Total Recovered*	*(%) Recovery*
10 mL	380	235	46	263	425	734	74
20 mL	380	450	43	305	321	669	67
10 mL	420	320	128	134	527	789	79
20 mL	420	450	134	202	369	703	70

The increase in hydrocarbon gas yield with increase in pressure to 450 bar observed at 420°C have been reported previously under anhydrous conditions (Hill et al., 1994, 1996). Hill et al. (1996) observed a slight decrease in gas yield from 90 to 210 bar and then an increase to 690 bar and then a decrease again at 2000 bar for oil cracking at 350 to 400°C, whereas for coal pyrolysis (Hill et al., 1994), both gas yield and VR increased with pressure to 690 bar before decreasing at 2000 bar at 300 and 340°C. The increase in gas yield (enhanced hydrocarbon generation) observed in this study at 420°C resulted from the promotional effect of water as the effect of 450-bar pressure attained became less dominant at 420°C. The increase in gas yield with pressure observed for oil cracking and coal pyrolysis (Hill et al., 1994, 1996) might be caused by the effect of pressure being less significant under high pressure–confined anhydrous conditions when compared with high-pressure hydrous conditions reported previously (Michels et al., 1995b; Landais et al., 1994). Michels et al. (1995b) and Landais et al. (1994) attributed the higher retardation effect observed under high-pressure hydrous conditions to the nature of the pressurizing medium (pyrolyzed sample in contact with pressurized water). The retardation of VR observed in this study at 430 to 500 bar is a result of the rock being in contact with pressurized water, whereas the increase in VR to 690 bar observed by Hill et al. (1994) might be caused by the fact that the pressure operating on the coal samples inside the gold bag has been reduced by the hydrocarbons generated inside the bag as well as the absence of water inside the bag.

The differences obtained at the three temperatures reflect the relative effects that temperature and pressure have on endothermic reactions from both thermodynamic (equilibrium) and kinetic standpoints. The extent to which any reaction will proceed at a given temperature and pressure is controlled by the Gibbs free energy (G in equation 1), which when negative favors the products and the reaction proceeds, whereas a positive value favors the reactants, that is, no reaction occurs. The Gibbs free energy is given by

$$\Delta G = \Delta H - T.\ \Delta S \tag{1}$$

where ΔG is the Gibbs free energy (kJ/mol), ΔS is entropy (J/mol K), ΔH is enthalpy (kJ/mol), and T is the absolute temperature. The effect of increasing the temperature at any pressure increases the $T.\ \Delta S$ term in equation 1, resulting in the increased probability that the $T.\ \Delta S$ term becomes larger than the enthalpy, with the result that G is negative and the reaction proceeds. The entropy (S) term in equation 1 is a measure of the disorder of the system, and the pressure reduces the mobility in the system, thereby reducing the entropy (Carr et al., 2009). Increasing the pressure at any temperature therefore reduces the $T.\ \Delta S$ term, increasing the probability that the G term will either be less negative or even positive, and if positive then the reaction cannot proceed.

From a kinetic standpoint, the rate at which a reaction proceeds is controlled by the activation energy in the Arrhenius first-order kinetic model. From the transition state theory, the activation energy is given by:

$$\text{Ea} = H^{\ddagger} + \text{R.T} \tag{2}$$

where $H^{\ddagger}$ is the activation enthalpy, which is given by:

$$\Delta H^{\ddagger} = \Delta U^{\ddagger} + \Delta \text{pV}^{\ddagger} \tag{3}$$

where $\Delta U^{\ddagger}$ is the internal energy at activation (kJ/mol), and $\Delta \text{pV}^{\ddagger}$ is the work (kJ/mol) against the force of the water that occurs as the volume of the activated complex increases. Internal energy includes the potential, kinetic, and thermal energies of the activated complex.

Pressure increases the activation energy caused by the energy required to create space for the hydrocarbons against the pressure. As shown in panels A and B of Figure 7, the expansion of the activated complex (activated volume) involves the transfer of energy (pV work in equation 3) from the kerogen into the water column, which increases the activation energy above the values determined by conventional kinetics using either Rock-Eval or microscale sealed vessel (MSSV) pyrolysis.

Theoretical considerations and the experimental results presented in this study and in Carr et al. (2009) indicate that both temperature and pressure exert significant controls on the extent of endothermic reactions, such as hydrocarbon generation and maturation reactions. Increasing temperature produces an increase in total hydrocarbon yield and maturation, whereas increases in pressure have the opposite effect, that is, hydrocarbon generation yields and maturation values are lower as reaction rates are retarded.

In the transition state theory, reaction rates are controlled by the activation volume ($\Delta^{\ddagger}\text{V}^{\text{o}}$), which is the volume difference between the activated complex and reactant. If $\Delta^{\ddagger}\text{V}^{\text{o}}$ is negative, then the reaction rate constant increases with increasing pressure. Conversely, if volume increases when the activated complex is formed, then pressure should slow the reaction rates. Hill et al. (1996) using gold bags to study oil cracking calculated $\Delta^{\ddagger}\text{V}^{\text{o}}$ values of +47 cm^3/mol in the 90- to 483-bar range at 400°C, whereas between 345 and approximately 690 bar at 350 and 380°C, and 483 and 690 bar at 400°C, $\Delta^{\ddagger}\text{V}^{\text{o}}$ is -14 cm^3/mol. Raising the pressure to between 690 and 2000 bar at temperatures between 350 and 400°C, $\Delta^{\ddagger}\text{V}^{\text{o}}$ is +5 cm^3/mol. Al Darouich et al. (2006) also using gold bags to study oil cracking, cal $\ddagger\ \text{V}^{\text{o}}$ values in the range of 40 to 140 cm^3/mol. Given the

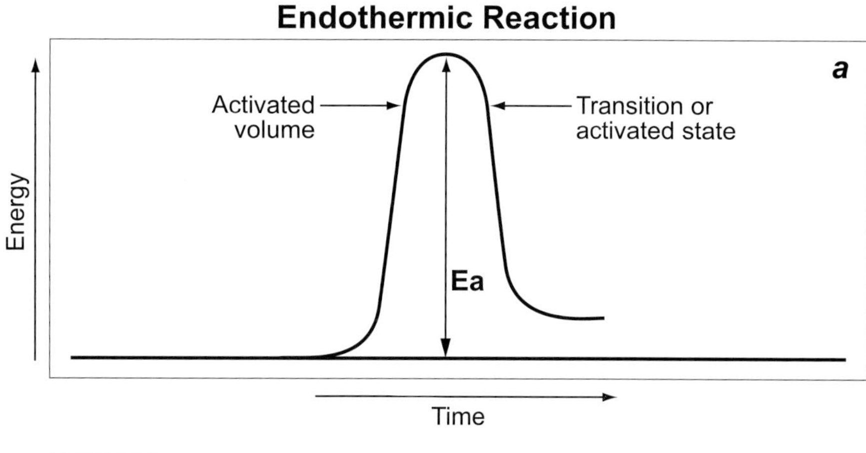

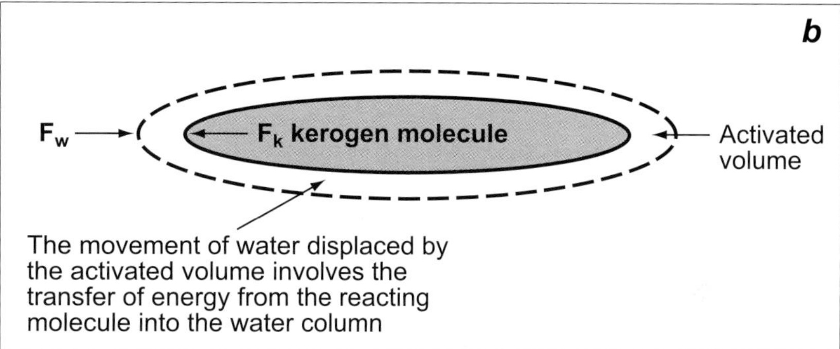

FIGURE 7. (A) Diagram illustrates the relationship between the activation energy, activation volume, and the transition or activated state for an endothermic reaction. (B) Schematic diagram shows the activation volume of the kerogen that forces the surrounding pore medium (water in this study) to move. Energy (thermal) must be transferred from the kerogen into the incompressible water to force the water to move away (pV work in equation 2), thereby enabling the kerogen to expend into its activated state. The higher the pressure, the more energy is transferred from the kerogen into the water column, and this energy is lost from the kerogen and the reaction is retarded.

relatively large size of the molecules in kerogen and petroleum, it would not be unrealistic to expect quite large changes upon activation, depending upon the exact nature of the activated complex. The situation is complicated by the fact that $\Delta^{\ddagger}V^{o}$ (like all volume parameters) is pressure dependent, and no theory exists to rationalize this variation.

To resolve this apparent ambiguity, we must also consider the complex thermodynamic effects in operation arising from equilibria based on competitive reactions. If the pseudo-equilibrium yield of methane for bitumen generation from kerogen is vastly reduced as shown by the reduced methane yields in the 310 and 320°C experiments reported by Carr et al. (2009), then the proportion of reactant (A_o) forming methane reduces correspondingly. Thus, taking the hypothetical case of a four-fold reduction in the methane yield at the point of maximum bitumen generation and a two-fold reduction from the lower reaction rate, then the thermodynamic effect would approximately give rise to approximately 50% less methane. Given these issues, the derivation of a kinetic model based on calculated activation volumes would appear to be exceedingly difficult, and a better approach might involve the modification by pressure of the activation energy and preexponential factors in the Arrhenius model using equations given in the transition state theory (see equations 2 and 3 and in Carr et al., 2009).

Hydrocarbon Generation: Time and Pressure Effects

At all three temperatures (350, 380, and 420°C), extending the heating time from 6 to 12 hr and then to 24 hr reduced the bitumen-plus-oil or oil yield and increased the gas yield at the same (or similar) pressure. In the 420°C experiments at high pressures (430–450 bar), for example, the gas yield increases from 120.77 mg/g after 6 hr to 137.48 mg/g after 12 hr, and to 167.40 mg/g after 24 hr. The difference between the 6- and 12-hr values is 16.71 mg/g, whereas that between 12 and 24 hr is 29.92 mg/g, indicating that although the duration of the experiment doubled, the yield has not doubled. As discussed previously, this is caused by the liquid product becoming more thermally stable as the product is composed more of oil caused by cracking of the bitumen with increase in residence time. Initially, this might be considered as being caused by the effect of the reaction nearing completion after 24 hr, but the presence of significant amounts of oil in the 420°C 450-bar sample indicates that the reactant has not been completely converted into products (gases).

The liquid product yields at 420°C also show the same pattern between 430 and 450 bar, with the bitumen-plus-oil yield decreasing from 351 mg/g for 6 hr to 296 mg/g for 12 hr and the oil yield also decreasing to 252 mg/g for 24 hr. The rate at which the bitumen-plus-oil phase is converted into gas during the 12 hr (difference between 24 and 12 hr experiments being 44 mg/g) does not seem to have doubled the rate at which it declined during the 6 hr between 12 and 6 hr (55 mg/g). The results again suggest that with increasing time, fixed temperature, and high water pressure as the reaction progresses, the rate of the reaction decreases because of enhanced thermal stability of the oil generated, which is less reactive than the starting kerogen and the intermediate bitumen. The results suggest that with increasing time at fixed temperature and water or supercritical water pressure, the rate may decrease because of the shift in the Gibbs free energy from negative toward zero, at which point the reaction is in equilibrium.

The results at 380°C show the same pattern of reduction in liquid product yield with increase in residence time, with the yields between 6 and 12 hr being greater than those between 12 and 24 hr at a similar pressure. Although the extension of the heating time allows increased amounts of bitumen and oil to be cracked to gas, pressure has a far more significant effect on oil than bitumen cracking. The small increase in the gas yield at 420°C for 24 hr is not considered to be caused by the reaction nearing completion because the combined bitumen and gas yields obtained at 320°C under 115 bar hydrous conditions are greater than 1000 mg/g (Carr et al., 2009), whereas at 420°C and 450 bar after 24 hr, the combined yields (bitumen plus oil and gas) obtained 420 mg/g. As the bulk of the mass in the bitumen and oil is converted to pyrobitumen or coke, the total organic carbon content of the residual rock increases from the minimum value obtained at the end of bitumen generation (see Carr et al., 2009) (Tables 7, 10). Clearly, the reactions at 450 to 500 bar at the three temperatures used in this study are still continuing, but they appear to be slowing down with increasing time because of the effects of pressure.

Vitrinite Reflectance

The VR data show four distinct trends, two of which are currently accepted as factors controlling the maturation of source rocks. First, increasing the temperature under either low-pressure (45–320 bar) hydrous conditions or high–water or supercritical water pressure conditions (450–500 bar) increases VR (Table 8; Figures 5, 6). Second, the effect of time also increases the VR, although some of the values such as the 380°C, 6 and 12 hr, 45- to 60-bar results are lower than the high-pressure values (Table 8; Figure 5). Third, pyrolysis under hydrous conditions always increases VR above the values obtained under nonhydrous conditions (Table 8; Figure 6). This increase in VR under hydrous conditions is consistent with the findings of Behar et al. (2003) who observed that the VR of lignite samples pyrolyzed under hydrous conditions was on average 0.2% R_o higher than when the lignites were pyrolyzed under nonhydrous conditions. The reason for this increase in VR under hydrous conditions is uncertain but might be caused by increased hydrocarbon yield that enhances structural reorganization (aromatization) of the vitrinite, thereby increasing the reflectance.

Fourth, and most significantly, high liquid or supercritical water (450–500 bar) pressure retards VR at 350 to 420°C, the VR values at 390 to 500 bar are lower than the values obtained under low-pressure hydrous (60–320 bar) conditions (Table 8; Figure 6). The 380°C values for 12 hr do not follow this trend possibly because of the change in appearance produced by pyrolysis and low abundances of vitrinite in the original source kerogen. The case of retardation of VR under water pressure is supported by the results obtained by the pyrolysis of two high-volatile coals (0.7% R_o) with high vitrinite contents at 350°C under 500 bar, which showed that the VR was always reduced relative to the coals pyrolyzed under low-pressure hydrous conditions (Uguna, 2007). The lower VR values under 450 to 500 bar pressure are probably related to the retardation of hydrocarbon generation in the same samples. The retention of volatiles (both hydrocarbon and nonhydrocarbon) in the kerogens pyrolyzed under 450- to 500-bar pressure as observed in the retardation of hydrocarbon generation in these samples prevents reorganization of the structure required to produce the polycyclic aromatic molecules that account for the higher reflectance in samples in the oil and gas windows (Carr, 1999).

Seewald and Eglinton (1994) and Seewald et al. (2000) investigated the function of aqueous fluids on vitrinite maturation by hydrous pyrolysis. Although pyrolysis using high-pH pore fluids yielded vitrinite with a reflectance of 1.06 to 1.19% R_o, pyrolysis with low-pH pore fluids gave vitrinite with a reflectance of 1.39% R_o. Seewald et al. (2000) examined the function of H^+ activity in the evolution of vitrinite maturation during aqueous pyrolysis. H^+ activity is much influenced by the physical state of water (vapor, liquid, supercritical). In addition, other problems with VR being retarded by water pressure exist. Khorsani and Michelsen (1994) found the evidence for the retardation of VR by pressure in geologic basins to be inconclusive and suggested that retardation was caused by reflectance differences produced by variations in both the chemical composition and structural heterogeneity of vitrinite. This issue about chemical differences between different types of vitrinite exists, as does the H^+ activity, and both contribute to variability of the individual values obtained from the different experiments. Overall, the chemical variability and H^+ activity are only secondary effects compared with the overall control exerted by the temperature-pressure-time conditions.

Implications for Hydrocarbon Generation and Maturation in Geologic Basins

Sealed pyrolysis vessels, such as MSSV, and gold bag and hydrous pyrolysis methods have been widely used to evaluate the hydrocarbon yields and the maturity obtained from source rocks in the laboratory to understand these systems. Clearly, the results generated from the laboratory cannot precisely simulate the processes as in geologic basins because of the necessity of using higher temperatures during short periods to provide the necessary data. The source rocks in geologic basins occur in relatively low temperature (<250°C) high–water pressure (>500 bar) environments, whereas the laboratory pyrolysis work reported in this study involves high-temperature (>350°C), low–water pressure environments. This means that laboratory pyrolysis studies will always favor enhanced hydrocarbon generation and maturation. Although hydrocarbon generation and maturation will still occur in the high-pressured conditions in geologic basins, the rates at which these reactions proceed are slower because of low temperatures when compared with laboratory conditions. As such, the rates at which these reactions proceed cannot be accurately modeled using an Arrhenius kinetic model, which ignores the effects of pressure.

Both thermodynamic and transition state theories, as well as the experimental results reported in this study, show that hydrocarbon generation and maturation are retarded under liquid water and supercritical water pressures. Clearly, either previously published kinetic models for hydrocarbon generation and maturation will have to be modified to include the effects of pressure, or entirely new models need to be written in which the activation energies and preexponential factors are modified by pressure. Another alternative might involve building databases for the activation volumes (see equation 5 in Carr et al., 2009) for the reactions under the range of source rock temperatures and pressures in geologic basins. The use of such pressure-controlled kinetic models will have implications for the predictions of volumes of hydrocarbon generated, timing of hydrocarbon generation, that is, the critical point in the petroleum system model, and the maturity in the deeper parts of the basins being modeled.

One of the effects of including pressure in the kinetic models used to predict hydrocarbon generation and maturation is that for a fixed transformation ratio or maturity, the thermal history used to model these values will have to be increased compared with kinetic models that do not include the effects of pressure. Incorporating the pressure effect into the kinetic models used in basin modeling will enable the models to include the elevated thermal histories predicted by crustal modeling for basin formation, but which are mostly ignored in petroleum system modeling with thermal calibration using the VR data and the Easy%R_o model of Sweeney and Burnham (1990). This approach results in models with considerable periods of constant heat flow during both subsidence and uplift events in the history of the basin (Carr, 2003; Carr and Scotchman, 2003), although the laws of physics and thermodynamics clearly indicate that such constant thermal histories cannot be correct.

Effect of Water Phase

The initial concern before conducting these experiments at higher maturity was the effect that supercritical water may have on the experimental results. However, the same result trend was observed under normal and supercritical water. Water initially promoted hydrocarbon generation and source rock maturation under low-pressure hydrous conditions, or low-pressure supercritical water conditions, before pressure retardation occurs. This indicates that the phase change going from liquid water to supercritical water is not significant. Thus, results obtained at lower temperatures (320 and 350°C) should be comparable to those obtained at 380 and 420°C.

CONCLUSIONS

The hydrocarbon generation yield data presented in this and previous study by Carr et al. (2009) show that hydrocarbon generation is retarded both by liquid and supercritical water at 450 to 500 bar. Maturation as measured by VR is also retarded at high (390–500 bar) pressures. The kinetic models used to model hydrocarbon generation or maturation in geologic basins must be modified to include the effects of pressure, or entirely new models must be developed in which the effects of temperature, pressure, and time can be integrated into the models. The effects of pressure will always increase the temperature required to produce a given transformation ratio.

Whether pressurized water or supercritical water is present does not seem to be a major factor in the rate of hydrocarbon generation from source rocks, although the presence of water at temperatures below 373°C means that most hydrocarbon generation in the source kitchens in geologic basins will involve water instead of supercritical water. However, because of the promotional effects of water on source rock maturation where maximum R_o is obtained at 150 to 300 bar pressure, the pressure range for observing retardation effects here has only been approximately 200 to 300 bar when conducting experiments close to 500-bar pressure. Much more work is required to investigate the effects of these factors on hydrocarbon generation and maturation, including the use of higher water pressures to extrapolate to the highest basin pressures required in basin modeling.

ACKNOWLEDGMENTS

We thank the Natural Environment Research Council (NERC grant no. NE/C507002/1), Statoil ASA, and Woodside Energy, Ltd., who have supported this work financially. This article reflects the views of the authors and is not indicative of the views or company policy of either Statoil or Woodside Energy. ADC publishes with the permission of the Executive Director, British Geological Survey (NERC). We thank Ron Hill, Ken Peters, and an anonymous reviewer.

REFERENCES CITED

Al Darouich, T., F. Behar, and C. Largeau, 2006, The effect on the thermal cracking of the light aromatic fraction of Safaniya crude oil – Implication for deep prospects: Organic Geochemistry, v. 37, p. 1159–1169.

Barth, T., B. J. Schmidt, and S. B. Nielsen, 1996, Do kinetic parameters from open system pyrolysis describe petroleum generation by simulated maturation?: Bulletin of Canadian Petroleum Geology, v. 44, no. 3, p. 446–457.

Behar, F., M. D. Lewan, F. Lorant, and M. Vandenbroucke, 2003, Comparison of artificial maturation of lignite in hydrous and nonhydrous conditions: Organic Geochemistry, v. 34, p. 575–600, doi:10.1016/S0146-6380(02)00241-3.

Braun, R. L., and A. K. Burnham, 1987, Analysis of chemical reaction kinetics using a distribution of activation energies and simpler models: Energy and Fuels, v. 1, p. 153–161, doi:10.1021/ef00002a003.

Brooks, J. D., W. R. Hesp, and D. Rigby, 1971, The natural conversion of oil to gas in sediments in the Cooper basin: APEA Journal, v. 11, p. 121–125.

Burnham, A. K., 1993, Relationship between hydrous and ordinary pyrolysis. Lawrence Livermore Laboratory Report UCRL-JC-114130.

Burnham, A. K., 1998, Comment on "Experiments on the role of water in petroleum formation" by M. D. Lewan: Geochimica et Cosmochimica Acta, v. 62, no. 12, p. 2207–2210, doi:10.1016/S0016-7037(98)00149-5.

Burnham, A. K., R. L. Braun, and A. M. Samoun, 1988, Further comparison of methods for measuring kerogen pyrolysis rates and fitting kinetic parameters: Organic Geochemistry, v. 13, p. 839–845.

Carr, A. D., 1999, A vitrinite reflectance kinetic model incorporating overpressure retardation: Marine and Petroleum Geology, v. 16, p. 353–377.

Carr, A. D., 2003, Thermal history model for the South Central Graben, North Sea, derived using both tectonics and maturation: International Journal Coal Geology, v. 54, p. 1–19, doi:10.1016/S0166-5162(03)00016-8.

Carr, A. D., and I. C. Scotchman, 2003, Thermal History modelling in the southern Faroe-Shetland Basin: Petroleum Geoscience, v. 9, p. 333–345, doi:10.1144/1354-079302-494.

Carr, A. D., C. E. Snape, M. Meredith, C. Uguna, I. C. Scotchman, and R. C. Davis, 2009, The effect of water pressure on hydrocarbon generation reactions: Some inferences from laboratory experiments: Petroleum Geoscience, v. 15, p. 17–26, doi:10.1144/1354-079309-797.

Dalla Torre, M., R. F. Mahlmann, and W. G. Ernst, 1997, Experimental study on the pressure dependence of vitrinite maturation: Geochimica et Cosmochimica Acta, v. 61, p. 2921–2928, doi:10.1016/S0016-7037(97)00104-X.

di Primio, R., and B. Horsfield, 2006, From petroleum-type organofacies to hydrocarbon phase prediction: AAPG Bulletin, v. 90, p. 1031–1058, doi:10.1306/02140605129.

Eglinton, T. I., S. J. Rowland, C. D. Curtis, and A. G. Douglas, 1986, Kerogen-mineral reactions at raised temperatures in the presence of water: Organic Geochemistry, v. 10, p. 1041–1052, doi:10.1016/S0146-6380(86)80043-2.

Erdmann, M., and B. Horsfield, 2006, Enhanced late gas generation potential of petroleum source rocks via recombination reactions. Evidence from the Norwegian North Sea: Geochimica et Cosmochimica Acta, v. 70, p. 3943–3956, doi:10.1016/j.gca.2006.04.003.

Hao, F., S. Youngchuan, L. Sitian, and Z. Qiming, 1995, Overpressure retardation of organic matter maturation and petroleum generation: A case study from the Yinggehai and Qiongdongnan Basins, South China Sea: AAPG Bulletin, v. 79, p. 551–562.

He, S., M. Middleton, A. Kaiko, C. Jiang, and M. Li, 2002, Two case studies of thermal maturity and thermal modelling within the overpressured Jurassic rocks of the Barrow Sub-basin, North West Shelf of Australia: Marine and Petroleum Geology, v. 19, p. 143–159, doi:10.1016/S0264-8172(02)00006-5.

Hesp, W., and D. Rigby, 1973, The geochemical alteration of hydrocarbons in the presence of water: Erdol Kohle-Erdgas, v. 26, p. 70–76.

Hill R. J., P. D. Jenden, Y. C. Tang, S. C. Teerman, and I. R. Kaplan, 1994, Influence of pressure on pyrolysis of coal: American Chemical Society Symposium, v. 570, p. 161–193.

Hill R. J., Y. Tang, I. R. Kaplan, and P. D. Jenden, 1996, The influence of pressure on the thermal cracking of oil: Energy Fuels, v. 10, p. 873–882.

Huijun, L., W. Tairana, M. Zongjin, and Z. Wencai, 2004, Pressure retardation of organic maturation in clastic reservoirs: A case study from the Banqiao Sag, Eastern China: Marine and Petroleum Geology, v. 21, p. 1083–1093, doi:10.1016/j.marpetgeo.2004.07.005.

Khorasani, G. K., and J. K. Michelsen, 1994, The effects of overpressure, lithology, chemistry and heating rate on vitrinite reflectance evolution, and its relationship with oil generation: APEA Journal, v. 34, no. pt. 1, p. 418–434.

Laidler, K. J., 1987, Chemical kinetics, 3rd ed. Prentice Hall Inc., New Jersey, p. 206–207.

Landais, P., R. Michels, and M. Elie, 1994, Are time and temperature the only constraints to the simulation of organic matter maturation?: Organic Geochemistry, v. 22, p. 617–630, doi:10.1016/0146-6380(94)90128-7.

Lewan, M. D., 1993, Laboratory simulation of petroleum formation: hydrous pyrolysis, *in* M. H. Engel and S. A. Macko, eds., Organic Geochemistry, p. 419–441.

Lewan, M. D., 1997, Experiments on the role of water in petroleum formation, Geochimica et Cosmochimica Acta, v. 61, p. 3691–3723.

McTavish, R. A., 1978, Pressure retardation of vitrinite diagenesis, offshore north-west Europe: Nature, v. 271, p. 648–650, doi:10.1038/271648a0.

McTavish, R. A., 1998, The role of overpressure in the retardation of organic matter maturation: Journal of Petroleum Geology, v. 21, p. 153–186, doi:10.1111/j.1747-5457.1998.tb00652.x.

Michels, R., P. Landais, R. P. Philp, and B. E. Torkelson, 1995a, Influence of pressure and the presence of water on the evolution of the residual kerogen during confined, hydrous and high pressure hydrous pyrolysis of Woodford Shale: Energy Fuels, v. 9, p. 204–215, doi:10.1021/ef00050a002.

Michels, R., P. Landais, B. E. Torkelson, and R. P. Philp, 1995b, Effects of effluents and water pressure on oil generation during confined pyrolysis and high-pressure hydrous pyrolysis: Geochimica et Cosmochimica Acta, v. 59, p. 1589–1604, doi:10.1016/0016-7037(95)00065-8.

Pitt, G. J., 1962, The kinetics of the evolution of volatile products from coal: Fuel, v. 41, p. 267–274.

Price, L. C., and L. M. Wenger, 1992, The influence of pressure on petroleum generation and maturation as suggested by aqueous pyrolysis: Organic Geochemistry, v. 19, p. 141–159, doi:10.1016/0146-6380(92)90033-T.

Schenk, H. J., and V. Dieckmann, 2004, Prediction of petroleum formation: The influence of laboratory heating rates on kinetic parameters and geological extrapolations: Marine and Petroleum Geology, v. 21, p. 79–95.

Seewald, J. S., and L. B. Eglington, 1994, Organic-inorganic interactions during vitrinite maturation: Constraints from hydrous pyrolysis experiment. Abstracts of 11th Annual Meeting of the Society for Organic Petrology, Jackson Hole 11, p. 91–93.

Seewald, J. S., L. B. Eglington, and Y. L. Ong, 2000, An experimental study of organic-inorganic interactions during vitrinite maturation: Geochimica et Cosmochimica Acta, v. 64, p. 1577–1591.

Stainforth, J. G., 2009, Practical kinetic modeling of petroleum generation and expulsion: Marine and Petroleum Geology, v. 26, p. 552–572, doi:10.1016/j.marpetgeo.2009.01.006.

Sweeney, J. J., and A. K. Burnham, 1990, Evaluation of a simple model of vitrinite reflectance based on chemical kinetics: AAPG Bulletin, v. 74, p. 1559–1570.

Uguna, C. N., 2007, Effect of water pressure on hydrocarbon generation, maturation and cracking, Ph.D. thesis University of Nottingham, p. 117–143.

Van Krevelen, D. W., C. Van Heeerden, and F. J. Huntjens, 1951, Physical chemical aspects of the pyrolysis of coal and related organic compounds: Fuel, v. 30, p. 253–259.

Zou, Y.-R., and P. Peng, 2001, Overpressure retardation of organic-matter maturation: A kinetic model and its application: Marine and Petroleum Geology, v. 18, p. 707–713, doi:10.1016/S0264-8172(01)00026-5.

Conference conveners Marek Kacewicz (left), Ken Peters (back), and Dave Curry (right) and The Geysers field trip organizer Noelle Schoellkopf.

The Conference Icebreaker reunited many old friends. Wally Dow (left) and Les Magoon (right), co-editors of AAPG Memoir 60 on "The Petroleum System—From Source to Trap" in 1994 share some time with Ken Peters, one of the conference organizers. Les Magoon presented an evening keynote lecture for the meeting on the 'History of the Petroleum System Concept'.

2

Haeseler, F., F. Behar, and D. Garnier, 2012, First stoichiometric model of oil biodegradation in natural petroleum systems: Part II: Application of the BioClass 0D approach to oils from various sources, *in* K. E. Peters, D. J. Curry, and M. Kacewicz, eds., Basin Modeling: New Horizons in Research and Application: AAPG Hedberg Series, no. 4, p. 39–50.

First Stoichiometric Model of Oil Biodegradation in Natural Petroleum Systems: Part II*: Application of the BioClass 0D Approach to Oils from Various Sources

Frank Haeseler, F. Behar, and D. Garnier
IFP Energies Nouvelles, Rueil Malmaison, France

ABSTRACT

The purpose of the this study was to apply the biodegradation model BioClass 0D proposed by Haeseler et al. (2010) to three nonbiodegraded oils representative of the three main types of source rocks (types I, II, and III of Tissot et al., 1974; Tissot and Welte, 1978). The oils are described by seven chemical classes: (1) the gas fraction is split into H_2S, CO_2, C_1, and C_2–C_4; (2, 3) the C_6–C_{14} fraction including saturates and aromatics; and (4–6) the C_{14+} fraction, including n-, iso-, cyclo-alkanes, aromatics, and nitrogen, sulfur, and oxygen–containing compounds (NSOs). This 0D model reconstructs the chemical evolution of the residual oil with increasing biodegradation and the amounts of products generated during either aerobic or anaerobic processes. Results show that with increasing biodegradation, the residual oil composition depends on the initial proportion of the different chemical classes. For instance, for the type I oil enriched in paraffins, the C_{14+} n-alkanes are still present when 60% of the original oil has already disappeared, whereas for the type II oil, the same chemical class disappears after only 30% of total hydrocarbon loss. During methanogenesis, the gas-oil ratio of the initial fluid significantly increases with increasing biodegradation. However, the volume of methane may be reduced because of its solubility in water or it may leak through the cap rock. The production of H_2S is always very low when sulfur minerals that can provide electron acceptors are absent. Results also show that the amount of water needed to provide the electron acceptors

*Part I of this chapter published in Organic Geochemistry, doi:10.1016/j.orggeochem.2010.05.019

DOI:10.1306/13311427H43462

depends strongly on the biological process responsible for oil biodegradation. The water ratio between aerobic biodegradation and methanogenesis might be as high as 10^7. Consequently, aerobic biodegradation may be limited compared with methanogenesis in petroleum systems in which the oil-water volume ratio varies during fluid history.

INTRODUCTION

Hydrocarbon biodegradation in nature or under laboratory conditions is based on successive reactions starting with aerobic catabolism and finishing with methanogenesis. This succession of reactions is based on the energy gain from highest to lowest being aerobic to methanogenesis. The first convincing evidence of biodegradation of oil in petroleum reservoirs was published by Williams and Winters (1969). At that time, aerobic bacteria were considered to be the most effective hydrocarbon degraders (Milner et al., 1977; Connan, 1984); however, the possibility of anaerobic degradation via sulfate reduction had not been dismissed. Jobson et al. (1979) simulated the two successive steps of aerobic and anaerobic biodegradation in the laboratory. Then, through complementary studies of contaminated aquifers and soils, the belief that aerobic processes dominate oil biodegradation in petroleum reservoirs was progressively replaced by the dominance of two anaerobic processes: sulfate reduction and methanogenesis (Connan et al., 1997; Magot et al., 2000; Orphan et al., 2000; Rozanova et al., 2001; Jackson and McInerney, 2002; Head et al., 2003; Röling et al., 2003; Aitken et al., 2004; Dolfing et al., 2008; Eschard and Huc, 2008; Jones et al., 2008). Literature data have clearly demonstrated that the aerobic process is the fastest, whereas the slowest process is methanogenesis (Rice and Claypool, 1981; Nealson and Rye, 2005). As suggested by Bailey et al. (1973a, b), over geologic time, both sulfate reduction and methanogenesis can result in significant biodegradation of hydrocarbons.

In naturally biodegraded oils, it is generally observed that most of the light C_6–C_{14} hydrocarbons and the C_{14+} n-alkanes are first eliminated, followed by the C_{14+} isoalkanes. But the light C_6–C_{14} hydrocarbons are not a monolithic class of compounds, and it is commonly noticed that the alkylcycloalkanes, naphthalenes, and phenanthrenes persist after the C_{14+} n-alkanes are gone. The C_{14+} cyclo-alkanes and aromatics are the most refractory, together with resins and asphaltenes. This means that the chemical degradation of oil follows a parallel series of reactions having different rates according to the sensitivity to biodegradation of the different chemical classes (Larter et al., 2006). Although qualitative scales for ranking biodegraded oils are available in literature (Peters and Moldowan, 1993; Wenger and Isaksen, 2002; Peters et al., 2005), it is still difficult to quantify hydrocarbon losses. Usually, the increasing concentrations of refractory compounds with increasing biological alteration were used as indicators of biodegradation levels. These compounds can be biomarkers (McCaffrey et al., 1996; Le Dréau et al., 1997; Grice et al., 2000) or metal-containing compounds such as Ni- or V-porphyrins (Sasaki et al., 1998, Magnier et al., 2001). Products generated during biodegradation, such as carboxylic acids, were also identified (Behar and Albrecht, 1984; Jaffé and Gallardo, 1993; Meredith et al., 2000, Fafet et al., 2008). More recently, new quantitative scales were proposed by measuring the isotopic enrichment of specific residual organic compounds (Wilkes et al., 2008; Asif et al., 2009). The total loss can also be determined by estimating individual losses of different chemical classes affected by biodegradation; the sum of these individual losses provides the total hydrocarbon loss. However, as two chemical classes may be degraded at the same time but at different relative rates, only minimum estimates of the total loss can be calculated (Behar et al., 2006).

An integrated model of biodegradation should include the following constraints:

1. Description of the different biological processes occurring during biodegradation from aerobic to methanogenic hydrocarbon biodegradation. This is a series of balanced stoichiometric equations in which a given hydrocarbon reacts with an electron acceptor and generates products, assuming that initial hydrocarbon reactant is totally consumed and no intermediate products remain in the system.
2. Description of the different chemical classes in a given oil with regard to their sensitivity to biodegradation. For each chemical class, the successive biological metabolic equation will be applied to a selected model compound.
3. Definition of the relative biodegradation sensitivity of each chemical class in a given oil (parameter can be tuned if proposed values are considered inappropriate for some oils)
4. Definition of the water volume in contact with oil because biological processes occur at the oil-water contact
5. Influence of temperature, which is a key parameter that influences the biodegradation rate

Table 1. Initial chemical composition of the three nonbiodegraded oils.

Oil	*Minas Oil*	*Safaniya Oil*	*Ardjuna Oil*
	Type I	*Type II*	*Type III*
°API	33.6	26.0	31.5
C_6–C_{14} saturates	12.2	20.6	20.0
C_6–C_{14} aromatics	1.4	3.6	6.7
C_{14+} n-alkanes	26.2	4.6	20.9
C_{14+} iso-alkanes	7.1	8.8	3.9
C_{14+} cyclo-alkanes	23.7	12.4	14.7
C_{14+} aro-alkanes	12.2	22.7	16.3
NSOs	17.4	27.3	17.6
Total	100.0	100.0	100.0

6. Influence of time. The biodegradation kinetic of the aerobic processes is the fastest. The time required for complete biodegradation depends on the biological process.
7. Influence of aqueous-phase chemistry such as salinity or mineralogy and the porous media (i.e., reservoir rock) with parameters such as irreducible water saturation. These parameters may have a strong influence on the biodegradation efficiency in a given reservoir rock.

In this workflow, the first three elements constitute the basis of the proposed BioClass 0D model (Haeseler et al., 2010). In this model, global stoichiometric biodegradation coefficients were proposed to describe each of the successive biological reactions from aerobic biodegradation to methanogenesis. Then, seven pseudocomponents were selected to represent the whole oil to which a specific biodegradation rate or biosensitivity was assigned. Consequently, the residual oil composition is computed with progressive biodegradation together with the individual loss per chemical class and the total hydrocarbon loss. It is also possible to determine the proportion of CO_2, H_2O, H_2S, and CH_4 generated during the different biological processes.

The aim of the study was to apply the BioClass 0D model to three nonbiodegraded oils derived from three main sources of organic matter, that is, types I, II, III (Tissot et al., 1974; Tissot and Welte, 1978). The evolution of the chemical compositions of each oil with increasing biodegradation was studied to test the impact of the initial proportion of a given chemical class on its residual composition. The gas volumes of H_2S, CO_2, and CH_4 generated during biodegradation are discussed in terms of gas-oil ratio (GOR) and acid gas risk. Finally, the volumes of water needed for each biological process are calculated to evaluate their impact on biodegradation efficiency in subsurface petroleum systems.

In a future work, the BioClass 0D model will be coupled to a dynamic model in a basin simulator to determine the extent of biodegradation expressed as cubic meters of biodegraded oil per million years. This integrated model will include the influence of temperature, time, water, hydrocarbon fluxes, and nature of the porous media.

EXPERIMENT

Samples

Three oils representative of types I, II, and III organic matter as defined by Tissot et al. (1974) were selected: type I Minas oil originates from the Sumatra Basin (Seifert and Moldowan, 1981); type II Safaniya oil comes from Saudi Arabia (Al Darouich et al., 2005); and type III Ardjuna oil comes from Sumatra (Noble et al., 1991). The chemical compositions of the three nonbiodegraded oils are reported in Table 1. The gas content associated to these oils is not known and thus not included. The Minas oil contains the highest proportion of C_{14+} n-alkanes and is depleted in both light hydrocarbons and C_{14+} aromatics. The highest content of NSOs, which are the sum of resins and asphaltenes (27.3 wt. %), is found in the marine Safaniya oil. The Ardjuna oil has similar proportions of C_6–C_{14} hydrocarbons (27 wt. %) and C_{14+} n- and iso-alkanes (25 wt. %), with 17.6 wt. % of nitrogen, sulfur, and oxygen–containing compounds (NSOs).

The three oils were subjected to a mathematical simulation of biodegradation using the BioClass 0D model to show how their compositions would behave if submitted to biological alteration in the subsurface.

Biodegradation Model

The conceptual model BioClass 0D for predicting both the global biodegradation yields and corresponding chemical changes in biodegraded petroleum fluids was thoroughly described by Haeseler et al. (2010). The petroleum fluid is represented by chemical classes and their sensitivity to biodegradation, for example, the C_6–C_{14} saturates, C_6–C_{14} aromatics, C_{14+} n + iso-alkanes, C_{14+} cyclo-alkanes, C_{14+} aromatics, and NSOs. Oil composition based on these chemical classes can be output from numerical basin models as described by Behar et al. (2008). Biodegradation comprises successive biological processes. This sequence starts with the aerobic degradation of hydrocarbons using oxygen as the electron acceptor until no more bioavailable oxygen remains in the system. Then, progressively, any

Table 2. Values of the relative biodegradation rates for the selected pseudocomponents (referred to as the K_{bio}).

Model Compound			
Chemical Class	***Chemical Structure***	C_xH_y	K_{bio}*
C_6–C_{14} saturates	Alkane (n + iso)	C_9H_{20}	34.6
	Alkyl cyclohexane	C_9H_{18}	34.6
C_6–C_{14} aromatics	Alkyl benzene	C_9H_{12}	21.2
C_{14+} saturates	n-alkene	$C_{18}H_{38}$	17.3
	iso-alkane	$C_{18}H_{39}$	10.4
	Alkyl decaline	$C_{18}H_{34}$	13.9
C_{14+} aromatics	Alkyl naphthalane	$C_{18}H_{24}$	2.6

*K_{bio} are normalized to 100.

available nitrates (denitrification) are used as electron acceptors for degradation followed by manganese and iron (metal reduction), sulfate (sulfate reduction), and finally, water (anaerobic methanogenesis). The net biodegradation reactions for the overall sequence are the following:

Aerobic Conditions:

$$C_xH_y + (x + 1/4y)O_2 \rightarrow xCO_2 + 1/2yH_2O, \quad (1)$$

Denitrifying Conditions:

$$C_xH_y + (2/3x + 1/6y)NO_3 \rightarrow xCO_2 + (1/3x + 1/12y)N_2 + 1/2yH_2O, \quad (2)$$

Sulfate-Reducing Conditions:

$$C_xH_y + (2/5x + 1/10y)SO_4 \rightarrow xCO_2 + (2/5x + 1/10y)H_2S + (-2/5x + 2/5y)H_2O, \quad (3)$$

Methanogenic Conditions:

$$C_xH_y + (x - 1/4y)H_2O \rightarrow (1/2x - 1/8y)CO_2 + (1/2x + 1/8y)CH_4 \quad (4)$$

In these four equations, the hydrocarbon is mineralized (totally biodegraded) to CO_2 and methane. This means that the generated polar compounds that are observed in nature during biodegradation, such as carboxylic acids or alcohols, are assumed to be converted to biogenic gas. They have been neglected because even for heavily biodegraded oils, the residual amount of these compounds does not exceed 1 or 2 wt. % (Behar and Albrecht, 1984; Fafet et al., 2008).

The successive biological hydrocarbon degradation reactions that occur in subsurface can be summarized as follows:

1. Hydrocarbon biodegradation uses the terminal electron acceptor with the highest redox potential as long as it is present in the environment and available to the microbes.
2. The hydrocarbon-degrading microbes will successively switch to the available electron acceptors with the highest possible redox potential until reaching methanogenic conditions.
3. The different hydrocarbon classes are all degraded in parallel but at different rates.

For each chemical class or model compound, a K_{bio} has been defined, which is a combination of accessibility and biosensitivity factors. The K_{bio} values proposed by Haeseler et al. (2010) were calibrated to fit the different biodegraded oils from the Carnaubais trend in the Potiguar Basin published by Behar et al. (2006) and validated on a series of biologically altered oils from the Williston Basin described by Bailey et al. (1973a, b). The K_{bio} values correspond to a relative biodegradation rate between the different hydrocarbon classes. The proposed K_{bio} values are reported in Table 2. The NSO compound class is assumed to be nondegradable, and the light saturates are the most sensitive to biodegradation followed by their aromatic homologs. In the C_{14+} fraction, the n-alkanes are half as sensitive as the light C_6–C_{14} saturates, and the aromatics are very refractory, but are still biodegradable. Each time a pseudocomponent is completely removed from the oil via biodegradation, the value of the corresponding K_{bio} becomes 0%, and the values of the others are recalculated so that their sum remains 100%.

It is assumed that whatever the biological process responsible for oil degradation, the successive sequence of compound class susceptibility remains the same. This is supported by observations of natural biodegradation series in various petroleum systems showing a similar evolution of the residual oil composition with increasing biodegradation: the light C_{14-} compounds and the C_{14+} n-alkanes are always most sensitive to biodegradation, whereas the C_{14+} iso- and cyclo-alkanes are more refractory. However, assigning a global K_{bio} to a given chemical class does not imply that all compounds in that class have the same biosensitivity. The K_{bio} values must be considered an average value between the most bioreactive and refractory compounds in a given chemical class.

In conclusion, starting with an initial composition and an arbitrary mass of nonbiodegraded oil, the BioClass 0D model calculates the residual masses of each class for a given biological process according to the stoichiometric equations 1 to 4. Then, it becomes trivial to determine

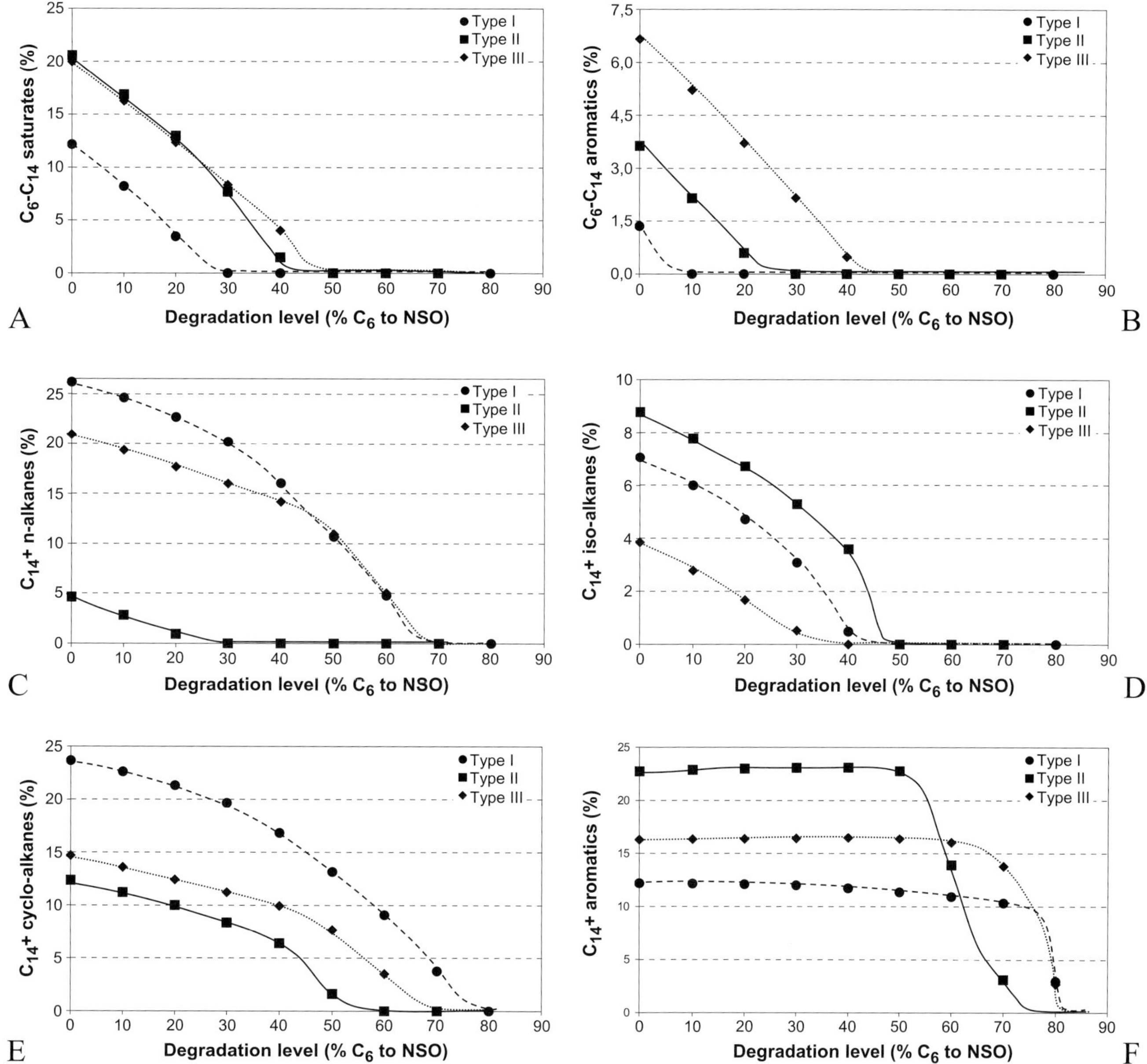

FIGURE 1. Relative content of biosensitive pseudocomponents as a function of the overall biodegradation yield of the oil (A = C_6–C_{14} saturates; B = C_6–C_{14} aromatics; C = C_{14+} n-alkanes; D = C_{14+} iso-alkanes; E = C_{14+} cyclo-alkanes; F = C_{14+} aromatics). The oil is considered to be the sum of all the liquid hydrocarbon fractions from C_6 to NSO compounds.

the relative chemical composition of the residual oil and the corresponding total hydrocarbon loss. The yields of the different metabolites produced during biodegradation, such as biogenic methane, H_2S, and CO_2, are also derived from these mass balance equations. Knowledge of the oil composition and the associated gases at each biodegradation step enables determination of the GOR. The API gravity can also be derived from the composition of the different hydrocarbon classes during increasing biodegradation by applying appropriate thermodynamic laws.

For application of the stoichiometric equations (1–4), calculations are done in moles but can be converted into kilograms or metric tons and recalculated in relative abundances.

RESULTS AND DISCUSSION

Evolution of Oil Composition during Biodegradation

The hydrocarbon compositions of the Minas, Safaniya, and Ardjuna oils are typical for oils originating from types I, II, and III source rocks, respectively. The initial compositions of these oils (Table 1) show that they significantly differ from each other

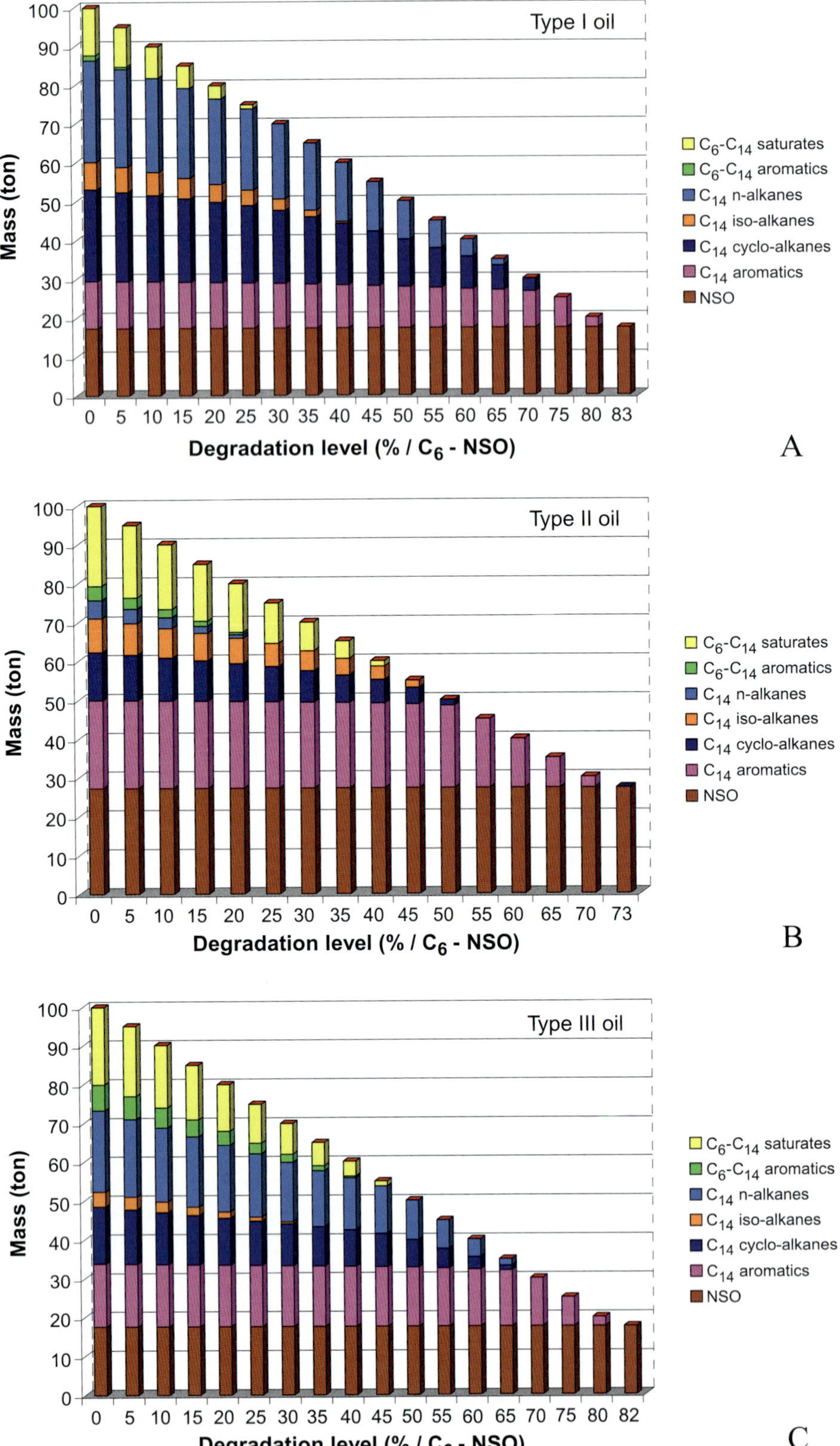

FIGURE 2. Masses of the pseudocomponents described by BioClass 0D as a function of the overall biodegradation yield of the oil (A = type I oil; B = type II oil; C = type III oil). The maximum biodegradation loss for each oil corresponds to 100 NSO, which is 83, 73, and 82%, respectively.

with regard to the proportions of the pseudocomponents chosen for their sensitivity to biodegradation. These oils with the posted initial compositions undergo biodegradation using the BioClass 0D model to evaluate eventual similarities or differences in relative chemical composition of the residual oils with increasing biodegradation.

The relative composition of each pseudocomponent is normalized with respect to the sum of all liquid oil components (from C_6 to NSO). These values are plotted

Table 3. Range of main disappearance of the six biodegradable classes.

	Wenger and Isaksen, 2002	*This Study**		
		Type I	*Type II*	*Type III*
C_6-C_{14} saturates	1 and 2	a	b	b and c
C_6-C_{14} aromatics	2 and 3	a	a	b and c
C_{14+} n-saturates	1 and 2	b	a	c
C_{14+} iso-alkanes	3	c and d	b and c	b
C_{14+} cyclo-alkanes	4+	d	c	c
C_{14+} aro-alkanes	4+	d	d	d

**For the present study, it is indicated in relation to the following quantitative scale (a = 0–20% of loss; b = 20–40% loss; c = 40–60% loss; d = 60–80% loss) and comparison with the qualitative scale of Wenger and Isaksen, 2002 (1 = slight; 2 = moderate; 3 = heavy; 4+ = severe).*

against the biodegradation yield in Figure 1. The biodegradation yields are determined from the residual oil masses corresponding to the indicated composition. For the three studied oils, the light fractions of both saturated and aromatic hydrocarbons (panels A and C of Figure 1 for the C_6–C_{14} saturates and C_6–C_{14} aromatics, respectively) appear to degrade linearly with respect to the level of degradation. In fact, this linearity only applies to these pseudocomponents characterized by the highest K_{bio} values and corresponds to degradation steps for which no pseudocomponent has been completely removed. For the C_{14+}-saturated hydrocarbons (panels C, D, and E of Figure 1, for C_{14+} n-alkanes, C_{14+} iso-alkanes, and C_{14+} cyclo-alkanes, respectively), the relative contents seem to follow a polynomial degradation curve with consistently decreasing proportions. The situation is significantly different for the C_{14+} aromatic hydrocarbons (Figure 1F), where relative abundance may even increase during early biodegradation (from 5 to 50%) because of the reduction of more easily degradable pseudocomponents. The significant reduction of the proportion of C_{14+} aromatic hydrocarbons in the oil mainly starts at the end of the oil biodegradation, in particular after the significant removal of the C_{14+} cyclo-alkanes. These results are in line with the K_{bio} values of the different pseudocomponents.

Because all hydrocarbon classes except the NSOs are biodegraded, the relative proportion of these polar compounds increases throughout biodegradation. The light fractions (C_6–C_{14} saturates and aromatics) may disappear during very early biodegradation (<10% for the type I oil) or at significantly higher biodegradation levels (>40% for types II and III oils). This clearly shows the importance of a thorough characterization of these light fractions and the possibility to use their abundance as a quick indicator of biodegradation level in altered oils. The parallel biodegradation scheme and the initial abundance of light hydrocarbons are responsible for the presence of C_6–C_{14} hydrocarbons and, in particular, the aromatic representatives that

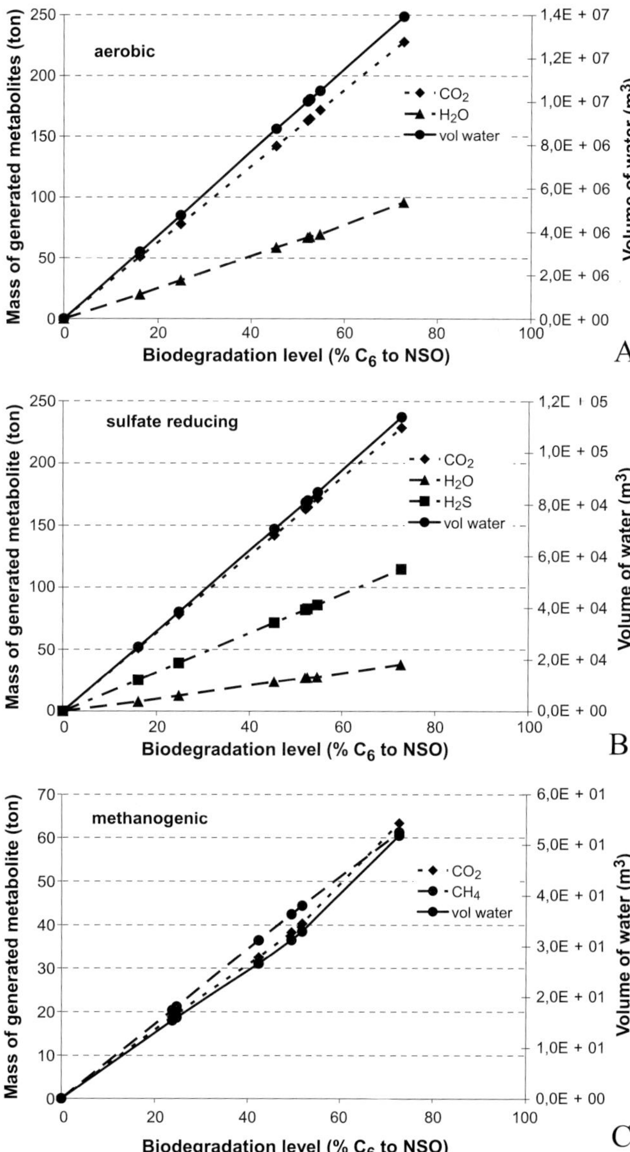

FIGURE 3. Quantities of metabolites generated during the exclusive (A) aerobic, (B) sulfate reducing, and (C) methanogenic biodegradation of type II oil. Also reported is the volume of water needed to provide the corresponding amounts of electron acceptors: dissolved oxygen, sulfate, and water as per equations 1, 3, and 4, respectively.

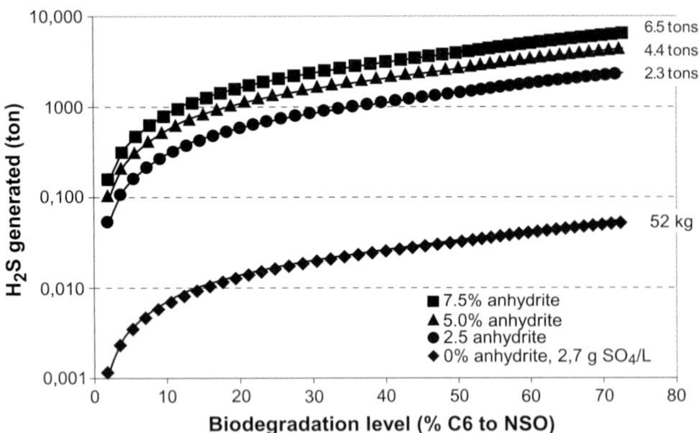

FIGURE 4. Masses of H_2S generated by BioClass 0D during the biodegradation of 100 tons as a function of biodegradation level. Calculations are based on successive alternations of sulfate reduction and methanogenesis with different sulfate scenarios (0% anhydrite in the rock and 2.7 g/L sulfate in the pore water, 7, 28, and 56 wt. % anhydrite in the reservoir rock considered as totally soluble in the pore water).

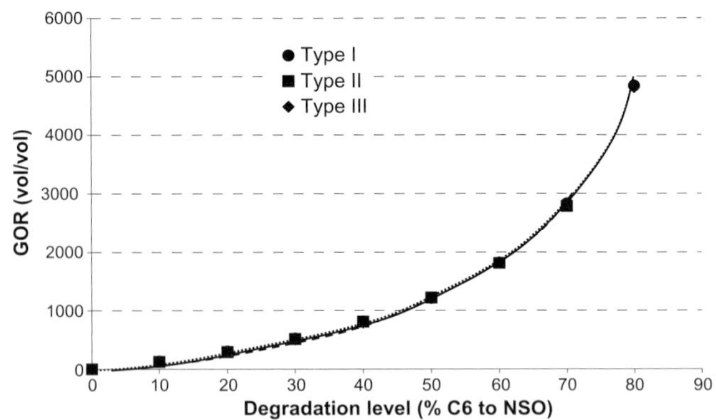

FIGURE 5. The gas-oil ratios (GORs) were determined assuming biodegradation-generated methane as the sole source of gas as calculated by the BioClass 0D model as a function of the biodegradation yield of the oil. Sulfate in the system: 2.7 g/L present in the water

can be expected in biologically altered oils that have lost as much as 40% of their mass.

Figure 1C shows that the C_{14+} normal alkanes can disappear during early biodegradation of the type II oil (≈20–25%), but they also can persist at much higher alteration levels (as much as 60–70%) for the types I and III oils. This compositional behavior can be attributed to both their K_{bio} and their initial abundance in the oil. Based on panels D and E of Figure 1, the other C_{14+} alkanes are completely degraded at biodegradation levels 35 to 50% for iso-alkanes and 55 to 75% for cyclo-alkanes.

The results of the biodegradation for the same three oil types can also be presented in the form of mass balances as shown in panels A and C of Figure 2. An arbitrary initial mass of oil was chosen at 100 tons. The total mass balances of the residual oil are plotted with increasing biodegradation extent (%) and are independent of the degradation reaction involved. As it was assumed that the NSO compounds are not biodegraded, the maximum biodegradation loss is 100 minus the initial proportion of NSOs. If these compounds are not considered as refractory, the total hydrocarbon loss may reach 100%. From these plots, it is possible to define a biosensitivity window for a given chemical class as a function of the total hydrocarbon loss: it is 0 to 45% for the C_6–C_{14} fraction and the C_{14+} iso-alkanes. It is a larger window, that is, 0 to 65%, for the C_{14+} n-alkanes because of their high proportion in paraffinic oils. The C_{14+} cyclo-alkanes start to be significantly degraded at more than 25 to 30% and are totally removed at more than 70%, the corresponding window is thus 30 to 70%. For the C_{14+} aromatics, biodegradation generally occurs above 50% degradation of the initial oil.

Results show that the scale of bioreactivity of the different chemical classes is dependent on their initial proportion in the oil. This is the result of assuming a parallel schema for the different chemical classes during oil biodegradation. For example, in the type I oil, the C_6–C_{14} aromatics are first totally removed, followed by the C_6–C_{14} saturates and the C_{14+} iso-alkanes. In a type II oil, the disappearance of the C_6–C_{14} occurs simultaneously with that of the C_{14+} n-alkanes. In the type III oil, the C_{14+} iso-alkanes are first degraded, followed by the C_6–C_{14} aromatics. For the more refractory classes, in the three oils, the C_{14+} cyclo-alkanes are first affected followed by the C_{14+} aromatics.

These trends can be compared with those observed on molecular groups on a scale ranging from level 1 to 10 proposed by Peters and Moldowan (1993) and Wenger and Isaksen (2002) and discussed in Peters et al. (2005). For that purpose, we proposed to split the biodegradation extent from Figure 2 into four main ranges: a = 0 to 20% and d = 60 to 100% and then to compare these ranges with the arbitrary scale (1 to 4+) of Wenger and Isaksen (2002). Results are reported in Table 3 presenting the ranges of disappearance for a given chemical class.

They show that according to the biodegradation scale proposed in this chapter and the different concentrations of the chemical classes in the three oils, the range of biological stability can be similar or larger than that proposed by Wenger and Isaksen (2002). For example, even if the biodegradation of the saturated C_6–C_{14} starts very early, this class of compounds can still be present when more than 40% of the hydrocarbons have been biodegraded. This chemical class is first removed in the type II oil, whereas it disappears later in the type III oil. These results demonstrate that the efficiency of disappearance by biodegradation for a given chemical class depends on both its biosensitivity and its initial concentration in the nonbiodegraded oil.

Evolution of Generated Products

Figure 3 shows the amounts of metabolites that would be generated given exclusive biodegradation by each of the aerobic, sulfate-reducing, or methanogenic biodegradation reactions using the type II oil. For this figure, only one electron acceptor has been provided to complete the oil biodegradation, which assumes that it is available to the hydrocarbon-degrading microorganisms until complete mineralization of the electron donor. As per equations 1, 3, and 4, these metabolites include the nonhydrocarbon gases (CO_2 and H_2S), water, and methane. In this figure, the volumes of water needed to provide the corresponding amounts of the different electron acceptors (dissolved oxygen, sulfate, and water, respectively) are also reported. The volumes of water were determined considering an oxygen concentration of 10 mg/L water corresponding to an oxygen saturation at 15°C, and a sulfate concentration of 2.7 g/L water, corresponding to the concentration of seawater at atmospheric conditions.

The simulations were completed using an initial arbitrary mass of 100 tons of oil. Therefore, the following data are presented for the metabolites as absolute masses expressed in tons and in cubic meters of water carrying the necessary amount of electron acceptors.

As shown in Figure 3A, aerobic biodegradation of all the oil (100 tons) generates 228 tons of CO_2 and 96 tons of H_2O, corresponding to incorporation of 224 tons of oxygen in the substrate. These 224 tons would need to be dissolved in 1.39×10^7 m^3 of water, considering a solubility of 10 mg/L. This clearly indicates the extremely large volume of water necessary to carry the electron acceptors for aerobic mineralization of petroleum in the subsurface.

Figure 3B shows the metabolites generated during sulfate reduction, assuming 2.7 g/L sulfate in the porewater. This concentration corresponds to normal seawater sulfate concentration. The mass of CO_2 generated is the same as for the aerobic biodegradation of the oil (228 tons). The mass of water generated as per equation 3 would be less (37.8 vs. 96 tons) than in the aerobic process for the mineralization of the 100 tons of the type II oil. During sulfate reduction, H_2S is also a major metabolite, and the stoichiometric model shows that 114.6 tons are generated. To achieve this biodegradation under sulfate-reducing conditions, a volume of 1.1×10^5 m^3 of water carrying the sulfate would be needed. This still requires large volumes of water providing electron acceptors but lower by two orders of magnitude than for aerobic biodegradation.

Figure 3C shows the metabolites generated during methanogenesis. This process clearly generates the smallest mass of metabolites than the others with 63.4 tons of CO_2 and 61.2 tons of CH_4 for the mineralization of 100 tons of type II oil. This biologic reaction needs a volume of only 51.8 m^3 of water to be completed.

These results show that CO_2 is generated by all of the processes (aerobic, sulfate reduction, and methanogenesis). As for methane, a significant proportion of this biogenic CO_2 may be lost by convection and diffusion because of its high aqueous solubility. Furthermore, the presence of high concentration of CO_2 may reduce the pH and induce diagenetic transformations of minerals in the porous media. The production of biogenic gas as pure methane may significantly increase the GOR, as indicated in Figure 4. The GOR is defined as the surface GOR and includes only the hydrocarbon gas. Figure 5 shows that its evolution is strictly linked to the biodegradation level, regardless of the oil used as source. This is a consequence of the stoichiometric methanogenesis equation in which the limiting factor is the electron acceptor. This equation strictly controls the mass of methane generated as a function of the hydrocarbon consumed, for example, the biodegradation level.

CONCLUSIONS

In a previous study, we developed a 0D model "BioClass 0D" that allows prediction of the residual chemical composition of an oil with increasing biodegradation and the corresponding hydrocarbon loss. In this 0D model, four successive stoichiometric equations are used for describing the aerobic and anaerobic processes. The reactants in these four equations are representative model compounds of the main chemical classes in the nonbiodegraded oil. Oil biodegradation is a series of parallel reactions in which relative biosensitivity rate, that is, K_{bio}, is assigned to the selected model compounds. It is assumed that whatever the biological process responsible for oil degradation, the successive sequence of oil degradation is the same. This is supported by observations of natural biodegradation series in various petroleum systems showing a similar evolution of the residual oil composition with increasing biodegradation: the light C_{14-} compounds and the C_{14+} n-alkanes are always most sensitive to biodegradation, whereas the C_{14+} iso- and cyclo-alkanes are more refractory. However, assigning a global K_{bio} to a given chemical class does not imply that all compounds in that class have the same biosensitivity. The K_{bio} values must be considered an average number between bioreactive and refractory compounds for a given chemical class. The amount and nature of the products generated during biodegradation depends on the biological process: CO_2 and H_2O under aerobic conditions, CO_2 and H_2S during sulfate reduction, and CH_4 and H_2O during methanogenesis.

This model was applied to three nonbiodegraded oils representative of the three main types of source rock organic matter. Results show that with increasing biodegradation, the residual oil composition depends on the initial proportion of the different chemical classes. For instance, for the type I oil that is enriched in paraffins, the C_{14+} n-alkanes are still present when 60% of the original oil has already disappeared, whereas for the type II oil, the same chemical class is observed only at less than 30% of total hydrocarbon loss. The computed residual composition of the oils follows the same sequence as that observed in natural series. Biosensitivity windows can be proposed for the global chemical classes as a function of the total hydrocarbon loss: it is 0 to 45% for C_6–C_{14} fraction and for the C_{14+} iso-alkanes, 0 to 65% for the C_{14+} n-alkanes, 30 to 70% for the C_{14+} cyclo-alkanes, and more than 50% for the C_{14+} aromatics.

BioClass 0D presently does not assume production of organic acids or other metabolites, which might accumulate in the oil and contribute for instance to an increase of the total acid number or NSO content. Such compounds are known to be generated, but no available data support their implementation within the stoichiometric hydrocarbon biodegradation equations. Another assumption considers that the NSOs are not subject to biodegradation, although this may not reflect reality. Aerobic biodegradation generates CO_2 and H_2O and requires very large volumes of water, providing the electron acceptor to allow significant biodegradation. The sulfate-reducing process generates CO_2, H_2O, and H_2S and requires either very large volumes of water or a sulfate supply provided from rock dissolution. Methanogenesis generates CO_2 and CH_4 and is the biological process requiring the most limited oil-water contact to allow hydrocarbon mineralization.

Knowing the amount of electron acceptors consumed during anaerobic biodegradation, it is possible to determine the amount of water required to provide these electron acceptors. For example, for the same biodegradation extent, the aerobic process needs a volume of water of 10^7 times more than for methanogenesis. The concentration of the electron acceptor, that is, dissolved oxygen in water, is only 0.01 g/L, whereas during methanogenesis, the water itself is the electron acceptor. The huge contrast of water volume is likely to have a significant impact on the key parameters controlling oil biodegradation.

All these considerations should be taken into account when elaborating in a future work the coupling of BioClass 0D with a dynamic model in a basin simulator. Indeed, during the history of the petroleum fluid from its primary expulsion, migration, reservoir filling, and accumulation, the oil-water ratio may vary significantly. This means that in addition to temperature, time, salinity, lithology, and reservoir geometry, an important factor should be the effective volume of water in contact with oil. This parameter may be effective as soon as the geological conditions are favorable for biodegradation.

ACKNOWLEDGMENTS

We thank Kenneth E. Peters for taking over the task of editor for this chapter and for his very accurate, precise, and helpful review of the manuscript. He was a real support to help us by improving this work. We also thank the anonymous reviewer who also helped us with a thorough review and pertinent comments to improve the article.

REFERENCES CITED

Aitken, C. M., D. M. Jones, and S. R. Larter, 2004, Anaerobic hydrocarbon biodegradation in deep subsurface oil reservoirs: Nature, v. 431, p. 291–294, doi:10.1038/nature02922.

Al Darouich, T., F. Behar, C. Largeau, and H. Budzinski, 2005, Separation and characterization of the C_{14-} aromatic fraction of Safaniya crude oil: Oil and Gas Science Technology, v. 60, p. 681–695, doi:10.2516/ogst:2005048.

Asif, M., K. Grice, and T. Fazeelat, 2009, Assessment of petroleum biodegradation using stable hydrogen isotopes of individual saturated hydrocarbon and polycyclic aromatic hydrocarbon distributions in oils from the Upper Indus Basin, Pakistan: Organic Geochemistry, v. 40, p. 301–311, doi:10.1016/j.orggeochem.2008.12.007.

Bailey, N. J. L., A. M. Jobson, and M. A. Rogers, 1973a, Bacterial degradation of crude oil: Comparison of field and experimental data: Chemical Geology, v. 11, p. 203–221, doi:10.1016/0009-2541(73)90017-X.

Bailey, N. J. L., H. R. Krouse, C. R. Evans,and M. A. Rogers, 1973b, Alteration of crude oil by waters and bacteria: Evidence from geochemical and isotope studies: AAPG Bulletin, v. 57, p. 1276–1290.

Behar, F., and P. Albrecht, 1984, Correlation between carboxylic acids and hydrocarbons in several crude oils. Alteration by biodegradation: Organic Geochemistry, v. 6, p. 597–604, doi:10.1016/0146-6380(84)90082-2.

Behar, F., H. L. de Barros Penteado, F. Lorant, and H. Budzinski, 2006, Study of biodegradation process along the Carnaubais trend: Potiguar Basin (Brazil), Part 1: Organic Geochemistry, v. 37, p. 1042–1051, doi:10.1016/j.orggeochem.2006.05.009.

Behar, F., F. Lorant, and L. Mazeas, 2008, Elaboration of a new compositional kinetic scheme for oil cracking: Organic Geochemistry, v. 39, p. 764–782, doi:10.1016/j.orggeochem.2008.03.007.

Connan, J., 1984, Biodegradation of crude oils in reservoirs, *in* J. Brooks and D. H. Welte, eds., Advances in

Petroleum Geochemistry 1: London, Academic Press, p. 299–335.

Connan, J., G. Lacrampe-Coulome, and M. Magot, 1997, Anaerobic biodegradation of petroleum in reservoirs: A widespread phenomena in nature, *in* Proceedings of the 18th International Meeting on Organic Geochemistry, September 22–26, 1997, Maastricht, The Netherlands, p. 5–6.

Dolfing, J., S. R. Larter, and I. M. Head, 2008, Thermodynamic constraints on methanogenic crude oil biodegradation: The International Society for Microbial Ecology Journal, v. 2, p. 442–452.

Eschard, R., and A.-Y. Huc, 2008, Habitat of biodegraded heavy oils: Industrial implications: Oil & Gas Science and Technology–Revue del l'Institut Français du Pétrole, v. 63, p. 587–607.

Fafet, A., F. Kergall, M. Da Silva, and F. Behar, 2008, Characterization of acid compounds in biodegraded oils: Organic Geochemistry, v. 39, p. 1235–1242, doi:10.1016/j.orggeochem.2008.03.008.

Grice, K., R. Alexander, and R. I. Kagi, 2000, Diamondoid hydrocarbon ratios as indicators of biodegradation in Australian crude oils: Organic Geochemistry, v. 31, p. 67–73, doi:10.1016/S0146-6380(99)00137-0.

Haeseler, F., F. Behar, D. Garnier, and P. Y. Chenet, 2010, First stoichiometric model of oil biodegradation in natural petroleum systems: Part I: The BioClass 0D approach, Organic Geochemistry, v. 41, p. 1156–1170, doi:10.1016/j.orggeochem.2010.05.019.

Head, I. M., D. Jones, and S. R. Larter, 2003, Biological activity in the deep subsurface and the origin of heavy oil: Nature, v. 426, p. 344–352.

Jackson, B. E., and M. McInerney, 2002, Anaerobic microbial metabolism can proceed close to thermodynamic limits: Nature, v. 415, p. 454–456, doi:10.1038/415454a.

Jaffé, R., and M. T. Gallardo, 1993, Application of carboxylic acid biomarkers as indicators of biodegradation and migration of crude oils from the Maracaibo Basin, Western Venezuela: Organic Geochemistry, v. 20, p. 973–984, doi:10.1016/0146-6380(93)90107-M.

Jobson, A. M., F. D. Cook, and D. W. S. Westlake, 1979, Interaction of aerobic and anaerobic bacteria in petroleum biodegradation: Chemical Geology, v. 24, p. 355–365, doi:10.1016/0009-2541(79)90133-5.

Jones, D. M., et al., 2008, Crude-oil biodegradation via methanogenesis in subsurface petroleum reservoirs: Nature, v. 451, p. 176–181, doi:10.1038/nature06484.

Larter, S., H. Huang, J. Adams, B. Bennett, O. Jokanola, T. Oldenburg, M. Jones, I. Head, C. Riediger, and M. Fowler, 2006, The controls on the composition of biodegraded oils in the deep subsurface: Part II: Geological controls on subsurface biodegradation fluxes and constraints on reservoir-fluid property prediction: AAPG Bulletin, v. 90, p. 921–938, doi:10.1306/01270605130.

Le Dréau, Y., F. Gilbert, P. Doumenq, L. Asia, J.-C. Bertrand, and G. Mille, 1997, The use of hopanes to track in situ variations in petroleum composition in surface sediments: Chemosphere, v. 34, p. 1663–1672, doi:10.1016/S0045-6535(97)00023-4.

Magnier, C., I. Kowalewski, B. Carpentier, A.-Y. Huc, E. Delamaide, and H. L. Penteado, 2001, Tentative quantification of the extent of biodegradation in oil fields using heavy metal content, *in* 20th International Meeting on Organic Geochemistry, September 10–14, 2001, Nancy, p. 413–414.

Magot, M., B. Ollivier, and B. K. C. Patel, 2000, Microbiology of petroleum reservoirs: Antonie van Leeuwenhoek, v. 77, p. 103–116, doi:10.1023/A:1002434330514.

McCaffrey, M. A., H. A. Legarre, and S. J. Johnson, 1996, Using biomarkers to improve heavy oil reservoir management: An example from the Cymric field, Kern County, California: AAPG Bulletin, v. 80, p. 898–913.

Meredith, W., S.-J. Kelland, and D. M. Jones, 2000, Influence of biodegradation on crude oil acidity and carboxylic acid composition: Organic Geochemistry, v. 31, p. 1059–1073, doi:10.1016/S0146-6380(00)00136-4.

Milner, C. W. D., M. A. Rogers, and C. R. Evans, 1977, Petroleum transformations in reservoirs: Journal of Geochemical Exploration, v. 7, p. 101–153, doi:10.1016/0375-6742(77)90079-6.

Nealson, K. H., and R. Rye, 2005, Evolution of metabolism, *in* W. H. Schlesinger, ed., Biogeochemistry, Amsterdam, Netherlands, Elsevier, p. 41–61.

Noble, R. A., C. H. Wu, and C. D. Atkinson, 1991, Petroleum generation and migration from Talang Akar coals and shales offshore NW Java, Indonesia: Organic Geochemistry, v. 17, p. 363–374, doi:10.1016/0146-6380(91)90100-X.

Orphan, V. J., L. T. Taylor, D. Hafenbradl, and E. F. Delong, 2000, Culture-dependant and culture-independent characterization of microbial assemblages associated with high-temperature petroleum reservoirs: Applied and Environmental Microbiology, v. 66, p. 700–711, doi:10.1128/AEM.66.2.700-711.2000.

Peters, K. E., and J. M. Moldowan, 1993, The biomarker guide: Interpreting molecular fossils in petroleum and ancient sediments: Englewood Cliffs, New Jersey, Prentice Hall, 363 p.

Peters, K. E., C. C. Walters, and J. M. Moldowan, 2005, The biomarker guide: Cambridge, Cambridge University Press, 1155 p.

Rice, D. D., and G. E. Claypool, 1981, Generation, accumulation, and resource potential of biogenic gas: AAPG Bulletin, v. 65, p. 5–25.

Röling, W. F. M., I. M. Head, and S. R. Larter, 2003, The microbiology of hydrocarbon degradation in subsurface petroleum reservoirs: Perspectives and prospects: Research in Microbiology, v. 154, p. 321–328, doi:10.1016/S0923-2508(03)00086-X.

Rozanova, E. P., I. A. Borzenkov, A. L. Tarasov, L. A. Suntsova, C. L. Dong, S. S. Belyaev, and M. V. Ivanov, 2001, Microbiological processes in a high-temperature oil field: Microbiology, v. 70, p. 102–110, doi:10.1023/A:1004809308305.

Sasaki, T., H. Maki, M. Ishihara, and S. Harayama, 1998, Vanadium as an internal marker to evaluate microbial degradation of crude oil: Environmental Science Technology, v. 32, p. 3618–3621, doi:10.1021/es980287o.

Seifert, W. K., and J. M. Moldowan, 1981, Paleoreconstruction by biological markers: Geochimica Cosmochimica Acta, v. 45, p. 783–794, doi:10.1016/0016-7037(81)90108-3.

Tissot, B., B. Durand, J. Espitalié, and A. Combaz, 1974, Influence of the nature and diagenesis of organic matter in formation of petroleum: AAPG Bulletin, v. 58, p. 499–506.

Tissot, B. P., and D. H. Welte, 1978, Petroleum formation and occurrence: Berlin, Springer-Verlag, 538 p.

Wenger, L. M., and G. H. Isaksen, 2002, Control of hydrocarbon seepage intensity on level of biodegradation in sea bottom sediments: Organic Geochemistry, v. 33, p. 1277–1292, doi:10.1016/S0146-6380(02)00116-X.

Wilkes, H., A. Vieth, and R. Elias, 2008, Constraints on the quantitative assessment of in-reservoir biodegradation using compound-specific stable carbon isotopes: Organic Geochemistry, v. 39, p. 1215–1221, doi:10.1016/j.orggeochem.2008.02.013.

Williams, J. A., and J. C. Winters, 1969, Microbial alteration of crude oil in the reservoir, *in* Symposium of Petroleum Transformations in Geological Environments, American Chemical Society National Meeting, Preprints 14, p. 22–31.

3

Schneider, F. J. S., J. A. Noya, and C. Magnier, 2012, Model of low-maturity generation of hydrocarbons applied to the Carupano Basin, Offshore Venezuela, *in* K. E. Peters, D. J. Curry, and M. Kacewicz, eds., Basin Modeling: New Horizons in Research and Applications: AAPG Hedberg Series, no. 4, p. 51–69.

Model of Low-Maturity Generation of Hydrocarbons Applied to the Carupano Basin, Offshore Venezuela

Frederic Jean Simon Schneider
Beicip-Franlab/Petróleo de Venezuela S.A. Estudio y Formación Acelerada de Iintegradores, Venezuela

Caroline Magnier
Institut Français du Pétrole Energies Nouvellas, Venezuela

Jose A. Noya
Petróleo de Venezuela S.A. Exploracion, Venezuela

ABSTRACT

The Carupano Basin is located in northeastern offshore Venezuela. This area is characterized by the interaction between the Caribbean and the South American Plate. It has two structural highs, the Los Testigos High located in the northern limit of the basin and the Patao High, which is between the Caracolito subbasin and the Paria subbasin. In the latter are located the main gas fields.

The generated gas is characterized by low maturity, and it has been attributed to biogenic processes because of its carbon isotopic signature. Nevertheless, gas compositions show that a thermogenic signature predominates with an increase of gas maturity from the east to the west, where condensates were found associated with gas.

To understand the origin of the gas, the total organic carbon $(TOC)_{SR}$ methodology was used to define continuous TOC profiles from sonic and resistivity wirelogs. As a result, we have shown that the whole column from the Eocene to the Pliocene consists of a poor source rock, except the middle Miocene that could be considered as a good source rock. The average TOC content of the middle Miocene can reach values around 2.5%. The kerogen is mostly type III continental–derived organic matter.

The thermal calibration and the basin modeling study shows that the bottom of the Paria subbasin has reached the oil window, whereas the bottom of the Caracolito subbasin has reached the gas window. Nevertheless, simulations of fluid-flow migration conducted using default type III kinetic parameters were not able to fill any of the known fields.

DOI:10.1306/13311428H43463

We conclude that the default kinetic parameters used for basin modeling are not able to reproduce the nature of these fluids. Indeed, in our study, the main part of the fields drainage area is in a low-maturity domain where the vitrinite reflectance (R_o) is less than 0.6%, but the kinetic parameters used were calibrated with kerogen samples for which R_o was taken to be approximately 0.6%.

Considering default type III kerogen as a starting point and using observed natural data such as gas compositions, a new set of kinetic parameters were derived to account for low gas maturity. This modified type III kerogen differs from the previous one by a 13% increase of the hydrogen index. The simulations conducted with this modified type III scheme allowed us to reproduce quite well the filling of the fields, as well as the composition of the hydrocarbons.

INTRODUCTION

The Carupano Basin is located in the northeastern Venezuela offshore area, close to the world-class petroliferous basins of Venezuela Oriental (Summa et al., 2003). This area is characterized by the complex interaction between the Caribbean and South American plates (Molnar and Sykes, 1969; Sykes et al., 1982; Burke et al., 1984; Speed, 1985; Eva et al., 1989; Mann et al., 1990; Pindell and Barret, 1990; Avé Lallemant, 1997; James, 2000). It is composed of two structural highs, the Los Testigos High located in the northern limit of the basin and the Patao High, which separates the Caracolito subbasin and the Paria subbasin (Figures 1, 2).

The main gas and condensate fields (Rio Caribe, Mejillones, Patao, and Dragon) are located in the Patao High (Figure 2). They were discovered as a result of the exploration campaign by Lagoven S.A. between 1979 and 1983 (Chalot et al., 1992). The gas reserves are estimated to be around 14 tcf (Liendo, 2008).

GEODYNAMIC SETTING

In Venezuela, the main geologic elements are the stable Guyana Shield to the south, the Miocene Andes Mountains to the northwest, and the Miocene–Pliocene right-lateral strike-slip Bocono/El Pilar fault system (Gonzalez de Juana et al., 1980). The latest element, in its eastern part—the Pilar, el Coche fault system—is part of the complex Caribbean-South America plate boundary (Molnar and Sykes, 1969).

Before the Jurassic, the Caribbean did not exist and the North and South American continental plates were united. The basic interpretation of the geologic history of northern South America consists of (Eva et al., 1989; Di Croce, 1995; Lugo and Mann, 1995) (1) a Triassic to Late Jurassic rifting event (breakup of Pangea); (2) a Late Jurassic–Oligocene passive margin phase; and (3) an early Miocene to present active margin phase.

The Tertiary evolution of the northeastern Venezuela offshore, which includes the Carupano Basin, is dominated by three main phases: (1) an Eocene to Oligocene extension, (2) an Oligocene to middle Miocene inversion, and (3) a late Miocene to present transtensional phase.

The Eocene to Oligocene extension is interpreted as an arc-normal extension associated with a retreating subduction boundary (Ysaccis, 1998) during the Paleogene convergence between North and South America (Pindell et al., 1991, 1998).

The Oligocene to middle Miocene inversion is related to the oblique collision between the Caribbean and the South American crusts (Dewey and Pindell, 1985, 1986; Speed, 1985). Many studies indicate that during the Neogene, the plate boundary zone evolved by right-oblique convergence (Speed, 1985; Pindell and Barret, 1990; Russo and Speed, 1992; Algar and Pindell, 1993; Avé Lallemant, 1997; Speed and Smith-Horowitz, 1998). The resulting deformed foreland comprises a Neogene fold and thrust belt and piggybacked foreland basins limited at its southern edge by blind thrusting and incipient folding (Roure et al., 1994; Parnaud et al., 1995).

The late Miocene to present transtension event began 12 m.y. ago (Algar and Pindell, 1993; Pindell, 1993; Pindell et al., 1998). The contemporary velocity of the Caribbean relative to the South American plate in eastern Venezuela is 2 cm/yr toward the east (Molnar and Sykes, 1969; Pérez et al., 2001; Weber et al., 2001). The motion is concentrated on right-slip faults, such as the El Pilar and Coche faults (Franke et al., 1996; Weber et al., 2001).

BASIN DESCRIPTION

Stratigraphic Framework

The sedimentary column of the Carupano Basin area is composed, from bottom to top, of the Mejillones Complex, the Tigrillo Formation, the Caracolito Formation,

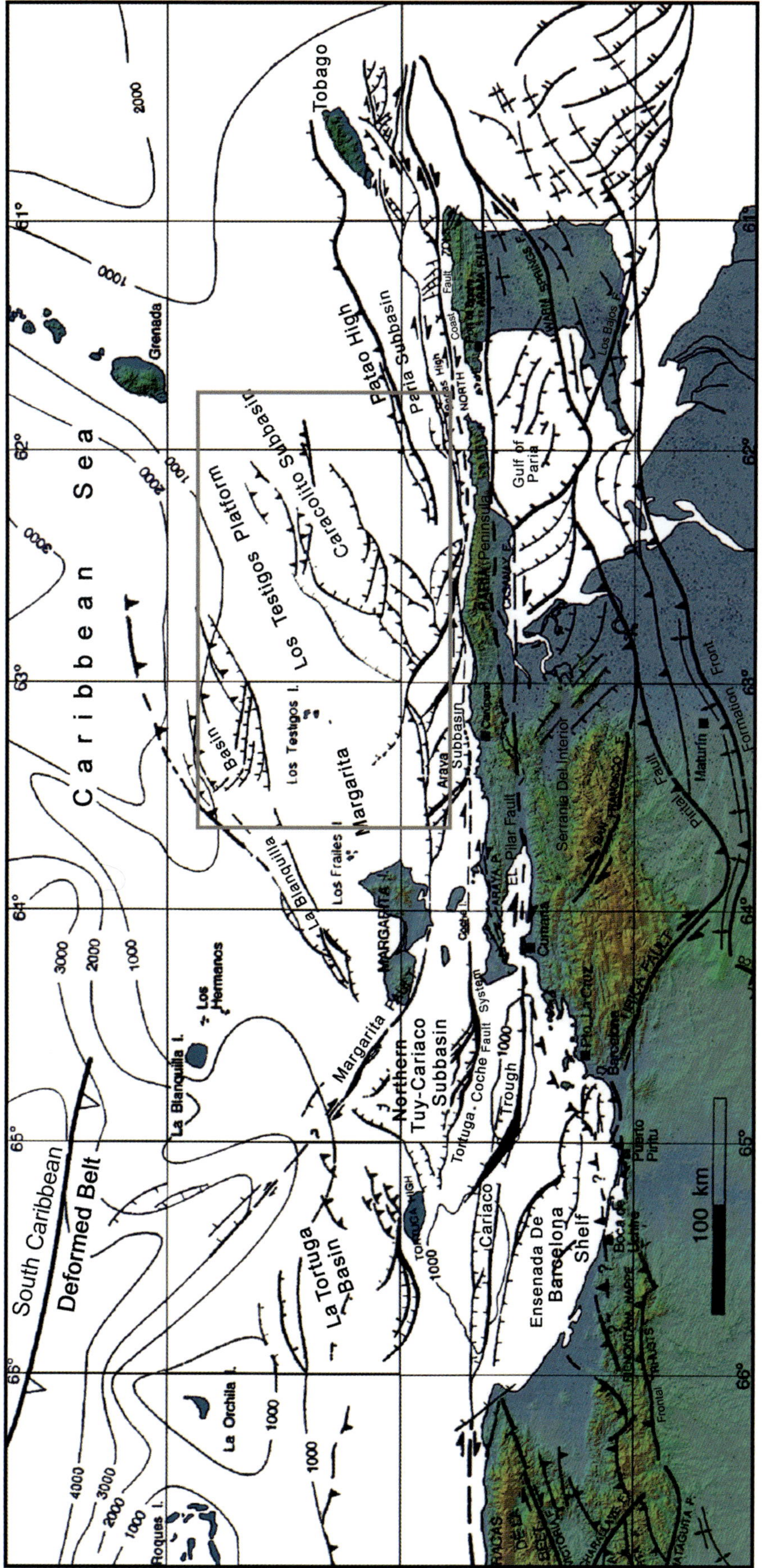

FIGURE 1. Structural setting of the studied area (modified from Ysaccis, 1998). The rectangle is the location of Figure 2.

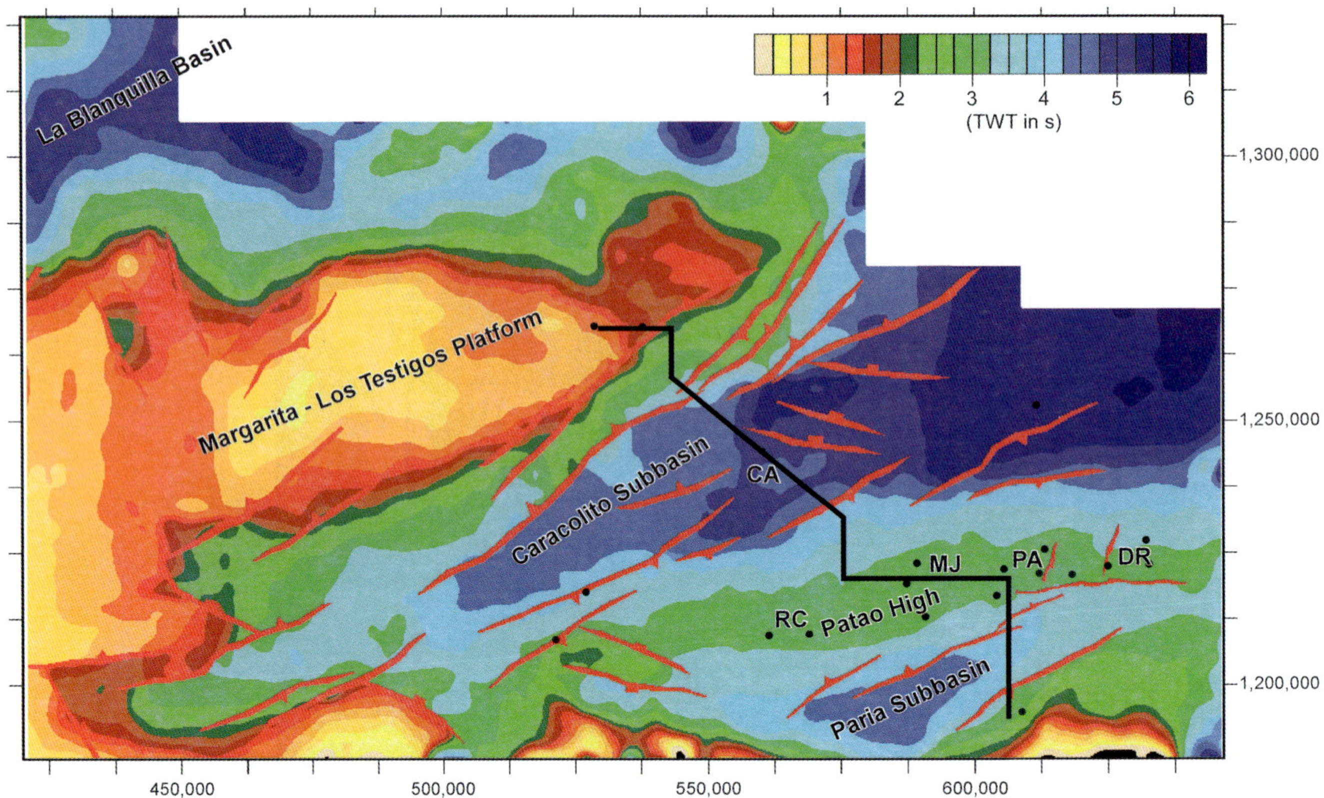

FIGURE 2. Simplified two-way traveltime (TWT) structural map at the top of the basement (mainly top Cretaceous) in the studied area showing the main structural elements: the Patao High, the Margarita-Los Testigos platform, the Paria subbasin, and the Caracolito subbasin. The broken line is the trace of the section presented in Figure 4. The dots indicate the well locations. RC = Rio Caribe; MJ = Mejillones; PA = Patao; DR = Dragon; CA = Caracolito.

the Tres Puntas Formation, the Cubagua Formation, and the Cumana Formation (Figure 3).

The Mejillones Complex is mainly composed of shales, limestones, sandstones interstratified with basalt layers deposited in a deep-marine environment. The age of these deposits is Early to Late Cretaceous. These rocks do not show evidence of metamorphism.

The Mejillones Complex correlates landward with the Lower Cretaceous Sucre Group (Barranquin and El Cantil formations) and the Upper Cretaceous Guayuta Group (Querecual and San Antonio formations). These Cretaceous strata were deposited on the north-facing passive margin of continental South America (Maresch, 1974; James, 1990).

The Tigrillo Formation is mainly composed of sandstones and shales and was deposited during the early and middle Eocene in a pelagic environment. The late Eocene is commonly absent in this area. This late Eocene unconformity is contemporaneous with the landward (southern) upper Eocene thin bioclastic limestones of the Tinajitas Member and the Peñas Blancas Formation, which are of shallow-marine provenance (Sageman and Speed, 2003).

The Caracolito Formation is mainly composed of shales and silty shales with some intercalated sandstones (turbidites). It was deposited in a bathyal environment during the Oligocene. This formation correlates with the landward Areo Formation (Castro and Mederos, 1985).

The Tres Puntas Formation is mainly composed of shales with some interstratified sandstones (turbidites) deposited in a bathyal environment at its base to a neritic environment in its upper part. The age of this formation is early to middle Miocene. The late Miocene is commonly absent in this area.

The lower part of the Cubagua Formation was deposited in bathyal environment and is mainly composed of shales with some fine-grained interstratified sandstones that constitute the reservoir rock for the known gas accumulations. The upper part, deposited in an inner neritic environment, contains a great variety of tropical shallow-marine sediments. The age of this formation is mainly Pliocene.

The Cumana Formation contains a great variety of tropical shallow-marine sediments deposited during the early to middle Pleistocene.

Structural Framework

The Carupano Basin is located on the southern margin of the Caribbean Plate. Two subbasins can be distinguished, Caracolito and Paria, separated by the

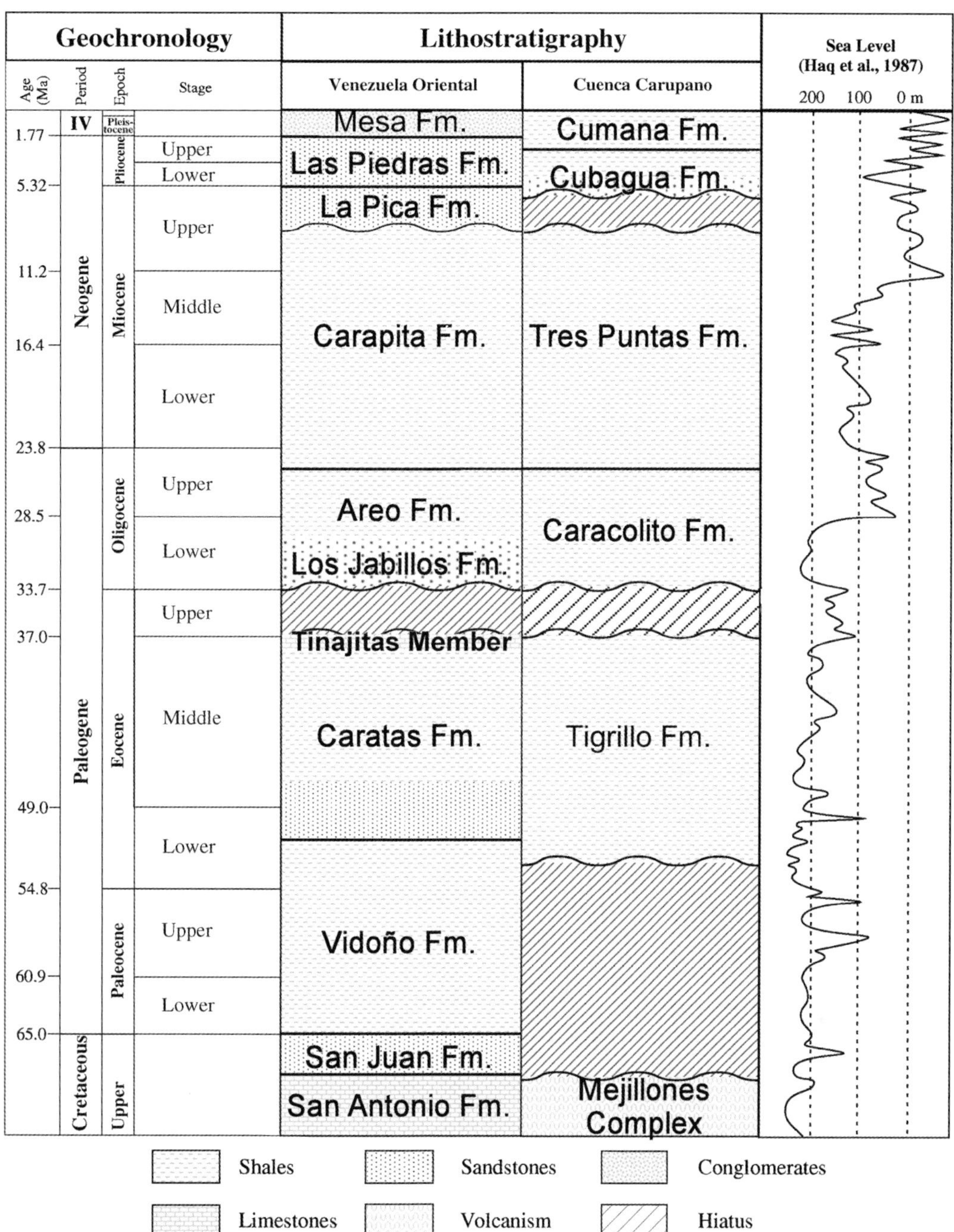

FIGURE 3. Simplified Cenozoic stratigraphic chart of the studied area (modified from Castro and Mederos, 1985). Only the dominant lithology is drawn for each formation.

Patao High (Figures 1, 2). The main depocenter of the Carupano Basin, the Caracolito subbasin formed during the Eocene in the forearc related to the Atlantic oceanic plate subduction below the Caribbean Plate. Part of this subbasin has been inverted during the Oligocene as shown by the anticline structures sealed by the upper Oligocene deposits (Figure 4).

On the Patao High where gas accumulations are located, the Miocene lies directly above deformed, but nonmetamorphosed, Cretaceous rocks.

Thermal Gradient

Horner-corrected bottomhole temperature data from 25 wells were used to determine the distribution of the present-day thermal gradient in this area. The map shown in Figure 5 was built from the well data, using an inverse square distance spatial interpolator followed by linear diffusion smoothing. The results indicate that the thermal gradient increases westward from 0.9°F/100 feet (~16°C /km) to 1.4°F/100 feet (~25°C/km). This gradient evolution, as shown in

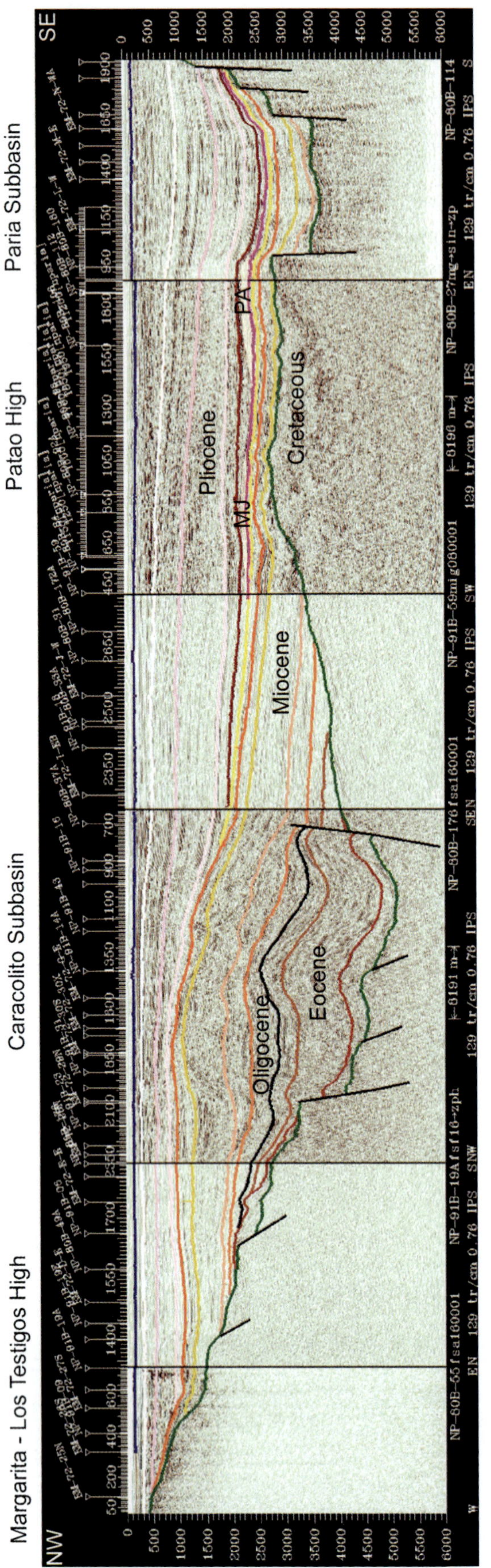

Figure 4. Composite northwest-southeast time section along the broken line showing the main structural elements: the Los Testigos high, the Caracolito subbasin, the Patao High, and the Paria subbasin. See location in Figure 1. MJ = Mejillones; PA = Patao.

Figure 5, does not show any obvious correlation with the main structural features.

Hydrocarbons

Dragon, Patao, and Mejillones are gas fields, whereas Rio Caribe is a wet gas-condensate field. In the C_1-C_2 to C_1-C_5 plot (Pixler, 1969), which compares the gas composition with typical compositions of dry and oil-associated gases, the Dragon, and Patao gases plot well above the dry thermal gas area on the plot (Verbanac and Duma, 1982); the Mejillones gas plots in the dry gas area. In this Pixler plot, the Rio Caribe gases fall in gas-condensate zone consistent with thermal gas (Table 1; Figure 6).

In the ethane-isobutane versus ethane-propane diagram (Prinzhofer and Battani, 2003), the gas compositions of the three fields that contain measurable amounts of isobutane (e.g., Rio Caribe, Mejillones, and Patao) plot in the low-maturity gas zone with no effect of biodegradation. Furthermore, the plot gives a rough maturity trend from Patao/Mejillones to Rio Caribe (Table 1; Figure 7). Note that in this diagram, any mixing with pure methane results in no displacement.

The low thermal maturity of the gases is also supported by their low carbon isotopic signature for methane. Indeed, the methane $\delta^{13}C$ values for two samples from Patao and Mejillones fields are -62.4‰ and -62.1‰ (Table 1).

The westward increase of the maturity correlates with the decrease of the dryness of the gas from Dragon to Rio Caribe as well as the gas composition that shows a westward increase of the C_{2+} hydrocarbon content (Table 1; Figure 8).

Although no doubt exists about the thermogenic origin of gases from the western field, one can question the origin of the gas from the eastern fields. The high methane content associated with the low carbon isotopic ratio might suggest a biogenic origin for the gases. Nevertheless, the relative concentrations of C_{2+} gaseous hydrocarbons in relation to $\delta^{13}C$ concentration in methane indicate that the gases from Patao and Mejillones plot in an intermediate zone, which includes both the biogenic and the thermogenic gas zones (Table 1) (Schoell, 1983). The same observation can be made in the Bernard diagram, where these two gases are in the mixed zone (Table 1; Figure 9). Furthermore, the results of the studies by Janezic (1979) and Coleman (1979) strongly suggest that C_2-C_4 hydrocarbons are not biogenic.

In summary, the proven gas is characterized by low maturity and its composition seems to show that a westward increase of maturity exists. However, mixing of mature hydrocarbons with biogenic methane cannot be excluded at this point. More data and especially a complete set of isotopic measurements will be necessary to test the mixing hypothesis.

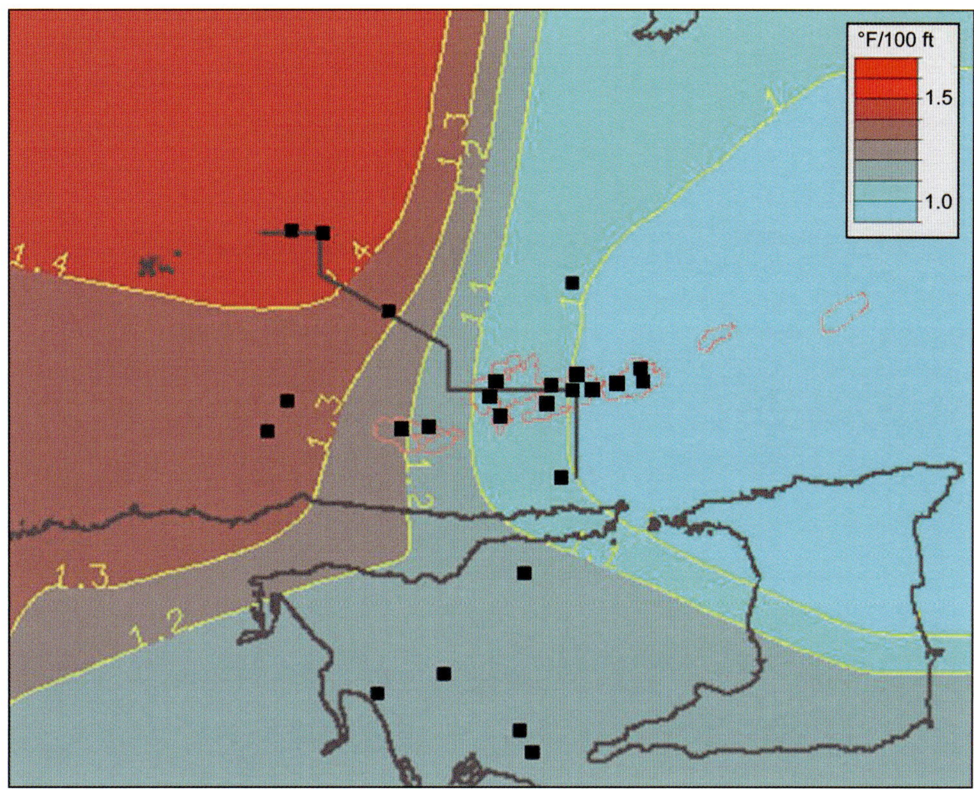

Figure 5. Present-day thermal gradient map of the studied area. Squares are control well locations.

Table 1. Average molar gas composition for the four known accumulations in the study area and methane $\delta^{13}C$ for the Patao and the Mejillones accumulations.

	RC (23)*		*MJ (9)*		*PA (36)*		*DR (17)*	
	E	*σ*	*E*	*σ*	*E*	*σ*	*E*	*σ*
C_1	83.36	2.78	94.96	1.89	98.26	0.52	99.29	0.22
C_2	8.23	1.12	0.39	0.12	0.10	0.02	0.09	0.02
C_3	4.23	0.94	0.29	0.10	0.06	0.02	0.06	0.03
iC_4	0.60	0.15	0.08	0.04	0.02	0.02		
C_4	1.18	0.32	0.09	0.04	0.00	0.00		
iC_5	0.29	0.09	0.04	0.02	0.00	0.00		
C_5	0.25	0.09	0.03	0.02	0.00	0.00		
C_5+	0.41	0.23						
N_2	0.52	0.20	3.86	1.96	1.34	0.57	0.40	0.16
CO_2	0.93	0.17	0.17	0.04	0.22	0.14	0.18	0.08
H_2S					0.01	0.02		
	100.01		99.91		100.00		100.02	
C_2+	15.19		0.92		0.18		0.15	
C_2/C_3	1.94		1.37		1.67		1.64	
C_2/iC_4	13.62		4.83		5.52			
$\delta 13C_1$ (‰)			−62.1		−62.4			

**RC = Rio Caribe; MJ = Mejillones; PA = Patao; DR = Dragon; E = average; σ = standard deviation.*

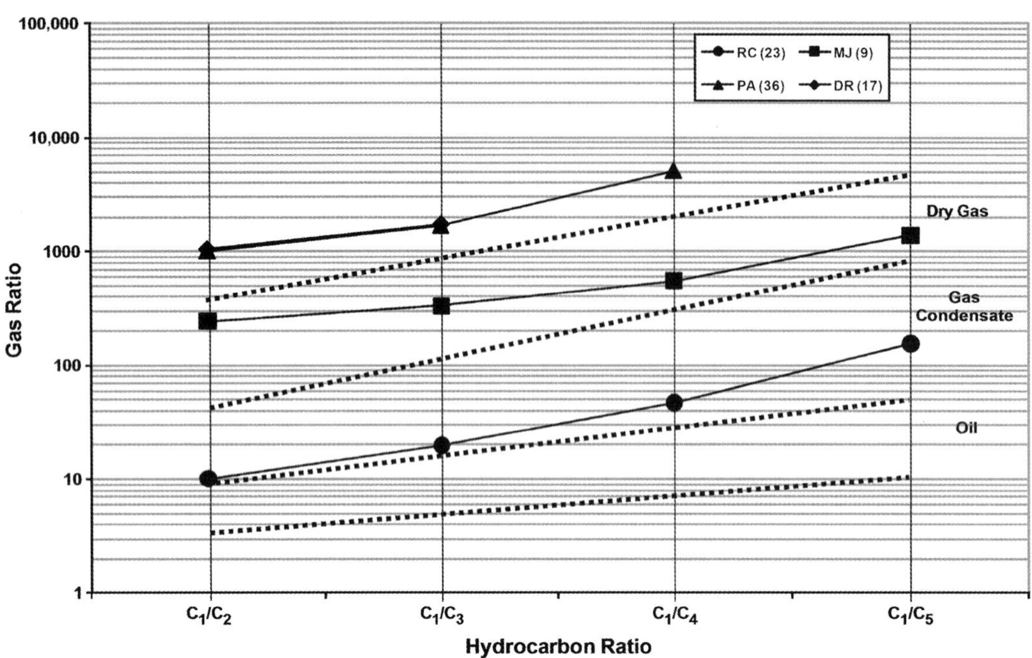

FIGURE 6. Pixler plot (Pixler, 1969) modified from Verbanac and Duma (1982) for different gases from the Patao High. RC = Rio Caribe; MJ = Mejillones; PA = Patao; DR = Dragon.

SOURCE ROCKS

Total Organic Carbon

To understand the origin of the gas, the first step was the definition of the source rock. For this purpose, the TOC_{SR} methodology (see Appendix) was used to define continuous TOC profiles from sonic and resistivity wireline logs (Figure 10). The method was applied to all of the wells that have sonic and resistivity logs. As a result, the whole column from the Eocene to the Pliocene could be considered as a source rock with average TOC content in the range of 0.5 to 1.2%, with a maximum averaged value of 2.5% for the upper part of the middle Miocene. However, considering that expulsion can be reached if the TOC content is greater than a critical value (Lewan, 1987), only the middle Miocene can be considered as a potential source rock.

Type

Kerogen type can be determined from hydrogen and oxygen index values determined from pyrolysis S2

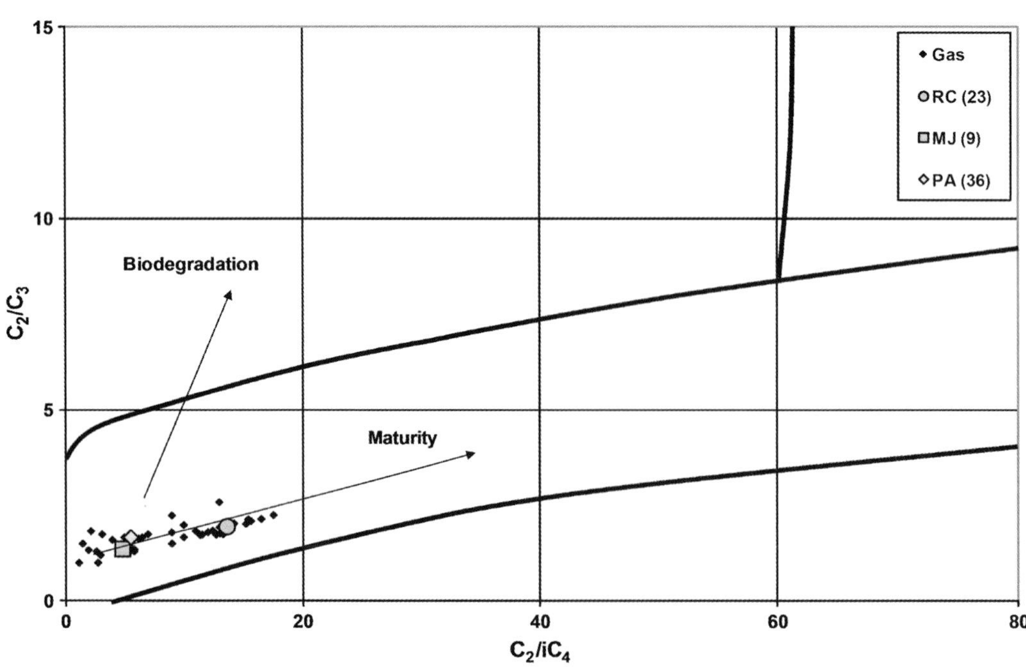

FIGURE 7. Ethane/isobutane versus ethane/propane diagram (Prinzhofer et al., 2000; Prinzhofer and Battani, 2003) showing that the gases are of low maturity and seem to belong to the same maturity trend. RC = Rio Caribe; MJ = Mejillones; PA = Patao.

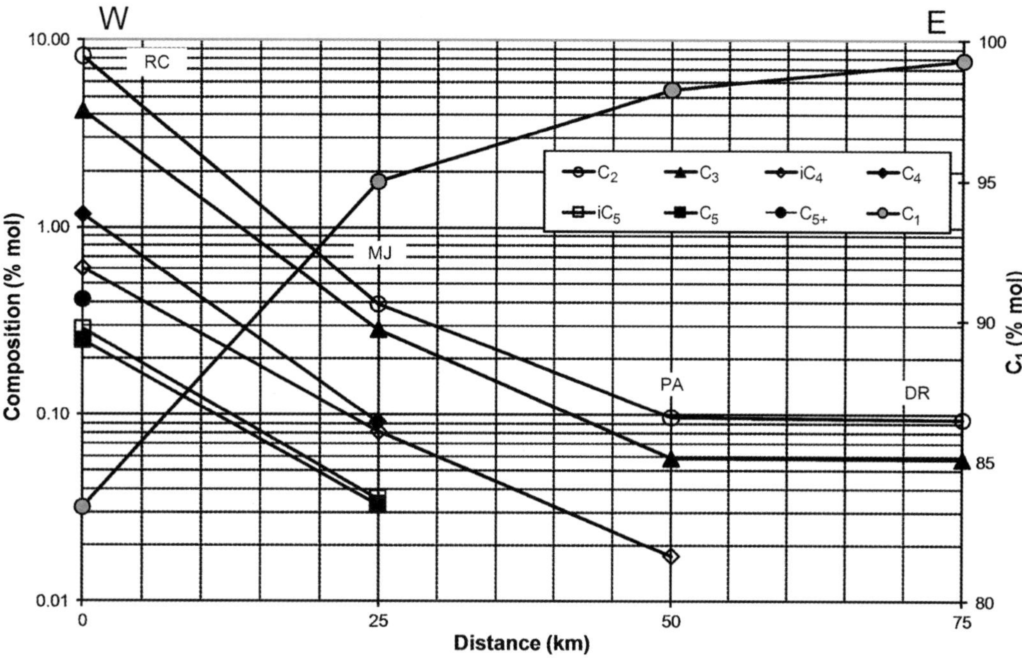

FIGURE 8. Chemical characterization of the gas showing that the westward decrease of the dryness correlated with the increase of the C_{2+} contents. The x origin has been arbitrarily fixed at Rio Caribe (RC) field. MJ = Mejillones; PA = Patao; DR = Dragon.

and S3 peaks and organic carbon data. When plotted on a van Krevelen–type diagram, most of the middle Miocene samples fall on or below the type III kerogen line (Table 2). The same result can be seen in the hydrogen index (HI) versus T_{max} diagram, where the samples fall in the type III–type IV domain (Figure 11). The samples from the upper part of the middle Miocene have higher HI than those coming from the lower part. Furthermore, they plot in the low-maturity (<0.6% vitrinite reflectance [R_o]) domain. In conclusion, the geochemical data show that the organic matter is mostly type III except in the middle Miocene, where part of the organic matter could be marine type II.

Maturity

In the Patao High, the measured maturity parameters (T_{max}, $\%R_o$) indicate that the top of the oil-generation zone was not reached in the section penetrated by the wells. None of the samples analyzed have pyrolysis T_{max} values more than 435°C or values of more than 0.6% R_o (Table 2).

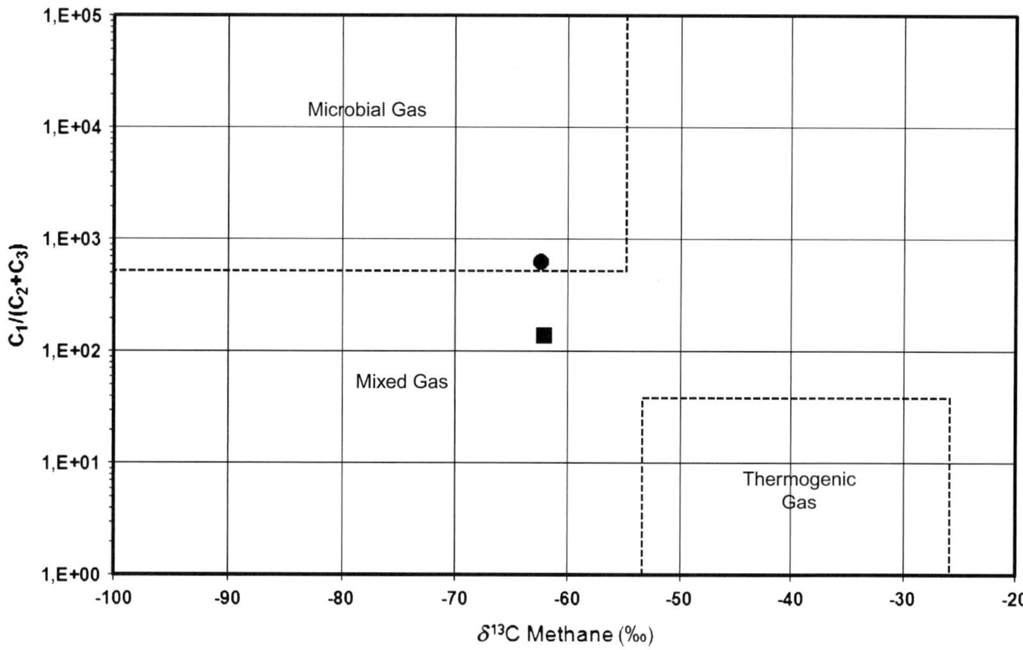

FIGURE 9. Bernard diagram (Bernard et al., 1978) adapted from Whiticar (1994) for the two gases with methane isotope analyses.

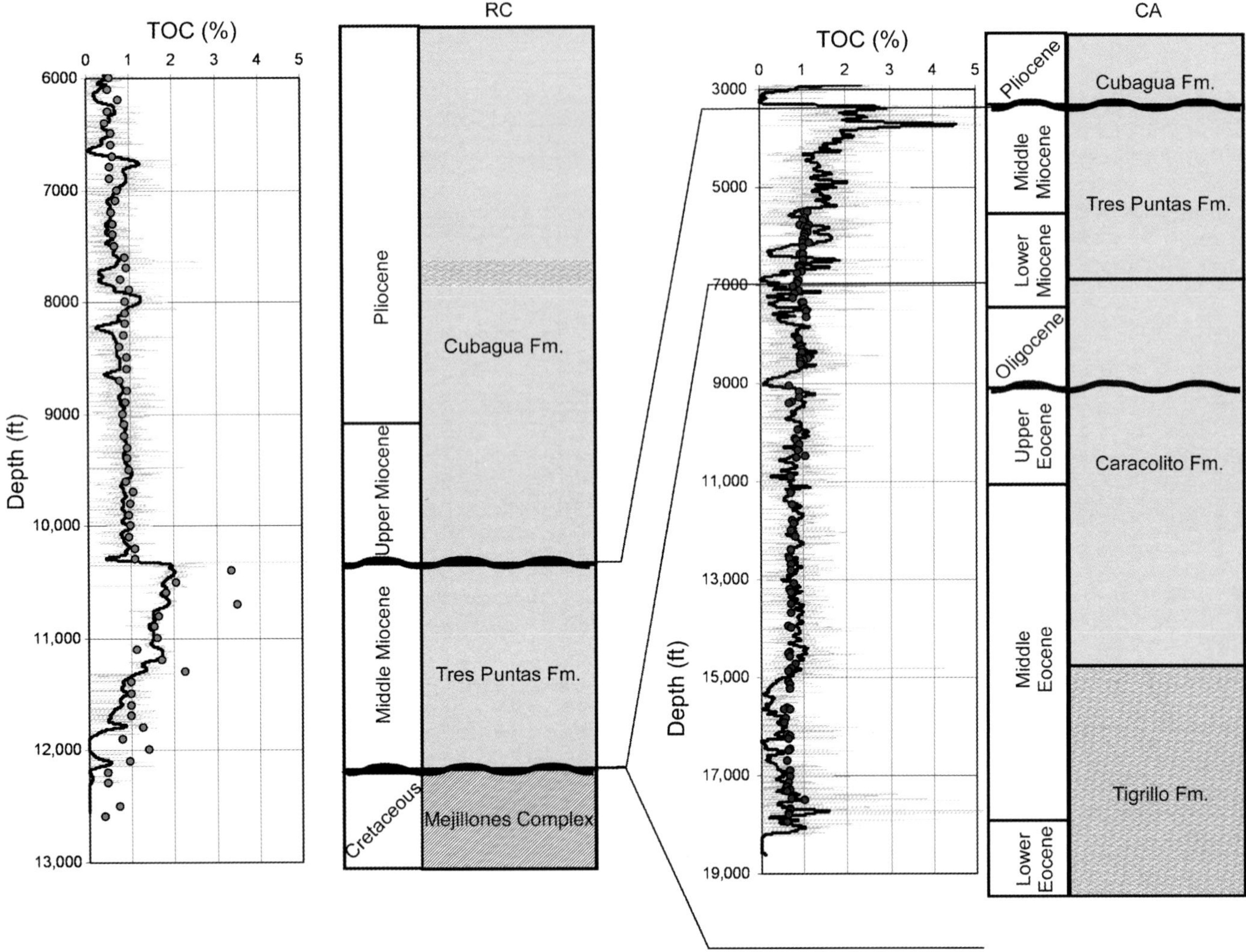

FIGURE 10. Continuous total organic matter (TOC) profiles calculated using the TOC_{SR} method. The calibration is performed on wells with TOC data, and then the method is applied on wells with no TOC data, but with sonic and resistivity logs. Dots are measurements, the gray curve is the TOC calculated each half foot, and the dark continuous curve is the TOC averaged for each 10-ft window. CA = Caracolito; RC = Rio Caribe; PA = Patao.

Table 2. Averaged Rock-Eval and vitrinite reflectance measurements for samples from the Tres Puntas Formation.

	TOC (%)		*Tmax °C*		*HI (mg/g TOC)*		*OI mg/g TOC)*		*% R_o*	
	E	*σ*	*E*	*σ*	*E*	*σ*	*E*	*σ*	*E*	*σ*
CA* (17)	0.99	0.11	428	3	93.9	27.9	35.6	12.6	0.40	0.01
RC (29)	1.34	0.68	428	3	160.6	76.6	32.3	7.6	0.46	0.03
DR (15)	0.91	0.19	429	3	84.3	44.9	131.7	126.3	0.39	0.03

**CA = Caracolito; RC = Rio Caribe; DR = Dragon; E = average; σ = standard deviation; TOC = total organic carbon; HI = Hydrogen Index; OI = Oxygen Index.*

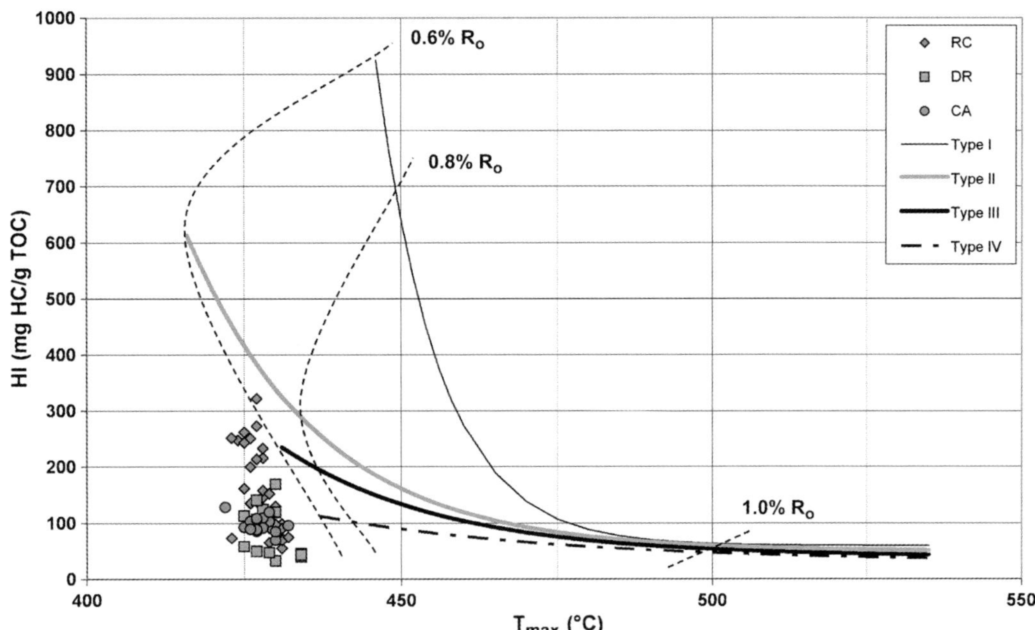

FIGURE 11. Hydrogen index (HI) versus T_{max} plot for the Tres Puntas Formation showing that the kerogen is mainly terrigeneous type III and immature. The evolution line for the different kerogen types and isomaturity lines were calculated using the default kinetic parameters from Temis Suite 2007. RC = Rio Caribe; DR = Dragon; CA = Caracolito; TOC = total organic carbon; R_o = vitrinite reflectance.

PETROLEUM GENERATION AND MIGRATION MODELING

Petroleum System

The gas in the four discovered fields is contained in the lower part of the Cubagua Formation (upper Miocene to lower Pliocene). As defined before, the source rock is supposed to be the upper part of the Tres Puntas Formation (middle Miocene). Nevertheless, we did not exclude the possibility that the Cretaceous might contain a type II source rock because of the lack of information.

First, the thermal calibration was performed. Then, a first set of simulations was conducted using default kinetic parameters for types III and II kerogens because no determination of the kinetic parameters for rock samples from the study area was possible.

Thermal Modeling

Twenty wells with temperature data were used to conduct the one-dimensional (1-D) thermal calibrations in Genex. The result of these 1-D calibrations is a present-day heat flow map.

The simulations show that the observed maturity level based on R_o data in five wells is well reproduced using a constant heat flow during geologic time. Thus, no need to vary the heat flow through time exists. Then, three-dimensional (3-D) simulations were performed with Temis using the heat flow map previously calculated. Ten key wells from the 20 used initially were extracted from the 3-D simulation to check that the thermal calibration was still valid.

The results of the simulation (Figure 12) show that the bottom of Paria subbasin has reached the oil window (>0.6% R_o), whereas the bottom of the Caracolito subbasin has reached the gas window (>2.0% R_o). Furthermore, the middle Miocene only reached the oil window in the southern part of the studied area (Figure 12).

Petroleum Generation

The simulations were done with default types III and II kerogens (Behar et al., 1997; Lorant and Behar, 2002) and were not able to reproduce the observed fields and their hydrocarbon compositions (Tables 3, 4).

The results show that the middle Miocene source rocks do not contribute to much accumulated hydrocarbons because these source rocks are not thermally mature enough to have expelled hydrocarbons. Furthermore, the default type III kerogen generates heavy products during the beginning of its maturity process (Table 4).

The Cretaceous source rocks with type II organic matter, however, may only contribute some liquid hydrocarbons in the eastern part of the basin (e.g., Patao and Dragon fields). These results indicate a maturity increase from west to east, which is in complete disagreement with the observations. We conclude that the Cretaceous could not be the source rock because of this maturity pattern.

DISCUSSION

At this point, we have shown that only the middle Miocene can be considered as efficient source rocks (e.g.,

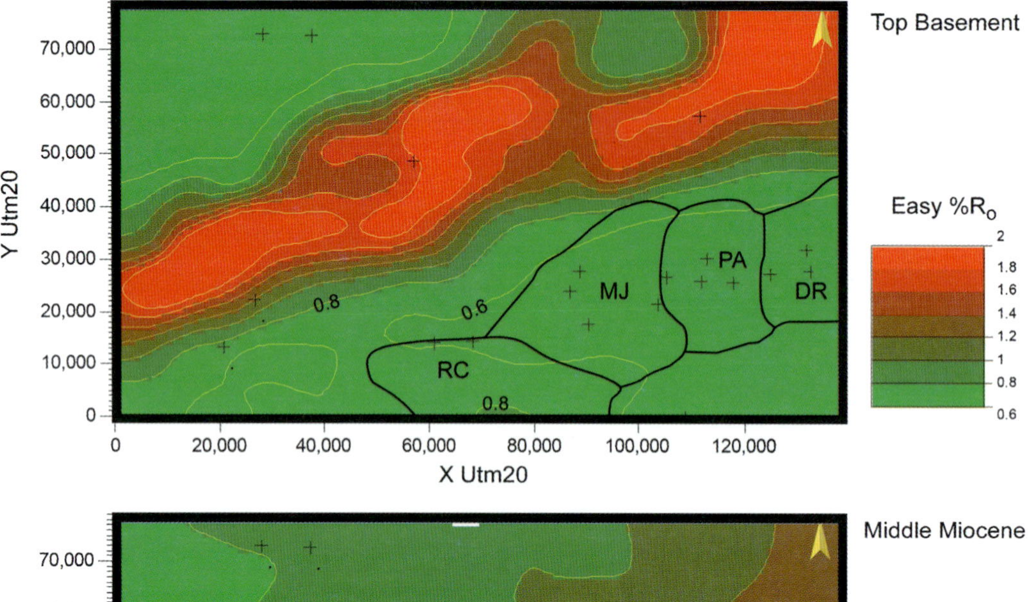

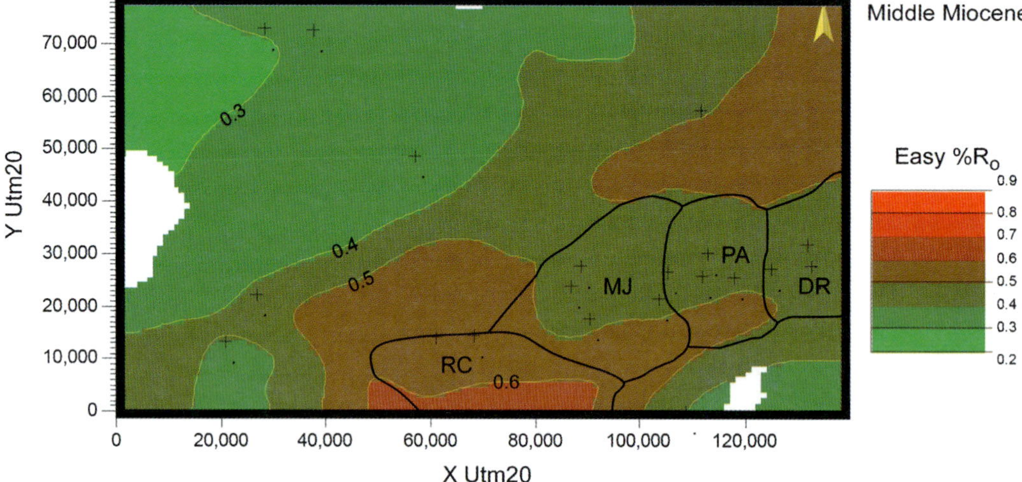

FIGURE 12. Maturity map (Easy %R_o; Sweeney and Burnham, 1990) calculated at the top of the basement and for the middle Miocene. The black contours represent the limits of the drainage area for each of the fields.

TOC >1%). These source rocks never reached the condensate and gas window in the area that includes the drainage areas of the known reservoirs.

Even if we consider older source rocks, such as the lower Eocene or the Cretaceous, which have not been drilled completely, the only zone where these rocks could have reached the condensate or gas window is the deeper part of the Caracolito subbasin. Nevertheless, what has been produced in the Caracolito subbasin never reached the drainage areas of the studied accumulations.

In this article, we did not consider diffusion fractionation that might generate isotopically light dry gas from normal thermal gas (Pernaton et al., 1996; Prinzhofer and Pernaton, 1997). In the same way, we did not consider the model developed by Vandre et al (2007) for the Nile Delta. In that model, very dry methane fields, such as the Dragon field, of possible microbial origin may coexist with condensate, such as Rio Caribe field, of thermogenic origin. These mixed fields can be explained by microbial alteration of preexisting thermogenic hydrocarbons in shallow reservoirs (<80°C). As a confirmation,

Table 3. Results of the three-dimensional simulation with default types II and III kerogens.

	Cretaceous (MM m^3)	*Miocene (MM m^3)*	*C1*	*C2*	*C3–C5 (% mol)*	*C5+*
RC*	0.00	0.0				
MJ	0.01	0.0	21.6	0.0	23.9	54.5
PA	3.58	0.0	22.3	7.7	20.9	49.1
DR	3.29	0.0	20.7	7.4	21.8	50.1

**RC = Rio Caribe; MJ = Mejillones; PA = Patao; DR = Dragon.*

Table 4. Initial and modified kinetic parameters used in this study for the type III kerogen.

			C14+			*C6–C13*				
E	***HI****	***NSO****	***Aromatics***	***cSat***	***nSat***	***Aromatics***	***Saturates***	***C3–C5***	***C2***	***C1***
Kcal/mol	***mg/g C***	***%***	***%***	***%***	***%***	***%***	***%***	***%***	***%***	***%***
Type III (Temis Suite 2008)								HI = 235.0 mg/g C		
								A = 3.1E + 15 1/s		
40										
42										
44										
46										
48										
50										
52										
54	1.2	49.3	12.7	1.4	3.3	16.7	16.7	0.0		
56	26.6	69.1	17.7	2.0	4.6	2.3	2.3	0.8	0.8	0.4
58	81.1	66.5	17.1	1.9	4.4	2.6	2.6	1.9	1.9	1.2
60	47.4	48.1	12.4	1.4	3.2	7.8	7.8	6.6	6.6	6.1
62	21.5	24.6	6.3	0.7	1.6	8.9	8.9	15.2	15.2	18.7
64	25.2	4.7	1.2	0.1	0.3	0.6	0.6	9.4	9.4	73.7
66	15.9							7.9	7.9	84.2
68	5.4							11.1	11.1	77.8
70	5.6							13.4	13.4	73.2
72	4.1							2.4	2.4	95.1
74	1.0									100.0
Modified type III (this work)								HI = 271.5 mg/g C		
								A = 3.1E + 15 1/s		
40	0.0								0.0	100.0
42	2.0								0.0	100.0
44	2.9							0.0	0.0	100.0
46	3.2							0.0	0.0	100.0
48	3.0							0.0	0.0	100.0
50	3.8	4.0	2.0	0.5	1.5	1.0	1.0	30.0	20.0	40.0
52	10.0	8.0	3.0	1.0	2.0	1.0	1.0	48.0	36.0	0.0
54	12.8	40.0	10.0	1.5	4.0	2.5	2.5	23.0	16.5	0.0
56	26.6	69.1	17.7	2.0	4.6	2.3	2.3	0.8	0.8	0.4
58	81.1	66.5	17.1	1.9	4.4	2.6	2.6	1.9	1.9	1.2
60	47.4	48.1	12.4	1.4	3.2	7.8	7.8	6.6	6.6	6.1
62	21.5	24.6	6.3	0.7	1.6	8.9	8.9	15.2	15.2	18.7
64	25.2	4.7	1.2	0.1	0.3	0.6	0.6	9.4	9.4	73.7
66	15.9							7.9	7.9	84.2
68	5.4							11.1	11.1	77.8
70	5.6							13.4	13.4	73.2
72	4.1							2.4	2.4	95.1
74	1.0									100.0

**HI = Hydrogen Index.*
***NSO = Resins and Asphaltenes.*

the C_2-C_3 versus C_2-iC_4 diagram (Figure 7) did not show any evidence of bacterial alteration.

As a consequence, the simplest way to explain the accumulations is to assume that the hydrocarbons were generated from the middle Miocene source rocks within the drainage areas, which are located in a low-maturity domain where R_o is less than 0.6% (e.g., the diagenetic zone).

The following question arises: how could that be possible?

The first point is that the kinetic parameters used were calibrated with kerogen samples with approximately 0.6% R_o (Behar et al., 1992, 1997). It appears that these parameters are not suitable for reproducing what occurred in the diagenetic zone (<0.6% R_o).

Dieckmann (2005) introduced an extended kinetic model in which the frequency factor is variable for each of the activation energies. This model predicts hydrocarbon generation at a wider range of temperatures and, in particular, it predicts generation at lower temperatures. Nevertheless, if it has been calibrated with a kerogen having 0.6% R_o, it cannot predict what has been generated by the kerogen for less than 0.6% R_o.

To account for the generation of significant volumes of hydrocarbons from Tertiary sediments, hydrocarbon generation at lower than standard temperatures has been considered. For example, Snowdon (1980) proposed that the occurrence of resinitic material in potential Tertiary source rocks would explain generation at low maturities. According to Clayton (1992), who studied the evolution of the organic matter since deposition, 20 to 30% of the deposited carbon is lost during diagenesis through biogenic, decarboxylation, or early thermogenic reactions. Roughly one half of this carbon may produce mainly methane.

Modified Kinetic Parameters

As mentioned before, kinetic parameters are determined in the laboratory using samples that are at the limit between diagenesis and catagenesis (~0.6% R_o). Let us consider a synthetic kerogen that is characterized as having a constant partial HI of 1 mg HC/g TOC for each of the activation energies from 34 to 74 kcal/mol. We performed a numerical maturity experiment on this kerogen to place it at a maturity level equivalent to 0.6% R_o. The results (Figure 13) show that all of the partial potentials have been totally consumed for activation energies less than 50 kcal/mol. For activation energies between 50 and 54 kcal/mol, the partial potentials were partially consumed. As a matter of fact, for the initial type III kerogen, the lowest activation energy that has a positive partial potential is 54 kcal/mol (Figure 13; Table 4).

If we consider the initial type III kerogen as a starting point and if we use observed natural data such as gas compositions, a new set of kinetic parameters can be derived to account for low maturities in the gas fields. These new kinetic parameters were determined by trial-and-error 3-D simulations. The objective was to reproduce the gas compositions in the four structures. The derived type III differs from the previous one by a 13% increase of the HI from 235 to 271 mg HC/g TOC (Figure 13; Table 4).

The simulations conducted with new fitted type III parameters enabled us to reproduce quite well the filling of the fields, as well as the general composition and state of the hydrocarbons (Figure 14).

CONCLUSIONS

The current view is that most of the large dry gas accumulations in shallow reservoirs are microbial in origin and are generated from the methanogenic pathway. However, isotopically light (more negative) methane is generally interpreted as biogenic gas, although this is also a common feature of all low-temperature gas formation processes. Recent studies have shown that isotopically light dry gas is not necessarily of microbial origin. Early thermal processes can create large amounts of dry gas that could be depleted in ^{13}C (Cramer et al. 1999; Rowe and Muehlenbachs, 1999).

In our simulations, most of the gas is generated at a very low maturity: 0.4 less than $\%R_o$ less than 0.6. In this domain, it has been stated that the thermal C-C–bound breakage is unlikely to be significant without catalysis. Because microbial enzymes are the most efficient low-temperature catalysts known, we could describe our kinetic model as a microbially enhanced thermal model.

The main conclusion of this study is that more effort should be devoted to the determination of kinetic parameters for source rocks to account for these phenomena, which occur in the diagenetic zone (<0.6% R_o). This chapter shows that modeling petroleum systems with kinetics parameters determined on samples at the end of the diagenetic zone (~0.6% R_o) can lead to an underestimation of the amount of hydrocarbons produced.

Determination of the compositional kinetic parameters for kerogen from the middle Miocene would help to improve the calibration. Indeed, even if the kerogen is obviously too mature to perform the calibration of the lowest activation energies, the results could be used as a better starting point than the default type III used in this study.

A possible future study might attempt to develop a calibration methodology using basin modeling to tune kinetic parameters.

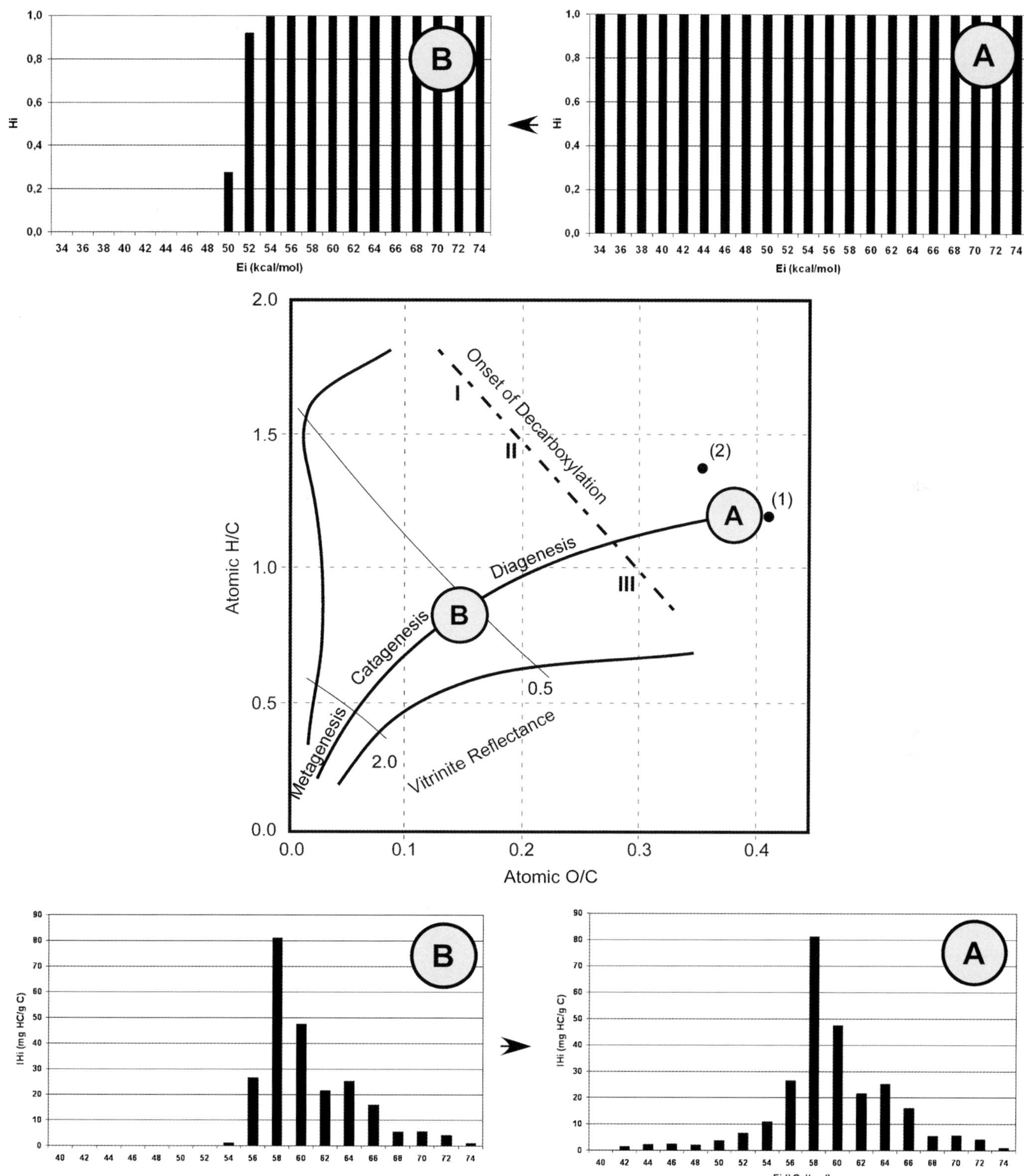

FIGURE 13. Evolution path of the organic matter since deposition showing that part of the hydrocarbons are produced within the diagenetic zone (from Clayton, 1992). This figure illustrates the methodology to determine new kinetic parameters and thus account for what occurred in the diagenetic zone that was not considered when these kinetic parameters were determined in the laboratory: (1) Fresh peat type III, (2) west Africa and Peru margin type II, (A) organic matter during deposition, (B) resulting kerogen at 0.6% R_o.

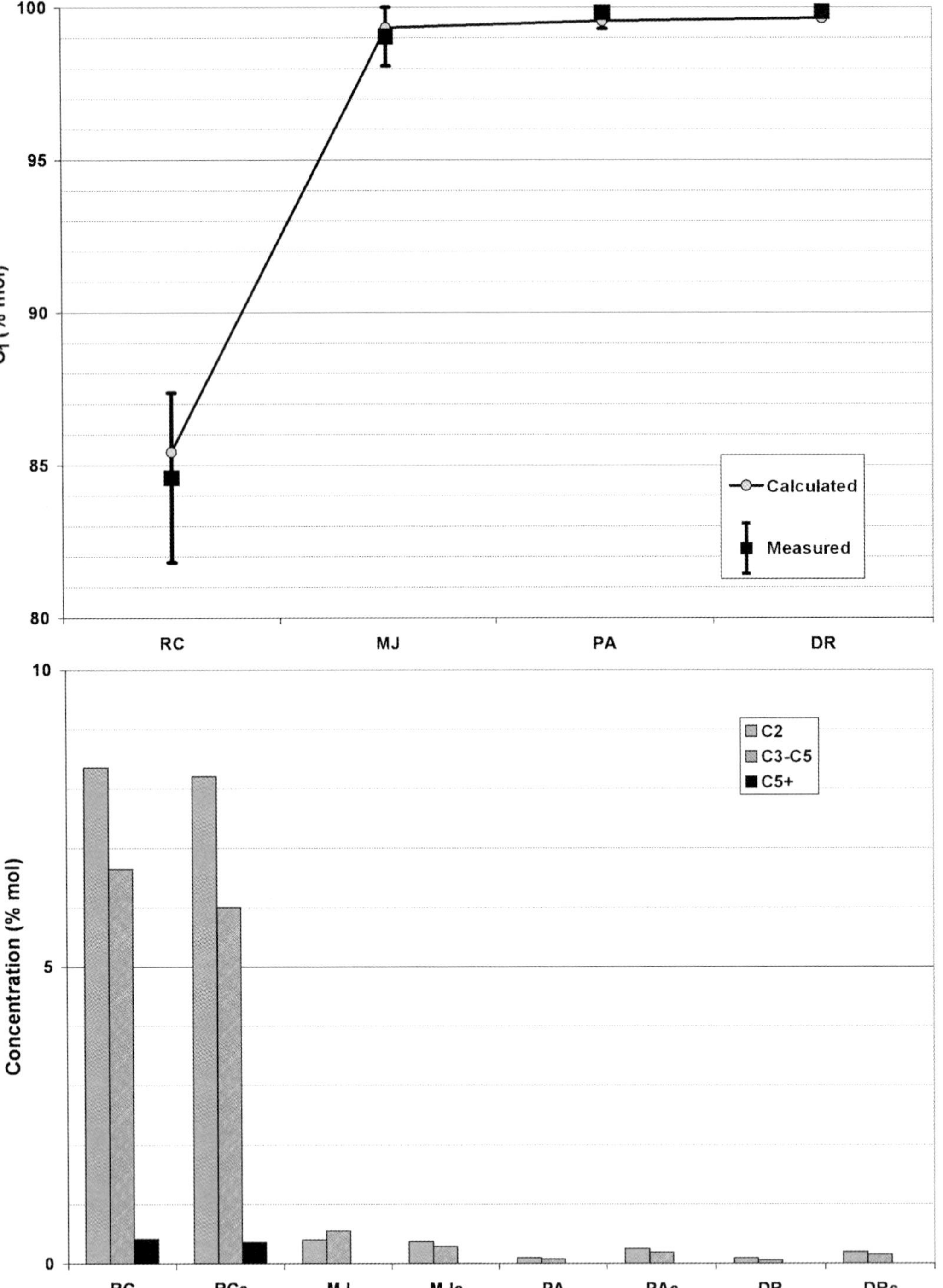

FIGURE 14. Comparison between the measured and the calculated gas compositions of the four fields. RC, MJ, PA, and DR are, respectively, the measurements from Rio Caribe, Mejillones, Patao, and Dragon; RCc, MJc, Pac, and DRc are the corresponding calculated compositions.

ACKNOWLEDGMENTS

We thank PDVSA for permission to publish the results; Michael Erdmann and an anonymous reviewer for making many constructive comments on a previous version of the manuscript; and Kenneth Peters for improving the quality of the manuscript.

APPENDIX: TOC_{SR} METHOD

Estimation of the organic content of potential source rocks is an important factor in the determination of the petroleum potential of sedimentary basins. In the Carbolog method (Carpentier et al., 1991) the organic content of petroleum source rocks is determined using the physical properties of the organic matter, which are characterized by a high sonic transit time (Δt) and by a high resistivity (R).

The Carbolog method uses a $\left(\frac{1}{\sqrt{R}}, \Delta t\right)$ diagram to derive, after calibration, the TOC from transit time and resistivity. The method assumes that (1) the transit time of the sediment is given as a linear combination of the individual transit times (generalization of the Wyllie formula), and (2) the resulting resistivity is given by the Archie formula with a cementation factor of 2.

Under these assumptions, a porous medium without organic matter is located in this diagram on a matrix-water line. Furthermore, the matrix-water line is supposed to be superimposed on the matrix-clay line. The main advantage of this method is its simplicity.

This method is designed for quite homogeneous matrix compositions having slight variations in the clay content and pore water salinity. This method seems to work properly for well-compacted sediments. However, it does not work correctly in young sediments, such as Miocene and Pliocene deposits from offshore Venezuela. This is mainly because of the fact that when porosity is high enough, the sonic transit time of the sediment cannot be expressed as a linear function of the sonic transit time of each of the components. When porosity is high enough, the Wyllie linear equation (Wyllie et al., 1956) is no longer valid and should be replaced, for instance, by the Raymer, Hunt, and Gartner (RHG) formula (Raymer et al., 1980).

The general concept of the modified method, here called TOC_{SR}, differs from the Carbolog method because it uses (1) Vsh to select only the shaly part of the sedimentary column, (2) the nonlinear RHG sonic-porosity relationship, and (3) a variable cementation factor. These differences from the initial Carbolog method increase the accuracy of TOC determination and allow the use of the tool in many more environments.

REFERENCES CITED

Algar, S. T., and J. L. Pindell, 1993, Structure and deformation history of the Northern Range of Trinidad and adjacent areas: Tectonics, v. 12, p. 814–829, doi:10.1029/93TC00673.

Avé Lallemant, H., 1997, Transpression, displacement partitioning, and exhumation in the eastern Caribbean/South American plate boundary zone: Tectonics, v. 16, p. 272–289, doi:10.1029/96TC03725.

Behar, F., S. Kressmann, J. L. Rudkiewicz, and M. Vandenbroucke, 1992, Experimental simulation in a confined system and kinematic modeling of kerogen and oil cracking: Organic Geochemistry, v. 19, p. 173–189, doi:10.1016/0146-6380(92)90035-V.

Behar, F., M. Vandenbroucke, Y. Tang, F. Marquis, and J. Espitalié, 1997, Experimental cracking of kerogen in open and closed systems: determination of kinetic parameters and stoichiometric coefficients for oil and gas generation: Organic Geochemistry, v. 26, p. 321–339, doi:10.1016/S0146-6380(97)00014-4.

Bernard, B. B., J. M. Brooks, and W. M. Sackett, 1978, Light hydrocarbons in recent Texas continental shelf and slope sediments: Journal of Geophysical Research, v. 83, p. 289–291, doi:10.1029/JB083iB01p00289.

Burke, K., C. Copper, J. F. Dewey, P. Mann, and J. L. Pindell, 1984, Caribbean tectonics and relative plate motions, *in* W. E. Bonini, R. B. Hargraves, and R. Shagam, eds., The Caribbean-South America plate boundary and regional tectonics: Geological Society of America Memoir 162, p. 31–63.

Carpentier, B., A. Huc, and G. Bessereau, 1991, Wireline logging and source rocks. Estimation of organic carbon by the CARBOLOG method, The Log Analyst, May–June, 1991, p. 279–297.

Castro, M., and A. Mederos, 1985, Litoestratigrafía de la Cuenca de Carupano, VI Congreso de Geología de Venezuela, Memoir 1, p. 201–225.

Chalot, J. P., R. Gómez, and F. Gonzalez, 1992, Exploration/development strategy for Rio Caribe field, SPE 23652, SPE Latin America Petroleum Engineering Conference, March 8–11, 1992, Caracas, Venezuela, p. 91–102.

Clayton, C., 1992, Source volumetrics of biogenic gas generation, *in* R. Vially, ed., Bacterial Gas: Paris, Editions Technip, p. 191–204.

Coleman, D. D., 1979, The origin of drift-gas deposits as determined by radiocarbon dating of methane, *in* R. Berger and H. E. Seuss, eds., Radiocarbon dating: Proceedings of the ninth international radiocarbon rating conference, 1976: Berkeley, University of California Press, p. 365–387.

Cramer, B., H. S. Poelchau, P. Gerling, N. V. Lopatin, and R. Littke, 1999, Methane release from groundwater—The source of natural gas accumulations in northern West Siberia: Marine and Petroleum Geology, v. 16, p. 225–244, doi:10.1016/S0264-8172(98)00085-3.

Dewey, J. F., and J. L. Pindell, 1985, Neogene block tectonics of eastern Turkey and northern South America: Continental applications of the finite difference method: Tectonics, v. 4, p. 71–83, doi:10.1029/TC004i001p00071.

Dewey, J. F., and J. L. Pindell, 1986, Neogene block tectonics of eastern Turkey and northern South America: Continental applications of the finite difference method, Reply: Tectonics, v. 5, p. 703–705, doi:10.1029/TC005i004p00703.

Di Croce, J., 1995, Eastern Venezuela Basin: Sequence stratigraphy and structural evolution: Ph.D. thesis, Rice University, Houston, Texas, 225 p.

Dieckmann, V., 2005, Modeling petroleum formation from heterogeneous source rocks: The influence of frequency factors on activation energy distribution and geological prediction: Marine and Petroleum Geology, v. 22, p. 375–390, doi:10.1016/j.marpetgeo.2004.11.002.

Eva, A., K. Burke, P. Mann, and G. Wadge, 1989, Four-phase tectonostratigraphic development of the southern Caribbean: Marine and Petroleum Geology, v. 6, p. 9–21, doi:10.1016/0264-8172(89)90072-X.

Franke, M., R. Russo, and W. Ambeh, 1996, Structure of the crust and upper mantle in northeastern South America: EOS: Transactions of the American Geophysical Union, v. 77, p. 182–183, doi:10.1029/96EO00124.

Gonzalez de Juana, C., J. M. Iturralde de Arozena, and X. Picard Cadillat, 1980, Geologia de Venezuela y de sus cuencas petroliferas, Tomos I y II, 1030 p. (Ediciones FONINVES, Caracas).

Haq, B., J. Hardenbol, and P. Vail, 1987, Chronology of fluctuating sea levels since the Triassic: Science, v. 235, no. 4793, p. 1156–1167, doi:10.1126/science.235.4793.1156.

James, K., 2000, The Venezuelan hydrocarbon habitat: Part 2, Hydrocarbon occurrences and generated-accumulated volumes: Journal of Petroleum Geology, v. 23, p. 133–164, doi:10.1111/j.1747-5457.2000.tb00488.x.

James, K. H., 1990, The Venezuelan hydrocarbon habitat, *in* J. Brooks, ed., Classic petroleum provinces: Geological Society (London) Special Publication 50, p. 9–35.

Janezic, G. G., 1979, Biogenic light hydrocarbon production related to near surface geochemical prospecting for petroleum (by title only): AAPG Bulletin, v. 63, p. 403.

Lewan, M. D., 1987, Petrographic study of primary petroleum migration in the Woodford Shale and related rock units, *in* B. Doligez, ed., Migration of hydrocarbons in sedimentary basins: Paris, Editions Technip, p. 113–130.

Liendo, J., 2008, Gas para todo el mundo, Dinero 237: Venezuela, Petróleo con reservas.

Lorant, F., and F. Behar, 2002, Late generation of methane from mature kerogens: Energy and Fuels, v. 16, p. 412–427, doi:10.1021/ef010126x.

Lugo, J., and P. Mann, 1995, Jurassic–Eocene tectonic evolution of Maracaibo Basin, Venezuela, *in* A. Tankard, S. Suarez, and H. Welsink, eds., Petroleum basins of South America: AAPG Memoir 62, p. 699–725.

Mann, P., C. Schubert C., and K. Burke, 1990, Review of Caribbean neotectonics, *in* G. Dengo and J. Case, eds., The Caribbean regions: Geological Society of America, the Geology of North America, v. H, p. 307–338.

Maresch, W. V., 1974, The plate tectonic origin of the Caribbean Mountain System of northern South America: Discussion and proposal: Geological Society of America Bulletin, v. 85, p. 669–682, doi:10.1130/0016-7606(1974)85<669:PTOOTC>2.0.CO;2.

Molnar, P., and L. Sykes, 1969, Tectonics of the Caribbean and Middle America regions from focal mechanisms and seismicity: Geological Society of America Bulletin, v. 80, p. 1639–1684, doi:10.1130/0016-7606(1969)80[1639:TOTCAM]2.0.CO;2.

Parnaud, F., Y. Gou, J.-C. Pascual, I. Truskowski, O. Gallago, and H. Passalaqua, 1995, Petroleum geology of the central part of the eastern Venezuela basins, *in* A. Tankard, S. Suarez, and H. Welsink, eds., Petroleum basins of South America: AAPG Memoir 62, p. 741–756.

Pérez, O., R. Bilham, R. Bendick, N. Hernández, M. Hoyer, J. Velandia, C. Moncayo, and M. Kozuch, 2001, Velocidad relativa entre las placas del Caribe y Sudamérica a partir de observaciones dentro del sistema de posicionamiento global (GPS) en El norte de Venezuela: Interciencia, v. 26, no. 2, p. 69–74.

Pernaton, E., A. Prinzhofer, and F. Schneider, 1996, Reconsideration of methane signature as a criterion for the genesis of natural gas: Influence of migration on isotopic signature: Revue de l'Institut Français du Pétrole, v. 51, no. 5, p. 635–651.

Pindell, J., and S. Barret, 1990, Geological evolution of the Caribbean region: A plate tectonics perspective, *in* G. Dengo and J. Case, eds., The Caribbean region: Geological Society of America: The Geology of North America, v. H, p. 405–432.

Pindell, J. L., 1993, Regional synopsis of Gulf of Mexico and Caribbean evolution, *in* J. L. Pindell and R. F. Perkins, eds., Mesozoic and Early Cenozoic development of the Gulf of Mexico and Caribbean region: A context for hydrocarbon exploration: Gulf Coast Section. SEPM Foundation, 13th Annual Research Conference, p. 251–274.

Pindell, J. L., J. P. Erikson, and S. T. Algar, 1991, The relationship between plate motions and sedimentary basin development in Northern South America: From a Mesozoic passive margin to a Cenezoic eastwardly progressive transpressional orogen, *in* K. A. Gillezau, ed., Transactions of the Second Geological Conference of the Geological Society of Trinidad and Tobago, Port-of-Spain, 1990, p. 191–202.

Pindell, J. L., R. Higgs, and J. F. Dewey, 1998, Cenozoic palinspatic reconstruction, paleogeographic evolution, and hydrocarbon setting of the northern margin of South America, *in* J. L. Pindell and C. Drake, eds., Paleogeographic evolution and nonglacial eustasy, northern South America: SEPM Special Publication 58, p. 45–86.

Pixler, B. O., 1969, Formation evaluation by analysis of hydrocarbon ratios: Journal of Petroleum Technology, v. 21, p. 665–670.

Prinzhofer, A., and A. Battani, 2003, Gas isotopes tracing: An important tool for hydrocarbons exploration: Oil and Gas Science Technology–Revue de l'Institut Français du Pétrole, v. 58, no. 2, p. 299–311.

Prinzhofer, A., M. R. Mello, and T. Takaki, 2000, Geochemical characterization of natural gas: A physical multivariable approach and its applications in maturity and migration estimate: AAPG Bulletin, v. 84, p. 1152–1172.

Prinzhofer, A., and E. Pernaton, 1997, Isotopically light methane in natural gas: Bacterial imprint or diffusive fractionation?: Chemical Geology, v. 142, p. 193–200, doi:10.1016/S0009-2541(97)00082-X.

Raymer, L. L., E. R. Hunt, and J. S. Gardner, 1980, An improved sonic transit time-to-porosity transform: Society of Petrophysicists and Well Log Analysts 21st Annual Logging Symposium, p. 1–12.

Roure, F., J. O. Carnavalli, Y. Gou, and T. Subieta, 1994, Geometry and kinematics of the north Monagas thrust belt, Venezuela: Marine and Petroleum Geology, v. 11, p. 347–359, doi:10.1016/0264-8172(94)90054-X.

Rowe, D., and A. Muehlenbachs, 1999, Low-temperature thermal generation of hydrocarbon gases in shallow shales: Nature, v. 398, p. 61–63, doi:10.1038/18007.

Russo, R. M., and R. C. Speed, 1992, Oblique collision and tectonic wedging of the South American continent and Carribean terrane: Geology, v. 20, p. 447–450, doi:10.1130/0091-7613(1992)020<0447:OCATWO>2.3.CO;2.

Sageman, B. B., and R. C. Speed, 2003, Upper Eocene limestones, associated sequence boundary, and proposed Eocene tectonics in eastern Venezuela, *in* C. Bartolini, R. T. Buffer, and J. Blickwede, eds., Circum Gulf of Mexico and the Caribbean, AAPG Memoir 79, p. 1–17.

Schoell, M., 1983, Genetic characterization of natural gases: AAPG Bulletin, v. 67, p. 2225–2238.

Snowdon, L. R., 1980, Resinite—A potential petroleum source in the upper Cretaceous/Tertiary of the Beaufort-Mackenzie Basin, *in* A. D. Miall, ed., Facts and principles of world petroleum occurrence: Canadian Society of Petroleum Geologists Memoir 6, p. 509–520.

Speed, R. C., 1985, Cenozoic collision of the Lesser Antilles arc and continental South America and the origin of the El Pilar fault: Tectonics, v. 4, p. 41–69, doi:10.1029/TC004i001p00041.

Speed, R. C., and P. Smith-Horowitz, 1998, The Tobago terrane: International Geology Review, v. 40, p. 805–830, doi:10.1080/00206819809465240.

Summa, L. L., E. D. Goodman, M. Richardson, I. O. Norton, and A. R. Green, 2003, Hydrocarbon systems of Northeastern Venezuela: Plate through molecular scale analysis of the genesis and evolution of the Eastern Venezuela Basin: Marine and Petroleum Geology, v. 20, p. 323–349, doi:10.1016/S0264-8172(03)00040-0.

Sweeney, J. J., and A. K. Burnham, 1990, Evaluation of a simple model of vitrinite reflectance based on chemical kinetics: AAPG Bulletin, v. 74, p. 1559–1570.

Sykes, L. R., W. R. McCann, and A. L. Kafka, 1982, Motion of Caribbean plate during last 7 million years and implications for earlier Cenozoic movement: Journal of Geophysical Research, v. 87, p. 10656–10676, doi:10.1029/JB087iB13p10656.

Vandre, C., B. Cramer, P. Gerling, and J. Winsemann, 2007, Natural gas formation in the western Nile Delta (Eastern Mediterranean): Thermogenic versus microbial mechanisms: Organic Geochemistry, v. 38, p. 523–539, doi:10.1016/j.orggeochem.2006.12.006.

Verbanac, B., and P. Duma, 1982, Hydrocarbon ratios in gas mixtures as indices of reservoir saturation and producing capacity: Geologijai I Geofizika, v. 33, p. 185–192.

Weber, J. C., T. H. Dixon, C. DeMets, W. B. Ambeh, G. Jansma, G. Mattioli, J. Saleh, G. Sella, R. Bilham, and O. Perez, 2001, GPS estimate of relative motion between the Carribean and South American plates, and geologic implications for Trinidad and Venezuela: Geology, v. 29, p. 75–78, doi:10.1130/0091-7613(2001)029<0075:GEORMB>2.0.CO;2.

Whiticar, M. J., 1994, Correlation of natural gases with their sources, *in* L. B. Magoon and W. G. Dow, eds., The petroleum system: From source to trap: AAPG Memoir 60, p. 261–283.

Wyllie, M. R. J., A. R. Gregory, and L. W. Gardner, 1956, Elastic wave velocities in heterogeneous and porous media: Geophysics, v. 21, p. 41–70, doi:10.1190/1.1438217.

Ysaccis, B. R., 1998, Tertiary evolution of the northeastern Venezuela offshore: Ph.D. thesis, Rice University, Houston, Texas, 285 p.

SECTION 2

Seismic/Pressure/Gravity Magnetics Revisited

4

Brink, H.-J., D. Gajewski, M. Baykulov, and M.-K. Yoon, 2012, Moho, Basin dynamics, salt stock family development, and hydrocarbon system examples of the North German Basin revisited by applying seismic common reflection surface processing, *in* K. E. Peters, D. J. Curry, and M. Kacewicz, eds., Basin Modeling: New Horizons in Research and Applications: AAPG Hedberg Series, no. 4, p. 71–86.

Moho, Basin Dynamics, Salt Stock Family Development, and Hydrocarbon System Examples of the North German Basin Revisited by Applying Seismic Common Reflection Surface Processing

Heinz-Juergen Brink, Dirk Gajewski, Mikhail Baykulov, and Mi-Kyung Yoon

University of Hamburg, Institute of Geophysics, Wave Inversion Technology, Germany

ABSTRACT

Proper two-dimensional and three-dimensional basin modeling relies on accurate seismic processing and interpretation, correct depth conversion of the identified sedimentary layers, reliable modeling of the thermal history of the basin, and understanding of the regional geodynamic setting. Seismic reprocessing using the common reflection surface (CRS) stack technique allows revised interpretation of the structural setting and the evolution of salt plugs in the area of the Glueckstadt Graben, located near the center of the North German Basin (NGB). Reprocessing of seismic data also provides an alternative view of the geodynamic origin of the basin. Reprocessing of data clearly demonstrates the capabilities of the CRS technique to improve the quality of low-fold data. The images display a considerably improved signal-to-noise ratio and much more detail than the common midpoint processing (CMP) of the 1980s. Moreover, a velocity model consistent with the data was built and used to perform prestack and poststack depth migrations. The image of a Jurassic salt plug indicates tectonics similar to observations in the Allertal region at the northern fringe of the inverted Lower Saxony Basin, where overthrusting plays a major role in the evolution of salt structures. Consequently, shortening of the Mesozoic strata was included in the revised interpretation. The reprocessing also provided new insights into the petroleum systems in this area, indicating possible new exploration targets. The results may lead to a new geologic understanding of the area. Instead of a two-story salt plug, steep reverse faults and associated salt structures similar to the features along the Allertal lineament may best explain the investigated seismic line. Furthermore, CRS processing leads to a new view of the shape of

DOI:10.1306/13311429H43464

the Moho in the center of the NGB. This view supports the assumption that the origin of the NGB may be more related to metamorphic processes during basin initiation than to crustal stretching.

INTRODUCTION

Proper basin modeling relies on accurate seismic processing and interpretation, correct depth conversion of the identified sedimentary layers, reliable modeling of the thermal history of the basin, and appropriate understanding of the regional geodynamic setting. Burial subsidence analysis classifies the effects of the developing sedimentary load and the basin forming and modifying tectonics as a function of time and is a key method for geodynamic evaluation. Required oil and gas generation modeling results can be obtained by application of standard lithospheric stretching models (McKenzie, 1978; Wernicke, 1981) that include isostatic processes and heat-flow modifications because of extension and thinning of the crust. This procedure is justified when an identified uplift of the crust-mantle boundary (Moho) by reflection or refraction seismic occurs below an investigated basin. However, in some basins such as Moho, uplift does not exist with the expected shape. Instead, a flat Moho appears to be present (Ilchenko, 1996; Lobkovsky et al., 1996; DEKORP-BASIN Research Group, 1999). Furthermore, quite often, a thickened lower crust with high rock densities and velocities is observed beneath a deep sedimentary basin (Lobkovsky et al., 1996; Mooney et al., 2001; Yegorova et al., 2007). In those cases, basin evolution and the development of the thickened lower crust appear to be related. Thickening by metamorphism of crustal rocks during times of increased heat flow may therefore play a dominant role for basin evolution (Brink, 2005a, b).

By revisiting seismic data from the North German Basin (NGB), this study resulted in reinterpretation of the shape of the Moho and thus changed the input parameters for basin modeling. Because the sedimentary fill of this basin was structurally modified by strong halokinetic processes, reinterpretation of reprocessed seismic data also led to alternative subsidence histories and structural evolution models for areas previously characterized by poor seismic imaging. An improved understanding of the general salt dynamics within the NGB may result from this study. Finally, the results of this work suggest the need to revisit some hydrocarbon plays.

Deep seismic reflection data obtained by industry were revisited with the aim to improve the data quality and study the influence of deep-rooted processes on formation and evolution of the NGB. Moho depth, Moho relief, Moho diversity, and possible metamorphic complexes within the lower crust and their effects on seismic signatures were some of the key issues to understand the basin evolution. Seismic events below the base Permian as well as scattered seismic reflections within salt features were topics for further hydrocarbon-related studies (Baykulov et al., 2008; Brink et al., 2008).

To provide more information on deep processes that influence basin formation, the data sets were also reprocessed, with an emphasis on structures in the lower crust (Yoon et al., 2008a, b). The reprocessing was performed using the common reflection surface (CRS) stack method (Tygel et al., 1997; Müller, 1999; Mann, 2002). In contrast to the classical common midpoint (CMP) stack scheme, the CRS stack sums more traces and improves the signal-to-noise ratio. The CRS stacking operator locally describes the reflection response of a curved rock interface in an inhomogeneous medium (Figure 1). The CRS stack itself provides a simulated zero-offset section and a set of three kinematic wave field attributes for each sample of the seismic traces. These attributes are related to the location, orientation, and local curvature of interfaces in the subsurface (Mann et al., 1999). The parameters can be used for subsequent processing, for example, velocity model building, multiple suppression (Zaske et al., 1999; Dümmong and Gajewski, 2008), as well as for prestack data enhancement by creation of CRS supergathers (Baykulov and Gajewski, 2008, 2009). The CRS stack method appears to be very well suited to perform optimized imaging of low-fold data from complex subsurface structures. Meanwhile, it is being successfully applied by the hydrocarbon industry for processing reflection data from sedimentary basins (Trappe et al., 2001; Bergler et al., 2002). First applications of the CRS stack method to crustal reflection data (4 s two-way traveltime [TWT]) from the Donbas Foldbelt in Ukraine (Menyoli et al., 2004) demonstrate that this technology is suitable for imaging complex overthrust environments.

MOHO OF THE NORTH GERMAN BASIN REVISITED

The reprocessing of seismic reflection data from the NGB (Figure 2) with comparably small offsets and low-fold coverage using the CRS stack method gave improved seismic images (Yoon et al., 2008a, b). The comparison with the results from conventional CMP stacking shows that the reprocessed sections provide a higher signal-to-noise ratio and enhanced visibility of reflections at all time levels (Figure 3). The CRS stack sections provided clearer structural images of the sedimentary part

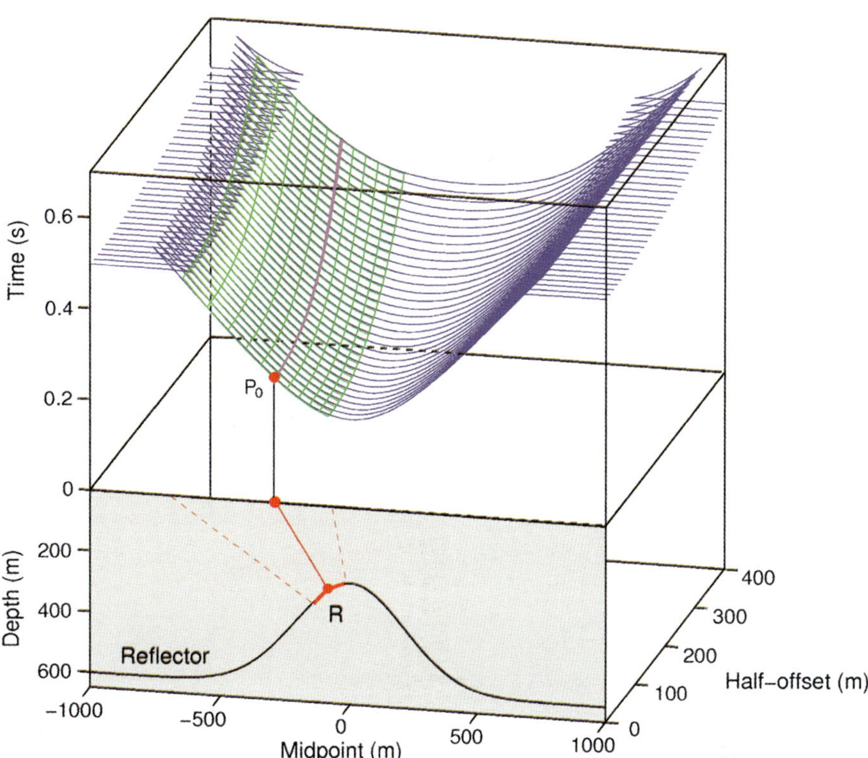

FIGURE 1. Common reflection surface (CRS) stacking surface for a constant velocity medium in a three-dimensional midpoint-offset-time domain modified from Müller (1999). The blue curves are the common-offset time responses of the curved subsurface reflector. The stacking surface results from approximating the true reflector by a reflector segment, R, which locally has the same curvature as the true reflector. The CRS stack sums the data along the green surface that coincides locally with the common-offset time response of the reflector and assigns the result to the point P0. Defined by the width of stacking surface in midpoint direction, the CRS might involve more traces during the stacking compared with the conventional common midpoint (CMP) stack (magenta line), thus, increasing the signal-to-noise ratio of stacked section.

down to 3 s TWT. The images of the salt domes were also improved and gave additional details on the internal salt structure. Within the center of the Glueckstadt Graben that is associated with a Triassic rift, which coincides with the center of the NGB, the Moho appears flat and about 1.5 s TWT later than previously interpreted (Dohr et al., 1989; Brink et al., 1990; Scheck-Wenderoth and Lamarche, 2005). In the deeper part of the section, weak and diffuse reflections that were almost invisible in the conventional CMP stack sections were enhanced. The strong reflections above the Moho events of the Glueckstadt Graben, previously interpreted as Moho reflections, may represent impedance contrasts caused by the high-density body that was found in a gravity study in the same area (Yegorova et al., 2007). According to Brink (2005a, b), this high-density body may result from metamorphism of the lower crustal rocks during the first stages of basin evolution and the related thermal event (Figure 4).

NORTH GERMAN BASIN EVOLUTION REVISITED

Some authors link the origin of the Southern Permian Basin and the NGB to rift processes in the crust (transtensional pull-apart basin; Bachmann and Grosse, 1989); others relate it mainly to convergent wrench tectonics, slab detachment, and thermal thinning of the mantle lithosphere caused by destabilization of the crust-mantle boundary during the Stephanian/Early Permian tectonomagmatic pulse (Ziegler, 1988, 1990; Van Wees et al., 2000; Neumann et al., 2004; Obst et al., 2004; Ziegler et al., 2004). According to Scheck and Bayer (1999), the Southern Permian Basin subsided in response to thermal relaxation of the lithosphere and sedimentary loading during late Early Permian to Middle Jurassic times, overprinted by the development of Triassic grabens and Late Cretaceous to Tertiary inversion tectonics.

The NGB has been an area of major subsidence since at least Late Proterozoic. This long-lasting process was episodically interrupted by orogenic phases, which are related to collisions of continents, terranes, and microplates during plate tectonic activity. During the Phanerozoic, important orogenic phases involving the development of accretionary wedges in the study area include the Caledonian (~420 Ma) and Variscan (~300 Ma) events.

Before the development of the North German (Permian) Basin, the underlying central European accreted crust (Norton and Johnson, 2001) became altered by magmatic intrusions and volcanic extrusions. This important thermal event, an increase of heat flow during Late Stephanian and Early Permian, is reported by many authors (Ziegler, 1990).

The rifting that occurred during this event comprises a set of Early Rotliegende (Autunian) roughly north-south–oriented grabens, more or less orthogonal to the basin axis (Gast, 1988). This differs from model predictions; for crustal stretching, the development of a graben system or pull-apart basin should be parallel

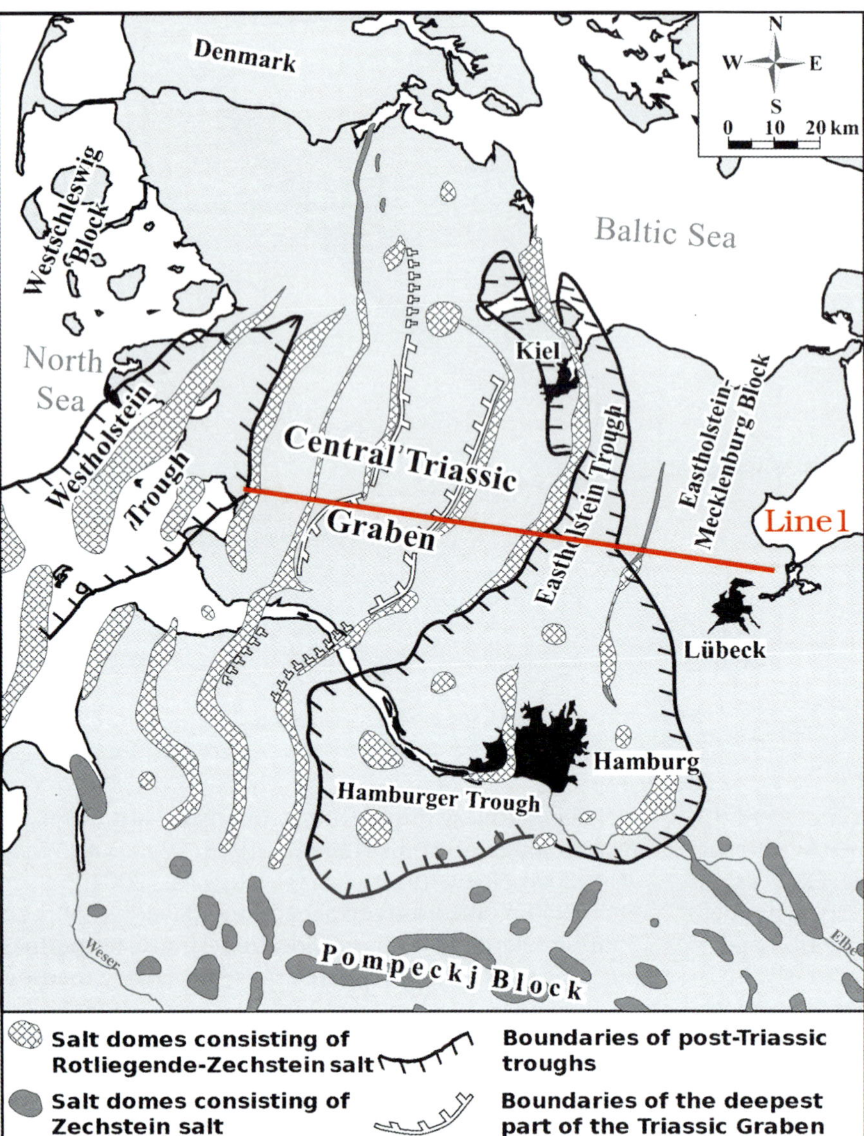

Figure 2. Map of northern Germany displaying major geologic units and cities, as well as the distribution of salt-stock families and salt plugs. The Central Triassic Graben is also known as the Glueckstadt Graben. Seismic line 1 is also called TON 8106 Extended.

to the axis of the basin before further basin evolution (McKenzie, 1978; Wernicke, 1981; Bachmann and Grosse, 1989; Ziegler, 1988, 1990).

The thermal anomaly during the Stephanian/Early Permian with a heat flow increase of about 10% is estimated to have lasted for approximately 25 m.y. (~300–275 Ma). For this period a substantial sedimentary record is missing from the North German Basin. During this event, metamorphic processes must have taken place in the middle and lower crusts, resulting in increased rock density and decreased rock volume (Brink 2005a, b; Brink, 2009) and in the development of an initial topographic depression about 250 m (820 ft) in depth (Figure 4). This topographic depression was subsequently filled with erosional products from the nearby Variscan Orogen. According to this model, the load of these sediments and the ongoing metamorphic event in the middle and lower crust became the driving forces for the developing sedimentary basin. According to Brink (2005a), the volume reduction processes resulting from metamorphism can explain roughly 30% of the subsidence; the remaining 70% is isostatically related to the sedimentary load. Instead of tectonic stretching accompanied by uplift of the Moho as commonly assumed, in this case, the volume reduction of the middle crust governs the basin evolution. As the middle crust thinned, the lower crust thickened by a lesser degree, which resulted in the growth of a high-density body in the lower crustal rocks directly underlying the evolving basin. Because metamorphic processes with a similar effect may also occur at the crust-mantle boundary, the Moho boundary below a basin may therefore vary during basin development. These processes can be quantitatively investigated, which leads to an evaluation of the basin-forming tectonics as a function of time, not divided into rift and postrift phases (Brink, 2005a).

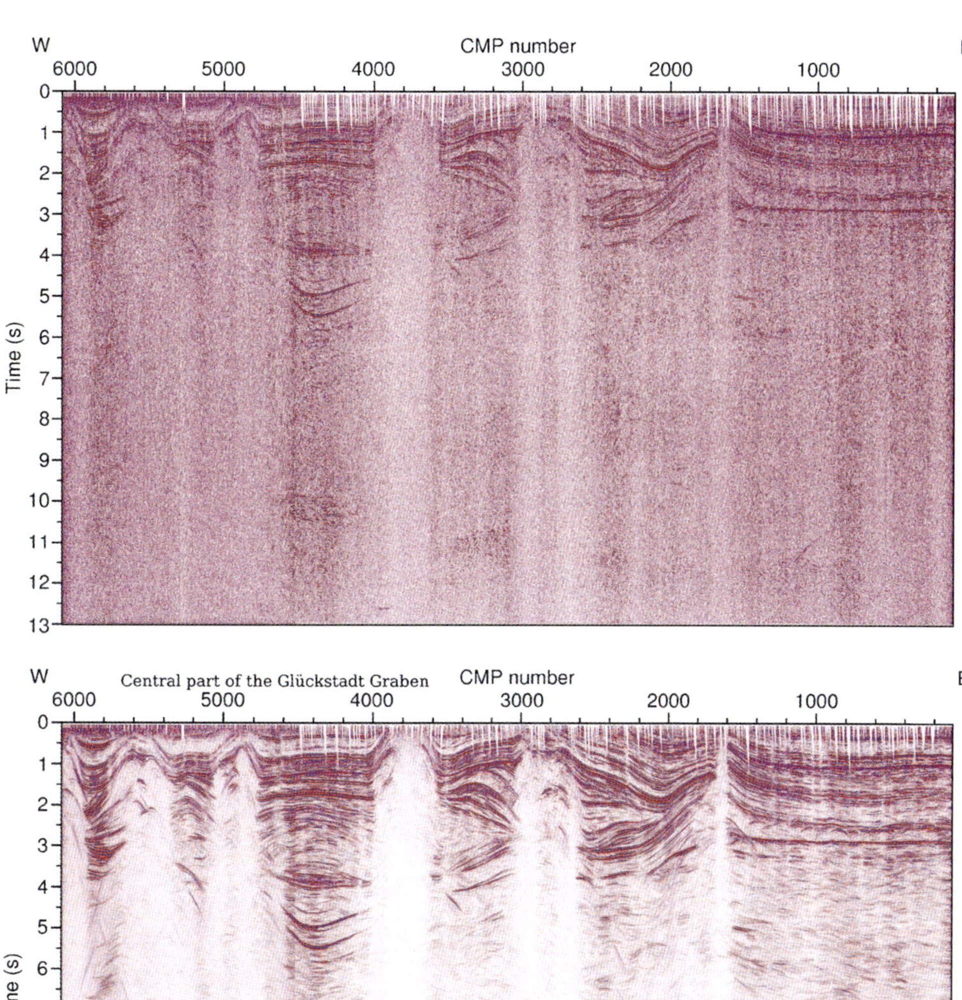

FIGURE 3. [top] The common midpoint (CMP) stack section of line 1 (TON 8106 Extended). The sedimentary cover is well imaged, whereas in the deeper part of the crustal reflections are hardly visible. [bottom] CRS stack section of line 1 (TON 8106 extended) with interpretation. The observed reflections in the time section indicate a high-density crustal structure below the graben center (see also Yoon et al., 2008a, b).

In the NGB, the present-day base of the post-Silurian Phanerozoic sediments is estimated to be at a depth of 10 to 15 km (6.2–9.3 mi). A thick sequence of Proterozoic and Lower Phanerozoic metasediments are assumed to underlie these sediments. The Proterozoic and Lower Phanerozoic rocks were imbricated during the Caledonian orogeny in an accretionary wedge (Norton and Johnson, 2001), which is thought to have merged with the microplates of eastern Avalonia. These deeply buried and metamorphically transformed Proterozoic and Lower Phanerozoic rocks, penetrated by magmatic intrusions, now form the middle and lower layers of the crust.

Metamorphic processes in the lower continental crust during basin development have already been qualitatively studied by Falvey (1974) and Haxby et al. (1976). According to Falvey, the rise in lithospheric temperature causes greenschist facies rocks in the continental crust to be metamorphosed to amphibolite facies accompanied by an increase in density. This causes a slight thinning of the crust, which to some extent compensates the thermal uplift. As the continental lithosphere cools after spreading starts, subsidence below the initial level occurs because of the denser and slightly thinner crust.

SALT DYNAMICS WITHIN THE NORTH GERMAN BASIN REVISITED

Salt plugs and their structural development represent a dominant component of the NGB (Figures 5, 6).

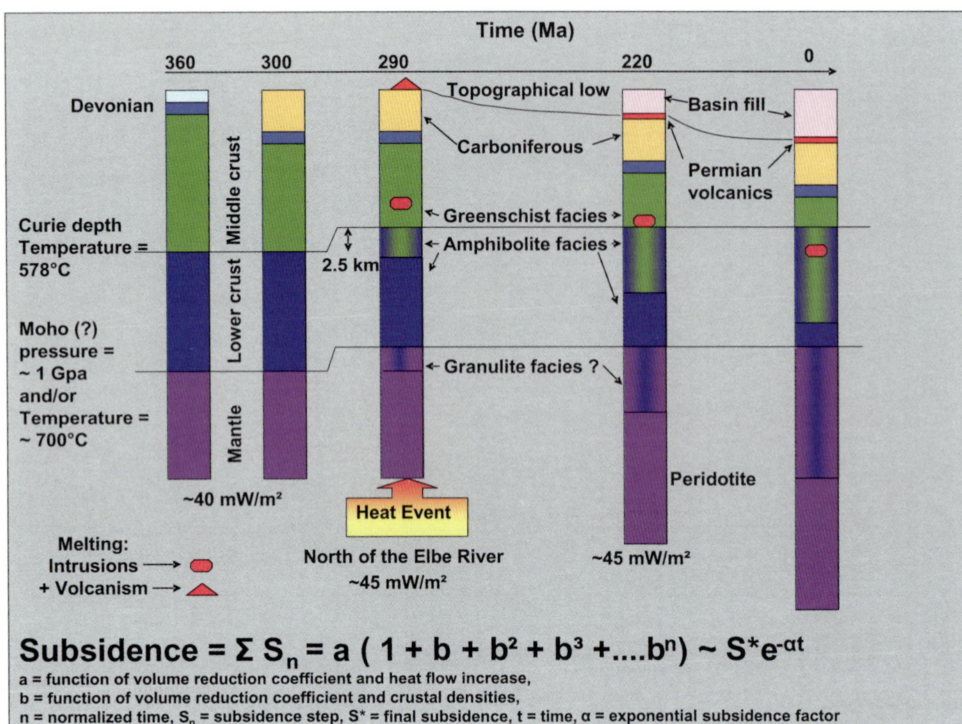

FIGURE 4. Relationship between metamorphism and subsidence processes of a young (accretionary) crust under the influence of a heat-flow anomaly (Brink, 2005a).

Many commercial products are linked to the effects of salt tectonics. Salt plugs are a source of raw material for different salts, and many oil accumulations are closely associated with the development of salt plugs (Boigk, 1981). Moreover, gas deposits were explored at the base of salt plugs (Pasternak, 2006). The functional dependence of the individual northern German salt structures and salt-stock families (Figure 7) was outlined by Sannemann (1968), and their history was described by Jaritz (1973). Trusheim (1957) introduced the term "halokinesis." In the explanations of the formation of salt plugs and the development of salt-plug families, these authors presumed density instabilities for the North German area between underlying light salt and overlying heavy sediments to be the major cause of buoyancy-driven halokinetic processes.

This assumption was already theoretically investigated by Hunsche (1978) and, using analog experiments, by Heye (1978). New general concepts of salt tectonics (Hudec and Jackson, 2007) and concepts specific for northwest Germany exist (Mohr et al., 2005). However, for the greater Glueckstadt Graben and Eastholstein

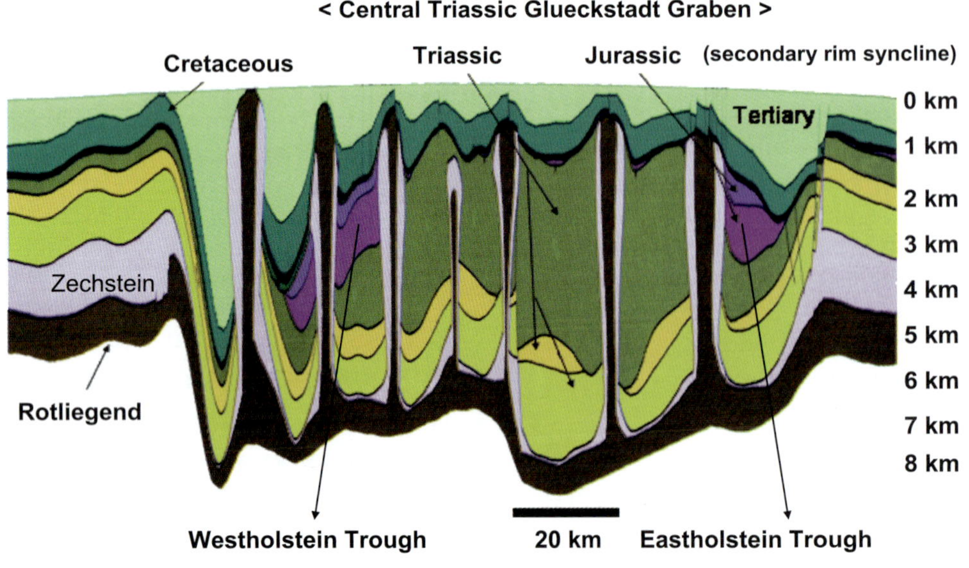

FIGURE 5. West-east cross section in Schleswig-Holstein with a salt-stock family (Baldschuhn et al., 2001).

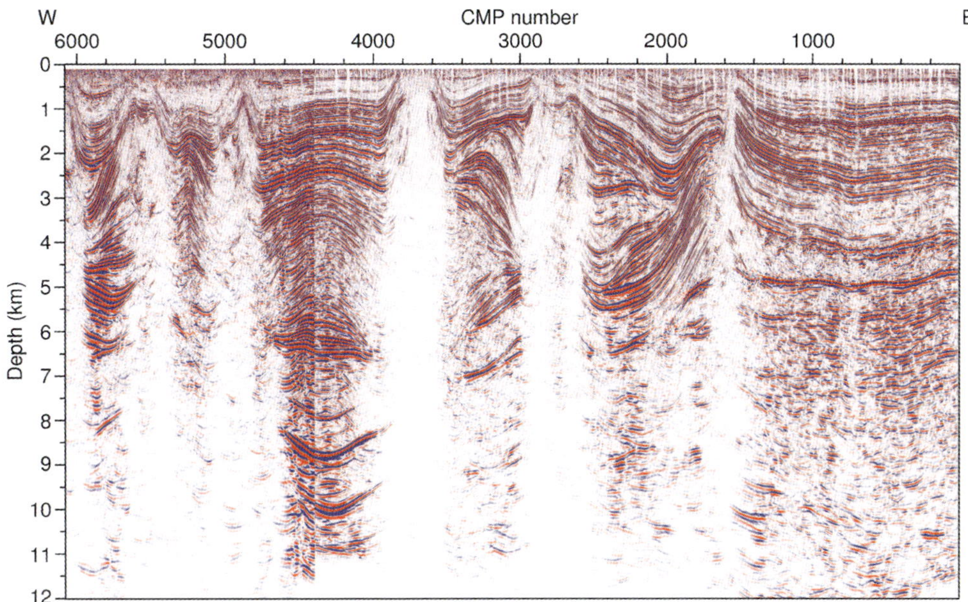

FIGURE 6. Pre–stack depth migration (PreSDM) of the common reflection surface (CRS) supergathers of seismic line 1 (TON 8106 Extended) along the central and eastern part of the west-east cross section in Schleswig-Holstein.

Trough areas, the evolution of salt-stock families was not yet revisited. Processing and interpretation of seismic reflection lines, gravity data, and density logs in boreholes raised doubts about the validity of the original concept and its value for the early stages (i.e., Lower Triassic Buntsandstein) of the halokinesis in Northern Germany. The following observations reported by Brink (1984, 1986, 1987) and Brink et al. (1992) are challenging the more or less accepted concept:

1. Within the Northwest German Basin, compressive "flower structures" were observed in seismic data at locations where previously salt plugs were assumed. This observation was further verified by drilling.
2. At the northern termination of the Glueckstadt Graben, a positive residual gravity anomaly at a location of a presumed salt plug required a reinvestigation by modern seismic data and was finally proved as an Early Triassic flower structure (Figure 8).
3. A salt plug in the Glueckstadt Graben center, which was interpreted to be a member of the so-called "two-story" salt structures with a core of Rotliegende salt (Early Permian) and flanks of Zechstein salt (Late Permian) (see Baldschuhn et al., 2001), was redefined as a Late Triassic (Keuper) salt structure.
4. Halokinesis during the Early Triassic (Buntsandstein) was certainly not governed by buoyancy because the rock density of the young and uncompacted Buntsandstein layer was most likely lower than the density of the underlying Permian salt.

Because of these doubts concerning the role of the primary domes in the Glueckstadt Graben area, a second-generation salt dome was revisited by Baykulov et al. (2008).

The Jurassic rim syncline of this salt stock is part of the Eastholstein Trough (see Figure 5), which contains a commercial petroleum system based on the Lower Jurassic–Liassic Posidonia shale as source rock and the Middle Jurassic Dogger sandstones as reservoir rocks. Because the position of this trough coincides with gravity and magnetic highs, its development is related to processes at greater depths (Brink, 1984). Therefore, in the chronological sequence and the spatial arrangement of the salt structures of the Glueckstadt Graben, different and time depending tectonic events may have combined with halokinetic process to produce the complex salt-plug distribution.

The reprocessed seismic depth image of the salt structure (Figure 6, CMP number 2500–3000) of the Eastholstein Trough contains significant internal reflections, which appear to be different from the image of salt plugs commonly observed in the NGB. Instead, it shows significant similarities to features imaged by a seismic line crossing the Allertal regional tectonic lineament close to the city of Celle in the Lower Saxony Basin in the south of the Pompeckj Block (compare Betz et al., 1987; Mohr et al., 2005). In these images, salt structures and compressive overthrusts are tectonically associated. Both features developed predominantly during inversion of the Lower Saxony Basin in Late Cretaceous to early Tertiary times and define the northern fringe of the Lower Saxony Basin. The similar structural style supports the assumption of a similar tectonic evolution within the Eastholstein Trough and the salt structures therein. However, the similarity does not include the same tectonic orientation and timing. As observed at the Allertal lineament, a complex three-dimensional tectonic setting with extensional, compressional, and transverse features can be expected. An interpretation of the reprocessed seismic

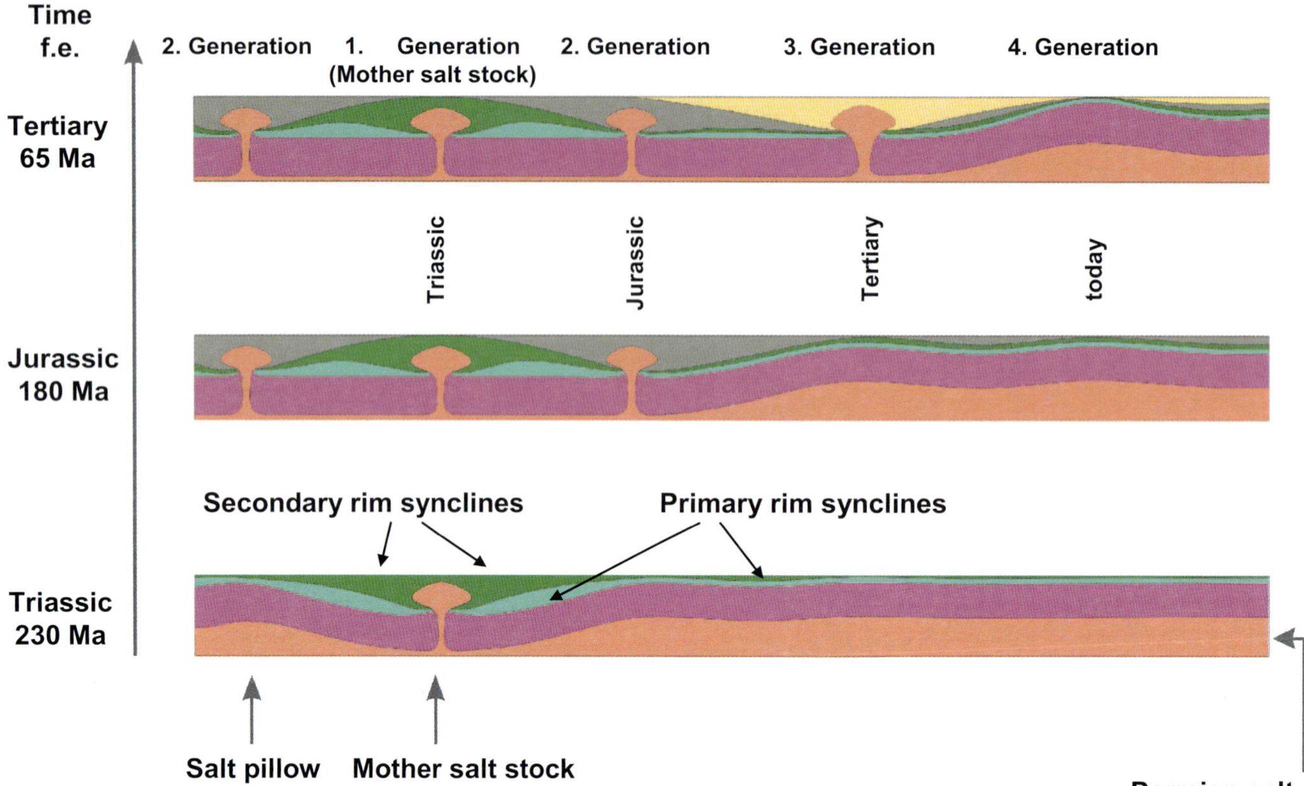

FIGURE 7. Concept of the evolution of a salt-stock family (modified from Sannemann, 1968). From a single salt stock (first-generation mother salt stock), salt migration spreads to adjacent areas because the unstable equilibrium of dense overburden and less dense salt has been disturbed. At a certain distance, new salt accumulation and salt-pillow formation with primary rim synclines take place. The salt pillow in turn is transformed into a diapir with secondary rim synclines (second generation), thereby creating a situation favoring the formation of another salt pillow (third generation), and so on.

FIGURE 8. Triassic flower structure at the position of a previously assumed salt stock (from Brink et al., 1992).

depth section based on the concept of compressive deformation is displayed in Figure 9 [top]. In such a structural setting with transverse movements and lateral salt flow, two-dimensional (2-D) balancing techniques are of limited value and have not been applied. Because one interpreted seismic line is not enough for a regional interpretation of the overall tectonics and the evaluation of a conjugate shear system, processing and interpretation of more seismic lines are required.

Two west-vergent steep faults are interpreted to occur below a salt body of unknown age (Rotliegende, Zechstein, Keuper(?)). This limited distribution of shale- and/or anhydrite-rich salt within the feature will not result in a well-defined Bouguer gravity low. A salt-rich feature was modeled by Yegorova et al. (2007) in a 2-D gravity study, based on work by Maystrenko et al. (2005), resulting in a poor fit of observed and modeled data. This poor fit may be caused by an inadequate shape of the investigated salt bodies, especially of the reprocessed feature, by an inadequate density distribution within the Mesozoic strata and below, by the complex three-dimensionality of the feature, or a combined effect. The present interpretation does not mean that the entire approximately 100-km (~62-mi)-long salt structure is similarly composed. However, the evidence of compressional tectonics is striking, and shortening of the Mesozoic strata by a significant amount is possible.

Tectonic processes that lead to shortening within the upper layers also had an accompanying effect within the basement. A reverse fault within the basement may have developed that was decoupled from the Mesozoic by the Permian salt layers. Alternately, metamorphic processes within the crustal rocks may have resulted in shrinkage of rock volumes caused by the pressure increase. Uplift of basement rocks could certainly account for the observed positive magnetic and gravimetric anomalies. However, such uplift is not supported by the seismic data.

On the other hand, according to Petrini and Podladchikov (2000), pressure gradients twice the lithostatic pressure can be achieved in regions exposed to horizontal shortening. This may result in metamorphism of greenschist facies into eclogite facies if conditions of the middle crust are assumed. The density of the affected rocks will then increase from about 2.7 g/cm^3 to about 3.4 g/cm^3, and the rock volumes will decrease. According to gravimetric and magnetic modeling (Scheibe et al., 2005), the presence of rocks with densities as high as 3.3 g/cm^3 within the lower and middle crust appears to be possible at depths less than 12 km (<7.5 mi). The magnetic anomaly is certainly caused by rocks that lie above the Curie depth. In such a case, no mafic intrusions into the lower and middle crust are required to generate the observed potential field anomalies. Intrusions as possible causes for the presence of the observed potential field anomalies were discussed by Brink (2005a, b) and Yegorova et al. (2007). However, because the gravimetric and magnetic highs are roughly coincident with the location of the Jurassic Eastholstein Trough, a volume reduction within deeper crustal levels and subsequent burial of the overlying strata appear to be more realistic than thrusting of the basement or the rise of magmatic intrusions. The burial of the trough took place before the Kimmerian inversion during the Late Jurassic, when large amounts of Jurassic to Triassic strata were eroded from adjacent areas. According to the presented assumptions, the burial and the inversion were accompanied by reverse faulting of the Mesozoic strata caused by the shortening of the crust, associated with the movement of salt. During the Tertiary, inversion processes occurred again and the structure became further overprinted.

This structural interpretation is supported by the seismic velocity distribution within the feature (Figure 9 [bottom]). A low-velocity zone at greater depth (3.5–5.5 km [2.2–3.4 mi]) between the two west-vergent steep faults ($x = 62–68$ km [39–42 mi]) is concordant with the interpreted dipping layers of base Jurassic to top Middle Triassic Muschelkalk. The magnitude of velocities indicates rather low-velocity sediments instead of a high-velocity salt body. The high-velocity body at a depth of about 3 km (1.9 mi) within the center of the salt complex and the thrust features may be part of the overlying salt body (second option, salt outline dotted). In this case, a high content of anhydrite may result in relatively higher velocities. Because light salt and heavy anhydrite are balancing each other, the resulting gravity effect may be as insignificant as observed. The consistency of the velocity model with the data is confirmed by the flatness of the prestack depth migrated CRS supergathers (Figure 10).

HYDROCARBON PLAYS REVISITED

Within the North German Basin, two major effective source rocks are widely distributed: Lower Jurassic (Liassic) Posidonia Shale (Binot et al., 1993) and Carboniferous coals (Cornford, 1998) and black shales, respectively (Gerling et al., 1999). Within the Eastholstein Trough, the Posidonia Shale charged the Dogger sandstone structures with petroleum below the Lower Cretaceous unconformity at the flank of the salt structure and at distal positions of the secondary rim syncline. Within the realm of the salt structure, seismic reflections have been remarkably recorded. In the CRS results, they show a reponse similar to the Posidonia Shale of the Eastholstein Trough. This source rock may therefore also be present below the remaining salt body in the upper central part of the salt structure (Figure 9, dotted black line). Whether Dogger reservoirs exist in a

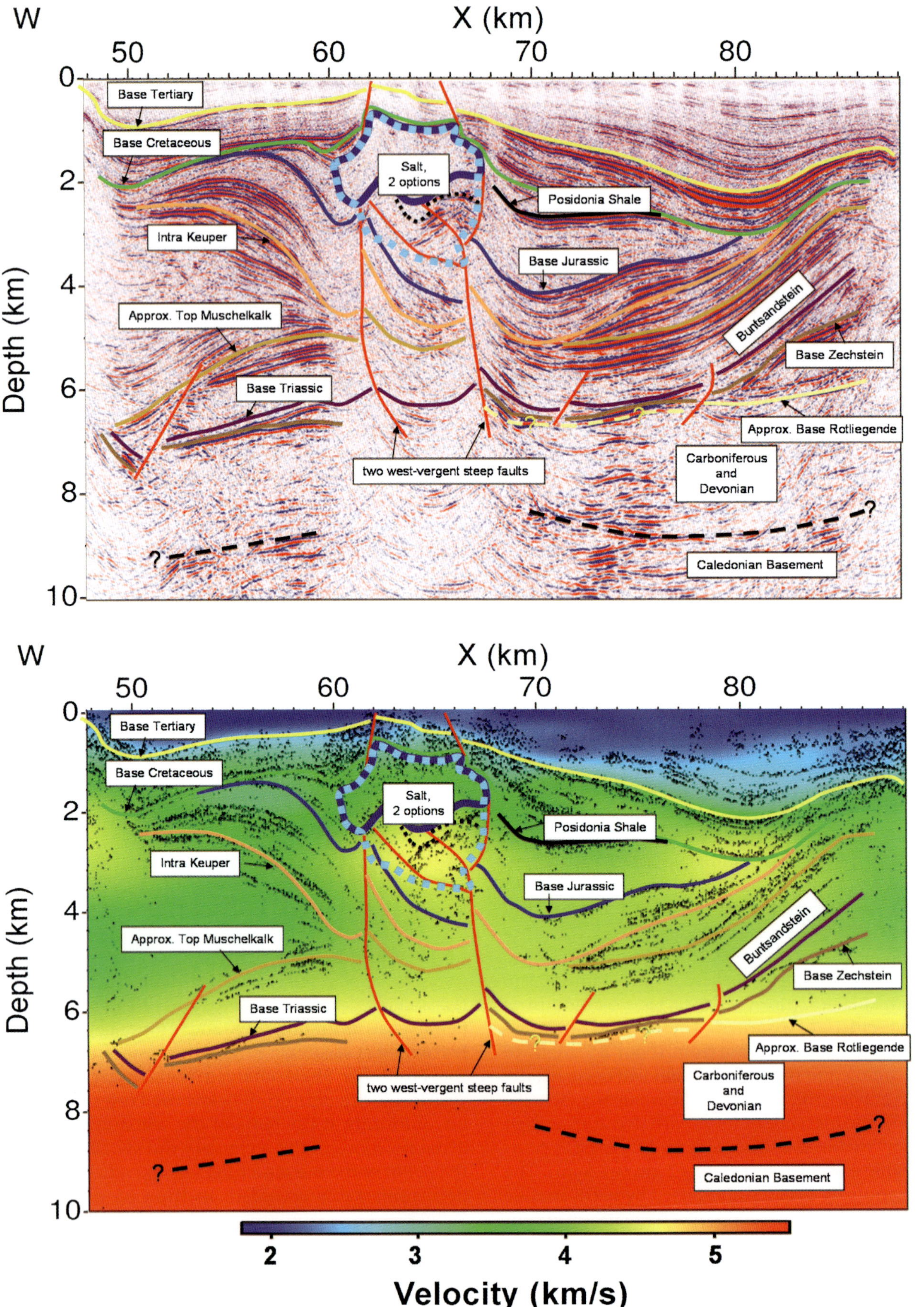
W
X (km)
E
50
60
70
80
0
2
4
6
8
10
Depth (km)
Base Tertiary
Base Cretaceous
Salt,
2 options
Posidonia Shale
Intra Keuper
Base Jurassic
Approx. Top Muschelkalk
Buntsandstein
Base Zechstein
Base Triassic
Approx. Base Rotliegende
two west-vergent steep faults
Carboniferous
and
Devonian
Caledonian Basement
W
X (km)
50
60
70
80
0
2
4
6
8
10
Depth (km)
Base Tertiary
Base Cretaceous
Salt,
2 options
Posidonia Shale
Intra Keuper
Base Jurassic
Approx. Top Muschelkalk
Buntsandstein
Base Zechstein
Base Triassic
Approx. Base Rotliegende
two west-vergent steep faults
Carboniferous
and
Devonian
Caledonian Basement
2
3
4
5
Velocity (km/s)

Figure 10. Common image gather after pre-stack depth migration of the common reflection surface (CRS) supergathers of seismic line TON 8106 Extended.

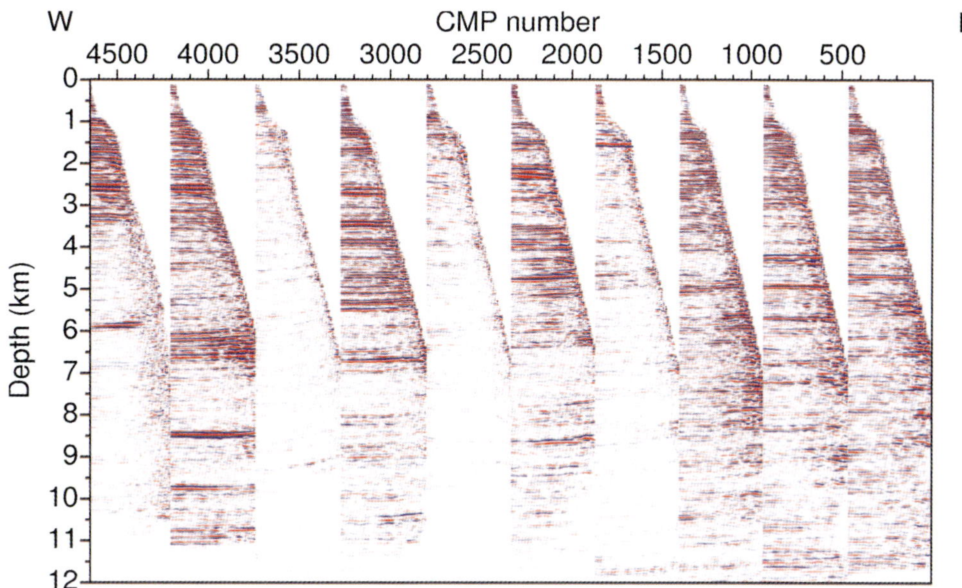

structurally high position within the structure is even more speculative.

Upper Carboniferous coals have not yet been penetrated by drilling within the area of Schleswig-Holstein. However, the Schleswig Z1 well, which tested the Triassic flower structure in the northern part of the Glueckstadt Graben, verified the presence of Lower Carboniferous black shales (Rodon and Littke, 2005). This observation and magnetotelluric (MT) data (Hoffmann et al., 2005, 2008) support the assumption that electrically conductive and highly coalified Carboniferous source rocks may be widely present within the Glueckstadt Graben (Figure 11). The MT data were recorded at 10 locations along an east-west–oriented profile across the total width of the graben. The central Glueckstadt Graben is characterized by a good conductor at a depth of approximately 10 km (6.2 mi). This conductor is missing on the adjacent graben flanks. For the first time, significant pre-Permian seismic reflections have been well imaged in shallower parts of the graben (Figure 9). These events could indicate the presence of coals and black shales within the Carboniferous and Devonian strata. In the center of the graben at very deep levels, these reflections were already identified in the images obtained by earlier processing (Bachmann and Grosse, 1989; Dohr et al., 1989; Brink et al., 1992). However, within the deepest parts of the graben, this potential source rock was overmature since Triassic times. This may not be the case below the eastern flank of the Eastholstein Trough, where the pre-Permian strata subsided still during the Tertiary and may then have entered and passed the gas-generation window. The exact timing depends on the amount of Permian salt, which migrated away from this position into the eastern third-generation salt plug, leaving a Tertiary rim syncline behind.

A third important and new observation from the reprocessed seismic line is the presence of faults (Figure 9), which cut through the Lower Triassic Buntsandstein sequence at a position where Rotliegende and Zechstein salts are very thin or absent (x = 70–80 km [43–50 mi], 6.5-km [40-mi] depth). This observation supports the existence of migration pathways from the mature Carboniferous source rock via the faults through the tight Lower Buntsandstein into the overlying sandstones of the Middle Buntsandstein as a necessary requirement for charging post-Permian closures (Brink, 2002; Karnin et al., 2006). At the fringes of the Eastholstein Jurassic Trough, Buntsandstein structures have been mapped (Figure 12) (Baldschuhn et al., 2001). One of the main risks of these structures is possible discharge by leakage along the flanks of adjacent salt features. Buntsandstein structures that were drilled in Schleswig-Holstein are mainly anticlinal closures above Permian salt pillows. Under these circumstances, the Permian salt acts as a perfect seal, and migration of hydrocarbons was severely hampered.

Figure 9. [top] Part of seismic line TON 8106 Extended (line 1), poststack depth migration of the common reflection surface (CRS) stack with interpretation. Solid black line = Posidonia Shale; dotted black line = inferred Posidonia Shale; thick solid blue line = salt body option 1; thick dotted light blue line = salt body option 2. [bottom] Part of seismic line TON 8106 extended (line 1), tomographic velocity distribution of CRS-processed line with interpretation (Brink et al., 2008; Baykulov et al., 2008).

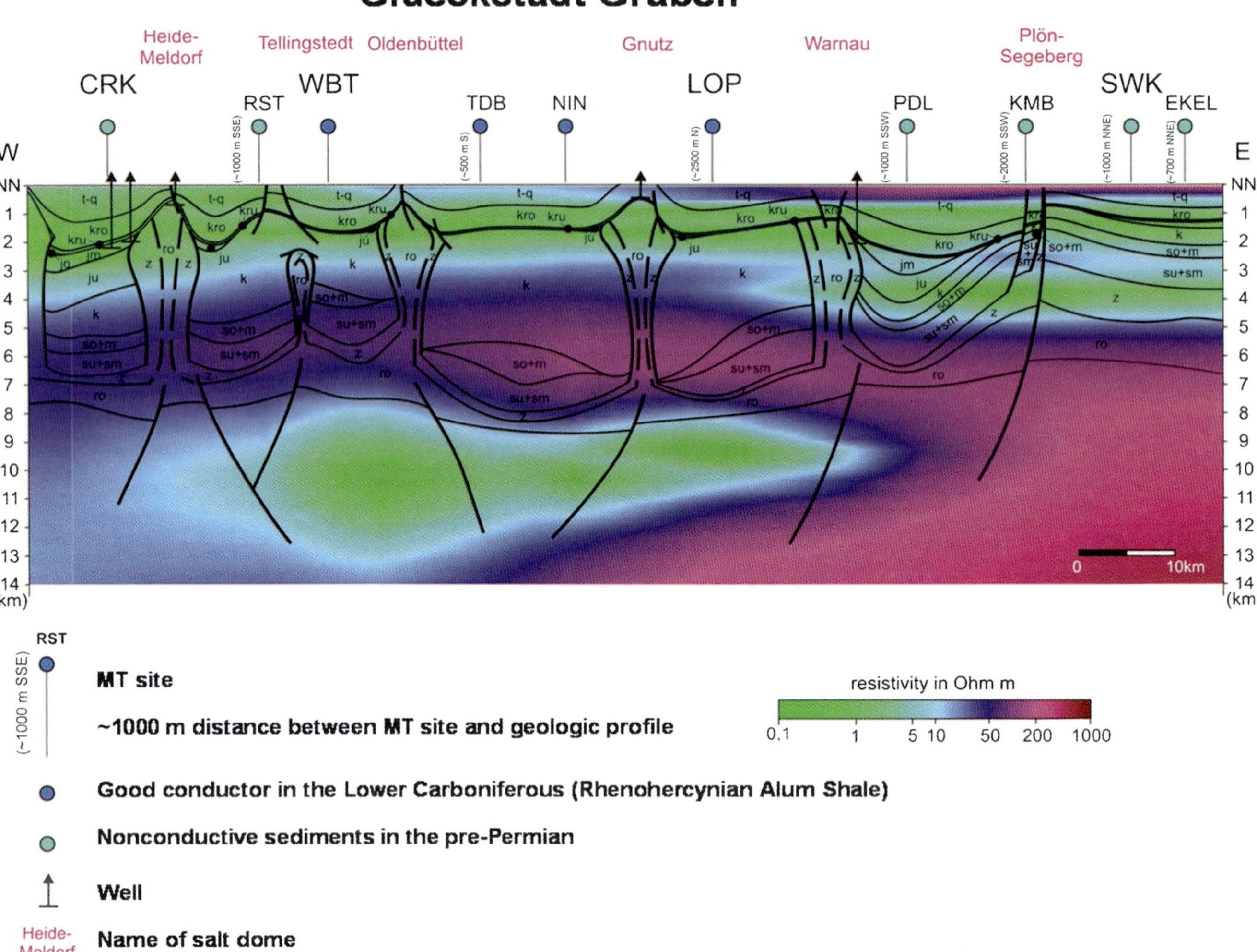

FIGURE 11. Integration of the Glueckstadt Graben two-dimensional magnetotelluric model into a geologic and tectonic cross section (redrawn and completed from Baldschuhn et al., 2001): q-t = Quarternary-Tertiary; kro = Upper Cretaceous; kru = Lower Cretaceous; jm = Dogger (Middle Jurassic); ju = Lias (Lower Jurassic); k = Keuper (Upper Triassic); so + m = Upper Buntsandstein and Muschelkalk (Upper Lower and Middle Triassic); sm + su = Middle and Lower Buntsandstein (Lower Triassic); z = Zechstein; ro = Upper Rotliegende (Hoffmann et al., 2008).

CONCLUSIONS

Seismic reprocessing using the CRS stack technique allows revised interpretation of the structural setting and the evolution of salt plugs in the area of the Glueckstadt Graben, NGB, and the geodynamic origin of the basin. The reprocessing of the data clearly demonstrates the capabilities of the CRS technique to improve the quality of low-fold data. The images display a considerably improved signal-to-noise ratio and many more details than the CMP processing of the 1980s. Moreover, a velocity model consistent with the data was built and used to perform prestack and poststack depth migrations. Prestack depth migrated CRS supergathers allowed reliable quality control of the velocity model used for migration, which was not possible by conventional processing. The image of the Jurassic salt plug indicates tectonics similar to observations in the Allertal region at the northern fringe of the inverted Lower Saxony Basin, where overthrusting plays a major role in the evolution of the salt structures. Consequently, shortening of the Mesozoic strata was included in the revised interpretation. The reprocessing also provided new insights into the petroleum systems in this area, indicating possible new exploration targets. The results may lead to a new geologic understanding of the area. Instead of a two-story salt plug, steep reverse faults and associated salt structures similar to the features along the Allertal lineament may best explain the investigated seismic line. Furthermore, CRS processing leads to a new view of the shape of

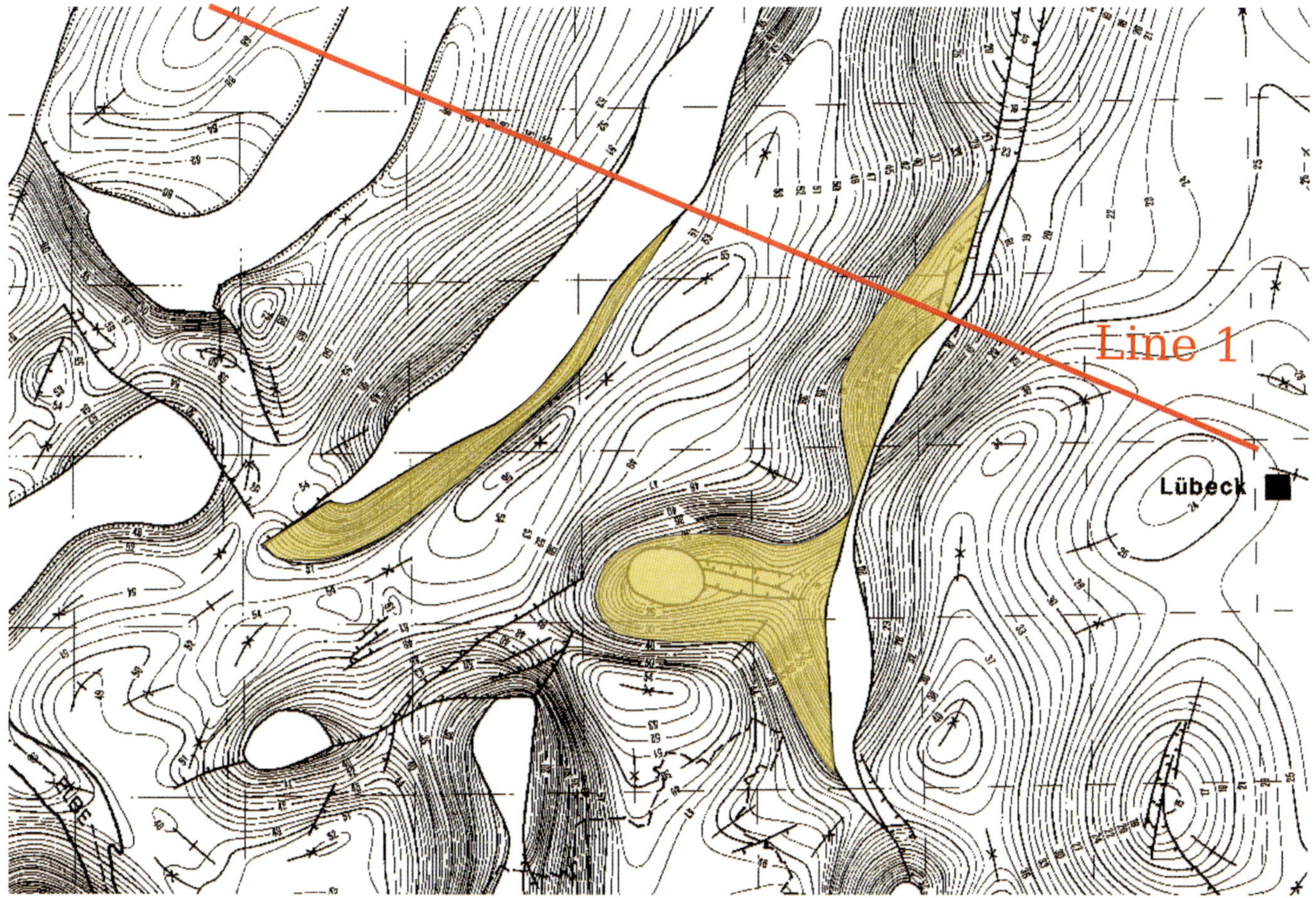

FIGURE 12. Early Triassic Buntsandstein structures (gold) near the common reflection surface (CRS)–processed seismic profile line 1 (red line, Figure 2), within the Eastholstein Trough (from Baldschuhn et al., 2001). Spacing between dashed grid lines: 10 km (6.2 mi).

the Moho in the center of the NGB. This view supports the assumption that the origin of the NGB may be more related to metamorphic processes during basin initiation than to crustal stretching.

ACKNOWLEDGMENTS

We thank the German Society of Petroleum and Coal Science and Technology (DGMK, project 577-1) and the German Science Foundation (DFG) for the support within the priority program SPP 1135 (Dynamics of Sedimentary Systems Under Varying Stress Conditions: The Example of the Central European Basin-System, Project Ga 350/12-1, 2). The seismic data were released through DGMK by the German oil and gas industry. We thank the Wave Inversion Technology (WIT) consortium for supporting the development of the CRS and NIP-wave tomography software. We acknowledge the supporting comments of two anonymous reviewers and the very valuable and constructive review of Ken Peters.

REFERENCES CITED

Bachmann, G. H., and S. Grosse, 1989, Struktur und Entstehung des Norddeutschen Beckens–geologische und geophysikalische Interpretation einer verbesserten Bouguer-Schwerekarte: Niedersächsische Akademie der Geowissenschaften: Veröffentlichung, v. 2, p. 23–47.

Baldschuhn, R., F. Binot, S. Fleig, and F. Kockel, 2001, Geotektonischer Atlas von Nordwest-Deutschland und dem deutschen Nordsee-Sektor-Strukturen, Strukturentwicklung, Paläogeographie: Geologisches Jahrbuch, Band A 153, 3 CD-ROM, p. 3–95.

Baykulov, M., and D. Gajewski, 2008, Seismic data enhancement with common reflection surface (CRS) stack method (abs.): Society of Exploration Geophysicists Expanded Abstracts 27, 2596, doi:10.1190/1.3063882.

Baykulov, M., and D. Gajewski, 2009, Prestack seismic data enhancement with partial common-reflection-surface (CRS) stack: Geophysics, v. 74, p. V49–V58, doi:10.1190/1.3106182.

Baykulov, M., H.-J. Brink, D. Gajewski, and M.-K. Yoon, 2008, Revisiting the structural setting of the Glueckstadt

Graben salt-stock family, North German Basin: Tectonophysics, v. 470, no. 1–2, p. 162–172, doi:10.1016/j.tecto.2008.05.027.

Bergler, S., P. Hubral, P. Marchetti, A. Cristini, and G. Cardone, 2002, 3D common-reflection-surface stack and kinematic wavefield attributes: The Leading Edge, v. 21, no. 10, p. 1010–1015, doi:10.1190/1.1518438.

Betz, D., F. Führer, G. Greiner, and E. Plein, 1987, Evolution of the Lower Saxony Basin: Tectonophysics, v. 137, p. 127–170, doi:10.1016/0040-1951(87)90319-2.

Binot, F., P. Gerling, W. Hiltmann, F. Kockel, and H. Wehner, 1993, The petroleum system in the Lower Saxony Basin, *in* A. M. Spencer, ed., Generation, accumulation and production of Europe's hydrocarbons: European Association of Petroleum Geoscientists Special Publication 3, p. 121–139.

Boigk, H., 1981, Erdöl und Erdölgas in der Bundesrepublik Deutschland, Enke-Verlag (Stuttgart), 330 p.

Brink, H.-J., 1984, Die Salzstockverteilung in Nordwestdeutschland: Geowissenschaften in unserer Zeit, 2, Jahrgang, v. 5, p. 160–166.

Brink, H.-J., 1986, Salzwirbel im Untergrund Norddeutschlands: Geowissenschaften in unserer Zeit, 4, Jahrgang, v. 3, p. 81–86.

Brink, H.-J., 1987, Salzwirbel oder instabile Dichtebeschichtung? Geowissenschaften in unserer Zeit, 5, Jahrgang, v. 4, p. 144–145.

Brink, H.-J., 2002, Halbwertszeiten im Kohlenwasserstoffhaushalt: Erdöl, Erdgas, Kohle, v. 2, p. 58–62.

Brink, H.-J., 2005a, The evolution of the North German Basin and the metamorphism of the lower crust: International Journal of Earth Sciences (Geol Rundsch), v. 94, p. 1103–1116, doi:10.1007/s00531-005-0037-7.

Brink, H.-J., 2005b, Liegt ein wesentlicher Ursprung vieler großer Sedimentbecken in der thermischen Metamorphose ihrer Unterkruste?-Das Norddeutsche Permbecken in einer globalen Betrachtung: Zeitschrift der Deutschen Gesellschaft für Geowissenschaften (ZDGG), v. 156, no. 2, p. 275–290, doi:10.1127/1860-1804/2005/0156-0275, Schweizerbart'sche Verlagsbuchhandlung, Stuttgart.

Brink, H.-J., 2009, Mantle plumes and the metamorphism of the lower crust and their influence on basin evolution: Marine and Petroleum Geology, v. 26, no. 4, p. 606–614, doi:10.1016/j.marpetgeo.2009.02.002.

Brink, H.-J., D. Franke, N. Hoffmann, W. Horst, and O. Oncken, 1990, Structure and evolution of the North German Basin: The European Geotraverse: Integrative studies, results from the 5th Earth Science Study Centre. European Science Foundation, Strasbourg, p. 195–212.

Brink, H.-J., H. Dürschner, and H. Trappe, 1992, Some aspects of the late- and post-Variscan development of the NW German Basin: Tectonophysics, v. 207, p. 65–95, doi:10.1016/0040-1951(92)90472-I.

Brink, H.-J., M. Baykulov, D. Gajewski, and M.-K. Yoon, 2008, Der Salzstock des ostholsteinischen Juratroges–eine seismische Re-Interpretation: DGMK-Tagungsbericht 2008 (CD-ROM).

Cornford, C., 1998, Source rocks and hydrocarbons of the North Sea, *in* K. W. Glennie, ed., Petroleum geology of the North Sea: Basic concepts and recent advances: London, Blackwell Scientific Publishers, p. 376–462.

DEKORP-BASIN Research Group, 1999, The deep crustal structure of the Northeast German Basin: New DEKORP-Basin '96 deep-profiling results: Geology, v. 27, p. 55–58, doi:10.1130/0091-7613(1999)027<0055:DCSOTN>2.3.CO;2.

Dohr, G., H. Dürschner, and H. A. K. Edelmann, 1989, Exploration geophysics in Germany: First Break, v. 7, no. 5, p. 153–172.

Dümmong, S., and D. Gajewski, 2008, A multiple suppression method via CRS attributes (abs.): Society of Exploration Geophysicists Expanded Abstracts, doi:10.1190/1.3063869.

Falvey, D. A., 1974, The development of continental margins in plate tectonic theory: Australian Petroleum Exploration Association Journal, v. 14, p. 95–106.

Gast, R., 1988, Rifting im Rotliegenden Niedersachsens: Geowissenschaften, v. 6, no. 4, p. 115–122.

Gerling, P., F. Kockel, and P. Krull, 1999, Das Kohlenwasserstoff-Potential des Präwestfals im norddeutschen Becken-Eine Synthese: Hamburg, DGMK-Forschungsbericht, v. 433, 107 p.

Haxby, W. F., D. L. Turcotte, and J. M. Bird, 1976, Thermal and mechanical evolution of the Michigan Basin: Tectonophysics, v. 36, p. 57–75, doi:10.1016/0040-1951(76)90006-8.

Heye, D., 1978, Experimente mit viskosen Flüssigkeiten zur Nachahmung von Salzstrukturen: Geologisches Jahrbuch, v. E12, p. 31–51.

Hoffmann, N., H. Jödicke, and L. Horejschi, 2005, Regional distribution of the Lower Carboniferous Culm and Carboniferous limestone facies in the North German Basin: Derived from magnetotelluric soundings: Zeitschrift der Deutschen Gesellschaft für Geowissenschaften (ZDGG), v. 156, no. 2, p. 323–339, doi:10.1127/1860-1804/2005/0156-0323, Schweizerbart'sche Verlagsbuchhandlung, Stuttgart.

Hoffmann, N., L. Hengesbach, B. Friedrichs, and H.-J. Brink, 2008, The contribution of magnetotellurics to an improved understanding of the geological evolution of the North German Basin: Review and new results: Zeitschrift der Deutschen Gesellschaft für Geowissenschaften (ZDGG), v. 159, no. 4, p. 591–606, doi:10.1127/1860-1804/2008/0159-0591, Schweizerbart'sche Verlagsbuchhandlung, Stuttgart.

Hudec, M. R., and M. P. A. Jackson, 2007, Terra infirma: Understanding salt tectonics: Earth Science Reviews, v. 82, p. 1–28, doi:10.1016/j.earscirev.2007.01.001.

Hunsche, U., 1978, Modellrechnungen zur Entstehung von Salzstockfamilien: Geologisches Jahrbuch, v. E12, p. 53–107.

Ilchenko, T., 1996, The Dniepr-Donets Rift: Deep structure and evolution from DSS profiling: Tectonophysics, v. 268, p. 83–98, doi:10.1016/S0040-1951(96)00221-1.

Jaritz, W., 1973, Zur Entstehung der Salzstrukturen Nordwestdeutschlands: Geologisches Jahrbuch, v. A10, p. 3–77.

Karnin, W. D., R. Gast, C. Bärle, B. Clever, M. Kühn, and J. Sommer, 2006, Play types, structural history and distribution of Middle Buntsandstein gas fields in NW Germany: Observations and their genetic interpretation: Zeitschrift der Deutschen Gesellschaft für Geowissenschaften (ZDGG), Schweizerbart'sche Verlagsbuchhandlung, Stuttgart, v. 157, no. 1, p. 121–134, doi:10.1127/1860-1804/2006/0157-0121.

Lobkovsky, L. I., A. T. Ismail-Zadeh, S. S. Krasovsky, P. Y. Kuprienko, and S. Cloetingh, 1996, Gravity anomalies and possible formation mechanism of the Dnieper-Donets Basin: Tectonophysics, v. 268, p. 281–292, doi:10.1016/S0040-1951(96)00223-5.

Mann, J., 2002, Extensions and applications of the common-reflection-surface stack method: Ph.D. thesis, University of Karlsruhe, Logos Verlag, Berlin, p. 2–52, ISBN 3-8325-0008-1.

Mann, J., G. Höcht, R. Jäger, and P. Hubral, 1999, Common reflection surface stack: An attribute analysis (abs.): 61st Meeting of European Association of Geoscientists and Engineers Extended Abstracts, session P140.

Maystrenko, Y., U. Bayer, and M. Scheck-Wenderoth, 2005, Structure and evolution of the Glueckstadt Graben due to salt movements: International Journal of Earth Sciences (Geol. Rundschau), v. 94, no. 5–6, p. 799–814, doi:10.1007/s00531-005-0003-4.

McKenzie, D. P., 1978, Some remarks on the development of sedimentary basins: Earth Planetary Science Letters, v. 40, p. 5–32.

Menyoli, E., D. Gajewski, and C. Hübscher, 2004, Imaging of complex basin structures with the common reflection surface (CRS) stack method: Geophysical Journal International, v. 157, p. 1206–1216, doi:10.1111/j.1365-246X.2004.02268.x.

Mohr, M., P. A. Kukla, J. L. Urai, and G. Bresser, 2005, Multiphase salt tectonic evolution, *in* NW Germany: Seismic interpretation and retro-deformation: International Journal of Earth Sciences (Geol. Rundschau), v. 94, p. 917–940, doi:10.1007/s00531-005-0039-5.

Mooney, W. D., R. G. Coleman, P. R. Reddy, I. Artemieva, M. Billien, J.-J. Leveque, and S. T. Detweiler, 2001, A review of new seismic constraints of crust and mantle structure from China and India coupled with seismic QS and temperature estimates for the upper mantle, 23rd Seismic Research Review: Worldwide Monitoring of Nuclear Explosions poster, sponsored by National Nuclear Security, Administration Office of Nonproliferation Research and Engineering, Office of Defense Nuclear Nonproliferations, contract no. DE-A104 98AL79758.

Müller, T., 1999, The common reflection surface stack—Seismic imaging without explicit knowledge of the velocity model: Ph.D. thesis, University of Karlsruhe, Der Andere Verlag, Bad Iburg, p. 9–36, ISBN 3-93436-606-6.

Neumann, E. R., M. Wilson, M. Heeremans, E. A. Spencer, K. Obst, M. J. Timmerman, and L. Kirstein, 2004, Carboniferous–Permian rifting and magmatism in southern Scandinavia, the North Sea and northern Germany: A review, *in* M. Wilson, E.-R. Neumann, G. R. Davies, M. J. Timmerman, M. Heeremans, and B. T. Larsen, eds., Permo-Carboniferous magmatism and rifting in Europe: Geological Society, London, Special Publications 223, p. 11–40.

Norton, I. O., and C. A. Johnson, 2001, Sedimentary basin development on accretionary crust: Exploration significance, *in* L. Moresi, R. D. Müller, and B. Hobbs, eds., Conference Volume American Geophysical Union Chapman Conference on Exploration Geodynamics, Perth, Australia, p. 137.

Obst, K., Z. Solyom, and L. Johansson, 2004, Permo–Carboniferius extension-related magmatism at the SW margin of the Fennoscandian Shield, *in* M. Wilson, E.-R. Neumann, G. R. Davies, M. J. Timmerman, M. Heeremans, and B. T. Larsen, eds., Permo–Carboniferous magmatism and rifting in Europe Geological Society (London) Special Publication 223, p. 259–288.

Pasternak, M., 2006, Exploration and production of crude oil and natural gas in Germany in 2005: Erdöl, Erdgas, Kohle, v. 122, no. 7/8, p. 260–272.

Petrini, K., and Y. Podladchikov, 2000, Lithospheric pressure-depth relationship in compressive regions of thickened crust: Journal of Metamorphic Geology, v. 18, p. 67–77, doi:10.1046/j.1525-1314.2000.00240.x.

Rodon, S., and R. Littke, 2005, Thermal maturity in the Central European Basin system (Schleswig-Holstein area): Results of 1-D basin modeling and new maturity maps: International Journal of Earth Sciences (Geol. Rundschau), v. 94, no. 5–6, p. 815–833, doi:10.1007/s00531-005-0006-1.

Sannemann, D., 1968, Salt-stock families in northwestern Germany, *in* J. Braunstein and G. D. O'Brien, eds., Diapirism and Diapirs: A Symposium: AAPG Memoir 8, p. 261–270.

Scheck, M., and U. Bayer, 1999, Evolution of the Northeast German Basin: Inferences from a 3-D structural model and subsidence analysis: Tectonophysics, v. 313, p. 145–169, doi:10.1016/S0040-1951(99)00194-8.

Scheck-Wenderoth, M., and J. Lamarche, 2005, Crustal memory and basin evolution in the Central European Basin System: New insights from a 3-D structural model: Tectonophysics, v. 397, p. 143–165, doi:10.1016/j.tecto.2004.10.007.

Scheibe, R., K. Seidel, M. Vormbaum, and N. Hoffmann, 2005, Magnetic and gravity modeling of the crystalline basement in the North German Basin: Zeitschrift der Deutschen Gesellschaft für Geowissenschaften (ZDGG), v. 156, no. 2, p. 291–298, doi:10.1127/1860-1804/2005/0156-0291.

Trappe, H., G. Gierse, and J. Pruessmann, 2001, The common reflection surface (CRS) stack: Structural resolution in time domain beyond the conventional NMO/DMO stack: First Break, v. 19, p. 625–633.

Trusheim, F., 1957, Über Halokinese und ihre Bedeutung für die strukturelle Entwicklung Norddeutschlands: Zeitschrift der Deutschen Geologischen Gesellschaft, v. 109, p. 111–151.

Tygel, M., T. Müller, P. Hubral, and J. Schleicher, 1997, Eigenwave based multiparameter traveltime expansions (abs.): 67th Annual International Meeting of Society of Exploration Geophysicists Expanded Abstracts, p. 1770–1773, doi:10.1190/1.1885776.

Van Wees, J.-D., R. A. Stephenson, P. A. Ziegler, U. Bayer, T. McCann, R. Dadlez, R. Gaupp, M. Narkiewicz, F. Bitzer, and M. Scheck, 2000, On the origin of the Southern

Permian Basin, Central Europe: Marine and Petroleum Geology, v. 17, p. 43–59, doi:10.1016/S0264-8172(99)00052-5.

Wernicke, B., 1981, Low-angle normal faults in the basin and range province: Nappe tectonics in an extending orogen: Nature, v. 291, p. 645–647, doi:10.1038/291645a0.

Yegorova, T., Y. Maystrenko, U. Bayer, and M. Scheck-Wenderoth, 2007, The Glueckstadt Graben of the North-German Basin: New insights into the structure from 3-D and 2-D gravity analyses: International Journal of Earth Sciences (Geol. Rundschau), v. 97, no. 5, p. 915–930, doi:10.1007/s00531-007-0228-5.

Yoon, M.-K., M. Baykulov, S. Dümmong, H.-J. Brink, and D. Gajewski, 2008a, New insights into the crustal structure of the North German Basin from reprocessing of seismic reflection data using the common reflection surface stack: International Journal of Earth Sciences (Geol. Rundschau), v. 97, no, 5, p. 887–898, doi:10.1007/s00531-007-0252-5.

Yoon, M.-K., M. Baykulov, S. Dümmong, H.-J. Brink, and D. Gajewski, 2008b, Reprocessing of deep seismic reflection data from the North German Basin with the common reflection surface stack: Tectonophysics, (2009) Deep seismic profiling of the continents and their margins, v. 472, no. 1–4, p. 273–283, doi:10.1016/j.tecto.2008.05.010.

Zaske, J., S. Keydar, and E. Landa, 1999, Estimation of kinematic wave front characteristics and their use for multiple attenuation: Journal of Applied Geophysics, v. 42, no. 3–4, p. 333–346, doi:10.1016/S0926-9851(99)00044-0.

Ziegler, P. A., 1988, Evolution of the Arctic-North Atlantic and the western Tethys: AAPG Memoir 43, p. 1–197.

Ziegler, P. A., 1990, Geological atlas of western and central Europe: Shell Internationale Petroleum Maatschappij B.V., ISBN 90-6644-125-9: Geological Society Publishing House.

Ziegler, P. A., M. E. Schumacher, P. Dezes, J. D. Van Wees, and S. Cloetingh, 2004, Post-Variscan evolution of the lithosphere in the Rhine Graben area: Constraints from subsidence modeling, *in* M. Wilson, E.-R. Neumann, G. R. Davies, M. J. Timmerman, M. Heeremans, and B. T. Larsen, eds., Permo–Carboniferous magmatism and rifting in Europe: Geological Society (London) Special Publication 223, p. 289–317.

5

Monnier, F., P.-Y. Chenet, J.-M. Laigle, F. Lorant, S. Pegaz-Fiornet, and A. Al-Khamiss, 2012, Prediction of fluid compositional heterogeneities in fields using local grid refinement: Example from the Jurassic of Northern Kuwait, *in* K. E. Peters, D. J. Curry, and M. Kacewicz, eds., Basin Modeling: New Horizons in Research and Applications: AAPG Hedberg Series, no. 4, p. 87–100.

Prediction of Fluid Compositional Heterogeneities in Fields Using Local Grid Refinement: Example from the Jurassic of Northern Kuwait

Frédéric Monnier, Pierre-Yves Chenet, and Jean-Marie Laigle

Beicip-Franlab, Rueil-Malmaison, France

Francois Lorant and Sylvie Pegaz-Fiornet

IFP Energies Nouvelles, Rueil-Malmaison, France

Awatif Al-Khamiss

Kuwait Oil Company (K.S.C.), Ahmadi, Kuwait

ABSTRACT

Applying basin modeling technology to predict high-resolution fluid distribution and properties, taking into account the local high resolution of the sediment properties in fields and prospects has been a growing need for the past 10 yr. To minimize simulation time, local grid refinement (LGR) techniques have been introduced. The main interest of LGR is to gain computing time and memory with respect to classical methods, such are Tartan gridding. With LGR, it is possible to define local areas with high resolution in a regional model. The LGR approach gives a more detailed picture of individual fields or prospects while using models of reasonable size. The models incorporate various regional elements of the petroleum system that include source rock and seal, for instance, to obtain a detailed understanding of local processes such as trap filling history.

To validate the LGR approach, a benchmark is performed. It aims at comparing the different refinement methods: (1) a high-resolution grid, (2) an LGR grid, (3) a Tartan grid, and (4) windowing. To test the behavior of and the results produced by LGR, this method is applied to a real case study from northern Kuwait. It illustrates the coupling between LGR and compositional three-dimensional Darcy flow modeling to predict the distribution of hydrocarbon composition and properties in local reservoir rock areas where accumulations are predicted. The LGR approach efficiently fills the gap between conventional basin modeling and reservoir modeling. Its application in northern Kuwait provides useful guidelines to predict API gravities, gas-oil ratio, and oil-water contact depth estimates in new prospects.

DOI:10.1306/13311430H43465

INTRODUCTION

Classical basin simulation is designed to account for large-scale processes, both in time and space, associated with petroleum systems. Applying this technology to predict high-resolution fluid distribution and properties, taking into account the local high resolution of the sediment properties in fields and prospects has been a growing need for the past 10 yr.

Because of the high degree of coupling between porous medium compaction, heat transfer, and multiphase fluid fluxes, computation time has been an issue for basin simulators. The number of cells in the grid is a strong limiting factor to building high-resolution models. Performing a multiphase multicomponent fluid-flow simulation with a generalized Darcy formulation in a volumetric grid with millions of cells is a challenging problem, requiring high-performance computing solutions (Requena et al., 2005; Agelas et al., 2007).

So far, to account for local high-resolution fluid properties, refining a grid locally was possible by generating a Tartan grid in x and y directions. In the z direction, the grid refinement (split layering) covers the entire domain of the block. However, the resulting grid still contains a lot of cells in areas where refinement is unnecessary, leading to a long computing time.

To overcome this situation and to represent better data availability (regional vs. prospect/structure of interest), a local grid refinement (LGR) function was developed. This technology, routinely applied to reservoir simulation (Ding and Lemonnier, 1993), allows to ascribe lateral and vertical grid refinement in selected areas where justified by data availability or where local property distributions need to be predicted. Outside of the local fine grid area, where less data are available, the resolution is kept coarse (Thibaut et al., 2005).

A case study of northern Kuwait illustrates the coupling between LGR and compositional three-dimensional (3-D) Darcy flow modeling to predict the distribution of hydrocarbon composition and properties in local reservoir rock areas where accumulations are suspected. The LGR approach gives a more detailed picture of individual fields or prospects while using a model of reasonable size for incorporating the various components of the petroleum system that include the regional source rock, seal, and hydrocarbon migration pathways from kitchen to trap.

GRID REFINEMENT TECHNIQUES

High-resolution grids consuming large computing resources are impractical for basin simulations in operational conditions, therefore, basin models need to be upscaled. However, depending on the data availability or the target of the study, it may be desirable to preserve a high resolution in some parts of the model to capture the interfacial exchange of fluids between the regional model and locally refined areas of interest. To achieve this objective, various techniques may be implemented (Figure 1).

Tartan Grid

A Tartan grid, compared with a regular Cartesian grid, has variable cell sizes in x and y directions. This is the easiest approach to apply to have a high mesh resolution in the areas of interest. Because of a constraint of regularity, the mesh remains dense in the lateral continuity of areas of interest and requires additional and unnecessary computing resources.

Windowing

When applying the windowing approach, a regional simulation with a coarse grid is performed, and some properties (transmissivities and overpressures) are extracted along the boundary of local areas of interest. These properties are then used as input boundary conditions for the local simulations. Each local simulation is managed as an individual basin model. At the end, this method requires $n + 1$ different simulations to model n locally refined areas. For the prediction of fluid flux–related variables, it is mandatory to compute accurately water flux through the interface between the local and regional models. This requires to upscale overpressures and transmissivities with precision. With the windowing approach, the overpressure extracted from the regional block may fail to account for the heterogeneities of the local block. For hydrocarbon migration computation, the flux conservation is not guaranteed.

Local Grid Refinement

In the LGR approach, the model grid is more detailed locally, and the grid is kept coarse elsewhere, without any lateral overmeshing. Compared with windowing, the LGR offers advantages. First, the LGR technology allows for keeping refined meshes in one or more particular stratigraphic intervals instead of having a refined definition all over the sedimentary column, which is one drawback of windowing. This drastically reduces the number of cells managed in the regional model while keeping a high xy resolution in local areas and in stratigraphic intervals (z) simulated simultaneously. Second, whereas a classical windowing implementation would require one simulation per local area in addition to the mandatory regional simulation, the LGR approach only requires a single simulation for a

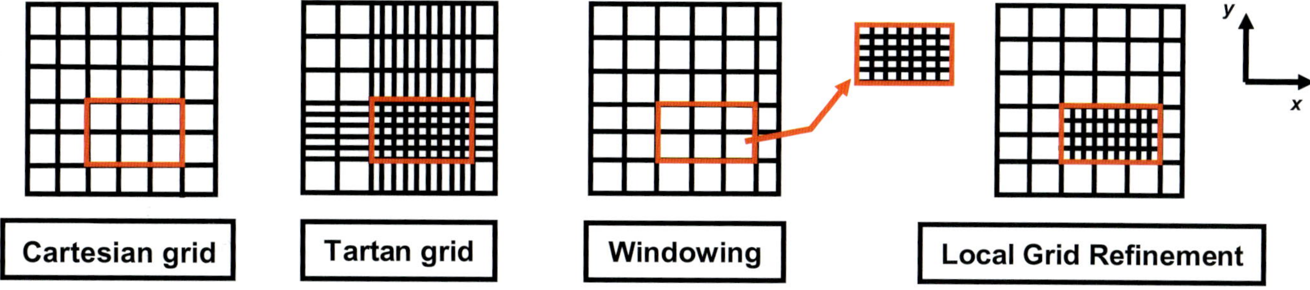

FIGURE 1. Different approaches of refinement.

case with multiple local areas. Moreover, if a behavior interaction occurs between local areas through time, it would be accounted for. Third, flux conservation for pressure and hydrocarbon fluid migration is ensured, whereas it cannot be guaranteed with windowing. Indeed, with windowing, the input flux at boundaries between local zones and the regional model does not account for the heterogeneities of the local grid as both models are run separately. In comparison, the LGR technique computes with precision water and hydrocarbon fluid fluxes through the interface between local and regional grids interactively.

Local Grid Refinement Calculator Main Features

The LGR calculator simultaneously solves the equations of mass conservation of solids and fluids, compaction, and Darcy's law for two-phase flow using a finite volume spatial discretization scheme (Schneider and Wolf, 2000; Schneider et al., 2000). For the transport equations, two different schemes can be used: a fully implicit or an IMPIMS (Implicitly Pressure-Implicitly Saturation) scheme (Wolf et al., 2009). The nonlinearities are treated with a Newton method, and a linear system is solved at each Newton iteration. The LGR simulator is parallelized and takes advantage of recent improvements on the convergence of the Newton algorithm both in terms of robustness and computing time (Agelas et al., 2007). Different kinds of preconditioners for the linear system can be used: incomplete LU factorization or algebraic multigrid (AMG) (Willien et al., 2009).

Contrary to standard grid refinement techniques applied in reservoir simulation (Reme et al., 1998), the LGR approach requires that the geometry of the local areas be correctly reconstructed through time. Depending on the grid size, gaps and/or overlaps develop through time in *z* direction as shown in Figure 2. Consequently, a step of backstripping (Schneider et al., 2000), which aims at inverting the geometry through geologic time, is performed for each local area accounting for the evolution of the full sedimentary column. In addition, the geometry of the coarse grid is updated with the fine

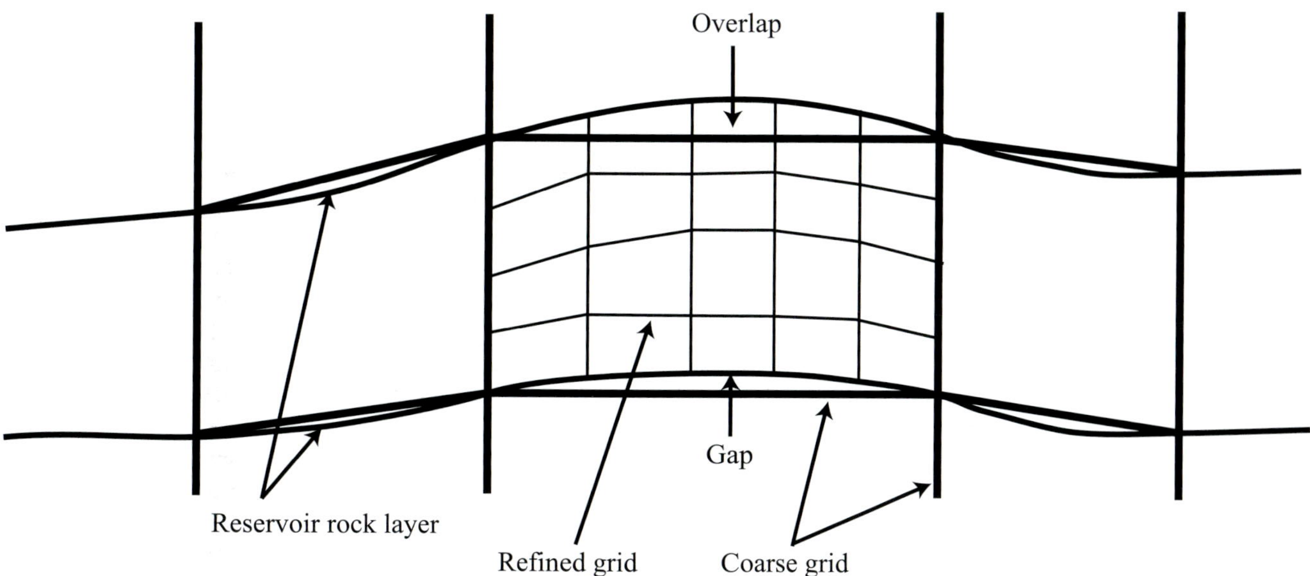

FIGURE 2. Local and coarse meshes for local grid refinement (LGR).

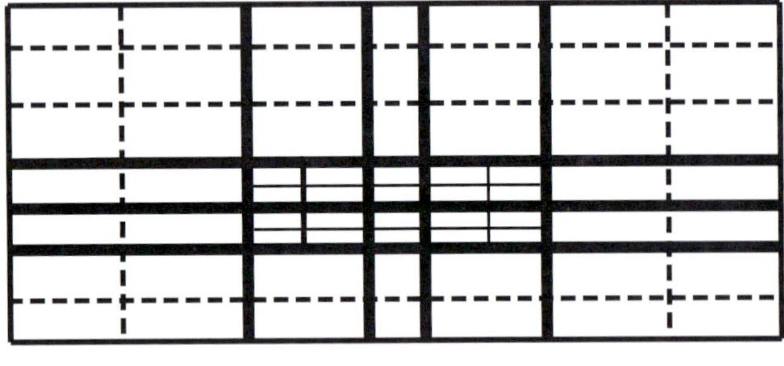

FIGURE 3. Constraints of the coarse grid on the local grid.

geometry of each local area at the interface over the whole column. This procedure, which windowing cannot do, takes into account the finest description of the geometric data available. To guarantee a reliable geometry computation, the following constraints on the mesh have to be respected: (1) each local grid must be a refinement of the coarser one, and (2) all nodes of the coarse grid contiguous to the locally refined zone must be shared by the local grid as well (Figure 3).

Validation of the Local Grid Refinement Approach

To validate the LGR approach, a benchmark was performed. It aims to compare the results between the different methods of refinement, shown in Figure 4, which are (1) a reference high-resolution grid, (2) a Tartan grid, (3) windowing, and (4) an LGR grid.

The test study covers an area of 70 × 100 km (43 × 62 mi) in northern Kuwait. The reference model is a high-resolution grid with a mesh size of 200 × 200 m (656 × 656 ft). It includes 45 layers and features overpressure below a thick evaporitic formation, that is, the Gotnia Formation. The Tartan, windowing, and LGR

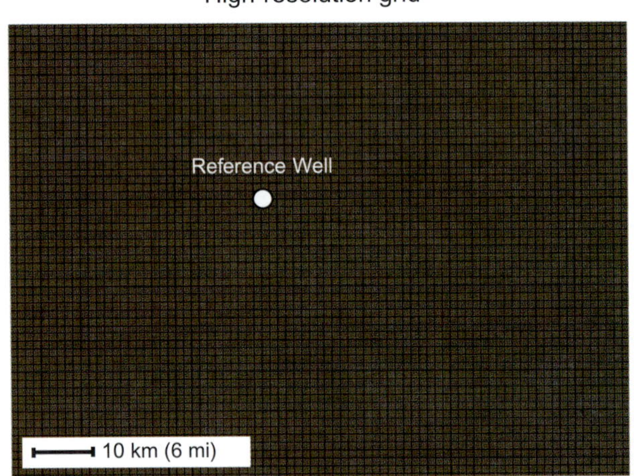

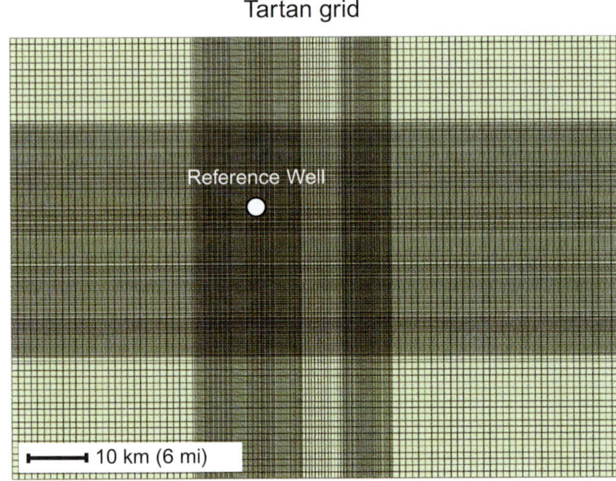

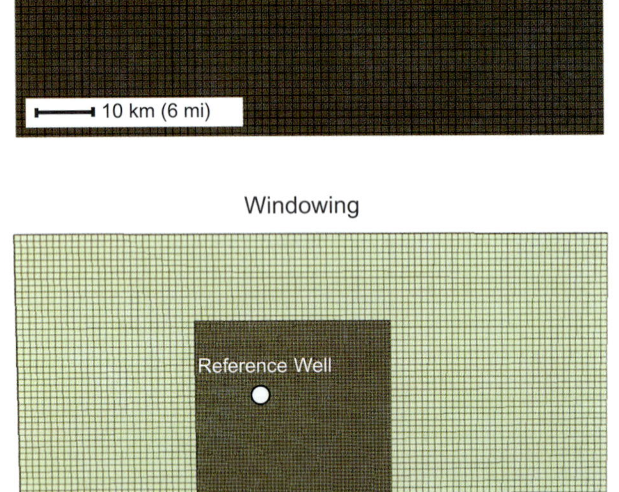

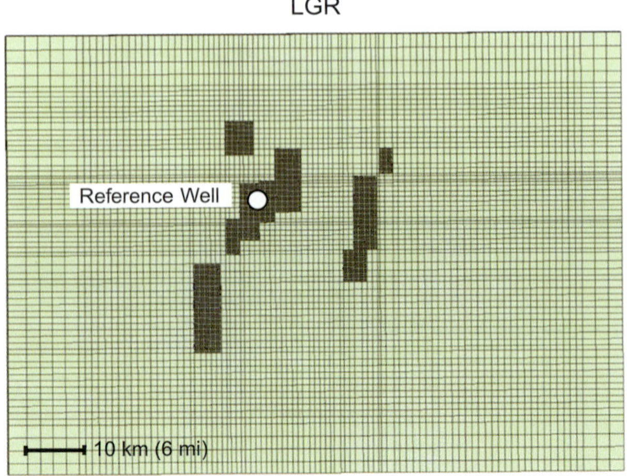

FIGURE 4. Illustration of the grids and location of the reference well used for benchmarking the four types of refinements.

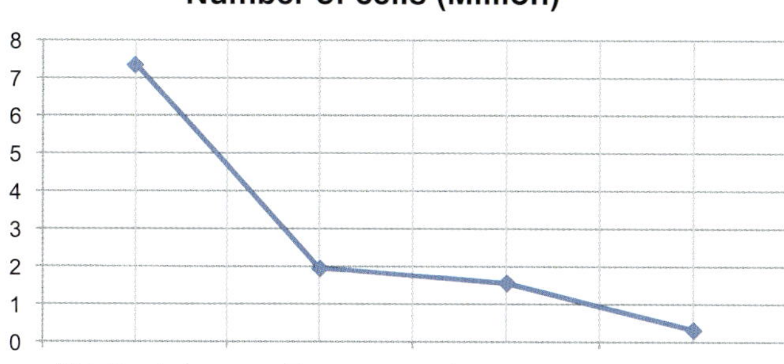

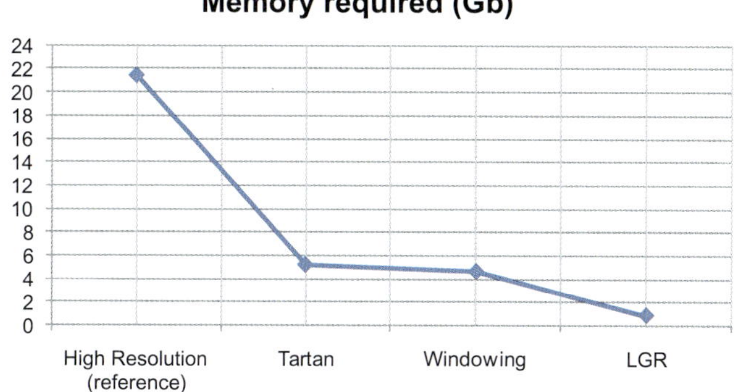

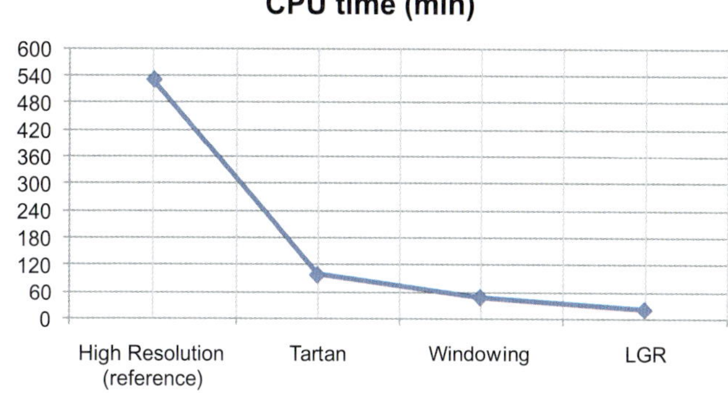

	High Resolution (reference)	Tartan	Windowing	LGR
Number of cells (Millions)	7.33	1.94	1.56	0.32
Memory required (Gb)	21	5	5	1
CPU time (min)	529	98	49	22
Speed up	1.0	5.4	10.8	24.1

FIGURE 5. Results of the benchmark comparing the four refinement methods. The local grid refinement (LGR) offers the best gain in computing time and memory use.

grids are derived from this reference model. Each of these grids focuses on the central part of the model where structures are examined. The resolution of 200 × 200 m (656 × 656 ft) from the reference high-resolution model has been kept in the local area of interest in all models. To capture all the structures, the Tartan grid and the windowing model cover a large area of 50 × 50 km (31 × 31 mi), whereas the LGR allows defining precisely the shape of the structures. For LGR, nine local areas were so defined. A pressure simulation was performed on each of these grids using a Linux cluster of 4 Intel dual-core processors cadenced at 2.9 GHz.

The ability of LGR to consider only refinement inside given *xy* areas and on a particular stratigraphic interval (*z*) reduces the 7.3 million initial cells of the reference high-resolution model down to 320,000 cells for the LGR model. The reductions with the Tartan grid and windowing reach 1.94 million and 1.56 million of cells, respectively (Figure 5). The added value of LGR is clearly demonstrated by a gain of 24-fold of computing time compared with the high-resolution reference solution while preserving the initial resolution for sediment properties definition and output variable values in the areas of interest. The inherent drawback of this solution is the poor definition of the result outside of the areas of interest. Figure 6 displays pressure as a function of depth at the reference well location (see Figure 4). It shows an excellent match between high-resolution grid of reference and LGR. The windowing result is less accurate mainly because of the nonconservation of fluxes as discussed in Section 2.2.

The gain in computer time observed in pressure simulation, exemplified in Figure 5, is also expected in full compositional hydrocarbon migration computation. Obviously, the computing time will increase depending on the hydrocarbon compositional complexity required (number of hydrocarbon classes). The Tartan case, with its larger number of cells is necessarily going to be slower. Windowing also has a larger number of cells, but because the regional and refined simulations are

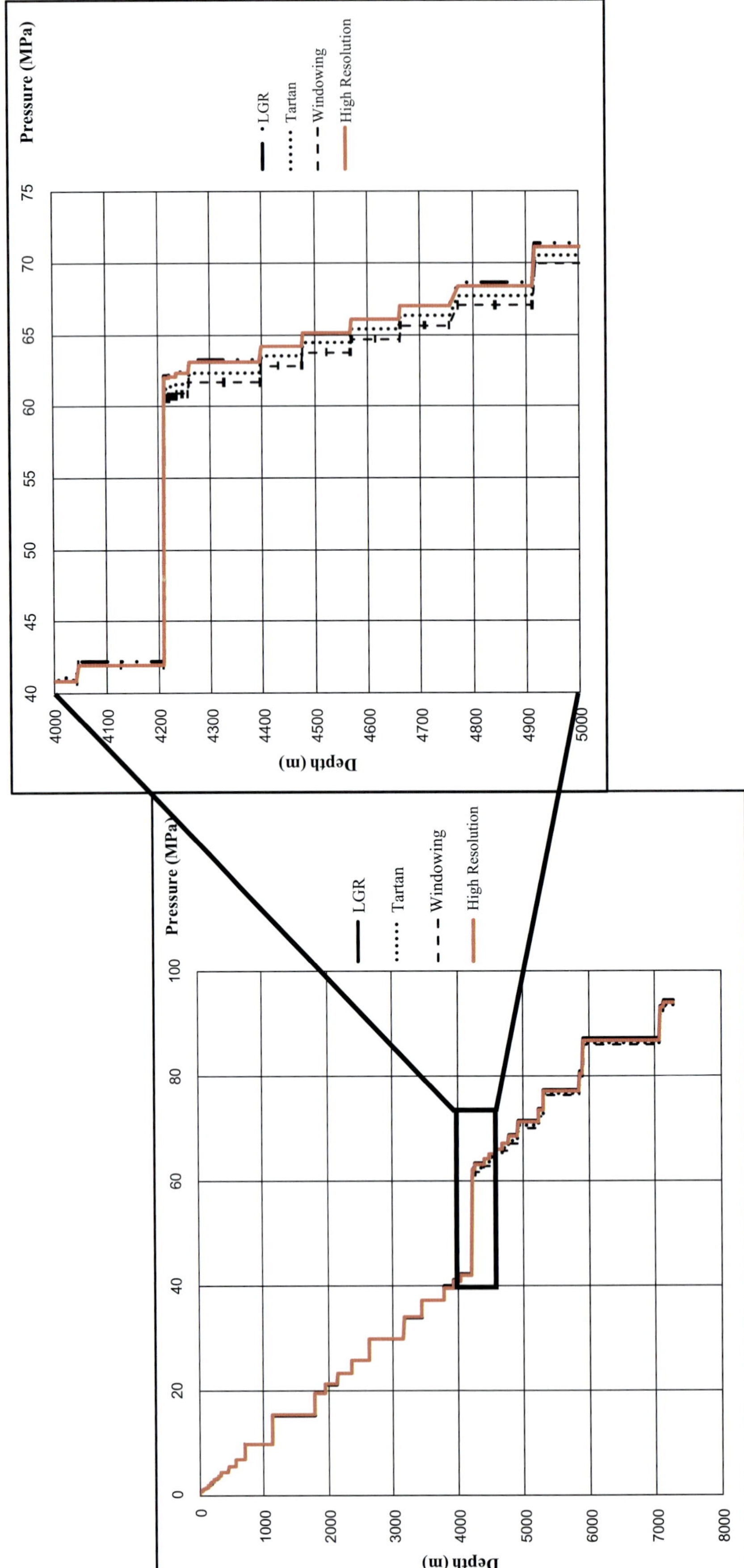

FIGURE 6. Pressure-depth plots of the benchmark simulations comparing the four refinement methods at the reference well location (Figure 4). The local grid refinement (LGR) fits best the high-resolution model.

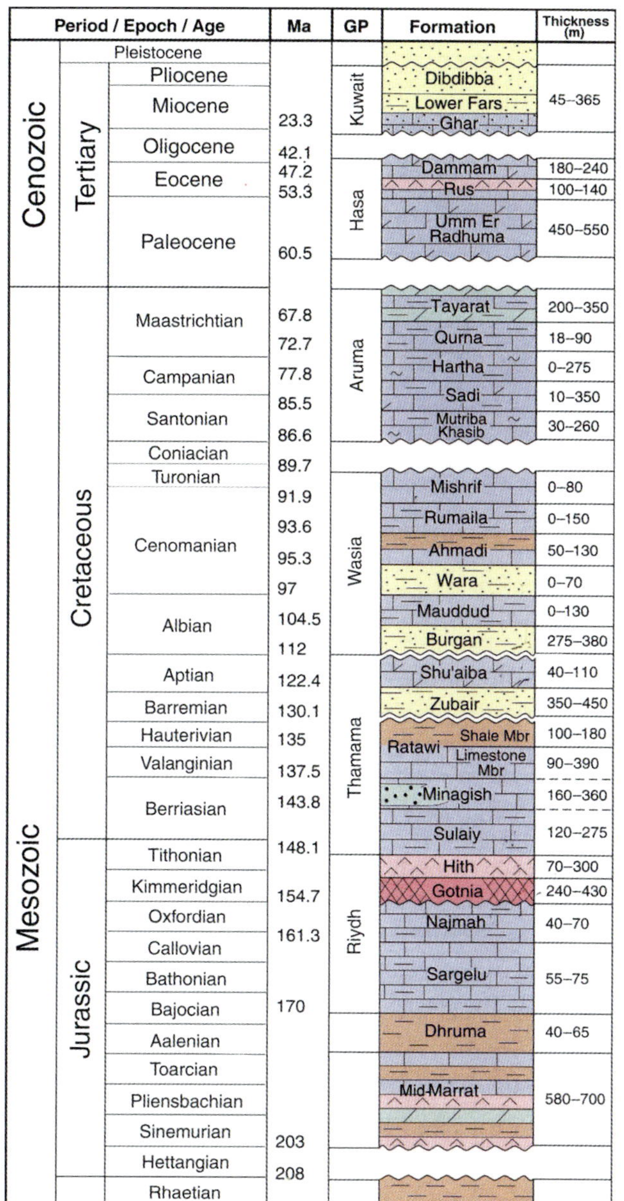

FIGURE 7. Stratigraphic section for onshore Kuwait (modified from Carman, 1996).

not coupled, the total computing time for both may actually be shorter than for Tartan or LGR grids. Yet, the severe drawback of windowing resides in that it does not guarantee fluid flux conservation. An LGR application in a real case of northern Kuwait is presented below.

LOCAL GRID REFINEMENT APPLICATION

The Study Area

The northern Kuwait study area is located onshore Kuwait and encompasses the northeastern region of the country. The area includes several producing oil fields, such as Sabriyah, Rhaudatain, and Umm Gudair. These fields produce from Jurassic carbonate and Cretaceous carbonate and siliciclastic reservoir rocks and contain oil with API gravities varying from 40° in the Jurassic to 28 to 35° in the Cretaceous. The gas-oil ratio (GOR) is above 2000 SCF/bbl (356 m^3/m^3) in the Jurassic, and below 1000 SCF/bbl (178 m^3/m^3) in the Cretaceous. Moderate overpressure exists in the Lower Cretaceous (1000–1500 psi [6895–10,340 kPa]), and strong overpressure exists in the Jurassic (>4500 psi [31,000 kPa]).

In 2009, northern Kuwait Jurassic oil potential was investigated through modeling techniques to plan delineation wells before development. In this respect, a 3-D petroleum system modeling study was undertaken using the LGR technique.

The Jurassic Mid-Marrat Formation, which hosts accumulations currently being delineated for development in the near future, is the main focus of the study. A 3-D integrated regional model of the entire onshore Kuwait, followed by a subregional model of northern Kuwait was initially developed. The latter was used as a basis for LGR implementation. This model uses a Darcy multiphase flow formalism, with a six-component kinetic scheme to provide a sufficiently accurate evaluation of oil API gravity and GOR.

General Context of the Petroleum System

The petroleum system of northern Kuwait developed itself in the context of the Arabian platform, which remained relatively stable since Paleozoic time. The main tectonic events are limited to mild vertical movements (with the possible exception of the Hercynian uplift phase), with a general progressive subsidence of an existing paleotopography acquired since the Paleozoic. This subsidence has slightly accelerated during the Middle to Upper Jurassic incipient extensional phases and has stopped or even slightly reverted during specific Cretaceous and early Tertiary compressionlike phases. These phases never led to detectable angular unconformities at seismic scale.

Since Paleozoic time, sediments deposited on a platform environment, with continental or nearshore clastics during the Paleozoic, inner-shelf carbonates and evaporites during Triassic (Khuff) and Lower and Middle Jurassic, outer-shelf evaporites during the Upper Jurassic (Gotnia and Hith evaporites), and shelf-type deposits during the Cretaceous and part of the Tertiary, and then back to continental deposits (Figure 7).

In northeastern Kuwait, the overall structural framework consists of a north-south arch developing since the Paleozoic, corresponding to the northern extension of the Burgan arch. This arch is subdivided in

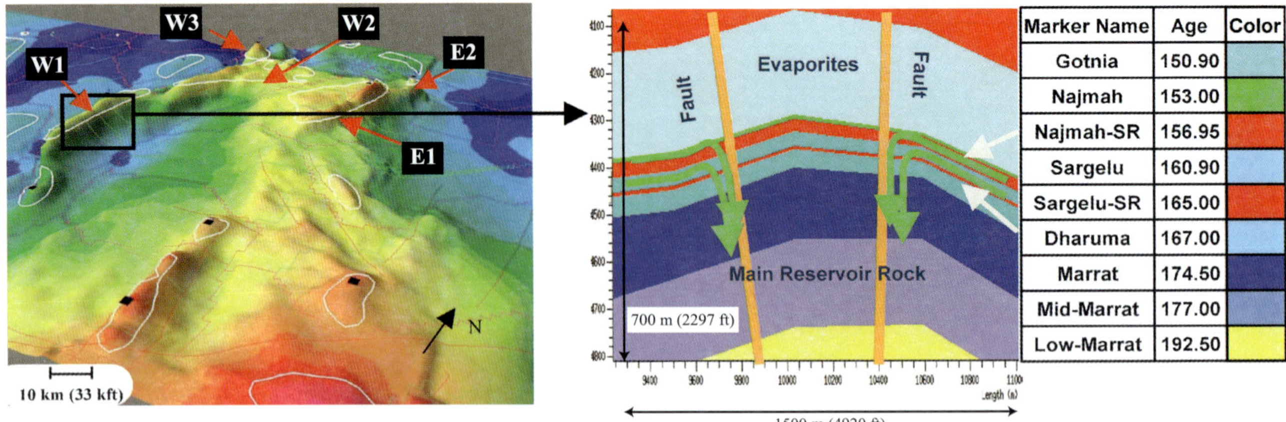

Marker Name	Age	Color
Gotnia	150.90	
Najmah	153.00	
Najmah-SR	156.95	
Sargelu	160.90	
Sargelu-SR	165.00	
Dharuma	167.00	
Marrat	174.50	
Mid-Marrat	177.00	
Low-Marrat	192.50	

Structural map of top Mid-Marrat reservoir showing the limits of the drainage areas (red lines) and the extension of closures (white lines) where LGR has been applied to the fields.

Hydrocarbons migrate from Najmah source rock with high Pc through faults with low Pc toward the main reservoir rock Mid-Marrat.

FIGURE 8. Description of the study area.

two branches near Sabriyah and Rhaudatain fields, respectively. The subsiding areas are located offshore to the east toward the Bubiyan island and to the west.

The oil fields are located on highs of the arch branches. The arch displays numerous faults oriented north-south (+-30°) in the Jurassic. The faults active through time affect the regional evaporitic seal, that is, the Late Jurassic Gotnia and Hith formations.

Petroleum System

The main Jurassic reservoir rocks are the Mid-Marrat Formation and the fractured carbonates of the Sargelu and Najmah formations sealed by the Gotnia-Hith evaporites (see Figure 7). In the Cretaceous sedimentary section, the main reservoir rocks are the Minagish, Burgan, and Mauddud formations sealed by shales (Ratawi shales for instance for the Minagish reservoir).

The main source rock is the Najmah shale (marine shale), with a high total organic content ([TOC] >6%). The Lower Cretaceous Makhul Formation with a TOC less than 2% may have contributed to Cretaceous accumulations but to a limited extent. The Silurian Qusaiba source rock is suspected to have contributed to the oil accumulations, although its existence has

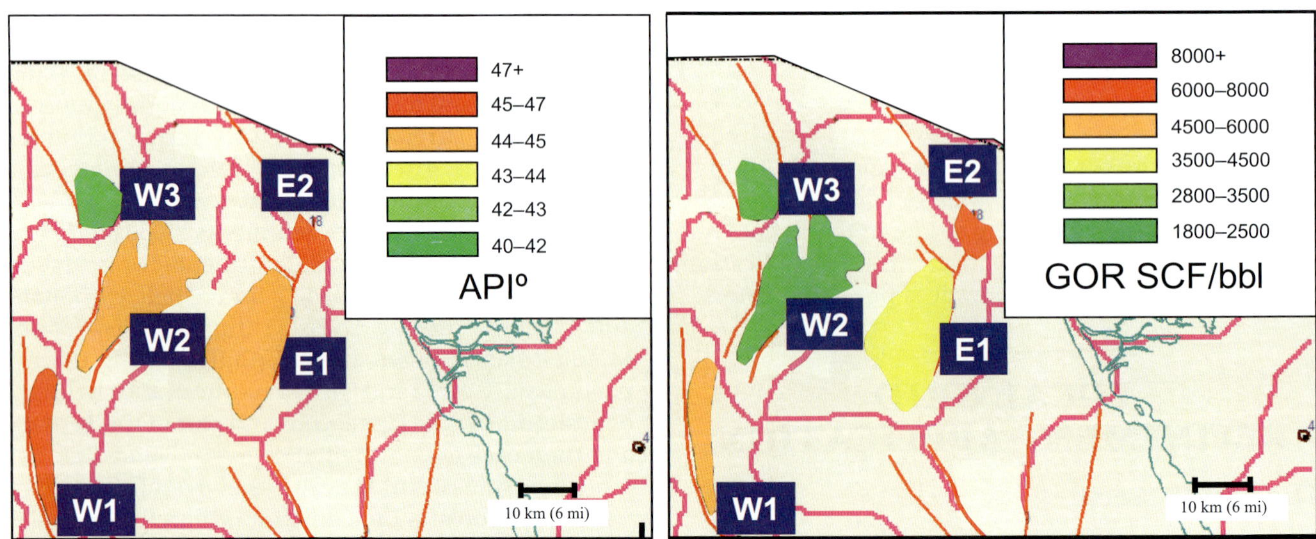

FIGURE 9. Average API gravity and gas-oil ratio (GOR) maps in the study area.

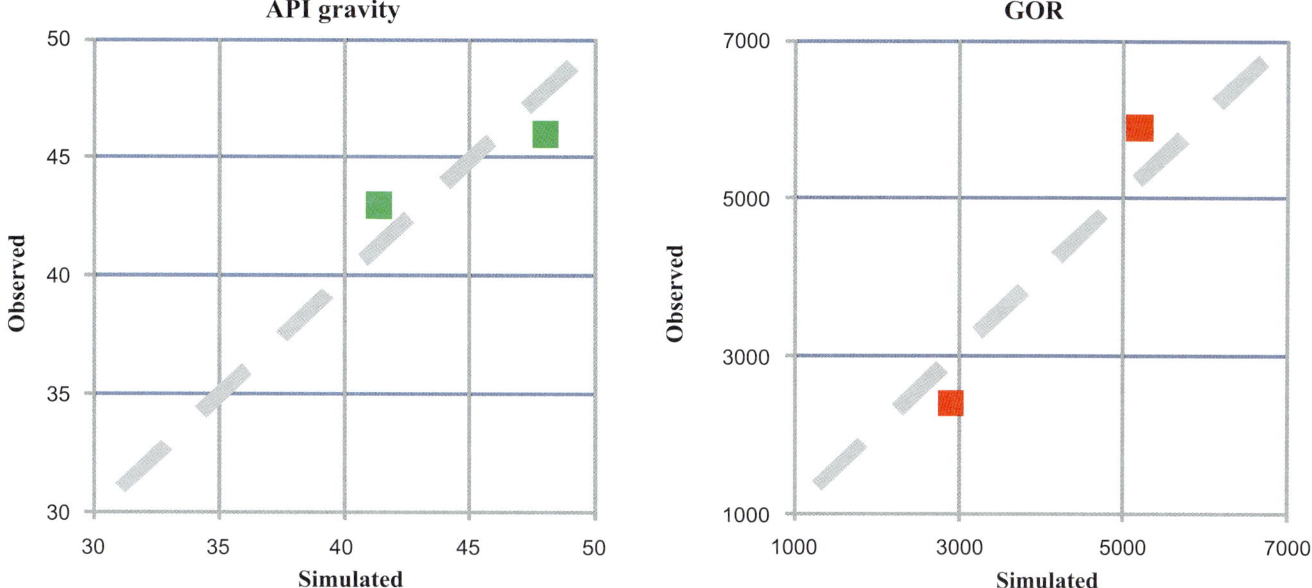

FIGURE 10. Calibration of the northern Kuwait three-dimensional model between observed and simulated API gravity and gas-oil ratio (GOR) in Mid-Marrat accumulations.

never been proven in Kuwait. Gas chromatography and biomarker signatures of oil are consistent with the Najmah oil type (Chenet et al., 2006).

The faults crossing the Gotnia-Hith evaporitic seal play a significant role through time in controlling hydrocarbon migration pattern and pressure differential developing above and below the seal (see Figure 8). This study suggests that the faults act first as conduits to fluid flow upward through the Gotnia-Hith evaporites, then as permeability barriers since approximately middle Eocene. Consequently, oil migrating from the Najmah source rock in the early phase of hydrocarbon generation through the faults across the Gotnia-Hith evaporites helped by developing overpressure-fed Cretaceous accumulations. Then, while the Najmah source rock continues generating, the faults closed and oil overfilling the thin reservoirs of the Najmah and Sargelu formations fed downward Mid-Marrat accumulations.

As suggested by the 3-D petroleum system simulation of northern Kuwait, the Jurassic accumulations underwent significant secondary cracking during the later phase of burial, leading to the observed high API

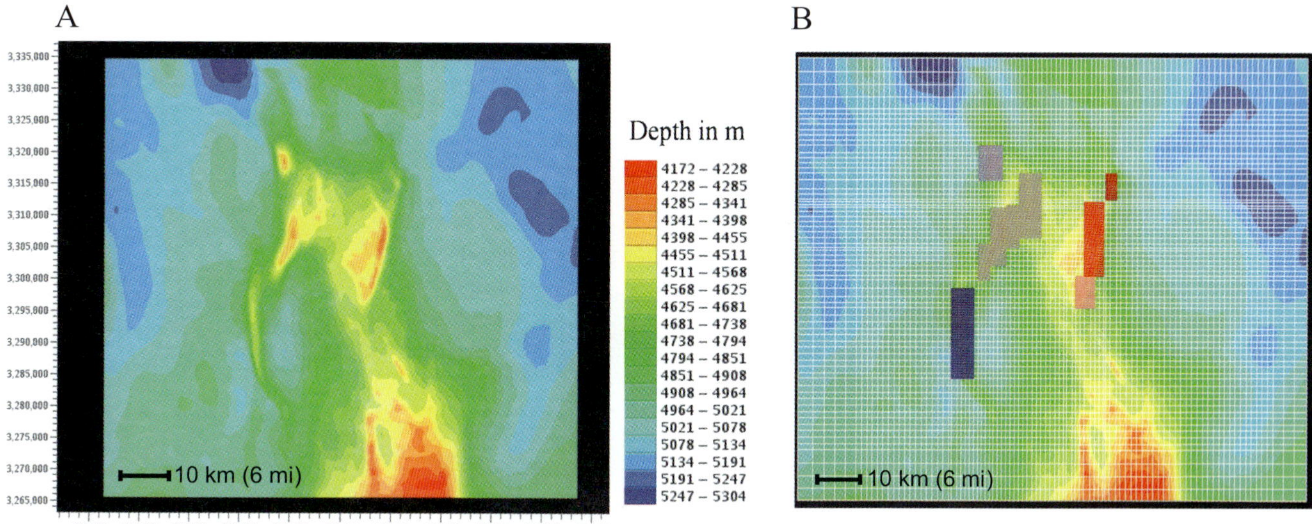

FIGURE 11. (A) Map view of structure on Mid-Marrat reservoir layer over the study area. The scale indicates the variation of structure depth. (B) Map view of the regional grid and local grid refinement (LGR) zones at Mid-Marrat Formation level over the structures of interest (colors differentiate the LGR zones).

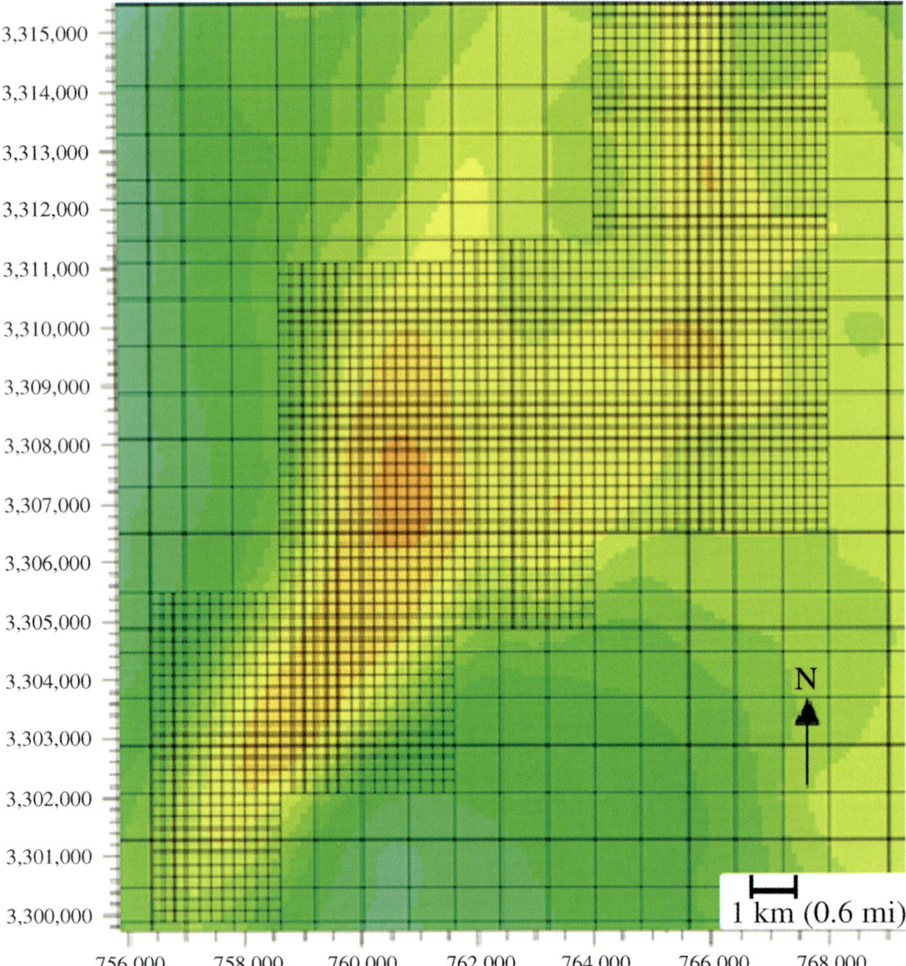

FIGURE 12. Detail showing where local grid refinement (LGR) was applied in *x* and *y* directions on Mid-Marrat Formation.

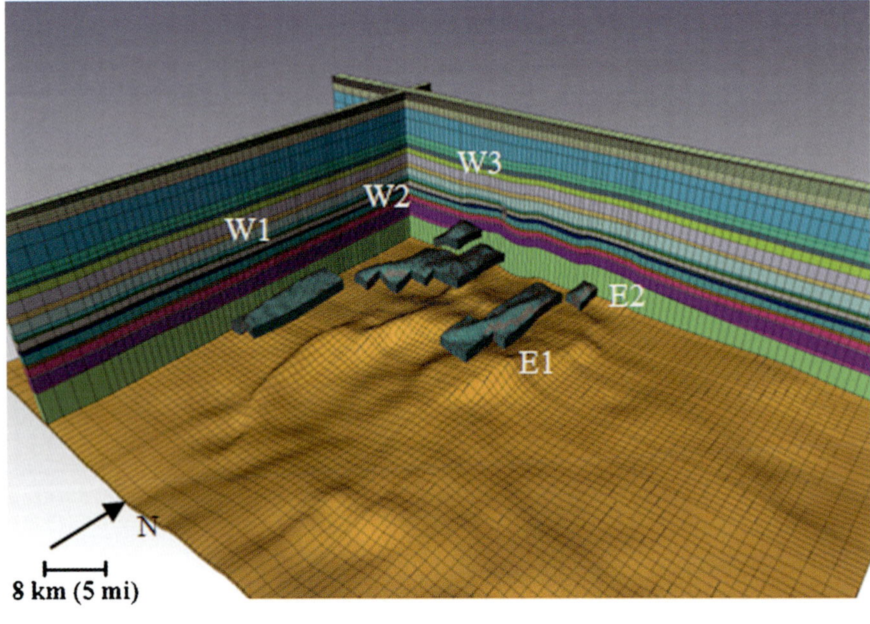

FIGURE 13. Three-dimensional regional view of local grid refinement (LGR) areas: W1,W2, W3, E1, and E2.

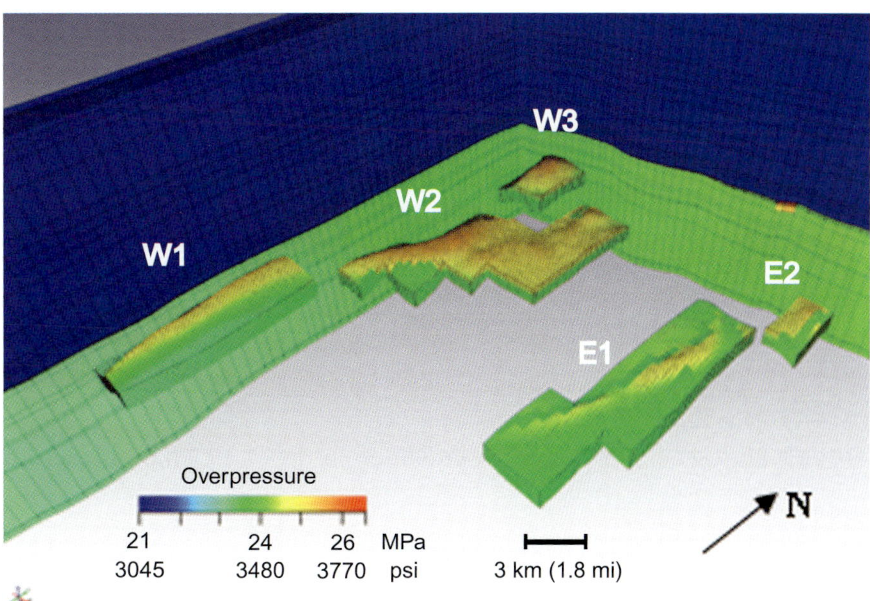

FIGURE 14. Overpressure regime below Gotnia evaporites.

gravities (>38°) and high GOR comprised between 2000 and 10,000 SCF/bbl or 356 and 1781 m^3/m^3 (Figure 9). The calibration of the 3-D model between observed and simulated API gravity and GOR in Mid-Marrat accumulations is shown in Figure 10. The oil accumulated in the Cretaceous escaped secondary cracking conditions and consequently kept its original gravities of 25 to 35° API and GOR less than 1000 SCF/bbl (178 m^3/m^3).

Local Grid Refinement Application to the Northern Kuwait Block

The objective of simulating the northern Kuwait block was to test the behavior of and the results produced by LGR technology in a real case scenario. The focus is limited to traps at the level of the Jurassic Mid-Marrat reservoir known to be prospective in the area. The model was refined to field-scale grid size in the areas of interest while the regional grid was loosened. The LGR zone was generated and populated in facies and source rock properties. A fine resolution grid of 200 × 200 m (656 × 656 ft) was applied to all the structures of interest at the level of the Mid-Marrat reservoir (177 Ma). Nine fine-scale grids were defined within the regional model. Elsewhere, the model was upscaled 5 times near the LGR zones and up to 10 times further away.

Vertically, the Mid-Marrat Formation was refined by a factor 4 to improve the description of the filling history of the existing fields and assess the depth of the oil-water contact (OWC). With this level of refinement, a cell in the loosest part of the regional model is approximately 100 times larger than a cell in the LGR zone (Figures 11, 12).

RESULTS

A regional view of the LGR model and the location of the structures of interest is displayed in Figure 13. The simulation provides results that demonstrate the

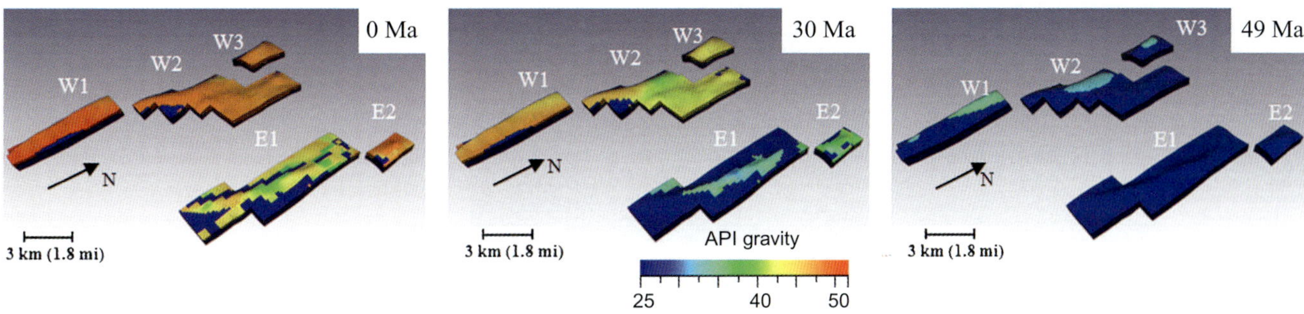

FIGURE 15. The API gravity distribution in the Mid-Marrat reservoir rock through time.

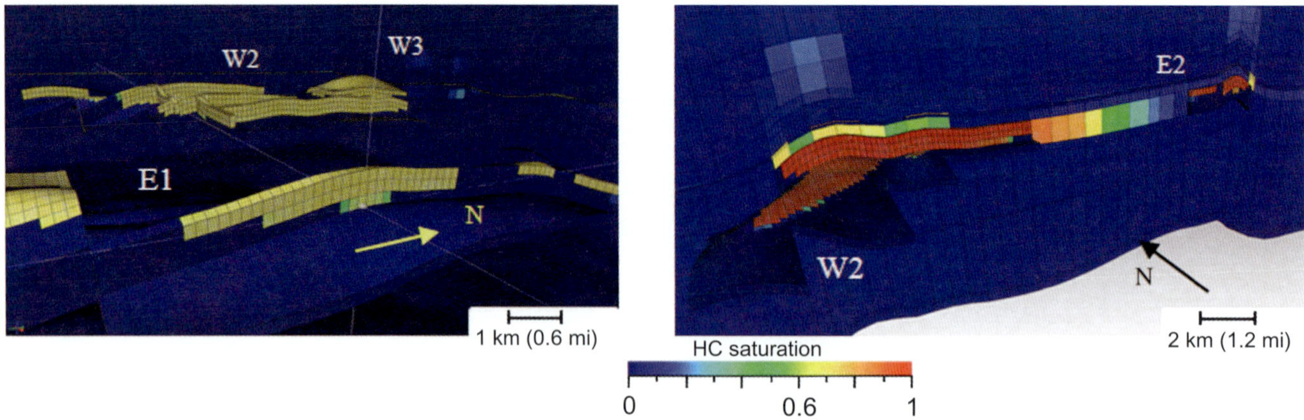

FIGURE 16. Detail of hydrocarbon saturation in the Mid-Marrat reservoir rock.

added value of the LGR approach for field delineation and development.

Delineation Wells Estimating Fracturing Risks

The detailed mapping of the overpressure allows to identify precisely the areas where the fracturing conditions are the riskiest when designing the field development plan. The W structures display the highest overpressure. In W2, the eastern extension of the structure compared with the western part seems safer to drill (Figure 14).

API Gravity

The structures located at various depths are supposed to communicate through the Mid-Marrat reservoir layer. The model, however, shows differences in oil quality (API gravity), suggesting specific infilling histories and compartmentalizations inducing gravity segregation. The E structures are filled up later than the W structures, yet E2, W3, and W2 display similar oil qualities (Figure 15).

Oil-Water Contact

The 3-D snapshots of Figure 16 show the coarse grid used as a framework and the refined grids corresponding to each field and prospect. The position of the oil-water contact (OWC) can be precisely modeled. Given that, in this particular case, the cells have an average thickness of 100 m (328 ft) in the refined areas, the vertical location of the OWC can be computed with an uncertainty of about 50 m (164 ft).

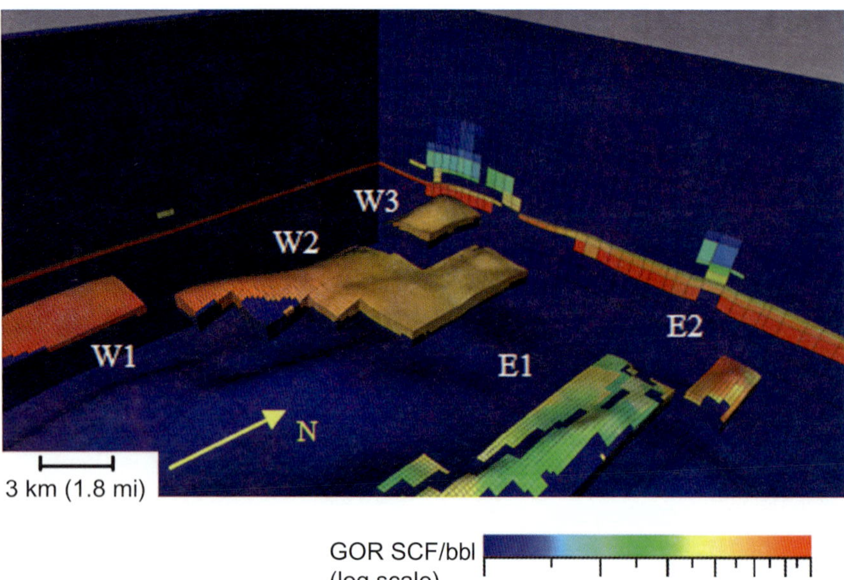

FIGURE 17. Detail of gas-oil ratio (GOR) in the Mid-Marrat reservoir rock.

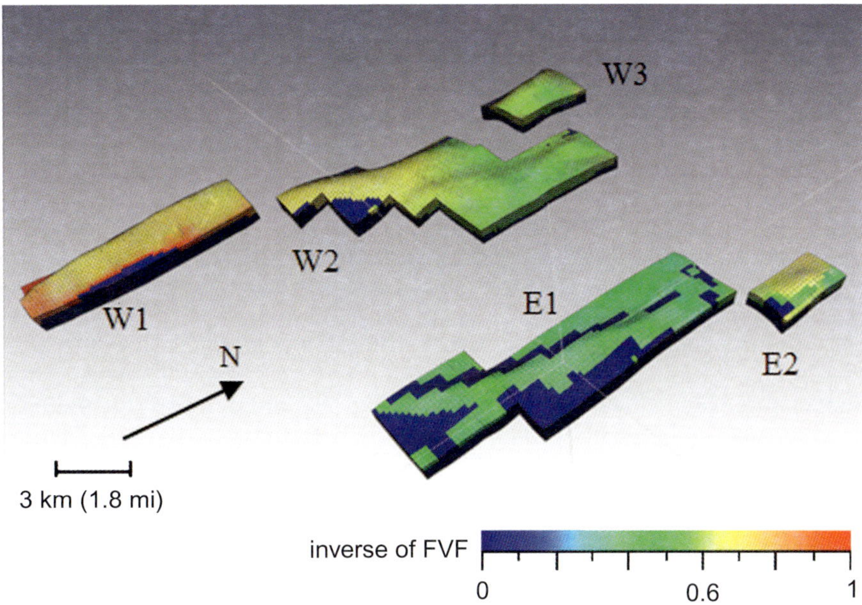

FIGURE 18. Detail of formation volume factors (FVFs) in the Mid-Marrat reservoir rock.

Gas-Oil Ratio and Formation Volume Factors

The calculation of GOR and formation volume factor can readily be used for a reservoir model initialization. As fields are probably connected at the level of the Mid-Marrat Formation, reservoir simulation can be made combining the various productions of the fields based on a more accurate description of the initial conditions. For instance, the E2 structure, although close to E1, displays a much higher GOR (Figures 17, 18).

CONCLUSIONS

The added value of LGR compared with the other methods of refinement, that is, Tartan gridding and windowing, resides in gaining computing time and memory while precisely describing the infilling processes of the accumulation areas. Moreover, the fluxes at the interface between refined local and regional grids of the model are computed accurately, and the feedback of local mesh heterogeneities on the fluxes at regional scale is accounted for. With LGR, performing a multiphase multicomponent fluid-flow simulation with a generalized Darcy formulation with precision in local areas of a regional block is now feasible within a reasonable computing time.

The LGR approach efficiently fills the gap between conventional basin modeling and reservoir modeling. Its application in northern Kuwait provides useful guidelines to predict API gravities, GOR, and OWC depth estimates in new prospects. The model shows differences in oil quality (API gravity) suggesting specific infilling histories and compartmentalizations inducing gravity segregation. The prediction of field compartmentalization and estimated position of the OWC facilitates implementation of delineation and development wells.

ACKNOWLEDGMENTS

We thank the Kuwait Oil Company for letting us present this work as a poster at the Hedberg Conference 2009 and publish it in the Hedberg AAPG special publication. We also thank the reviewers Leslie Magoon and Frederic Schneider for their constructive comments.

REFERENCES CITED

Agelas, L., I. Faille, S. Wolf, and S. Requena, 2007, High performance computing Darcy compositional single-phase fluid-flow simulations (abs.): Proceedings of the AAPG Hedberg Conference Basin Modeling Perspectives: Innovative developments and novel applications, The Hague, Netherlands, http://www.searchanddiscovery.com/abstracts/html/2007/hedberg/short/agelas.htm (accessed May 19, 2011).

Carman, G. J., 1996, Structural elements of onshore Kuwait, GeoArabia: Middle East Petroleum Geosciences, v. 1, no. 2, p. 239–266.

Chenet, P. Y., N. Bianchi, M. Pernelle, A. Prinzhofer, F. Lorant, S. K. Bhattacharya, A. Al-Khamiss, and M. Al-Hajeri, 2006, Oil and gas fingerprinting in onshore Kuwait: Implications for the petroleum system (abs.): Proceedings of the GEO 2006 Middle East Conference and Exhibition, Manama, Bahrain, http://www.searchanddiscovery.com/abstracts/pdf/2006/geo/index.htm?q=%2BtextStrip%3Achenet#C (accessed May 19, 2011).

Ding, Y., and P. A. Lemonnier, 1993, Development for dynamic local grid refinement in reservoir simulation: Proceedings of the 12th SPE symposium on reservoir simulation, New Orleans, Louisiana, paper SPE 25279, http://www.onepetro.org/mslib/app/Preview.do?paperNumber=00025279&societyCode=SPE (accessed May 19, 2011).

Reme, H., G. Å. Øye, G. E. Fladmark, and M. Espedal, 1998, Application of local grid refinements for simulation of fluid flow through faulted and fractured reservoirs: http://citeseerx.ist.psu.edu/viewdoc/summary?doi=10.1.1.43.8436 (accessed December 10, 2010).

Requena, S., M. Thibaut, S. Pegaz, and L. Agelas, 2005, Pushing the limits of 3-D basin modeling with linux clusters (abs.): Proceedings of the 67th European Association of Geoscientists & Engineers Conference and Exhibition: Petroleum systems, Madrid, Spain, http://www.earthdoc.org/detail.php?pubid=1149 (accessed May 19, 2011).

Schneider, F., and S. Wolf, 2000, Quantitave HC potential evaluation using 3-D basin modeling: Application to Franklin structure, Central Graben, North Sea, U.K.: Marine and Petroleum Geology, v. 17, p. 841–856, doi:10.1016/S0264-8172(99)00060-4.

Schneider, F., S. Wolf, I. Faille, and D. Pot, 2000, A 3-D basin model for hydrocarbon potential evaluation: Application to Congo offshore: Oil & Gas Science and Technology, v. 55, no. 1 , p. 3–13, doi:10.2516/ogst:2000001.

Thibaut, M., Y. Caillabet, E. Flauraud, and F. Schneider, 2005, Integrating local grid refinement methods for basin modeling (abs.): Proceedings of the 67th European Association of Geoscientists & Engineers Conference and Exhibition: Petroleum systems, Madrid, Spain, http://www.earthdoc.org/detail.php?pubid=1150 (accessed May 19, 2011).

Willien, F., I. Chetvchenko, R. Masson, P. Quandalle, L. Agelas, and S. Requena, 2009, AMG preconditioning for sedimentary basin simulations in Temis calculator: Marine and Petroleum Geology, v. 26, p. 519–524, doi:10.1016/j.marpetgeo.2009.01.014.

Wolf, S., I. Faille, S. Pegaz-Fiornet, and F. Willien, 2009, A new innovating scheme to model hydrocarbon migration at basin scale: A pressure-saturation splitting (abs.): Proceedings of AAPG Hedberg Conference Basin and Petroleum System Modeling: New horizons in research and applications, Napa, California, http://www.searchanddiscovery.com/abstracts/html/2009/hedberg/abstracts/short/wolf.htm (accessed May 19, 2011).

Steam rises from fumaroles in the Sulfur Springs thermal area of The Geysers geothermal field near a pipe carrying superheated steam from a producing well.

SECTION 3

Structural and Tectonic Processes

6

Gibson, R., 2012, A methodology to incorporate dynamic salt evolution in three-dimensional basin models: Application to regional modeling of the Gulf of Mexico, *in* K. E. Peters, D. J. Curry, and M. Kacewicz, eds., Basin Modeling: New Horizons in Research and Applications: AAPG Hedberg Series, no. 4, p. 103–118.

A Methodology to Incorporate Dynamic Salt Evolution in Three-Dimensional Basin Models: Application to Regional Modeling of the Gulf of Mexico

Richard Gibson[1]

BP America, Houston, Texas, U.S.A.

ABSTRACT

Construction of three-dimensional (3-D) basin models in areas of detached salt tectonics poses difficult challenges but is necessary to simulate the 3-D thermal effects of salt and correctly model subsalt burial histories. Over much of the offshore northern Gulf of Mexico Basin, a regional salt canopy detaches shallow structures, formed by growth and subsequent collapse of allochthonous salt sheets, from subsalt structures formed mostly in response to movement of deep (autochthonous) salt. Dynamic simulation modeling of this system requires (1) understanding the evolution of salt distribution and thickness through time, (2) a methodology to incorporate thickness changes within the simulation model, and (3) geometric solutions to account for the fact that allochthonous salt occurs at various stratigraphic levels across the basin.

Twenty regionally mapped horizons, including top and base of allochthonous salt and a composite weld representing areas of collapsed salt canopy, were used to build a regional Gulf of Mexico numerical simulation model. Salt isopach maps for sequential stages of the basin evolution were derived by vertical backstripping using "regionals" constructed to approximate the predeformation geometry of selected horizons. For a given horizon, the salt thickness changes since horizon deposition is represented by the difference between a mapped horizon and its regional. Simple rules were applied to partition the derived salt thicknesses between the allochthonous and autochthonous salt levels. To model the climb of the salt canopy across stratigraphy, the basin model was divided into subsalt and suprasalt parts containing horizons of equivalent age separated by an intervening allochthonous salt layer. The thickness of both the allochthonous and autochthonous salt layers were altered through time using the salt isopachs.

[1]*Present address:* Apache Corp., Tulsa, Oklahoma.

DOI:10.1306/13311431H43466

The resulting simulation model reasonably represents the large-scale structural evolution of the northern Gulf of Mexico Basin, including (1) progressive southward displacement and evacuation of salt along the Louann level, (2) basinward stratigraphic climb and progressive welding of salt within the canopy, (3) seaward progradation of depositional systems throughout the Mesozoic and Cenozoic, and (4) Miocene uplift and erosion of the onshore part of the basin. The methodology outlined can be adapted to assist in building basin models in other structurally complex basins.

INTRODUCTION

Natural deformation in sedimentary basins is commonly disharmonic with depth because of the movement of material along gently dipping detachment surfaces. The resulting structural geometries pose significant challenges for traditional basin modeling approaches because of the constraints imposed by vertical backstripping throughout the model evolution. Simple backstripping of such systems generates geometric artifacts caused by the structurally impacted thicknesses between horizons (Figure 1). When doing basin modeling for the purpose of evaluating petroleum system elements, these artifacts impact the geometry of horizons back through time and, thus, the inferred burial histories used to determine source rock maturity evolution, petroleum migration pathways, and reservoir quality prediction.

Complex structurally impacted burial histories are routinely simulated in one-dimensional simulation models by thickening and thinning of individual layers through time (McBride et al., 1998b; Gibson et al., 2004). Similar approaches can be applied in two-dimensional (2-D) or three-dimensional (3-D) models if the thickness changes can be constrained. Software developments in recent years allow the use of preexisting structural restorations as templates or direct input to 2-D models (Schneider et al., 2002a, b; Lampe et al., 2006), and Baur et al. (2009) extended this capability to 3-D with an application to the Bolivian thrust belt. These latter approaches require that a 2-D section or 3-D volume be divided into individual fault blocks that move separately in the basin model. Because of the requirement for a restored 2-D section or 3-D volume as input, use of these advanced tools is time and resource intensive.

Salt basins provide a particular challenge for basin modelers. Salt-related structures can include simple diapirs, listric growth fault systems, thin-skinned contractional fold and thrust belts, salt nappes and canopies, and a wide variety of welds produced by emplacement and subsequent evacuation of salt. Salt movement occurs in the vertical and all lateral directions, and the resulting structures generally cannot be adequately described in two dimensions. Complex 3-D structural geometries, combined with the extreme thermal conductivity contrast between salt and other sediments (Mello et al., 1995), necessitate the use of 3-D basin modeling to correctly simulate the multidirectional flow of heat and fluid in these systems. Although 3-D structural restorations of salt systems can be done (Rivero et al., 2005; Grando et al., 2008), they are typically restricted to small areas and are not done routinely enough to be used as standard basin model inputs.

In this chapter, a simple methodology is developed to describe variations in subsurface salt geometry through time and implement these variations into basin models. The workflow described is meant to allow the basin modeler to realistically simulate the geometric evolution using standard basin modeling software without doing a full 3-D structural restoration. This approach does not replace restoration as it has no ability to test or validate a structural interpretation. The current study focuses on a regional-scale model of the northern Gulf of Mexico Basin built to underpin an exploration framework for the area. However, the approaches described here can be applied to other salt basins and adapted to work in both extensional and contractional basins.

GULF OF MEXICO OVERVIEW

The Gulf of Mexico Basin formed as a result of rifting of the southern margin of North America during the mid-Mesozoic (Sawyer et al., 1991; Pindell and Kennan, 2001). The basement consists of Paleozoic rocks of the Ouachita-Appalachian orogenic belt and other terranes sutured to North America during the Carboniferous (Woods et al., 1991). Rifting began in the late Triassic and culminated during the Jurassic. Terrigenous sediments were deposited in early rift basins and overlain by regionally extensive salt deposits (Louann Salt), probably during the Callovian (Salvador, 1991). The original thickness of salt is unknown, although its areal extent can be reasonably established based on sedimentary onlap relationships around the basin margin (Figure 2). Oceanic spreading began either during or soon after salt deposition and resulted in the formation of a crescent-shaped area of oceanic crust between the southern margin of North America and the Yucatan block, which moved counterclockwise out of the present-day Gulf of Mexico

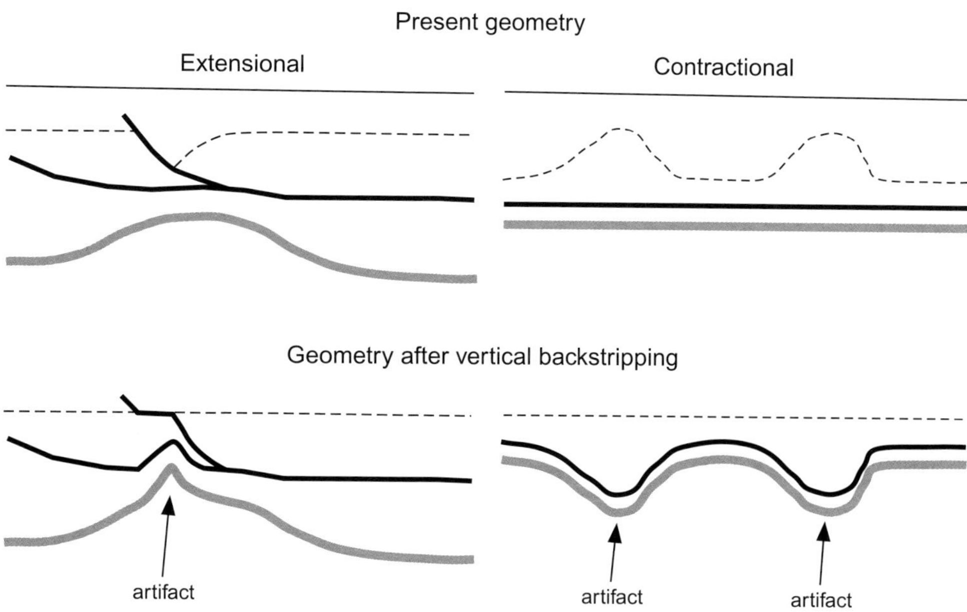

FIGURE 1. Schematic cross sections illustrating common geometric artifacts created by vertically backstripping extensional and contractional structures with low-dipping detachments (bold black lines). Errors are propagated downward so that they impact the geometry of deep layers (e.g., bold gray line) that may be source rocks, deep carrier beds, or target reservoirs.

around a pole of rotation near present-day Cuba (Pindell and Kennan, 2001).

Since the Late Jurassic, the Gulf of Mexico has been a continuous site of sediment deposition, capturing much of the drainage from the southern half of the North American continent. Carbonate deposition, interrupted by several brief periods of siliciclastic influx, predominated during the Cretaceous sea level highstand and led to the development of a prominent shelf margin by the end of the Early Cretaceous (McFarlan and Menes, 1991; Sohl et al., 1991). Major siliciclastic input began in the early Paleogene with deposition of the Wilcox Group (onshore) and its deep-water equivalent strata, mostly via deltas located west and northwest of the basin center (Galloway et al., 2000). Major Oligocene deltas, including the Frio, Vicksburg, and Hackberry, continued to supply sediment to the west and northwestern parts of the basin. Beginning in the early Miocene, marginal uplift of the northwestern basin margin caused drainages to shift eastward, and subsequent deltaic input has been provided mostly by river systems draining into the central Gulf of Mexico through present-day Louisiana (Galloway et al., 2000). The Tertiary depositional setting was mostly progradational, with the shelf margin migrating basinward as much as 280 km (175 mi) since the Late Cretaceous (Figure 2A). Throughout the Tertiary, the eastern Gulf of Mexico (Florida Platform) remained mostly starved of siliciclastic input, and the relict Cretaceous carbonate platform has been progressively drowned.

During basin filling, the Louann Salt reacted in a dramatic manner to a combination of sediment loading and basinward tilting. Evidence of salt movement can be found in the earliest postsalt sediments (Oxfordian Norphlet) as thickness variations and rafting caused by basinward sliding (MacRae and Watkins, 1996; White et al., 1999). Throughout the Cretaceous and Paleogene, salt movement along its original stratigraphic (autochthonous) level created regional and local thickness variations (Figures 3, 4) within the overlying sediment packages (Hall, 2002). Autochthon-sourced salt diapirs and massifs formed by passive downbuilding (White et al., 1999; Johnson et al., 2006). Beginning at least as early as the late Paleogene, salt diapirs and massifs in the downdip part of the basin began to expand into allochthonous sheets and canopies (Diegel et al., 1995). Much of the original autochthonous salt was transferred into the allochthons throughout the Neogene. The detailed geometries, timing, and relative roles of autochthonous and allochthonous salt movement during this period have been discussed by numerous authors (Diegel et al., 1995; Peel et al., 1995; Rowan, 1995, 2002; McBride et al., 1998a; Hall, 2002). Continued sedimentation on top of the allochthonous salt caused many of the early canopies to be welded out as the salt continued to move basinward and higher in the stratigraphic section.

As seen today, the trail of allochthonous salt movement is recorded by a complex linked weld system. For the purposes of this study, this weld system was simplified so that it could be represented as a single surface constructed from many individual weld segments. The philosophy used was to merge the segments together, giving priority to the shallowest weld at any location. The end result is a composite weld surface that regionally climbs from deep in the sedimentary section, near the present-day coastline, to within the Pliocene–Pleistocene section in deep water, where it is continuous with the landward end of the present-day allochthonous salt canopy (Figure 2B).

Sediments deposited on top of the Louann Salt or on the later formed allochthonous salt canopies slid toward the basin center using the salt as the main detachment(s). Much of the central part of the basin south of the Cretaceous shelf edge and north of the present-day salt canopy is dominated by extensional structures (Figure 2A). This domain includes (1) normal faults that root down into the salt autochthon, (2) shale-detached fault systems

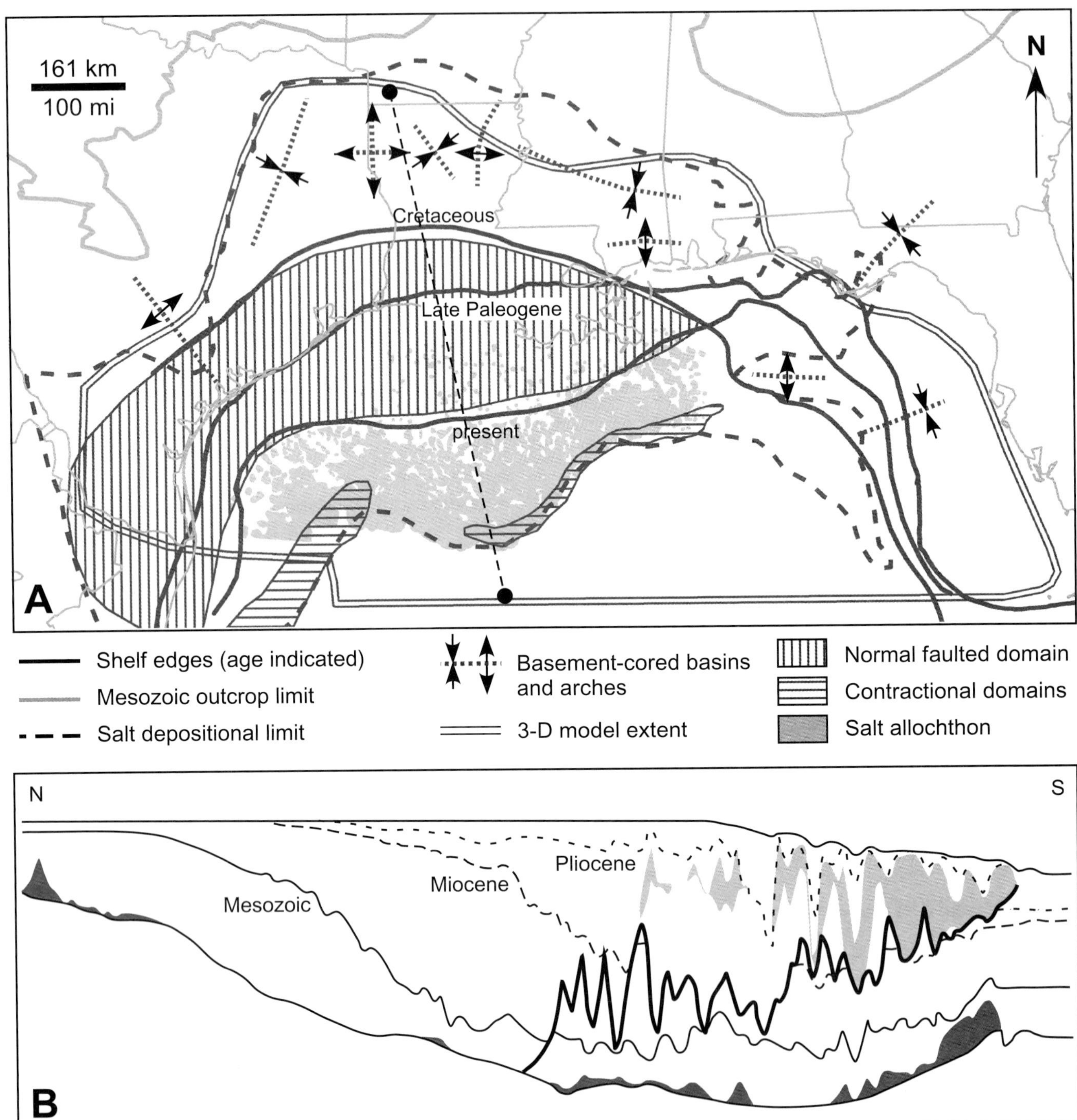

FIGURE 2. (A) Simplified elements map of the northern Gulf of Mexico showing various major features and provinces discussed in text. The salt allochthon is from Hall (2002); salt depositional limit is modified from Salvador (1991); shelf edges are from Galloway et al. (2000) . The extent of the three-dimensional model is indicated by a double black line. A dotted line with circle terminations is the position of the cross section in Figures 2B, 6C, 7, 10, and 12. (B) Simplified regional cross section showing geometry of three selected age horizons, autochthonous (dark gray) and allochthonous (light gray) salt, and regional composite weld system (thick black line) tracking the southward climb of salt through the stratigraphic section.

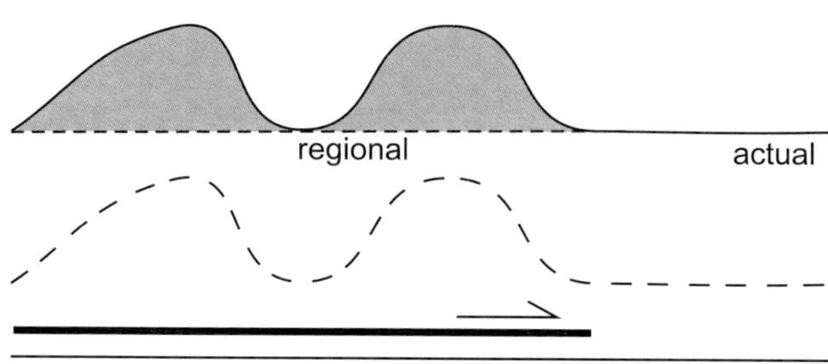

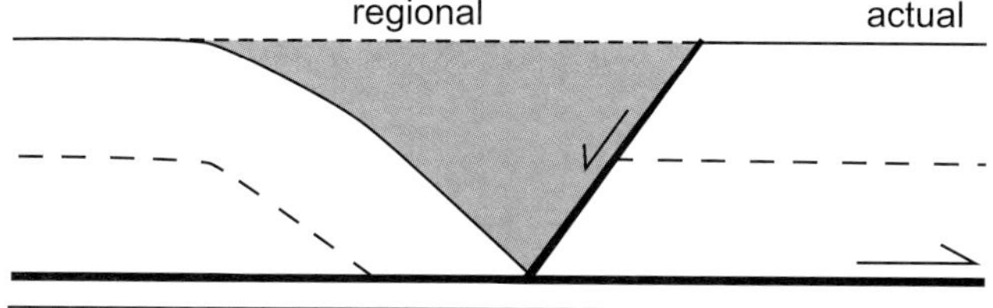

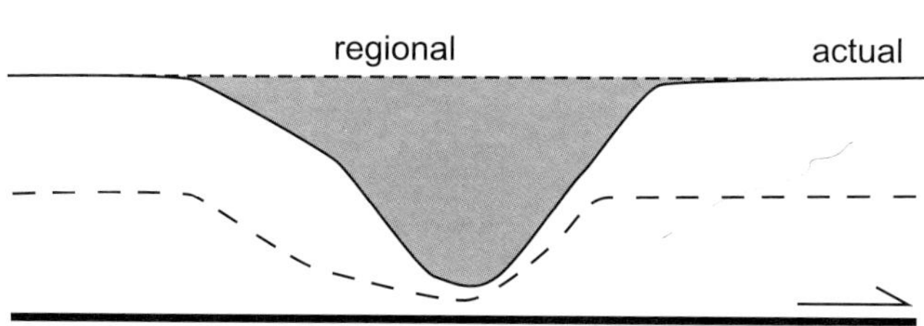

FIGURE 3. Definition of regional relative to actual horizons for contraction, extension, and deflation (e.g., salt withdrawal). Gray areas indicated material added (in contraction) or removed from the section by lateral or vertical movements during deformation since the actual horizon was deposited.

that accommodate expansion of the Paleogene sediment wedges (mostly located onshore Texas), and (3) arrays of normal faults that accommodate subsidence and possible extension above and adjacent to the allochthonous salt (Diegel et al., 1995; Rowan et al., 1999). In contrast, contractional fold and thrust belts (Weimer and Buffler, 1992; Trudgill et al., 1999) comprise much of the downdip end of the system (Figure 2A). The amount of shortening in these fold belts is relatively minor compared with the width of the entire basin (e.g., 6–9 km [3.7–5.6 mi] in Perdido fold belt; Trudgill et al., 1999), and a significant fraction of the total shortening needed to balance the updip extension may be hidden in cryptic contractional features among complex salt structures (figure 3 of Rowan, 2002) combined with local closure of salt diapirs.

BASIN MODELING CHALLENGES

The goal of this study was to create a numerical simulation model of the northern Gulf of Mexico that realistically restores back through time with minimal artifacts (Figure 1). The area of interest includes all of the U.S. offshore plus the onshore parts of the basin from Alabama westward to south Texas. Encompassed within this 1.1 million km^2 area is most of the Louann Salt Basin preserved on the northern margin of the Gulf of Mexico (Figure 2A).

The nature of available simulation software places limitations on how one can model complex geologic systems. Specifically, all displacements between grid nodes (or cells) are constrained to be vertical. Using traditional basin modeling tools, apparent thickness changes caused by contraction or extension must be approximated by changing the thicknesses of specific layers during the model evolution. Defining the nature of these thickness changes is a major challenge. The following section describes the method used to handle these complexities.

Figures 2B and 5A illustrate one feature of the Gulf of Mexico that poses the most significant challenge for simulation model construction: the composite weld that records the past position of the salt canopy occurs at different stratigraphic positions across the basin, starting out within the oldest strata toward the north and climbing updip and upsection basinward. As a result, the same stratigraphic age horizon occurs above the composite weld in some areas and below the weld in others. This feature poses a modeling challenge because horizons within numerical simulation models are constrained to existing throughout the model extent, and the intervening layers must have either zero or a finite positive thickness (i.e., horizons cannot cross). The work-around used to model this situation is described in subsequent sections.

Lateral Translations

At the scale of the modeling presented here, the Gulf of Mexico Basin is pinned at the updip and downdip ends beyond the depositional limit of salt where there has been no lateral translation. In between, I have assumed that most of the deformation can be approximated by vertical movements in response to salt inflation and/or withdrawal, and that lateral movements can be neglected on the scale of the entire basin. This is clearly untrue in detail as demonstrated for the contractional Perdido and Mississippi Fan fold belts (Weimer and Buffler, 1992; Trudgill et al., 1999) and farther updip within the extensional domains (Diegel et al., 1995; McBride et al., 1998a). However, the lateral translations are small compared with the length of a basin-scale cross section and can be neglected for the purpose of producing a reasonable basin-scale model. Clearly, capturing these translations

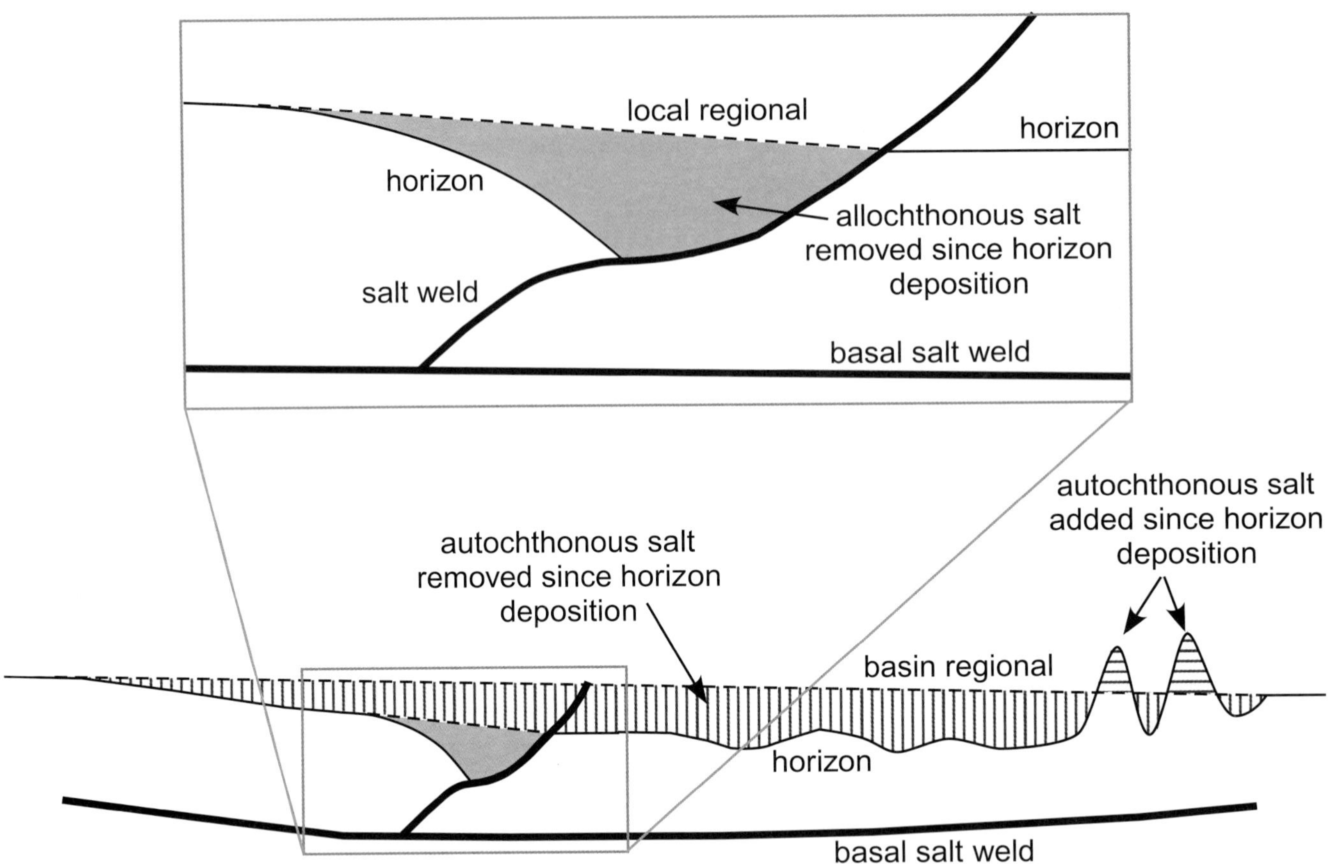

FIGURE 4. Schematic illustration showing the use of multiple regionals for estimating salt volume changes in the Gulf of Mexico. Various patterns indicate areas of salt thickness change. See text for discussion.

becomes more important when building local basin models to describe detailed prospect-scale structural histories.

Estimating Salt Distribution and Thickness Through Time

A key challenge for producing geometrically reasonable evolutionary models in a salt basin is being able to describe the changes in thickness of mobile layers. To simulate deformation in the Gulf of Mexico, multiple lithologic intervals need to change thickness through time, both thickening and thinning, sometimes in succession. Most notably, these include the autochthonous and allochthonous salt layers, but areas also exist where other parts of the sedimentary section effectively thin because of localized extreme extension (rafting) or thicken because of local thrust duplication. For the purposes of this chapter, only the salt is considered. Building these thickness changes into the basin model requires description of the thickness of impacted layers as maps for the key time steps.

The methodology used in this study is based on the concept of a structural "regional," defined by Marshak and Woodward (1988) as the elevation of a bedding surface before deformation, commonly reflecting its original depositional attitude. Contraction or inflation of material lifts the actual horizons above regional, whereas extension or withdrawal of material drops horizons below regional (Figure 3). In the case where structures are created solely by salt inflation and/or deflation, the area (or volume in 3-D) between the actual horizon and the regional approximates the amount of salt added or removed since deposition of the horizon. Extension combined with salt removal will cause the removed salt thickness to be overestimated because part of the created space is caused solely by extension.

To quantify the salt thickness using this approach in the Gulf of Mexico, two sets of regionals need to be defined and used for each horizon that will be input to the basin model (Figure 4). One of these, the "local regional", is used to estimate the salt thickness change associated with the allochthonous salt. The other, the "basin regional," is used to estimate the salt volume changes within the original (autochthonous) salt level.

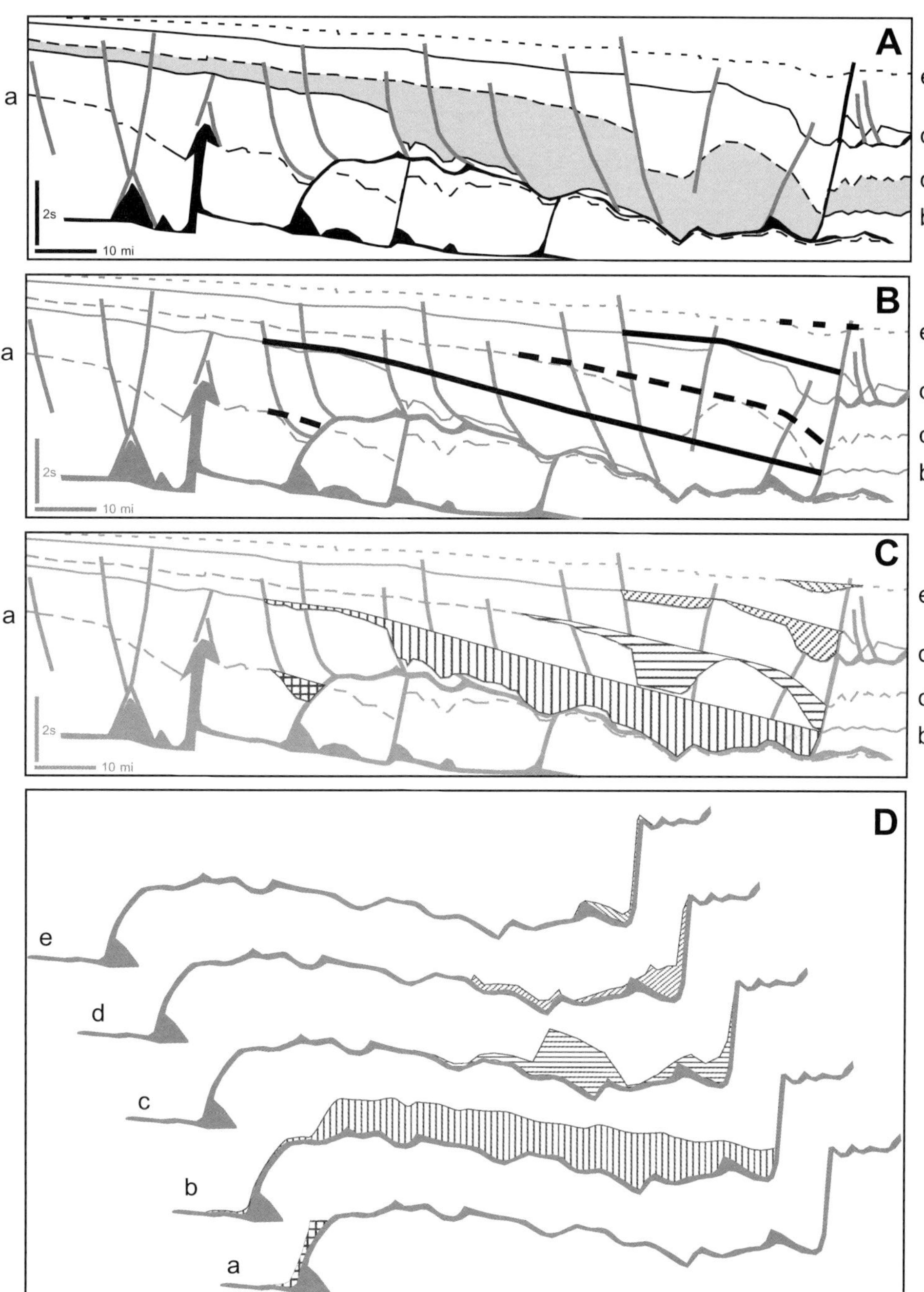

FIGURE 5. Cross section across the Louisiana shelf to illustrate the use of local regionals to derive the salt evolution. (A) Original cross section simplified and redrawn from Diegel et al. (1995). Heavy black areas and lines are salt bodies and welds; line types and shading indicate correlative stratigraphic horizons. Note how weld climbs stratigraphically toward the south (right) and individual horizons exist above or below the main weld at different places in the section. (B) Section with local regionals (heavy black lines) added for five horizons. (C) Section with patterns indicating areas of salt removed since deposition of the five horizons. (D) Amounts of salt removed for each time step superimposed on the weld surface to show the implied growth (steps a, b) and subsequent collapse (c–e) of the salt canopy.

The local regional is constructed by interpolating between preserved pieces of a stratigraphic horizon on either side of an area of salt withdrawal above a partially or completely welded salt allochthon. As shown in Figure 5, the local regional may connect across withdrawal basins that are entirely above a salt weld (c–e in Figure 5B) or connect pieces of the same horizon on opposite sides of the weld (a, b in Figure 5B). The difference between the regional and the actual horizon (merged with the weld) is inferred to be the salt thickness change since deposition of that horizon. In panels C and D of Figure 5, the salt thickness change is shown for five horizons. For the oldest horizon (a), salt withdrawal was localized in a narrow area corresponding to the salt feeder connecting up from the autochthon. Since horizon b deposition, a broad zone of salt withdrawal existed, corresponding to an area of deposition on top of a wide salt canopy. During subsequent periods (c–e), areas of salt withdrawal gradually decreased as the canopy collapsed, creating accommodation space for deposition of younger sediment.

Construction of the basin regional requires taking a basin-scale view and defining areas where horizons were not influenced by salt movement. The horizons in these areas are used to constrain the geometry of the basin

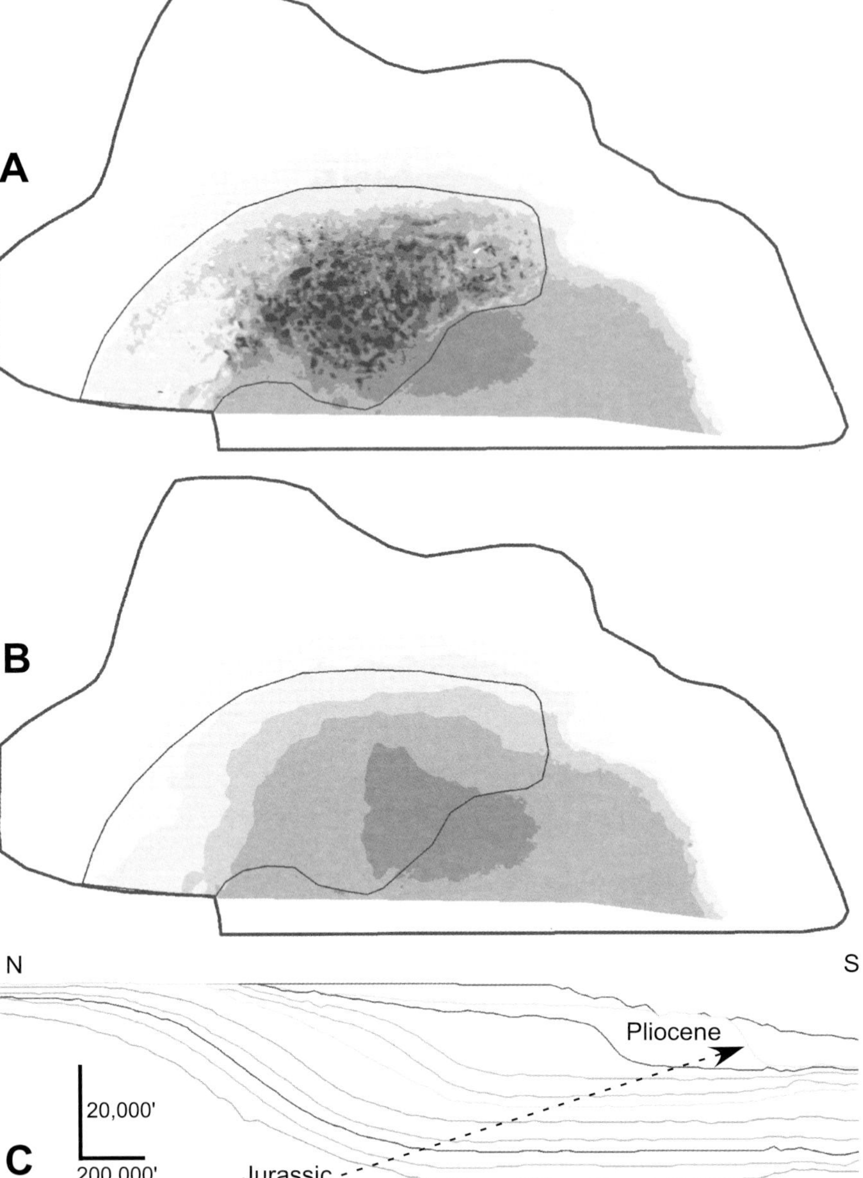

Figure 6. Methodology for producing basin regionals. (A) Miocene depth map (dark areas are deepest) with dashed polygon showing area impacted by salt movement. (B) Same map after deletion of the area within the dashed polygon and regridding. (C) Cross section (location in Figure 2A) showing final geometry of basin regionals. Dashed arrow indicates age progression from Jurassic through Pliocene.

regional. In the case of the Gulf of Mexico, one of these areas is the present-day abyssal plain, which was beyond the original depositional edge of salt (heavy dashed line in Figure 2A). The other area exists along the updip edge of the basin. For each horizon, a line is present north of which minimal deformation occurred other than regional tilting. The basin regional is a surface (or line in 2-D) that connects these two areas and approximates what a depositional surface would look like if salt movement had not occurred. From a practical point of view, constructing these surfaces requires (1) starting with a mapped horizon surface, (2) deleting the part that was impacted by salt movement, and (3) interpolating through the deleted area to create a smooth surface (Figure 6). The interpolation needs to be sufficiently controlled to assure reasonable horizon shapes and avoid crossing surfaces. In this case, pseudodata were used during the interpolation to constrain the regridding and impose a simple shelf-slope-basin profile to the surfaces (Figure 6C).

Use of the local and basin regionals to reconstruct salt thicknesses is done as a series of grid operations in the following sequence for each basin model input horizon (Figure 7). Grid operations are conducted using a combination of the Earthvision and Trinity software packages. First, the horizon surface is combined with the weld to create a composite surface with no holes (dashed line in Figure 7A). The allochthonous salt thickness change is calculated as the difference between this surface and the local regional (horizontal ruled area in Figure 7A). Once this is done, the autochthonous salt thickness change can be calculated as the difference between the local regional and the basin regional (vertically ruled area

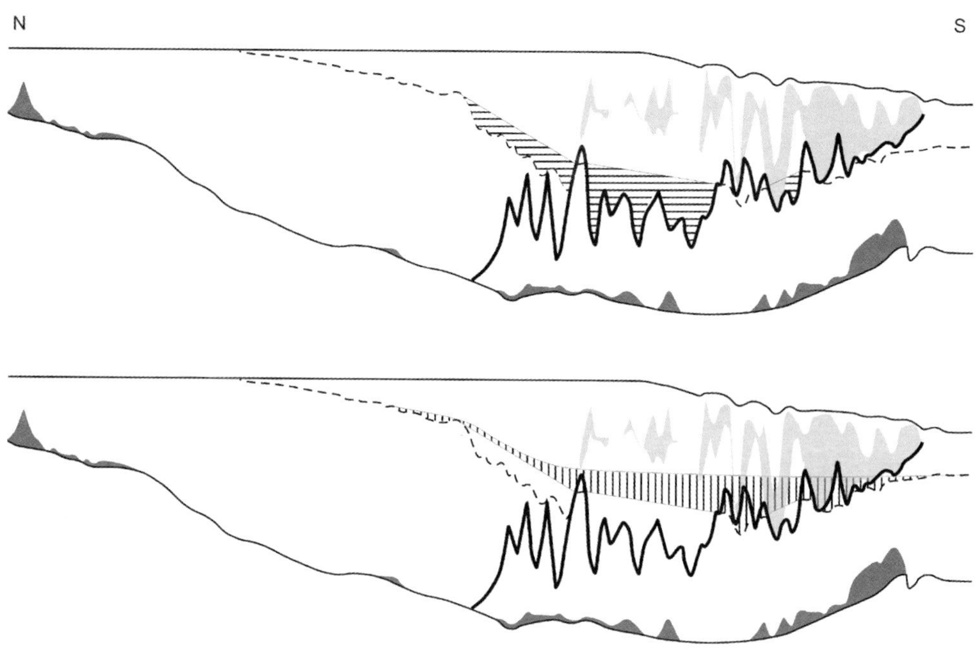

FIGURE 7. Regional cross section (location in Figure 2A) showing sequential application of local and basin regionals to derive thickness change of allochthonous salt (horizontally ruled) and autochthonous salt (vertically ruled) since Miocene horizon deposition. See Figure 2 for other symbols.

in Figure 7B). The salt thickness changes calculated during these steps are relative to present day. Therefore, areas in which salt has been withdrawn since horizon deposition will yield positive salt thickness changes, reflecting the fact that the actual horizon has dropped below the local and/or basin regional (Figures 3, 7). In areas of contraction and/or inflation, where horizons are elevated above regional, the salt thickness change values will be negative relative to present day, indicating that the salt was thinner at the time of horizon deposition (Figure 3).

To complete the process of estimating the allochthonous or autochthonous salt thickness at the time a horizon was deposited, it is necessary to add back any salt that remains today in a position that it likely occupied during horizon deposition. Therefore, the total allochthonous salt thickness at each time step is the thickness change plus any preserved allochthonous salt situated structurally below the local regional. Similarly, the total autochthonous salt thickness is the sum of the autochthonous salt thickness change and any salt remaining today below the basin regional. In areas of contraction, the present-day thickness and thickness change values will be opposite in sign, resulting in salt thinning back through time.

Clearly, numerous factors can contribute to errors in the final results of this process. First and foremost are errors associated with the interpretation of horizons, which are likely given the poor quality of seismic data in most areas underlain by shallow salt. Second, the regionals are interpolations with limited constraints across areas where the original horizon geometries were modified. This is especially true for construction of the basin regionals, which require interpolation across very long distances with sparse constraints. In some areas where both salt layers are welded out, it is impossible to unambiguously separate subsidence associated with allochthonous and autochthonous salt withdrawal. In areas such as these, I have made a choice, typically preferring to attribute subsidence in the shallow section to allochthonous salt withdrawal. Extension caused by slip on faults above salt welds (either autochthon or allochthon) will tend to cause overestimation of the removed salt thickness. Finally, the estimated salt thicknesses were not corrected for the fact that the surrounding sediment compacts during burial whereas the salt does not. This choice was made because I felt that the impact would be secondary to other uncertainties previously discussed. The lack of decompaction will tend to cause an underestimation of the salt paleothicknesses.

Despite these sources of error, the salt restoration process yields what appear to be realistic results, illustrated on selected maps in Figure 8. Throughout the Mesozoic to the present day, the autochthonous salt has progressively shrunk, whereas the allochthonous salt has appeared and grown in extent. During the Cretaceous, the autochthonous salt covered most of the study area but was already thin and patchily distributed along the northern basin margin. Allochthonous salt was restricted to small patches that represent the earliest canopies overlying feeders connecting to the autochthon. Throughout the Paleogene, the northern margin of thick salt migrated southward, and a large allochthon appeared in the western Gulf of Mexico. Thinning of this early allochthon and the autochthon occurred during the Neogene as isolated salt canopies grew across much of the deep-water central Gulf of Mexico. Since the mid-Neogene, the isolated deep-water canopies coalesced into the regionally extensive salt allochthon that we observe today. Most of

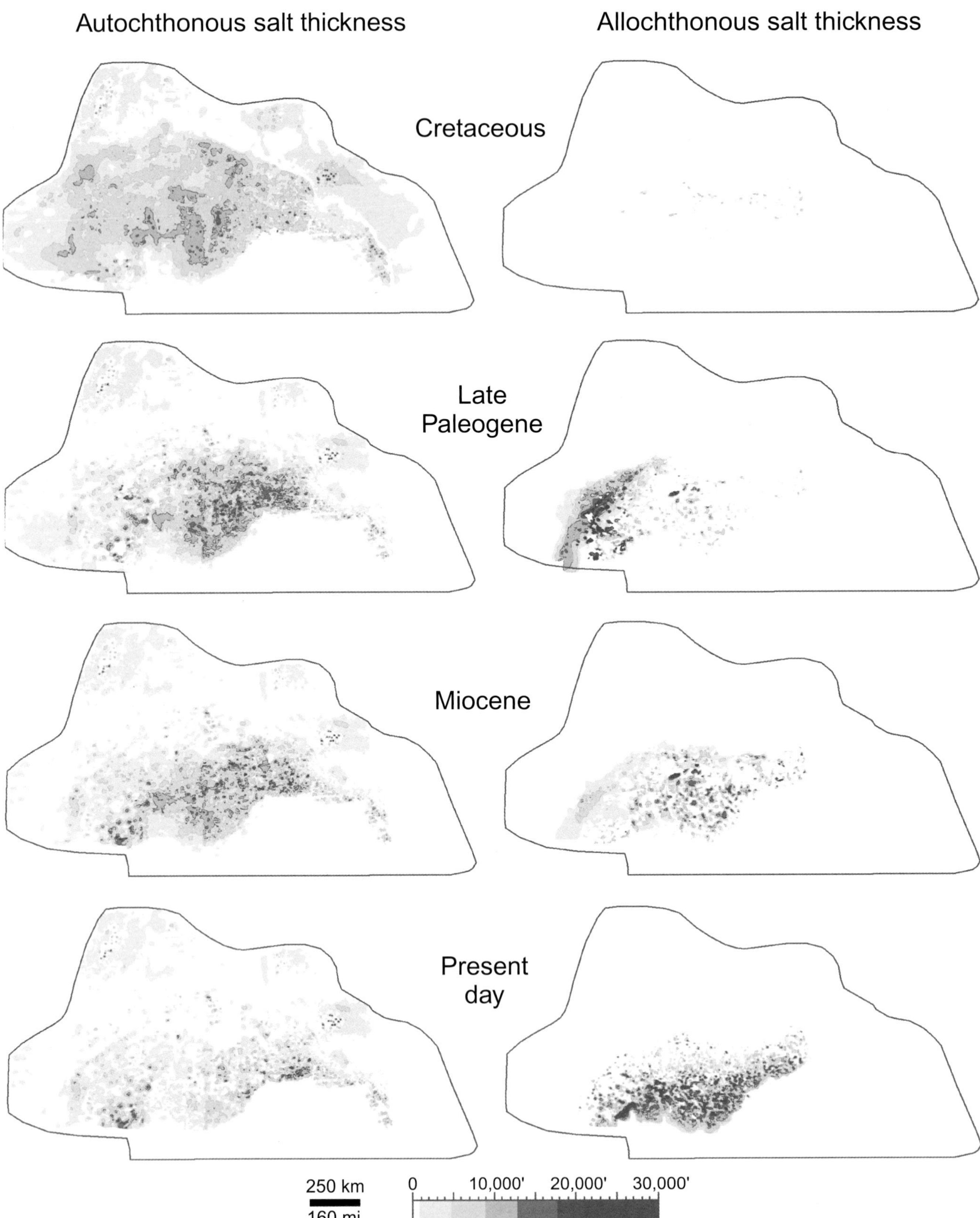

FIGURE 8. Maps of autochthonous (left) and allochthonous (right) salt thickness at four times during the Gulf of Mexico evolution using the methodology described in the text. Note the progressive thinning and areal shrinkage of the autochthonous salt, and the growth and basinward movement of the allochthonous salt through time.

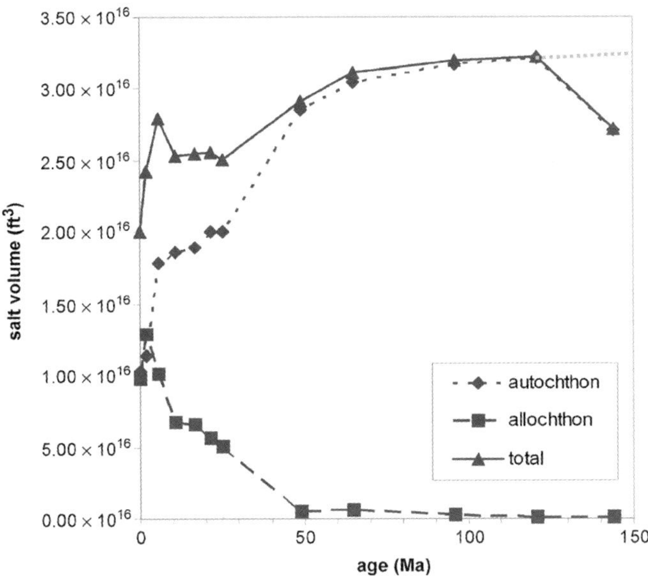

FIGURE 9. Plot showing salt volume change implied by the restoration. Final volume is 30 to 40% less than the starting volume. Gray dotted line is extrapolation of volumes back toward the time of salt deposition.

the autochthonous salt is thin or welded out, with the exception of thick pillows associated with the frontal fold belts along the southern edge of salt.

Because the salt restoration process is a purely geometric exercise done without using any volumetric constraints, one check on the validity of the results is to examine the salt volumes through time implied by the salt thickness maps. The total salt volume should not increase forward in time to the present day. Salt volumes for each time step were calculated in Trinity from the salt isopachs. As shown in Figure 9, the autochthon salt volume generally decreases through time, slowly until the Eocene (~50 Ma) and more rapidly after that. The allochthonous salt volume changes in the opposite way. The total salt volume generally decreases, ending with 30 to 40% less than the Mesozoic values.

Although the overall patterns in Figure 9 appear realistic, two anomalies need to be discussed. First of all, both the autochthonous and total salt volumes for the oldest time step (Late Jurassic) are lower than the corresponding values for several of the later time steps. Clearly, this is not physically possible. The most likely explanation for this is that the study area does not extend sufficiently far onshore to encompass the area completely unaffected by salt movement. Examination of Figure 2A shows this to be the case where the northern limit of the model is south of the Louann Salt depositional limit in eastern Louisiana and Mississippi. As a result, the regional surface is at too low an elevation and the salt volume at the updip end of the basin is underestimated. A linear extrapolation based on the values for subsequent time steps (120–65 Ma) can be used to correct the Late Jurassic salt volume and estimate the original depositional volume (~9.4×10^{14} m^3 [~3.4×10^{16} ft^3]).

The second anomaly is the abrupt increase of both the allochthonous and total salt volumes at approximately 5 Ma. At present, the cause of this is not entirely clear, although one might suspect that it is associated with errors in constructing the regionals for this time step. Large-scale transfer of salt from the autochthon to allochthon at this time created a complex salt and minibasin geometry, making it difficult to reliably select locations at which to pin the local regionals. Even moderate errors in the geometric construction of the regionals can translate to large volumetric errors.

Constructing the Basin Model

Now that a series of salt thickness maps exists, the challenge becomes constructing the basin model. As shown in Figure 2A, any single chronostratigraphic horizon occurs both above and below the allochthonous salt, thereby crossing the allochthonous salt layer. Because crossing horizons cannot be handled by basin modeling software, a geometric work-around was applied in which the model was effectively split into two parts, separated by the allochthonous salt layer.

The way this is done is illustrated in Figure 10. For each horizon that crosses the weld, two horizons were created by merging parts of the horizon with the weld and the top or base salt. The part of the horizon above the weld/salt is combined with the top salt, allochthonous salt weld, and surface bathymetry in that order of preference. Similarly, the part of the horizon below the weld/salt is merged with the base salt, allochthonous salt weld, and autochthonous salt weld, in that order. In this way, two horizons are created such that the intervening thickness is zero between the hanging wall and footwall cutoffs at the allochthonous salt weld (Figure 10). Once this process is applied to all surfaces that cross the weld, a complete stratigraphic sequence exists, consisting of duplicate horizons above and below the weld.

These duplicate age horizons require the use of simulation software that permits multiple horizons to have the same age while occupying different positions within the stratigraphic column. Because it allows this functionality, the Petromod System was chosen as the modeling tool for this project. The layer structure used in the model is shown schematically in Figure 11. For each horizon that exists above the allochthonous weld/salt, a corresponding age horizon exists below with a different horizon name. When the model is run, the age-corresponding layers appear at the appropriate times, regardless of their position within the stratigraphic column. The allochthonous and autochthonous salt layers are bounded by top and base surfaces, which are the same where the salt layer

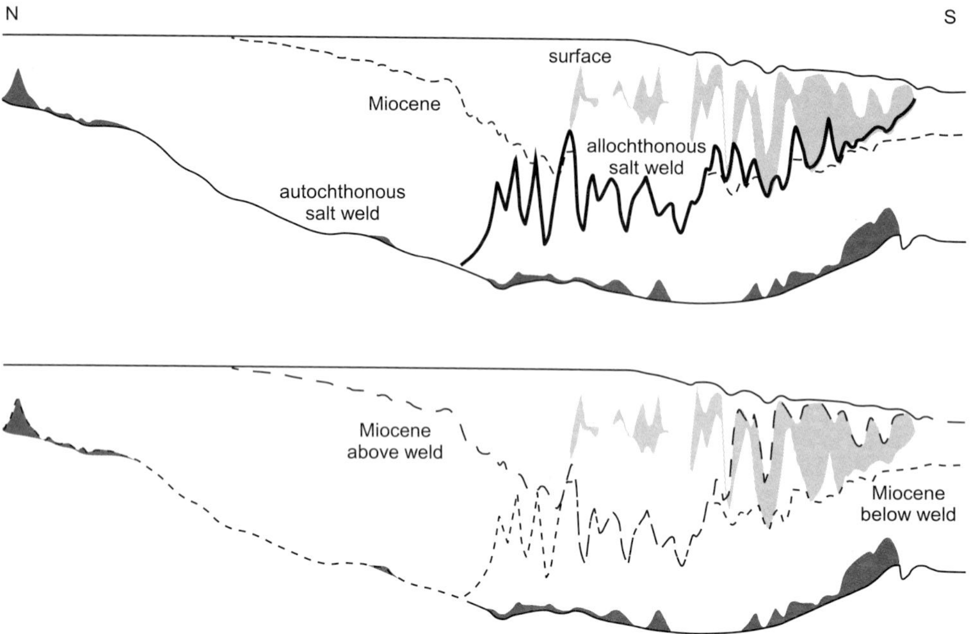

FIGURE 10. Regional cross section (location in Figure 2A) showing how a mapped horizon, weld, and top and/or base salt was adapted for inclusion in the simulation model. Lower section shows the same horizon after being split into two horizons, one above and one below the weld. See text for discussion.

is welded out. The salt isopachs derived in the previous section are then used to change the thickness of the salt layers throughout the model evolution (Figure 11).

RESULTING MODEL GEOMETRY

Some results extracted from the 3-D model after simulation are shown in Figures 12 and 13. The geometry of the cross sections in Figure 12 demonstrates the overall validity of the approach used for the model construction. The progressive change in shape of the base of the model reflects the overall sediment loading and subsidence of the basin. Although local restoration artifacts remain at some time steps, they are minor given the overall structural complexity. In general, these artifacts are restricted to instances where the two salt layers undergo dramatic thickness change at the same time. One notable feature in the Paleogene and Miocene restoration steps is the abrupt northward deepening of the base of the model about midway along the section, which corresponds to the transition from an area of thick salt northward into an expanded siliciclastic section. This is probably not a real geologic feature but is most likely a result of doing the salt thickness estimation without considering decompaction of the surrounding sediments. Incorporating decompaction would increase the salt thickness estimates and minimize this artifact.

The key elements of the Gulf of Mexico evolution are well illustrated by cross sections (Figure 12) and a

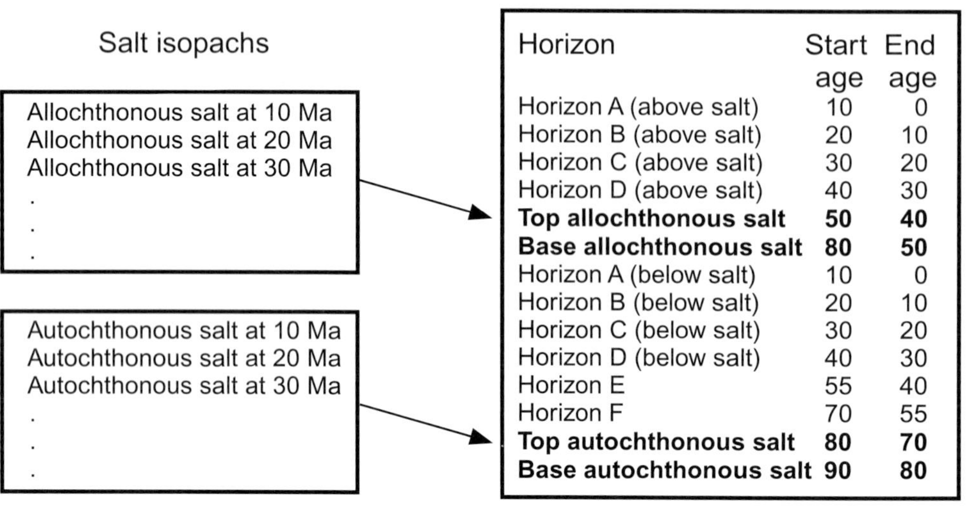

FIGURE 11. Schematic illustration of layer sequence used in the construction of the three-dimensional model when multiple horizons of the same age exist. The stratigraphic section is duplicated above and below the allochthonous salt layer. Salt thickness maps are applied to the salt layers to change the thickness back through time.

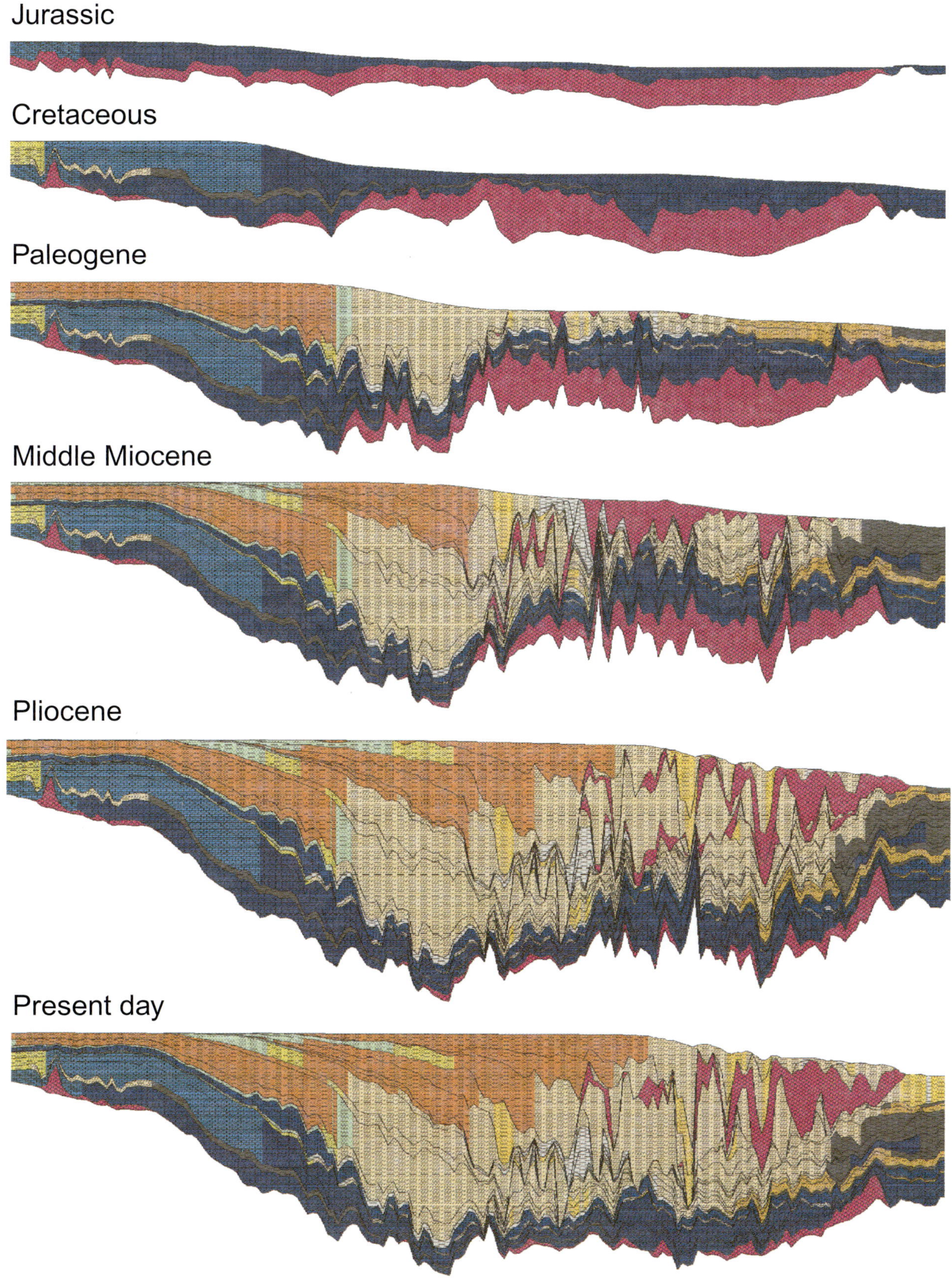

FIGURE 12. Two-dimensional extractions (location in Figure 2A) from the three-dimensional model at various time steps showing the progressive geometric evolution. Colors indicate lithology as follows: pink = salt; blue = carbonates and siliciclastic-poor units; green, yellow, orange, brown, and gray = siliciclastic-dominated units.

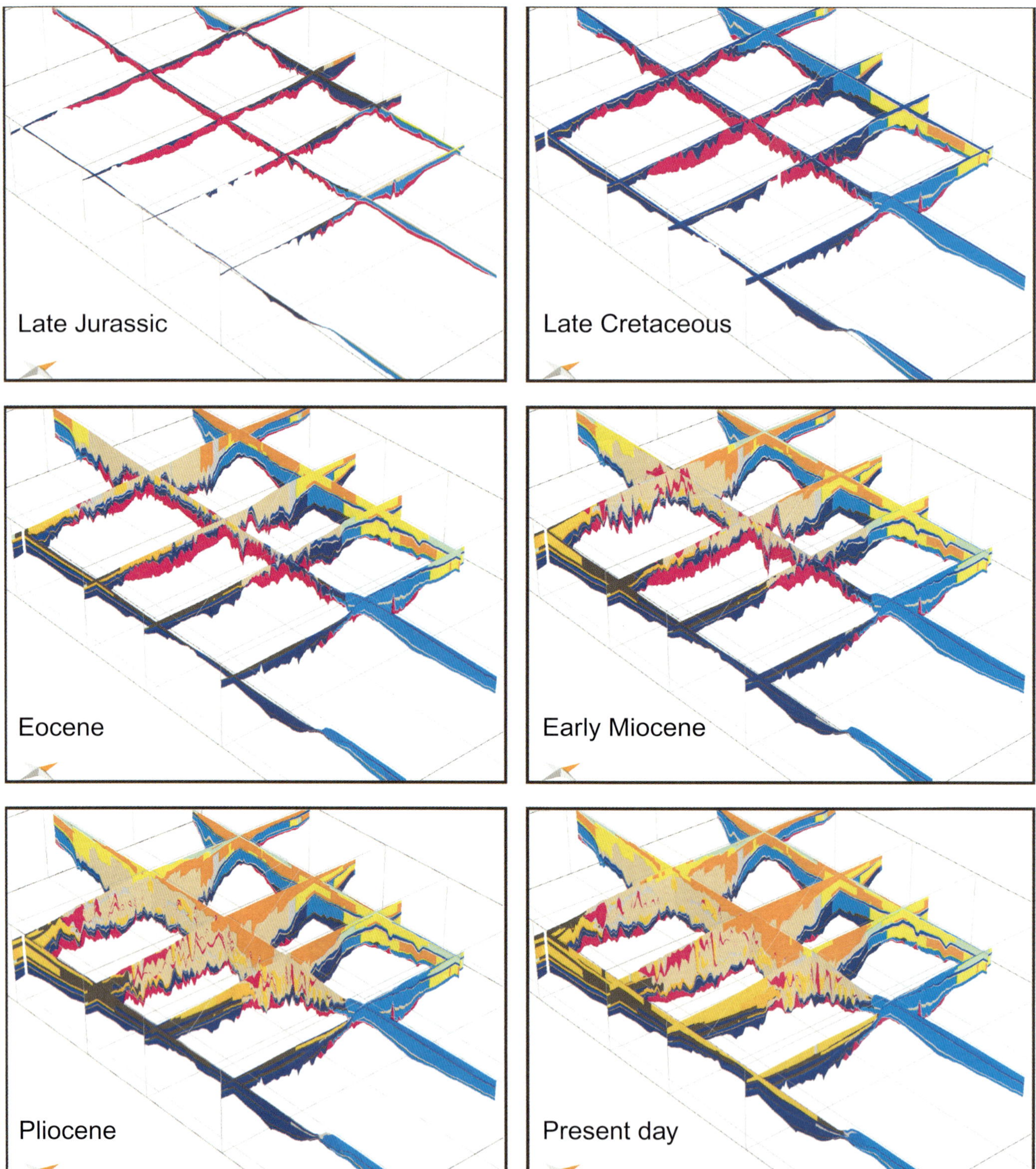

Figure 13. Fence diagram illustrating geometric evolution of the model in three dimensions. Orange arrow points north. Pink = salt; blue = carbonates and siliciclastic-poor units; green, yellow, orange, brown, and gray = siliciclastic-dominated units.

fence diagram cut through the model (Figure 13). Salt began to mobilize in the Mesozoic so that much of the original autochthonous salt was squeezed out from beneath the developing carbonate shelf during the Cretaceous. This salt migrated laterally into a thick massif that underlay the slope and deep-water part of the basin. Subsequent loading by Paleogene siliciclastic sediments, focused in the western Gulf of Mexico, continued to drive the autochthonous salt seaward and initiated the first salt canopy, consisting of isolated upward-widening salt bodies. Neogene siliciclastic deposition was focused farther eastward and further shrunk the area of thick autochthonous salt by driving more salt into the allochthonous canopy. The early formed isolated salt bodies merged into an extensive salt sheet in response to Neogene deposition. Sediment deposition on top of the salt allochthon, especially during the Pliocene–Pleistocene, remobilized salt and formed secondary minibasins. Minor uplift during the Neogene resulted in some erosion of stratigraphy along the northern basin margin and accentuated the basinward tilt of beds.

SUMMARY

Basin simulation in areas of complex thin-skinned structural geology has long been hampered by the inability of vertical backstripping algorithms to reproduce realistic burial histories, thus creating serious geometric artifacts in paleo–time steps. In this chapter, I have outlined a simple approach to derive the apparent thickness evolution of ductile layers by creating regionals and using them to estimate volume changes through time. Using standard 3-D basin modeling tools, this technique was applied to create input for a regional simulation model of the Gulf of Mexico, where the mobility of multiple salt layers generated extremely complex structural geometries. Although the approach is not a rigorous 3-D structural restoration and the final product has some flaws, the resulting model provides an overall realistic view of the basin evolution and a sufficiently accurate framework on which to base regional petroleum system evaluations.

ACKNOWLEDGMENTS

This project would not have been possible without the contributions of many individuals currently or previously involved in BP Gulf of Mexico efforts, including R. Barrett, W. Bunting, M. Casey, T. Fitzpatrick, J. Gosses, K. Hargrove, W. Hart, T. Heyn, D. Jordan, J. Kirkova-Pourciau, A. Leroy, R. Kasino, S. Krueger, J. Laird, G. Lyman, K. Meisling, D. Muller, B. Nguyen, G. Pfau, D. Phillips, R. Priem, T. Searcy, J. Stephens, M. Steuer, J. Turner, and many others. S. Paulson and C. Gong were early adopters of the model-building approach described and helped significantly improve the methodology. R. Tscherny, formerly of Integrated Exploration Systems (IES), was critical in the initial workflow development because of his knowledge of PetroMod. The final article benefited from reviews by S. Krueger, C. Yeilding, K. Peters, and C. Lampe. Finally, I thank the management of BP America and Gulf of Mexico Exploration for permission to publish this work.

REFERENCES CITED

Baur, F., M. Di Benedetto, T. Fuchs, C. Lampe, and S. Sciamanna, 2009, Integrating structural geology and petroleum systems modeling: A pilot project from Bolivia's fold and thrust belt: Marine and Petroleum Geology, v. 26, p. 573–579, doi:10.1016/j.marpetgeo.2009.01.004.

Diegel, F. A., J. F. Karlo, D. C. Schuster, R. C. Shoup, and P. R. Tauvers, 1995, Cenozoic structural evolution and tectono-stratigraphic framework of the northern Gulf Coast continental margin, *in* M. P. A. Jackson, D. G. Roberts, and S. Snelson, eds., Salt tectonics: A global perspective: AAPG Memoir 65, p. 109–151.

Galloway, W. E., P. E. Ganey-Curry, X. Li, and R. T. Buffler, 2000, Cenozoic depositional history of the Gulf of Mexico Basin: AAPG Bulletin, v. 84, p. 1743–1774.

Gibson, R. G., L. I. P. Dzou, and D. F. Greeley, 2004, Shelf petroleum system of the Columbus Basin, offshore Trinidad, West Indies: I. Source rock, thermal history, and controls on product distribution: Marine and Petroleum Geology, v. 21, p. 97–108, doi:10.1016/j.marpetgeo.2003.11.003.

Grando, G., Z. Schleder, R. Shackleton, G. Seed, T. Buddin, K. McClay, and F. Borraccini, 2008, 3-D structural evolution of the salt controlled Frampton anticline, Atwater Valley fold belt, deep-water Gulf of Mexico (abs.): AAPG Annual Convention, San Antonio, Texas, http://www.searchanddiscovery.com/abstracts/html/2008/annual/abstracts/409767.htm (accessed May 23, 2011).

Hall, S. H., 2002, The role of autochthonous salt inflation and deflation in the northern Gulf of Mexico: Marine and Petroleum Geology, v. 19, p. 649–682, doi:10.1016/S0264-8172(02)00025-9.

Johnson Jr., J. R., M. A. Meylan, and D. F. Ufnar, 2006, Evolution of a salt diapir within the Mississippi Salt Basin, U.S.A.: Interaction of salt migration and sediment deposition: Gulf Coast Association of Geological Societies Transactions, v. 56, p. 309–322.

Lampe, C., K. Kornpihl, S. Sciamanna, T. Zapata, G. Zamora, and R. Varadé, 2006, Petroleum systems modeling in tectonically complex areas: A 2-D migration study from the Neuquen Basin, Argentina: Journal of Geochemical Exploration, v. 89, p. 201–204, doi:10.1016/j.gexplo.2005.11.041.

MacRae, G., and J. S. Watkins, 1996, Desoto Canyon salt basin: Tectonic evolution and salts structural styles, *in* J. O. Jones and R. L. Freed, eds., Structural framework of the northern Gulf of Mexico: Gulf Coast Association of Geological Societies Special Publication, v. 46, p. 53–61.

Marshak, S., and N. B. Woodward, 1988, Introduction to cross section balancing, *in* S. Marshak and G. Mitra, eds., Basic methods of structural geology: Englewood, Cliffs, New Jersey, Prentice-Hall, 446 p.

McBride, B. C., P. Weimer, and M. G. Rowan, 1998a, The evolution of allochthonous salt systems, northern Green Canyon and Ewing Bank (offshore Louisiana), northern Gulf of Mexico: AAPG Bulletin, v. 82, p. 1013–1036.

McBride, B. C., P. Weimer, and M. G. Rowan, 1998b, The effect of allochthonous salt on the petroleum systems of northern Green Canyon and Ewing Bank (offshore Louisiana), northern Gulf of Mexico: AAPG Bulletin, v. 82, p. 1083–1112.

McFarlan Jr., E., and L. S. Menes, 1991, Lower Cretaceous, *in* A. Salvador, ed., The geology of North America: The Gulf of Mexico Basin: Geological Society of America Decade of North American Geology, v. J, p. 181–204.

Mello, U. T., G. D. Karner, and R. N. Anderson, 1995, Role of salt in restraining the maturation of subsalt source rocks: Marine and Petroleum Geology, v. 12, p. 697–716, doi:10.1016/0264-8172(95)93596-V.

Peel, F. J., C. J. Travis, and J. R. Hossack, 1995, Genetic structural provinces and salt tectonics of the Cenozoic offshore U.S. Gulf of Mexico: A preliminary analysis, *in* M. P. A. Jackson, D. G. Roberts, and S. Snelson, eds., Salt tectonics: A global perspective: AAPG Memoir 65, p. 153–175.

Pindell, J., and L. Kennan, 2001, Kinematic evolution of the Gulf of Mexico and Caribbean: GCS-SEPM Foundation 21st Annual Research Conference, *in* R. H. Fillon, N. C. Rosen, and P. Weimer, eds., Petroleum Systems of Deep Water Basins, p. 193–220.

Rivero, C., F. Bilotti, R. Boyce, M. Strickler, and T. Clarke, 2005, 3D decompaction and restoration of salt-cored fold structures, deep-water Gulf of Mexico (abs.): AAPG Annual Convention, Calgary, Alberta, Canada, http://www.searchanddiscovery.com/abstracts/html/2005/annual/abstracts/rivero.htm (accessed May 23, 2011).

Rowan, M. G., 1995, Structural styles and evolution of allochthonous salt, central Louisiana outer-shelf and upper slope, *in* M. P. A. Jackson, D. G. Roberts, and S. Snelson, eds., Salt tectonics: A global perspective: AAPG Memoir 65, p. 199–228.

Rowan, M. G., 2002, Salt-related accommodation in the Gulf of Mexico deep water: Withdrawal or inflation, autochthonous or allochthonous?: Gulf Coast Association of Geological Societies Transactions, v. 52, p. 861–869.

Rowan, M. G., M. P. A. Jackson, and B. D. Trudgill, 1999, Salt-related fault families and fault welds in the northern Gulf of Mexico: AAPG Bulletin, v. 83, p. 1454–1484.

Salvador, A., 1991, Triassic-Jurassic, *in* A. Salvador, ed., The geology of North America: The Gulf of Mexico Basin: Geological Society of America Decade of North American Geology, v. J, p. 131–180.

Sawyer, D. S., R. T. Buffler, and R. H. Pilger Jr., 1991, The crust under the Gulf of Mexico Basin, *in* A. Salvador, ed., The geology of North America: The Gulf of Mexico Basin: Geological Society of America Decade of North American Geology, v. J, p. 53–72.

Schneider, F., H. Devoitine, I. Faille, E. Flauraud, and F. Willien, 2002a, Ceres 2-D: A numerical prototype for HC potential evaluation in complex area: Oil & Gas Science and Technology, v. 54, p. 607–619, doi:10.2516/ogst:2002041.

Schneider, F., J. L. Faure, and F. Roure, 2002b, Methodology for basin modeling in complex area: Examples from Eastern Venezuelan and Canadian Foothills (abs.): AAPG Hedberg Conference, Deformation history, fluid flow reconstruction and reservoir appraisal in foreland fold and thrust belts, Palermo-Mondello, Sicily, Italy, http://www.searchanddiscovery.com/abstracts/pdf/2002/hedberg_sicily/ (accessed May 23, 2011).

Sohl, N. F., R. E. Martinez, P. Salmeron-Urena, and F. Soto-Jaramillo, 1991, Upper Cretaceous, *in* A. Salvador, ed., The geology of North America: The Gulf of Mexico Basin: Geological Society of America Decade of North American Geology, v. J, p. 205–244.

Trudgill, B. D., M. G. Rowan, J. C. Fiduk, P. Weimer, P. E. Gale, B. E. Korn, R. L. Phair, W. T. Gafford, G. R. Roberts, and S. W. Dobbs, 1999, The Perdido fold belt, northwestern deep Gulf of Mexico, Part 1: Structural geometry, evolution, and regional implications: AAPG Bulletin, v. 83, p. 88–113.

Weimer, P., and R. T. Buffler, 1992, Structural geology of the Mississippi Fan fold belt, deep Gulf of Mexico: AAPG Bulletin, v. 76, p. 225–251.

White, G. W., S. J. Blanke, and C. F. Clawson II, 1999, Evolutionary model of the Jurassic sequences of the east Texas Basin: Implications for hydrocarbon exploration: Gulf Coast Association of Geological Societies Transactions, v. 49, p. 488–498.

Woods, R. D., A. Salvador, and A. E. Miles, 1991, Pre-Triassic, *in* A. Salvador, ed., The geology of North America: The Gulf of Mexico Basin: Geological Society of America Decade of North American Geology, v. J, p. 109–129.

7

Lampe, C., K. J. Bird, T. E. Moore, R. A. Ratliff, and B. Freeman, 2012, Modeling^3: Integrating structural modeling, fault property analysis, and petroleum systems modeling—An example from the Brooks Range foothills of the Alaska North Slope, *in* K. E. Peters, D. J. Curry, and M. Kacewicz, eds., Basin Modeling: New Horizons in Research and Applications: AAPG Hedberg Series, no. 4, p. 119–136.

Modeling^3: Integrating Structural Modeling, Fault Property Analysis, and Petroleum Systems Modeling—An Example from the Brooks Range Foothills of the Alaska North Slope

Carolyn Lampe

UCON Geoconsulting, Cologne, Germany

Kenneth J. Bird and Thomas E. Moore

U.S. Geological Survey, Menlo Park, California, U.S.A.

Robert A. Ratliff

Landmark Software and Services, Highlands Ranch, Colorado, U.S.A.

Brett Freeman

Badley Geoscience, Lincolnshire, United Kingdom

ABSTRACT

Seismic interpretation and various modeling techniques, including structural modeling, fault-seal analysis, and petroleum systems modeling, have been combined to conduct an integrated study along a tectonically complex compressional cross section in the Brooks Range foothills of the Alaska North Slope. In the first approach, relatively simple models have been developed to show the interaction and codependency of various parameters such as changing geometry over time in a compressional regime, character and timing of faults with respect to sealing or nonsealing quality, thermal and maturity evolution of the study area, as well as petroleum generation, migration, and accumulation over time, with respect to the geometry changes and the fault properties. Modeling results show that a comprehensive understanding of all aspects involved in basin evolution is crucial to understand the petroleum systems, to be able to reproduce what is observed in the field, and to ultimately predict what can be expected from a prospect area. This integrated approach allows a better understanding of the complex petroleum systems of the Brooks Range foothills.

DOI:10.1306/13311432H43467

INTRODUCTION

The North Slope of Alaska, including the adjacent continental shelves of the Beaufort and Chukchi Seas, is thought to hold significant amounts of undiscovered petroleum resources (Houseknecht and Bird, 2006). Most known petroleum accumulations involve structural or combination structural-stratigraphic traps related to closures along the Barrow Arch, a regional basement high, yet this geologically complex region also includes prospective strata within passive-margin, rift, and foreland-basin sequences. Both extensional and compressional structures provide substantial exploration targets in the shelf and turbidite rock sequences of the Jurassic through the Tertiary. Although the area is well explored in the greater Prudhoe Bay, little exploration has occurred elsewhere, with only a few wells and seismic surveys providing insight into the complex geology.

A seismic section, approximately 50 km (31 mi) long, trending north-south across the Brooks Range foothills in the southern National Petroleum Reserve-Alaska (NPRA) and including the 3414 m (11,200 ft) exploratory Awuna-1 well, is the basis for an integrated comprehensive modeling study. Various investigative methods are used and combined to assess the complex geology and geohistory of the foothills. These include seismic reprocessing and interpretation, palinspastic reconstruction of the structural architecture, estimation of fault rock properties, and finally, two-dimensional (2-D) basin and petroleum systems modeling using TecLink technology.

To minimize technical uncertainty and drilling risk, an improved understanding of the geologic system is required. For the first time, structural, fault, and petroleum systems modeling have been combined and fully integrated to assess the processes and risks involved in the structural and geologic development of the study area.

GEOLOGIC SETTING

The study area (gray rectangle in Figure 1) lies in the Colville foreland basin between the Brooks Range in the south and the Barrow Arch in the north. The area is characterized by compressional tectonics, related to Tertiary deformation of the Cretaceous and Tertiary Brookian fold and thrust belt, which is the northernmost continuation of the back-arc fold and thrust belt of the Cordilleran orogen. Uplift of the Brooks Range, tectonic loading by the northward-moving deformation front, and extensive flysch and molasse deposits shed into the rapidly subsiding foreland basin in the early Cretaceous and resulted in strata more than 10 km (~33,000 ft) thick in the southern part of the foreland basin, thinning northward toward the Barrow Arch. The stratigraphy can be subdivided into four major stratigraphic sequences, where the provenance of the sediments was controlled by major tectonic events (Lerand, 1973; Bird and Molenaar, 1987; Moore et al., 1994): (1) the slightly metamorphosed and highly deformed Proterozoic to Devonian sedimentary and igneous rocks of the Franklinian sequence, which is referred to as "basement" in this study; (2) Mississippian to Triassic northerly derived passive margin siliciclastic and carbonate rocks of the Ellesmerian sequence; (3) the rift-related Jurassic to Lower Cretaceous siliciclastic rocks of the Beaufortian sequence; and (4) the Lower Cretaceous to Cenozoic siliciclastic rocks of the Brookian sequence, the thick flysch and molasse units derived from the Brooks Range to the south.

Four source rocks specific to the study area are considered (Figure 2; Table 1), namely, (1) the Middle and Upper Triassic Shublik Formation, which consists of a predominantly organic carbon–rich mixture of shale, marl, and limestone deposited under upwelling conditions; (2) the directly overlying Lower Jurassic K1 sequence of the Kingak Shale (Houseknecht and Bird, 2004), which represents bottomset facies of rift-related southward-prograding clinoforms; (3) the Lower Cretaceous informally named pebble shale unit (PBS), a thin (~30 m [~100 ft]) organic-rich transgressive shale; and (4) the Hue–Gamma Ray Zone (Hue-GRZ), representing the distal condensed bottomset facies of eastward-prograding Brookian clinoforms. The Shublik Formation and the K1 member are separated from the PBS and Hue-GRZ north of the study area by the regional Lower Cretaceous Unconformity (LCU) (Bird, 1994; 2001); however, in the study area, the LCU is represented by a conformable contact between the uppermost (Lower Cretaceous) Kingak Shale and PBS.

The Awuna well fails to penetrate any of the four source rocks, which is why the source rock properties—total organic carbon (TOC) and hydrogen index (HI)—are derived and extrapolated from key wells in the NPRA and the greater Alaska North Slope (Peters et al., 2003) (Table 1). In the study area, the original TOC of the Shublik Formation is estimated to be 4.1 wt. %, with an estimated original HI of 1000 mg hydrocarbon (HC)/g TOC. The K1 member has a TOC of 5.0 wt. %, with an HI of 400 mg HC/g TOC. The PBS and Hue-GRZ formations feature TOC values of 2.1 wt. % and HI values of 200 and 197 mg HC/g TOC, respectively. Primary oil and gas reactions for source rock kinetics are based on custom measurements of thermally immature Hue Shale and Shublik source rocks (Lillis et al., 1999) and measurements by Masterson (2001) for the PBS and Kingak Shale. Secondary cracking is assigned based on GFZ/GeoS4 (Secondary Reaction T2 P10 Crack from the PetroMod library), applying a narrow distribution of activation energies for destruction of oil instead of a single activation energy.

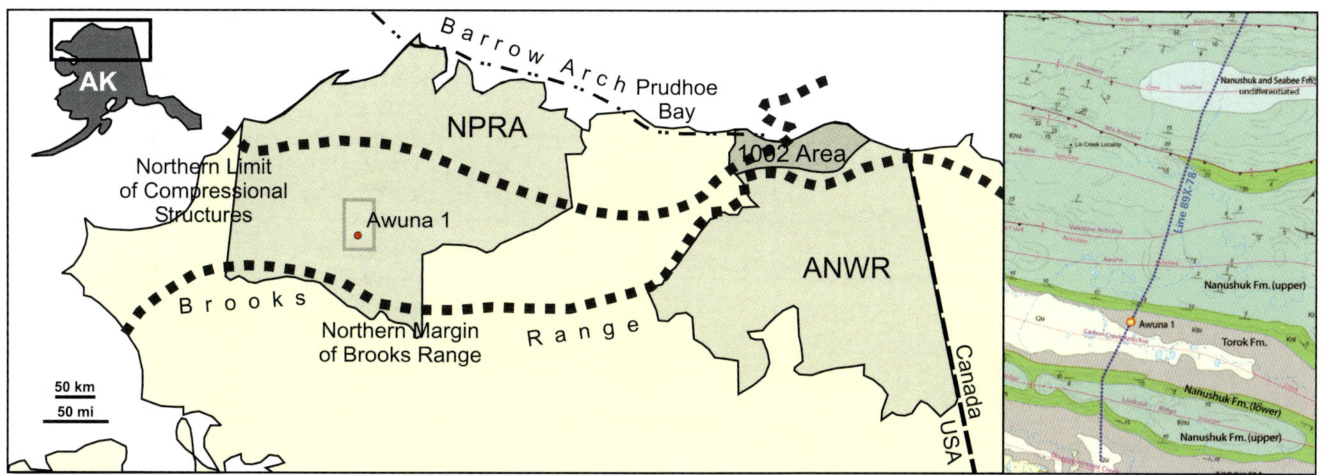

FIGURE 1. Location of the study area (gray rectangle) on the Alaska North Slope. The position of the 89X-78 seismic line and the Awuna-1 well within the gray rectangle is shown in detail on the right. NPRA = National Petroleum Reserve-Alaska; ANWR = Arctic National Wildlife Refuge.

The entire Brookian sequence (Torok and Nanushuk formations) represents both the overburden and the main reservoir rock in the area. Because of simplifications necessary for the modeling, these two formations are assumed to be more or less homogeneous with a mixed percentage of sand, silt, and/or shale (Table 1). In reality, the lowermost Torok Formation probably contains tongues of Hue Shale-type source rock (Creaney and Passey, 1993; Bird, 2001). Based on wells in the vicinity of the study area, the lower Torok is likely more sand rich than the upper part and may include locally confined lenticular turbidite sand bodies that could provide stratigraphic traps (Houseknecht and Schenk, 2007). However, these are unaccounted for in this model.

Reprocessing and Interpretation of Seismic Data

The analyzed seismic section (line 89X-78, Figure 1) was acquired in 1978 as part of the 1974–1982 government exploration program of NPRA, during which more than 21,000 line-km (~31,000 mi) of 2-D seismic data was collected (Gryc, 1988). Stacked data provided by the seismic contractor for this line were migrated, and then the depth was converted using posted stacking velocities. As with most foothills seismic lines, the time section is displayed with a sloping datum, which for this line ranged from sea level in the north to nearly 396 m (1300 ft) in the south. The depth section is displayed

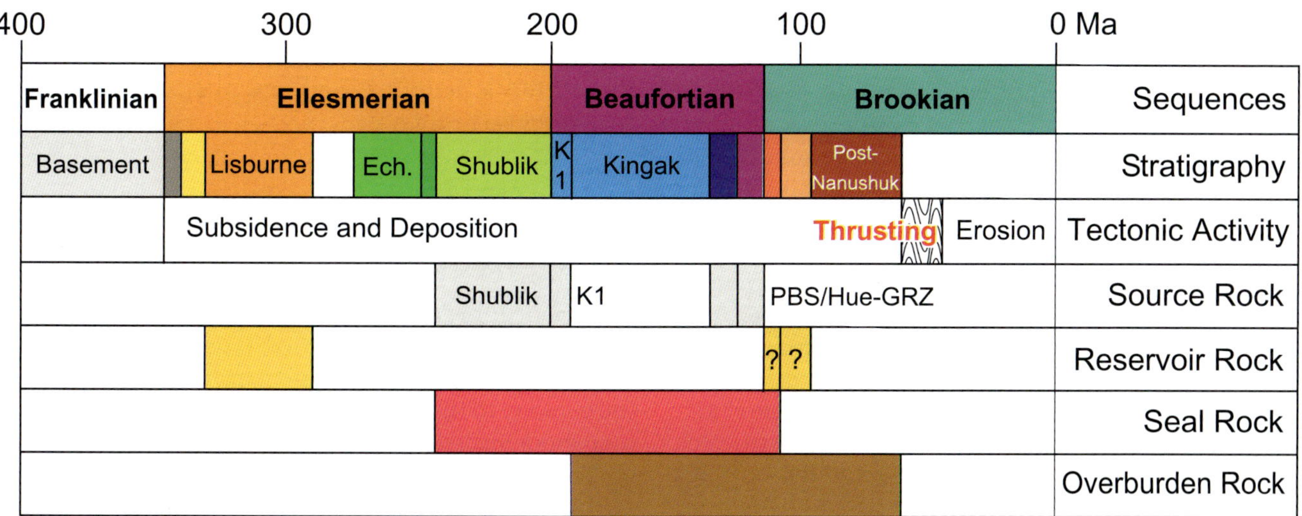

FIGURE 2. Simplified events chart specific for the study area depicting the stratigraphy and the major petroleum systems (source rock, reservoir rock, seal rock, overburden rock), as well as the main tectonic events. PBS = pebble shale unit; Hue-GRZ = Hue–Gamma Ray Zone.

Table 1. Rock units, sequences, depositional ages, lithology mixtures, and source rock properties used for all models to determine decompaction and compaction (structural and petroleum systems modeling), fault properties (fault seal analysis), and thermal history in combination with petroleum migration (petroleum systems modeling).*

Rock Units	*Sequences*	*Depositional Age (Ma)*		*Lithology ss/sltst/sh (%)*	*TOC (wt. %)*	*HI (mg HC/g TOC)*
		From	*To*			
Deformation, uplift, and erosion		60	0			
Post-Nanushuk		94	60	40/40/20		
Nanushuk Formation	Brookian	107	94	30/30/40		
Torok (upper, gray unit)		111	107	19/01/80		
Torok (middle, blue unit)		111.2	111	20/05/75		
Torok (lower, green unit)		112	111	30/10/60		
Hue (GRZ) Shale		122	112	0/5/95	2.1	197
Pebble Shale Unit	Beaufortian	133	122	0/10/90	2.1	200
Lower Cretaceous unconformity		133	133			
Kingak Shale		200	133	0/10/90	5.0	400
Shublik Formation		245	200	20 lst/80 sh	4.1	1000
Ivishak Formation		250	245	0/40/60		
Echooka Formation		275	250	20/30/50		
Permian unconformity	Ellesmerian	290	275			
Lisburne Group		335	290	50 lst/40 dol/10 sh		
Kayak Shale		345	335	10 lst/10 slt/80 sh		
Kekiktuk Formation		350	345	50/30/20		
Basement	Franklinian	>400	350	Argillite		

**ss = sandstone; sltst = siltstone; sh = shale; lst = limestone; dol = dolomite; TOC = total organic carbon; HI = hydrogen index; HC = hydrocarbon; GRZ = Gamma Ray Zone.*

with a horizontal datum of 396 m (1300 ft) above sea level. Initially, six horizons—top basement, top Shublik, LCU, top Torok, and two internal Torok—were interpreted based on regional seismic reflectors and outcrop geology (Figure 3).

For modeling purposes (both structural and petroleum systems modeling), the initial interpretation of the seismic section (six horizons) had to be subdivided to account for a total of 15 layers between the top of the basement and the youngest strata before uplift and erosion (Table 1). Poor seismic resolution at depth made mapping even of the primary six horizons difficult, especially in the southern half of the line. The additional horizons required by the model (Figure 4) were added to the section in a "layer-cake" manner using reasonable thicknesses derived from regional relations (Moore et al., 1994; Bird, 2001). Eroded units have been reconstructed based on the regional sonic-log analysis of Burns et al. (2007), combined with vitrinite reflectance and apatite fission-track data. According to this, maximum erosion occurred in the south of the section with an estimated paleo-overburden of 7.2 km (23,600 ft).

The cross section indicates that most of the geologic time is represented by the Ellesmerian and Beaufortian sequences below the LCU, comprising roughly 240 m.y. Yet, most of the section's thickness is in the Brookian sequence, namely, in the lower Torok Formation, deposited within a period of only 0.8 m.y., and the entire Torok representing only 5 m.y.

Palinspastic Reconstruction of Structural Architecture

Application of structural modeling to the present interpretation of the 2-D section is designed to improve the structural interpretation and to palinspastically restore the geology to its undeformed configuration. The structural modeling requires a kinematically reasonable model of the deformation and results in iterative improvements of the structural interpretation through use of conservation (balancing) of area and line lengths between the undeformed and deformed geometries. Once balanced, the structural model can be portrayed as a series of

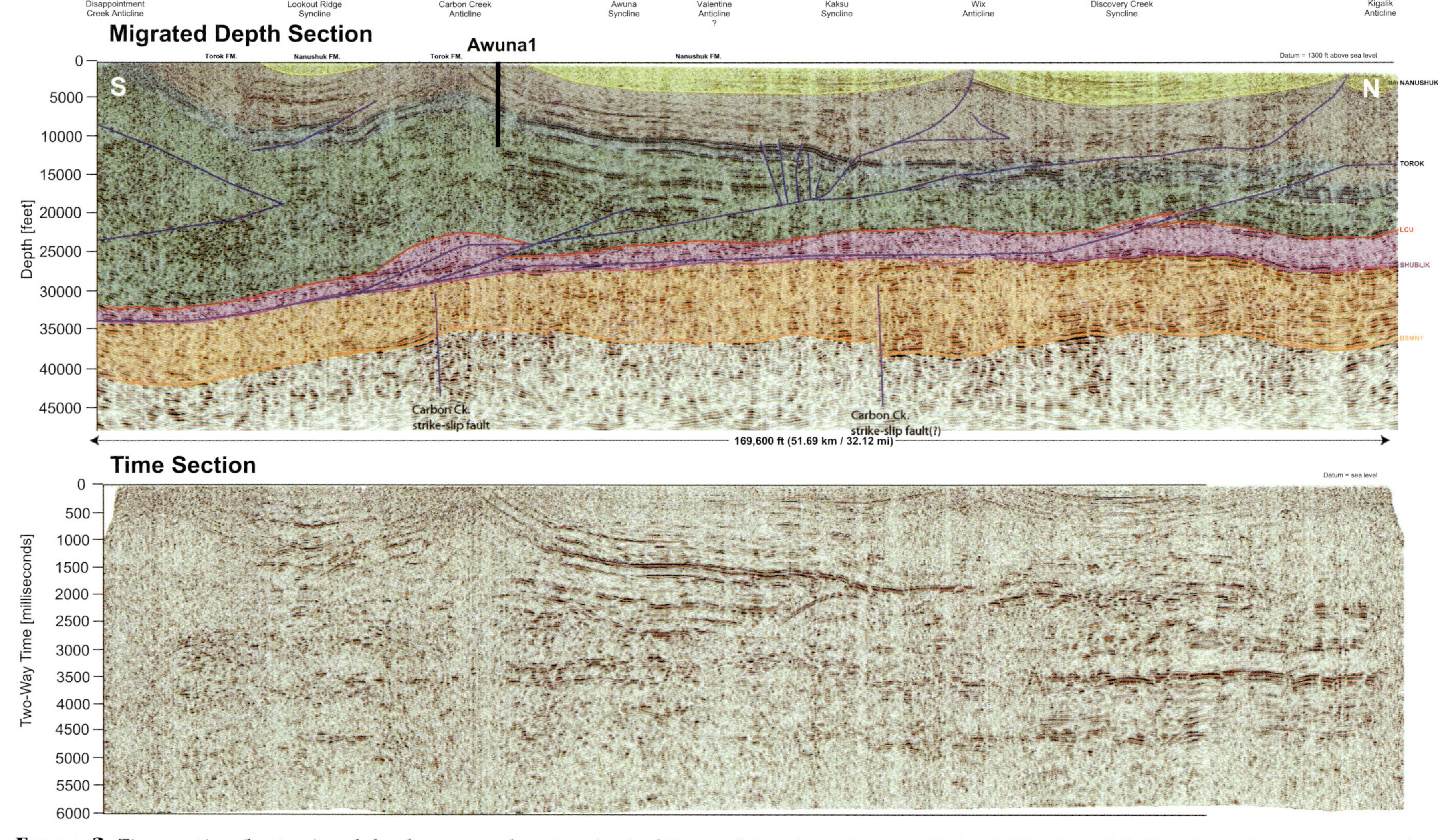

FIGURE 3. Time section (bottom) and depth-converted section (top) of National Petroleum Reserve-Alaska (NPRA) line 89X-78 with preliminary structural-stratigraphic interpretation as basis for the two-dimensional analyses. Depth section is not vertically exaggerated.

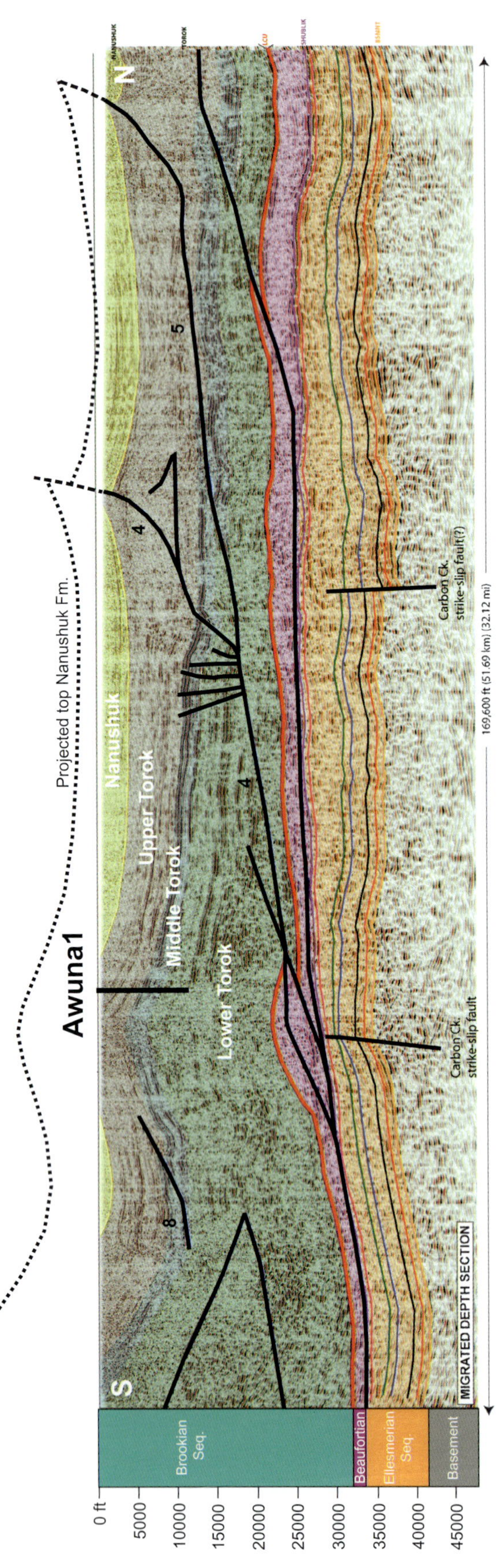

FIGURE 4. Interpreted seismic section showing the layers used in the models (see also Figure 2; Table 1), the faults (black), and the downdip correlative horizon of the Lower Cretaceous unconformity ([LCU] red), as well as the estimated erosional overburden (dashed horizon).

quantitatively constrained paleomodels that can be integrated into petroleum systems analysis.

Palinspastic restoration of the 2-D section, including the interpretation of several paleosections, has been performed using Geo-Logic Systems LithoTect software package. The reconstruction tracks movement through time and shows periods of deposition and subsequent erosion to account for the possible impact of overburden and compaction during burial history of the foreland basin. Understanding the evolution of the relative position of rock blocks involved in faulting allows establishing the geometric and temporal controls on potential HC migration pathways.

Structural modeling is an iterative process where initial interpretations are expected to require modification in the balancing process. Changes may be necessary to improve the interpretation by modifying either structural or stratigraphic aspects. The interpretation shown in Figure 3 (top) and Figure 4 is unbalanced, and incremental modeling highlighted some issues that can be manipulated in a straight restoration. However, for proper balancing, minor revisions of the original interpretation were necessary, such as modifications of the fault trajectories or the hanging-wall/footwall cutoffs (Figure 5). In addition, the post-Nanushuk layer has been added (Table 1). It is not present in the present-day section, yet accommodates some of the erosion and represents the geologic period during which deformation occurred. The revisions show nicely how structural modeling can improve a cross section and are ultimately helpful for better understanding of the structure.

Regarding timing and age of deformation, regional fission-track data from a transect about 60 km (37 mi) to the east across the same structures indicate the same cooling history, which is consistent with a simple model (Cole et al., 1997). However, the data are complicated, with retention of old provenance ages and indications that cooling may have been either protracted from about 60 to 45 Ma or episodic with cooling at 60 and 45 Ma—or anything in between. This leaves room for many complexities in the movement history as long as the vertical throw is minimal.

In the absence of other information, the simplest explanation has been to only restore the structures identified on the cross section. The sequential restoration implies that each fault accommodates its final displacement and then becomes inactive, and then another fault takes over (Figure 6). This is an important assumption for the subsequent fault seal analysis.

Decompaction of the strata during backstripping and paleosection balancing has been taken into account using standard Sclater and Christie (1980) function mixtures corresponding to the lithologies assigned to the model (Table 1).

The "flying wedge" (paleosection 6 in Figure 6) that comes in from the south is the tip end of a triangular

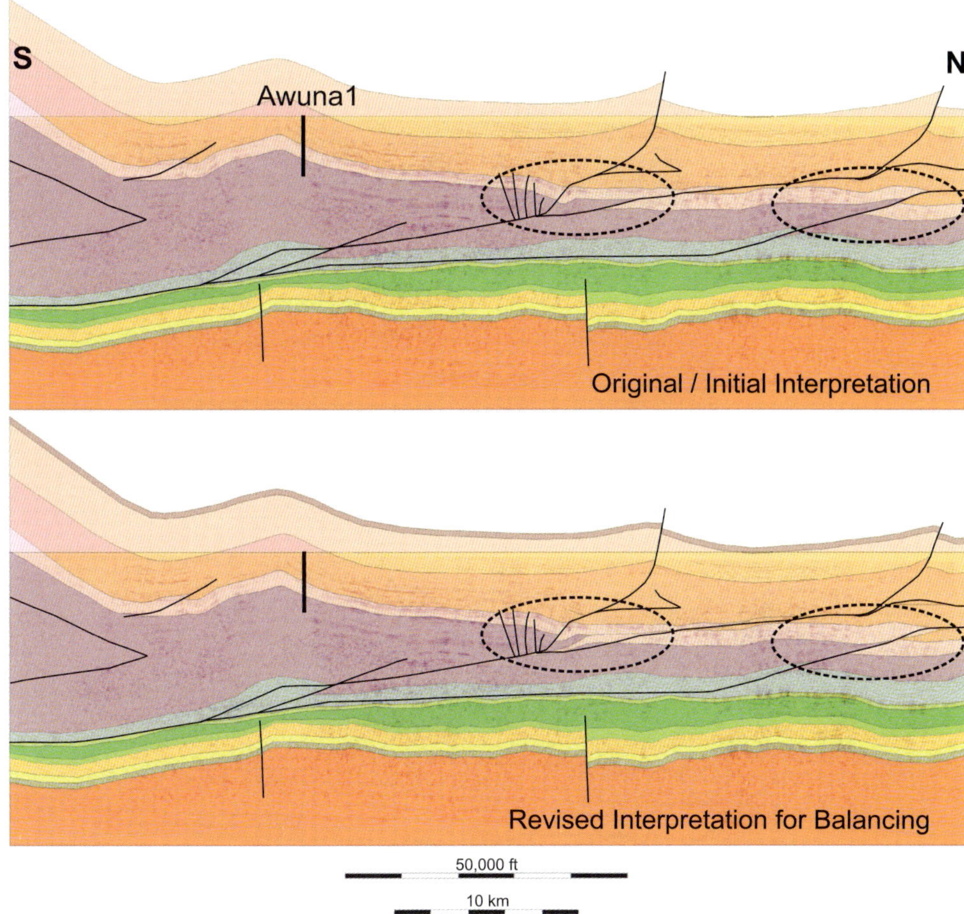

FIGURE 5. Comparison of the original interpretation and the revised interpretation of the present-day situation used for structural balancing.

zone (passive-roof duplex) that can be traced along the length of the Brooks Range foothills (Potter and Moore, 2003). Little stratigraphic control would provide an idea about how much shortening is represented in the duplex, and the seismic data from that area are typically very noisy probably because of steep southward dips that are characteristic. Moreover, south of the Awuna well, the duplex seems to have been influenced by the main trace of the predeformational Carbon Creek strike-slip fault, which lies beneath the southern part of the section. For these reasons, this part of the section is shown as a rootless wedge.

Estimation of Fault Rock Properties

Faults modify the connectivity of reservoir or carrier rock units both geometrically (as block boundaries separating individual fault blocks) and by acting as physical barriers to flow, hence, fault interpretation needs to be a primary task when considering petroleum systems modeling. Lithologic information (derived from well data or outcrop), a thorough stratigraphic interpretation of the seismic section, and the evolution of displacement through time provide the means to determine the sealing potential of faults using quantitative parameters such as the shale gouge ratio (SGR) (Yielding et al., 1997). In this chapter, lateral and vertical fault correlation, fault relationships, fault-surface extent, fault displacement, and growth histories have been addressed using Badley Geoscience TrapTester technology. Ultimately, the migration model is conditioned with estimates of the capillary properties of the faulted rocks through time.

The Vshale log for the Awuna well has been used to guide the prediction of fault rock properties. Vshale is a generic term indicating the fraction of shaly material in a given volume of rock. It can be the result of a petrophysical study or of field-core and laboratory analyses and is used as an estimate of clay content. The well transects ad hoc upper, middle, and lower parts of the Torok Formation of the Brookian sequence (Figure 4). According to the log of the Awuna well, the transected rocks are shale prone. For the three Torok intervals, the average shale content of each zone is greater than 50%. This implies that for any fault displacement greater than the thickness of the sandy layers, the SGR will also be in excess of 50%. To test structural and stratigraphic sensitivities, the Vshale curve has been adjusted to make it 10% more sandy, but this adjustment is insignificant to the outcome. The basic conclusion is that all the

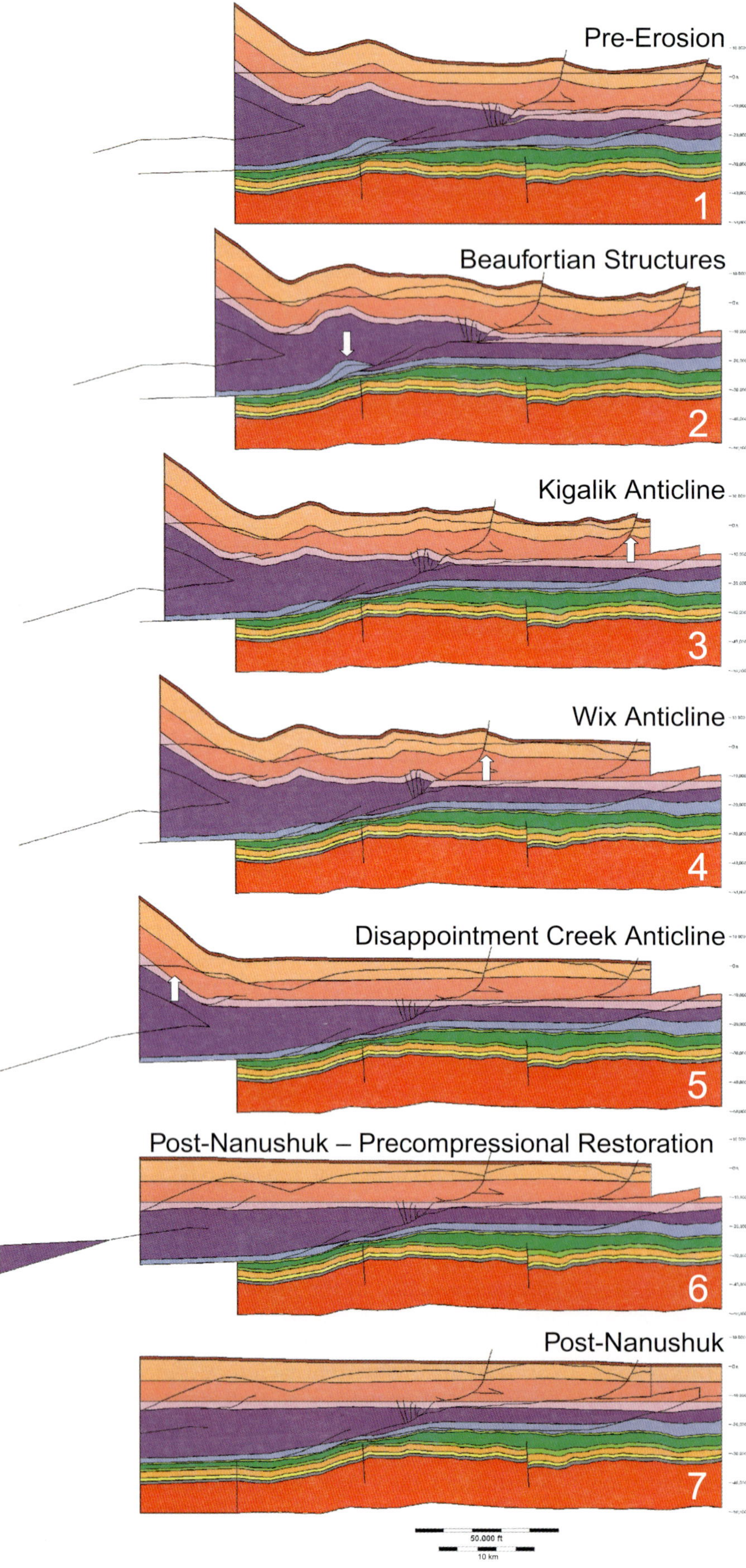

FIGURE 6. Incremental evolution of the cross section. Paleosections 1 through 6 are bed length and area balanced. Paleosection 7 assumes the undeformed initial geometry before faulting and folding. Deformation occurred over a period of 15 m.y. between 60 and 45 Ma. Uplift and erosion are assumed for the remaining 45 m.y. until the present day. White arrows depict respective structures formed during deformation.

faults will be completely sealing (capillary entry pressure of 0.7–0.8 MPa [102–116 psi]) at displacements in excess of 30 m (100 ft). Thirty meters is about one tenth the average displacement on the major faults.

The displacement along the major faults in the section is remarkably uniform between 300 and 600 m (1000–2000 ft). The sequential restoration implies the simplification that each thrust fault becomes inactive after maximum displacement, at that point the next fault takes over. The restored faults move in sequence from south to north, and the displacement is complete at the end of each sequence (compare with Figure 6). Hence, the capillary structure of each fault is also "complete" at the end of each sequence.

To determine the fault properties, a triangle plot (Figure 7A) was used, depicting the sequence-throw juxtaposition. This requires the Vshale log as well as some knowledge of the fault displacement. Fault sealing potential is considered to be related to the shale percentage within the part of the sequence that has moved past a point on the fault surface. This is termed "shale gouge ratio" (Yielding et al., 1997). The triangle plot is interpreted by reading off the SGR for a given layer-by-layer juxtaposition at a given displacement (Figure 7B). The SGR values are then converted to threshold pressures (fault capillary pressure [FCP], capillary entry pressure [CEP]; Bretan and Yielding, 2005), which are used to determine the seal capacity on the fault trace for particular fluid densities (oil and/or gas).

In general, when the displacements are of the same order as, or less than, the thickness of the flow units, cross-fault flow potential is relatively high. As displacement increases, the expected composition of the fault zones approaches the average composition of the section or well so the faults tend to be more prone to seal. Based on in-situ measurements and laboratory experiments, an SGR of 50% corresponds to a minimum entry pressure of 0.8 MPa (116 psi) (Bretan and Yielding, 2005). In the Awuna well, sand units occupy only 1 or 2% of the sequence, and the SGR is extremely high at very low displacements. The notion of cross-fault flow at any but the smallest offset is tenuous, and geometric connectivity has disappeared by the time the displacement is in excess of 30 m ([100 ft] i.e., one tenth the average displacement) (Figure 8).

Assuming that the Awuna log is representative for the upper part of the sequence, very little opportunity for cross-fault flow occurs. The CEPs of as much as 0.01 or 0.02 MPa (1.45 or 2.9 psi) rapidly increase to 0.7 or 0.8 MPa (102 or 116 psi) once displacement has exceeded 30 m (100 ft). In terms of the faulting sequence, the major fault 8 will be a barrier to flow before fault 4 becomes active (Figure 4). Likewise, fault 4 will be a barrier before fault 5 is active. In practical terms, this means that the upper part of the central fault block is isolated early in the structural history. Based on wells in the vicinity of the study area, the deeper part of the lower Torok Formation may be more sand prone than the middle and upper parts but also includes Hue Shale–type low-permeability layers, that is, the faults in the deeper part of the sequence are likely to also have CEPs in excess of 0.8 MPa (116 psi).

Two-Dimensional Basin and Petroleum Systems Modeling with TecLink Technology

The PetroMod TecLink technology is used to integrate and combine the results of the structural and fault property analyses in a classic petroleum systems model. In this study, a 2-D, pressure-volume-temperature (PVT)–controlled, multicomponent, three-phase petroleum migration model demonstrates the integration of geometry changes, fault properties, thermal history, source rock maturation, and HC migration and accumulation through time to understand the geologic evolution and the possible petroleum distribution in the study area.

Teclink Technology

The time framework for a basin and petroleum systems model is generally provided by the depositional and erosional ages of the strata (Figure 9). The strata in the study area were deposited from 400 Ma (basement) to 60 Ma (post-Nanushuk Formation), with the Nanushuk Formation (94 Ma) representing the youngest strata in the study area at present day. These depositional ages are referred to as "event steps" and are the basis for a past to present-day forward modeling simulation (Hantschel and Kauerauf, 2009). During event stepping, compaction is calculated based on initial rock porosity, compressibility, and permeability, as well as fluid viscosity. Present-day thickness and time of deposition and erosion are also taken into account. This approach is suitable for most processes dominated by normal deposition and erosion or in extensional basins (Broichhausen, 2004). However, in case of complex paleothickness or paleogeometry variations, including for example, salt movement, mud-diapirism, or complex extensional and compressional movement, kinematic, palinspastic, and structural restorations are required that account for movement through time. Reconstructed intermediate paleogeometries act as a geometric guide during forward modeling and are directly implemented into the modeling process as paleosteps. Seven such paleosteps are used in the model (Figure 6). Instead of applying regular event stepping all the way from the oldest to present-day geometry, event stepping is calculated only up to the first paleogeometry (post-Nanushuk geometry at 60 Ma, paleosteps 6 and 7, respectively, in Figure 6), upon which paleostepping takes over, that is, the time framework is

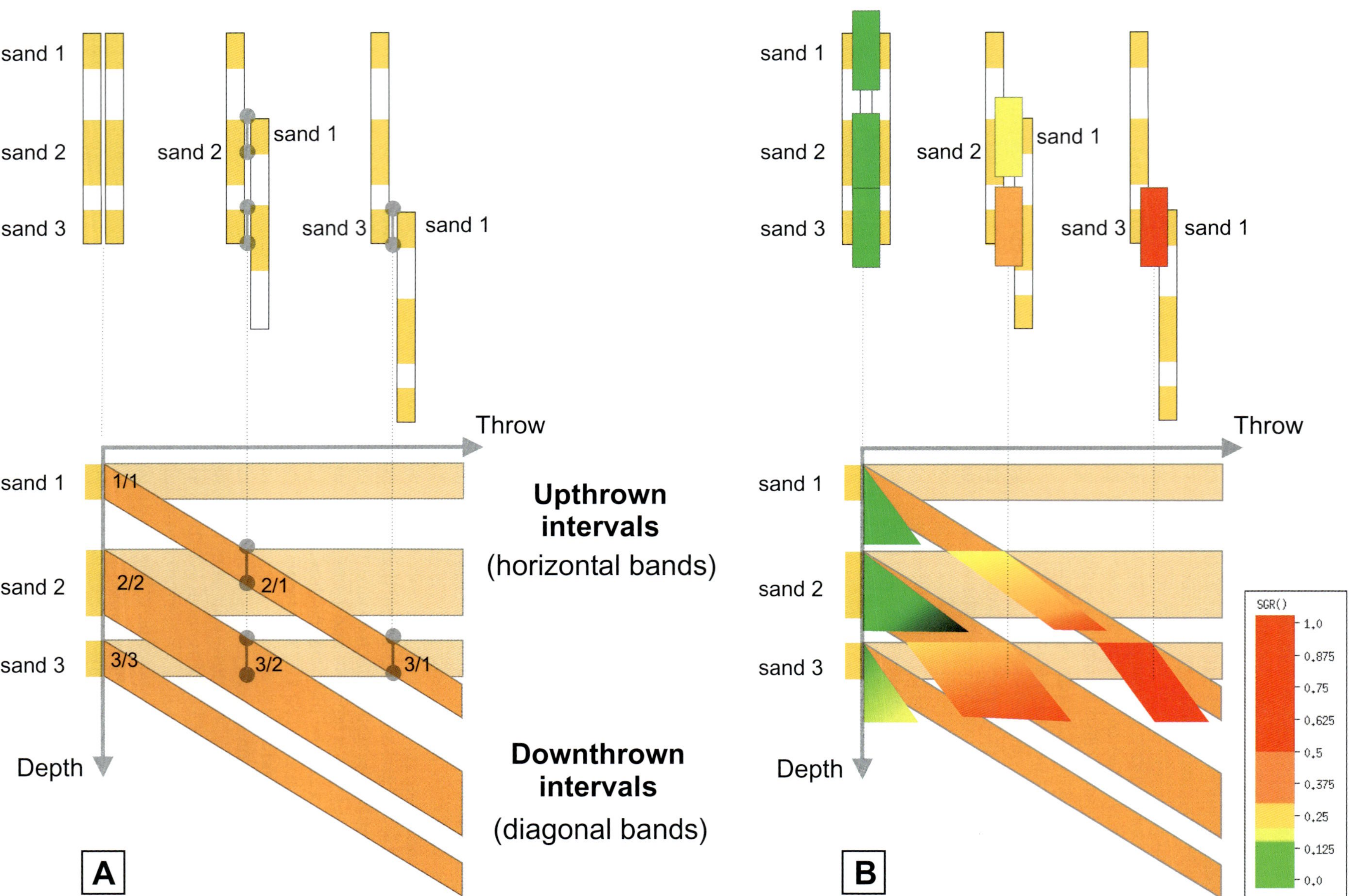

Figure 7. (A) Triangle plot showing various types of sand juxtaposition with respect to fault displacement (throw). 1/1 = self-juxtaposed intervals; 3/2 = intervals juxtaposed against different intervals (e.g., sand 3 in footwall against sand 2 in hanging wall). (B) Triangle plot showing the increase of shale gouge ratio (SGR) with increasing fault displacement (throw). The SGR (~fault-zone clay content) is predicted for each throw increment and displayed on the sand-on-sand juxtapositions. The SGR scale: green = low SGR, sands are likely to leak; red = high SGR, sands are likely to seal.

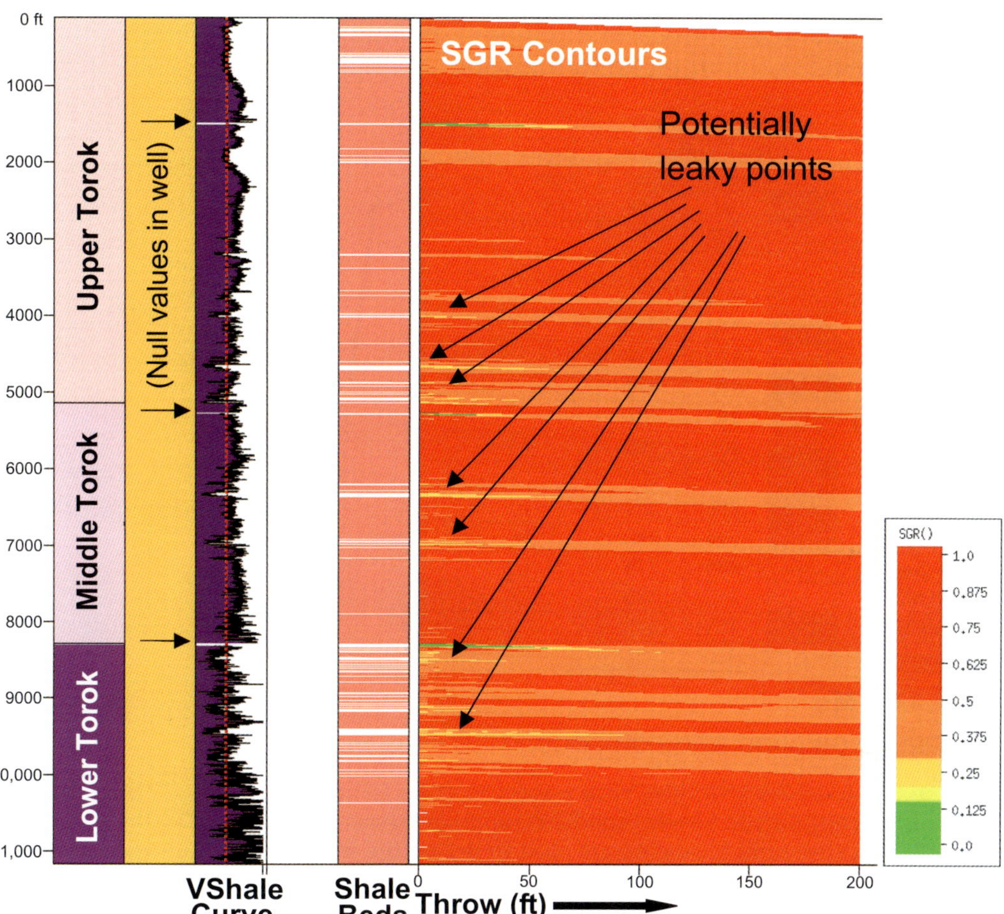

FIGURE 8. Sequence-throw juxtaposition diagram constructed from the Awuna well. Note that the Vshale curve plots predominantly shale (to the right of the 50% shale, red dotted line). See also light red intervals to the left of the throw diagram, which depict shale. The throw diagram is contoured for shale gouge ratio (SGR). Note that the faults become virtually sealing with fault displacement (throw) in excess of 30 m (100 ft). The SGR scale: green = low SGR, sands are likely to leak; red = high SGR, sands are likely to seal.

now determined by the number of paleosteps (geometry snapshots in time) instead of by the deposition of the strata (Figure 9).

In addition to discretizing time not only with event, but also with paleosteps, a model that has undergone thrusting (doubling of strata) has to be subdivided in a way that avoids multiple depth values. Here, each thrust sheet represents an individual geometric unit or fault block without any vertically repetitive strata. Although these fault blocks essentially represent individual submodels, they are combined to a comprehensive model during simulation of each time step. Thermal, pressure, and migration properties are calculated for the entire model considering mutual and overlapping effects at the block boundaries.

A detailed description of the event versus paleostepping technology, as well as the fault-block concept, can be found in Lampe et al. (2006), Baur et al. (2009), and Hantschel and Kauerauf (2009).

Model Calibration

Basal heat flow is determined from one-dimensional calibrations of vitrinite reflectance data across the entire North Slope (Lampe et al., 2003), with special emphasis on the Awuna well. Present-day heat flow is highly variable, with relatively low values in the south near the foothills of the Brooks Range to high values near the north coast. The observed pattern is similar to that measured by Deming et al. (1992) and is consistent with forced convection by a topographically driven groundwater flow system. The model is calibrated using a conductive thermal background heat flow of 60 mW/m^2 with a superimposed cooling effect by recharging groundwater coinciding with the formation of the Brooks Range 120 m.y. ago, resulting in a minimum heat flow of 40 mW/m^2 at 60 Ma, which is slowly rising during subsequent deformation, uplift, and erosion to about 53 mW/m^2 at the present day. Changing paleo–water depths through time and related sediment surface temperatures are taken into account. The resulting calibration for temperature and vitrinite reflectance is shown in Figure 10.

The present-day temperature distribution in the model suggests equilibrium conditions in that the isotherms run horizontally across the structures (Figure 11, top). However, vitrinite reflectance indicates that it was at steady state (maximum burial and temperature) before deformation. Highly mature source rock was brought up to cooler regions during deformation, hence, the

Event-Stepping
Using depositional ages of formations (time)

Rock Units	Deposition (Ma) from	Deposition (Ma) to	Erosion (Ma) from	Erosion (Ma) to
Post-Nanushuk	94	60	45	0
Nanushuk	107	94		
Upper Torok	111	107		
Middle Torok	111.2	111		
Lower Torok	112	111.2		
Hue-GRZ	122	112		
PBS	133	122		
Kingak	193	133		
K1	200	193		
Shublik	245	200		
Ivishak	250	245		
Echooka	275	250		
Lisburne	335	275		
Kayak	345	335		
Kekiktuk	350	345		
Basement	400	350		

400 – 60 Ma
normal deposition and compaction

Paleostepping
Using ages of deformation (geometry)

	Paleosections	Age (Ma)
	Present Day	0
1	Preerosion	45
2	Beaufortian Structures	48
3	Kigalik Anticline	51
4	Wix Anticline	54
5	Dissappointment Anticline	57
6	Post–Nanushuk	60

60 - 45 Ma
paleosections with predefined geometry

45 – 0 Ma
erosion

FIGURE 9. Model discretization with respect to time. Depositional ages of the individual formations are used for the classic deterministic forward modeling approach (event stepping). During deformation, predefined geometries derived from structural balancing are used to define the model evolution (paleostepping). The numbers (1–6) in the paleostepping table correspond to the individual balanced sections in Figure 6. PBS = pebble shale unit; Hue-GRZ = Hue–Gamma Ray Zone.

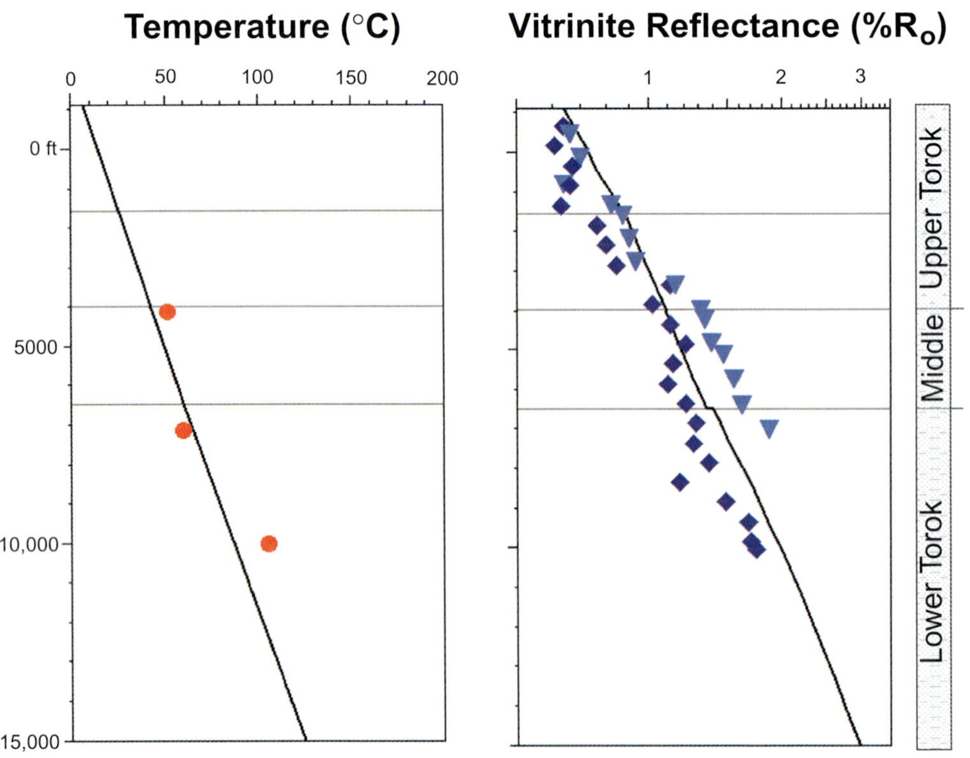

FIGURE 10. Temperature and vitrinite reflectance along the Awuna well. Dots = measured data; lines = calculated data. The measured vitrinite reflectance shows two different sources of data: diamonds = Arco; triangles = U.S. Geological Survey.

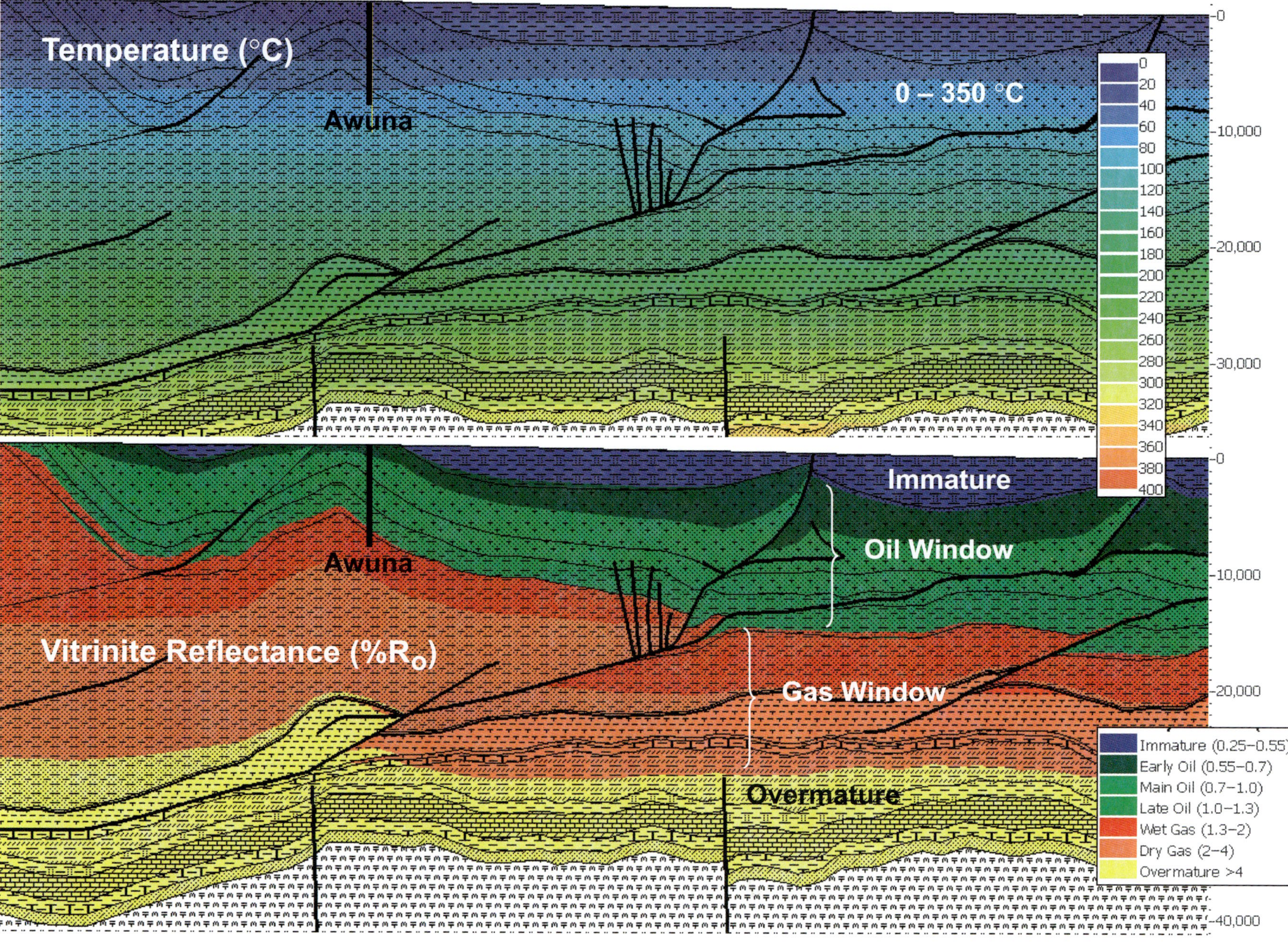

FIGURE 11. Calculated temperature (top) and vitrinite reflectance (bottom) for the cross section. Note that the isotherms run more or less horizontal across the structures whereas the level of maturity is offset by the thrusting. Depth in feet, distance in meters.

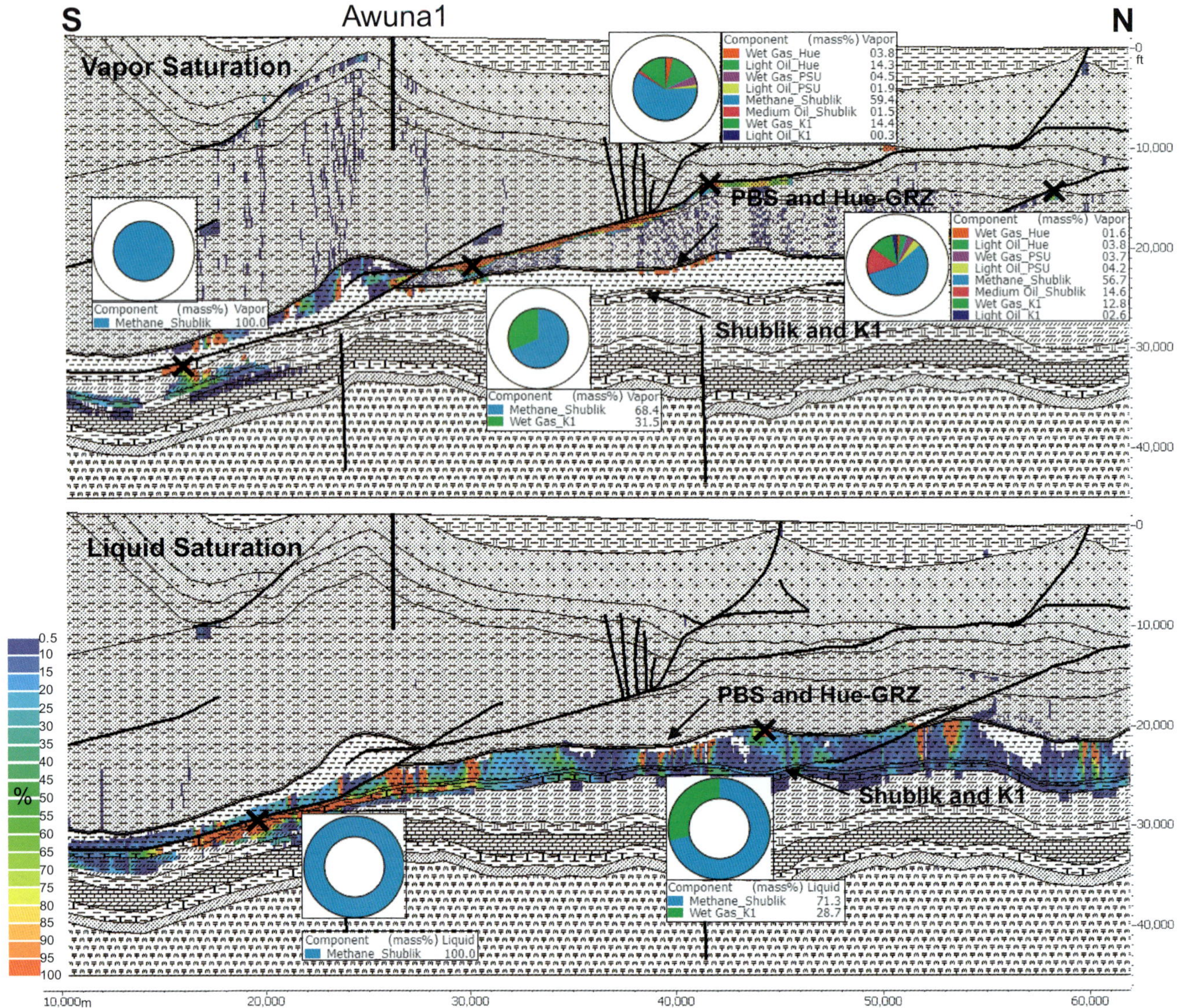

FIGURE 12. Vapor (top) and liquid (bottom) saturation as well as petroleum composition for the cross section. The black Xs depict the position of the analyses. The source rocks (pebble shale unit [PBS] and Hue–Gamma Ray Zone [Hue-GRZ] above the Lower Cretaceous unconformity [LCU] as well as the pre-LCU Shublik and K1) are indicated for reference. Liquid saturation is mostly confined to the overpressured pre-LCU strata whereas vapor saturation dominates the Brookian sequence and is trapped against the impermeable faults. In the southern (left) part of the section, downward expulsion from the Shublik Formation into the slightly more permeable Ivishak and Echooka formations can be observed. Depth in feet, distance in meters.

thrusting shows up in the vitrinite reflectance pattern (Figure 11, bottom). This pattern nicely highlights the importance of structural modeling before maturity and/or migration modeling.

Migration and Petroleum Systems

Migration modeling was performed, applying the hybrid method (Hantschel and Kauerauf, 2009), which combines Darcy flow for low-permeability rocks with buoyancy-driven flow path migration in high-permeability strata. As most of the strata are low-permeability rocks, that is, they do not qualify as good reservoir or carrier rocks (Figure 2), Darcy flow dominates the migration and can be depicted as saturations (i.e., percentage of fill for a given cell, Figure 12) and flow vectors (Figure 13).

Liquid saturation is confined to the Shublik and Kingak formations, which are subject to significant overpressure of as much as 105 MPa (15,229 psi) excess hydraulic pressure caused by the mostly shaly lithology, which remains undercompacted because of the rapid deposition of the Brookian sequence. The vapor saturation occurs dominantly in the shallower Brookian

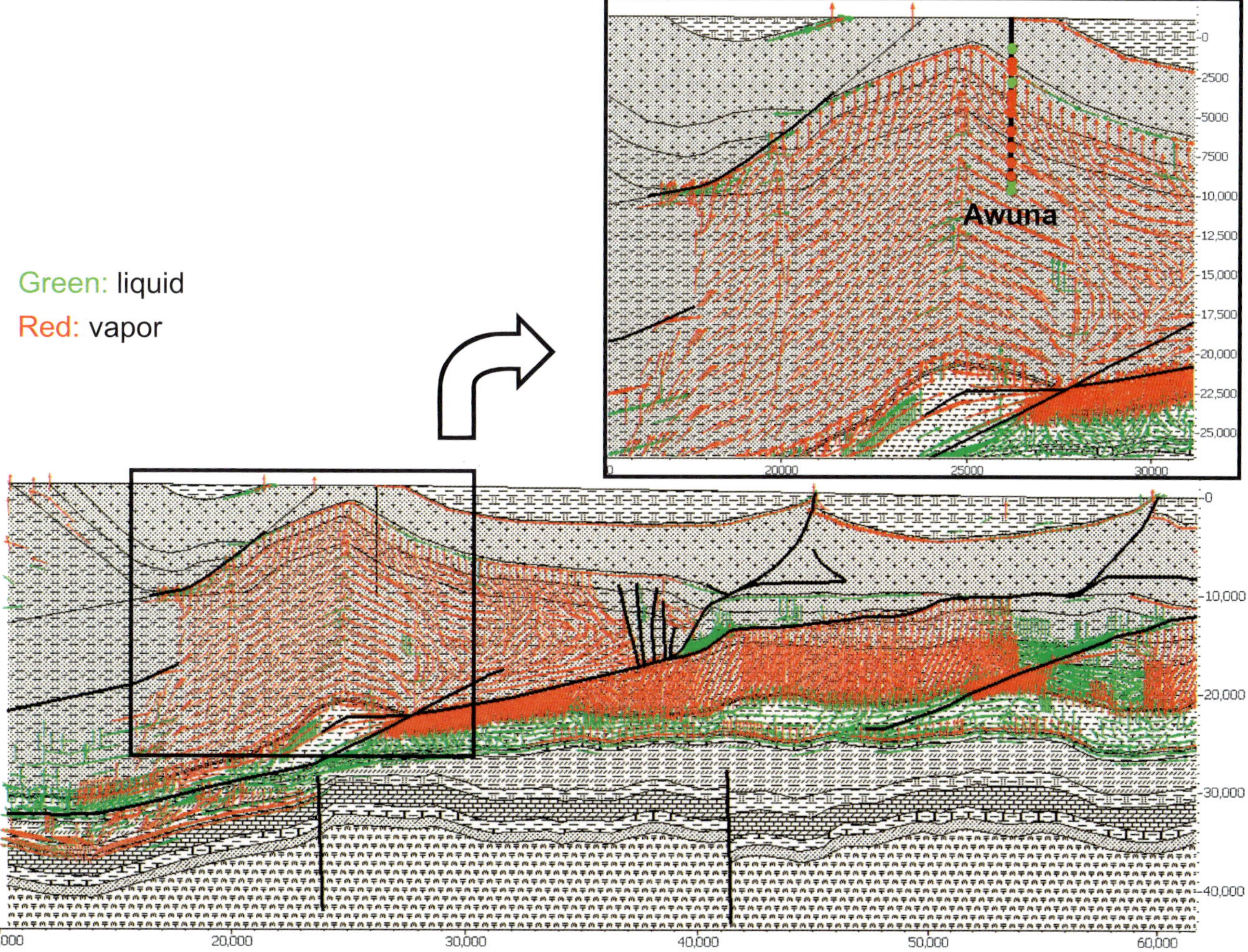

FIGURE 13. Present-day flow vectors derived from Hybrid migration in the two-dimensional section and in the Awuna well vicinity. The calculated hydrocarbon occurrences coincide well with the observed oil and gas shows in the Awuna well (green and red dots). Depth in feet, distance in meters.

strata, which is generally more permeable so that modeled excess hydraulic pressure only reaches as much as 0.6 MPa (87 psi). The mostly impermeable faults act as barriers for flow and mostly confine the petroleum to the deeper parts of the basin. However, some leaking occurs in the crests of the anticlines (southern part of the section) and where the thrust faults reach the surface (northern part of the section), which is consistent with observations in the study area (Figure 13).

The composition of the petroleum indicates extensive secondary cracking of the deep K1 and Shublik source rocks to wet gas and methane in the deeper overpressured pre-LCU sequences where the petroleum is in liquid phase (Figure 12, bottom). Only in the Brookian sequence can primary oil and gas from all four source rocks be found, where it is mostly in vapor phase (Figure 12, top). As to be expected, the thick and rich Shublik source rock contributes about 60 to 100% of the petroleum in the system, depending on the strata in which it is trapped.

Hydrocarbon indications in the Awuna well, summarized by Hayba et al. (2002), consist of 34 oil and 98 gas occurrences. Oil occurrences are mostly fluorescence with some slight oil staining and some dead oil and solid bitumen. Gas occurrences are mostly methane (C_1) mud gas readings greater than 20,000 ppm or ethane (C_2) mud gas readings greater than 2000 ppm. Surface gas seeps may be common but are unidentified unless gas bubbles are observed in bodies of water. One such seep, Pahron Lake gas seep, is known to occur a few miles north of the study area (Whittington and Keller, 1949). The geologic map of this area shows a fault nearby (Mull et al., 2006). The present-day flow vectors (Figure 13) of the model show a mixture of liquid (green) and vapor (red) petroleum moving through the system. These petroleum shows coincide mostly with the HC occurrences, both liquid and vapor, in the Awuna well and together with the HC seeps at the surface validate the model results.

Figure 14 shows the event chart for the petroleum systems in the study area. Trying to assess the petroleum

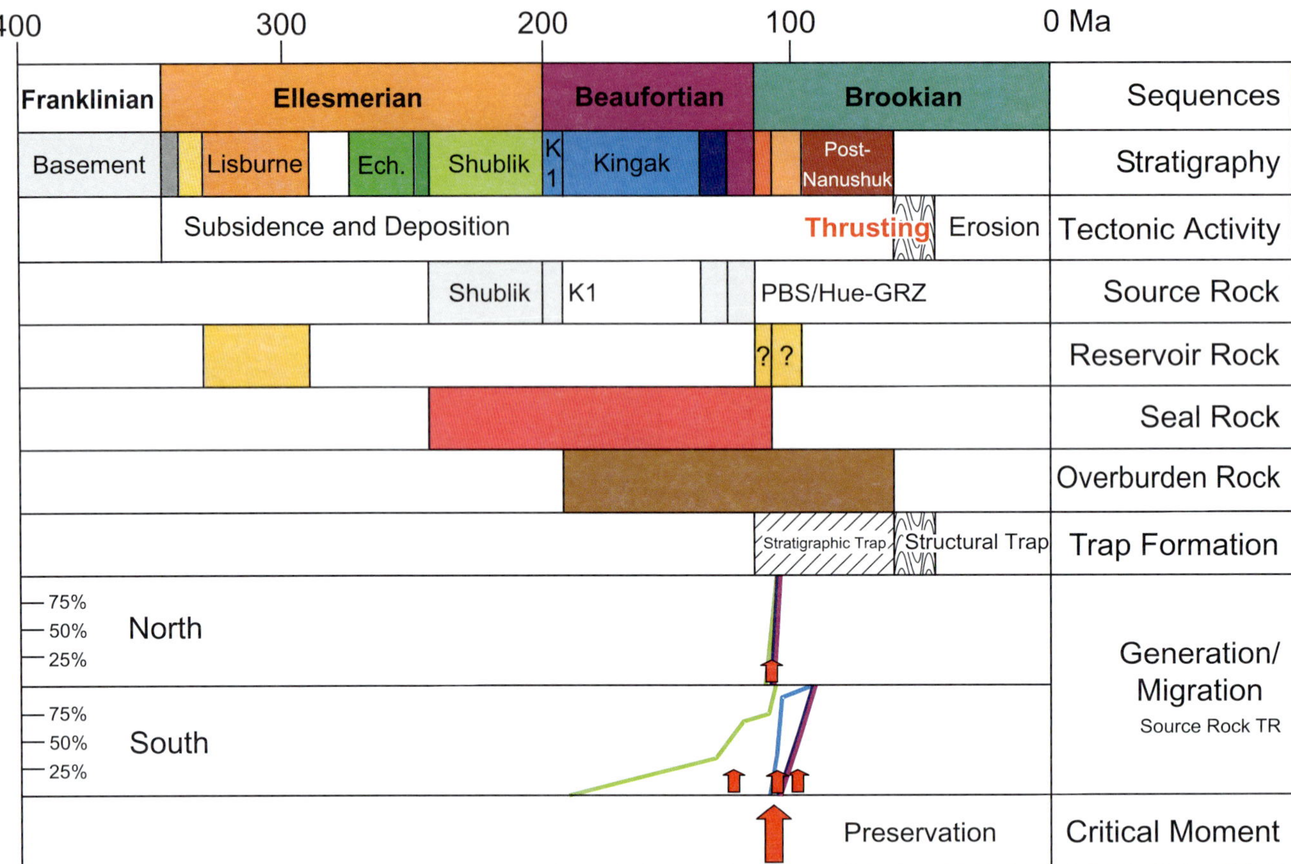

FIGURE 14. Events chart for the study area. The generation-migration curve colors refer to the individual source rock color to show which source rock became mature at what point in time. The locations have been picked at the edges of the model, the far north and south, respectively. They show source rock transformation ratio (TR) between 0 (no organic material has been converted to kerogen) and 100% (all organic material has been converted to kerogen). Note that petroleum generation and migration is confined to a very short time interval because of the rapid deposition of the Brookian sequence. Trap formation occurred largely after petroleum generation, and a defined regional seal is absent. PBS = pebble shale unit; Hue-GRZ = Hue–Gamma Ray Zone.

potential of the foothills fold and thrust belt, the source rock maturation is confined to a very narrow time window, governed by the rapid deposition of the Torok Formation. Subsequent thrusting has little effect on the source rocks as they were spent by the time the study area got deformed. Structural traps formed after the critical moment. In addition, the modeled reservoir rocks (Torok and Nanushuk formations) lack seals, although they may have more or less sealing intervals intercalated or may feature confined sandstone reservoirs encased in shale, respectively. In summary, the generated petroleum had plenty of time and opportunity to escape. However, the resource assessment conducted by the U.S. Geological Survey (Potter and Moore, 2003) postulates the presence of stratigraphic traps that may be charged, and some may have even survived structural deformation in the broad synclinal areas. The assessment also postulated that some stratigraphic traps will be breached in the deformation, and the released HCs will either be lost to the surface or be secondarily trapped in structures.

CONCLUSIONS

Integrating palinspastic reconstruction work, fault seal analysis, and petroleum systems modeling enables the simulation of dynamic processes and allows accurate assessment and prediction of the extent, timing, and potential of petroleum systems in tectonically complex areas such as the Colville foreland basin.

This multidisciplinary approach allows a better understanding of the basin geometry, the function of faults and their possible effects on the migration pattern and timing, as well as the temperature, maturation, and migration history through time. At the same time, feedback between the individual modeling processes can be ensured while the superior functionality of the respective tools or method is retained. This ultimately increases confidence in the model results.

The presented study is a good demonstration of the degree of complexity one gets in doing structural modeling and fault seal analysis in combination with petroleum systems analysis, even in seemingly simple structural settings.

ACKNOWLEDGMENTS

We thank Oliver Schenk and Les Magoon for their detailed and thoughtful review of the manuscript.

REFERENCES CITED

Baur, F., M. Di Benedetto, T. Fuchs, C. Lampe, and S. Sciamanna, 2009, Integrated structural geology and petroleum systems modeling: A pilot project from Bolivia's fold and thrust belt: Marine and Petroleum Geology, v. 26, no. 4, p. 573–580, doi:10.1016/j.marpetgeo.2009.01.004.

Bird, K. J., 1994, The Ellesmerian(!) petroleum system, North Slope of Alaska, U.S.A., *in* L. B. Magoon and W. Dow, eds., The petroleum system: From source to trap: AAPG Memoir 60, p. 339–358.

Bird, K. J., 2001, Alaska: A twenty-first century petroleum province, *in* M. W. Downey, J. C. Threet, and W. A. Morgan, eds., Petroleum provinces of the twenty-first century: AAPG Memoir 74, p. 137–165.

Bird, K. J., and C. M. Molenaar, 1987, Stratigraphy of the northern part of the Arctic National Wildlife Refuge, northeastern Alaska, *in* K. J. Bird and L. B. Magoon, eds., Petroleum geology of the northern part of the Arctic National Wildlife Refuge, Northeastern Alaska: U.S. Geological Survey Bulletin, v. 1778, p. 37–59.

Bretan, P., and G. Yielding, 2005, Using buoyancy pressure profiles to assess uncertainty in fault seal calibration, *in* P. Boult and J. Kaldi, eds., Evaluating fault and cap rock seals: AAPG Hedberg Series 2, p. 151–162.

Broichhausen, H., 2004, Mudstone compaction and overpressure calculation in a 3-D petroleum systems analysis tool: Development of the software and application to the North Sea: Ph.D. thesis, Shaker Verlag, Aachen, 264 p.

Burns, W. M., D. O. Hayba, E. L. Rowan, and D. W. Houseknecht, 2007, Estimating the amount of eroded section in a partially exhumed basin from geophysical well logs, an example from the North Slope, *in* P. Haeussler and J. Galloway, eds., Studies by the U.S. Geological Survey in Alaska, 2005: U.S. Geological Survey Professional Paper 1732-D, p. 18, http://pubs.usgs.gov/pp/pp1732/pp1732d/ (accessed November 19, 2010).

Cole, F., K. J. Bird, J. Toro, F. Roure, P. B. O'Sullivan, M. Pawlewicz, and D. G. Howell, 1997, An integrated model for the tectonic development of the frontal Brooks Range and Colville Basin 250 km west of the Trans-Alaska Crustal Transect: Journal of Geophysical Research, v. 102, no. B9, p. 20,685–20,708.

Creaney, S., and Q. R. Passey, 1993, Recurring patterns of total organic carbon and source rock quality within a sequence-stratigraphic framework: AAPG Bulletin, v. 77, no. 3, p. 386–401.

Deming, D., J. H. Sass, A. H. Lachenbruch, and R. F. de Rito, 1992, Heat flow and subsurface temperature as evidence for basin-scale ground-water flow, North Slope Alaska: Geological Society of America Bulletin, v. 104, p. 528–542, doi:10.1130/0016-7606(1992)104<0528:HFASTA>2.3.CO;2.

Gryc, G., ed., 1988, Geology and exploration of the National Petroleum Reserve in Alaska, 1974 to 1982: U.S. Geological Survey Professional Paper 1399, 940 p.

Hantschel, T., and A. Kauerauf, 2009, Fundamentals of basin and petroleum systems modeling: Heidelberg, Springer Verlag, 476 p.

Hayba, D. O., K. J. Bird, and C. Garrity, 2002, Chapter 5, Subsurface oil and gas indications in and adjacent to the NPRA, *in* D. W. Houseknecht, ed., National Petroleum Reserve, Alaska (NPRA) core images and well data: U.S. Geological Survey Digital Data Series DDS-75 (4 CD-ROM).

Houseknecht, D. W., and K. J. Bird, 2004, Sequence stratigraphy of the Kingak Shale (Jurassic–Lower Cretaceous), National Petroleum Reserve in Alaska: AAPG Bulletin, v. 88, no. 3, p. 279–302, doi:10.1306/10220303068.

Houseknecht, D. W., and K. J. Bird, 2006, Oil and gas resources of the Arctic Alaska Petroleum Province: U.S. Geological Survey Professional Paper 1732-A, http://pubs.usgs.gov/pp/pp1732/pp1732a/ (accessed November 19, 2010).

Houseknecht, D. W., and C. J. Schenk, 2007, Outcrops of turbidite facies in the Torok Formation: Reservoir analogs for the Alaska North Slope, *in* T. Nilsen, R. D. Shew, G. S. Steffens, and J. R. J. Studlick, eds., An atlas of deepwater outcrops: Models and analogs: AAPG Studies in Geology 56, p. 373–377.

Lampe, C., K. E. Peters, L. M. Magoon, K. J. Bird, and P. G. Lillis, 2003, The Shublik's petroleum systems of the Alaskan North Slope: A numerical journey from source to trap: U.S. Geological Survey Open-File Report 03-326, 3 sheets, http://geopubs.wr.usgs.gov/open-file/of03-326/ (accessed November 19, 2010).

Lampe, C., K. Kornpihl, S. Sciamanna, T. Zapata, and G. Zamora, 2006, Petroleum systems modeling in tectonically complex areas: A 2-D migration study from the Neuquen Basin, Argentina: Journal of Geochemical Exploration, v. 89, p. 201–204, doi:10.1016/j.gexplo.2005.11.041.

Lerand, M., 1973, Beaufort Sea, *in* R. G. McGrossan, ed., The future petroleum provinces of Canada: Their geology and potential: Canadian Society of Petroleum Geologists Memoir 1, p. 315–386.

Lillis, P. G., M. D. Lewan, A. Warden, S. M. Monk, and J. D. King, 1999, Identification and characterization of oil types and their source rocks: Chapter OA (oil analyses), *in* ANWR assessment team, the oil and gas resource potential of the 1002 Area, Arctic National Wildlife Refuge, Alaska: U.S. Geological Survey Open-File Report 98-34, CD-ROM-ANWR, OA-1–OA-101.

Masterson, W., 2001, Petroleum filling history of central Alaska North Slope fields: Ph.D. thesis, University of Texas at Dallas, Dallas, Texas, 222 p.

Moore, T. E., W. K. Wallace, K. J. Bird, S. M. Karl, C. G. Mull, and J. T. Dillon, 1994, The geology of Alaska, *in*

The geology of North America: Geological Society of America, Boulder, Colorado, v. G1, p. 49–140.

Mull, C. G., D. W. Houseknecht, G. H. Pessel, and C. P. Garrity, 2006, Geologic map of the Lookout Ridge quadrangle, Alaska: U. S. Geological Survey Scientific Investigations map 2817-C scale: 1:250,000.

Peters, K. E., K. J. Bird, M. A. Keller, P. G. Lillis, and L. M. Magoon, 2003, Distribution, richness, quality, and thermal maturity of source rock units on the North Slope of Alaska: U.S. Geological Survey Open-File Report 03-328, 3 sheets, http://geopubs.wr.usgs.gov/open-file/of03-328/ (accessed November 19, 2010).

Potter, C. J., and T. E. Moore, 2003, Brookian structural plays in the National Petroleum Reserve-Alaska: U.S. Geological Survey Open-File Report 03-266, http://pubs.usgs.gov/of/2003/of03-266/ (accessed November 19, 2010).

Sclater, J., and P. Christie, 1980, Continental Stretching: An explanation of the post-mid-Cretaceous subsidence of the central North Sea Basin: Journal of Geophysical Research, v. 85, p. 3711–3739, doi:10.1029/JB085iB07p03711.

Whittington, C. L., and S. A. Keller, 1949, Preliminary reports on the Carbon Creek anticline and on the upper Meade River, Alaska: U.S. Geological Survey Preliminary Report 23, 16 p.

Yielding, G., B. Freeman, and D. T. Needham, 1997, Quantitative fault seal prediction: AAPG Bulletin, v. 81, p. 897–917.

8

Nelskamp, S., J. D. van Wees, and R. Littke, 2012, Structural evolution, temperature, and maturity of sedimentary basins in the Netherlands: Results of combined structural and thermal two-dimensional modeling, *in* K. E. Peters, D. J. Curry, and M. Kacewicz, eds., Basin Modeling: New Horizons in Research and Applications: AAPG Hedberg Series, no. 4, p. 137–156.

Structural Evolution, Temperature, and Maturity of Sedimentary Basins in the Netherlands: Results of Combined Structural and Thermal Two-Dimensional Modeling

Susanne Nelskamp and Jan D. van Wees

TNO, Geo-Energy, Utrecht, The Netherlands

Ralf Littke

Institute for Geology and Geochemistry of Petroleum and Coal, RWTH Aachen University, Aachen, Germany

ABSTRACT

Structural modeling combined with basin modeling was used to demonstrate the influence of the structural history on hydrocarbon generation. For this purpose, a tectonic setting in the Netherlands was selected that shows large-scale tectonic inversion, associated erosion, and later subsidence. On the basis of a 300-km (186-mi) two-dimensional section that crosses the main tectonic features of this setting, a structural model consisting of 21 paleosections was created. The results generated by the structural model show that the Late Cretaceous inversion affected the basins the most, whereas the erosion in the Jurassic had the strongest influence on the structural highs. This can be seen from the amount of erosion associated with these erosion phases. Using the structural model as input for the basin model allowed the temperature and maturity of the sediments to be calculated. A temperature profile at 2000-m (6562-ft) depth along the section shows that the present-day temperature distribution is also strongly influenced by the inversion. In the inverted basins, highly conductive layers, such as overcompacted sediments or salt, are closer to the surface, which results in higher temperatures than in the noninverted. Finally, the timing of hydrocarbon generation from the Posidonia Shale source rock was found to be related to the structural history within the basin. In strongly inverted parts of the basin, present-day burial is insufficient to restart hydrocarbon generation, but in less inverted parts, hydrocarbon generation resumed during the Tertiary.

DOI:10.1306/13311433H43468

INTRODUCTION

Basin modeling is widely used for the study of temperature and maturity evolution. In structurally complex areas, however, basin modeling alone is not sufficient to describe the processes that lead to the maturation of source rocks and the generation of hydrocarbons. Using the results of structural modeling as a basis for basin modeling gives more detailed information about the influence of structural evolution on the temperature and maturity of a basin. Structural modeling reveals lateral variability in tectonic evolution, inferred from vertical motions and erosion patterns. Basin models then constrain these results to observed maturity and present-day temperature trends.

A structurally complex but well-documented setting in the Netherlands was chosen to illustrate how the structural evolution affects the petroleum system through time. In the Netherlands, detailed one-dimensional (1-D) model information (Nelskamp et al., 2008) and country-wide maps of the important stratigraphic layers and structural elements (TNO, 2004a) are available. This excellent database provided the input data for structural modeling and subsequent basin modeling. In this chapter, these methods were used to improve heat-flow (temperature) history, petroleum generation, subsidence, and amount and timing of erosion. For this purpose, a two-dimensional (2-D) section was selected in the Netherlands onshore region, trending approximately perpendicular to the axes of the most important structural elements of the Dutch onshore basins.

The Netherlands is situated north of the Variscan deformation front; to the southwest, it is bordered by the London-Brabant Massif, which is also the southwestern margin of the Southern Permian Basin and the Central European Basin System (Littke et al. 2008a, b). During the Mesozoic, the Netherlands became subdivided into several different basins in response to the breaking up of the Pangea (Ziegler, 1988, 1990). The main structures in this basin system are the West Netherlands Basin (WNB), the Central Netherlands Basin (CNB), the Lower Saxony Basin (LSB), the Broad Fourteens Basin (BFB), the Roer Valley Graben, the Vlieland Basin, and the Dutch Central Graben (Figure 1). They are bordered by a series of highs or platform areas, such as the Zeeland Platform bordering the London-Brabant Massif in the west, the Zandvoort Ridge, separating the WNB and CNB, the Peel Block as the continuation of the Zandvoort Ridge to the southeast, the Texel-Ijsselmeer High, situated between the CNB and the Vlieland Basin, the Cleaver Bank High west of the Central Graben, and the Friesland Platform between the CNB, the Vlieland Basin, and the LSB.

A detailed description of the geologic history of the entire Netherlands can be found in Wong et al. (2007). The individual structures have also been studied extensively. The Mesozoic evolution of the WNB and the BFB has been well studied (Van Wijhe, 1987; Hooper et al., 1995; de Jager et al., 1996; van Balen et al., 2000; Verweij and Simmelink, 2002; Verweij et al., 2003). The southern region of the Netherlands, namely, the Roer Valley Graben and the adjacent highs, has been examined in detail (Zijerveld et al., 1992; Geluk et al., 1994; Gras, 1995; Gras and Geluk, 1999; van Balen et al., 2002; Michon et al., 2003). The German (central) part of the LSB was studied recently by Petmecky et al. (1999), Senglaub et al. (2005, 2006), and Munoz et al. (2007). Geluk published several studies dealing with the Permian–Triassic evolution of the Netherlands (Geluk et al., 1996; Geluk and Röhling, 1997; Geluk, 2000) and summarized them in his thesis (Geluk, 2005). Whereas the Upper Carboniferous was the focus of hydrocarbon exploration since the early 1950s (Thiadens, 1963; van Buggenum and den Hartog Jager, 2007), the more deeply buried Paleozoic sequences have only recently attracted more attention as source as well as reservoir rocks (van Balen et al., 2000; Kombrink, 2008).

This chapter introduces the methods and data input as used in the study. In the second part, the results of the structural modeling integrating 1-D modeling results from Nelskamp et al. (2008) and others are presented, and their implications for our understanding of the tectonic evolution of the area are discussed. Subsequently, the structural model is used as input for basin modeling, and maturity, temperature evolution, and hydrocarbon generation are presented.

METHODS AND DATA INPUT

Input for Structural Modeling

The 2-D section was put together using data from two different sources (Figure 2). The Late Permian and younger part had previously been published as section EE′ in the onshore geologic atlas (TNO, 2004a) compiled from the systematic subsurface mapping conducted by the Dutch Geological Survey (TNO-NITG). As the 2-D section is the result of detailed mapping and not based on a seismic section, the fault and layer geometries could not be checked against seismic data. To be able to calibrate the temperature and maturity with measurements from Paleozoic rocks, rocks of Carboniferous age were added to the model on the basis of the base Permian subcrop map (published in van Buggenum and den Hartog Jager, 2007) and estimates of thickness of the individual Paleozoic layers calculated from sparse well data. In the original section, the faults were drawn to extend into Pre-Permian rocks, so in the section studied, we extended them into the rocks of the Paleozoic age. The geometry of these faults in the Paleozoic is unknown, and it was assumed that the inclination and

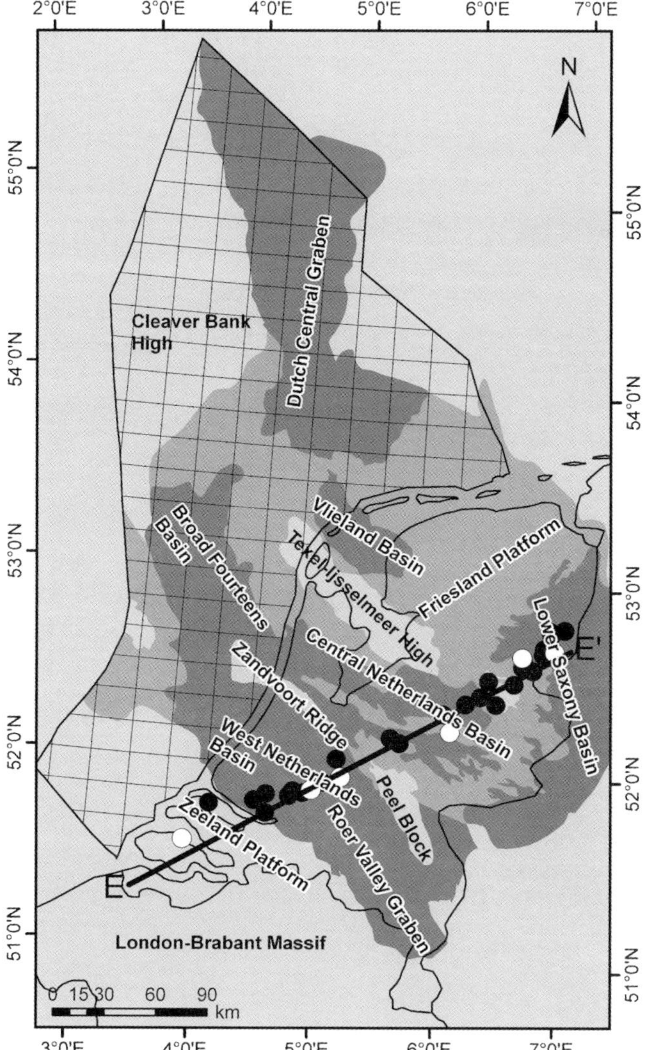

FIGURE 1. Overview map of the Netherlands showing the major structural features. EE′ is the studied two-dimensional transect crossing the West Netherlands Basin and the Central Netherlands Basin and reaching the margin of the Lower Saxony Basin. Black dots represent the position of wells included in this study, and white dots represent the position of the six wells for which the vitrinite reflectance calibration is shown in Figure 11A.

direction they displayed in the Mesozoic continue in the Paleozoic.

The part of the Netherlands studied lies on the margin of the Southern Permian Basin. With the exception of the northeasternmost part (LSB), the Zechstein deposits in this area do not contain thick salt layers. In contrast to the salt-dominated basins farther north, a decoupling of tectonic movements or faults in salt is not seen. Another peculiarity of the Dutch subsurface is the steepness of the faults. The major basins in the Netherlands developed under a transtensional regime (de Jager, 2007) governed by strike-slip movements that commonly resulted in steep faults, as can be seen on the section.

We used structural information from the section and the results of the 1-D modeling (Nelskamp et al., 2008), plus other detailed studies on the area (de Jager et al., 1996; van Balen et al., 2000; TNO, 2004a—detailed list in introduction) to make a structural reconstruction. A total of 21 paleosections were created with 2-DMove of Midland Valley, using Move On Fault Restorations and Flexural Slip Unfolding/Restore and also decompaction with and without isostasy (see also Midland Valley, 2008).

Conceptual Model

A conceptual model was established to define the input parameters for each numerical model. The conceptual model consists of depositional, nondepositional, and erosional events that have absolute ages. These events must be continuous. Because the stratigraphic information on the 2-D line is less detailed than the 1-D well reports, the conceptual model for structural reconstruction purposes was simplified. The timing and thickness of erosions was based on the 1-D modeling results. In cases where erosion occurred within a layer (e.g., sub-Herzynian and Laramide phases in the Cretaceous), the erosion was assumed to have occurred after that layer was deposited, as it was impossible to distinguish between pre-erosion and post-erosion sediments. The conceptual 2-D model is shown in Figure 3. The timing of the sections is based on the Geological Time Scale 2004 (Gradstein et al., 2004) (Figure 4).

Lithology

Lithological information was derived from the descriptions in Van Adrichem Boogaert and Kouwe (1993), paleogeographic maps of Ziegler (1990), and simplified interpolations from the lithologic descriptions in well reports. In numerical basin modeling, the lithology defines properties, such as porosity, permeability, chemical and mechanical compaction, and thermal conductivity. Several user-defined lithologies were created in PetroMod by mixing the standard lithologies (Table 1).

Input for Numerical Basin Modeling

The modeling in this study was performed with the PetroMod v10 program of IES/Schlumberger, Aachen. For the principles on which this software is based, see Poelchau et al. (1997), Yalcin et al. (1997), and Hantschel and Kauerauf (2009).

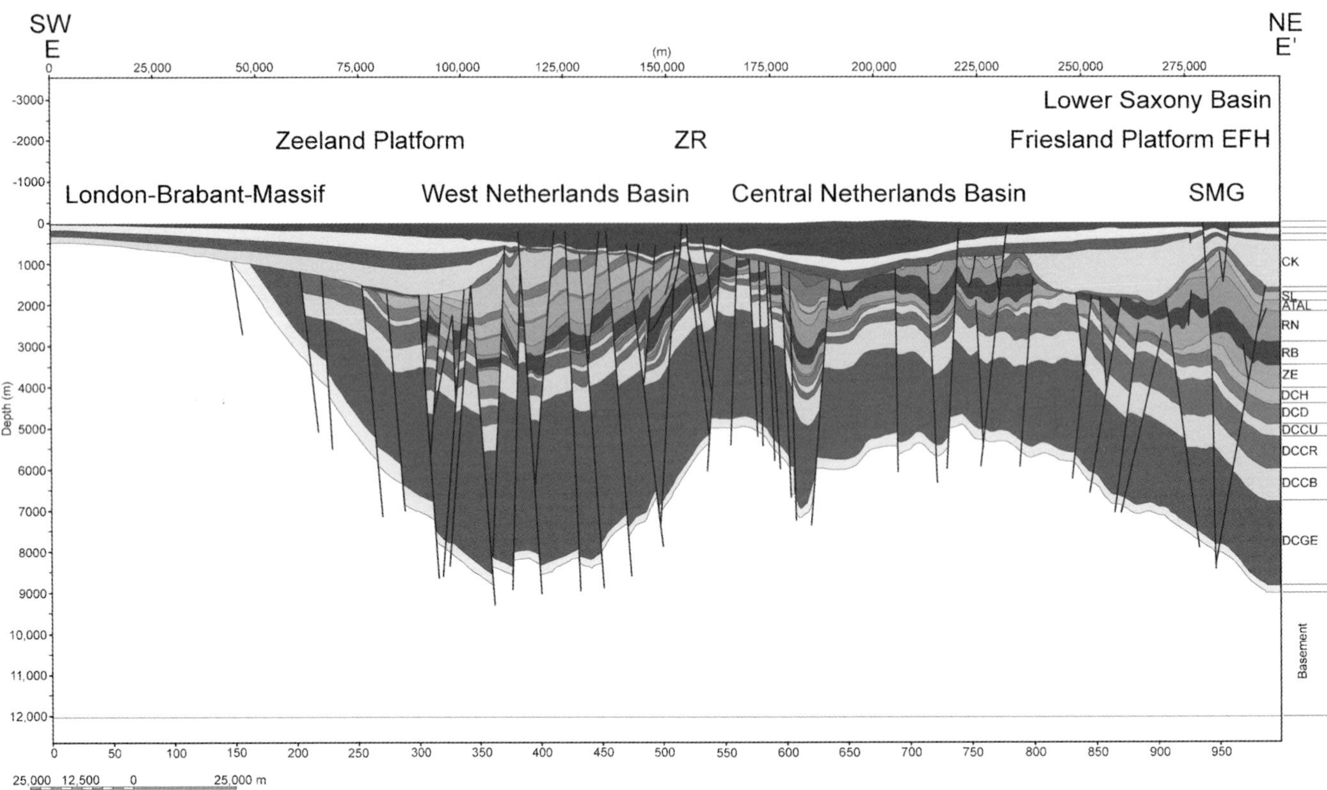

FIGURE 2. The present-day cross-section EE′ (modified from TNO, 2004a). Vertical exaggeration is 1:10. EFH = Emmen-Fehndorf High; SMG = Schoonebeek-Meppen Graben; ZR = Zandvoort Ridge; CK = Chalk Group; SL = Schlieland Group; ATAL = Aalburg and Sleen Formation; RN = Upper Germanic Trias Group; RB = Lower Germanic Trias Group; ZE = Zechstein Group; DCH = Hunze Subgroup; DCD = Dinkel Subgroup; DCCU = Maurits Formation; DCCR = Ruurlo Formation; DCCB = Baarlo Formation; DCGE = Epen Formation.

Boundary Conditions

To calculate the temperature trend of a 2-D section, the boundary conditions of the base and top must be defined. The temperature at the base of the model is defined by the basal heat flow (HF), whereas at the top, the sediment-water interface temperature (SWIT) is used. When estimating the SWIT, the paleowater depths have to be taken into account. What defines the HF through the model is the temperature difference between the base and the top, together with the physical properties of the lithologies.

Basal Heat Flow

The basal HF determines how much energy is introduced into the system from below. For the present-day situation, it can be calibrated using present-day well-temperature measurements. The calibration of paleo-HF values is less straightforward, especially if insufficient calibration data are available. In this study, the present-day HF applied was similar to the average values for continental sedimentary basins (Allen and Allen, 2005). We used paleo-HF of 65 mW/m^2, which was kept constant through time. Toward the LSB, the HF was slightly diminished, and a value of 55 mW/m^2 was applied. In Nelskamp et al. (2008), we discussed the effect of different paleo-HF scenarios on the calibration of the 1-D models and showed that the calibration data do not indicate whether HF was greater in the past. Therefore, the past HF cannot be constrained.

Sediment-Water Interface Temperature

The upper boundary condition is set by the SWIT, which is a function of surface temperature and water depth. The SWIT was calculated using the integrated PetroMod tool based on research done by Wygrala (1989). This tool generates a global surface temperature curve on the basis of the continent in which the study area is located and its latitude. To calculate the SWIT, the paleowater depth must be defined.

Paleowater Depth

The paleowater depth of the 2-D model was defined partly by the isostatic reconstruction during the decompaction and partly on the basis of paleogeographic

Table 1. Lithology parameters as defined in PetroMod.

Lithology	*Density (kg/m^3)*	*Initial Porosity*	*Minimum Porosity*	*Compressibility Max (1E-7/kPa)*	*Compressibility Min (1E-7/kPa)*	*Thermal Conductivity at 20°C (W/m/K)*	*Thermal Conductivity at 100°C (W/m/K)*	*Heat Capacity at 20°C (kcal/kg/K)*	*Heat Capacity at 100°C (kcal/kg/K)*	*Radioactive Heat at 0% Porosity ($microW/m^3$)*
100Chalk	2680.0	0.70	0.01	4613.3	41.6	2.90	2.62	0.203	0.234	0.60
100Sand	2720.0	0.41	0.01	274.7	11.5	3.95	3.38	0.204	0.236	0.70
100Shale	2700.0	0.70	0.01	4032.7	40.3	1.64	1.69	0.206	0.238	2.03
75Shale_25Sand	2705.0	0.63	0.01	3093.2	33.1	2.04	2.01	0.205	0.237	1.70
70Marl_30Shale	2700.0	0.56	0.01	1754.4	25.9	1.88	1.87	0.204	0.235	1.43
70Shale_30Lime	2712.0	0.64	0.01	3077.9	34.1	1.97	1.94	0.204	0.236	1.53
70Shale_20Lime_10Silt	2710.0	0.65	0.01	3096.5	34.3	1.89	1.89	0.206	0.238	1.59
70Shale_20 Sand_10Lime	2708.0	0.62	0.01	2962.8	32.5	2.08	2.03	0.205	0.237	1.60
60Shale_35Sand_5Coal	2652.0	0.60	0.01	6886.1	56.4	2.05	2.06	0.211	0.243	1.47
60Shale_30Evap_10Lime	2785.0	0.47	0.01	2504.6	26.2	2.61	2.47	0.197	0.228	1.32
60Shale_20Silt_20Lime	2712.0	0.63	0.01	2796.8	32.4	1.94	1.92	0.207	0.239	1.49
50Shale_50Sand	2710.0	0.55	0.01	2153.7	25.9	2.55	2.39	0.205	0.237	1.37
50Anhd_35 Dolo_15Shale	2866.5	0.23	0.01	693.2	9.9	4.47	3.82	0.193	0.222	0.48
50Shale_35 Silt_15Sand	2710.0	0.60	0.01	2420.2	29.3	2.02	1.99	0.210	0.242	1.46
50Shale_25Sand_25Lime	2715.0	0.58	0.01	2297.5	28.0	2.38	2.26	0.204	0.236	1.29
50Sand_25Silt_25Shale	2715.0	0.52	0.01	1404.6	21.1	2.69	2.49	0.208	0.240	1.10
50Shale_24Sand_24Silt_ 2Coal	2687.6	0.60	0.01	4079.1	39.3	2.07	2.04	0.210	0.243	1.42
40Anhy_40 Dolo_20Salt	2852.0	0.15	0.01	100.9	4.4	5.39	4.45	0.195	0.226	0.16
40Lime_40Dolo_20Shale	2752.0	0.48	0.01	1247.5	20.4	3.04	2.75	0.204	0.235	0.67
40Salt_40 Anhy_20Shale	2824.0	0.15	0.01	806.5	8.1	4.87	4.14	0.195	0.226	0.47
40Sand_30Shale_30Lime	2720.0	0.53	0.01	1574.7	22.6	2.79	2.56	0.203	0.235	1.00
40Shale_20Sand_20Silt_ 20 Lime	2716.0	0.57	0.01	2045.2	26.6	2.31	2.20	0.207	0.239	1.22

<table>
<tr><th>Name</th><th colspan="2">Section age (Ma)</th><th colspan="5">Lithology (with fractions in %)</th></tr>
<tr><td>Present-day</td><td colspan="2">0</td><td colspan="5"></td></tr>
<tr><td>Base NU</td><td colspan="2">23.03</td><td colspan="5">50Shale_50Sand</td></tr>
<tr><td>Base NM</td><td colspan="2">37.2</td><td colspan="2">50Shale_50Sand</td><td>50Shale_35Silt_15Sand</td><td colspan="2">50Sand_25Silt_25Shale</td></tr>
<tr><td>Base NL</td><td colspan="2">58.7</td><td colspan="2">50Shale_50Sand</td><td>50Shale_35Silt_15Sand</td><td colspan="2">50Sand_25Silt_25Shale</td></tr>
<tr><td>Uplift CK</td><td colspan="2">80</td><td colspan="5"></td></tr>
<tr><td>Pre-uplift CK</td><td colspan="2">83.5</td><td colspan="5"></td></tr>
<tr><td>Base CK</td><td colspan="2">99.6</td><td colspan="5">100Chalk</td></tr>
<tr><td>Base KN</td><td colspan="2">140.2</td><td colspan="3">40Sand_30Shale_30Lime</td><td colspan="2">70Marl_30Shale</td></tr>
<tr><td>Base SL</td><td colspan="2">155.7</td><td colspan="2">50Shale_50Sand</td><td>70Shale_20Sand_10Lime</td><td colspan="2">70Shale_30Lime</td></tr>
<tr><td>Uplift ATBR</td><td colspan="2">157</td><td colspan="5"></td></tr>
<tr><td>Pre-uplift ATBR</td><td colspan="2">161.2</td><td colspan="5"></td></tr>
<tr><td>Base ATBR</td><td colspan="2">166</td><td colspan="5">70Shale_20Lime_10Silt</td></tr>
<tr><td>Pre-uplift AT</td><td colspan="2">167.7</td><td colspan="5"></td></tr>
<tr><td>Base ATWD</td><td colspan="2">175.6</td><td colspan="5">70Shale_20Lime_10Silt</td></tr>
<tr><td>Base ATPO</td><td colspan="2">183</td><td colspan="5">100Shale</td></tr>
<tr><td>Base ATAL</td><td colspan="2">203.6</td><td colspan="5">60Shale_20Silt_20Lime</td></tr>
<tr><td>Base RN</td><td colspan="2">245</td><td colspan="3">50Shale_25Sand_25Lime</td><td colspan="2">60Shale_30Evap_10Lime</td></tr>
<tr><td>Base RB</td><td colspan="2">251</td><td colspan="3">100Sand</td><td colspan="2">40Shale_20Sand_20Silt_20Lime</td></tr>
<tr><td>Base ZE</td><td colspan="2">256.1</td><td>100Shale</td><td>60Shale_30Evap_10Lime</td><td>50Anhy_35Dolo_15Shale</td><td>40Anhy_40Dolo_20Salt</td><td>40Salt_40Anhy_20Shale</td></tr>
<tr><td>Base RO</td><td colspan="2">270</td><td colspan="5">100_Sand</td></tr>
<tr><td>Uplift DC</td><td colspan="2">290</td><td colspan="5"></td></tr>
<tr><td>Pre-uplift DC</td><td colspan="2">305</td><td colspan="5"></td></tr>
<tr><td></td><td colspan="2">Layer age (Ma)</td><td colspan="5"></td></tr>
<tr><td></td><td>from</td><td>to</td><td colspan="5"></td></tr>
<tr><td>DCH</td><td>309</td><td>305</td><td colspan="5">50Shale_50Sand</td></tr>
<tr><td>DCD</td><td>310</td><td>309</td><td colspan="5">50Shale_50Sand</td></tr>
<tr><td>DCCU</td><td>312</td><td>310</td><td colspan="5">60Shale_35Sand_5Coal</td></tr>
<tr><td>DCCR</td><td>313</td><td>312</td><td colspan="5">60Shale_35Sand_5Coal</td></tr>
<tr><td>DCCB</td><td>314</td><td>313</td><td colspan="5">50Shale_24Sand_24Silt_2Coal</td></tr>
<tr><td>DCGE</td><td>326.5</td><td>314</td><td colspan="5">75Shale_25Sand</td></tr>
<tr><td>CL</td><td>355</td><td>326.5</td><td colspan="5">40Lime_40Dolo_20Shale</td></tr>
</table>

FIGURE 3. Conceptual two-dimensional model. In the case that two or more lithologies are listed for one layer, they represent the lateral variation and approximate position of the lithology along the model (left in the table equals southwest, right equals northeast). NU = Upper North Sea Group; NM = Middle North Sea Group; NL = Lower North Sea Group; CK = Chalk Group; KN = Rijnland Group; SL = Schieland Group; ATBR = Brabant Formation; ATWD = Werkendam Formation; ATPO = Posidonia Shale Formation; ATAL = Aalburg and Sleen Formation; RN = Upper Germanic Trias Group; RB = Lower Germanic Trias Group; ZE = Zechstein Group; RO = Upper Rotliegende Group; DCH = Hunze Subgroup; DCD = Dinkel Subgroup; DCCU = Maurits Formation; DCCR = Ruurlo Formation; DCCB = Baarlo Formation; DCGE = Epen Formation; CL = Carboniferous Limestone Group.

reconstructions and lithology. Studies have shown that in the study region, the water depth was rarely greater than 200 m (656.2 ft), that is, it is mostly shallow water or continental deposits (TNO, 2000, 2002, 2004b; de Lugt et al., 2003).

RESULTS AND DISCUSSION

Structural Evolution

In the following section, the results of the detailed analysis of the tectonic structures are described, starting from the present-day situation and going back in time, reflecting the sequence followed in the structural reconstruction.

Present-Day Situation

The main tectonic features can be seen on the present-day section (Figure 2). Starting in the southwest, the London-Brabant Massif is a distinct feature, with pre-Carboniferous rocks close to the surface and Tertiary rocks thinning toward the massif. On the Zeeland Platform, Carboniferous rocks are present, dipping toward the WNB and becoming more deeply buried toward the northeast. At the transition to the WNB, the combined Upper Cretaceous–Cenozoic succession is at its thickest and the westernmost pre–Upper Cretaceous–Mesozoic sediments of the section are mapped. The WNB was tilted and eroded on its northeastern margin, resulting in an erosionally thinned Jurassic succession on its northeastern flank. The transition from the WNB to the Zandvoort Ridge is marked by a large fault zone that was reactivated during the Late Cretaceous and along which most of the inversion of the WNB occurred. The Zandvoort Ridge— situated between the WNB and CNB— is a pronounced Jurassic erosion structure. On the Zandvoort Ridge, Upper Cretaceous and Cenozoic sediments overlie mostly Triassic sediments. The CNB is marked by the disappearance of Upper Cretaceous sediments. At the present day, it consists of a series of fault blocks with very different amounts of uplift and erosion. The Cenozoic succession is thickest above the CNB. On the Friesland Platform, thick Upper Cretaceous sediments occasionally directly overlie Permian or even Carboniferous sediments. The CNB and the Friesland Platform have Namurian to Lower Westphalian sediments subcropping beneath the Mesozoic succession. The studied part of the LSB in the northeasternmost part of the study area has a thinner Upper Cretaceous and Jurassic succession and thicker Zechstein and Triassic sediments than the other basins. The 2-D section only marginally transects the LSB. The structural high at the transition of the Friesland Platform to the LSB is the mildly inverted Schoonebeek-Meppen Graben, a part of the LSB, whose bounding faults were reactivated during pulses of inversion. The line ends on a former structural high, the Emmen-Fehndorf High, which experienced subsidence during inversion. The part of the LSB is effectively reduced to this small structural high of the Schoonebeek-Meppen Graben, which clearly differs from the center of the LSB, where much more Cretaceous inversion and subsequent erosion occurred (as much as 6 km [3.7 mile]; Senglaub et al., 2005, 2006). Locally, the present-day Carboniferous sediments can be deeper than 6000 m (19,685 ft) and attain thicknesses exceeding 4000 m (13,123 ft; TNO, 2004a).

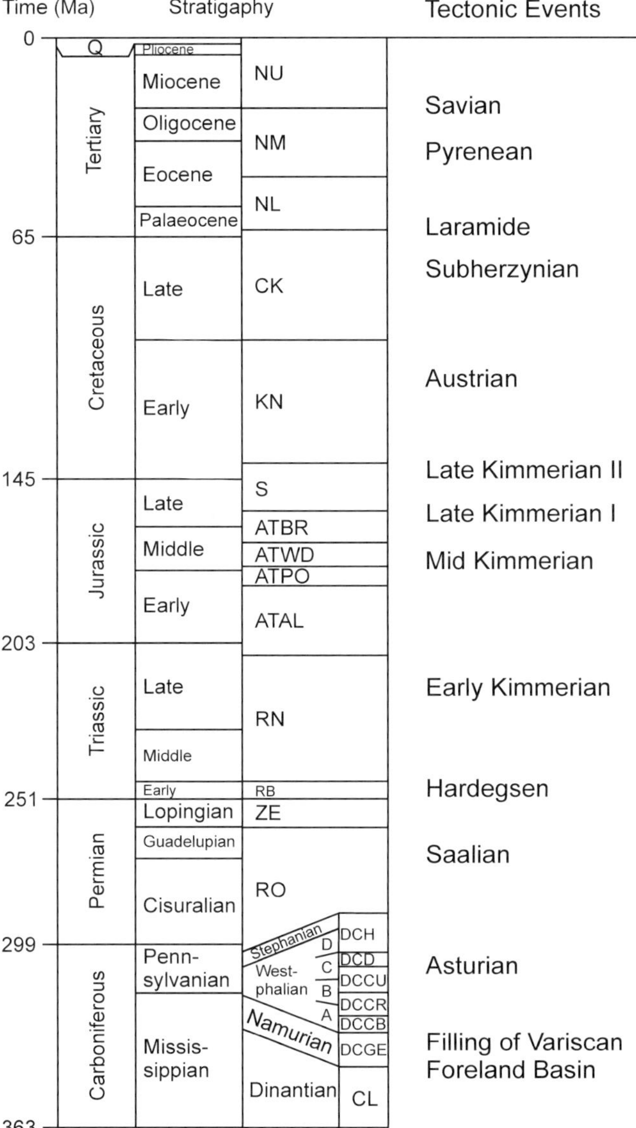

Figure 4. Stratigraphic table with subdivisions from the Dutch nomenclature after van Adrichem Boogaert and Kouwe (1993) and main tectonic events. NU = Upper North Sea Group; NM = Middle North Sea Group; NL = Lower North Sea Group; CK = Chalk Group; KN = Rijnland Group; SL = Schlieland Group; ATBR = Brabant Formation; ATWD = Werkendam Formation; ATPO = Posidonia Shale Formation; ATAL = Aalburg and Sleen Formation; RN = Upper Germanic Trias Group; RB = Lower Germanic Trias Group; ZE = Zechstein Group; RO = Upper Rotliegende Group; DCH = Hunze Subgroup; DCD = Dinkel Subgroup; DCCU = Maurits Formation; DCCR = Ruurlo Formation; DCCB = Baarlo Formation; DCGE = Epen Formation; CL = Carboniferous Limestone Group (for detailed information on these stratigraphic units, see van Adrichem Boogaert and Kouwe [1993]).

Cenozoic

Quaternary sediments consist of marine, fluvial, lacustrine, and glacial deposits. During the Tertiary, mostly siliciclastic sediments were deposited, commonly interrupted by uplift and small-scale erosion or nondeposition caused by compressional tectonic movements related to the Alpine orogeny (Pyrenean and Savian phases). In this study, no tectonic events were included during the Tertiary because the Tertiary inversion pulses are beyond the resolution of the 2-D model.

Post-Cretaceous Inversion

At the end of the Cretaceous, the sub-Hercynian and the Laramide tectonic phases caused inversion and erosion of the basins. Figure 5 is a section restored to the Late Cretaceous–Cenozoic boundary, after major inversion. During the inversion, basins were eroded while sediments of Late Cretaceous age accumulated on former structural highs (London-Brabant Massif, Zeeland Platform, Zandvoort Ridge, and Friesland Platform).

Eroded thicknesses were established on the basis of this reconstruction, and the 1-D modeling results were presented in Nelskamp et al. (2008). Worum (2004) identified an asymmetric inversion pattern in the WNB with similar maximum erosion thicknesses in the northeast. Van Balen et al. (2000) described the WNB as a north-bounded half graben during the Triassic, which was subdivided into various subunits with differential subsidence in the Jurassic. However, no thickening of Triassic sediments toward the northeast can be seen in the section we studied, and the remaining Jurassic sediments have a rather uniform thickness, with values decreasing toward the basin boundaries. Therefore, a symmetric basin geometry for the WNB was assumed to

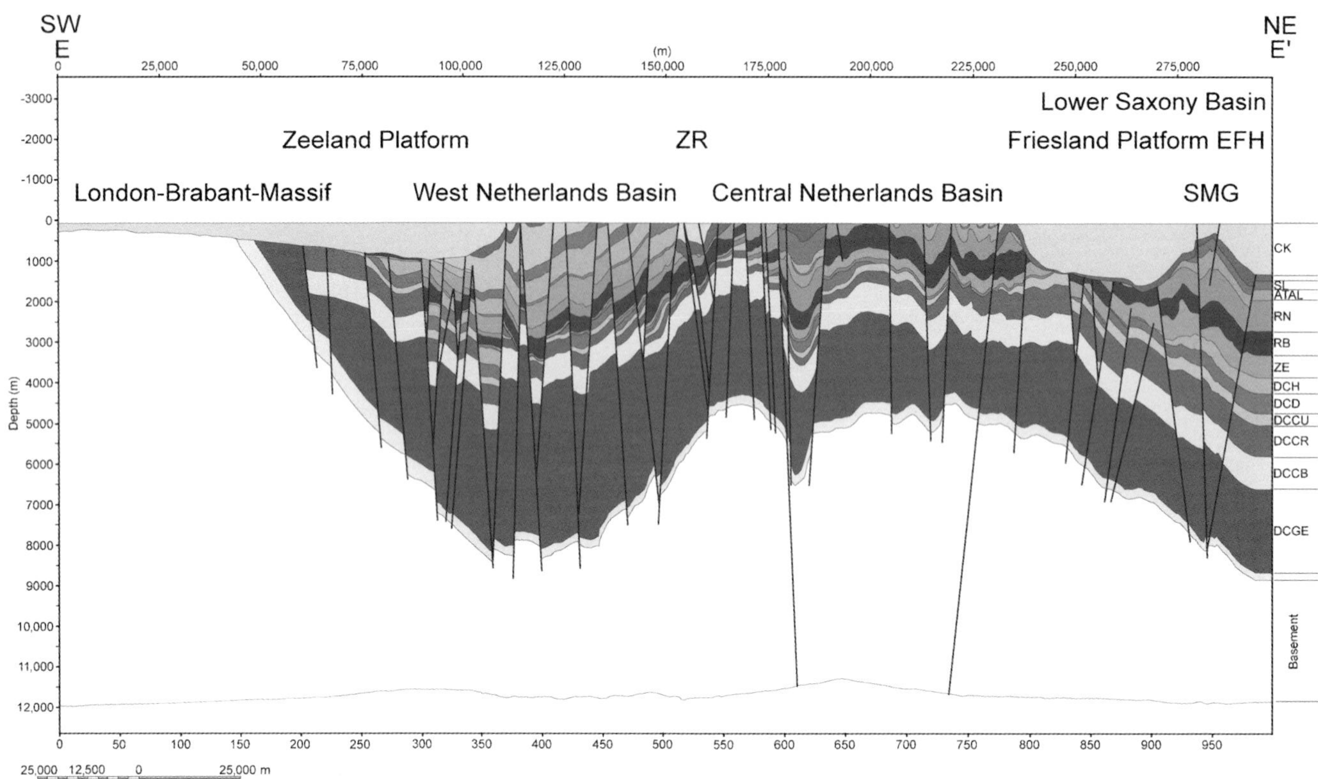

Figure 5. Reconstruction of the tectonic setting after the Late Cretaceous inversion. Vertical exaggeration is 1:10. EFH = Emmen-Fehndorf High; SMG = Schoonebeek-Meppen Graben; ZR = Zandvoort Ridge; CK = Chalk Group; SL = Schlieland Group; ATAL = Aalburg and Sleen Formation; RN = Upper Germanic Trias Group; RB = Lower Germanic Trias Group; ZE = Zechstein Group; DCH = Hunze Subgroup; DCD = Dinkel Subgroup; DCCU = Maurits Formation; DCCR = Ruurlo Formation; DCCB = Baarlo Formation; DCGE = Epen Formation.

have existed before the inversion. The same assumption was made for the CNB. The minimum thickness of the layers was extrapolated from the basin margins or from where the layer was not eroded.

Before the Late Cretaceous Inversion

Figure 6 shows the situation before the inversion in the Late Cretaceous in the basins but after the deposition of Upper Cretaceous sediments on the highs. The basin geometry of the CNB is restored, and the tilting of the WNB is undone. The eroded thicknesses estimated for the eastern part of the WNB exceed 1500 m (4921 ft); this compares with the estimate of 1100 to 1700 m (3609–5577 ft) obtained by Worum (2004) using the interval velocity and a geometric reconstruction for the WNB. The maximum estimate for eroded sediments in the CNB is 2500 m (8202 ft). A thinned Upper Cretaceous layer of calcareous sediments was assumed to have been deposited over the entire area before the inversion. This section— restored to the time immediately before the inversion— shows the geometry at the time of deepest burial for the severely eroded basins (WNB ~500 m [1640 ft] more than the present day, CNB ~1200 m [3937 ft] more than the present day). The depths of the Mesozoic succession in the WNB and the CNB are similar. The base of the Carboniferous section is deeper in the WNB because of the previously mentioned Namurian subcrop in the CNB, whereas a Late Westphalian to Stephanian subcrop exists in the WNB. At the beginning of the Cretaceous, marine influences led to the deposition of claystones, siltstones, and sandstones in the basin areas.

The Late Jurassic Erosion

During the Late Jurassic, mostly fluvial continental sedimentation occurred, concentrated in the basins. The Middle to Late Jurassic uplift of the Central North Sea Dome led to deep erosion in the CNB and the Dutch part of the LSB (Wong, 2007). This erosion removed sediments of Jurassic, Triassic, and Permian age from the platform areas and— to a lesser extent— also from the basin margins. From the thick open marine sediments deposited in the Early Jurassic, it was inferred that this period was one of steady subsidence. During the Toarcian, a short period of anoxic conditions prevailed, during which the organic-rich Posidionia Shale

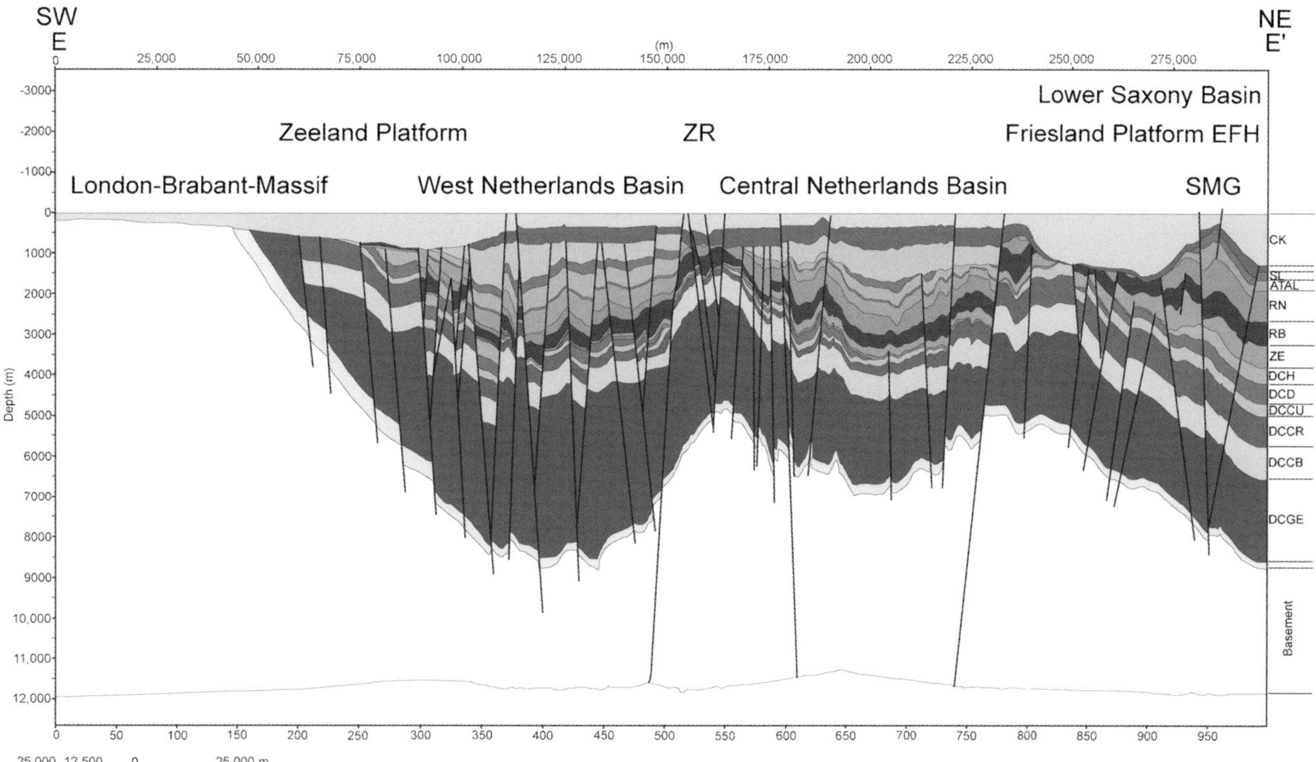

FIGURE 6. Reconstruction of the Late Cretaceous inversion. The estimated eroded thicknesses have been interpolated from the surrounding areas and the one-dimensional models. Vertical exaggeration is 1:10. EFH = Emmen-Fehndorf High; SMG = Schoonebeek-Meppen Graben; ZR = Zandvoort Ridge; CK = Chalk Group; SL = Schlieland Group; ATAL = Aalburg and Sleen Formation; RN = Upper Germanic Trias Group; RB = Lower Germanic Trias Group; ZE = Zechstein Group; DCH = Hunze Subgroup; DCD = Dinkel Subgroup; DCCU = Maurits Formation; DCCR = Ruurlo Formation; DCCB = Baarlo Formation; DCGE = Epen Formation.

Formation was deposited. It is likely that Jurassic sediments, including Posidonia Shale, were deposited everywhere, although less thickly on the highs. Triassic and Permian sediments were restored, becoming thicker toward the former center of the Permian–Triassic Southern Permian Basin. The London-Brabant Massif in the southwest was a stable high or platform during the Permian to Jurassic; evidence for this is the thinning of these sediments toward the LBM (Wong, 2007). Fission-track measurements from the southern border of the LBM yield ages of 209 to 146 Ma, indicating continuous uplift and probably erosion during the entire Jurassic (Vercoutere and van den Houte, 1993). On the basis of vitrinite reflectance measurements and conodont color alteration indices (Helsen 1995), it can be inferred that thick Carboniferous sediments exist on the London-Brabant Massif before the erosion (Figure 7A, B).

Late Triassic

In the Upper Triassic section, the beginning of structural segmentation can be seen (Figure 8). During the Late Triassic, basin differentiation started, and thicker sediments were deposited in the basins than on the surrounding highs (Geluk, 2007). Clearly visible is the thickening of the Upper Triassic succession in the WNB, CNB, and LSB. The underlying Lower Triassic and Zechstein layers become thicker northeastward, but not in the basins. During the Late Permian (Zechstein), periodic marine incursions resulted in the deposition of several series of claystones, carbonates, and evaporites that thicken from the London-Brabant Massif, where no evaporites were deposited, toward the basin center in the northeast. In the study area, sediments of Rotliegende age are found only in the area of the CNB, which was the only basin active as early as Rotliegende times (Duin et al., 2006).

After the Late Carboniferous–Early Permian Erosion

At the end of the Carboniferous, wrench tectonics created several structural features (Campine Basin, Roer Valley Graben etc.; Duin et al., 2006) and led to the erosion of Carboniferous sediments in several areas. The Hunze Formation (Westphalian D possibly up to Stephanian) was eroded in most of the Netherlands, except in the Campine Basin (future Zeeland Platform— WNB

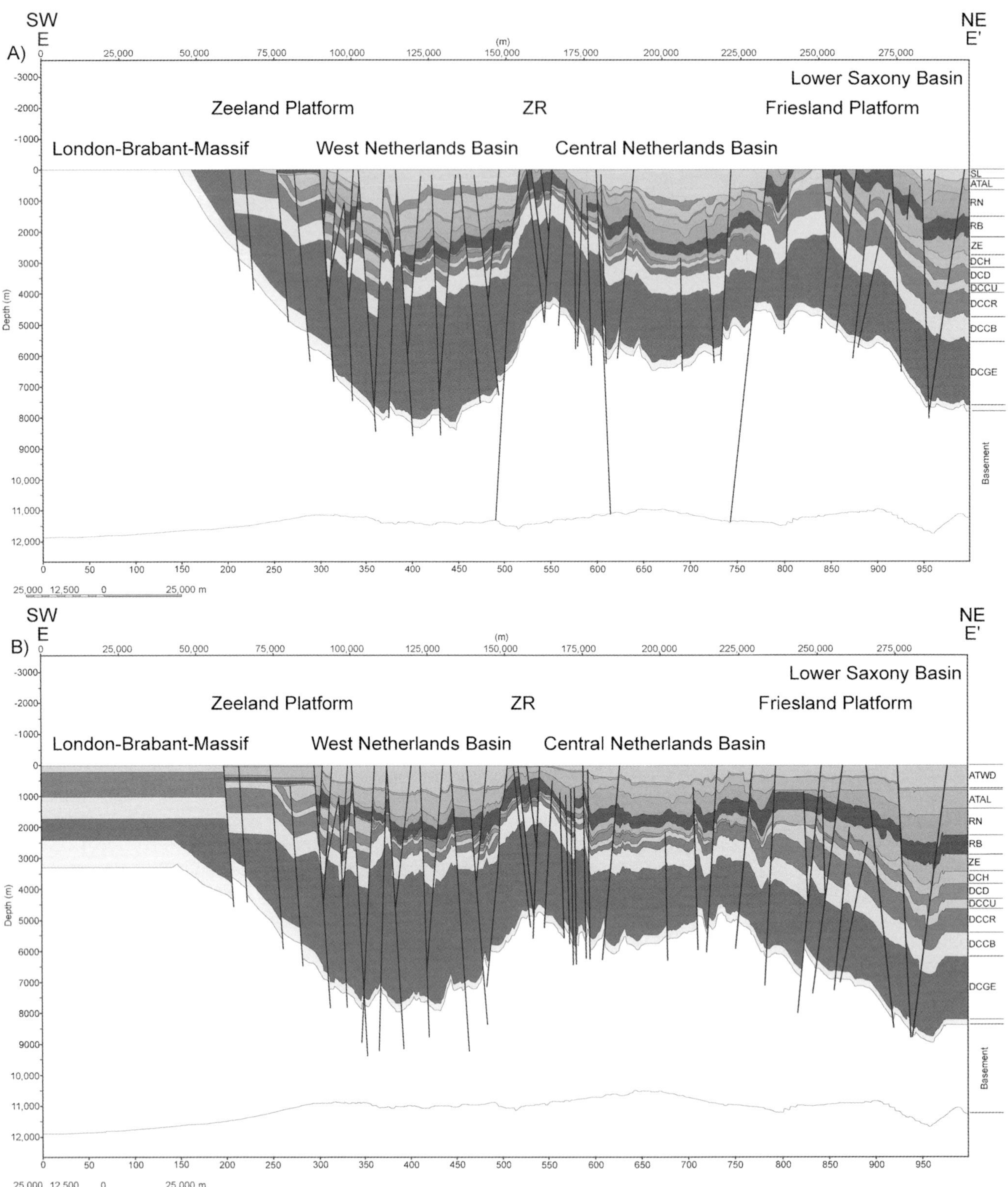

FIGURE 7. (A, B) Reconstruction of the Late Jurassic erosion: (A) situation after erosion, (B) reconstructed geometry prior to erosion. Vertical exaggeration is 1:10. ZR = Zandvoort Ridge; SL = Schlieland Group; ATAL = Aalburg and Sleen Formation; RN = Upper Germanic Trias Group; RB = Lower Germanic Trias Group; ZE = Zechstein Group; DCH = Hunze Subgroup; DCD = Dinkel Subgroup; DCCU = Maurits Formation; DCCR = Ruurlo Formation; DCCB = Baarlo Formation; DCGE = Epen Formation; ATWD = Werkendam Formation.

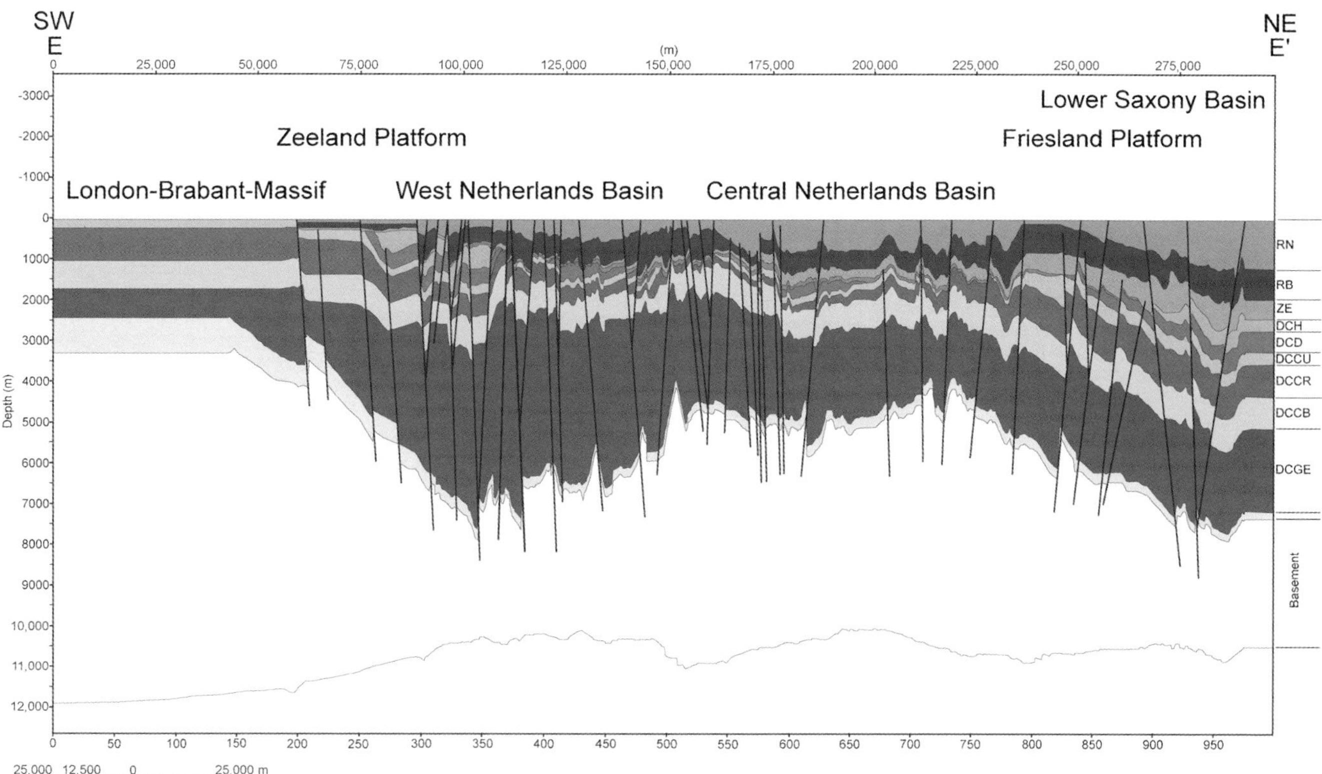

FIGURE 8. Reconstructed geometry during the Late Triassic. Vertical exaggeration is 1:10. RN = Upper Germanic Trias Group; RB = Lower Germanic Trias Group; ZE = Zechstein Group; DCH = Hunze Subgroup; DCD = Dinkel Subgroup; DCCU = Maurits Formation; DCCR = Ruurlo Formation; DCCB = Baarlo Formation; DCGE = Epen Formation.

transition) and the Ems Low. Erosion was most severe in the central part of the WNB, the Zandvoort Ridge, and the Friesland Platform, already contouring the two latter future structural elements (Figure 9). The major large fault zones of the Netherlands trending northwest-southeast also came into existence during that time.

Before the Late Carboniferous–Early Permian Erosion

Figure 10 shows the modeled depositional thicknesses of the Carboniferous sediments. The Hunze and Dinkel formations were assumed to be deposited in a more or less continuous thickness over the study area. The thickening of the Ruurlo Formation on the Friesland Platform toward the northeast is an artifact of reconstruction of the Late Jurassic erosion and probably does not mirror real geologic conditions. The Namurian Epen Formation is thickest in the area of the Campine Basin, which has been described as one of the more rapidly subsiding areas during the Namurian (Kombrink, 2008).

The structural reconstructions presented show artifacts of restoration errors in short wavelength folds visible in the deep layers. This is related to the simplified fault geometry in the upper part of the section and the added geometry of the Carboniferous layers and cannot be avoided. In addition, strike-slip movements are common, resulting in out-of-plane motions.

Basin Modeling Results

The results of the structural analysis were imported into PetroMod, which was then run with the previously mentioned boundary conditions. The results of the temperature, maturity, and hydrocarbon generation calculations are subsequently described.

Temperature and Maturity Distribution

The temperature history was calculated using HF through time, in agreement with 1-D modeling assumptions (Nelskamp et al., 2008). The HFs adopted were 65 mW/m^2 for the WNB and CNB and 55 mW/m^2 for the LSB. The slightly lower HF in the LSB is well constrained by vitrinite reflectance data (Figure 11A, B). The highest maturities beneath the Cenozoic subcrop at relatively shallow depth occur in the CNB and the London-Brabant Massif. In the WNB, the maturity follows the dip of the layers, a result of the tilting of the basin during the inversion. The fit of the vitrinite reflectance values in the WNB and CNB is far from perfect (Figure 11A).

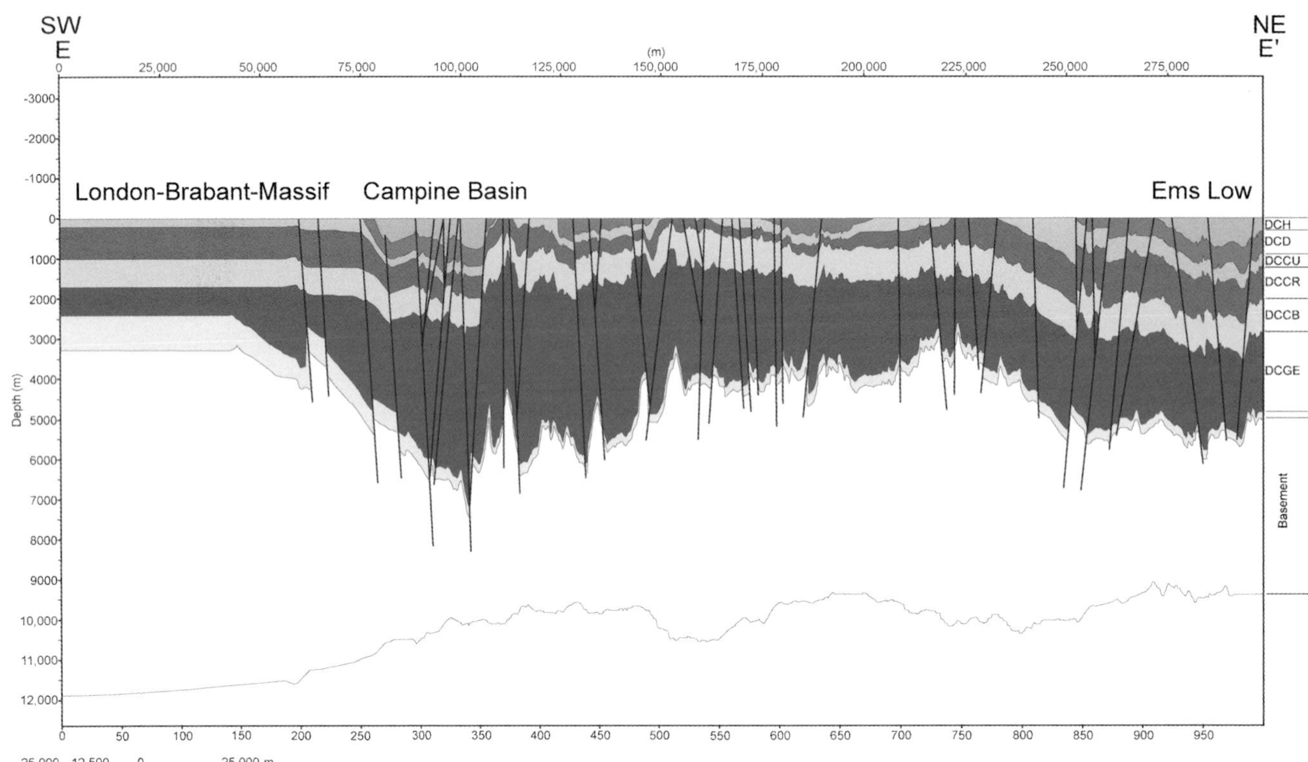

FIGURE 9. Reconstructed outcrop after the Later Carboniferous–Early Permian erosion. Vertical exaggeration is 1:10. DCH = Hunze Subgroup; DCD = Dinkel Subgroup; DCCU = Maurits Formation; DCCR = Ruurlo Formation; DCCB = Baarlo Formation; DCGE = Epen Formation.

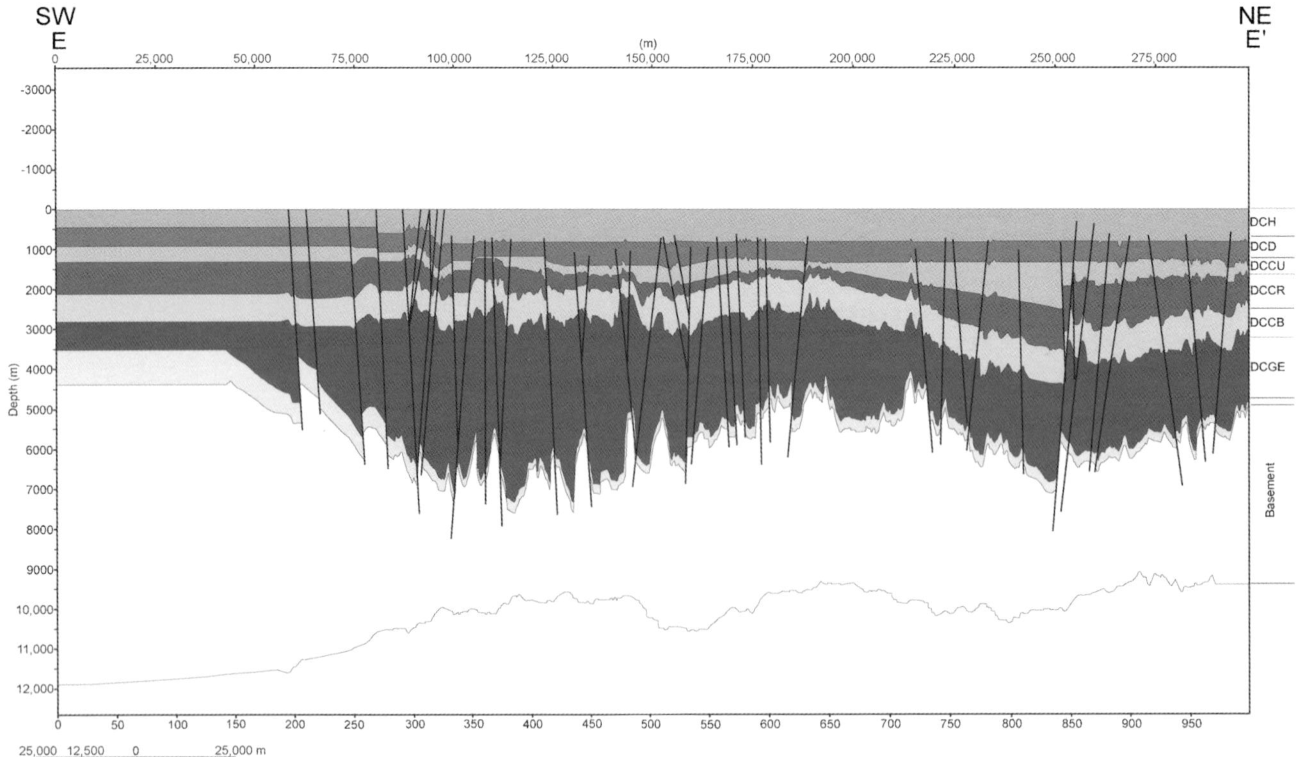

FIGURE 10. Reconstructed Carboniferous thicknesses. The thickening of the Maurits Formation on the Friesland Platform is an artifact of the reconstruction. Vertical exaggeration is 1:10. DCH = Hunze Subgroup; DCD = Dinkel Subgroup; DCCU = Maurits Formation; DCCR = Ruurlo Formation; DCCB = Baarlo Formation; DCGE = Epen Formation.

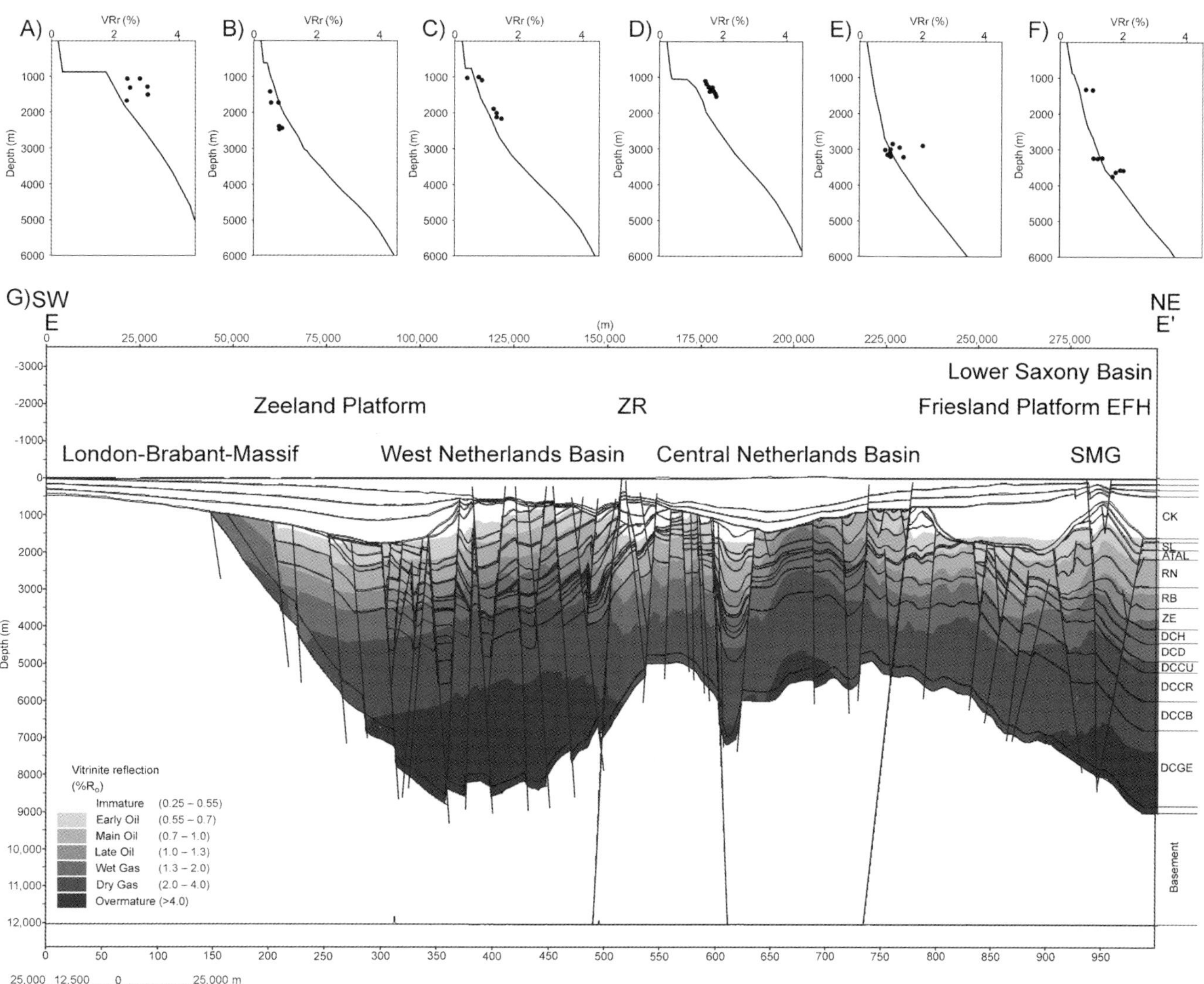

FIGURE 11. (A–G) Calibration of the maturity (A–F) in six wells along the transect (A = KTG-01; B = OTL-01; C = EVD-01; D = APN-01; E = COV-10; F = SCH-447) and (G) along the entire transect. EFH = Emmen-Fehndorf High; SMG = Schoonebeek-Meppen Graben; ZR = Zandvoort Ridge; CK = Chalk Group; SL = Schlieland Group; ATAL = Aalburg and Sleen Formation; RN = Upper Germanic Trias Group; RB = Lower Germanic Trias Group; ZE = Zechstein Group; DCH = Hunze Subgroup; DCD = Dinkel Subgroup; DCCU = Maurits Formation; DCCR = Ruurlo Formation; DCCB = Baarlo Formation; DCGE = Epen Formation.

Applying slightly higher HF does not resolve this problem. More erosion would have to be assumed to fit the values. However, the wells used for calibration do not lie directly on the 2-D section, and there may be spatial variations in eroded thickness or basal HF (Figure 1).

The present-day temperature distribution is shown in Figure 12A. Today, it is steady-state for conditions considered in the model. Figure 12B shows the present-day temperature at 2000-m (6562-ft) depth. The temperature increases toward the eastern boundary on the WNB, from about 80°C (176°F) on the Zeeland Platform to 95°C (203°F). On the Zandvoort Ridge, lower temperatures prevail, whereas in the western CNB, high temperatures of about 95°C (203°F) were again modeled. In the eastern, drastically inverted, part of the CNB, the temperature decreases to about 85°C (185°F) at the 2000-m (6562-ft) depth, stays at that value on the Friesland Platform, and falls slightly toward the LSB.

The temperature at the 2000-m (6562-ft) depth is influenced by the basal HF, the thermal conductivity of the sediments, and the burial history (in particular, compaction and porosity). In uplifted areas, overcompacted sediments with high bulk thermal conductivities are close to the surface, which increases the temperature. In contrast, in rapidly subsiding areas, recently deposited and undercompacted sediments can reduce temperatures because of their relatively low bulk thermal conductivity. In addition to these effects, the presence of very conductive layers (such as salt or sandstone) and poorly conductive layers (such as shale or coal) affects

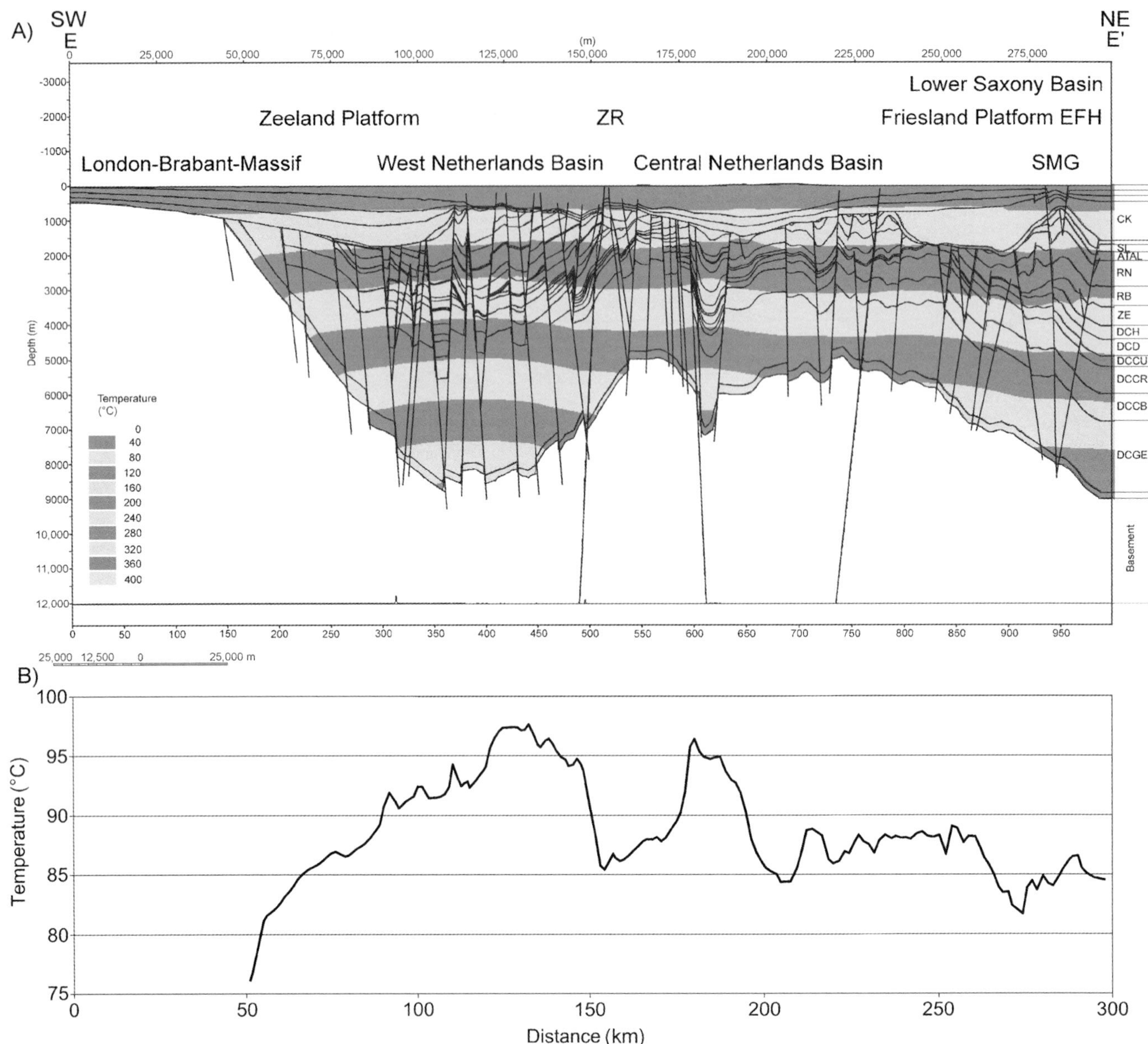

FIGURE 12. (A, B) Present-day simulated temperature distribution (A) along the section and (B) at the 2000-m (6562-ft) depth (B). Temperature gradient in (A) ranges from 0°C (32°F) to 400°C (752°F). EFH = Emmen-Fehndorf High; SMG = Schoonebeek-Meppen Graben; ZR = Zandvoort Ridge; CK = Chalk Group; SL = Schlieland Group; ATAL = Aalburg and Sleen Formation; RN = Upper Germanic Trias Group; RB = Lower Germanic Trias Group; ZE = Zechstein Group; DCH = Hunze Subgroup; DCD = Dinkel Subgroup; DCCU = Maurits Formation; DCCR = Ruurlo Formation; DCCB = Baarlo Formation; DCGE = Epen Formation.

temperatures to an extent that depends on their relative positions. In the WNB, the effects of the inversion and related temperature increase can be observed. In the western part, thick young layers (Chalk to present day) were deposited, resulting in low temperatures. In the eastern part of the basin, the inversion brought highly compacted older sediments close to the surface, resulting in relatively high temperatures. In the CNB, the situation is more complex. Here, thick Tertiary sediments were deposited, but because of prior inversion, overcompacted deep layers are directly juxtaposed below the Tertiary section. In addition, the Zechstein salt influences the thermal field. Model runs suggest that the temperature at 2000 m (6562 ft) in the CNB is mainly influenced by the salt: where the salt is deeper than 2000 m (>6562 ft), the temperature is higher; where the salt is at depths of less than 2000 m (<6562 ft), the temperature is lower. The influence of salt on temperature distribution within basins has also been discussed by Neunzert et al. (1996).

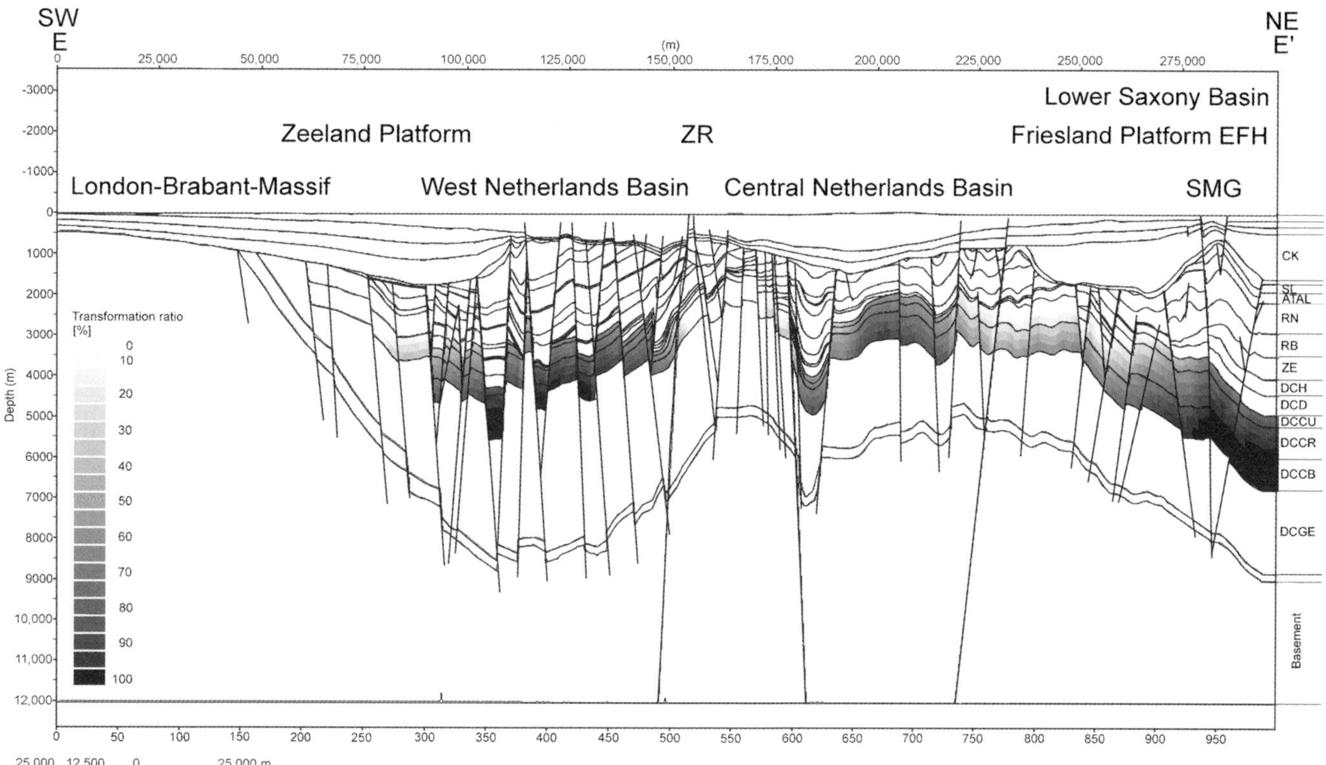

FIGURE 13. Transformation ratio (%) of the source rocks along the transect. White = no transformation, black = 100% transformation. EFH = Emmen-Fehndorf High; SMG = Schoonebeek-Meppen Graben; ZR = Zandvoort Ridge; CK = Chalk Group; SL = Schlieland Group; ATAL = Aalburg and Sleen Formation; RN = Upper Germanic Trias Group; RB = Lower Germanic Trias Group; ZE = Zechstein Group; DCH = Hunze Subgroup; DCD = Dinkel Subgroup; DCCU = Maurits Formation; DCCR = Ruurlo Formation; DCCB = Baarlo Formation; DCGE = Epen Formation.

Hydrocarbon Generation

The transformation ratio is a measurement of how much of the hydrocarbon potential has been realized from a source rock. In the Netherlands, the main gas source rocks are the Westphalian (Baarlo, Ruurlo, and Maurits formations; Figure 4) layers, and the main oil source rock is the Posidonia Shale Formation. Pepper and Corvi (1995) TIII–V kinetics were used for the Carboniferous source rocks, and the Dieckmann et al. (1998) TII (Posidonia) kinetics were used for the Posidonia Shale.

The highest transformation ratios for the Carboniferous rocks occur in the WNB, the CNB, and the LSB (Figure 13). The transformation ratio over time for six extraction points from the Maurits Formation (two per basin) is shown in Figure 14. The position of the extraction points of Figures 14–16 is shown in Figure 17. In the deep part of the LSB, hydrocarbon generation started in the Late Triassic, and more than 60% transformation occurred before the end of the Jurassic. Continuing burial since Late Cretaceous time resulted in renewed transformation until the Neogene. Near the margins of the LSB, generation started at the same time, but less potential was transformed. Since the erosion in the Jurassic, no hydrocarbons were generated. The Maurits Formation of the CNB generated the first hydrocarbons in the Jurassic. The Late Cretaceous uplift halted the generation in the eastern, strongly inverted, part of the CNB, but in the western part, hydrocarbon generation resumed during Tertiary. The transformation history of the eastern part of the WNB is similar to that of the eastern part of the CNB; it also started to generate hydrocarbons in the Jurassic and stopped transformation after the inversion. The western part of the WNB did not start to generate hydrocarbons until Late Cretaceous and experienced significantly less erosion during the inversion. This area continued to generate hydrocarbons from Carboniferous rocks until the present day (Figure 14).

Transformation ratio histories for three extraction points from the Baarlo Formation of the Zeeland Platform, Zandvoort Ridge, and Friesland Platform are shown in Figure 15. Generation of oil and gas from the Carboniferous source rocks on the Friesland Platform had already started in the Carboniferous, but was interrupted by the Late Carboniferous–Early Permian erosion. Deeper burial in Late Triassic–Early Jurassic triggered some generation but was briefly interrupted by the Late Jurassic uplift and erosion. Since then, there has been no further generation. Transformation of the hydrocarbon potential

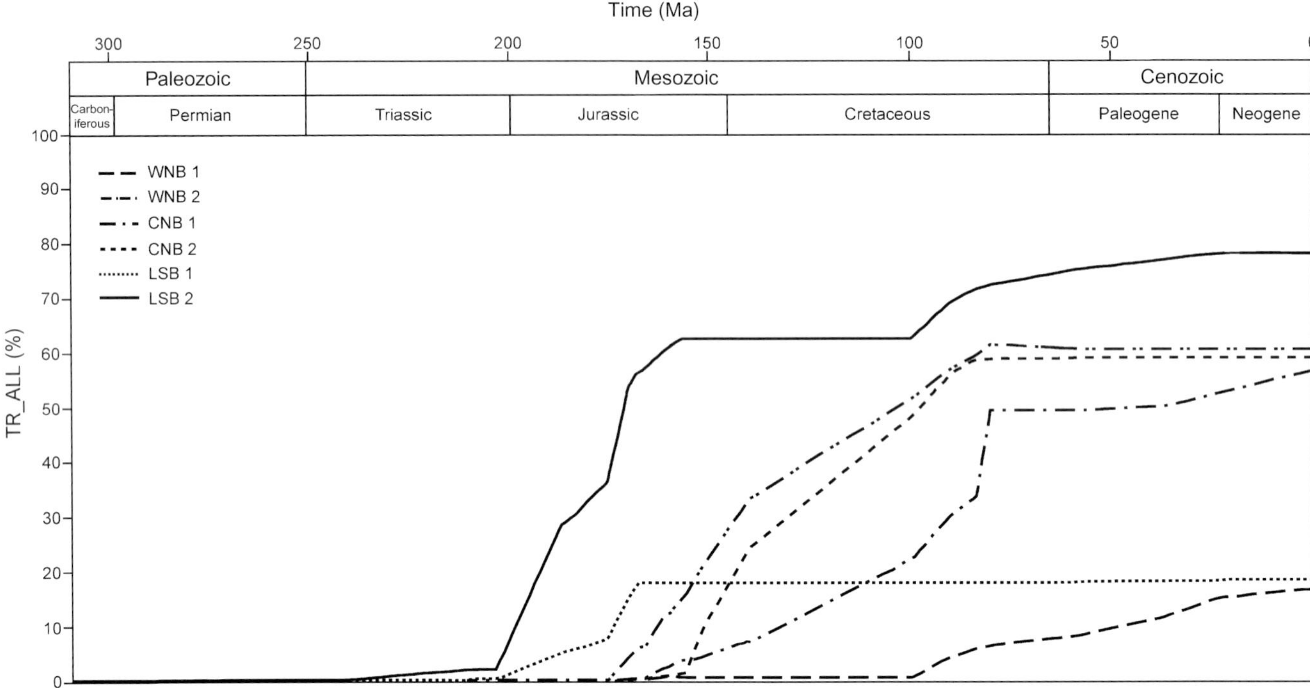

FIGURE 14. Transformation ratio (%) of the Maurits Formation (DCCU) at six different points along the transect. WNB = West Netherlands Basin; CNB = Central Netherlands Basin; LSB = Lower Saxony Basin. For position of the extractions, see Figure 17.

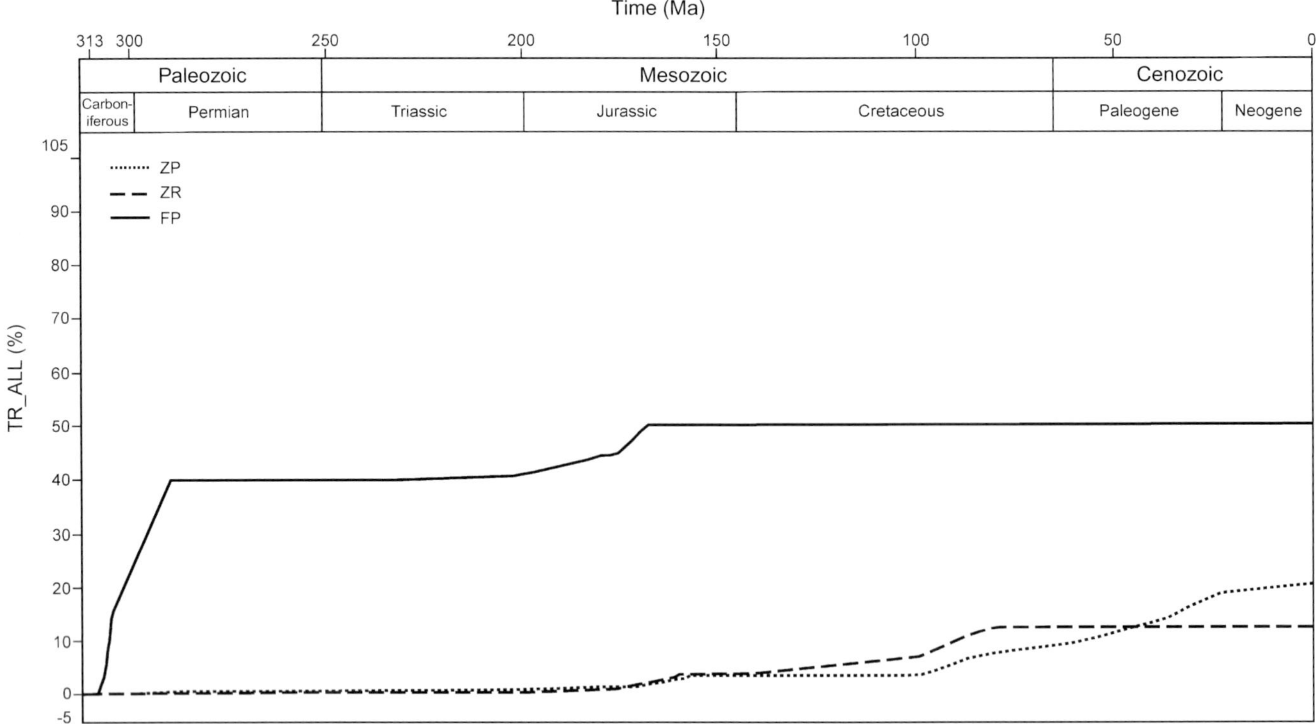

FIGURE 15. Transformation ratio (%) for three points from the Carboniferous Baarlo Formation (DCCB) from the highs. ZP = Zeeland Platform; ZR = Zandervoort Ridge; FP = Friesland Platform. For position of the extractions, see Figure 17.

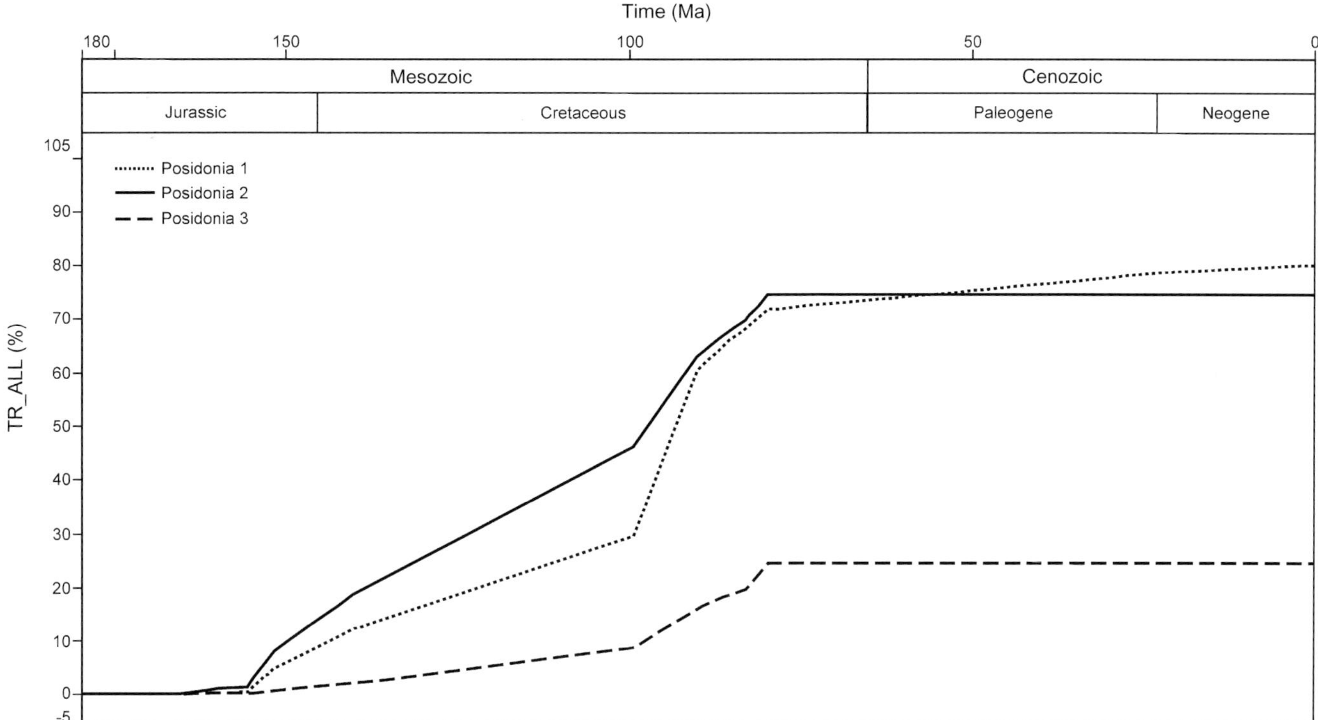

FIGURE 16. Transformation ratio (%) for the Posidonia Shale Formation (ATPO) from three points in the West Netherlands Basin. For position of the extractions, see Figure 17.

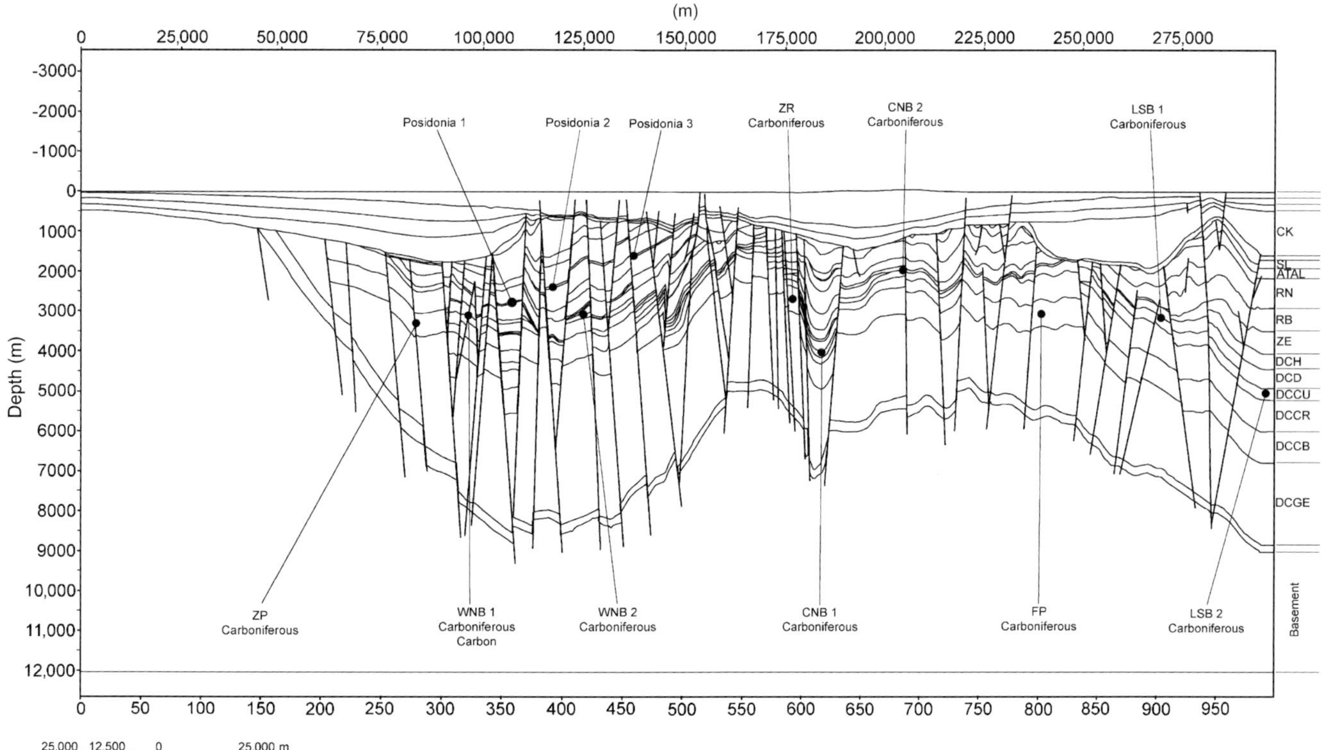

FIGURE 17. Position of the transformation ratio extractions shown in Figures 14–16. WNB = West Netherlands Basin; CNB = Central Netherlands Basin; LSB = Lower Saxony Basin; CK = Chalk Group; SL = Schlieland Group; ATAL = Aalburg and Sleen Formation; RN = Upper Germanic Trias Group; RB = Lower Germanic Trias Group; ZE = Zechstein Group; DCH = Hunze Subgroup; DCD = Dinkel Subgroup; DCCU = Maurits Formation; DCCR = Ruurlo Formation; ZR = Zandervoort Ridge; FP = Friesland Platform. DCCB = Baarlo Formation; DCGE = Epen Formation.

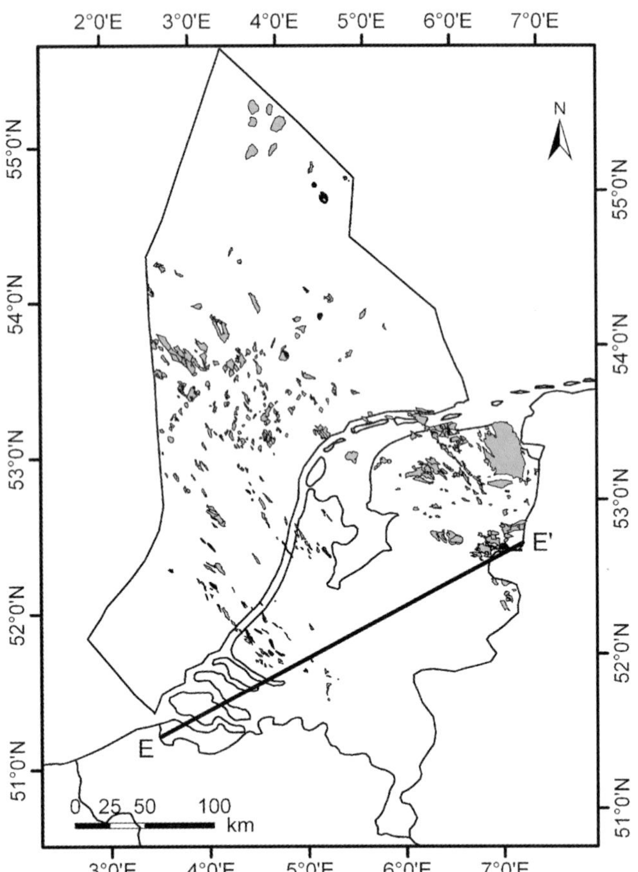

FIGURE 18. Gas (gray) and oil (black) accumulations in the Netherlands and position of the studied section EE′.

of the Carboniferous rocks on the Zandvoort Ridge and the Zeeland Platform started in the Early Jurassic. On the Zandvoort Ridge, the generation continued until the Late Cretaceous; on the Zeeland Platform, transformation continues today.

Three extractions were analyzed for the Posidonia Shale in the WNB, the only region in the study area where the formation is still present (Figure 16). The position of the extraction points can be seen in Figure 17. The Posidonia Shale started generation shortly after deposition in the Late Jurassic. The highest transformation ratio can be seen in the central extraction, representing the deeper burial in the center of the basin. At the beginning of the Late Cretaceous, all three areas experienced increasing transformation ratios, but the westernmost area shows the biggest increase, which is related to the deposition of a thicker Chalk layer in that region. The areas of the WNB affected by the inversion stopped transformation after the Late Cretaceous, whereas the western part of the basin continues to generate hydrocarbons today.

In the Netherlands, gas fields were found in the WNB and the LSB (Figure 18). In the CNB, the Carboniferous sediments have also generated a significant amount of their potential, but no gas accumulations have yet been found. Based on the timing of the generation, generation stopped after the Late Cretaceous inversion phase in the eastern part of the WNB and in the CNB, whereas the generation of hydrocarbons continued during the Cenozoic in areas with present-day gas reservoirs (Figure 14).

The present-day distribution of the Posidonia Shale is limited to the western part of the WNB, the CNB outside the study area, the German LSB, and the offshore basins (de Jager and Geluk, 2007). This is also reflected in the distribution of oil reservoirs in the Netherlands (Figure 18). In the study area, the Posidonia Shale is only present in the WNB. Oil generation started in the Late Jurassic and for most of the WNB continued until the Late Cretaceous. Again, only in the western part of the WNB has generation continued until the present day (Figure 16).

CONCLUSIONS

In structurally complex regions, tectonic evolution has a huge impact on temperature, maturity, and hydrocarbon generation. Structural reconstructions in combination with petroleum systems modeling create a good basis for the understanding of the effects that influence the thermal field and hydrocarbon distribution.

The present-day temperature at the 2000-m (6562-ft) depth in the Netherlands is controlled by basal HF and the position of layers of high or low conductivity either as a result of burial and uplift or because of lithology. High temperatures occur (1) in areas where uplift and erosion brought overcompacted sediments closer to the surface or (2) above thick Zechstein salt. Low temperatures occur where thick layers of sediments were deposited rapidly or where the Zechstein salt is shallower than 2000 m (6562 ft). The Late Cretaceous erosion phase had the most important influence on the oil and gas system in the basins. The drastic uplift and erosion at that time interrupted oil and gas generation in parts of the WNB and in the CNB, and in these areas, the subsidence and temperature never reattained the preinversion values. Furthermore, reactivation of faults, coupled with the evolution and modification of structures, possibly resulted in spilling from paleotraps and the loss of preinversion hydrocarbon accumulations. On the highs, the Late Carboniferous–Early Permian and the Jurassic erosion phases had the most influence on the oil and gas system. Early generation was interrupted by uplift and erosion, and only on the Zeeland Platform does present-day burial provide conditions for current generation. The occurrence of oil fields is further controlled by the distribution of the Posidonia Shale source rock. This source rock was eroded in most of the Netherlands during the Late Jurassic (on platform areas) or in the Late

Cretaceous. The present-day discovered oil accumulations coincide with areas where the Posidonia Shale is still present and hydrocarbon generation in the oil and gas-prone source rocks continued after the Late Cretaceous.

ACKNOWLEDGMENTS

This study was performed in the context of the special priority program 1135 of the German Research Foundation (DFG) on the CEBS (grant Li618/15). We wish to express our gratitude to this organization. In addition, we thank the reviewers Bill Kilsdonk and Kenneth Peters and one anonymus reviewer for their comments on a previous draft of this manuscript and Joy Burrough for editing the English.

REFERENCES CITED

Allen, P. A., and J. R. Allen, 2005, Basin analysis: Principles and applications: Oxford, Blackwell, 549 p.

de Jager, J., 2007, Geological development, *in* T. E. Wong, D. A. J. Batjes, and J. de Jager, eds., Geology of the Netherlands: Amsterdam, Royal Netherlands Academy of Arts and Science, p. 5–26.

de Jager, J., and M. C. Geluk, 2007, Petroleum geology, *in* T. E. Wong, D. A. J. Batjes, and J. de Jager, eds., Geology of the Netherlands: Amsterdam, Royal Netherlands Academy of Arts and Science, p. 241–264.

de Jager, J., M. A. Doyle, P. J. Grantham, J. E. Mabillard, H. E. Rondeel, D. A. J. Batjes, and W. H. Nieuwenhuijs, 1996, Hydrocarbon habitat of the West Netherlands Basin, Geology of gas and oil under the Netherlands, v. 1, Dordrecht, Kluwer Academic Publisher, p. 191–209.

de Lugt, I. R., J. D. van Wees, and T. E. Wong, 2003, The tectonic evolution of the southern Dutch North Sea during the Paleogene: Basin inversion in distinct pulses: Tectonophysics, v. 373, p. 141–159, doi:10.1016/S0040-1951(03)00284-1.

Dieckmann, V., B. Horsfield, H. J. Schenk, and D. H. Welte, 1998, Kinetics of petroleum generation and cracking by programmed-temperature closed-system pyrolysis of Posidonia Shale: Fuel, v. 77, p. 23–31, doi:10.1016/S0016-2361(97)00165-8.

Duin, E. J. T., J. C. Doornenbal, R. H. B. Rijkers, J. W. Verbeek, and T. E. Wong, 2006, Subsurface structure of the Netherlands: Results of recent onshore and offshore mapping: Netherlands Journal of Geoscience, v. 85, p. 245–276.

Geluk, M. C., 2000, Late Perminan (Zechstein) carbonate-facies maps, the Netherlands: Geologie en Mijnbouw/Netherlands Journal of Geosciences, v. 79, p. 17–27.

Geluk, M. C., 2005, Stratigraphy and tectonics of Permo–Triassic basins of the Netherlands and surrounding areas: Ph.D. thesis, University of Utrecht, Utrecht, the Netherlands, 171 p.

Geluk, M. C., 2007, Triassic, *in* T. E. Wong, D. A. J Batjes and J. de Jager, eds., Geology of the Netherlands: Amsterdam, Royal Netherlands Academy of Arts and Science, p. 85–106.

Geluk, M. C., and H. G. Röhling, 1997, High-resolution sequence stratigraphy of the Lower Triassic "Buntsandstein" in the Netherlands and northwestern Germany: Geologie en Mijnbouw, v. 76, p. 227–246.

Geluk, M. C., E. J. T. Duin, M. Dusar, R. H. B. Rijkers, M. W. van den Berg, and P. van Rooijen, 1994, Stratigraphy and tectonics of the Roer Valley Graben: Geologie en Mijnbouw, v. 73, p. 129–141.

Geluk, M. C., A. Plomp, T. H. M. van Doorn, H. E. Rondeel, D. A. J. Batjes, and W. H. Nieuwenhuijs, 1996, Development of the Permo–Triassic succession in the basin fringe area, southern Netherlands, Geology of gas and oil under the Netherlands: Dordrecht, Kluwer Academic Publisher, v. 1, p. 57–78.

Gradstein, F., J. Ogg, and A. Smith, 2004, A geologic time scale, 2004: Cambridge, Cambridge University Press, 589 p.

Gras, R., 1995, Late Cretaceous sedimentation and tectonic inversion, southern Netherlands: Geologie en Mijnbouw, v. 74, p. 117–127.

Gras, R., and M. C. Geluk, 1999, Late Cretaceous–Early Tertiary sedimentation and tectonic inversion in the southern Netherlands: Geologie en Mijnbouw, v. 78, p. 1–19.

Hantschel, T., and A. I. Kauerauf, 2009, Fundamentals of basin and petroleum systems modeling: Berlin, Springer, 476 p.

Helsen, S., 1995, Burial history of Paleozoic strata in Belgium as revealed by conodont color alteration data and thickness distributions: Geologische Rundschau, v. 84, no. 4, p. 738–747, doi:10.1007/s005310050036.

Hooper, R. J., L. S. Goh, F. Dewey, J. G. Buchanan, and P. G. Buchanan, 1995, The inversion history of the northeastern margin of the Broad Fourteens Basin, *in* A. J. Fleet, ed., Basin inversion: Geological Society (London) Special Publication 88, p. 307–317.

Kombrink, H., 2008, The Carboniferous of the Netherlands and surrounding areas: A basin analysis: Ph.D. thesis, University of Utrecht, Utrecht, no. 294, 184 p.

Littke, R., U. Bayer, D. Gajewski and S. Nelskamp, 2008a, Dynamics of complex intracontinental basins: Berlin Heidelberg, Springer, 520 p.

Littke, R., M. Scheck-Wenderoth, M. R. Brix, and S. Nelskamp, 2008b, Subsidence, inversion and evolution of the thermal field, *in* R. Littke, U. Bayer, D. Gajewski, and S. Nelskamp, eds., Dynamics of complex intracontinental basins: Berlin Heidelberg, Springer, p. 125–155.

Michon, L., R. T. von Balen, O. Merle, and H. J. M. Pagnier, 2003, The Cenozoic evolution of the Roer Valley rift system integrated at a European Scale: Tectonophysics, v. 367, p. 101–126, doi:10.1016/S0040-1951(03)00132-X.

Midland Valley, 2008, Move 2008.1 Tutorial, Midland Valley, 450 p.

Munoz, Y. A., R. Littke, and M. R. Brix, 2007, Fluid systems and basin evolution of the western Lower Saxony Basin, Germany: Geofluids, v. 7, p. 335–355, doi:10.1111/j.1468-8123.2007.00186.x.

Neunzert, G. H., R. Gaupp and R. Littke, 1996, Absenkungs- und Temperaturgeschichte paläozoischer und mesozoischer Formationen im Nordwestdeutschen Becken: Zeitschrift der deutschen Geolodischen Gesellschaft, v. 147, p. 183–208.

Nelskamp, S., P. David, and R. Littke, 2008, A comparison of burial, maturity and temperature histories of selected wells from sedimentary basins in the Netherlands: International Journal of Earth Sciences, v. 97, p. 931–953.

Pepper, A. S., and P. J. Corvi, 1995, Simple kinetic models of petroleum formation. Part I: Oil and gas generation from kerogen: Marine and Petroleum Geology, v. 12, no. 3, p. 291–319, doi:10.1016/0264-8172(95)98381-E.

Petmecky, S., L. Meier, H. Reiser, and R. Littke, 1999, High thermal maturity in the Lower Saxony Basin: Intrusion or deep burial?: Tectonophysics, v. 304, p. 317–344, doi:10.1016/S0040-1951(99)00030-X.

Poelchau, H. S., D. R. Baker, T. Hantschel, B. Horsfield, B. Wygrala, and D. H. Welte, 1997, Basin simulation and the design of the conceptual basin model: Petroleum and basin evolution: Berlin, Springer, p. 3–70.

Senglaub, Y., M. R. Brix, A. C. Adriasola, and R. Littke, 2005, New information on the thermal history of the southwestern Lower Saxony Basin, northern Germany, based on fission track analysis: International Journal of Earth Sciences, v. 94, no. 5–6, p. 876–896, doi:10.1007/s00531-005-0008-z.

Senglaub, Y., R. Littke, and M. R. Brix, 2006, Numerical modeling of burial and temperature history as an approach for an alternative interpretation of the Bramsche anomaly, Lower Saxony Basin: International Journal of Earth Sciences, v. 95, no. 2, p. 204–224, doi:10.1007/s00531-005-0033-y.

Thiadens, A. A., 1963, The Paleozoic of the Netherlands: Koninklijk Nederlands Geologisch Mijnbouwkundig Genootschap, Verhandelingen, Geologische Serie, v. 21, no. 1, p. 9–28.

TNO, 2000, Geological atlas of the subsurface of the Netherlands, Explanation to map sheet VI: Veendam-Hogeveen: Utrecht, Netherlands Institute of Applied Geoscience TNO, National Geological Survey.

TNO, 2002, Geological atlas of the subsurface of the Netherlands, Explanation to map sheets VII and VIII: Noordwijk-Rotterdam and Amsterdam-Gorinchem, Utrecht, 135 p.

TNO, 2004a, Geological atlas of the subsurface of the Netherlands–Onshore: Utrecht, Netherlands Institute of Applied Geosciences TNO, 103 p.

TNO, 2004b, Geological atlas of the subsurface of the Netherlands, Explanation to map sheet IX: Harderwijk-Nijmegen: Utrecht, 123 p.

van Adrichem Boogaert, H. A., and W. P. F. Kouwe, 1993, Stratigraphic nomenclature of the Netherlands, revision and update by RGD and NOGEPA: Mededelingen Rijks Geologische Dienst: Maastricht, Netherlands, Rijks Geologische Dienst: http://www.dinoloket.nl/nomenclator/nl/start/introduction/home.html (accessed December 1, 2010).

van Balen, R. T., F. van Bergen, C. de Leeuw, H. J. M. Pagnier, H. J. Simmelink, J. D. van Wees, and J. M. Verweij, 2000, Modeling the hydrocarbon generation and migration in the West Netherlands Basin, the Netherlands: Geologie en Mijnbouw/Netherlands Journal of Geosciences, v. 79, p. 29–44.

van Balen, R. T., J. M. Verweij, J. D. van Wees, H. J. Simmelink, F. van Bergen, and H. J. M. Pagnier, 2002, Deep subsurface temperatures in the Roer Valley Graben and the Peelblock, the Netherlands: New results: Geologie en Mijnbouw/Netherlands Journal of Geosciences, v. 81, p. 19–26.

van Buggenum, J. M., and D. G. den Hartog Jager, 2007, Silesian, *in* T. E. Wong, D. A. J. Batjes, and J. de Jager, eds., Geology of the Netherlands: Amsterdam, Royal Netherlands Academy of Arts and Science, p. 43–62.

van Wijhe, D. H., 1987, Structural evolution of inverted basins in the Dutch offshore: Tectonophysics, v. 137, p. 171–219, doi:10.1016/0040-1951(87)90320-9.

Vercoutere, C., and P. van den Houte, 1993, Post-Paleozoic cooling and uplift of the Brabant Massif as revealed by apatite fission track analysis: Geological Magazine, v. 130, no. 5, p. 639–646, doi:10.1017/S001675680002094X.

Verweij, J. M., and H. J. Simmelink, 2002, Geodynamic and hydrodynamic evolution of the Broad Fourteens Basin (the Netherlands) in relation to its petroleum systems: Marine and Petroleum Geology, v. 19, p. 339–359, doi:10.1016/S0264-8172(02)00021-1.

Verweij, J. M., H. J. Simmelink, R. T. van Balen, and P. David, 2003, History of petroleum systems in the southern part of the Broad Fourteens Basin: Geologie en Mijnbouw/Netherlands Journal of Geosciences, v. 82, p. 71–90.

Wong, T. E., 2007, Jurassic, *in* T. E. Wong, D. A. J Batjes, and J. de Jager, eds., Geology of the Netherlands: Amsterdam, Royal Netherlands Academy of Arts and Science, p. 107–126.

Wong, T. E., D. A. J Batjes, and J. de Jager, 2007, Geology of the Netherlands: Amsterdam, Royal Netherlands Academy of Arts and Science, 356 p.

Worum, G., 2004, Modeling of fault reactivation potential and quantification of inversion tectonics in the southern Netherlands: Ph.D. thesis, Vrije Universiteit Amsterdam, Amsterdam, No. 20040606, 152 p.

Wygrala, B. P., 1989, Integrated study of an oil field in the southern Po Basin, northern Italy: PhD thesis, Berichte Kernforschungsanlage Jülich 2313, 217 p.

Yalcin, N. M., R. Littke, R. F. Sachsenhofer, D. H. Welte, B. Horsfield, and D. R. Baker, 1997, Thermal histories of sedimentary basins, Petroleum and basin evolution: Berlin, Springer, p. 71–167.

Ziegler, P. A., 1988, Evolution of the Arctic, North Atlantic and western Tethys: AAPG Memoir 43, 196 p.

Ziegler, P. A., 1990, Geological atlas of Western and Central Europe, Shell Internationale Petroleum Maatschappij, Geological Society Publishing House, 239 p.

Zijerveld, L., R. Stephenson, S. Cloetingh, E. Duin, and M. V. van den Berg, 1992, Subsidence analysis and modeling of the Roer Valley Graben: Tectonophysics, v. 208, p. 159–171, doi:10.1016/0040-1951(92)90342-4.

Northeast view toward Clear Lake (upper right background) from crest of Mayacamas Mountains, The Geysers geothermal field, and the northwest flank of Cobb Mountain. Photo was taken standing on Franciscan/ophiolitic basement. Gathering lines from steam wells are in the foreground. The brush-covered terrain to the right (northeast) is serpentinite overlain by Clear Lake Volcanics.

SECTION 4

New Generation Methods/Unconventional Approaches

9

Derks, J. F., O. Swientek, T. Fuchs, A. Kauerauf, M. Al-Quattan, M. Al-Saeed, and M. Al-Hajeri, 2012, Three-dimensional basin and petroleum system model of the Cretaceous Burgan Formation, Kuwait: Model-in-model, high-resolution charge modeling, *in* K. E. Peters, D. J. Curry, and M. Kacewicz, eds., Basin Modeling: New Horizons in Research and Applications: AAPG Hedberg Series, no. 4, p. 159–174.

Three-Dimensional Basin and Petroleum System Model of the Cretaceous Burgan Formation, Kuwait: Model-in-Model, High-Resolution Charge Modeling

Jan Frederik Derks, Oliver Swientek, Thomas Fuchs, and Armin Kauerauf

IES GmbH, Aachen, Germany

Meshari Al-Quattan, Mariam Al-Saeed, and Mubarak Al-Hajeri

Kuwait Oil Company, Ahmadi, Kuwait

ABSTRACT

In basin and petroleum system modeling, the spatial resolution of models is too coarse to cover all of the relevant geologic processes that occur in reservoirs. Reservoir models are built on a static (time invariant) grid and cannot cover the charge history of a field or processes related to changing geologic structures through time. A new method to combine regional-scale petroleum system models and local reservoir- or prospect-scale models was developed and applied to an oil field in Kuwait. In the field, heavy oil zones occur at the original oil-water contact and also in stratigraphic and structural positions above it. Heavy oil occurs in the highly permeable Fourth Sand and Middle Third Sand of the Burgan Formation in the field. This study demonstrates that the heavy oil distribution in those layers can be explained by the petroleum charge history. An early charge from the Cretaceous Makhul Formation was replenished by the Cretaceous Kazhdumi Formation, the stratigraphic equivalent of the Burgan Sands. These sediments were deposited in the area of the Dezful Embayment of the Zargos Fold Belt. No Jurassic charge, breaking through the Gotnia evaporites, is needed to fill the structures of the Cretaceous Burgan reservoirs in the field. Well data and the results of the high-resolution petroleum system model covering the area of the oil field and describing the distribution of charge from the regional model to the field scale lead to the conclusion that the heavy oil zones are mainly the result of gravity segregation, although some influence of water washing cannot be excluded.

DOI:10.1306/13311434H43469

INTRODUCTION

One of the key decisions in basin and petroleum system modeling is the scale of the model. When deciding on the dimensions and resolution of a model, it is necessary to consider computation time and the resolution required to describe a particular process. The size of the model should be assessed before the model is built. The regional extensions of the petroleum system and the basin are the first input parameters, defined by the geologic setting. These elements might have to be modified in the process of modeling if the original input dimensions or parameters do not cover the entire system, but in most cases, the general dimensions are known in advance. The processes that are investigated via the petroleum system model should be understood and defined as precisely as possible to decide which resolution of the model will properly reflect these processes (Waples, 1994). The number of elements (cells) in the model required to fulfill the user's requirements generally represent the maximum number of elements in a project. In the end, the limitations of the simulation speed restrict the size of a model. The compromise between size and resolution of a model and the practical limits of computation power control the total number of elements and scale of the model. The properties of rocks and fluids have to be upscaled to describe regional processes (e.g., fluid flow) and petroleum distribution (e.g., pressure compartments) in a basin and petroleum system model (Hantschel et al., 2000).

The horizontal scale of the grid of such models is in most cases in the range of kilometers. Reservoir models have much higher resolutions in the range of 100 to 10 m (328 to 33 ft) but are built on a static grid that cannot cover dynamic processes over geologic time scales. To understand these dynamic processes in reservoirs related to changes in the geologic structures that affect the charge on a scale near that of reservoir models, it is necessary to cover the entire regional charge area of the petroleum system and the high-resolution reservoir area at the same time. This can be achieved by local refinement of a coarse grid (Caillabet et al., 2004) or by combining two models with different scales (model-in-model). The presented approach uses the model-in-model method; two models are built that communicate in respect to pressure, temperature, and fluid flow. The local model is part of the regional model but has higher resolution, commonly referred to as local grid refinement. The regional model covers the kitchen area and the migration pathways to the prospect or field. In the local model, the boundary conditions of the regional model are considered and the charge is transferred from the regional coarse grid to the local high-resolution grid.

The purpose of this study was to use model-in-model grid refinement to analyze the processes and charge history of a giant oil field in Kuwait that lead to heavy oil occurrence. In the first step, the charge area and history of the field were described using a regional-scale model. In the second step, multiphase kinetics were used to model the generation and expulsion of hydrocarbons from the main source rocks and their migration from the kitchen areas to the field. The distribution of the components on the high-resolution grid of the local model was correlated to the observed occurrence of heavy oil in wells. This approach provided explanations for the occurrence of heavy oil in the field.

MODEL-IN-MODEL GRID REFINEMENT

General Overview

Local grid refinement (LGR) is the generic term for a wide range of methods used to improve simulation accuracy in local areas of interest. The underlying grid is locally refined to represent data of different quality and to respect differences in resolution. The LGR drastically decreases the number of cells compared with a globally refined grid and thus decreases simulation time and requires less computer memory.

In general, one can distinguish between coupled and decoupled LGR methods. Decoupled methods combine at least two different-size grids: one regional coarse grid and at least one local fine grid. First, the regional coarse grid is simulated with globally defined boundary conditions, and then the local fine grid is simulated, taking its boundary conditions from the regional model. In the literature, decoupled LGR is also called telescopic mesh refinement (only coupled methods are called local grid refinement; Ward et al., 1987). A special case is the model-in-model or windowing approach (Hantschel and Kauerauf, 2009), where it is necessary to build two different models, a regional and at least one local model with different grid resolutions. "Coupled" means that the simulators can deal with variable grid spacing, and therefore, only one run has to be performed.

The LGR methods are well known and commonly used in reservoir simulators (Boyett et al., 1992). However, classic reservoir simulators are not able to account for timing effects, burial history, or structural changes. Although basin modeling has dealt with these time-related effects for many years, LGR methods were only recently implemented in basin modeling applications. With these new developments, it is now possible to refine models in specific areas from basin to prospect scale, enabling reservoir engineers to directly use basin modeling as an interpretation tool.

The Method

Local grid refinement methods were applied to petroleum system modeling in the past, but only with

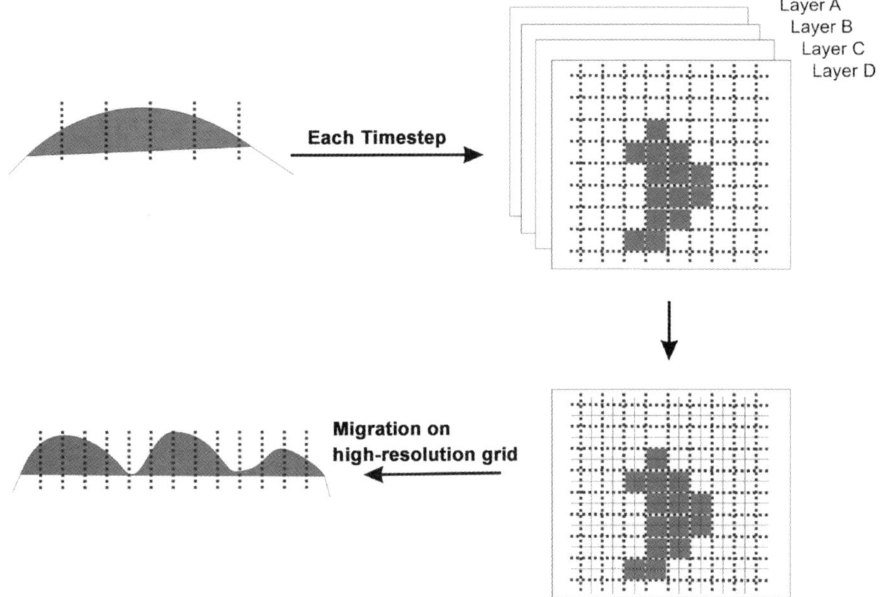

Figure 1. Process of map-based mass distribution on local refined grid in the model-in-model method. Area-yield saturations for each component are written in a separate file at each time step. During the model-in-model simulation, these saturations are projected on the refined high-resolution grid in the local model.

very limited numbers of grid cells. In these applications, a high-resolution grid was upscaled to lower resolution (Caillabet et al., 2004). In this study, the model-in-model approach with two independent grid resolutions of the models was applied for the first time in a basin and petroleum system model application. A regional model with a coarse grid resolution of 1 × 1 km (0.6 × 0.6 mi), covering an area of 69,500 km^2 (26,834 mi^2), was built and combined with a local model with a fine resolution of 50 × 50 m (164 × 164 ft), covering 1285 km^2 (496 mi^2). The regional model was simulated as usual, that is, no additional files were required, and the local model did not have to be specified in that part of the simulation. This has the advantage that the regional model has to be simulated only once, and several local models can be defined and simulated on the basis of the existing regional model. No restriction on the migration method exists. The simulation of the regional model is performed on the coarse grid, and overlays for temperature, pressure, and the masses per area for each component are generated. In the next step, the local model is calculated using the generated overlays of the regional model. For the temperature and pressure calculation, the simulator uses the values of the regional model at the borders of the local model and assigns these values as the new boundary conditions to the local model. The unrealistic "no-flow" boundary conditions at the sides of the model are thus replaced by fixed values for temperature and pressure (Hantschel and Kauerauf, 2009). Temperature and pressure equations are solved on the fine grid for each time step using these derived boundary conditions.

Applying the same concept of using values at the boundaries of the local model for a migration model is problematic. Problems arise when hydrocarbons leave the local model but not the regional model because they can reenter the area of the local model at a later time step in the simulation. A coupled LGR method is needed to address these problems. To solve this issue, we apply a migration method that we call in-situ injection: At the beginning of each event, all residual masses and saturations in the local model are cleared. The new masses are obtained by an in-situ projection of the mass overlays of the regional model (Figure 1). In detail, for each cell in the local model, the global X-Y position is determined and the corresponding component masses of the regional model at this position are adopted. To respect the different sizes of the cells, the area yield overlays are taken (mass per area) and scaled by the area of the cell. All mass-related fields such as saturations and total masses are updated, and the specified migration method is applied. Generation inside the source rocks, as well as secondary cracking within and outside the source rocks of the local model are switched off. This means that generation and secondary cracking effects are always calculated on the coarse grid of the regional model.

Some advantages of the model-in-model approach combined with our method of in-situ injections are present: The method is straightforward. No complex regridding algorithm has to be implemented, and the extension to a coupled migration mode is simple. However, the coupled mode extension requires extensive memory because two models need to be stored at the same time. In addition, different options for the regional and the local model can be used, for example, one can use different migration methods or apply a vertical refinement by adding additional layers, decreasing the maximum cell thickness or increasing the migration steps in the local model. One advantage of the model-in-model

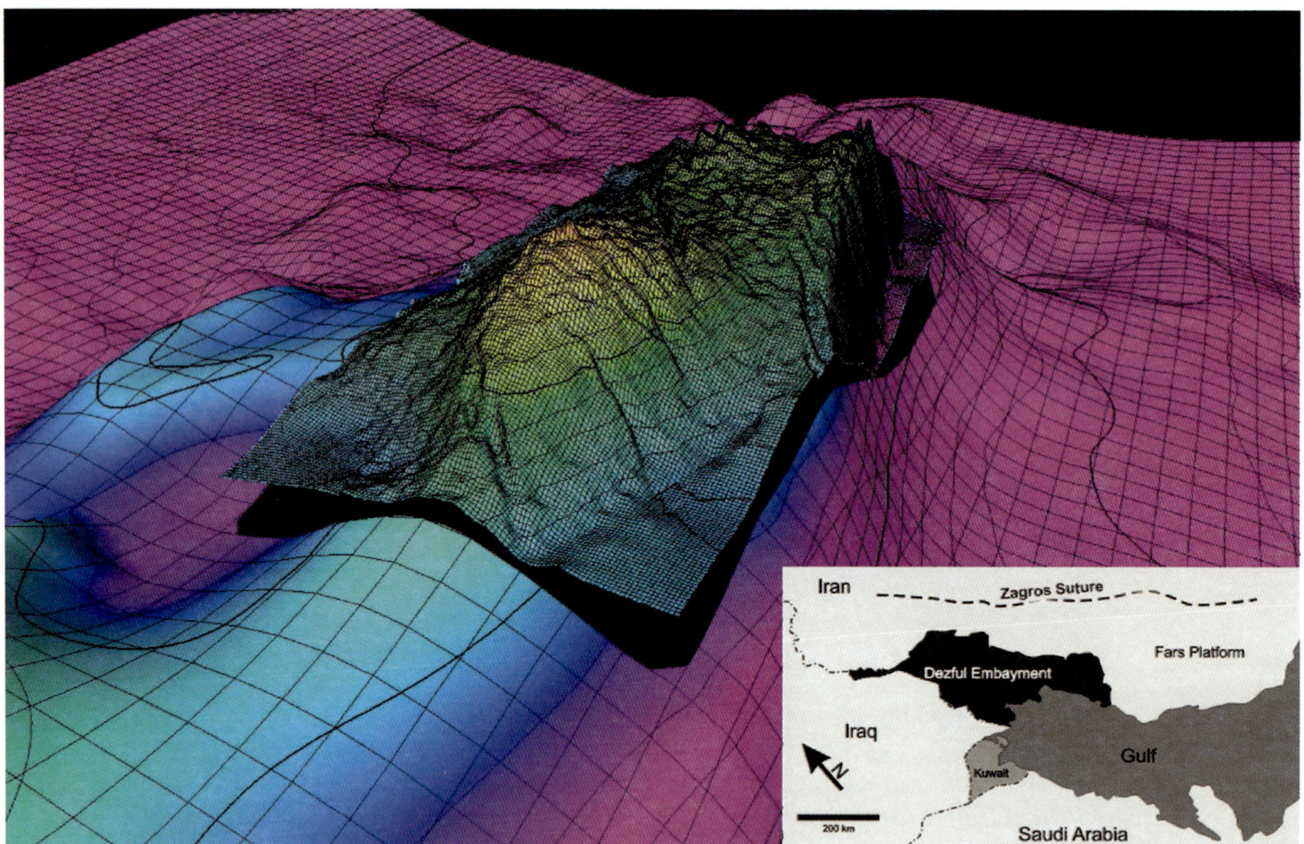

FIGURE 2. Local grid refinement of the study area in Kuwait with model-in-model approach: X, Y resolution of local model, 100 m (328 ft); X, Y resolution of regional model, 1200 m (3937 ft). Contours show depth in 50-m (164-ft) steps.

approach is the multiple use of a regional model for several locally refined models. However, this could also be seen as a disadvantage because two models need to be built and simulated separately. Small accumulations could be missed if a structure is too small to be recorded in the regional model, so no masses are stored in that area that could be transferred to the local model.

CASE STUDY

In this case study, the method of noncoupled model-in-model grid refinement was first applied to a large oil field of Kuwait to investigate the occurrence of heavy oil within the reservoirs of the Burgan Formation of that field. Two models were built. One regional model, covering the entire charge area of the field, was built in a first phase of the study. In a second phase, a local model was built (Figure 2), covering the field geometries in detail. Migration was modeled using the hybrid method, where Darcy migration is used for low-permeability layers and flowpath migration is used for high-permeability carrier beds in the same simulation run (Hantschel et al., 2000; Hantschel and Kauerauf, 2009). Transfer of migrated hydrocarbons from the regional model to the local model was done using the in-situ injection method.

Regional Geology

Kuwait is located on the Arabian Platform Margin between the Precambrian Arabian shield in the west and the Zagros Fold Belt in the east. Throughout its geologic history, the area was in a relatively stable shelf position characterized by continuous deposition of mainly carbonate platform sediments with intercalated clastic sequences (Figure 3) (Yousif and Nouman, 1997). Several hiati and unconformities reflect repeated phases of uplift and erosion. The most important unconformity is the Upper Eocene Dammam Formation, which is associated with the northeast tilting of the predefined structures in the region, caused by the Zagros orogeny (Al-Sulaimi and Mukhopadhyay, 2000). During the Late Jurassic, sabkha-like settings were established, forming large salt and anhydrite deposits (Ali, 1995). The general dip of the strata is very shallow directed to the northeast and is subdivided by anticlinal structures. The present-day geometries

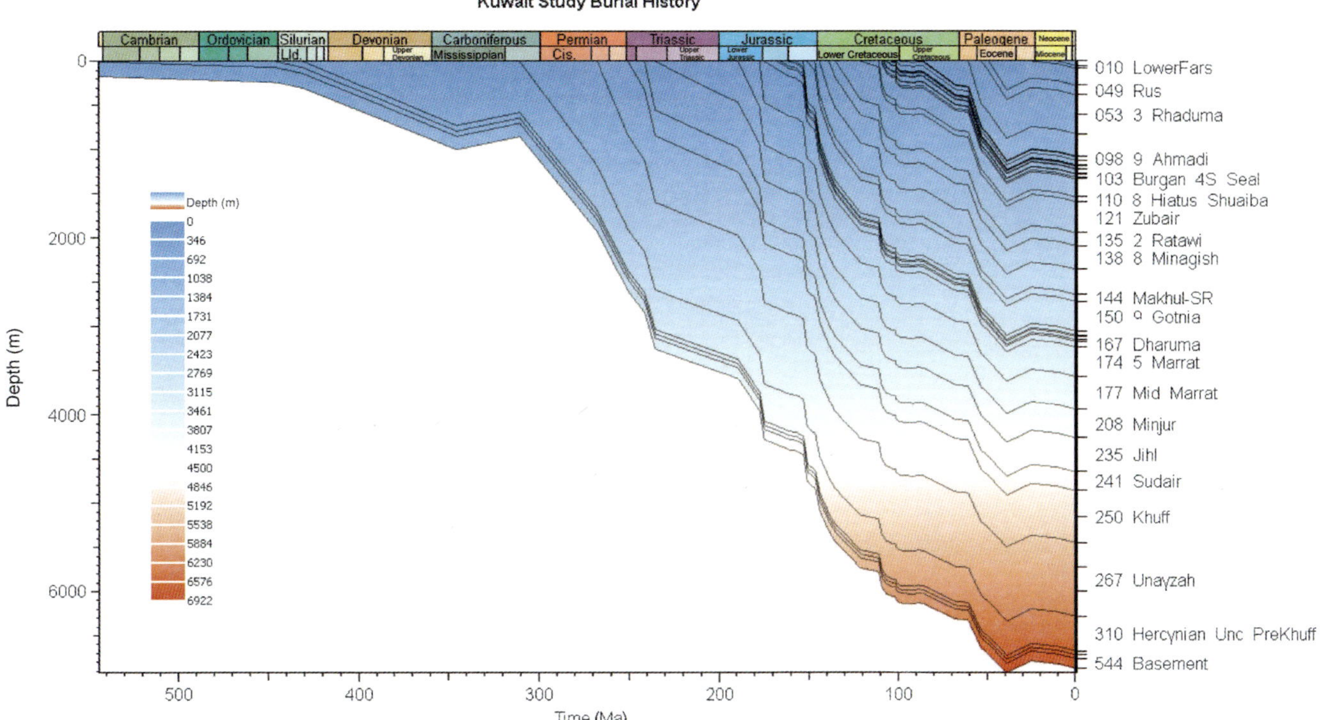

FIGURE 3. Burial history of the field, Kuwait case study.

mainly originated during the Jurassic to Cretaceous when the Kuwait Arch formed as a result of east-west compressional stresses along older Paleozoic predefined lineaments (Warsi, 1990; Carman, 1996). In younger times, structures like the Ahmadi Ridge were caused by the Zagros orogeny (Bou-Rabee, 1996; Ziegler, 2001) (Figures 4–6).

Heavy Oil Occurrence

The heavy oil distribution in the field is not always associated with the original oil-water contact (OOWC). The OOWC before start of production was at a similar level throughout the field (Kirby et al., 1998). Water washing, gravity segregation, and biodegradation can explain heavy oil associated with an OOWC. The occurrence of heavy oil in stratigraphically higher position, not related to the OOWC, needs to be explained considering the charge history of the field and the pressure-volume-temperature conditions in the past. The main reservoirs of the field are the sands of the Wasia Group that have a stacked geometry with five sand layers, separated by thin, more shaly seals in the Burgan Formation and the Mauddud Carbonate, which separates the Wara Formation from the underlying Burgan Formation (Figure 7). These seals limit vertical communication between the layers in the group. Heavy oil precipitation at permeability barriers was described in other fields (Carpentier et al., 2007), but heavy oil is not only associated with these barriers in the investigated field. The heavy oil zones occur mainly in two of the layers of the reservoir: the Fourth Sand and Middle Third Sand (Figure 8), which show the highest porosities and permeabilities in the field and are not directly connected. The heavy oil was encountered in several wells drilled during the long history of the field development and can be tracked on well logs. The heavy oil zones are characterized by a discontinuous change of the API gravity from an average value of 32° to values as low as to 10°. The oils are characterized by average sulfur contents of 2.5%. In general, the heavy oil zones are interpreted to result from gravity-induced separation of phases within the oil column (Kaufman et al., 2000).

MODEL SETUP AND BOUNDARY CONDITIONS

Regional Model Setup

In this study, 36 depth-converted seismic-derived maps were used for three-dimensional thermal and maturity history reconstructions and for three-dimensional migration modeling. The model covered an area of 69,500 km^2 (26,834 mi^2) including offshore and onshore parts of Kuwait, Iraq, Iran, and Saudi Arabia. The Burgan Layer was manually split into the subdivision of Wara, Mauddud, Third Sand Upper, Third Sand Middle, Third Sand Lower, and Fourth Sand according to the depth

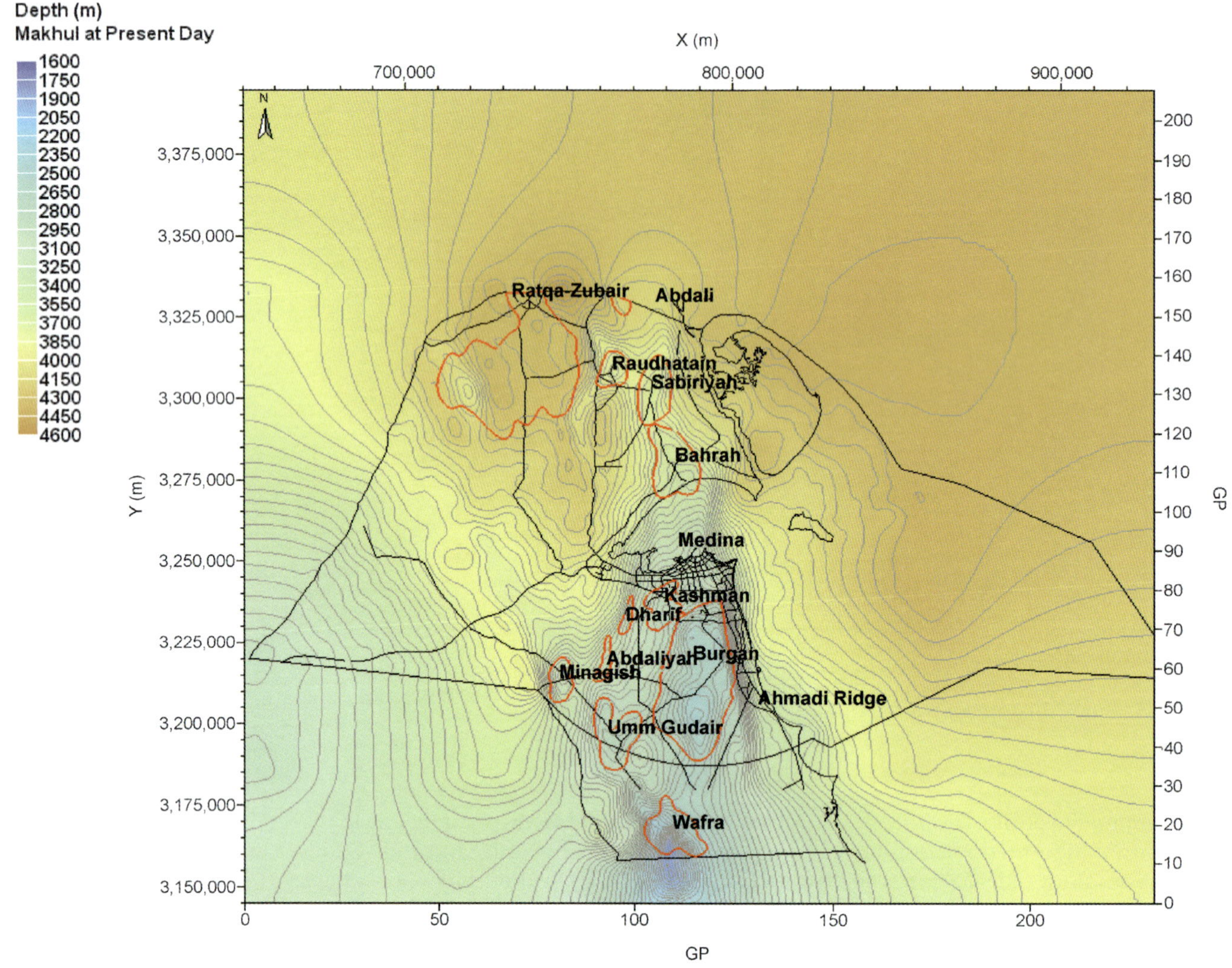

FIGURE 4. Depth map: Top Makhul. The regional model encompasses the entire area of Kuwait and adjacent countries. Political boundaries of the State of Kuwait and road map for orientation. Major oil fields are labeled. Contours are depth values in 50-m (164-ft) steps.

maps provided for the reservoir-scale model and well-pick information. The Burgan sands are vertically compartmentalized, and during production, the homogeneous OOWC developed into different OWC for the Third Sand and Fourth Sand (Kirby et al., 1998). However, the homogeneous OOWC might be a sign that a communication and equilibration between the different reservoir layers occurred on geologic time scales. To cover these observations in the model, a thin sealing layer was inserted for each sand layer to create a subdivision of the Burgan sands. The seal capacities were calibrated to the reported original column heights of the field. The Zubair layer was included in the map stack, derived from the Shuaiba map geometries. Where the maps did not cover the entire charge area of the field, the geometries were extrapolated. Additional horizon lines were created by seismic interpretation along two-dimensional seismic lines for the Partitioned Neutral Zone to guide the extrapolation. These lines were appended to the depth maps, and the remaining areas were extrapolated using minimum curvature extrapolation. By using this method, the southern charge area limitations of the field could be defined much more precisely than before. Additional minor corrections and editing (e.g., removal of crosscutting horizons, continuation of horizons) were conducted manually.

The most important faults, which could have a significant impact on migration routes, were also incorporated in the model. For the regional model, 120 faults were selected out of a data set of 461 fault surfaces, following the criteria listed below:

1. Major trends, visible in map view, should be represented by the selected faults.
2. Regional faults that are interpreted over a greater distance or that crosscut more layers are preferred.

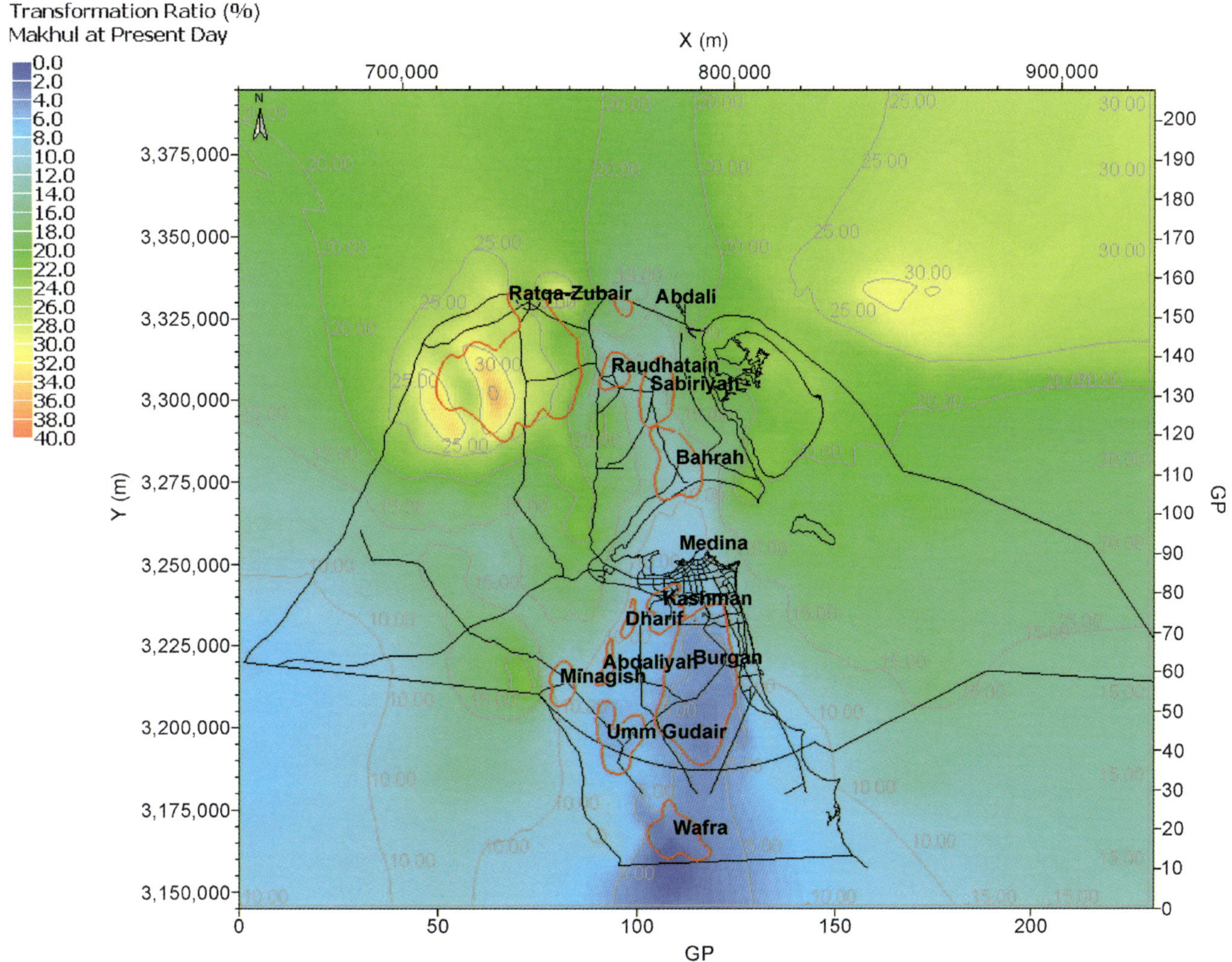

FIGURE 5. Present-day transformation ratio of the Cretaceous Makhul Formation source rock.

3. Faults that are continued by other faults are preferred and merged.
4. If more than two parallel fault planes are very near each other, every second fault plane is selected if they have the same depth and lateral extension.
5. If the distance between two faults is less than five grid points, then the larger fault is selected if no special geometry (e.g., closures, drainage area boundaries) is affected.

Fault modeling was part of the sensitivity analysis of the study. In the final scenario, the faults were assigned to be closed and open at specific compressional and extensional episodes in the tectonic history of Kuwait. This scenario was used for the final sensitivity analysis to evaluate the importance of the fault behavior.

The age assignment was based on stratigraphic information from the Kuwait Oil Company (KOC) and the International Stratigraphic Chart (International Commission on Stratigraphy, 2009). Most of the lithotypes assigned to the model are mixed lithologies from standard lithologies available in the modeling software. Data for the definition of the lithologies were taken from reports prepared for KOC, internal proprietary data of KOC, and special publications on depositional environments and facies distributions of the model area (Ziegler, 2001; Strohmenger et al., 2006). Lithologies were modified in a second stage of modeling using calibration data for porosity and permeability derived from well-log data. The facies assignment was done by digitizing facies maps provided by KOC and extrapolation of these maps to undefined areas. Additional data for well picks and lithology from internal reports and several publications were taken into account (Bordenave and Burwood, 1990; Warsi, 1990; Ali, 1995; Bou-Rabee, 1996; Carman, 1996; Al-Husseini, 1997; Yousif and Nouman, 1997; Ziegler, 2001; Konyuhov and Maleki, 2006; Strohmenger et al., 2006). In total, six erosion and hiatus surfaces were assigned to the model. The eroded amounts varied between nondeposition during the hiati and more than 600 m

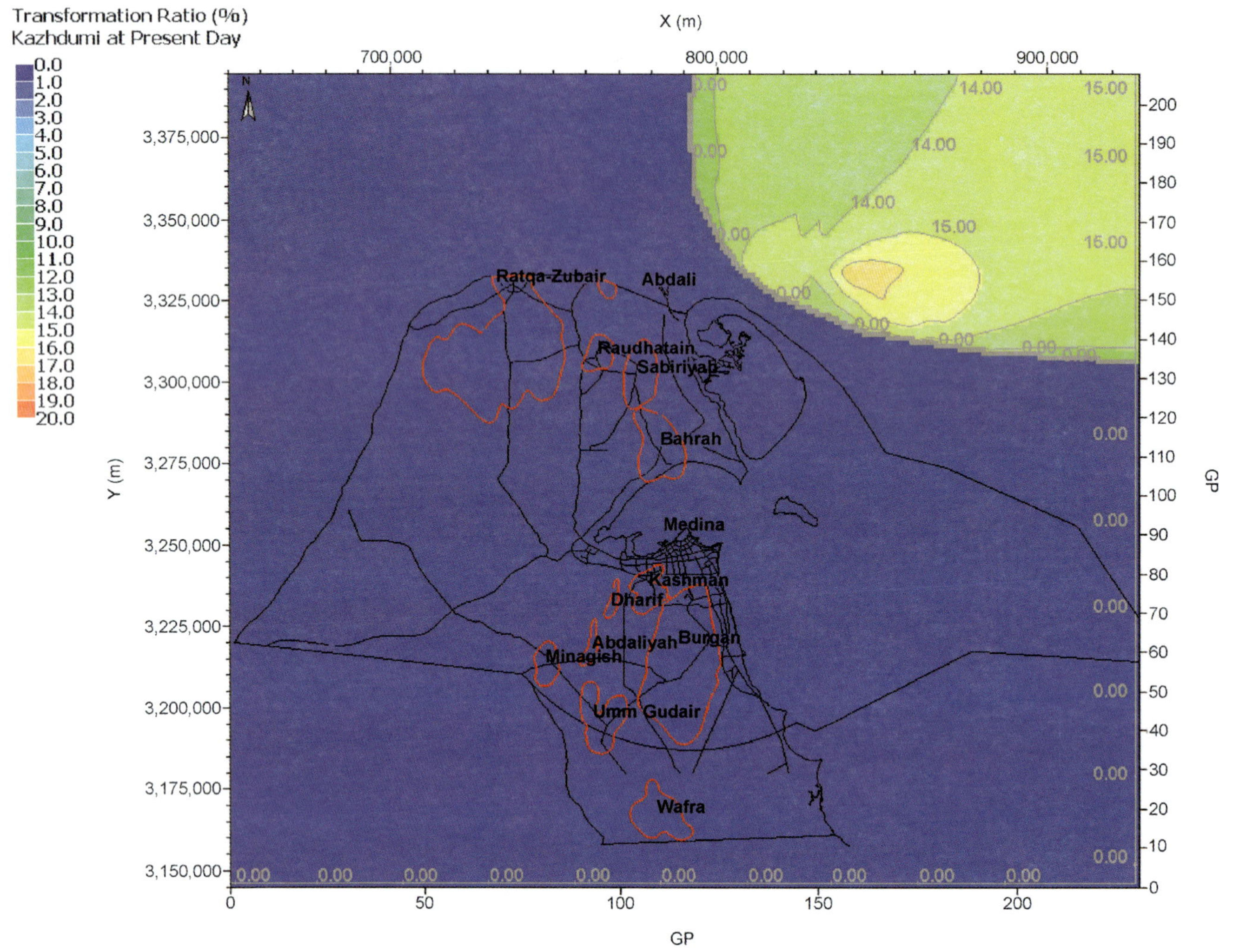

FIGURE 6. Present-day transformation ratio of the Cretaceous Kazhdumi Formation source rock.

(1969 ft) of missing section in some areas. These missing thicknesses were quite significant and could have a serious impact on source rock maturation. The major unconformities are as follows (Figure 3):

1. Hercynian Unconformity (uniform erosion).
2. Minjur–Upper Triassic Hiatus.
3. Shuaiba–Lower Cretaceous Hiatus.
4. Mishrif–Upper Cretaceous (minor uniform erosion).
5. Tayarat–Uppermost Cretaceous (uniform erosion).
6. Damman–Eocene (erosion map; 76–624 m [249–2047 ft] of eroded sediment).

Source Rock Definition

We believe that the main source rocks in Kuwait are the Cretaceous Makhul and Jurassic Najmah and Sargelu formations separated by the Gotnia evaporates, with an additional contribution of the Albian Kazhdumi Formation from the north and northeast. All source rocks contain marine type II kerogen, with only minor terrigenous input of organic matter. Source rock parameters were derived from internal KOC reports, well data, and the literature. The Makhul (Sulaiy) Formation source rock is an argillaceous limestone with an average total organic carbon (TOC) of 2.5% and a Rock-Eval hydrogen index (HI) of 350 mg HC/g TOC. In the model, TOC and HI distribution maps derived from well data were used as input. Pyrolysis data for the Makhul Formation source rock were available from an internal report of KOC, summarizing data for 37 samples from 15 wells throughout Kuwait. The measured TOC values were back calculated to original values using the method of Peters et al. (2005), assuming an original HI of 600 mg/g TOC and an initial production index of 0.02. Original TOC values from pyrolysis–gas chromatography measurements, literature, and new kinetic measurements were used to create TOC maps for the Makhul, Najmah, and Sargelu formations. In addition to these data, multicomponent

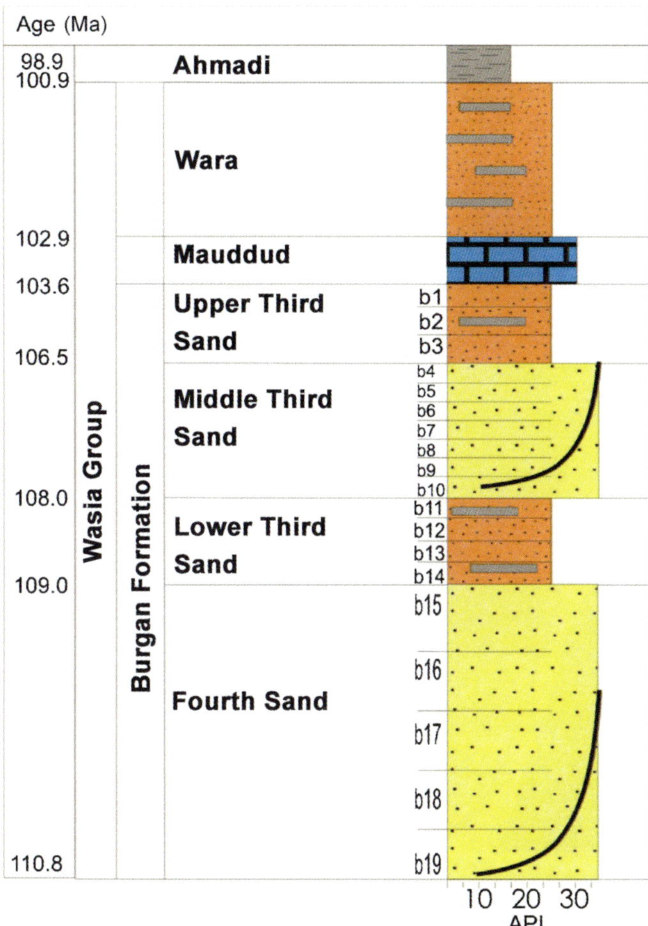

FIGURE 7. Schematic stratigraphic column of the Wasia Group. The Burgan Formation is subdivided into 19 layers. Heavy oil zones can be found mainly in the Third Sand and Fourth Sand of the Burgan Formation with gradually decreasing API values toward the original oil-water contact at the base of the layer.

kinetics for phase separation were measured on three of six samples from the Sargelu, Najmah, and the Makhul formations to better understand the maturation of the main source rocks in the area of the study. The measured kinetic reactions were calibrated against fluid compositional data of the known accumulations in the field derived from production fluid analysis. To determine the different contributions of the source rocks to the petroleum systems, source rock tracking was used. For each source rock, an individually flagged kinetic was assigned to quantify the amount of hydrocarbons from each source in every modeled accumulation.

The kerogen-hydrocarbon conversion of the two main Cretaceous source rocks, identified by source rock tracking, was modeled in the second step using kinetics that describe the conversion of kerogen to 14 hydrocarbon components, each following the phase kinetics approach of di Primio et al. (1998). The phase behavior of the 14-component system does not significantly change by adding more components to the system. The 14 components with groups of carbon numbers that comprise chain lengths of 10 carbon atoms (C_7–C_{15} in PK-P10, C_{16}–C_{25} in PK-P20 to C_{56+} in PK-P60) are therefore an optimum to describe the phase behavior of the fluids. A detailed description of the kerogen conversion allows the prediction of the phase behavior of the components and property prediction of the fluid phases in the model. The kinetic parameters for the Makhul source rock were measured on two core samples, whereas those for the Kazhdumi were assumed based on published data (Bordenave and Burwood ,1990; Beydoun et al., 1992). The core samples of the Makhul source rock were derived from a well in the field from a depth between 2638.6 and 2666.4 m (8656.8–8748 ft). The TOC values of the two samples were 2.37% in the shallow sample and 3.29% in the deeper sample. The HI values of the samples were 387 and 413 mg HC/g TOC. The kinetics measured by microscale sealed vessel pyrolysis show a relatively wide distribution of activation energies with a single peak from 55 to 56 kcal/mol. The measured T_{max} value of 434 to 437°C indicates immature to moderately mature samples. For the Kazhdumi source rock, only general assumptions on TOC and HI values were possible. An average TOC value of 5.7% and an original HI of 600 mg HC/g TOC based on literature data was used (Beydoun et al.,1992). For the description of the Kazhdumi source rock, a unimodal kinetic model with a relatively broad distribution of activation energies similar to the Makhul kinetic distribution, was chosen. The Kazhdumi-derived oils in the Dezful Embayment of the Zargos Fold Belt are described as sulfur rich (Bordenave and Burwood, 1990). Based on these data, sulfur-rich type II source rock kinetics were selected (Figure 9).

Boundary Conditions

Sediment-water interface temperatures ([SWIT] as upper boundary condition for temperature calculations) and paleowater depth (PWD) were obtained using the SWIT and PWD trends provided by KOC for all of Kuwait. The trends were edited, considering erosion events. In addition, a comparison of the depositional environment (Al-Husseini, 1997; Ziegler, 2001) and the assigned PWD and SWIT values was made. The whole area of the study was part of the same tectonic environment, and the assumed water depth was never much deeper than 50 m (164 ft), based on the sedimentary record. A single trend for the entire study can be applied without losing information.

Six heat-flow maps for different times were digitized based on internal studies of KOC and used for the initial simulation. After this initial run, wells with calibration data (temperature and vitrinite reflectance [$\%R_o$]) were added to the model, and an uncertainty analysis

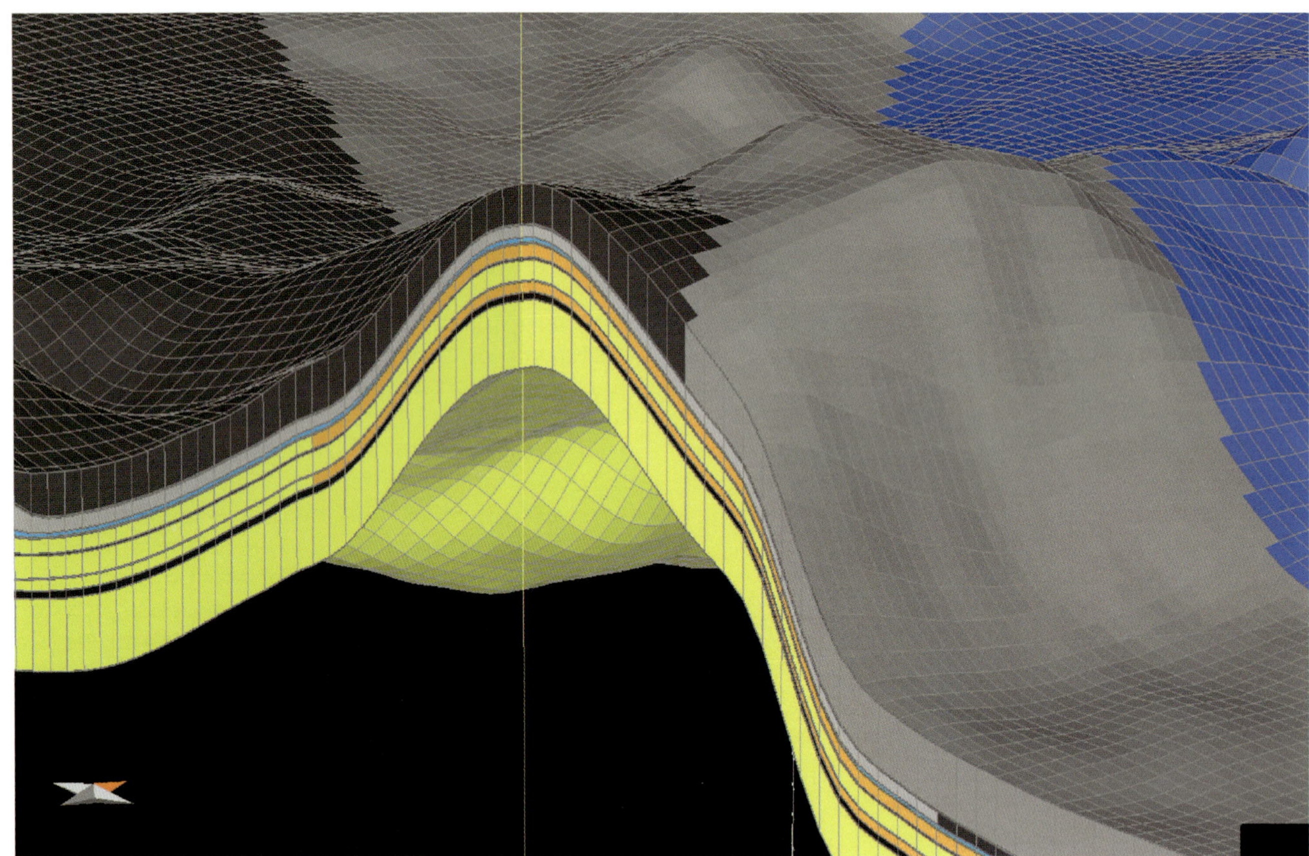

FIGURE 8. Cut cube section through a three-dimensional cube of the Wasia Group in the area of the study for the regional coarse resolution model. Layers shown are described in Figure 7. Dark-gray to gray: shaly lithologies; yellow and orange: sand and sand-shale lithologies, respectively.

was performed to calibrate the heat-flow maps. After calibration, the measured temperatures and vitrinite reflectance matched to the modeled data. Because of its structural setting, thermal boundary conditions of Kuwait are believed to have been relatively constant through geologic time with a slightly decreasing heat flow from 75 to 80 mW/m^2 in the lower Paleozoic to present-day values less than 50 mW/m^2.

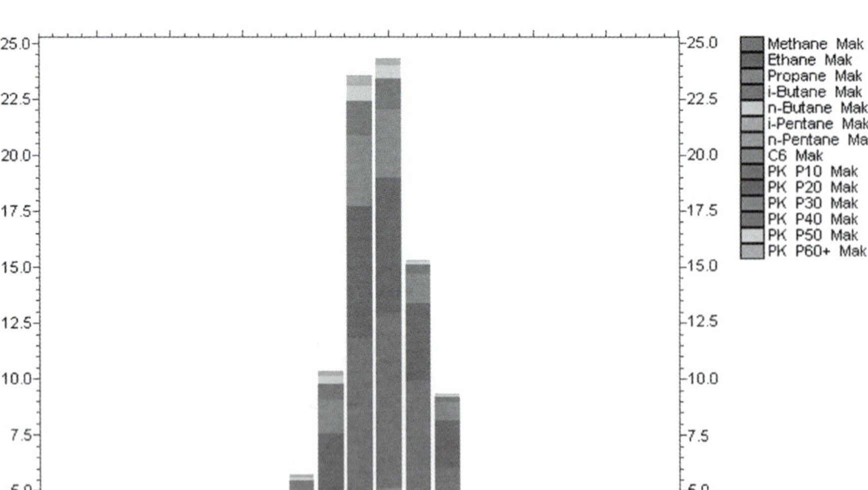

FIGURE 9. Makhul source rock: Phase kinetic activation energies of the 14 components. Kinetics were measured on thermally immature core samples from a well in the oil field. A = 1.53E + 14.

LOCAL MODEL SETUP AND BOUNDARY CONDITIONS

The local model was based on geometry maps of 50 × 50 m (164 × 164 ft) horizontal resolution and sampled to a grid of 100 × 100 m (328 × 328 ft) horizontal x and y dimensions. The total dimensions of the local model covered an area of 1285 km^2 (496 mi^2). The maximum thickness of a layer was defined as 400 m (1312 ft). If the thickness of the layer was less than 400 m (<1312 ft), no subdivision was made. The number of maps and the stratigraphic subdivision was equal to the regional model. Facies definitions and assignment were done using the input data of the regional model. Boundary conditions for heat flow, sediment-water interface temperature, and water depth were applied according to the maps, generated for the regional model. Temperature, pressure, porosity, and fluid property calibration was done using proprietary data of KOC from more than 300 production wells of the field.

RESULTS

Petroleum System

The regional model demonstrated clearly that the uppermost significant source rock in the petroleum system of the field is the Lower Cretaceous Makhul Formation, which overlies the Gotnia evaporites. The main kitchen areas for this source rock are in the north and offshore areas of Kuwait. The Makhul source rock provides the main charge for the field. The expulsion of hydrocarbons started around 105 m.y. ago and continues today. In the kitchen areas, the Makhul is producing mature oil in the main oil window. Because of the low heat-flow regime, the extent of maturation is relatively low, resulting in a relatively low transformation ratio at the present day (Figure 5). Migration follows a series of fill and spill paths from the kitchen areas in the north to the field. Once the main seals were deposited and had gained a certain capillary entry pressure caused by compaction, accumulations formed in the model around 95 Ma in the lower sands of the Burgan Formation.

However, the amount of hydrocarbons in the field can only be explained by a charge of at least one additional source rock because the Makhul Formation alone cannot provide sufficient charge in the model using the source rock parameters that were assumed from the available well data. Several studies (Kaufman et al., 2000; Internal report provided by KOC) show small differences in geochemical data that suggest that the oil in the field could have originated from one or more source rocks having similar geochemical composition, but geochemical fingerprinting is reported to be difficult as oil mixing occurred during the production history of the field (Kaufman et al., 2000; Kirby et al., 1998). Earlier petroleum system modeling studies of Kuwait by KOC suggest additional charge from the Najmah and Sargelu source rocks below the Gotnia halite/anhydrite succession (Al-Khamis et al., 2006). The Gotnia consists of four salt layers and intercalated anhydrites up to a thickness of 300 m (984 ft). We think that the ductile behavior of the salt prevents potential faults (cutting the Gotnia) from acting as conduits for migration. An additional proof for the sealing capacity of the Gotnia is the overpressure that can be found directly below this layer, which cannot be explained by open fault behavior (Ali, 1995). Therefore, we do not believe that the Najmah and Sargelu source rocks contribute to the Cretaceous petroleum system.

The second important source rock for the field area is the Cretaceous Kazhdumi Formation, which is the stratigraphic equivalent of the Burgan Formation to the northeast in the Zagros Foredeep of the Dezfull Embayment. The Kazhdumi Formation consists of bituminous marls and shaly limestones and is a proven source rock for several oil fields in the Zagros Fold Belt (Bordenave and Burwood, 1990; Beydoun et al., 1992). It acts as a kitchen area for parts of the charge for the Burgan sands by lateral migration from the northeast. The Kazhdumi reached peak expulsion around 39 Ma and continues to contribute hydrocarbons to the petroleum system at the present day while still in the early oil window (Figure 6).

Oil from the Makhul and the Kazhdumi source rocks mixed in the reservoir. The main reservoirs in the field are the sands of the Burgan Formation, which are subdivided into the four main layers of the Fourth Sand, Lower Third Sand, Middle Third Sand, and Upper Third Sand (Figure 7). The Burgan Formation was modeled using 19 horizons. The thinner Mauddud carbonate layer separates the overlying Wara clastic reservoir from the underlying Burgan Formation. The main regional seal of the Wasia Group, including the Burgan, Mauddud, and Wara formations, is the Ahmadi Shale. Within the Burgan Formation, thinner shaly seals limit vertical communication between the reservoir layers. The Burgan Sands are subject to a strong lateral water drive (Kaufman et al., 2000). The aquifer supports the pressure depletion of the reservoirs because of hydrocarbon production for more than 50 yr. Production has resulted in only a minor pressure decrease. The lateral water drive can be explained in the model by pressure equilibration between the deeper, more compacted areas of the Burgan Formation in the west of Kuwait and the shallower reservoirs of the field.

Heavy Oil in the Model

By using the high-resolution geometries of the local model in combination with the charge history of the regional model, it was possible to track the distribution of the generated components from the kitchen areas to the field and the remigration of these components within the field. The first charge can be associated with the initial maturation of the Makhul source rock. The hydrocarbon mixture migrates as a single liquid phase. The available pore volume and the sealing capacity, which was calibrated against the reported column height of the oil field before onset of production (t_0), control the distribution of the components in the reservoirs.

The largest amount of heavy components occurs in the Fourth Sand and Third Sand reservoir layers. The second charge derived from the Kazhdumi source rock delivered early mature petroleum into the system and mixing occurred within the reservoir of the field. The modeled amount of heavy components rises to a maximum during the second oil charge. The relative amount of heavy components in the mixture after the second charge remains stable until present day (Figure 10). The availability of heavy oil components in these high-porosity and high-permeability areas of the model can explain the main occurrence of the heavy oil zones in the Third Sand and Fourth Sand from the distribution of the components in these layers. However, the presence of these heavy components alone does not explain the mechanisms involved in their precipitation.

Paleo-Accumulation Analysis

The positions of wells with heavy oil zones were compared with the distribution of accumulations in the reservoirs through geologic time on the high-resolution grid of the local model. Present-day heavy oil wells tend to occur mainly at the edges of paleo-accumulations. In the northern part of the field, the complex charge history resulted in many small-scale fill and spill paths caused by the local geometries of the layers and many separate accumulations in the geologic past. Proprietary data of KOC show low amounts of water-soluble components in degraded oils from the field. However, clear geochemical evidence for water washing was not found. The strong water drive in the area of the field (Kaufman et al., 2000) could cause water washing at the paleo-OWC of the accumulations, leading to a relative increase of heavy oil components in these areas caused by the solution of light hydrocarbons in the range of C_4 to C_7, including benzene and toluene. The process is described in Jones and Smith (1965), Bailey et al. (1973), and Palmer (1984). When the filling of the reservoirs continues, these zones could be flushed by new charge and might then be found above the new OWC.

Biodegradation

Biodegradation can result in the removal of lighter hydrocarbons by bacterial consumption (Connan, 1984; Larter et al., 2003; Wilhelms et al., 2004). Biodegradation is the most obvious explanation for the formation of heavy oil in the field because of the contact with the strong aquifer and low reservoir temperatures around 50°C. However, present-day biodegradation cannot be considered as the only major cause of alteration within the oil column because degradation is not only occurring at the OWC but occurs also at certain levels within the reservoir, far from the OOWC. An internal study by KOC on the geochemical properties of two oils from the oil field shows an unresolved complex mixture with nondegraded n-alkanes on top in the whole oil chromatograms, indicative of mixing of oils. In addition, the $\delta^{13}C$ isotope values are lighter (more negative) than could be expected from the observed rank of biodegradation. This observation suggests an additional charge of nondegraded oil in a later stage of the history of the field after biodegraded accumulations were already present.

Repeated uplift of the area of the oil field indicates that the reservoirs were buried deeper than the present day during the geologic past. Calibration of the heat-flow regime shows that the reservoirs of the field were heated to above 70°C. In biodegradation models, the rate of biodegradation is modeled as a function of temperature and charge history of the accumulation. A temperature of 80°C is considered as the upper temperature limit for microbial activity (Wilhelms et al., 2001). The degradation rates start to be affected at temperatures around 60°C. Based on these degradation models, only limited degradation rates during the deeper burial of the reservoirs in the Oligocene and only minor reactivation of microbial activity until the present day can be observed in the simulation. At the time of deepest burial of the accumulation around 39 Ma, the biodegradation rate becomes zero, then slowly rises to values of 7 to 9 kg/m^2 per Ma until 10 Ma, continuing to rise to 36 kg/m^2 per Ma at the present day.

The Cretaceous aquifer in areas surrounding the oil field is deep and hotter than 80°C. The differential burial induces a pressure difference in the aquifer and leads to a water movement toward the dome structure of the field. The heating of the aquifer by passing deeper areas at the present day could explain the prevention of new contamination before the water reaches the field reservoirs.

Gravity Segregation

In addition to the heavy oil wells positioned at the edges of paleo-accumulations, some wells contained heavy oil at the crest of the field, where huge hydrocarbon columns accumulated. Heavy oil in these wells cannot be explained by a paleo-OWC at the edge of an accumulation. The most probable cause for the occurrence of heavy oil in these cases is gravity segregation. Gravity segregation leads to the separation of components having different densities (Schulte, 1980; Hirschberg, 1988). In the highly permeable reservoir layers of the oil

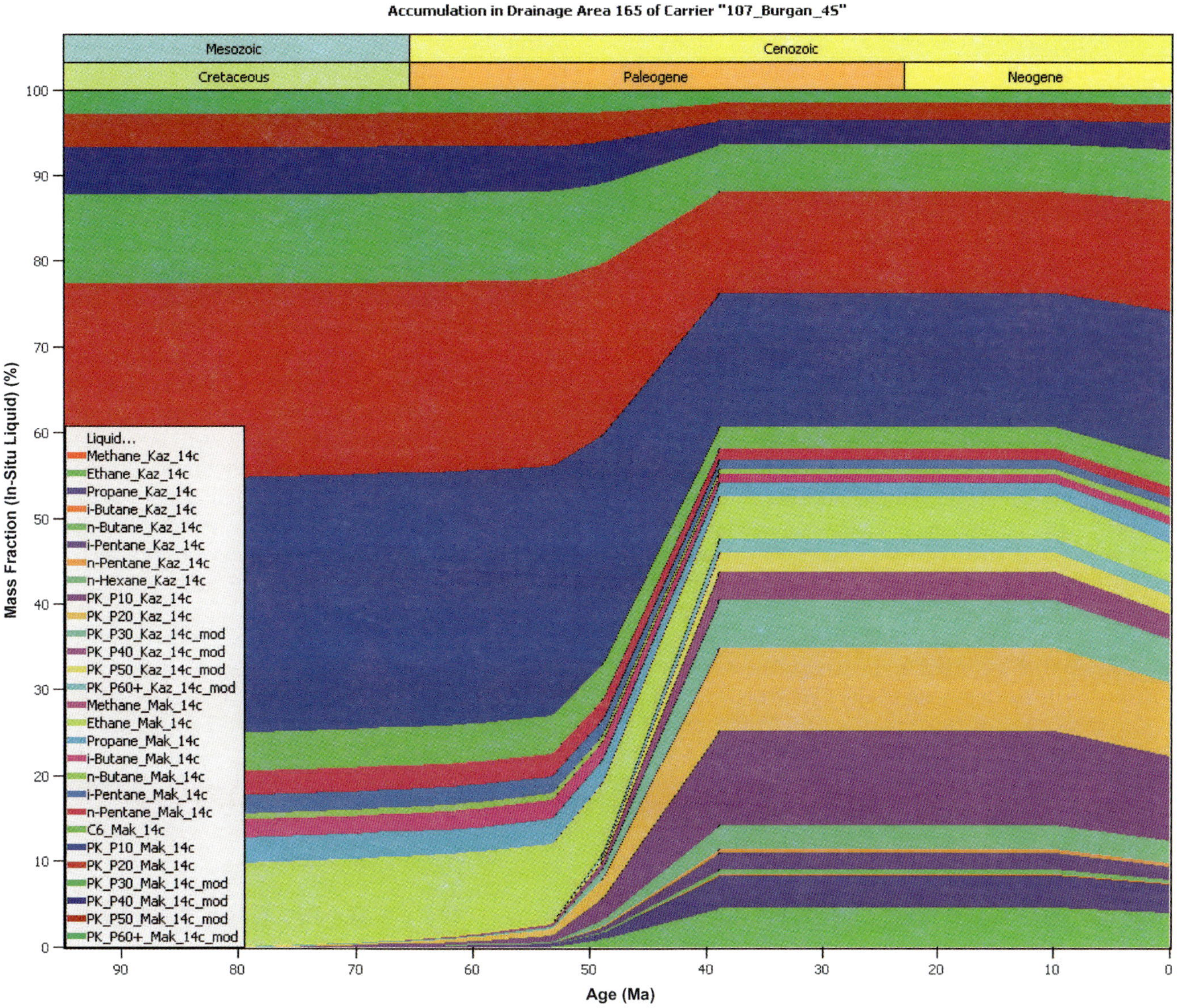

FIGURE 10. Charge history of the Burgan Formation in the oil field. Source rock tracking demonstrates the impact of the second charge from the Cretaceous Kazhdumi source rock beginning around 50 Ma. Liquid components tracked are given in the legend. PK-labeled component groups comprise chain lengths of 10 carbon atoms (C_7–C_{15} in PK-P10, C_{16}–C_{25} in PK-P20 to C_{56+} in PK-P60).

field, the mobility of the compounds is high enough and the length of time during which an oil column existed is sufficient for gravity segregation to occur (Jedaan et al., 2007). The limited communication among the individual reservoir horizons leads to separate systems located on top of each other and prevented equilibration of the hydrocarbon composition throughout the oil column.

DISCUSSION

The distribution of heavy oil components can be described by tracking the C_{56+} hydrocarbons (PK P60) from the kitchen areas described by the regional model to the field and the remigration of the components within the oil field in the local model. This group contains the heaviest components in the model and represents average properties of typical heavy oil components like long-chained alkanes, asphaltenes, and resins. It can be shown that the amount of these components is highest in the Third Sand and Fourth Sand of the Burgan Formation of the field. The reason for this is that the highest porosity and permeability occur in these layers, and the largest liquid accumulations can be found in these reservoirs. In the field, most of the heavy oil identified in the wells occurs in these two highly permeable layers. These complex heavy components are the first products of kerogen conversion and can be found mainly in the early mature oils that are produced. These oils migrate first in the carriers with the highest permeabilities and thus represent the initial charge to the reservoirs. This model can explain the distribution of the components necessary for the formation of heavy oil in the field.

Gravity segregation is the most probable process that accounts for precipitation of a separate heavy oil phase in the Burgan Formation of the field, although water washing may also contribute. This does not include the shallower Tayarat Formation, where clearly biodegraded oils can be found (M. Bhattacharya, 2009, personal communication).

In the simulation of the model, the filling history for the accumulations in the field and the temperature history were used to evaluate the risk of biodegradation in the reservoirs of the Burgan and Wara formations. The potential biodegradation rate of the accumulations was compared using methods described by Larter et al. (2003) and Wilhelms et al. (2004) and further developed to a quantification method by the University of Aachen, Germany. This method was applied in the PetroMod application that was used in the study. The results demonstrate that the heavy oil in the Cretaceous reservoirs of the field cannot be explained by biodegradation alone. Although present-day temperatures and a strong aquifer favor conditions for microbial activity, the burial history of the field limits the biodegradation rates in the model.

A strong water drive can cause the depletion of water-soluble components at the OWC. This leads to a relative increase of the amount of insoluble components, resulting in a higher density of the remaining hydrocarbon mixture. The initial water drive in the oil field is reported by reservoir engineers of KOC as an edge water drive with limited or no bottom component. Many heavy oil zones that have been identified in wells correspond to the edges of paleo-accumulations in the model. These areas were later flushed by additional charge and tilting of the OWC and are now in positions above the OOWC.

In the crest areas, very large oil columns were developed early in the filling history of the field. These areas in the Third Sand and Fourth Sand have high porosities and permeabilities that enabled the separation of components in a connected oil column by gravity. The initial filling of the reservoirs occurred between 90 and 100 Ma and continues until present day, giving separation processes enough time to have a significant effect.

CONCLUSIONS

The most probable reason for the heavy oil zones above the OOWC in the model of the Cretaceous Burgan Formation in Kuwait is gravity segregation in highly permeable reservoir zones with huge oil columns and a long preservation time. In addition, geochemical data and the position of the wells at the edges of the paleo-accumulations suggest that water washing affected these paleo-accumulations. Additional studies are needed to strengthen this latter hypothesis. By using the model-in-model approach in petroleum system modeling, it was possible to combine the charge history of the field, including lateral long-distance migration, with the high-resolution geometries of the field in the local model to investigate each paleo-accumulation in much higher detail than previously possible. The model results were combined with well data that identified heavy oil zones. It is now possible to combine large mega-regional petroleum system models with locally refined models without losing information because of different cell sizes. In the future, this integrated approach in combination with classic reservoir modeling applications will help to better identify reservoir-scale processes that directly affect the production and development of reservoirs.

ACKNOWLEDGMENTS

We thank the Kuwait Oil Company and Schlumberger for permission to publish this article. We also thank K. Peters, K. Weissenburger, and an anonymous reviewer for their helpful comments and suggestions.

REFERENCES CITED

Ali, M. A., 1995, Gotnia salt and its structural implications in Kuwait, *in* M. I. Al-Husseini, ed., Middle East petroleum geosciences: GEO.94, Gulf PetroLink, Bahrain, v. 1, p. 133–143.

Al-Husseini, M. I., 1997, Jurassic sequence stratigraphy of the western and southern Arabic Gulf: GeoArabia, v. 2, no. 4, p. 361–382.

Al-Khamis, A., S. Bhattacharya, M. Al-Hajeri, M. Rao, P. Chenet, N. Bianchi, M. Pernelle, and A. Prinzhofer, 2006, Petroleum systems in onshore Kuwait: From Paleozoic deep targets to Cretaceous (abs.): Geo 2006 Abstracts, Manama, Bahrain, http://www.searchanddiscovery.com/abstracts/pdf/2006/geo/index.htm (accessed May 23, 2011).

Al-Sulaimi, J., and A. Mukhopadhyay, 2000, An overview of the surface and near-surface geology, geomorphology and natural resources of Kuwait: Earth Science Reviews, v. 50, p. 227–267, doi:10.1016/S0012-8252(00)00005-2.

Bailey, N., H. Krouse, C. Evans, and M. Rogers, 1973, Alteration of crude oil by waters and bacteria: Evidence from geochemical and isotope studies: AAPG Bulletin, v. 57, p. 1276–1290.

Beydoun, Z. R., M. W. Huges Clarke, and R. Stoneley,

1992, Petroleum in the Zagros Basin: A Late Tertiary foreland basin overprinted onto the outer edge of a vast hydrocarbon-rich Paleozoic–Mesozoic passive-margin shelf, *in* R. W. Macqueen and D. A. Leckie, eds., Foreland basins and fold belts: AAPG Memoir 55, p. 309–339.

Bordenave, M. L., and R. Burwood, 1990, Source rock distribution and maturation in the Zargos Orogenic Belt: Provenance of the Asmari and Bangestan reservoir oil accumulations: Organic Geochemistry, v. 16, p. 369–387, doi:10.1016/0146-6380(90)90055-5.

Bou-Rabee, F., 1996, The tectonic and depositional history of Kuwait from seismic reflection data: Journal of Petroleum Geology, v. 19, p. 183–198, doi:10.1111/j.1747-5457.1996.tb00424.x.

Boyett, B. A., M. S. El-Mandouh, and R. E. Ewing, 1992, Computational methods in geosciences: Society for Industrial and Applied Mathematics, p. 15–28.

Caillabet, Y., E. Flauraud, and F. J. S. Schneider, 2004, Local grid refinement methods for basin modeling (abs.): AAPG International Conference Abstracts 2004, Cancun, Mexico, http://www.searchanddiscovery.com/documents/abstracts/2004intl_cancun/index.htm#c (accessed May 23, 2011).

Carman, G. J., 1996, Structural elements of onshore Kuwait: GeoArabia, v. 1, no. 2, p. 239–266.

Carpentier, B., H. Arab, E. Pluchery, and J.-M. Chautru, 2007, Tar mats and residual oil distribution in a giant oil field offshore Abu Dhabi: Journal of Petroleum Science and Engineering, v. 58, p. 472–490, doi:10.1016/j.petrol.2006.12.009.

Connan, J., 1984, Biodegradation of crude oils in reservoirs, *in* J. Brooks and D. Welte, eds., Advances in Petroleum Geochemistry: London, Academic Press, v. 1, p. 299–335.

di Primio, R., V. Dieckmann, and N. Mills, 1998, PVT and phase behavior analysis in petroleum exploration: Organic Geochemistry, v. 29, p. 207–222, doi:10.1016/S0146-6380(98)00102-8.

Hantschel, T., and A. Kauerauf, 2009, Fundamentals of basin and petroleum systems modeling: Berlin, Springer, 476 p.

Hantschel, T., A. Kauerauf, and B. Wygrala, 2000, Finite element analysis and ray tracing modeling of petroleum migration: Marine and Petroleum Geology, v. 17, p. 815–820, doi:10.1016/S0264-8172(99)00061-6.

Hirschberg, A., 1988, The role of asphaltenes in compositional grading of a reservoirs fluid column, SPE paper no. 13171: Journal of Petroleum Technology, v. 40, p. 89–94.

International Commission on Stratigraphy, 2009, International Stratigraphic Chart 2009: http://www.stratigraphy.org/upload/ISChart2009.pdf (accessed March 15, 2010).

Jedaan, N. M., A. Al Abdulmalik, D. Dessort, V. L. N. De Groen, C. J. Fraisse, and E. Pluchery, 2007, Characterization, origin and repartition of tar mat in the Bul Hanine field in Qatar: International Petroleum Technology Conference 2007, v. 3, p. 1882–1893.

Jones, T. S., and H. M. Smith, 1965, Relationships of oil composition and stratigraphy in the Permian basin of west Texas and New Mexico, *in* A. Young and J. E. Galley, eds., Fluids in subsurface environments: AAPG Memoir 4, p. 101–224.

Kaufman, R. L., C. S. Kabir, B. Abdul-Rahman, R. Quttainah, H. Dashti, J. M. Pederson, and M. S. Moon, 2000, Characterizing the Greater Burgan field with geochemical and other field data: Society of Petroleum Engineering: SPE Reservoir Evaluation & Engineering, v. 3, p. 118–126.

Kirby, R. H., B. S. Carr, J. Al-Humoud, A. I. Safar, D. Al-Matar, and W. Naser, 1998, Characterization of a vertically compartmentalized reservoir in a supergiant field, Burgan Formation, Greater Burgan field, Kuwait, Part 1: Stratigraphy and water encroachment (abs.), SPE 49214: Society of Petroleum Engineering Annual Technical Conference Abstracts 1998, New Orleans, p. 509–520.

Konyuhov, A. I., and B. Maleki, 2006, The Persian Gulf Basin: Geological history, sedimentary formations, and petroleum potential: Lithology and Mineral Resources, v. 41, p. 344–361, doi:10.1134/S0024490206040055.

Larter, S. R., A. Wilhelms, I. M. Head, M. Koopmans, A. C. Aplin, R. di Primio, C. Zwach, M. Erdmann, and N. Telnaes, 2003, The controls on the composition of biodegraded oils in the deep subsurface-Part 1: Biodegradation rates in petroleum reservoirs: Organic Geochemistry, v. 34, p. 601–613, doi:10.1016/S0146-6380(02)00240-1.

Palmer, S., 1984, Effect of water washing on C_{15+} hydrocarbon fraction of crude oils from northwest Palawan, Philippines: AAPG Bulletin, v. 68, p. 137–149.

Peters, K., C. Walters, and J. Moldowan, 2005, The biomarker guide, 2d ed.: New York, Cambridge University Press, 1155 p.

Schulte, A., 1980, Compositional variations within a hydrocarbon column due to gravity: Society of Petroleum Engineering, American Institute of Mining Metallurgical and Petroleum Engineers, SPE paper no. 9235, SPE Annual Technical Conference and Exhibition, September 21–24, 1980, Dallas, Texas, 10 p.

Strohmenger, C. J., P. E. Patterson, G. Al-Sahlan, T. M. Demko, R. W. Wellner, G. G. McCrimmon, and N. Al-Ajmi, 2006, Sequence stratigraphy and reservoir architecture of the Burgan and Mauddud formations (Lower Cretaceous), Kuwait, *in* P. M. Harris and L. J. Weber, eds., Giant hydrocarbon reservoirs of the world: From rocks to reservoir characterization and modeling: AAPG Memoir 88, p. 213–245.

Waples, D., 1994, Modeling of sedimentary basins and petroleum systems, *in* L. B. Magoon and W. G. Dow, eds., The petroleum system: From source to trap: AAPG Memoir 60, p. 307–322.

Ward, D. S., D. R. Buss, J. W. Mercer, and S. S. Hughes, 1987, Evaluation of a groundwater corrective action at the Chem-Dyne hazardous waste site using a telescopic mesh refinement modeling approach: Water Resources Research, v. 23, p. 603–617, doi:10.1029/WR023i004p00603.

Warsi, W. E. K., 1990, Gravity field of Kuwait and its relevance to major geological structures: AAPG Bulletin, v. 74, p. 1610–1622.

Wilhelms, A., S. R. Larter, I. Head, P. Farrimond, R. di Primio, and C. Zwach, 2001, Biodegradation of oil in uplifted basins prevented by deep-burial sterilization: Nature, v. 411, p. 1034–1037, doi:10.1038/35082535.

Wilhelms, A., M. Erdmann, and S. R. Larter, 2004, Easy Fest versus BDI uncertainties in predrill prediction of biodegradation degree in subsurface petroleum reservoirs: AAPG Annual Meeting 2004, Dallas, Texas, http://www.searchanddiscovery.com/abstracts/html/2004/annual/abstracts/Wilhelms.htm (accessed May 23, 2011).

Yousif, S., and G. Nouman, 1997, Jurassic geology of Kuwait: GeoArabia, v. 2, p. 91–110.

Ziegler, M. A., 2001, Late Permian to Holocene paleofacies evolution of the Arabian Plate and its hydrocarbon occurrences: GeoArabia, v. 6, p. 445–504.

10

Verweij, H. M., M. Souto Carneiro Echternach, N. Witmans, and R. Abdul Fattah, 2012, Reconstruction of basal heat flow, surface temperature, source rock maturity, and hydrocarbon generation in salt-dominated Dutch Basins, *in* K. E. Peters, D. J. Curry, and M. Kacewicz, eds., Basin Modeling: New Horizons in Research and Applications: AAPG Hedberg Series, no. 4, p. 175–195.

Reconstruction of Basal Heat Flow, Surface Temperature, Source Rock Maturity, and Hydrocarbon Generation in Salt-Dominated Dutch Basins

Hanneke M. Verweij[1], Mônica Souto Carneiro Echternach, Nora Witmans, and Rader Abdul Fattah

TNO Built Environment and Geosciences–Geological Survey of the Netherlands, Utrecht, the Netherlands

ABSTRACT

A rapidly growing demand for improved understanding of the Dutch subsurface exists because of the need for alternative energy supplies, such as geothermal energy, as well as for finding and producing more oil and gas in this mature area for petroleum exploration. We use basin modeling to integrate the wealth of new data and information that are increasingly available on the Dutch subsurface. In addition, we develop different approaches to improve the basin modeling results.

Here, we present novel approaches to reconstruct the surface and bottom thermal boundary conditions for basin modeling. The first approach involves assessment of Tertiary sediment-water interface temperatures from information on local and global climate changes that was recently discovered using geobiological and geochemical techniques. The second approach involves multiple one-dimensional probabilistic tectonic heat-flow modeling to calculate the basal heat-flow history and construct paleo–heat-flow maps.

This chapter presents modeling results for the Terschelling Basin and southern part of the Dutch Central Graben that demonstrate the effect of incorporating the tectonic heat-flow boundary condition and detailed knowledge of Tertiary climate changes on source rock maturity and hydrocarbon generation. The simulation results show a marked difference in generated hydrocarbon volumes and a shift in the timing of Tertiary generation compared with simulations using a default surface temperature boundary condition based on paleolatitudes of the research area.

[1]Also at the Politecnico di Torino, Utrecht, the Netherlands.

DOI:10.1306/13311435H43470

In the complex salt-dominated case study area, more detailed knowledge on the timing of oil generation from Jurassic source rocks and gas generation from Carboniferous source rocks is critical to evaluate prospectivity. Recently gathered, analyzed, and mapped data and information for the southern part of the Dutch Central Graben and Terschelling Basin were used as input for a full three-dimensional (3-D) reconstruction of the burial history, including temperature, source rock maturity, and timing of hydrocarbon generation. Simulation results show that the combination of new data, new surface and basal thermal histories, and 3-D basin modeling improved the understanding of the burial, thermal, and maturity history of the area and provided more detail on the timing of hydrocarbon generation from the Jurassic and Carboniferous source rocks. The 3-D modeling results revealed significant lateral variations in maturity and hydrocarbon generation history within each source rock related to their structural positions.

INTRODUCTION

The Netherlands is considered to be a mature area for petroleum exploration. Significant amounts of gas have been discovered and produced during the last 50 yr, and yet there remain large volumes of gas to be found and developed (EBN, 2009). Finding and producing more oil and gas before the aging infrastructure is abandoned is a high priority, but also an increasing focus is on alternative energy supplies, such as geothermal energy. This has promoted a rapidly growing demand for more detailed understanding of the Dutch subsurface. A wealth of data and information has been gathered in ongoing detailed mapping programs of subareas of offshore Netherlands (Duin et al., 2006, Van Gessel et al., 2008). We use basin modeling approaches to integrate these data and to evaluate the interdependencies of the different processes that affect rocks and fluids during geologic history. Special attention is paid to processes and conditions affecting hydrocarbon potential as well as present-day geothermal conditions. Two important boundary conditions that influence basin modeling results are the histories of basal heat flow and sediment-water interface temperature (SWIT). Here, we present recently developed approaches to determine these thermal boundary conditions and show examples of their application.

The study area encompasses the salt-dominated and highly overpressured Terschelling Basin and southern part of the Dutch Central Graben located in offshore Netherlands (Figure 1). A critical factor to evaluate the prospectivity of this area is the timing of hydrocarbon generation from the gas-prone Carboniferous and oil-prone Jurassic source rocks in relation to the timing of fault activity, salt movement, and the development of overpressures. This chapter shows results of three-dimensional (3-D) basin modeling of the burial history and the history of temperature, source rock maturity, and timing of hydrocarbon generation in the study area from 320 Ma to the present day.

GEOLOGIC SETTING

The recently published books and articles by De Mulder et al. (2003), TNO-NITG (2004), Duin et al. (2006), and Wong et al. (2007), and references therein, provide an extensive overview of the regional geologic and petroleum geologic setting of the Netherlands. The following description of the geologic setting is based on these publications and on the more detailed results of recently completed mapping of the studied area, in particular Verweij and Witmans (2009).

The dominant features of the present-day structural framework and lithostratigraphic buildup of the area (Figure 2) reflect its complex geodynamic history. Important phases of this geodynamic history include the Saalian phase of uplift and erosion, Late Permian–Triassic rifting phases, Middle Kimmerian phase of erosion, Late Jurassic–Early Cretaceous rifting and sub-Hercynian inversion phase. The main structural elements in the area are the Dutch Central Graben and the Terschelling Basin, as well as the adjacent Central Offshore Platform, Vlieland High, Ameland Block, and the northwestern extension of the Lauwerszee Trough that reaches the southern border of the studied area (Figure 1). Some of these structures are of Carboniferous origin and regained importance during the Mesozoic and Cenozoic.

The presence of Zechstein evaporites (Figures 1, 2) greatly influenced the post-Permian structural and sedimentary development of the area. The estimated original depositional thickness of the Zechstein Group was approximately 540 m (~1772 ft), whereas its present-day thickness ranges from approximately 5000 m (~16,404 ft) in salt structures to only a few meters in salt withdrawal areas. Approximately 30 identified salt structures follow the structural grain of the area (Figure 1). The timing of salt movement is related to repeated phases of active fault movement (see also Remmelts, 1996). Halokinesis started as early as Early to Middle Triassic. Extensive Late Jurassic halokinesis occurred in the Dutch Central

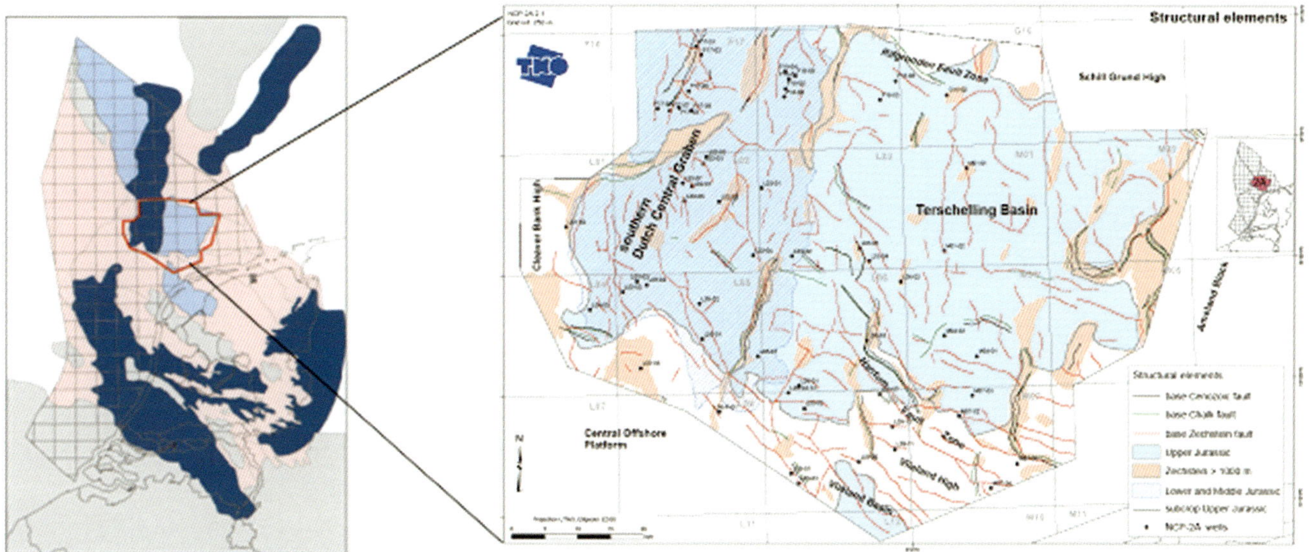

FIGURE 1. The Netherlands onshore and offshore and location of the study area (blue = basins and grabens; pink = platforms; gray = highs). The inset shows the structural elements and salt structures in the study area; the southern part of the Dutch Central Graben and the Terschelling Basin are the main structural elements.

Graben during the Callovian, Oxfordian, and Kimmerian and in the Terschelling Basin during the Portlandian and Ryazian. The salt structures located along the boundary between the graben and the Terschelling Basin and those within the Terschelling Basin were reactivated during the Late Cretaceous–Early Tertiary tectonic phase. The salt structures in the Terschelling Basin remained active during the Tertiary and Quaternary. The Zechstein salt was and still is a factor in preservation of high overpressures in the study area (Verweij et al., 2009a).

The main source rocks for gas are the coal measures of the Carboniferous Limburg Group, which underlie the entire area. These source rocks contain type III kerogen and include the Baarlo, Ruurlo, and Maurits formations. The occurrence of the Maurits Formation is probably restricted to the southwestern part of the area (according to the subcrop map in Verweij and Witmans, 2009). Coal layers in the Central Graben Subgroup (Schieland Group) are considered to be secondary source rocks for gas. The Jurassic Posidonia Shale Formation is the main source rock for oil (De Jager and Geluk, 2007). The distribution area of this type II source rock was greatly reduced by erosion shortly after its deposition during the Jurassic. Erosional remnants occur only in the Dutch Central Graben. Possible additional source rocks occur in the Aalburg and Sleen formations of the Jurassic Altena Group (De Jager and Geluk, 2007).

Proven reservoirs for gas are the Lower Slochteren Member of the Upper Rotliegend Group in the southernmost part of the study area, sandstone units of the Main Buntsandstein Subgroup, such as the Lower Volpriehausen Sandstone Member and Detfurth Member, Solling Fat Sandstone Member of the Upper Germanic Trias Group, and sandstone units of the Schieland Group (Terschelling Sandstone Member). The proven reservoir for oil is the Jurassic Friese Front Formation of the Schieland Group. The lateral continuity of the sandstone reservoir units of the Triassic groups and also those of the Schieland Group were disrupted by the large Zechstein salt structures (apparent in thickness maps published in Verweij and Witmans, 2009).

The cap rock of the Lower Slochteren sandstone reservoir is the overlying Silverpit Formation, whereas the evaporites of the laterally extensive Zechstein Group form the regional top seal. The Triassic reservoirs are capped by Triassic shales. Salt withdrawal zones and/or active faults are thought to have provided or may still provide migration paths for gas of Carboniferous origin, allowing charging of the Triassic sandstone reservoirs. Hydrocarbon migration and accumulation probably occurred in overpressured pre-Zechstein units and Triassic units during part of geologic history (as indicated by present-day pressure conditions; Verweij et al., 2009a). Hence, timing of gas generation from Carboniferous source rocks, especially in relation to timing of salt deformation, fault activity, and overpressure generation and dissipation, is critical for the correct evaluation of the Rotliegend and Triassic plays.

The intra-Jurassic and Lower Cretaceous shales seal the Jurassic oil reservoirs (Figure 2). The present-day highly variable burial depth of the Posidonia source rocks reflects regional variations in burial history. This regional variation in burial history is, among other things, related to salt deformation and Late Cretaceous–Early Tertiary

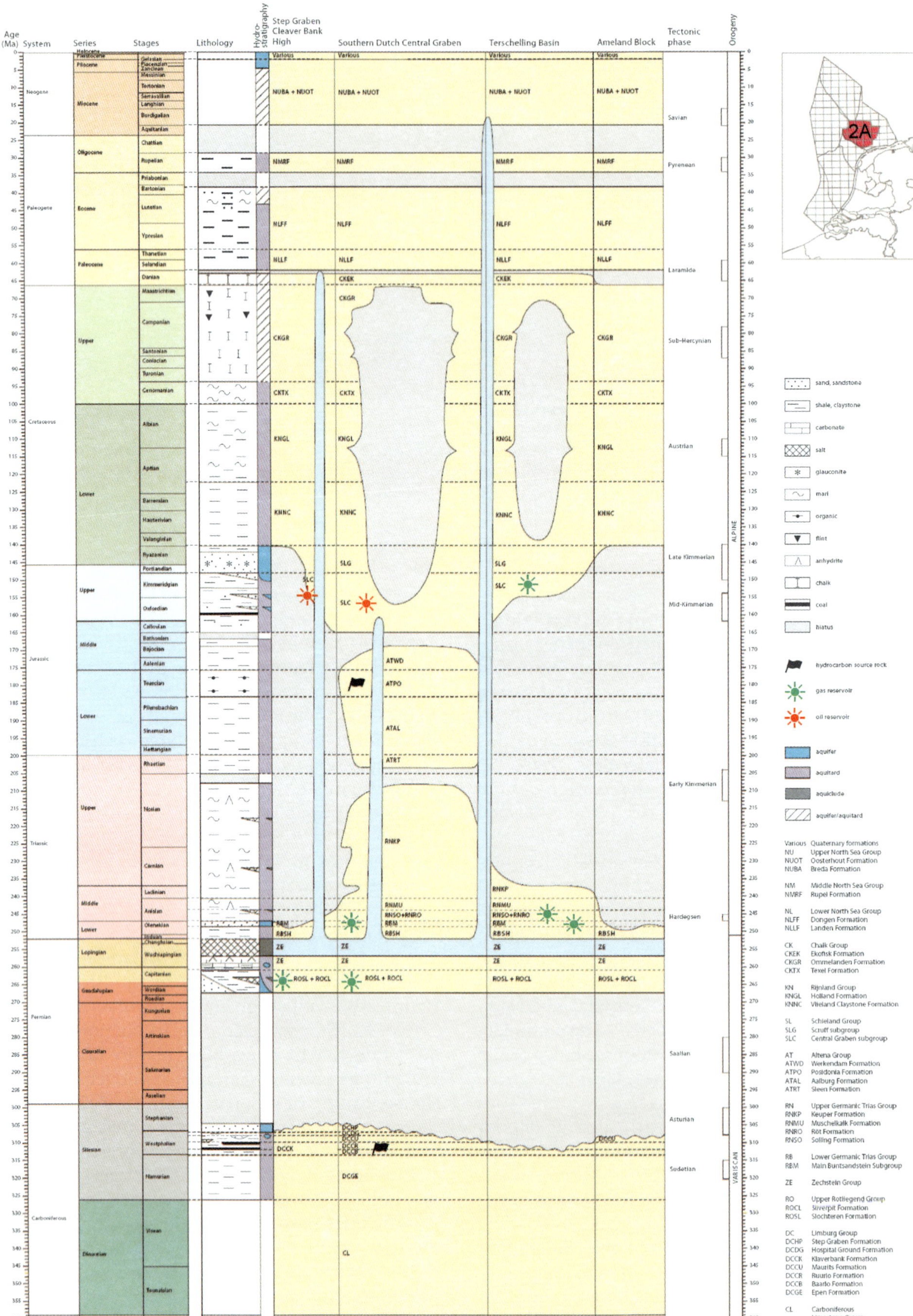

FIGURE 2. Tectonostratigraphic chart of the Dutch Central Graben and Terschelling Basin (from Verweij and Witmans, 2009).

basin inversion. As a consequence, more detailed knowledge on the regional variation in timing of oil generation is critical for useful evaluation of the prospectivity of the Jurassic play.

BASIN MODELING: INPUT AND BOUNDARY CONDITIONS

The basin modeling program PetroMod (version 10) of IES Integrated Exploration Systems was used to study the 3-D burial, temperature, maturity, and hydrocarbon generation history of the case study area. Basic data requirements for the modeling include present-day geometry, lithologic properties, quantified time sequence of events during geologic history, boundary conditions, and calibration data. The results of the detailed mapping of the Terschelling Basin and the southern part of the Dutch Central Graben (Verweij and Witmans, 2009) provided the present-day stratigraphic and structural framework of the sedimentary fill and properties of the rocks and fluids, including lithology and calibration data (temperatures, porosities, permeabilities, pressures, and vitrinite reflectance values) required for the numerical modeling. More detailed information on basin modeling is published in Verweij et al. (2009b).

Geologic Model

The basic 3-D stratigraphic model (depths and thicknesses) consists of 11 main stratigraphic groups from the Upper North Sea Group to Upper Rotliegend Group. This model was expanded below the Upper Rotliegend Group with the Step Graben and Hospital Ground, Maurits, Ruurlo, and Baarlo formations of the Limburg Group. The model was refined with the Posidonia Shale source rock and additional reservoir units (Jurassic Terschelling Sandstone Formation and Triassic Solling Formation, Lower Volpriehausen Sandstone Member and Lower Detfurth Sandstone Member) and Triassic seals (evaporites of the Röt Formation). The refined model includes 29 layers plus basement. Lithologic compositions based on information from Van Adrichem Boogaert and Kouwe (1993-1997) were assigned to each layer. Lithofacies changes were assigned to the bottom layer of the Lower Germanic Trias Group and to the Upper Rotliegend Group (Table 1; Figure 3).

The model includes two source rock intervals: gas-prone coals of the Baarlo, Ruurlo, and Maurits formations (kerogen type III; total organic carbon [TOC] = 2.0, 2.0, and 5.0 wt. %, respectively; hydrogen index [HI] = 250 mg HC/TOC) and the oil-prone Posidonia Shale Formation (kerogen type II; TOC = 9.0 wt. %, HI = 600 mg HC/g TOC). The source rock characteristics of the Posidonia Shale Formation are based on measured values in the southern Dutch Central Graben (Verweij and Witmans, 2009). These values are slightly lower than average values for the Netherlands published by De Jager and Geluk (2007). No measured data were available for the Carboniferous source rocks in the study area. The original TOC and HI values used in the modeling were estimated from measured data from surrounding areas in our in-house database and regional knowledge (Van Balen et al., 2000). In the model, pore water has constant density and is incompressible, solid rock is incompressible, and salt is impermeable.

The geologic history incorporates the three main phases of erosion identified in the area: the Saalian, Middle Kimmerian, and sub-Hercynian phases, and additional periods of nondeposition (Table 1). The 3-D thicknesses of eroded sediments were estimated for each period using stratigraphic information from well-log interpretations and initial 1-D basin modeling at well locations. This resulted in 11 erosion maps (corresponding to the 11 stratigraphic layers affected by erosion and published in Verweij et al., 2009b).

Boundary Conditions

Paleowater Depth

Paleowater depths (PWDs) were allowed to vary in time, but were kept constant over the entire area at a certain time (Figure 4).

Sediment-Water Interface Temperature

Two scenarios were used for the paleo-SWITs; both scenarios are representative of the whole area. The first was calculated using an integrated PetroMod tool that considers the PWD and the evolution of ocean surface temperatures through time, depending on paleolatitude of the area (Figure 4). Later, the Tertiary history of the SWITs was reconstructed in more detail to investigate the effect of recently discovered warming and cooling trends in the Tertiary on source rock maturity and hydrocarbon generation in the studied region.

Cenozoic climate history has been studied in detail worldwide the past few years, including the reconstruction of paleo–sea surface and continental temperatures using geochemical and geobiological techniques. These studies revealed long time trends of global warming (e.g., Paleocene–early Eocene warming) and cooling (e.g., Eocene), as well as geologically brief episodes of globally elevated temperatures superimposed on these long-term trends (Sluijs et al., 2006, Zachos et al., 2008). The most well-known hyperthermal is the Paleocene–Eocene Thermal Maximum, at approximately 55.5 Ma, that was marked by a 5 to 8°C warming with a duration of 170,000 yr (Sluijs et al., 2006; Sluijs and Brinkhuis, 2008). Tertiary climate reconstructions are now becoming

Table 1. Model stratigraphy, 30 layers and assigned lithologic composition, and timing and duration of periods of sedimentation, erosion, and nondeposition.

Layer	*Deposition Age (Ma)*		*Erosion Age (Ma)*		*Lithology*
	From	*To*	*From*	*To*	
1 Upper North Sea Group-top	1.81	0	0	0	75% Sand 25% Shale
2 Upper North Sea Group-bottom	5.33	1.81	0	0	75% Sand 25% Shale
3 Middle North Sea Group	20.43	14.8	0	0	75% Sand 25% Shale
4 Lower North Sea Group	56.6	30.4	0	0	100% Shale
5 Chalk Group-top	80	61.7	0	0	100% Chalk
6 Chalk Group-bottom	99	83.5	83.5	80.5	100% Chalk
7 Rijnland Group	140	99	80.5	80	60% Shale 20% Silt 20% Marl
8 Schieland Group-Scruff subgroup	145	140	0	0	75% Shale 25% Silt
9 Schieland Group-top	146.8	145	0	0	75% Shale 25% Silt
10 Terschelling Sandstone Formation	148	146.8	0	0	100% Sand
11 Schieland Group-bottom	154	148	0	0	75% Shale 25% Silt
12 Posidonia Shale Formation	183	176	173	172	100% Shale
13 Altena Group	203.6	183	172	162	75% Shale 25% Silt
14 Upper Germanic Trias Group-top	241	203.6	162	157	50% Shale 25% Silt 25% Limestone
15 Upper Germanic Trias Group-salt	243	241	157	156	100% Salt
16 Upper Germanic Trias Group-bottom*	245	243	156	154	50% Shale 25% Silt 25% Limestone; 100% Sand
17 Lower Germanic Trias Group-top	247.6	246.2	0	0	75% Shale 25% Silt
18 Lower Detfurth Sandstone Member	247.8	247.6	0	0	100% Sand
19 Lower Germanic Trias Group-middle	248.6	247.8	0	0	75% Shale 25% Silt
20 Lower Volpriehausen Sandstone Member	249	248.6	0	0	100% Sand
21 Lower Germanic Trias Group-bottom	254	249	0	0	75% Shale 25% Silt
22 Zechstein Group	258	254	0	0	100% Salt
23 Upper Rotliegend Group-top	260.85	258	0	0	75% Shale 25% Silt
24 Upper Rotliegend Group-middle*	266.17	260.85	0	0	75% Shale 25% Silt; 25% Sand 25% Silt 25% Salt 25% Lime
25 Upper Rotliegend Group-bottom*	267.5	266.17	0	0	75% Shale 25% Silt; 100% Sand
26 Step Graben/Hospital Ground formations	308	300	300	292	60% Sand 20% Silt 18% Shale 2% Coal
27 Maurits Formation	310	308	292	288	80% Shale 15% Sand 5% Coal
28 Ruurio Formation	312	310	288	280	78% Shale 20% Sand 2% Coal
29 Baario Formation	316.5	312	280	278	48% Shale 25% Silt 25% Sand 2% Coal
30 Basement	320	316.5	0	0	Basement

**Lithofacies maps used: more than one lithology per layer as indicated.*

available for the Netherlands (Donders et al., 2009) and surrounding areas (Mosbrugger et al., 2005). We used the results of these studies in combination with published paleoclimate data from other parts of the world (Sluijs et al., 2006; Pearson et al., 2007; Zachos et al., 2008) to tentatively compile different scenarios for the evolution of SWIT during the Tertiary in the Dutch North Sea area. The reconstructions involved the following main adaptations: the conversion of published deep sea temperatures from Zachos et al. (2008) to sea surface temperatures by increasing the deep sea temperatures 10°C (the correction of 10°C was estimated from the difference in Eocene temperature of deep sea water and sea surface temperature in the tropics published by Pearson et al., 2007, and the latitudinal Eocene temperature gradient published by Greenwood and Wing, 1995); conversion of the published sea surface temperatures for New Jersey (Sluijs et al., 2006) to SWITs for North Sea

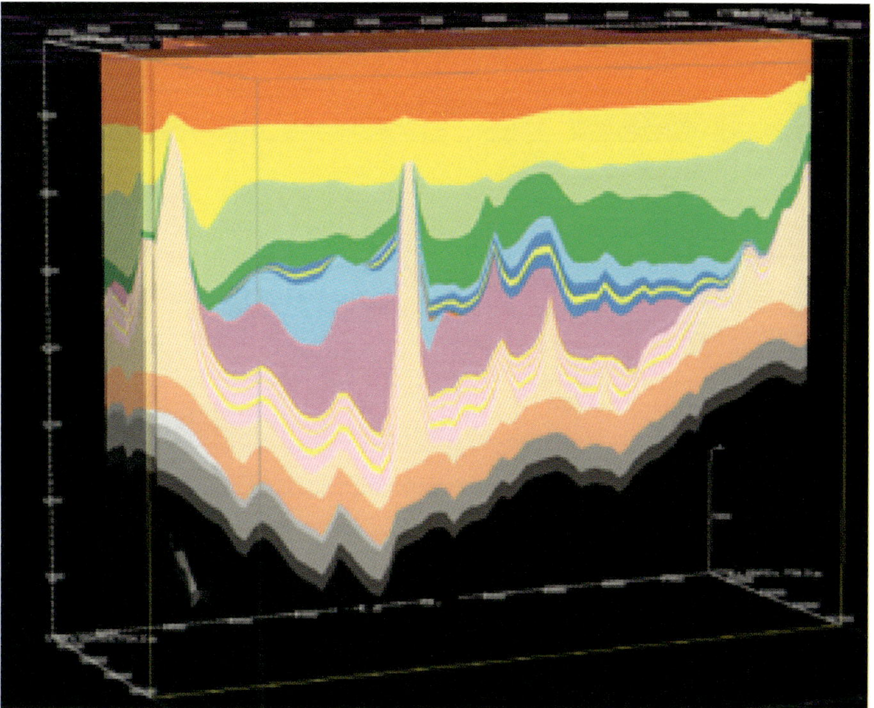

FIGURE 3. Present-day three-dimensional geometric model of the southern Dutch Central Graben (left side of figure) and Terschelling Basin.

latitudes (correction for latitude, -5°C); and corrections for water depths. The reconstructed SWIT trend applied in the study area (Figure 5) clearly deviates from the default SWIT especially in relation to the Paleocene–Eocene thermal maximum and the middle Miocene climate optimum and subsequent rapid cooling.

Basal Heat Flow

The basal heat-flow boundary condition was kept constant at 60 mW/m^2 in the initial modeling scenarios. Later scenarios used a variable basal heat-flow boundary condition (Figure 4) estimated in-house by Abdul Fattah et al. (2008) using probabilistic tectonic heat-flow modeling (Van Wees et al., 2009). The modeling used the reconstructed history of sedimentation, uplift, and erosion at two well locations, representative of the western and eastern parts of the area. The heat-flow history in the western and eastern parts differs because of differences in tectonic history. The reconstructed basal heat-flow histories at the two well locations were used to generate heat-flow maps for the whole area for all event times. A heat-flow map reflects the different basal heat-flow histories in the western and eastern parts of the area, that is, the southern Dutch Central Graben and Terschelling Basin, respectively.

Default Setups and Assumptions

The PetroMod default setups that were used in the simulations concern the lithology and mixed lithology and their associated default properties (thermal conductivity, radiogenic heat production, and heat capacity), default mechanical compaction equations, and default porosity-permeability relations. Measured porosities and permeabilities provided the basis for selecting the proper compaction and porosity-permeability relations. The best fits with measured reservoir porosities were obtained for one-dimensional (1-D) and 3-D burial history simulations assuming hydrostatic conditions. Because the porosity history exerts an important influence on the calculated bulk thermal conductivities and as a consequence on temperature evolution of the basin, we decided to run the 3-D simulations of temperature and maturity assuming hydrostatic conditions. Present-day temperature data and vitrinite reflectance data were used to calibrate the 1-D and 3-D models.

The salt movement tool in PetroMod was used to simulate the movement of Zechstein salt, taking the calculated original thickness of the Zechstein Group as original depositional thickness. This original constant thickness of the Zechstein Group over the whole area was calculated from the present-day volume of the Zechstein Group to be 500 to 600 m (1640–1969 ft). Salt movement was set to start in the Triassic (in accordance with regional observations by Remmelts, 1996).

The simulations of the thermal and maturity history started with a thermal model, applying default PetroMod lithology–related thermal conductivities (Sekiguchi, 1984 model), heat capacities and radiogenic heat production, and a constant basal heat flow of 60 mW/m^2 as bottom boundary condition, zero heat flow across lateral boundaries, a transient heat flow through the sediments, and the default history of SWITs. A second scenario for the

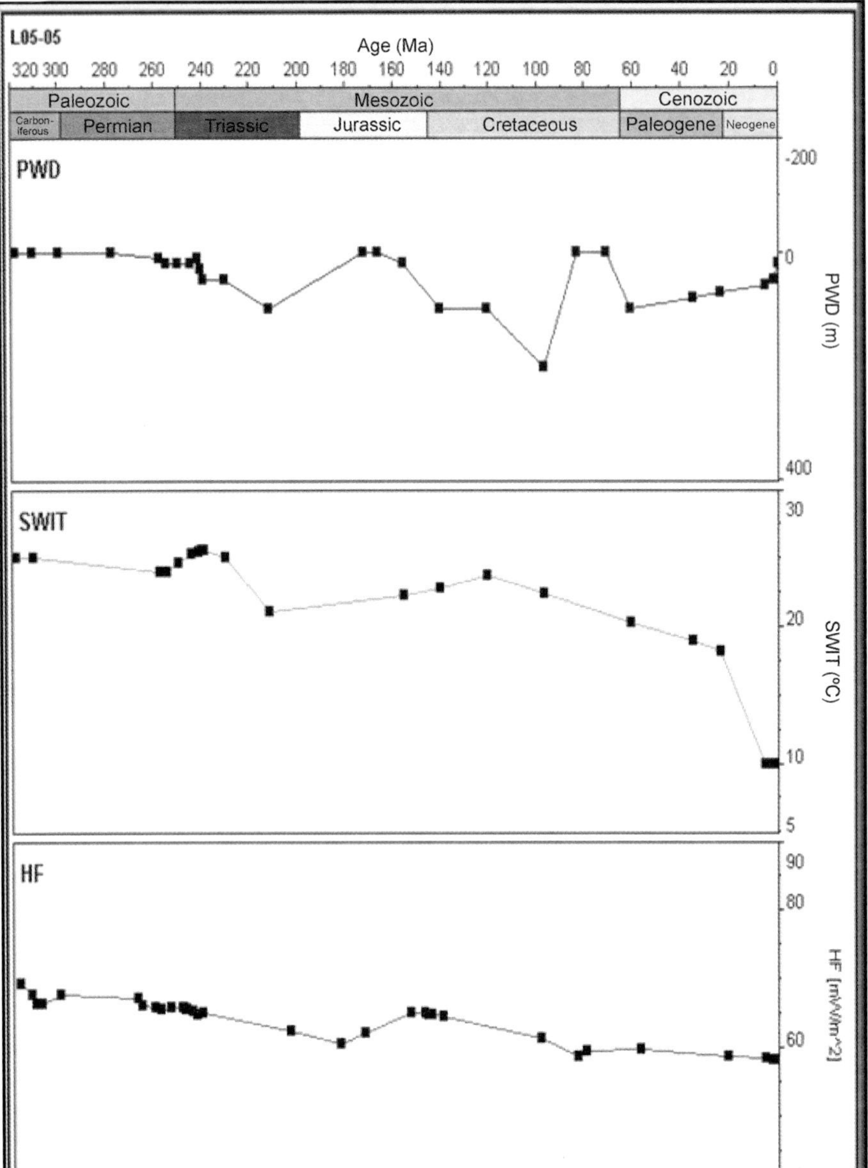

FIGURE 4. Boundary conditions for three-dimensional basin modeling: history of water depth (paleowater depth = PWD), sediment-water interface temperature (SWIT), and basal heat flow (HF). The PWD and SWIT are representative for the whole area; HF shown in this figure is representative of the western part of the area (Dutch Central Graben).

thermal modeling incorporated the results of a tectonic reconstruction of the basal heat-flow boundary condition.

The maturity modeling is based on the Sweeney and Burnham (1990) kinetic model. The calculations of the transformation ratios are based on the Burnham (1989) T2 kinetic model for the Posidonia Shale Formation and the Burnham (1989) T3 kinetic model for the Carboniferous source rocks.

RESULTS

Burial History

The similarities and differences in the burial history of the Dutch Central Graben and the Terschelling Basin are illustrated by the simulated burial histories at well locations F17-05 and M01-02, respectively (Figures 6, 7). Three phases of rapid subsidence and sedimentation occur in the Dutch Central Graben and the Terschelling Basin as well as in the adjacent platform and highs during the Late Carboniferous, Late Permian–Early Triassic (~270–245 Ma), and Pliocene–Quaternary. These burial histories are similar to those described for other Dutch subbasins (De Jager, 2003, 2007). Figures 6 and 7 show that in the Dutch Central Graben and the Terschelling Basin, Saalian uplift and erosion are significant and the Carboniferous and Posidonia source rocks are at maximum burial depth at the present day. The main differences in burial history between the graben and the basin occur during Jurassic–Cretaceous.

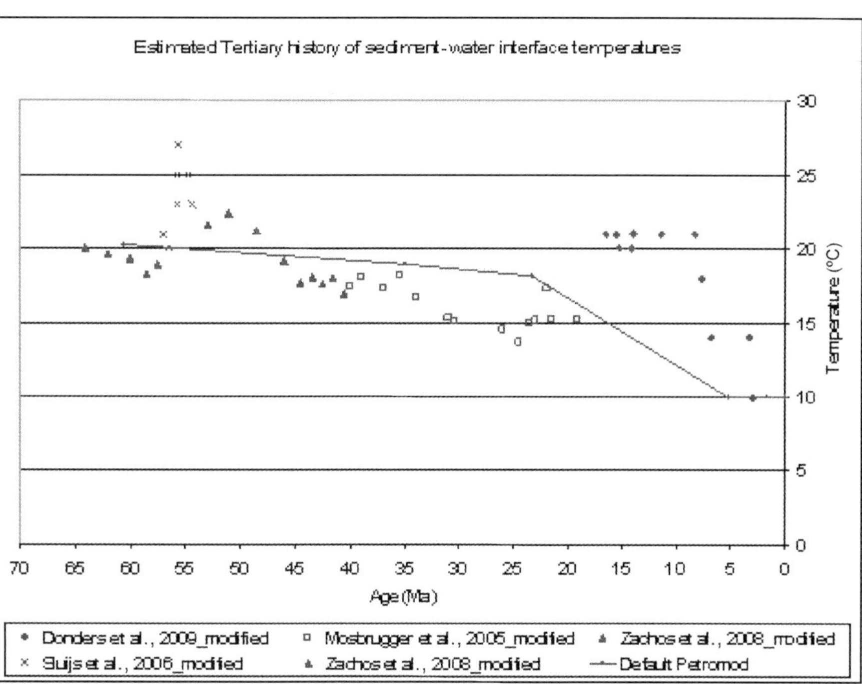

FIGURE 5. Crossplot showing a scenario for the Tertiary history of sediment-water interface temperatures (SWITs) in the Dutch North Sea based on recently discovered warming and cooling trends compared with paleotemperatures calculated by PetroMod Version 10 (IES Integrated Exploration Systems) from the paleolatitude of the study area and paleowater depth (= default PetroMod). Donders et al. (2009): mean annual surface temperatures northwest Europe (including southeast Netherlands); corrected for paleowater depth; Mosbrugger et al. (2005): mean annual surface temperatures northwest Europe; corrected for paleowater depth; Zachos et al. (2008): deep sea temperatures; converted to sea surface temperatures by increasing the deep sea temperatures by 10°C; Sluijs et al. (2006): sea surface temperatures for New Jersey; converted to SWITs for North Sea latitudes (correction for latitude: -5°C; correction for water depth; -2°C).

Tectonic Subsidence

The tectonic subsidence was reconstructed by Abdul Fattah et al. (2008) at six well locations using probabilistic tectonic heat-flow modeling (Van Wees et al., 2009): F17-03, L02-05, and L05-03 in the Dutch Central Graben; M04-03 and L03-03 in the Terschelling Basin; and L09-10 on the Vlieland High. The reconstruction was based on the same time sequence of events, lithologies, and PWD and SWITs used for the PetroMod simulations, whereas important lithology-dependent compaction and thermal parameters were made consistent as much as possible with the default parameter values incorporated in PetroMod. Tectonic subsidence is calculated from the decompacted burial history at the wells. This involves removing the impact of decompacted sediments, isostatic balancing, and the paleowater depth from the subsidence curve of the well. The resulting tectonic subsidence is called the observed tectonic subsidence because it is obtained from the observed thicknesses and lithology parameters for the formations. The observed tectonic subsidence history (Figure 8) reveals two phases of rapid subsidence at all well locations, namely, during the Late Carboniferous and Late Permian–Early Triassic (~270–245 Ma). In addition, Figure 8 shows a Late Jurassic–Early Cretaceous phase of rapid tectonic subsidence in the Terschelling Basin that was also apparent in the burial history of the basin (Figure 7). The main regional difference in tectonic subsidence history is the amount of uplift at approximately 160 Ma between the wells F17-03, L02-05, and L05-03 located in the Dutch Central Graben and the wells M04-03 and L03-03 located in the Terschelling Basin and on the Vlieland High (L09-10). Abdul Fattah et al. (2008) performed forward modeling of the tectonic subsidence at two well locations (L05-03 and L09-10) representing the two different tectonic histories. A uniform stretching model was used for the forward tectonic modeling, except during the Jurassic uplift phase, which was defined as an underplating event. Figures 9 and 10 show the best fit tectonic subsidence and the associated crustal stretching factors (d). The crustal stretching factors for the Saalian and the Late Cretaceous uplift phases are less than 1, indicating that the crust undergoes shortening as a result of tectonic inversion. The crustal stretching factor for the Middle–Late Jurassic uplift phase $d = 1$ (and a subcrustal stretching $b > 1$) reflects the fact that uplift was set to be caused by thermal uplift and mantle upwelling (underplating) rather than tectonic inversion (Abdul Fattah et al., 2008).

Thermal History

Basal Heat-Flow History

The history of basal heat flow was reconstructed by Abdul Fattah et al. (2008) for L05-03 and L09-10 wells, representative of the western area (Dutch Central Graben) and the eastern area (Terschelling Basin and Vlieland High), respectively. The tectonic model used to predict the basal heat flow (described by Van Wees et al., 2009) incorporates the effects of sediment infill and changes

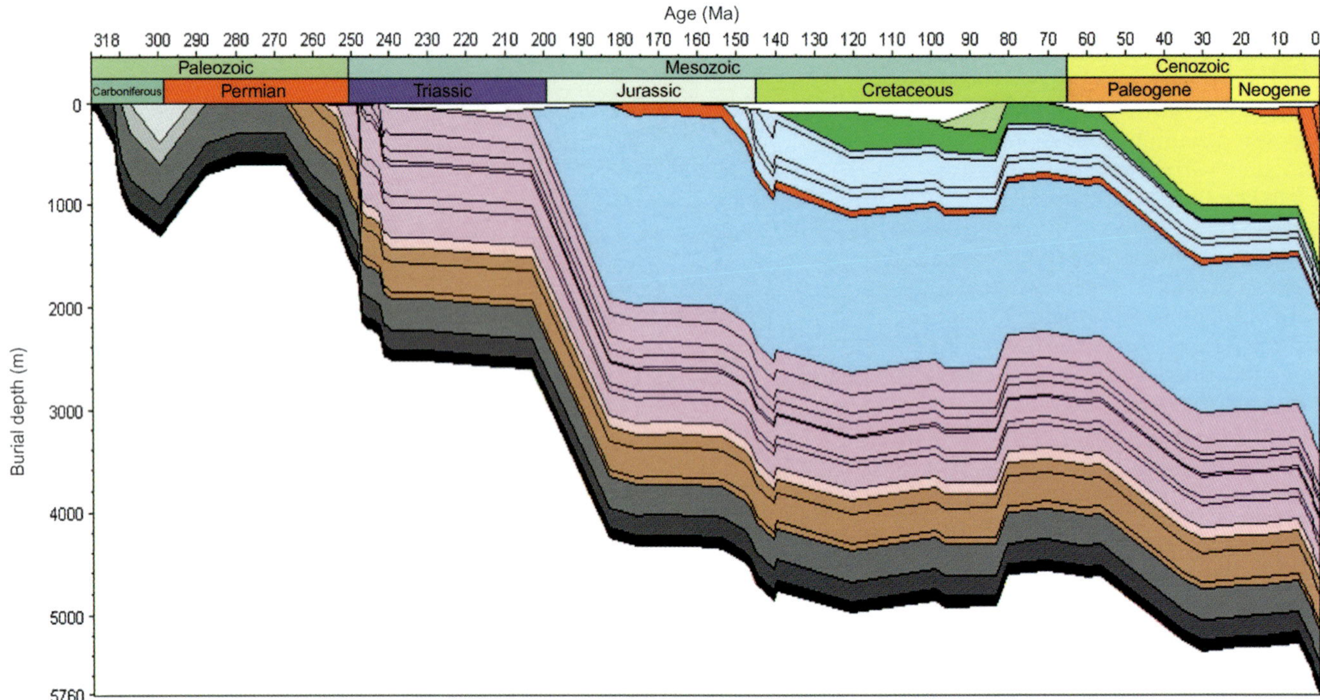

FIGURE 6. One-dimensional burial history at well F17-05 in the Dutch Central Graben. Note the presence of the complete Altena Group (dark blue) and the Posidonia Shale Formation (red) and the absence of the Chalk Group (light green) because of Late Cretaceous uplift and erosion and absence of Carboniferous Maurits Formation (light gray) because of Saalian erosion (movement of Zechstein salt is not included here).

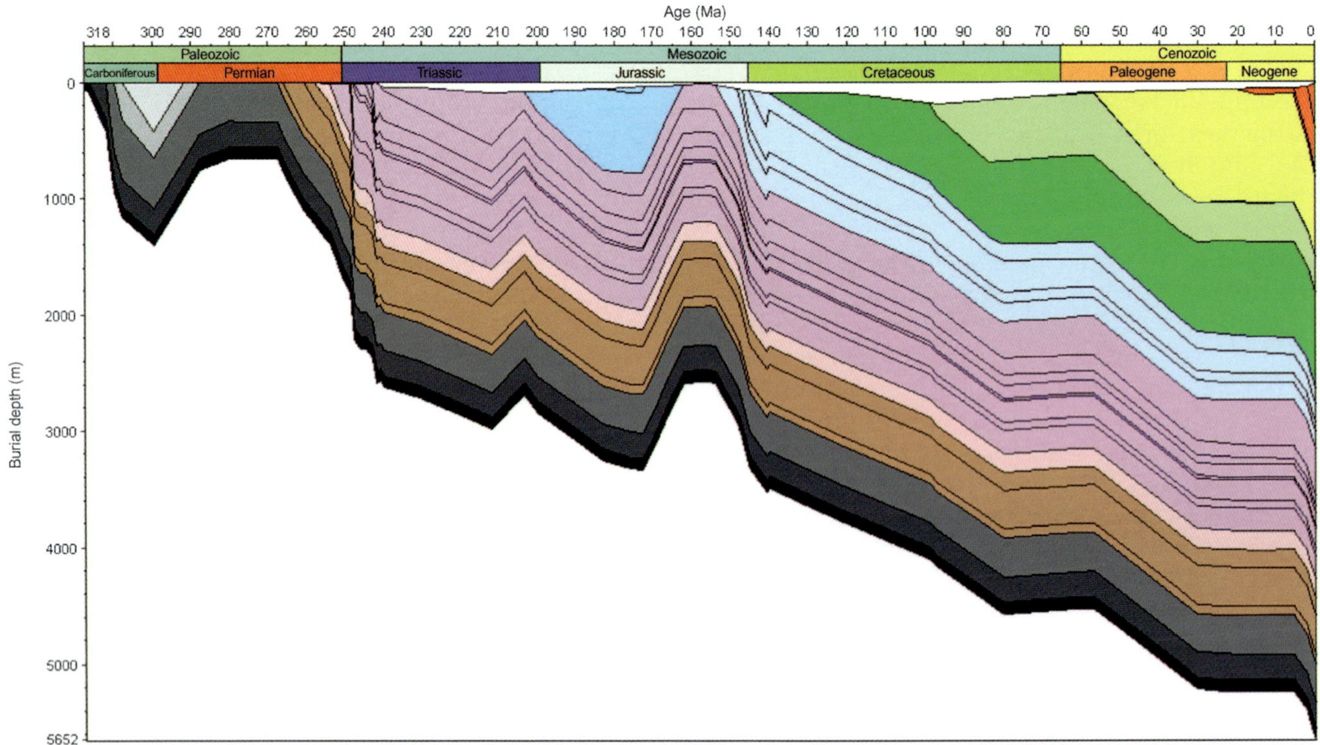

FIGURE 7. One-dimensional burial history at well M01-02 in the Terschelling Basin. Note the complete erosion of the Altena Group (dark blue) during Jurassic uplift and erosion and the absence of the Posidonia Shale Formation (movement of Zechstein salt is not included here).

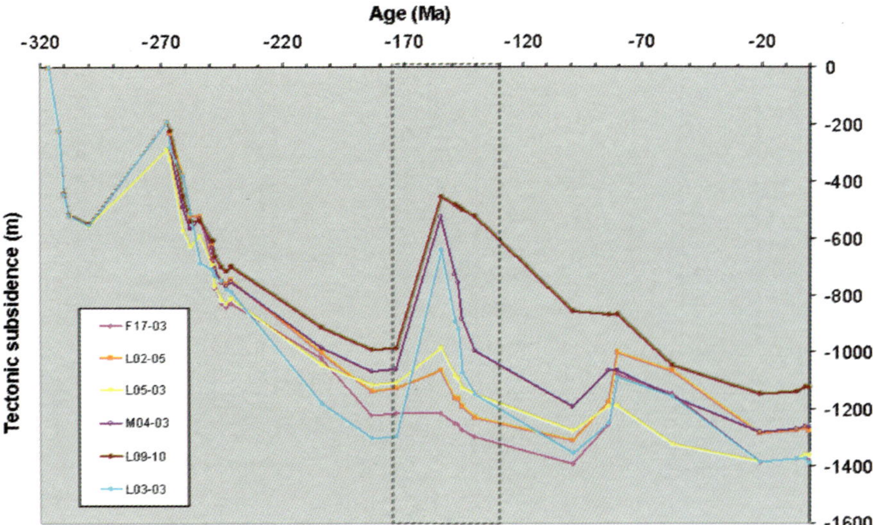

Figure 8. Tectonic subsidence at six well locations. Note the difference in amount of uplift around 160 Ma between the wells F17-03, L02-05, L05-03 located in the Dutch Central Graben and the wells M04-03, L03-03, L09-10 located in the Terschelling Basin and on the Vlieland High (L09-10) (modified from Abdul Fattah et al., 2008).

in heat production in the crust, in contrast to classical McKenzie models (McKenzie, 1978). The reconstructed tectonic subsidence and associated heat flow are shown in Figures 9 and 10.

Modeling results indicate that present-day basal heat flow at both wells is at its minimum, reaching values of 58 to 60 mW/m^2 with the slightly higher values in the eastern part of the area. The Middle–Late Jurassic uplift is associated with a peak in basal heat flow reaching 77 mW/m^2 in L09-10 and 65 mW/m^2 in L05-03. The Late Carboniferous and Early Triassic extensional tectonic phases (Figures 9, 10) result in a limited increase in basal heat flow. During these phases, the cooling effect of sediment deposition and the decrease in radiogenic heat production in the thinning crust produce considerably lower heat flow than expected from models neglecting these aspects. The Middle–Late Jurassic uplift phase was modeled as an underplating event. Underplating is associated with the active introduction of heat from the mantle and as a result significantly increases heat flow (basal heat-flow peaks in Figures 9, 10). The variable basal heat-flow boundary condition reconstructed from tectonic forward modeling produces a generally higher heat flow during history in comparison with the constant boundary condition of 60 mW/m^2.

Evaluation of the influence of these two different boundary conditions on the 1-D and 3-D simulation of temperature and maturity history (using the default SWIT boundary condition) indicated that no to only very slight differences in simulated present-day temperature and vitrinite reflectance values exist and that the simulation results agree with measured temperature and vitrinite reflectance data. The incorporation of time-dependent changes of basal heat flow into the boundary conditions resulted in somewhat higher calculated paleotemperatures and paleomaturities (vitrinite reflectance values), mostly in the deeper parts of the basin, and especially in the Terschelling Basin and on the Vlieland High during the Jurassic periods of underplating-related increased basal heat flow. The main differences in simulated

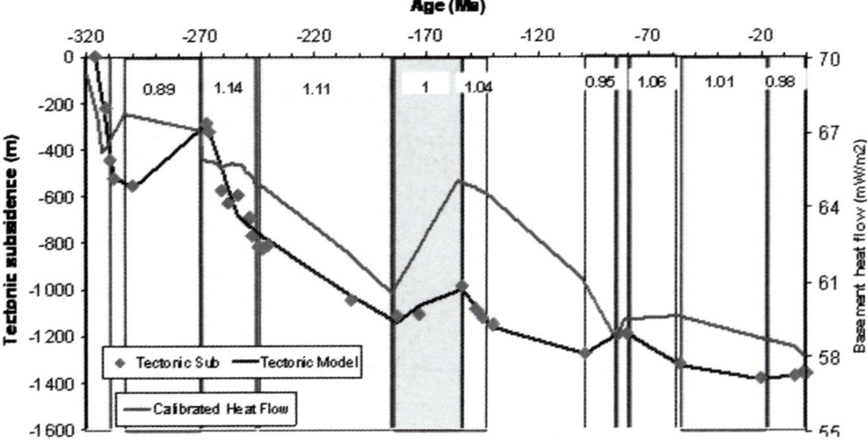

Figure 9. Modeled tectonic subsidence with associated crustal stretching factors and observed tectonic subsidence at well L05-03 (Dutch Central Graben) and associated calibrated basal heat flow (from Abdul Fattah et al., 2008).

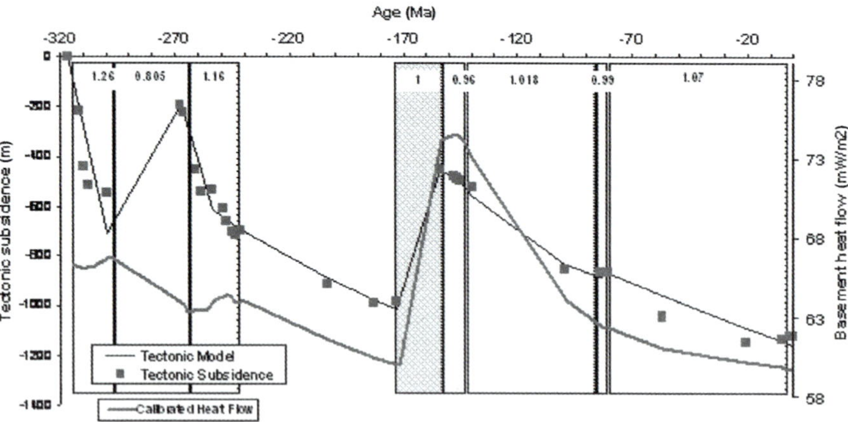

FIGURE 10. Modeled tectonic subsidence with associated crustal stretching factors and observed tectonic subsidence at well L09-10 (Terschelling Basin/Vlieland High) and associated calibrated heat flow (from Abdul Fattah et al., 2008).

temperature occur from the Middle Jurassic to the end of the Cretaceous, and the differences in maturity appear with a time delay from the Early Cretaceous to the early Paleogene (Figure 11). Temperatures and maturities reach higher values for the variable basal heat-flow boundary condition. However, these differences disappear again during the Paleogene. The results suggest that the Middle–Late Jurassic peak in basal heat-flow influences the timing of maturation, but not the maximum maturity level of the source rock reached during geologic history.

Surface Temperature History

Figure 5 shows the selected trend of SWITs based on recently discovered warming and cooling trends in the Tertiary. This more detailed SWIT boundary condition was used in 1-D to 3-D simulations of source rock maturation and hydrocarbon generation in different structural settings in the Netherlands, including the Terschelling Basin and Dutch Central Graben, to investigate the effect on the prospectivity in the Dutch part of the North Sea. Figures 12 and 13 show preliminary results

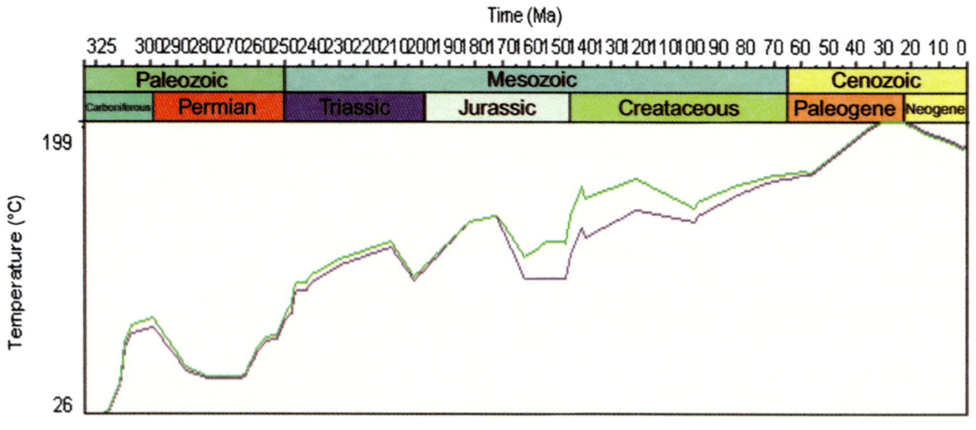

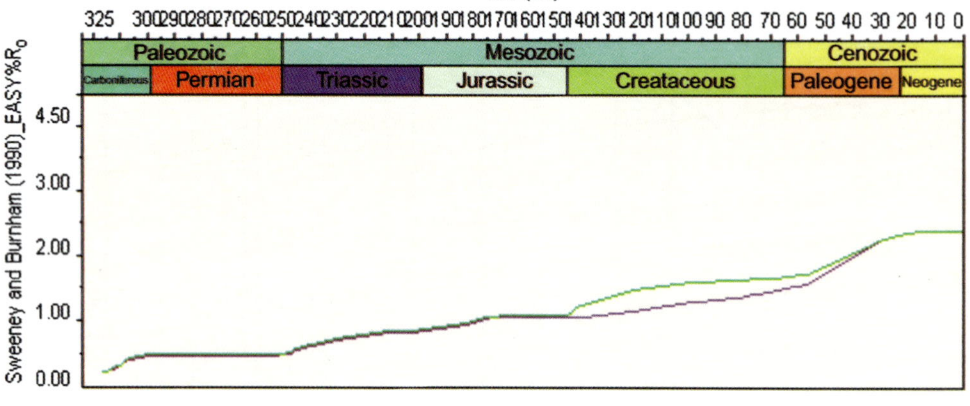

FIGURE 11. Simulated temperature and maturity history for deepest layer in well L09-10 for constant heat-flow boundary condition of 60 mW/m^2 (purple line) and for variable heat-flow boundary condition (green line). The level of maturity is indicated by the simulated vitrinite reflectance %R_o, using the Sweeney and Burnham (1990) kinetic model.

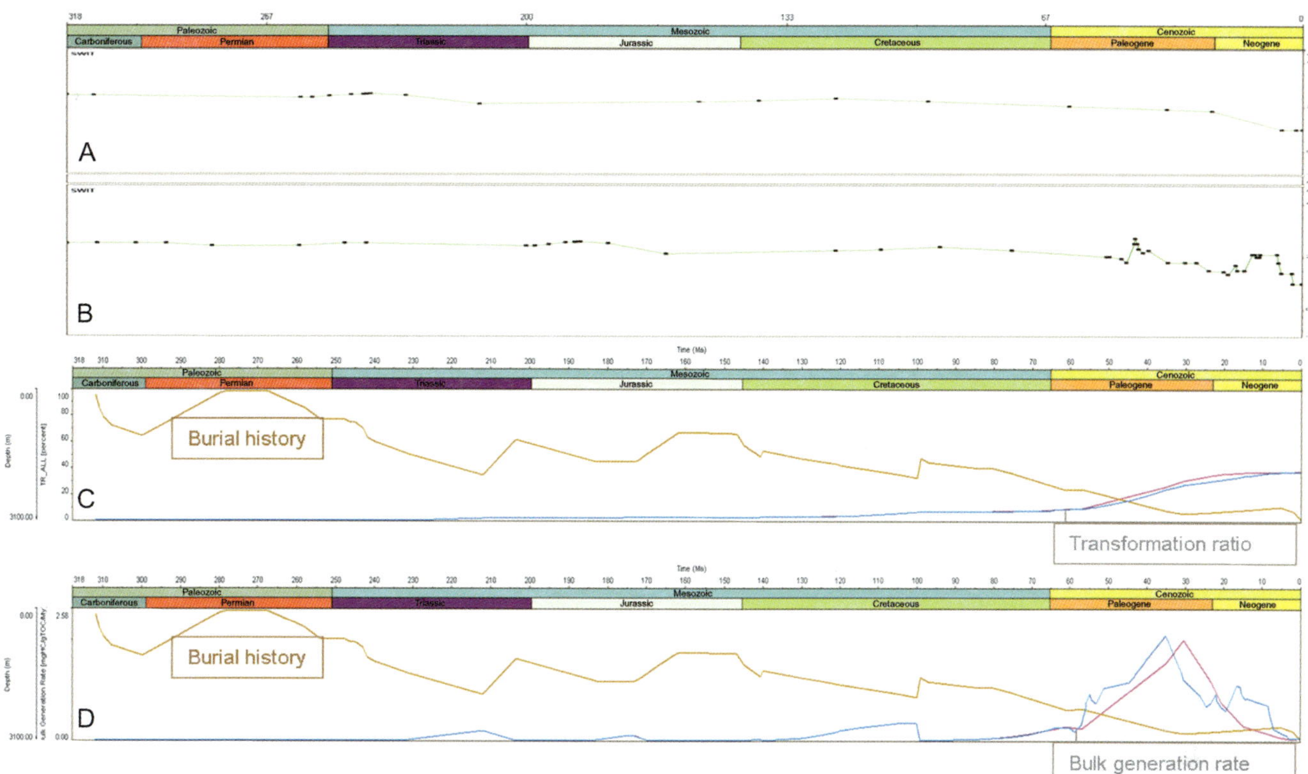

FIGURE 12. Results of three-dimensional basin modeling of the Terschelling Basin and Dutch Central Graben showing effect of applying different Tertiary sediment-water interface temperature (SWIT) boundary conditions; one-dimensional extraction at well M07-04. (A) Default PetroMod SWIT boundary condition; (B) User-defined SWIT boundary condition; (C, D) Burial history of the gas-prone Carboniferous source rock (Baarlo Formation) and comparison of the influence of different SWIT boundary conditions on timing and magnitude of the transformation ratio (C) and bulk generation rate (D) of the source rock (blue relates to user-defined SWIT).

for the study area. The simulated maturities for the Carboniferous (Figure 12) and Posidonia Shale (Figure 13) source rocks are both somewhat lower in Paleogene and the beginning of the Neogene, whereas the calculated maturity at the present day is approximately the same for both SWIT boundary conditions. A clear shift in timing of hydrocarbon generation toward the Neogene exists, especially in the Posidonia Shale source rock (Figures 12D, 13D). A decrease in SWIT in Late Paleogene and the increased SWITs during the Middle Miocene coincide with a period of only minor changes in burial depth of the source rocks. As a consequence, the changes in SWIT are clearly reflected in changes of the source rock temperatures and also in the hydrocarbon generation rates: Figures 12D and 13D show that the hydrocarbon generation in the Carboniferous and the Posidonia Shale source rocks increases in response to the Middle Miocene climate optimum.

Temperature, Maturity, and Hydrocarbon Generation History of Source Rocks

The results of the 3-D thermal and maturity simulations of the source rocks, presented later, are based on the tectonic heat-flow boundary conditions and on default PetroMod SWIT conditions. For steady-state PetroMod simulations, the regional variations in temperature in a stratigraphic unit at a given time mainly result from regional variations in depth of burial, basal heat flow, and bulk thermal conductivity of the sedimentary sequence. Here, we use transient simulations. These also incorporate the effects of paleoboundary conditions (e.g., paleosurface temperatures) and effects of rapid sedimentation or uplift on the temperature distribution at a given time. The lateral variations in thermal and maturity history of a source rock in this salt-dominated area may also depend on its position relative to salt structures, which have a high thermal conductivity.

Temperature, Maturity, and Hydrocarbon Generation History Posidonia Shale Formation

The present-day distribution of the Posidonia Shale Formation is restricted to the Dutch Central Graben. The time-dependent basal heat-flow input for the 3-D PetroMod modeling is constant for the whole Dutch Central Graben. As a consequence, regional variations

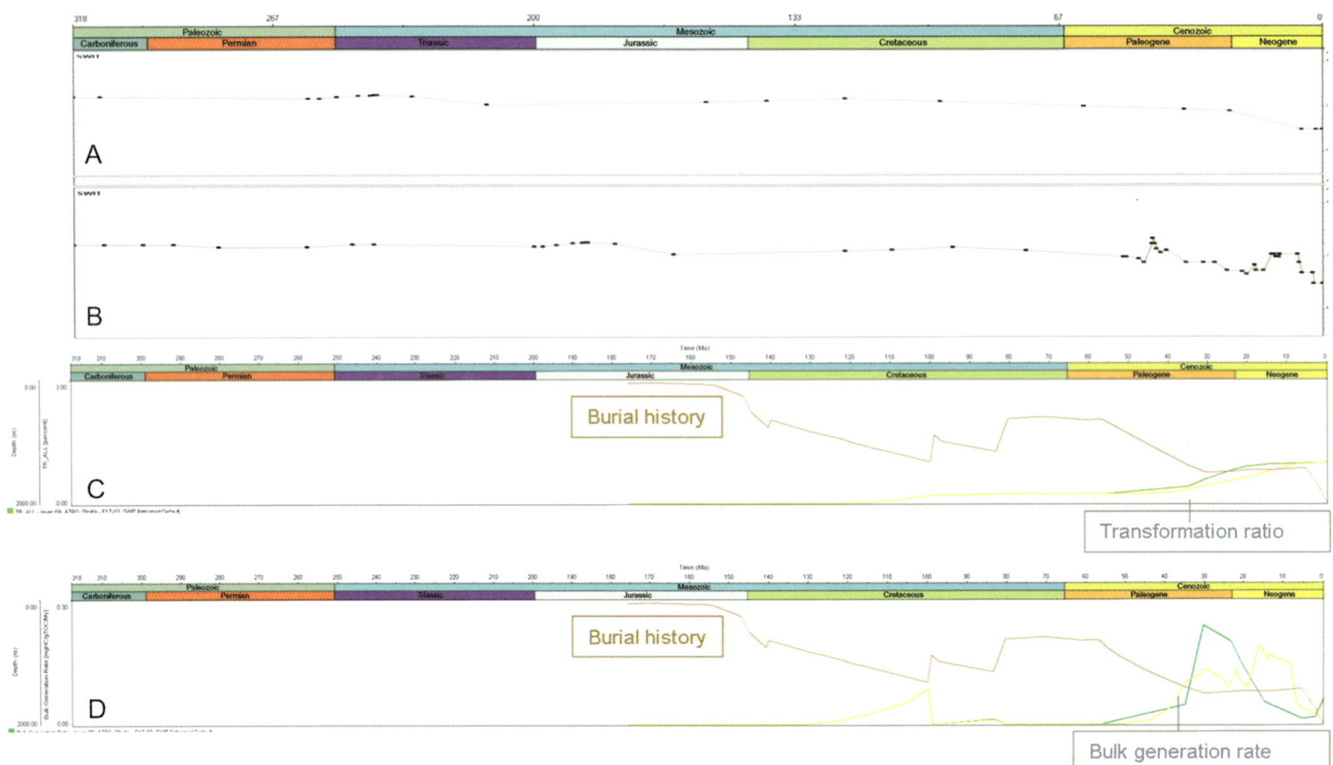

FIGURE 13. Results of three-dimensional basin modeling of the Terschelling Basin and Dutch Central Graben showing effect of applying different Tertiary SWIT boundary conditions; one-dimensional extraction at well F17-03. (A) Default PetroMod SWIT boundary condition; (B) User-defined SWIT boundary condition; (C, D) Burial history of the oil-prone Jurassic source rock (Posidonia Shale Formation) and comparison of the influence of different SWIT boundary conditions on timing and magnitude of the transformation ratio (C) and bulk generation rate (D) of the source rock (yellow relates to user-defined SWIT).

in calculated temperature history of the Posidonia Shale result mainly from differences in burial history and more locally from differences in position relative to salt structures.

Figure 14 shows the simulation results for three 1-D extractions of the 3-D basin modeling (Figure 15). The burial history of the Posidonia Shale Formation in the Central Graben shows large differences (Figure 14A) resulting in very different temperature and maturity histories depending on structural position. During the Latest Paleogene and subsequent Neogene, the temperatures in the source rock also decrease during increasing burial (Figure 14A, B). This is probably caused by the sharply decreasing SWITs in combination with the relatively low basal heat flow. Only in the recent past do temperatures in the source rock start to increase again. Present-day temperatures in the Posidonia Shale Formation are less than previous values.

The transformation ratio is the ratio of generated petroleum to the original petroleum potential of a source rock. The ratio indicates that in the southwestern part of block F17 (location 1), the Posidonia Shale Formation already started generating hydrocarbons during the Cretaceous, with generation rates reaching maximum values just before Late Cretaceous uplift (Figure 14C, D). The generation is resumed during Paleogene and practically stops at the end of the Paleogene. At location 2 (block L02), the maturity of the Posidonia increases gradually and the source rock does not generate hydrocarbons until the Paleogene and continues into the Neogene. In the inverted center of the Dutch Central Graben (location 1), the Posidonia did not reach a mature state for hydrocarbon generation (<0.55% R_o; Figure 14D).

The migration, charging, and preservation conditions of the hydrocarbons generated before the Late Cretaceous inversion were influenced to a greater or lesser extent by the inversion movements and the accompanying erosion of the inverted central parts of the graben. This is in contrast with the hydrocarbons generated in the Tertiary, that is, after the inversion movements. The 3-D simulations provide a detailed picture of the regional variations of present-day maturities in the Posidonia Shale Formation and the timing of hydrocarbon generation and, as such, represent a firm basis to further evaluate migration and petroleum charge (not treated here).

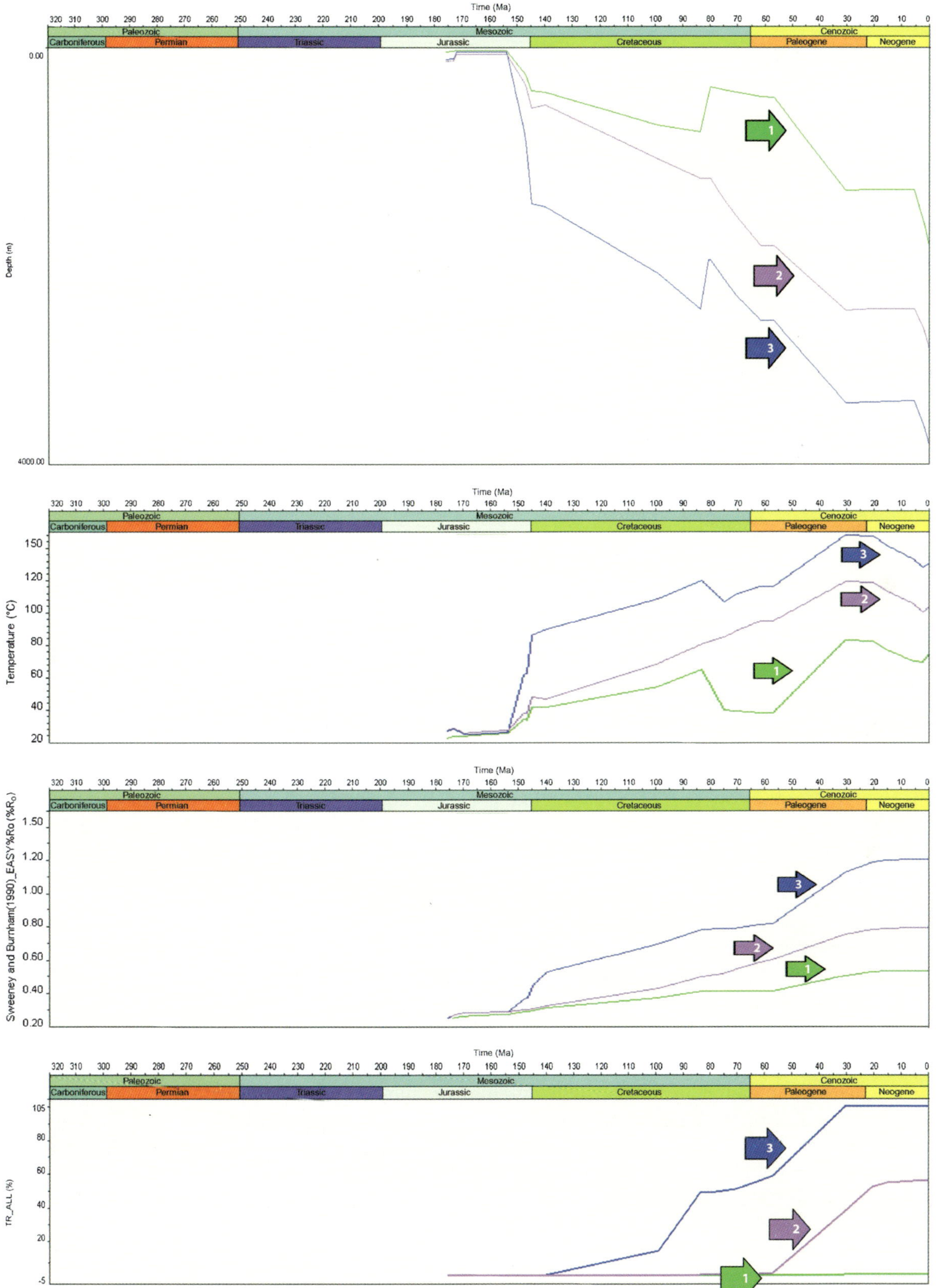

FIGURE 14. One-dimensional (1-D) extractions from three-dimensional simulated history of burial (A), temperature (B), maturity (C), and transformation ratios (D) of the Jurassic source rock (Posidonia Shale Formation) at the three selected locations (1) block F17 (inverted graben center), (2) block L02, and (3) southwestern part of block 17 (Figure 15 shows locations of 1-D extractions). The level of maturity (C) is indicated by the simulated vitrinite reflectance $\%R_o$, using the Sweeney and Burnham (1990) kinetic model.

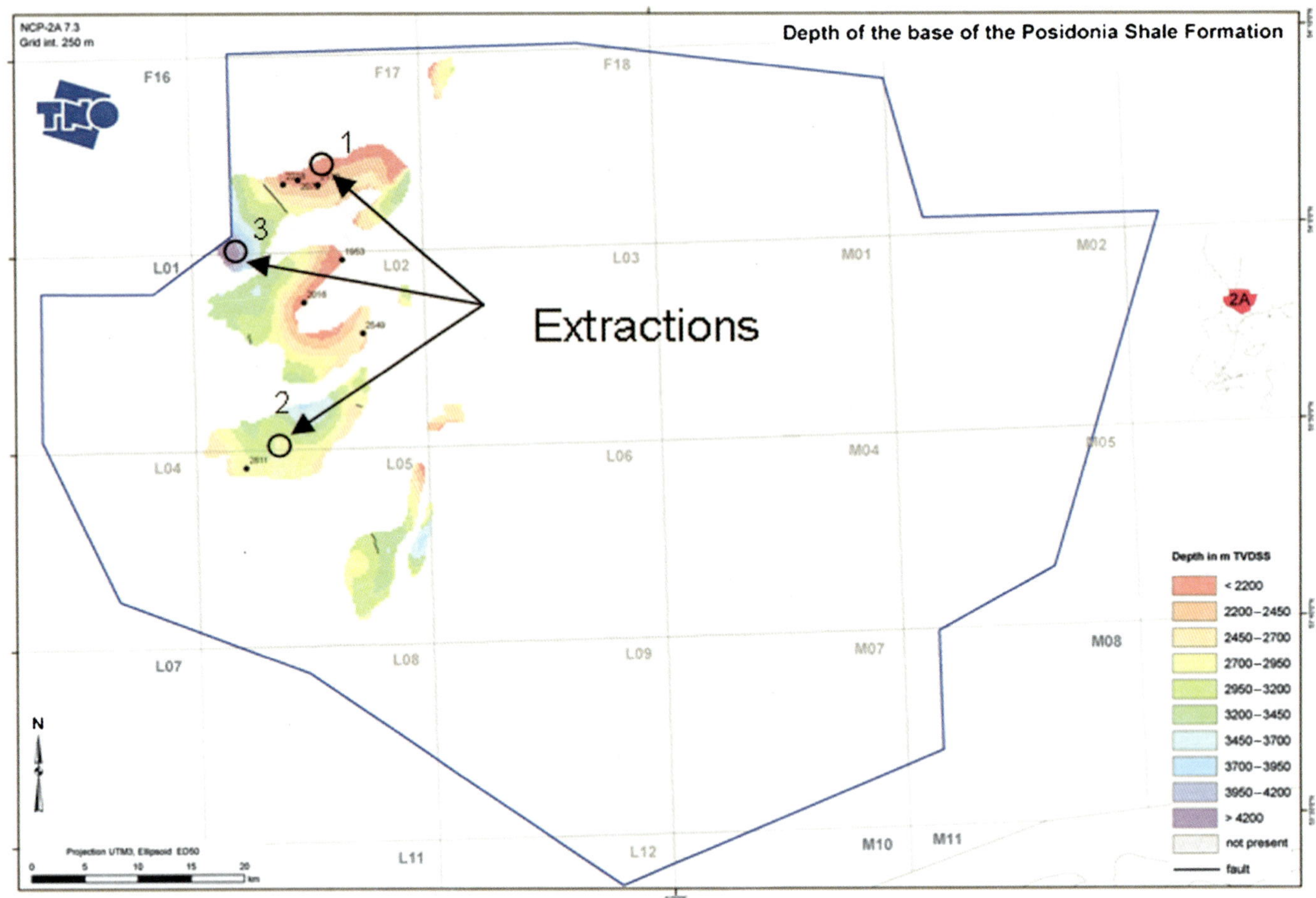

FIGURE 15. Present-day distribution and depth of the Posidonia Shale Formation. Locations of the selected one-dimensional extractions of the three-dimensional simulated histories of burial, temperature, maturity, and transformation ratio of the Posidonia Shale Formation: Location 1, block F17 (present-day depth Posidonia = 1745 m [5725 ft]); Location 2, block L02 (present-day depth Posidonia = 2762 m [9062 ft]); Location 3, southwestern part of block 17 (present-day depth Posidonia = 3768 m [12,362 ft]).

Temperature, Maturity, and Hydrocarbon Generation History of Carboniferous Source Rocks

The Carboniferous source rocks (Baarlo and Ruurlo formations) are present throughout the area. Lateral variations in calculated temperature history of these source rocks will mainly result from differences in burial history, position relative to salt structures, and in contrast to the Posidonia Shale Formation, also from differences in basal heat-flow history.

Results of four representative 1-D extractions from 3-D simulations of the Ruurlo Formation (Figures 16, 17) illustrate the temperature evolution in the Carboniferous source rocks in the Dutch Central Graben, Terschelling Basin, and Vlieland High.

The results of the 3-D simulation of the Carboniferous source rocks show that the temperature history mostly follows the burial history, except during the Late Paleogene to Pliocene. In addition, Middle Kimmerian uplift and erosion of the Terschelling Basin and Vlieland High are associated with decreasing temperatures of the source rocks despite increasing basal heat flow. The simulated present-day temperatures in the Carboniferous source rocks are lower than the maximum temperatures during previous burial. The simulated history of transformation rates for the Ruurlo and Baarlo formations in the graben and the Terschelling Basin reveals important information on the timing of hydrocarbon generation: (1) an initial phase of hydrocarbon generation before the Middle Kimmerian uplift in the Terschelling Basin, (2) a major phase of hydrocarbon generation in both the graben and the basin during the Late Jurassic and Early Cretaceous, and (3) only limited resumption of hydrocarbon generation in the graben and variable resumption of hydrocarbon generation in the Terschelling Basin after the sub-Hercynian uplift during the Paleogene (Figure 17D). Top view maps of the maturity distribution

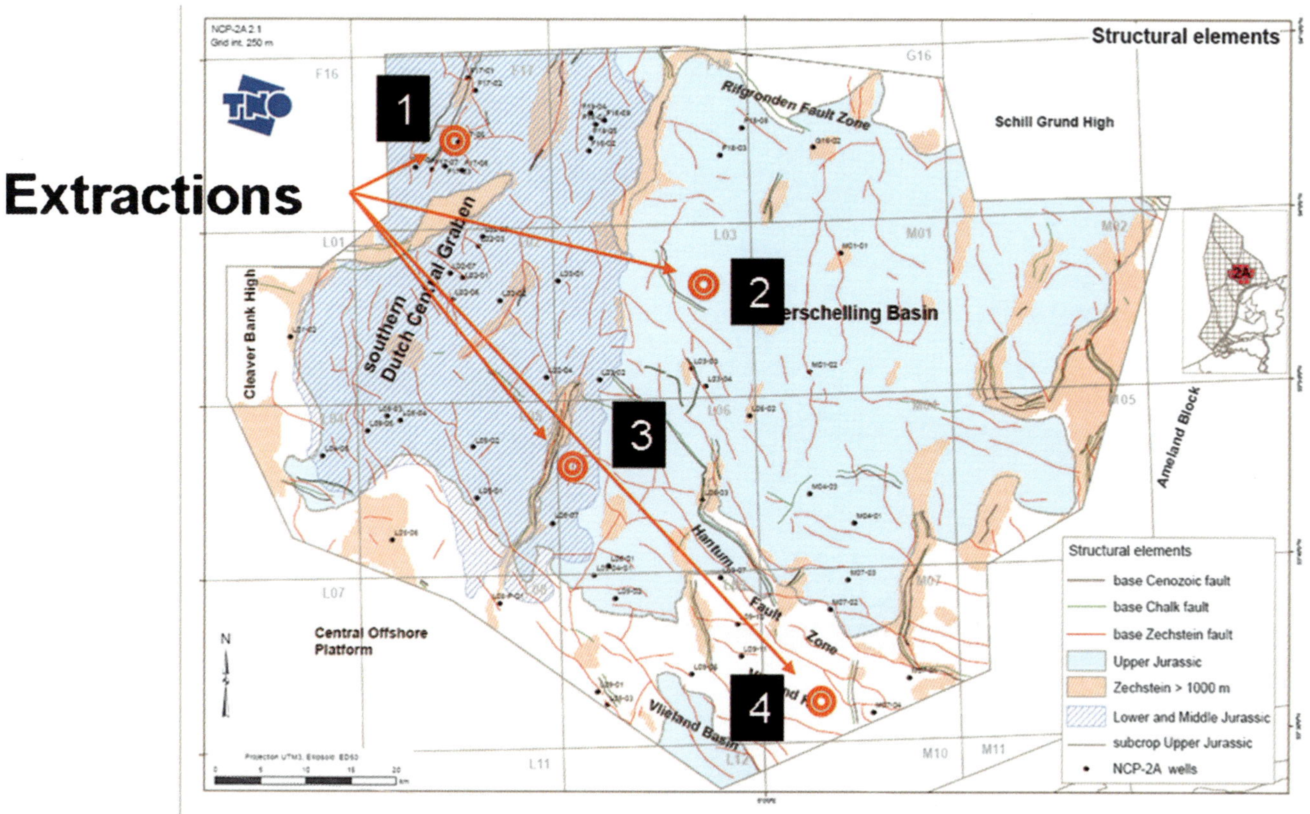

FIGURE 16. Location of four selected one-dimensional extractions of the three-dimensional simulated history of burial, temperature, maturity, and transformation ratio of the Carboniferous source rock (Ruurlo Formation): location 1, block F17 in Dutch Central Graben; locations 2 and 3, blocks L03 and L06 in the Terschelling Basin; location 4, block M07 on Vlieland High.

of the Ruurlo Formation before and after the sub-Hercynian uplift (Figures 18, 19) show a clear increase of source rock maturity along the boundaries of the study area. This indicates a Tertiary phase of hydrocarbon generation from Carboniferous source rocks on the Cleaver Bank High, Central Offshore Platform, Vlieland High, and Schill Grund High.

The calculated present-day variation of maturity at the top of the Maurits and Ruurlo formations shows that these source rocks are in the dry gas window in large parts of the graben and basin, whereas local wet gas conditions could be related to the simulated cooling effect below large salt structures. In addition, wet gas conditions occur at their borders and on the adjacent platform and highs (Figure 20).

CONCLUSIONS

In this study, we applied new approaches to reconstruct SWIT and basal heat-flow boundary conditions in more detail. The results show that these approaches provide a more refined reconstruction of temperature and maturity histories and associated hydrocarbon generation. These approaches to reconstruct the paleo-SWITs and basal heat flows can be applied in different geologic settings worldwide and during different geologic periods.

New data on the complex southern part of the Dutch Central Graben and Terschelling Basin were used as input for a full 3-D reconstruction of the burial and temperature history, and source rock maturity and timing of hydrocarbon generation of the study area. The combination of new data, new surface and basal thermal histories, and 3-D basin modeling improved understanding of the burial, thermal, and maturity history of the area and provided more detailed information on the timing of the main periods of hydrocarbon generation in the Posidonia Shale and Carboniferous source rocks. The 3-D modeling results revealed significant lateral variations in maturity and hydrocarbon generation history for each source rock related to the structural positions of each source rock. For example, the Posidonia Shale Formation already started generating hydrocarbons during the Cretaceous with maximum generation rates just before Late Cretaceous uplift west of the inverted center of the Dutch Central Graben. The generation resumed during the Paleogene and stopped at the end of

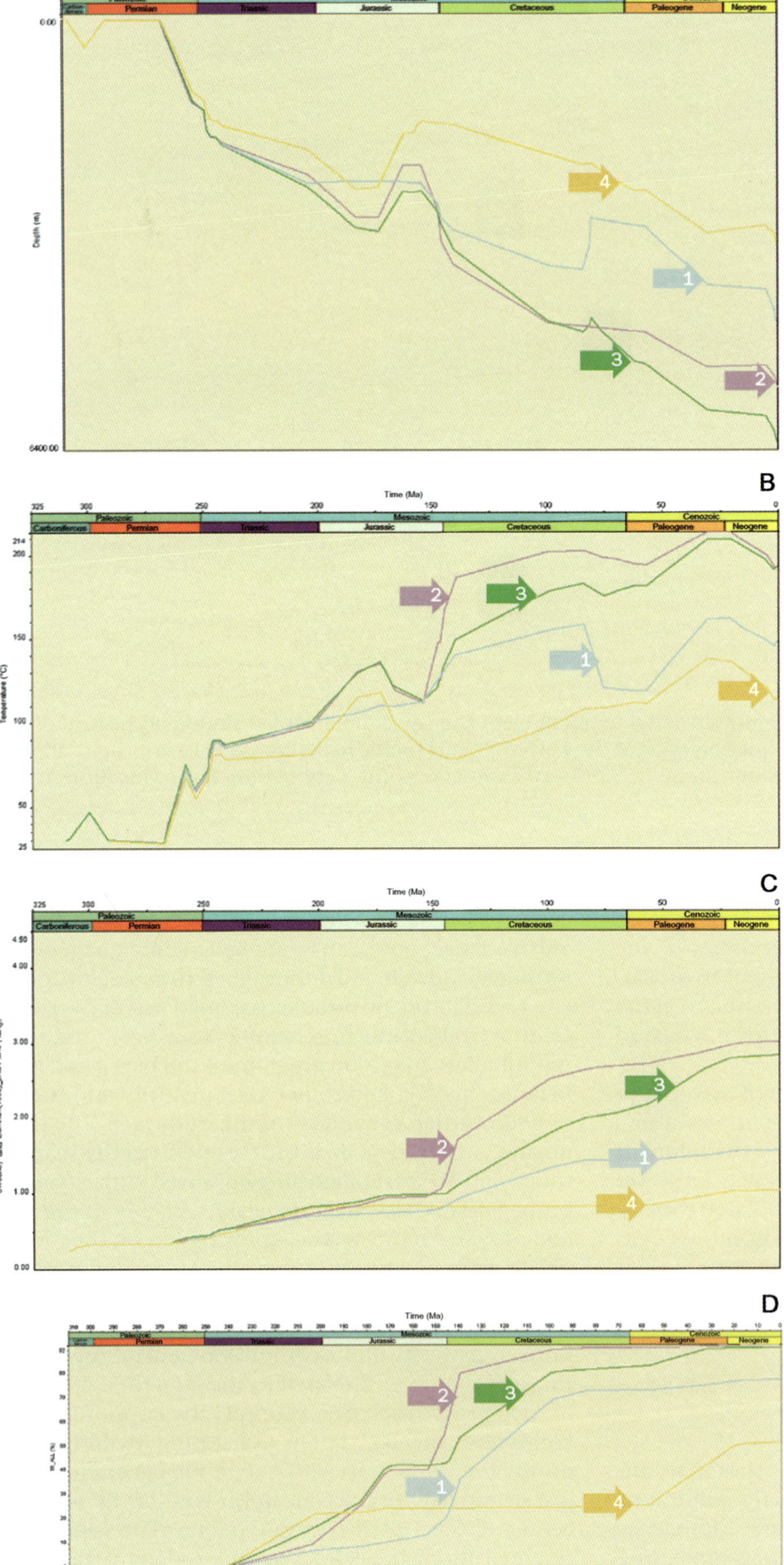

FIGURE 17. One-dimensional extractions from three-dimensional simulated history of burial (A), temperature (B), maturity (C), and transformation ratios (D) of the Ruurlo Formation at the four selected locations (see Figure 16) Location 1, block F17 in Dutch Central Graben (blue); locations 2 and 3, blocks L03 and L06 in the Terschelling Basin (purple and green, respectively); location 4, block M07 on Vlieland High (yellow). The level of maturity (C) is indicated by the simulated vitrinite reflectance $\%R_o$, using the Sweeney and Burnham (1990) kinetic model.

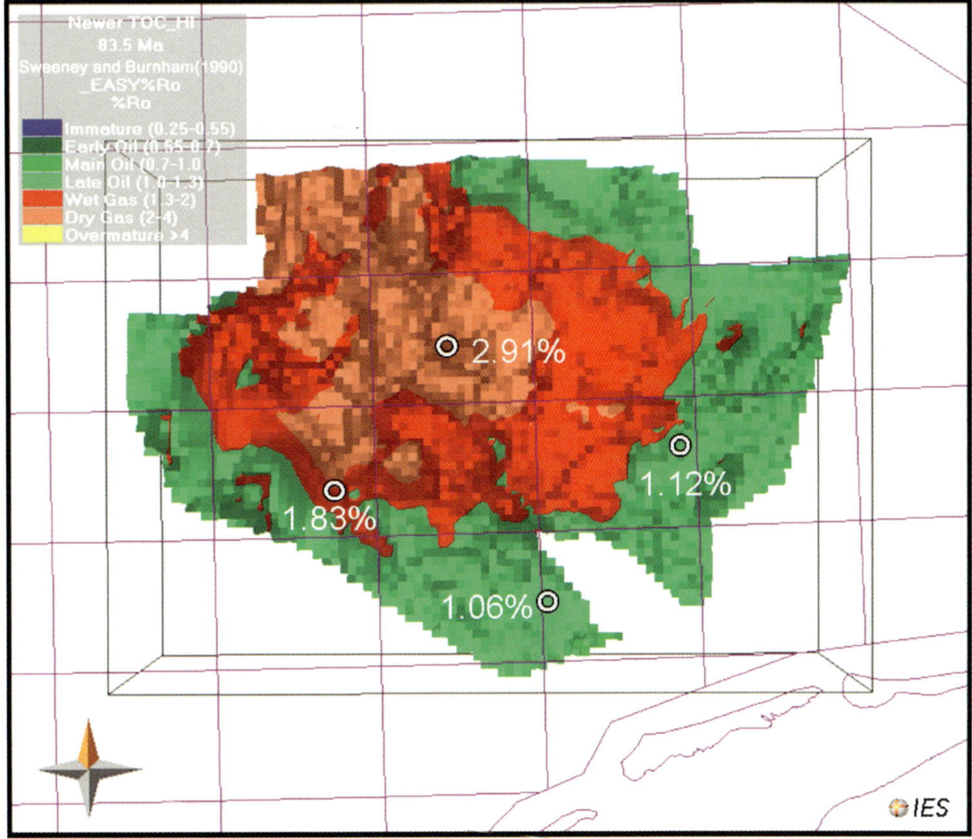

FIGURE 18. Top view: Distribution of calculated maturity (vitrinite reflectance [R_o]) at the top of the Ruurlo Formation just before sub-Hercynian uplift (83.5 Ma). The Sweeney and Burnham (1990) kinetic model was used to calculate the vitrinite reflectance $\%R_o$. TOC = total organic carbon; HI = hydrogen index.

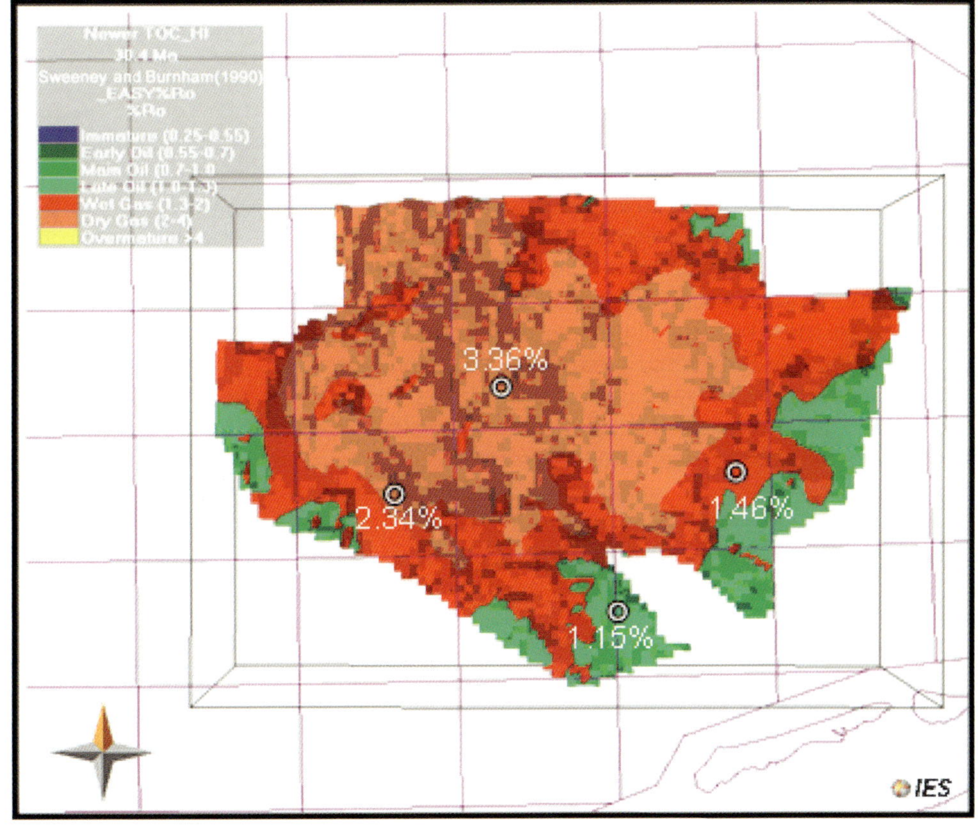

FIGURE 19. Top view: Distribution of calculated maturity (vitrinite reflectance [R_o]) at the top of the Ruurlo Formation after sub-Hercynian uplift (30.4 Ma). The Sweeney and Burnham (1990) kinetic model was used to calculate the vitrinite reflectance $\%R_o$. TOC = total organic carbon; HI = hydrogen index.

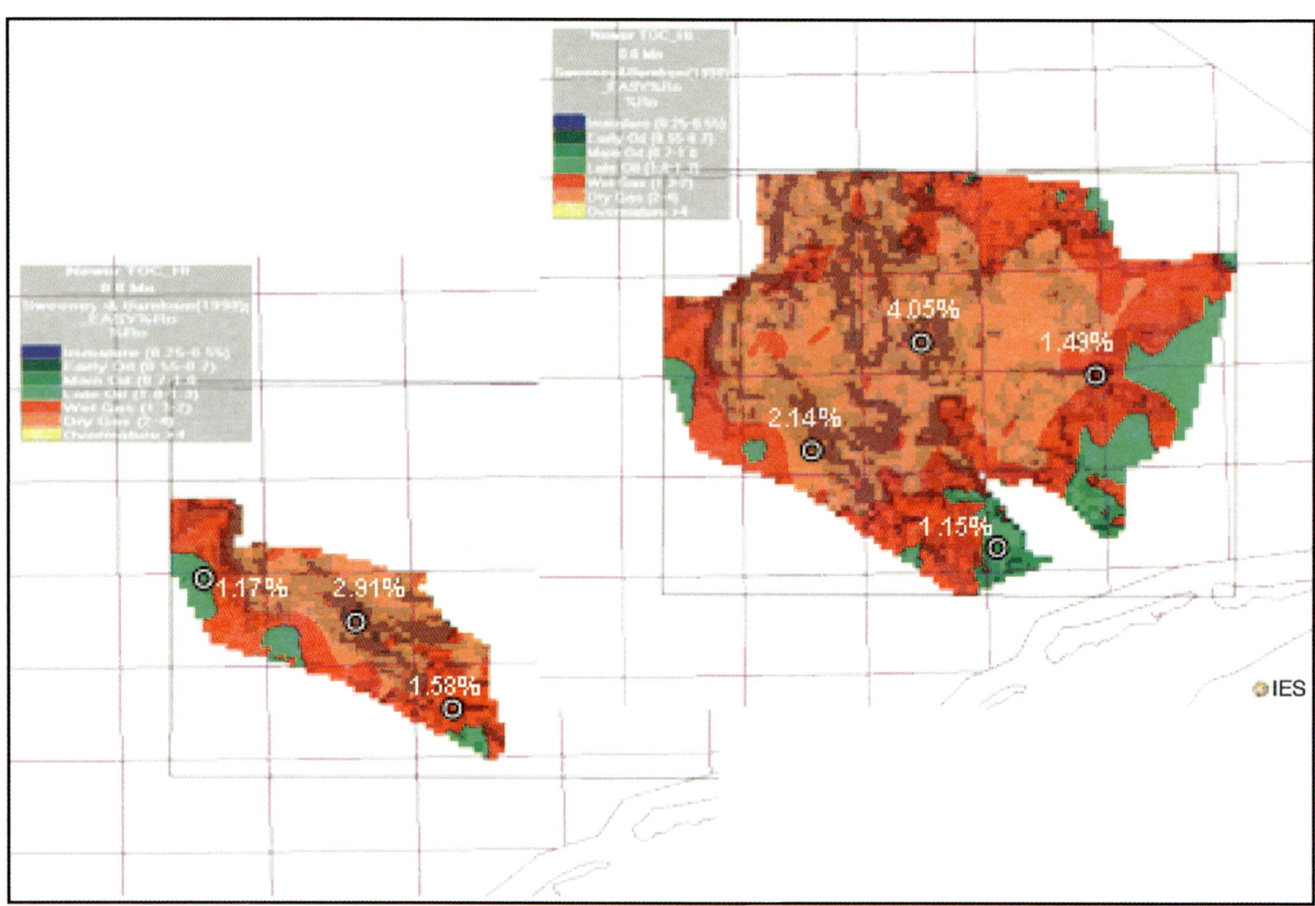

FIGURE 20. Top view: Distribution of calculated maturity (vitrinite reflectance [R_o]) at the top of the Maurits Formation and the Ruurlo Formation (left and right side of figure, respectively). Maturity of Ruurlo Formation varies between overmature (>4% R_o) in the deepest part of the Terschelling Basin to immature for gas generation (green) on the platform and highs adjacent to the Terschelling Basin and the Dutch Central Graben (according to Sweeney and Burnham, 1990, maturity classification). TOC = total organic carbon; HI = hydrogen index.

the Paleogene. In contrast, in the inverted center of the graben, the Posidonia Shale Formation did not reach maturity for hydrocarbon generation. The simulated history of transformation rates for the Carboniferous source rocks revealed an initial phase of hydrocarbon generation before the Middle Kimmerian uplift in the Terschelling Basin, a major phase of hydrocarbon generation in both the graben and the basin during the Late Jurassic and Early Cretaceous, and only limited resumption of hydrocarbon generation in the graben and variable resumption of hydrocarbon generation in the Terschelling Basin after the sub-Hercynian uplift during the Paleogene.

ACKNOWLEDGMENTS

We thank the reviewers Fokko van Hulten (EBN, The Netherlands) and Robert Ondrak (GFZ Potsdam, Germany), and the editor Ken Peters for their valuable comments.

REFERENCES CITED

Abdul Fattah, R., J.-D. van Wees , J. M. Verweij, and D. Bonte, 2008, Tectonic heat-flow modeling for the NCP2a area of the Dutch offshore, TNO Built Environment and Geosciences-National Geological Survey, the Netherlands, TNO report 2008-U-R0558/A (report is publicly available at TNO), 51 p.

Burnham, A. K., 1989, A simple kinetic model of petroleum formation and cracking, Lawrence Livermoore National Laboratory Report UCID 21665, 11 p.

De Jager, J., 2003, Inverted basins in the Netherlands, similarities and differences: Netherlands Journal of Geosciences–Geologie en Mijnbouw, v. 82-4, p. 355–366.

De Jager, J., 2007, Geological development, *in* Th. E. Wong, D. A. J. Batjes, and J. de Jager, eds, Geology of the Netherlands: Amsterdam, Royal Dutch Academy of Arts and Sciences, p. 5–26.

De Jager, J., and M. C. Geluk, 2007, Petroleum geology, *in* Th. E. Wong, D. A. J. Batjes, and J. de Jager, eds, Geology of the Netherlands: Amsterdam, Royal Dutch Academy of Arts and Sciences, p. 241–264.

De Mulder, F. J., M. C. Geluk, I. L. Ritsema, W. E. Westerhoff, and Th. E. Wong, 2003, De ondergrond van Nederland, Wolters-Noordhoff Groningen/Houten, The Netherlands, 379 p. (in Dutch).

Donders, T. H., J. W. H. Weijers, D. K. Munsterman, M. L. Kloosterboer-van Hoeve, L. K. Buckles, R. D. Pancost, S. Schouten, J. S. Sinninghe Damsté, and H. Brinkhuis, 2009, Strong climate coupling of terrestrial and marine environments in the Miocene of northwest Europe: Earth and Planetary Science Letters, v. 281, p. 215–225, doi:10.1016/j.epsl.2009.02.034.

Duin, E. J. T., J. C. Doornenbal, R. H. B. Rijkers, J. W. Verbeek, and Th. E. Wong, 2006, Subsurface structure of the Netherlands—results of recent onshore and offshore mapping: Netherlands Journal of Geosciences-Geologie en Mijnbouw, v. 85-4, p. 245–276.

EBN, 2009, Focus on Dutch gas, Energie Beheer Nederland BV (EBN), Utrecht, The Netherlands, EBN report, 16 p.

Greenwood, D. R., and S. L. Wing, 1995, Eocene continental climates and latitudinal temperature gradients: Geology, v. 23, no. 11, p. 1044–1048, doi:10.1130/0091-7613(1995)023<1044:ECCALT>2.3.CO;2.

McKenzie, D., 1978, Some remarks on the development of sedimentary basins: Earth and Planetary Science Letters, v. 40, p. 25–32, doi:10.1016/0012-821X(78)90071-7.

Mosbrugger, V., T. Utescher, and D. L. Dilcher, 2005, Cenozoic continental climate evolution of Central Europe: Proceedings of the National Academy of Science U. S. A., v. 102, no. 42, p. 14,964–14,969, doi:10.1073/pnas.0505267102.

Pearson, P. N., B. E. van Dongen, C. J. Nicholas, R. D. Pancost, S. Schouten, J. Singano, and B. S. Wade, 2007, Stable warm tropical climate through the Eocene Epoch: Geology, v. 35, p. 211–214, doi:10.1130/G23175A.1.

Remmelts, G., 1996, Salt tectonics in the southern North Sea, the Netherlands, *in* H. E. Rondeel, D. A. J. Batjes, and W. H. Nieuwenhuijs, eds, Geology of gas and oil under the Netherlands: Dordrecht, Kluwer, p. 143–158.

Sekiguchi, K., 1984, A method for determining terrestrial heat flow in oil basinal areas: Tectonophysics, v. 103, p. 67–79, doi:10.1016/0040-1951(84)90075-1.

Sluijs, A., and H. Brinkhuis, 2008, Rapid carbon injection and transient global warming during the Paleocene–Eocene thermal maximum: Netherlands Journal of Geosciences–Geologie en Mijnbouw, v. 87-3, p. 201–206.

Sluijs, A., et al., 2006, Subtropical Arctic Ocean temperatures during the Paleocene/Eocene thermal maximum: Nature, v. 441, no. 7093, p. 610–613, doi:10.1038/nature04668.

Sweeney, J. J., and A. K. Burnham, 1990, Evaluation of a simple model of vitrinite reflectance based on chemical kinetics: AAPG Bulletin, v. 74, p. 1559–1570.

TNO-NITG (Netherlands Institute of Applied Geoscience TNO–National Geological Survey), 2004, Geological atlas of the subsurface of the Netherlands: Onshore: Utrecht, The Netherlands, TNO-NITG, 101 p.

Van Adrichem Boogaert, H. A., and W. F. P. Kouwe, eds., 1993-1997, Stratigraphic nomenclature of the Netherlands, revision and update by RGD and NOGEPA: The Netherlands, Mededelingen Rijks Geologische Dienst, v. 50.

Van Balen, R. T., F. van Bergen, C. de Leeuw, H. Pagnier, E. Simmelink, J.-D. van Wees, and J. M. Verweij, 2000, Modeling the hydrocarbon generation and migration in the West Netherlands Basin, the Netherlands: Geologie en Mijnbouw/Netherlands Journal of Geosciences, v. 79, p. 19–26.

Van Gessel, S., H. Doornenbal, E. Duin, and N. Witmans, 2008, 3-D Subsurface mapping of the Dutch offshore: Results and progress, TNO built environment and geosciences: National Geological Survey, Utrecht, The Netherlands, Information on GeoEnergy June 2008, p. 12–17.

Van Wees, J. D., F. van Bergen, P. David, M. Nepveu, F. Beekman, S. Cloetingh, and D. Bonte, 2009, Probabilistic tectonic heat flow modeling for basin maturation: Assessment methods and applications, *in* H. Verweij, M. Kacewicz, J. Wendebourg, G. Yardley, S. Cloetingh, and S. Düppenbecker, eds., Thematic set on basin modeling perspectives: Marine and Petroleum Geology, v. 26, p. 536–551, doi:10.1016/j.marpetgeo.2009.01.020.

Verweij, J. M., and N. Witmans, 2009, Terschelling Basin and southern Dutch Central Graben mapping and modeling: Area 2A, TNO Built Environment and Geosciences-National Geological Survey, Utrecht, the Netherlands, TNO report TNO-034-UT-2009-05169, 65 p. (report is publicly available at www.nlog.nl).

Verweij, H., S. Nelskamp, M. Souto Carneiro Echternach, and E. Simmelink, 2009a, Overpressure generation and preservation in salt-dominated basins of the Netherlands offshore area, Abstract keynote lecture at the Geofluids VI conference on fluid evolution, migration and interaction in sedimentary basins and orogenic belts, e, April 14–17, 2009, Adelaide, Australia: Journal of Geochemical Exploration, v. 101, p. 108, doi:10.1016/j.gexplo.2008.12.051.

Verweij, J. M., M. Souto Carneiro Echternach, and N. Witmans, 2009b, Terschelling Basin and southern Dutch Central Graben, Burial history, temperature, source rock maturity and hydrocarbon generation: Area 2A, TNO Built Environment and Geosciences-National Geological Survey, Utrecht, The Netherlands, TNO report TNO-034-UT-2009-02065, 46 p. (report is publicly available at www.nlog.nl).

Wong, T. E., D. A. J. Batjes, and J. de Jager, eds, 2007, Geology of the Netherlands: Amsterdam, Royal Netherlands Academy of Arts and Sciences, 354 p.

Zachos, J. C., G. R. Dickens, and R. E. Zeebe, 2008, An Early Cenozoic perspective on greenhouse warming and carbon-cycle dynamics: Nature, v. 451, no. 7176, p. 279–283, doi:10.1038/nature06588.

Condensers cooled by water collected mercury volatilzed by heating of ore in the retort pictured earlier. Mercury mines in this area follow a northwest-southeast trend of hydrothermal alteration of serpentinite to silica carbonate rock, the common host rock for mercury mineralization. The mined hydrothermal trend delineates a former zone of fluid venting from The Geysers geothermal system. Boil-off of fluids from this hydrothermal system over the last ~0.5 million years resulted in shrinkage of the fluid system and migration to the northeast, where heat is now released in fumaroles and hot springs along a sub–parallel fault system at Big Sulfur Creek. The heat sources for The Geysers geothermal system are magmas associated with the Clear Lake (~0–2 Ma) and Sonoma (~2.5–8.0 Ma) volcanic fields.

11

Wolf, S., I. Faille, S. Pegaz-Fiornet, F. Willien, and B. Carpentier, 2012, A new efficient scheme to model hydrocarbon migration at basin scale: A pressure-saturation splitting, *in* K. E. Peters, D. J. Curry, and M. Kacewicz, eds., Basin Modeling: New Horizons in Research and Applications: AAPG Hedberg Series, no. 4, p. 197–205.

A New Efficient Scheme to Model Hydrocarbon Migration at Basin Scale: A Pressure-Saturation Splitting

Sylvie Wolf, Isabelle Faille, Sylvie Pegaz-Fiornet, Françoise Willien, and Bernard Carpentier

IFP Energies Nouvelles 92852 Rueil-Malmaison Cedex, France

ABSTRACT

Oil modeling in sedimentary basins commonly involves a great variety of complex physics, which leads to a fully coupled nonlinear set of partial differential equations. The most classical sequential time stepping is the fully implicit method, which computes pressure and oil saturation simultaneously. Although this method provides accurate solutions, it turns out to be computationally expensive. The aim of this chapter is to propose a new time-stepping strategy that allows computational cost to be minimized while preserving the accuracy of the numerical solution. The new approach is based on separating pressure from oil saturation and using a local time-step technique to calculate the oil saturation. An effective reduction of the central processing unit time is reached and is illustrated through several case studies.

INTRODUCTION

The modeling of multiphase flow in a sedimentary basin commonly involves a great variety of complex physics (Schneider and Wolf, 2000; Schneider et al., 2000; Mello, 2009). First, we have to describe the formation of the basin through geologic time, namely, the deposition and the erosion of the sediments. In relation to this, we have to handle phenomena such as sediment compaction and water expulsion using rheological laws. At this level, difficulties may arise because of overpressures. Second, we have to predict the quantity of hydrocarbons generated in deep source rocks at high temperatures. In this respect, first-order kinetic reactions are commonly involved to model the cracking of kerogen into hydrocarbon compounds. Last, it is critical to account for the migration of hydrocarbons toward reservoirs. This migration is classically modeled by Darcy's law, which answers specific objectives of a basin simulation, that is, oil trapping history and pressure prediction.

The numerical approximation of such a physical system leads to a fully coupled nonlinear set of partial differential equations. The most traditional sequential time stepping is the fully implicit method, which computes pressure and oil saturation simultaneously. Although this method provides accurate solutions, it turns out to be computationally expensive. On the one

DOI:10.1306/13311436H43471

hand, at each time step, we are faced with a very large system of nonlinear equations, to which the Newton iterations must be heavily applied (Scheichl et al., 2003; Willien et al., 2009). However, the time step itself chosen must be small enough if convection prevails in the problem under consideration. This occurs, in particular, in regions located on sandy layers where oil just passes through at a rapid velocity.

The aim of this chapter is to propose a new time-stepping strategy that allows computational cost to be minimized while preserving the accuracy of the numerical solution. Our approach relies on the key observation that pressure and temperature do not have a significant influence on the displacement of oil. Indeed, the latter is essentially driven by buoyancy forces. This motivates us to separate the calculation of pressure and porosity from that of oil saturation. This splitting can be achieved by expressing the oil Darcy velocity as a function of the total fluid velocity. In this approach, we now solve two small systems (pressure and saturation) instead of a single large one. This is advantageous, insofar as the central processing unit (CPU) time can be drastically decreased. Moreover, it can be shown that for an incompressible fluid, the new strategy does not create any additional error. In the range of geologic problems at issue, the pressure block always appears to be the more costly, but also the more stable with respect to large time steps. At this point, the saturation block remains as ill conditioned as in the full implicit scheme, but its cost is negligible compared with the pressure block. This motivates us to apply the local time-step technique to the saturation block to improve its stability while ensuring that the overall cost is lower up to a speed-up factor of five times that of the fully implicit scheme. This new approach has been implemented in the TEMIS software with local grid refinement capability to capture hydrocarbon traps more easily. It has been parallelized and extended to compressible fluid. We have performed several computations of two-dimensional (2-D) sections and three-dimensional (3-D) blocks from real case studies.

This chapter is outlined as follows. We first start by stating the continuous problem. Then, its space discretization is given. The next section is devoted to its time discretization, which leads to the formulation of the decoupled scheme called implicit pressure-implicit saturation (IMPIMS) scheme. Local time steps are introduced and last, numerical results are shown.

THE CONTINUOUS MODEL

In this section, we present the mathematical model that is used in the basin simulator software TEMIS (Schneider et al., 2000).

NOMENCLATURE

S_α	*Phase saturation (or phase volumetric fraction in the fluid)*
P_α	*Pore pressure of the phase*
σ_z	*Lithostatic load (or vertical stress)*
σ	*Mean effective stress*
ϕ	*Porosity of the porous medium*
$\vec{V}_s$	*Solid velocity*
$\vec{V}_\alpha$	*Mean velocity of the phase*
$\vec{U}_\alpha$	*Darcy velocity of the phase*
$\vec{U}_T$	*Total velocity*
Q_α	*Source term*
ρ_α	*Phase density*
ρ_s	*Solid density*
G	*Gravity acceleration*
β, γ	*Rheology parameters*
$\overline{\overline{K}}$	*Permeability tensor*
η_α	*Phase mobility*
P_c	*Capillary pressure*

The subscript α denotes the fluid phase (α = w for water, o for oil). The coordinate system is chosen such that the *z* axis is pointing vertically downward.

The equations are the conservation of mass for water and oil and the compaction of sediments, coupled with Darcy's law.

- Mass balance for water:

$$\frac{\partial}{\partial t}(\rho_w S_w \phi) + \mathrm{div}(\rho_w S_w \phi \overrightarrow{V_w}) = \rho_w q_w \tag{1}$$

- Mass balance for oil:

$$\frac{\partial}{\partial t}(\rho_o S_o \phi) + \mathrm{div}(\rho_o S_o \phi \overrightarrow{V_o}) = \rho_o q_o \tag{2}$$

- Vertical equilibrium:

$$\frac{\partial \sigma_z}{\partial z} = (\phi \rho_w + [1 - \phi]\rho_s) g \tag{3}$$

- Porous medium rheology (elastoviscoplastic law):

$$\frac{d\phi}{dt} = -\beta(\phi, \sigma)\frac{\partial \sigma}{\partial t} - \gamma(\phi, \sigma)\sigma \tag{4}$$

- Stress relation:

$$\sigma_z = \sigma + P_w \tag{5}$$

- Generalized Darcy's law for each phase:

$$\begin{aligned}\overrightarrow{U_\alpha} &= \phi\, S_\alpha(\overrightarrow{V_\alpha} - \overrightarrow{V_s}) \\ &= -\eta_\alpha \overline{\overline{K}}\,(\nabla P_\alpha - \rho_\alpha g\, \nabla z),\ \alpha \in \{w, o\}\end{aligned} \tag{6}$$

All in all, this gives rise to a system of 7 equations for 11 unknowns, which are the porosity, the vertical stress, the mean effective stress of the porous medium, and for each fluid phase (water and oil), its saturation, pressure, mean velocity, and Darcy velocity (ϕ, σ_z, σ, S_α, P_α, $\vec{V}_\alpha$, $\vec{U}_\alpha$).

The remaining terms appearing in the equations are parameters of the system, which depend on the unknowns defined above or on the temperature T (which is assumed to be given at each time step).

The source term for water q_w has a non-zero value in the layers that erode or deposit. It is tied to the sedimentation or erosion rate. The source term for oil, q_o, corresponds to the hydrocarbon generation during primary cracking.

To close the system, we further impose:

$$\begin{cases} S_w + S_o = 1 & (7) \\ P_o = P_w + P_c(\phi, S_o) & (8) \end{cases}$$

By substituting equations 5 to 8 with 1 to 4, we obtain a system of two partial differential equations in the four unknowns, S_o, P_w, σ_z, ϕ, which will be the primary unknowns of our system.

Given a bounded domain, $\Omega := \Omega(t)$ with boundary $\Gamma := \Gamma(t)$, the forward simulation consists of numerically solving this system in the domain Ω subject to suitable boundary conditions and initial conditions. The geometric history of the domain Ω is calculated during the backstripping process.

CELL-CENTERED FINITE VOLUME DISCRETIZATION IN SPACE

We are not interested here in modeling the water compaction. Therefore, we do not focus on the resolution of equations 3 and 4. In the following, we assume water and oil to be noncompressible fluids. We mesh the medium Ω into cells Ω_k of measure $|\Omega_k|$, outward normal $\vec{n}$, and boundary $\partial\Omega_k$. Then we divide the mass conservation equations 1 and 2 by the constant fluid density ρ_α and integrate them over each element Ω_k.

$$\int_{\Omega_k} \frac{\partial(S_\alpha \phi)}{\partial t} + \int_{\Omega_k} \mathrm{div}(S_\alpha \phi \overrightarrow{V_s}) = \int_{\Omega_k} q_\alpha \qquad (9)$$

Now using the given displacement velocity, $\vec{V}_s$, and exchanging the order of integration and differentiation, we can switch from the partial to the full-time derivative in the first term of equation 9. We obtain:

$$\frac{d}{dt}\int_{\Omega_k} (S_\alpha \phi) = \int_{\Omega_k} \frac{\partial(S_\alpha \phi)}{\partial t} + \int_{\Omega_k} \mathrm{div}\,(S_\alpha \phi \overrightarrow{V_s}) \qquad (10)$$

Substituting equation 10 with equation 9 and using Green's formula and the generalized Darcy's law (equation 6) yield:

$$\frac{d}{dt}\int_{\Omega_k} S_\alpha \phi + \int_{\partial\Omega_k} \overrightarrow{U_\alpha} \cdot \vec{n} = \int_{\Omega_k} q_\alpha \qquad (11)$$

To derive a complete space discretization, we resort to quadrature rules to approximate the integrals.

Mass Approximation

The mass term on the left side of equation 11 and the source terms on the right side contain only values of the unknowns at the center of the cell Ω_k. They are expressed as:

$$\int_{\Omega_k} S_\alpha \phi = |\,\Omega_k\,|\, S_{\alpha,k}\phi_k$$

$$\int_{\Omega_k} q_\alpha = |\,\Omega_k\,|\, q_{\alpha,k}$$

Flux Approximation

The next step consists of adding the conservation equation 11 and replacing fluid velocity by the total fluid velocity. The saturation in the mass term is therefore eliminated by substituting the closure equation 7:

$$\frac{d}{dt}\int_{\Omega_k} \phi + \int_{\partial\Omega_k} \overrightarrow{U_T} \cdot \vec{n} = \int_{\Omega_k} (q_w + q_o)$$

with the total fluid $\overrightarrow{U_T} = \overrightarrow{U_w} + \overrightarrow{U_o}$.

The final system to be solved is the following one, defined on each cell Ω_k

$$(F_k)\begin{cases} Rheology\ (\phi, \sigma, \sigma_z) = 0 \\ \frac{d}{dt}\int_{\Omega_k} \phi + \int_{\partial\Omega_k} \overrightarrow{U_T} \cdot \vec{n} = \int_{\Omega_k} (q_w + q_o) \\ \frac{d}{dt}\int_{\Omega_k} S_o \phi + \int_{\partial\Omega_k} \overrightarrow{U_o} \cdot \vec{n} = \int_{\Omega} q_o \\ S_w + S_o = 1 \end{cases}$$

where $\overrightarrow{U_o}$ is given by

$$\overrightarrow{U_o} = \frac{\eta_o}{\eta_w + \eta_o}\left(\overrightarrow{U_T} + \eta_w[\rho_o - \rho_w]\,\overline{\overline{K}}\,g\nabla z - \overline{\overline{K}}\,\eta_w \nabla P_c\right)$$

For each edge δ_k of the boundary of the cell Ω_k, we can write for each vector $\vec{U}$

$$\int_{\partial\Omega_k} \vec{U}\cdot\vec{n} = \sum_{\delta_k \epsilon \partial\Omega_k} \int_{\delta_k} \vec{U}\cdot\vec{n}$$

The flux terms have also been approximated using a midpoint rule on each interface δ_k of the mesh. For instance, the classical two-point scheme with harmonic average of the permeability that ensures the continuity of the water fluxes over δ_k.

The most widely used method in commercial reservoir simulators is still the implicit upwind scheme (Natvig and Lie, 2008). The robustness, simplicity, and stability of this scheme make it preferable to more sophisticated schemes for real basin models with large variations in flow speed and porosity. The upwind scheme can be written as:

$$F_{o,\delta_k} = \frac{\eta_{o,l_o}}{\eta_{o,l_o} + \eta_{w,l_w}}\left(F_{T,\delta_k} + \eta_{w,l_w} G_{o,\delta_k}\right)$$

where $F_{o,\delta}$ (resp. $F_{T,\delta}$) is an approximation of the oil flux $\int_\delta \vec{U}_o.\vec{n}$ (resp. the total flux $\int_\delta \vec{U}_T.\vec{n}$), and $G_{o,\delta}$ is an approximation of the capillary forces and the buoyancy forces:

$$\int_{\delta_k} \overline{\overline{K}}(\nabla P_c - [\rho_w - \rho_o]g\nabla z)\cdot\overrightarrow{n}$$

The mobilities are upwinded according to the sign of the fluxes:

$$(l_o, l_w) = \begin{cases} (1,1), \text{ if}: F_{T,\delta_k} \geq 0,\ G_{o\delta_k} \leq 0,\ F_{T,\delta_k} + \eta_{w,1}G_{o\delta_k} \geq 0. \\ (2,1), \text{ if}: F_{T,\delta_k} \geq 0,\ G_{o\delta_k} \leq 0,\ F_{T,\delta_k} + \eta_{w,1}G_{o\delta_k} < 0. \\ (1,1), \text{ if}: F_{T,\delta_k} \geq 0,\ G_{o\delta_k} > 0,\ F_{T,\delta_k} - \eta_{o1}G_{o\delta_k} \geq 0. \\ (1,2), \text{ if}: F_{T,\delta_k} \geq 0,\ G_{o\delta_k} > 0,\ F_{T,\delta_k} - \eta_{o1}G_{o\delta_k} < 0. \end{cases}$$

We rewrite the full system (Fk) into two subsystems:

$$(G_k)\begin{cases} Rheology(\phi, \sigma, \sigma_z) = 0 \\ \frac{d}{dt}\int_{\Omega_k}\phi + \int_{\partial\Omega_k}\overrightarrow{U_T}\cdot\overrightarrow{n} = \int_{\Omega_k}(q_w + q_o) \end{cases}$$

$$(H_k)\begin{cases} \frac{d}{dt}\int_{\Omega_k} S_o\phi + \int_{\partial\Omega_k}\overrightarrow{U_o}\cdot\overrightarrow{n} = \int_{\Omega_k} q_o \\ S_w + S_o = 1 \end{cases}$$

The subsystem (Gk) depends on the porosity and on the pressure via the total flux. It models the total fluid behavior. The subsystem (Hk) depends on the oil saturation and on the oil velocity; it models the oil behavior. This system is nonlinear in saturation and pressure. We use a convenient adaptation of Newton's method for its resolution. This is the aim of the next section.

TIME DISCRETIZATION

We use a fully implicit Euler time discretization. We define a sequence of time steps $t^{(n)} > 0$ with times $t^{(n+1)} = t^{(n)} + \Delta t^{(n)}$, for (n = 0, 1, . . .,) and where $u^{(n)}$ denotes the value of u at time $t^{(n)}$. Then, we approximate the derivatives by simple finite differences between $t^{(n)}$ and $t^{(n+1)}$ and evaluate the remaining terms at $t^{(n+1)}$:

$$\frac{d}{dt}\int_{\Omega_k} u = \frac{u_k^{(n+1)} \mid \Omega_k \mid^{(n+1)} - u_k^{(n)} \mid \Omega_k \mid^{(n)}}{\Delta t^{(n)}}$$

Fully Implicit Time Discretization

The numerical solution reduces to solving a system of nonlinear equations at each time step:

$$\begin{Bmatrix} G[R^{(n+1)}, S^{(n+1)}] = 0 \\ H[R^{(n+1)}, S^{(n+1)}] = 0 \end{Bmatrix}$$

for $R^{(n+1)}: = [P_k^{(n+1)}, \phi_k^{(n+1)}, \sigma_k^{(n+1)}]^T_{k=1,\ldots,N}$ and $S^{(n+1)}: = [S_{o,k}^{(n+1)}]^T_{k=1,\ldots,N}$.

We solve this system using the classical Newton method, with an initial guess $[R^o,S^o] = [R^{(n)},S^{(n)}]$ and a stopping criterion on the residuals $||G||<\varepsilon_G$ and $||H||<\varepsilon_H$ and the variations of the solution $||\, y^m - y^{m-1} \,||< \varepsilon_y$, leading in each Newton iteration m to a system of linear equations with the unknown y = (R,S):

$$\begin{pmatrix} \frac{\partial G}{\partial y}[y^m]\delta y^m = -G[y^m] \\ \frac{\partial H}{\partial y}[y^m]\delta y^m = -G[y^m] \end{pmatrix}$$

The Jacobian of this system requires all the partial derivatives $\frac{\partial G}{\partial R}, \frac{\partial G}{\partial S}, \frac{\partial H}{\partial R}, \frac{\partial H}{\partial S}$ to be evaluated.

Decoupled Time Discretization

The decoupled time discretization demands two resolution steps.

Step 1

The first step consists of solving the (G) subsystem while setting the S unknown at the fixed value $S^{(n)}$:

$$G_d(R) = G[R, S^{(n)}] = 0$$

Therefore $R_d^{(n+1)}$ is the solution of the Newton algorithm:

$$\frac{\partial G_d}{\partial R}(R_d^m)\delta R_d^m = -G_d(R_d^m)$$

Step 2

The second step consists of solving the (H) subsystem by injecting the previous solution of step 1:

$$H_d(S) = H[R_d^{(n+1)}, S] = 0$$

$S_d^{(n+1)}$ is the solution of the Newton algorithm:

$$\frac{\partial H_d}{\partial S}(S_d^m)\delta S_d^m = -G_d(S_d^m)$$

Fewer partial derivative evaluations are required compared with the fully implicit method.

Integrated Master Plan–Integrated Master Schedule: The Decoupled Scheme

Here, we apply this splitting process to the oil migration.

First, we solve the pressure $P_w^{(n+1)}$ by the resolution of the pressure block while setting the oil saturation at its value determined at time step $S_o^{(n)}$. Thus, the total flux depends on pressure only:

$$(G_k)\begin{cases} Rheology\ (\phi, \sigma, \sigma_z) = 0 \\ \frac{d}{dt}\int_{\Omega_k}\phi + \int_{\partial\Omega_k}\overrightarrow{U_T}\cdot\overrightarrow{n} = \sum_{\alpha\epsilon\{w,o\}}\int_{\Omega_k} q_\alpha \end{cases}$$

Second, we solve oil saturation, $S_o^{(n+1)}$, by injecting the known total velocity computed previously in the oil block:

$$(H_k)\begin{cases} \frac{d}{dt}\int_{\Omega_k} S_o\phi + \int_{\partial\Omega_k}\overrightarrow{U_o}\cdot\overrightarrow{n} = \int_{\Omega_k} q_o \\ S_w + S_o = 1 \end{cases}$$

Remark 1 (The implicit pressure–explicit saturation [IMPES] scheme). The IMPES scheme is a particular splitting time discretization. In the oil velocity, we take the oil saturation at the previous time step, $S_o^{(n)}$. It is less effective, as it uses smaller time steps to guarantee stability.

Remark 2 (Conservativity). In the case of noncompressible fluids, the fully implicit scheme and the decoupled scheme are both conservative regarding oil and water.

Remark 3 (Alternative). In the block system (Gk), we assign water properties to the total fluid, meaning that oil saturation is taken as equal to zero. We lose water conservation, but the decoupled scheme is more robust, thus leading to larger time steps.

Remark 4 (Compressibility). For compressible fluids, we add a compressibility factor in the pressure block, which has the following expression:

$$C_\alpha = S_\alpha^n \frac{1}{\rho_\alpha^n}\frac{d\rho_\alpha}{dP}(P - P^n)$$

Therefore, the pressure block has the following expression:

$$(G_k^\rho)\begin{cases} Rheology\ (\phi, \sigma, \sigma_z) = 0 \\ \frac{d}{dt}\int_{\Omega_k}\phi + \int_{\partial\Omega_k}\overrightarrow{U_T}\cdot\overrightarrow{n} + \sum_{\alpha\epsilon\{w,o\}}\int_{\Omega_k}\phi C_\alpha = \sum_{\alpha\epsilon\{w,o\}}\int_{\Omega_k} q_a \end{cases}$$

And the oil block has the following one:

$$(H_k^\rho)\begin{cases} \frac{d}{dt}\int_{\Omega_k}\rho_o S_o\phi + \int_{\partial\Omega_k}\rho_o\overrightarrow{U_o}\cdot\overrightarrow{n} = \int_{\Omega_k}\rho_o q_o \\ S_w + S_o = 1 \end{cases}$$

LOCAL TIME STEPS

The pressure block (G) is the elliptic part of the system. It is large and ill conditioned because of large heterogeneities. For an efficient and robust iterative solution of this system, it is therefore crucial to find good preconditioners. Sparse incomplete LU factorization with 0 level in fill, known as ILU(0) is not always efficient. Algebraic multigrid (AMG) methods are sometimes required but are very expensive (Scheichl et al., 2003; Willien et al., 2009).

In contrast, the oil block (H) is the hyperbolic or transport-dominated part of the system. ILU(0) preconditioners are very efficient, but sometimes, we need to reduce the time step to ensure stability conditions; also, the time step can be rejected because of nonphysical computed solutions. To avoid restarting the time step and inverting the pressure block too frequently, a local time step management has been implemented for saturation resolution. The decoupled scheme is very appropriate for that purpose.

During the time step $\Delta t^{(n+1)}$, we compute the pressure $P_w^{(n+1)}$ by solving the (G) system. Then, we divide this time step into smaller time steps $\delta t^{p,(n+1)}$:

$$\sum_{p=0}^{N}\delta t^{p,(n+1)} = \Delta t^{(n+1)}$$

During each local time step, $\delta t^{p,(n+1)}$, we resolve the (H) system assuming the two following hypotheses to be satisfied:

Hypothesis 1. Water balance is true at each local time step.

Hypothesis 2. Total flux is kept constant at each local time step. An adaptive time step control approach is used in which a small time step is taken at the beginning and then increased by a given multiplying factor until the maximum time step is reached.

NUMERICAL RESULTS

Central Processing Unit Time

The IMPIMS scheme has been implemented in a prototype version of the IFP basin calculator TEMIS. This prototype has been parallelized (Masson et al., 2004). The reference solution is given by the fully implicit scheme. The efficiency of the decoupled method is given by its computational cost. For the IMPIMS scheme, the total fluid is assumed to have water properties for pressure resolution. In 3-D case studies, local grid refinement (LGR) is possible to describe our zone of interest better. This LGR is fully integrated in the whole simulation (Monnier et al., 2009).

We have a very significant reduction of the CPU time, up to a factor of 5 with the decoupled scheme compared with the fully implicit scheme (Tables 1, 2). In Table 2, the column "water" refers to a simulation without any hydrocarbon.

Solutions

The solution for oil migration is computed along three 2-D sections (A, B, and C) and is displayed at present time for both the decoupled scheme and the fully implicit scheme (Figures 1–3). For the decoupled scheme, the total fluid for pressure resolution is assumed to have water properties. The decoupled scheme produces good matches with respect to the fully implicit method. The comparisons are made qualitatively because it is difficult to exhibit a criterion to assess the degree of similarity. The solution can differ significantly in places, but the main basin objectives are well represented. The traps and migration pathways are well modeled, whatever scheme is used. However, we note a larger discrepancy in section C (Figure 3). Hydrocarbons migrate much more through the cap rock for the decoupled scheme. In this particular case, reservoirs have a water saturation that is below the irreducible water saturation, which means that water does not move any more. This contradicts the hypothesis that the total flux is calculated with zero oil saturation. That, in turn, explains why oil migrates vertically because of bad values of the total flux. A better approximation of the total flux could reduce this leakage. However, the CPU time is drastically reduced up to a factor of 5.

A 3-D block has also been performed (Figures 4, 5). The reservoir has been refined to describe the hydrocarbons better. Figure 5 shows the LGR zone in the 3-D block. The mesh is refined simultaneously in the three directions x, y, z (Thibaut et al., 2005; Carpentier et al., 2009). The trap is well modeled by the decoupled scheme compared with the fully implicit scheme, as shown in Figure 5.

Table 1. Run times for oil simulations of two-dimensional sections on PC Linux 32 bits 2 processors.

Two-Dimensional Section	*Fully (min)*	*IMPIMS* (min)*	*Acceleration Factor*
197 × 54 layers	44	8.21	5.3
160 × 51 layers	10.65	5	2.1
470 × 31 layers	49.66	17	2.9
169 × 44 layers	16.30	6.16	2.6

**IMPIMS = implicit pressure–implicit saturation scheme; acceleration factor = the ratio between the run time with the fully implicit scheme over the run time with the decoupled scheme.*

Table 2. Run times for oil simulations of three-dimensional blocks on PC Linux 32 bits 2 processors.

Three-Dimensional Block (Zone LGR)*	*Water (min)*	*Fully (min)*	*IMPIMS (min)*	*Acceleration Factor*
Tartan Grid: 47 × 49 × 45 layers	14	258	28.56	9
19 × 18 × 43 layers (2 LGR zones)	3.46	153	25.48	6
42 × 47 × 31 layers (1 LGR)	10	160	39.56	4

**LGR = local grid refinement; water = refers to simulation without any oil; IMPIMS = implicit pressure–implicit saturation scheme; acceleration factor = the ratio between the run time with the fully implicit scheme over the run time with the decoupled scheme.*

CONCLUSIONS

A new time-stepping scheme based on separating pressure from oil saturation has been successfully developed for oil modeling in sedimentary basins. It is capable of yielding accurate solutions compared with those computed by the fully implicit scheme, in the sense that the main objectives of basin simulation (trap-filling history, oil trapping, and migration pathways) are well represented. The computational cost for the decoupled scheme is minimized. More specifically, we have reached an effective reduction of the CPU time up to a factor of 5 compared with the fully implicit method, thanks to the

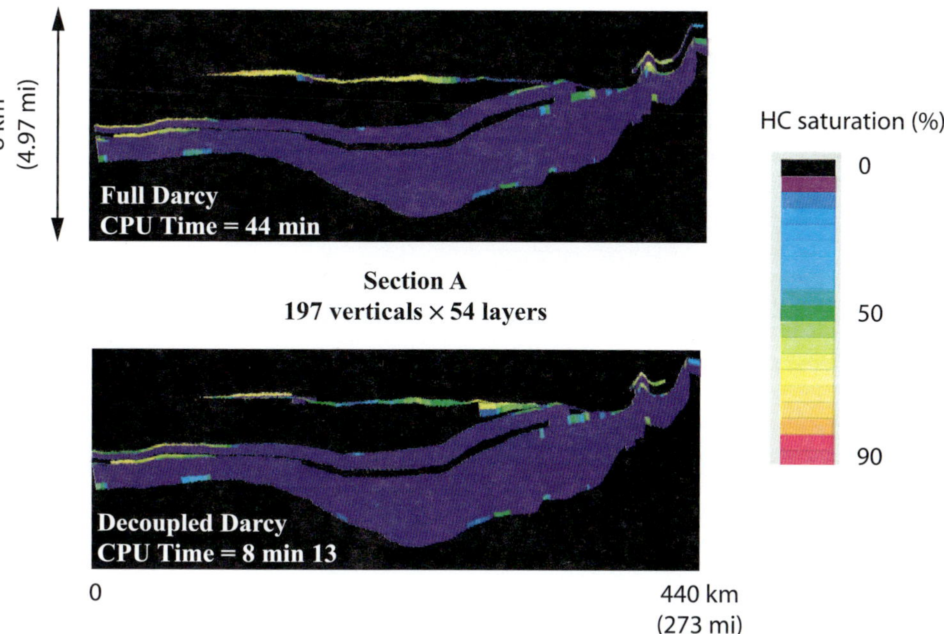

Figure 1. Section A. Comparison of oil saturation between the fully implicit scheme and the decoupled scheme. Results are very similar. CPU = central processing unit; HC = hydrocarbon.

possibility of using larger time steps within a guarantee of stability.

Beyond the efficiency, the work has clearly demonstrated that IMPIMS also exhibits more flexibility in introducing additional physical features. It now becomes easy to incorporate new phenomena such as mineral diagenesis, rock mechanisms, fluid fracturing, or oil biodegradation. This opens up the prospect of dealing with highly complex problems, which ultimately leads to a deeper insight into fluid flow. The next work will be to extend this decoupled scheme to compositional oil migration to improve the performance of oil simulations (Agelas et al., 2007).

ACKNOWLEDGMENTS

We thank Theresa Burke, Ken Peters, and an anonymous reviewer for their invaluable advice and relevant questions. We also thank Quang Huy Tran for his help in correcting the manuscript.

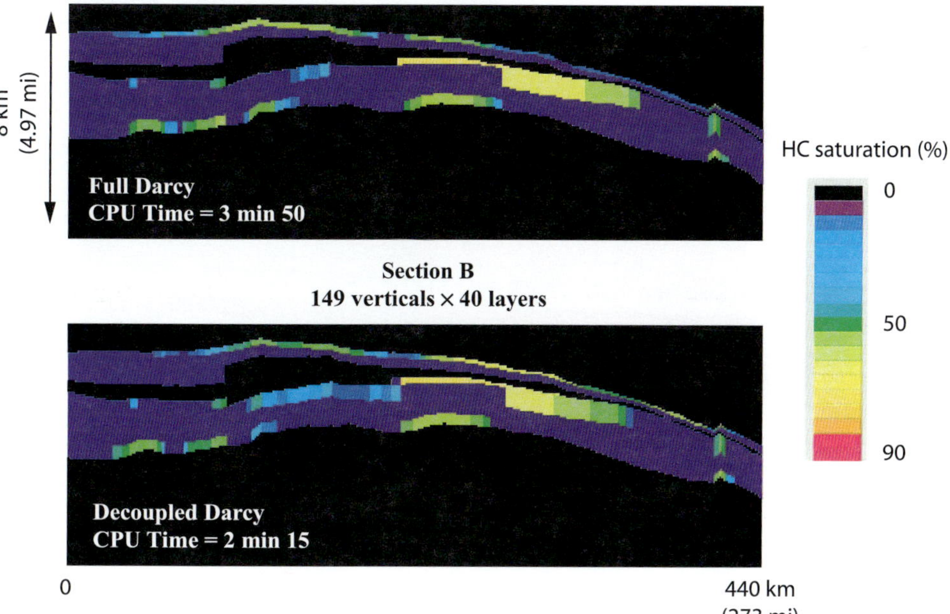

Figure 2. Section B. Comparison of oil saturation between the fully implicit scheme and the decoupled scheme. Results are very similar. CPU = central processing unit; HC = hydrocarbon.

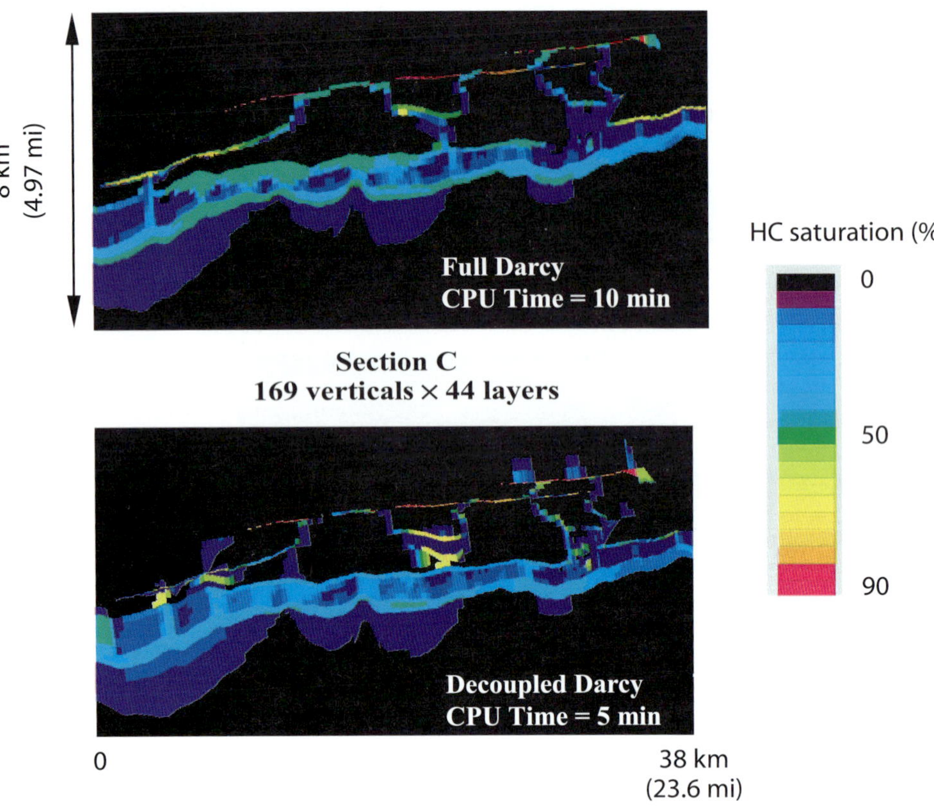

FIGURE 3. Section C. Comparison of oil saturation between the fully implicit scheme and the decoupled scheme. Migration pathways and oil traps are well captured. CPU = central processing unit; HC = hydrocarbon.

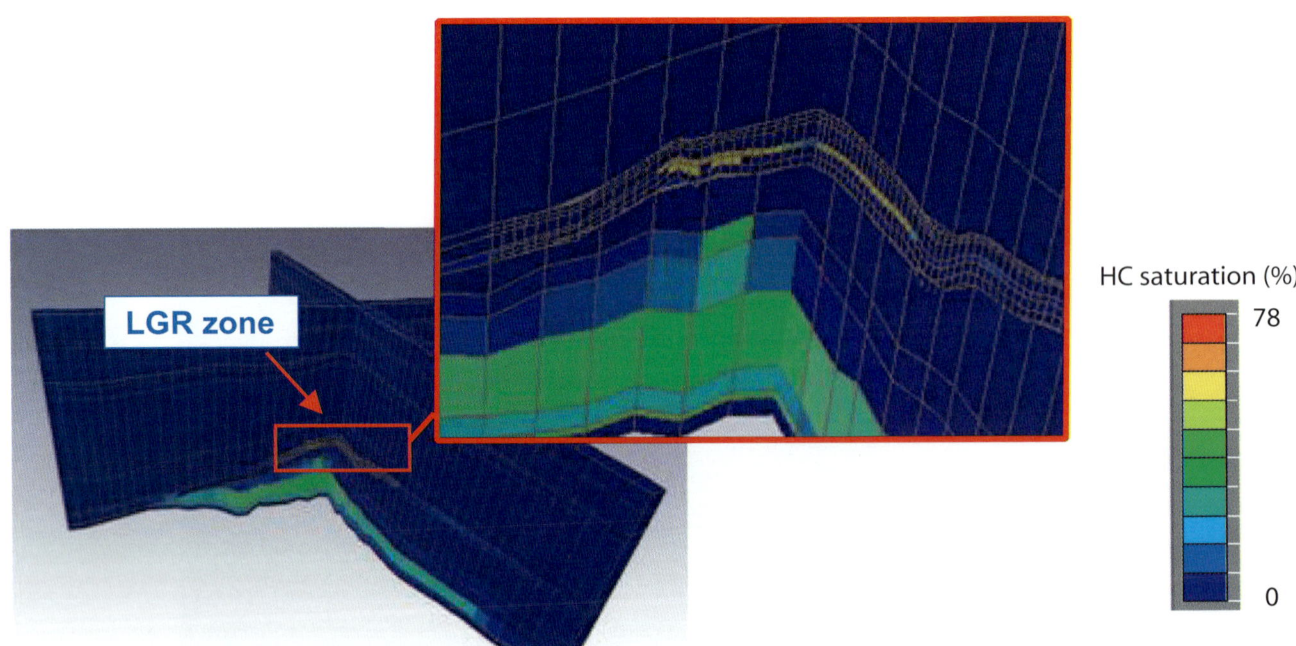

FIGURE 4. Oil saturation in two-dimensional cross sections of the three-dimensional block illustrating the local grid refinement (LGR) zone. HC = hydrocarbon.

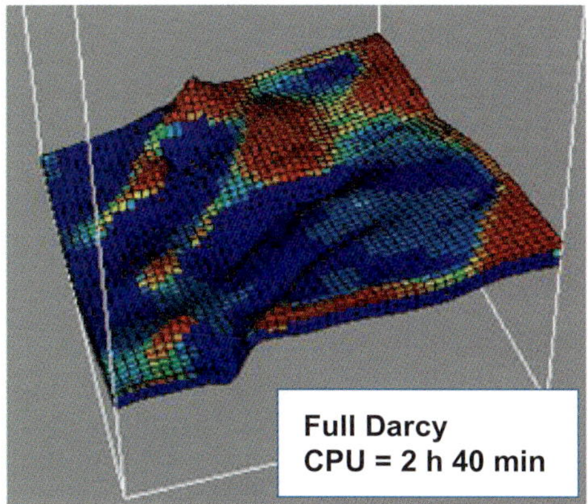

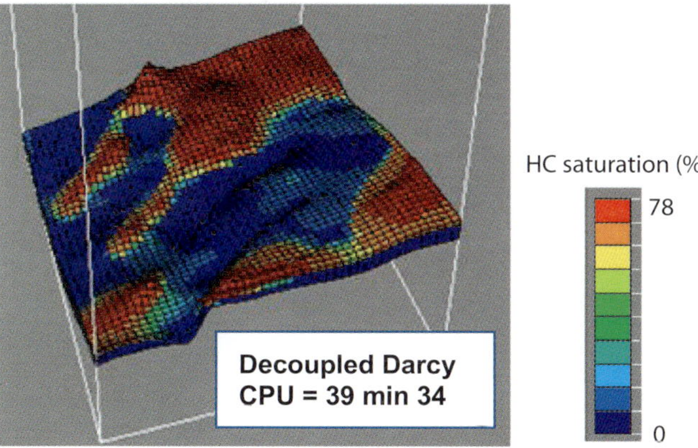

Figure 5. Oil saturation in the reservoir layer of the three-dimensional block. Comparison between the fully implicit scheme and the decoupled scheme. Cells with high saturations are well captured even if local discrepancies are present. CPU = central processing unit; HC = hydrocarbon.

REFERENCES CITED

Agelas, L., I. Faille, S. Wolf, and S. Requena, 2007, High performance computing Darcy compositional single phase fluid flow simulations (abs.): AAPG Hedberg Conference Basin Modeling Perspectives: Innovative developments and novel applications, The Hague, The Netherlands, http://www.searchanddiscovery.com/abstracts/html/2007/hedberg/short/agelas.htm (accessed May 24, 2011).

Carpentier, B., I. Kowalewski, S. Pegaz-Fiornet, and S. Wolf, 2009, 3-D Modeling, biodegradation and local grid refinement, a new approach to predict biodegradation at prospect scale (abs.): AAPG Hedberg Conference Basin and Petroleum System Modeling: New horizons in research and applications, Napa, California, http://www.searchanddiscovery.com/abstracts/html/2009/hedberg/abstracts/short/carpentier.htm (accessed May 24, 2011).

Masson, R., P. Quandalle, S. Requena, and R. Scheichl, 2004, Parallel preconditioning for sedimentary basin simulations: Lecture notes in computer science: Berlin, Springer, v. 2907, p. 59–81.

Mello, U., 2009, A control-volume finite-element method for three-dimensional multiphase basin modeling: Marine and Petroleum Geology, v. 26, p. 504–518, doi:10.1016/j.marpetgeo.2009.01.015.

Monnier, F., F. Lorant, P.-Y. Chenet, J.-M. Laigle, and A. Al-Khamiss, 2009, Prediction of fluid compositional heterogeneities in accumulations using TEMIS Suite® enhanced with local grid refinement (LGR): Example from the Jurassic of northern Kuwait (abs.): AAPG Hedberg Conference Basin and Petroleum System Modeling: New horizons in research and applications, Napa, California, http://www.searchanddiscovery.com/abstracts/html/2009/hedberg/abstracts/extended/monnier/monnier.htm (accessed May 24, 2011).

Natvig, J. R., and K. A. Lie, 2008, On efficient implicit upwind schemes: Proceedings of the 11th European Conference on the Mathematics of Oil Recovery, September 2008, Bergen, Norway, https://www.sintef.no/project/GeoScale/papers/ecmorxi-ro.pdf (accessed May 24, 2011).

Scheichl, R., R. Masson, and J. Wendebourg, 2003, Decoupling and block preconditoning for sedimentary basin simulations: Computational Geosciences, v. 7, no. 4, p. 295–318, doi:10.1023/B:COMG.0000005244.61636.4e.

Schneider, F., and S. Wolf, 2000, Quantitative HC potential evaluation using 3-D basin modeling: Application to Franklin structure, Central Graben, North Sea, United Kingdom: Marine and Petroleum Geology, v. 17, p. 841–856, doi:10.1016/S0264-8172(99)00060-4.

Schneider, F., S. Wolf, I. Faille, and D. Pot, 2000, A 3-D basin model for hydrocarbon potential evaluation: Application to Congo offshore: Oil & Gas Science and Technology, v. 55, no. 1, p. 3–13, doi:10.2516/ogst:2000001.

Thibaut, M., Y. Caillabet, E. Flauraud, and F. Schneider, 2005, Integrating local grid refinement methods for basin modeling (abs.): Proceedings of the 67th EAGE Conference and Exhibition, Madrid, Spain, http://www.earthdoc.org/detail.php?pubid=1150 (accessed May 24, 2011).

Willien, F., I. Chetvchenko, R. Masson, P. Quandalle, L. Agelas, and S. Requena, 2009, AMG preconditioning for sedimentary basin simulations in TEMIS calculator: Marine and Petroleum Geology, v. 26, p. 519–524, doi:10.1016/j.marpetgeo.2009.01.014.

SECTION 5

Uncertainty in Basin Modeling

12

Hicks, P. J. Jr., C. M. Fraticelli, J. D. Shosa, M. J. Hardy, and M. B. Townsley, 2012, Identifying and quantifying significant uncertainties in basin modeling, *in* K. E. Peters, D. J. Curry, and M. Kacewicz, eds., Basin Modeling: New Horizons in Research and Applications: AAPG Hedberg Series, no. 4, p. 207–219.

Identifying and Quantifying Significant Uncertainties in Basin Modeling

Paul J. Hicks Jr. and Carmen M. Fraticelli

ExxonMobil Exploration Co., Houston, Texas, U.S.A.

Jennifer D. Shosa

ExxonMobil Upstream Research Co., Houston, Texas, U.S.A.

Martine J. Hardy

ExxonMobil International Limited, Surrey, United Kingdom

Michael B. Townsley

ExxonMobil Technical Computing Co., Houston, Texas, U.S.A.

ABSTRACT

Basin models are used to address a variety of questions concerning oil and gas generation, reservoir pressure and temperature, and oil quality. A large number of input parameters are required for a basin model, and many are functions of both space and time. Examples include isopach thicknesses and ages, amount of eroded/missing section, rock properties (e.g., porosity, thermal conductivity), and heat flow and surface temperature boundary conditions. Most, if not all, of these model input parameters have associated uncertainties, and it can be difficult and time consuming to adequately quantify these uncertainties and propagate them through a basin model to assign error bars, probabilities, and risks to the output properties of interest.

In this chapter, we propose a workflow that allows a basin modeler to identify key input parameters and quantify and propagate uncertainties in these key input parameters through a model to evaluate the model results in light of a business question. We demonstrate this workflow using a hypothetical illustration in which uncertainties in key input parameters that control hydrocarbon generation, volumes, and timing are identified, quantified, and propagated through a basin model.

The workflow proposed in this chapter was designed to (1) identify the purpose(s) of the model; (2) develop a base-case scenario; (3) identify the input parameters whose uncertainty might affect the output property of interest; (4) perform screening simulations to identify the key input parameters; (5) evaluate the range of uncertainty in the key input parameters; (6) propagate the uncertainty in key input parameters through

DOI:10.1306/13311437H41527

the model to the output properties of interest and estimate ranges of uncertainty for these input parameters; and (7) iterate as needed to fine tune the input parameters and dependencies between input parameters, fine tune error bars and weights for calibration data, and improve the base-case scenario.

INTRODUCTION

Basin modeling is an increasingly important element of exploration, development, and production workflows. Problems addressed with basin models typically include questions regarding burial history, source maturation, hydrocarbon yields (timing and volume), hydrocarbon migration, hydrocarbon type and quality, reservoir quality, and reservoir pressure and temperature prediction for pre–drill analysis. As computing power and software capabilities increase, the size and complexity of basin models also increase. These larger, more complex models address multiple scales (well to basin) and problems of variable intricacy, making it more important than ever to understand how the uncertainties in input parameters affect model results.

Increasingly complex basin models require an ever-increasing number of input parameters with values that are likely to vary both spatially and temporally. Some of the input parameters that are commonly used in basin models and their potential effect on model results are listed in Table 1. For a basin model to be successful, the modeler must not only determine the most appropriate estimate for the value for each input parameter, but must also understand the range of uncertainty associated with these estimates and the uncertainties related to the assumptions, approximations, and mathematical limitations of the software. This second type of uncertainty may involve fundamental physics that are not adequately modeled by the software and/or the numerical schemes used to solve the underlying partial differential equations. Although these issues are not addressed in this chapter or by the proposed workflow, basin modelers should be aware of these issues and consider them in any final recommendations or conclusions.

Considering Uncertainty

Uncertainty is present in most, if not all, model inputs and calibration data. These uncertainties generate uncertainties in the model outputs. Sometimes, the resultant uncertainties are not significant enough to impact decisions based on the model results. Other times, these uncertainties can make the model results virtually useless in the decision-making process. Of course, a wide range of cases exist between these extremes, and this is where basin modelers commonly work. In these cases, the model results can be useful, but the uncertainties surrounding the model predictions can be difficult to fully grasp and communicate. Successful decisions based on models in which significant uncertainties exist require that the modeler (1) identify and quantify uncertainties in key input parameters, (2) adequately propagate these uncertainties from input through to output, particularly for three-dimensional models, and (3) clearly communicate this information to decision makers.

Several potential pitfalls can be avoided. Some of the most common are:

- not keeping the goal of the model in mind as it is built to direct and focus the appropriate levels of effort during the model construction
- not clearly identifying which uncertainties in input parameters will have significant impacts on the results and subsequent business decisions
- ignoring the uncertainties because not enough time or a lack of knowledge on how to adequately handle them exists
- working hard on issues that we are comfortable of working on, instead of on those that are truly important
- not recognizing feasible alternative scenarios

Why consider uncertainty? Is not a single deterministic case sufficient for analysis? Consider the simple case of estimating charge volume to a trap. The necessary minimum charge volume required for success (i.e., low charge risk) and a range associated with this minimum charge have been defined and are illustrated by the vertical black solid and dashed lines, respectively, in Figure 1. If a model predicts a charge volume greater than the minimum, then it might be said that little or no charge risk exists. Similarly, if a model predicts a charge volume less than the minimum, then we might say that a significant charge risk exists. These cases are illustrated in Figure 1 and are labeled "low risk" and "high risk," respectively. However, the perception of what is low risk and what is high risk can change greatly when the probability of an outcome is considered. In this example, the difference between low risk and high risk becomes less definitive, as indicated in Figure 1. Although this is a simplistic illustration, all of the key input parameters in a basin model have the potential to cause this degree of ambiguity in the final results. For that reason, estimates of the range of possible outcomes are as important to the final analysis as estimates of the most likely outcome.

Table 1. Some of the input parameters commonly used in basin models and their potential effects on model results.

Property	*Effect*
Depths (or isopachs)	Thickness of each unit controls the relative amount of each rock type (see bulk rock properties) and controls the depth and, therefore, temperature and maturity of each stratigraphic unit.
Ages	Ages of each surface and event control timing and thereby control the transient behavior in the model.
Bulk rock properties Stratigraphy/lithology Compaction curves Thermal conductivity Density, heat capacity Radiogenic contribution	The bulk rock properties control the thermal and fluid transport properties (thermal conductivity, density, heat capacity, radiogenic heat, and permeability) that control thermal and pressure evolution in the model.
Missing section (erosion)	The amount of missing section controls the burial history of sediments below the associated unconformity. The burial history controls bulk rock properties (primarily through the compaction history) and the temperature and pressure through time.
Fluid properties	Fluid properties control bulk rock properties in the porous rocks, migration rates, and affect seal capacity.
Source properties Thickness, original total organic carbon and hydrogen index Kinetics, retention/expulsion model	Source properties control the timing, rate, and fluid type for hydrocarbon generation and expulsion from the source rocks.
Bottom boundary condition Heat flow (or temperature)	The bottom boundary condition directly affects the temperatures throughout the model. The effects are roughly proportional to depth.
Top boundary Temperature, and water depth	Changes in the top temperature boundary condition affect the steady-state temperature through the model ~1:1. Changes in water depth can affect the top temperature boundary condition and the pressure at the mudline.
Calibration data Temperature, maturity, observed hydrocarbons including known accumulations Lithology, and rock properties	The quantity of good calibration data has a direct impact on the quality of the model and the reliability of predictions based on model outputs.

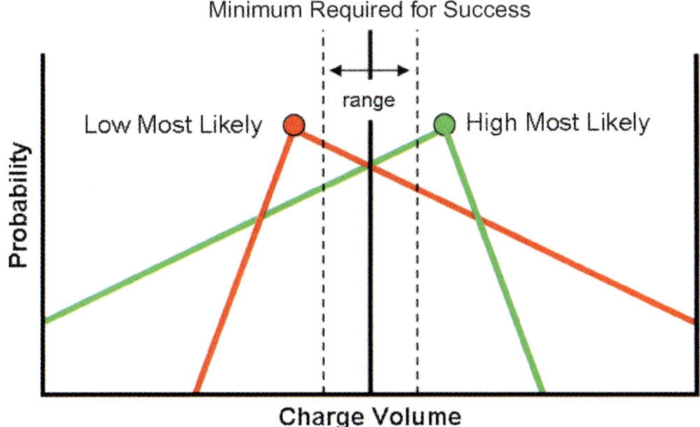

FIGURE 1. Illustration of the importance of considering uncertainty in an analysis. The "High Most Likely" case (green) has a most likely charge greater than the minimum and the "Low Most Likely" case (red) has a most likely charge less than the minimum. However, a consideration of the probability distributions (triangular distributions in this example) can alter our perception of what is "low risk" and what is "high risk."

Handling Uncertainty

The degree of effort required to adequately handle uncertainty will vary for different parameters and model objectives. In thinking about how to quantify uncertainty, it is commonly useful to divide input parameters into two groups: (1) scalars and (2) maps and volumes. Assigning and propagating uncertainties in scalar values is rather straightforward; assigning and propagating uncertainties in maps and volumes is far less straightforward.

Examples of scalars in a model are the age of a surface or an event or the coefficient in a rock property equation (e.g., compaction, thermal conductivity, or permeability). Probability distribution functions can be used to describe most likely values and the estimated range of scalar values. These distribution functions are used to describe the range and likelihood of values and then to propagate the described uncertainty through a basin model. Simple distributions such as uniform (rectangular), triangular, normal, and log normal are commonly used, but any distribution can be used as long as it honors the observed data while properly accounting for any dependencies.

Examples of maps and volumes include depth surfaces (including paleo–water depths), facies distributions, and basal heat-flow maps. In general, quantifying uncertainties in maps and volumes is much more difficult than quantifying uncertainty in scalars because they consist of multiple values and the spatial dependencies of these values are commonly not easily characterized. Varying each location in a map or volume independently of others is not an ideal solution as this does not consider spatial dependencies. Several approaches have been used in the industry to address the issues associated with maps and volumes. These approaches include simple additive and multiplicative factors, geostatistical approaches, forward modeling, and weighting of multiple scenarios.

Consider two common cases: (1) uncertainty in the depth to a surface generated during a two-dimensional seismic interpretation and (2) uncertainty in the distribution of lithologies within an isopach. Various approaches are available to address each case. In the first case, three primary sources of uncertainty are present: (1) picking the correct surface from the seismic and carrying it throughout the data, (2) interpolation of the surface between the data constraints, and (3) the time-to-depth conversion. In the second case, a different set of uncertainty issues present themselves. Uncertainty in the values of the facies properties and uncertainty in the spatial distribution of those values exist.

Options for handling uncertainty in the depth to a surface might include using a stochastic additive or multiplicative factor to handle uncertainties in interval velocities and interpolation uncertainties using a conditional geostatistical technique (such as Gaussian simulation) to handle estimation away from well control and/or generating multiple scenarios to handle uncertainty in interpretation. Options for handling uncertainty in the distribution of lithologies might include using a conditional geostatistical technique to handle estimation away from data control, forward modeling to handle extrapolation to areas with no data control, and generation of multiple scenarios to handle uncertainty in interpretation and understanding. The best approach in any given situation depends on the type of input parameter under consideration and the available data.

APPROACH

The workflow presented involves the following steps:

Step 1: Identify the purpose(s) of the model.

Step 2: Develop a base-case scenario.

- The objective of this step is to build a base-case model to be used in subsequent sensitivity analyses.

Step 3: Identify input parameters whose uncertainty might affect the output property of interest and estimate the preliminary ranges of uncertainty for these input parameters.

Step 4: Perform screening simulations to identify key input parameters.

- In the screening step, the parameters whose uncertainty has a significant effect on the output property of interest are identified. The output property of interest should be selected with care and in light of the overall objective of the modeling exercise. For example, given the same model, the key uncertainties are likely to be different depending on whether prediction of overpressure, hydrocarbon charge, or oil quality motivated the construction of the model. The method involves conducting a set of deterministic simulations where each parameter is independently varied from the minimum to the maximum value. The results are plotted using a tornado or similar plot to aid in identification of parameters whose uncertainty has the greatest effect on the desired result. In this evaluation, it is important to consider potential nonlinearities in parameter behavior and to reevaluate the notional base-case model created in step 2.

Step 5: Evaluate the range of uncertainty in key input parameters.

- Once the key parameters have been identified, uncertainties in these parameters are quantified and assigned. The quantification in this step is generally more rigorous than in step 3 (preliminary screening). Quantification typically involves selecting and populating a probability distribution (e.g., uniform, triangular, normal) for each of the key input parameters identified.

Step 6: Propagate the uncertainty in key input parameters through to the output properties of interest via Monte Carlo or similar analysis.

- The final step is a Monte Carlo simulation where values for the input parameters identified in step 4 are randomly selected from the distributions assigned in step 5. The key results from each realization are saved for subsequent evaluation.

Step 7: Iterate as needed to fine tune the input parameters and dependencies between input parameters, to fine tune error bars and weights for calibration data, and to improve the base-case scenario.

Although several approaches could be used to quantify uncertainty in models, the approach presented uses Monte Carlo simulation. Monte Carlo simulation has the advantage of being (1) able to handle any probability distribution function, (2) able to account for dependencies between variables, and (3) straightforward to implement. It is also typically straightforward to analyze the results of the Monte Carlo simulation. The disadvantages include that it may require a large number of realizations to adequately sample the possible solution space, and it may be difficult to adequately develop probability distribution functions or the required realizations, particularly for maps and volumes.

Hypothetical Example

A hypothetical example is presented to illustrate the approach described. Although the geology is synthetic, it was constructed with realistic basin modeling issues in mind. In this example, the traps of interest formed about 15 Ma. The primary question addressed by the model is, "What is the volume of oil charge to each of the traps during the last 15 m.y.?"

For the purposes of this illustration, the migration analysis has been simplified, and it has been assumed that a present-day map-based drainage analysis is sufficient. A map view of the key surface for the map-based drainage analysis is shown in Figure 2, and a cross section through the model is shown in Figure 3. A burial history curve at location X in Figure 2 is shown in Figure 4. Also shown in Figure 4 are three potential hydrocarbon source rocks, Upper Jurassic, Lower Cretaceous, and lower Miocene. The sources are modeled as uniformly distributed marine source rocks with some terrigenous input.

Step 1: Identify the Purpose of the Model

As previously stated, the purpose of this model is to estimate the volume of oil charge to individual traps during the last 15 m.y.

Step 2: Develop a Base-Case Scenario

The next step is to develop and calibrate a base-case scenario. Values for the selected parameters used in this example are listed in the "Most Likely" column of Table 2. In this hypothetical model, only one calibration point is present, so a match to the data is relatively straightforward, but is also nonunique (Figure 5). The

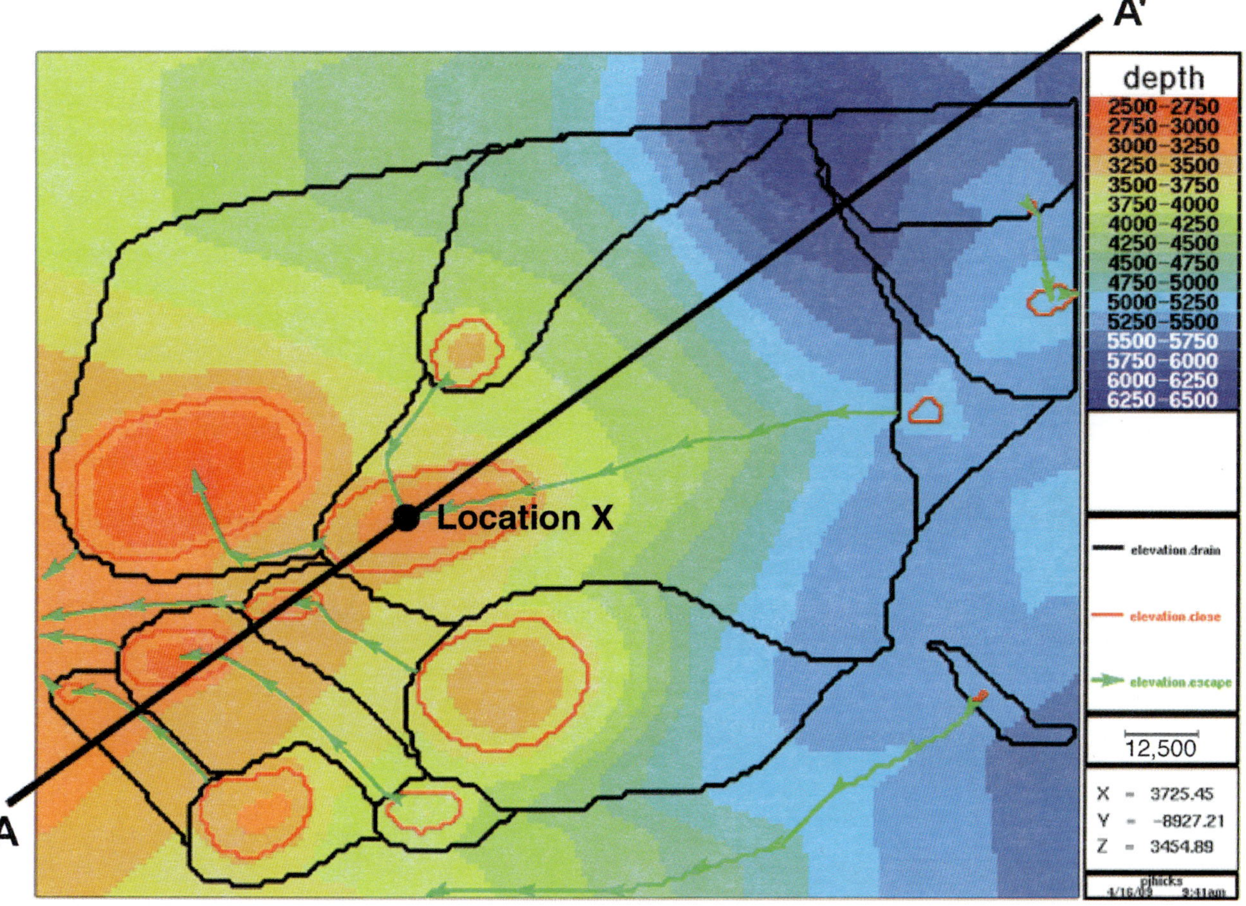

FIGURE 2. Present-day depth structure map (meters) of the key migration surface. The outlines of the drainage polygons are shown in black, the closures are outlined in red, and the escape paths are shown in green. The scale bar in the legend represents 12,500 m (41,010 ft).

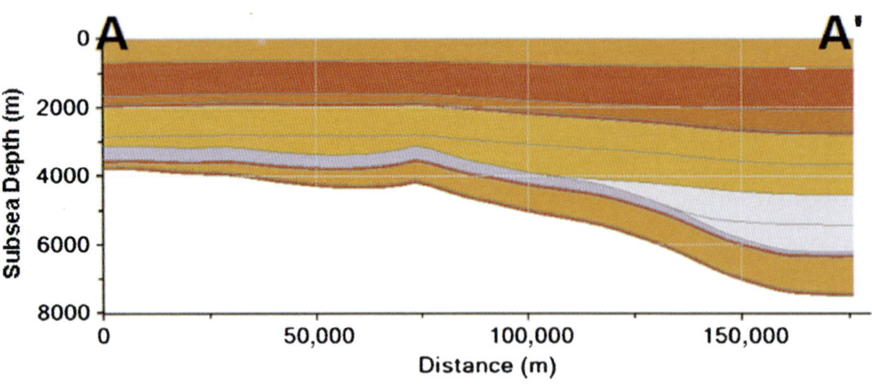

FIGURE 3. Cross section through the model along the line AA′ shown in Figure 2.

uncertainty around the single temperature measurement (±15°C) is indicated by the error bars.

Step 3: Identify and Estimate Uncertainty in Input Parameters

In addition to the uncertainties in source type, kinetics, and thickness, it is hypothesized that uncertainties in the depths, amount of missing (eroded) section, lithology (rock properties), shale grain conductivity, radiogenic heat contribution, surface temperature, and basal heat flow (magnitude and timing of extension) could significantly affect the outcome of the model. The estimations of uncertainty in these input parameters are listed in the "Minimum" and "Maximum" columns in Table 2.

Step 4: Perform Screening Simulations to Identify Key Input Parameters

Evaluating the sensitivity of results to individual parameters involves exploring the solution space by running a series of basin model simulations in which each parameter is set equal to the maximum value and then to the minimum value while all of the other parameters are held at their base-case value. This process results in 2N + 1 realizations, where N is the number of parameters for which ranges have been defined. In this example, uncertainties were defined for the surface temperature, the magnitude, age, and duration of the rifting event, the background heat flow, the shale conductivity and radiogenic heat generation, the lithology of the upper Miocene and Pliocene isopachs, the depths, the missing section, and the generative characteristics of all three source rocks. These uncertainties are summarized in the "Minimum" and "Maximum" columns of Table 2.

The simulation results are summarized in a tornado chart (Figure 6). The yields are plotted on a log scale to more clearly examine the low-yield (high-risk) cases. Analysis of this plot provides a good opportunity to think about the problem. What properties are important? How important are they? Are there any surprises? The basin modeler should spend some time evaluating the behavior of each parameter to make sure

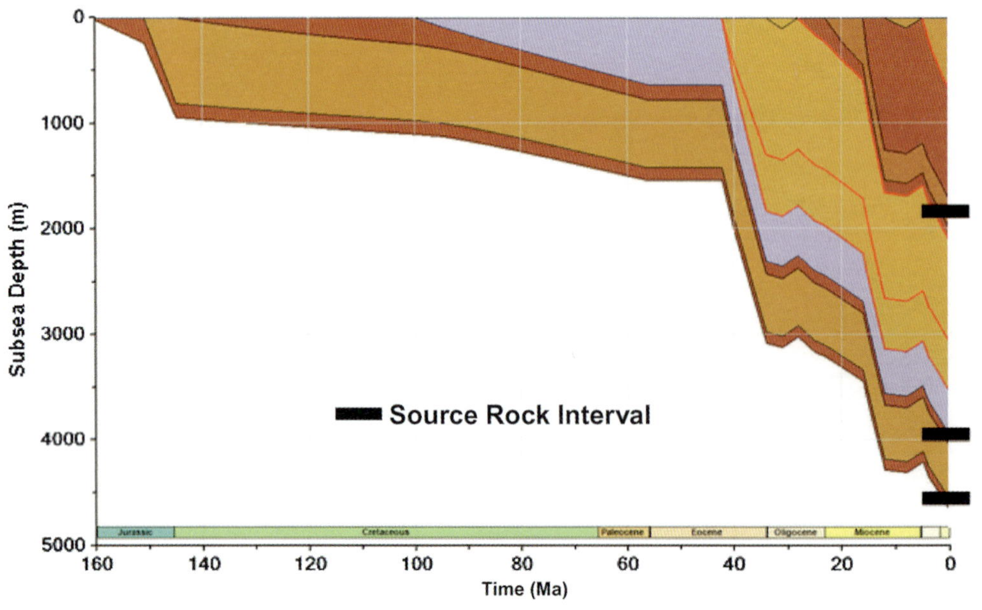

FIGURE 4. Burial history curve for location X represented by the dot in Figure 2. Source rocks are in the middle of each indicated isopachs.

Table 2. The input parameters for the hypothetical example.*

Property	*Minimum*	*Most Likely*	*Maximum*
Surface temperature (°C)	10	20	25
Age of rift (Ma)	40	50	60
Duration of rift (m.y.)	5	10	20
Background heat flow (mW/m^2)	40	50	60
Magnitude of extension (γ, 0–1)	0.1	0.64	0.9
Shale conductivity (W/m.K)	1.87	2.34	2.81
Shale heat generation (mW/m^3)	1.0	2.1	4.1
Lithology of upper Miocene and Pliocene–Pleistocene isopachs	10% less shale; 10% more sand	Base	30% more shale; 30% less sand
Depths	10% shallower	Base	10% deeper
Missing section			
Upper Miocene (m)	10	100	1000
Lower Oligocene (m)	10	100	1
Source properties (all 3 sources)			
Location within isopach	10% from top	Middle	90% from top
Net thickness (m)	15	30	60
oHI (mg-HC/g oTOC)**	500	600	700
oTOC (%)**	4	4	4
OMT lower Miocene source (type II %/type III %)**	50/50	90/10	100/0
OMT other 2 sources (type II %/type III %)**	80/20	90/10	100/0

**The base case used the "Most Likely" values. The "Minimum" and "Maximum" values were used in the screening step and as bounds on the Monte Carlo distributions.*

***oHI = original hydrogen index of the source rock; oTOC = original total organic carbon content of the source rock; OMT = organic matter type of the source rock.*

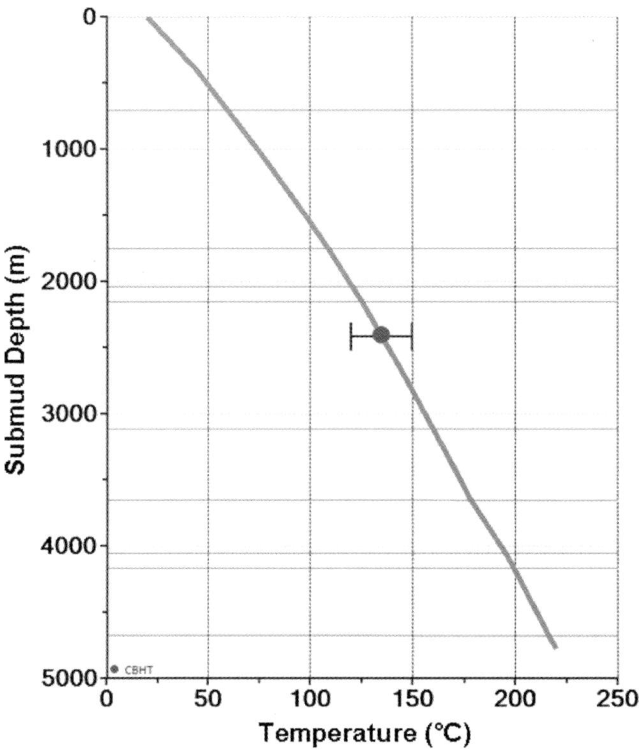

FIGURE 5. Thermal profile calibrated to corrected bottom-hole temperature.

it is understood and makes geologic sense. A limitation of this process is that it does not account for dependencies between input parameters. Thus, the modeler should also give potential dependencies some thought. Examples include a positive correlation between the source rock total organic carbon and hydrogen index, between the mudline temperature and paleo–water depth, between the ages and thickness of isopachs and the timing and magnitude of extension, and between the stratigraphy and paleo–water depths.

Modelers should realize that although it is possible in this sort of analysis for some of these scenarios to be inconsistent with the calibration data, a mismatch on its own is not sufficient reason to narrow the range of values for one of these variables. A particular value of one parameter can cause a mismatch with the data because the value of another parameter is incorrect. If both values were set appropriately, then the model results might be consistent with the calibration data. These interdependency issues will be discussed in more detail later.

The parameters in a tornado plot are sorted by decreasing the range of the net yield resulting from the range given to each input parameter. By constructing the plot in this way, the input parameter uncertainties producing the greatest range in model results are at the top. Twenty-six different parameters were varied in this

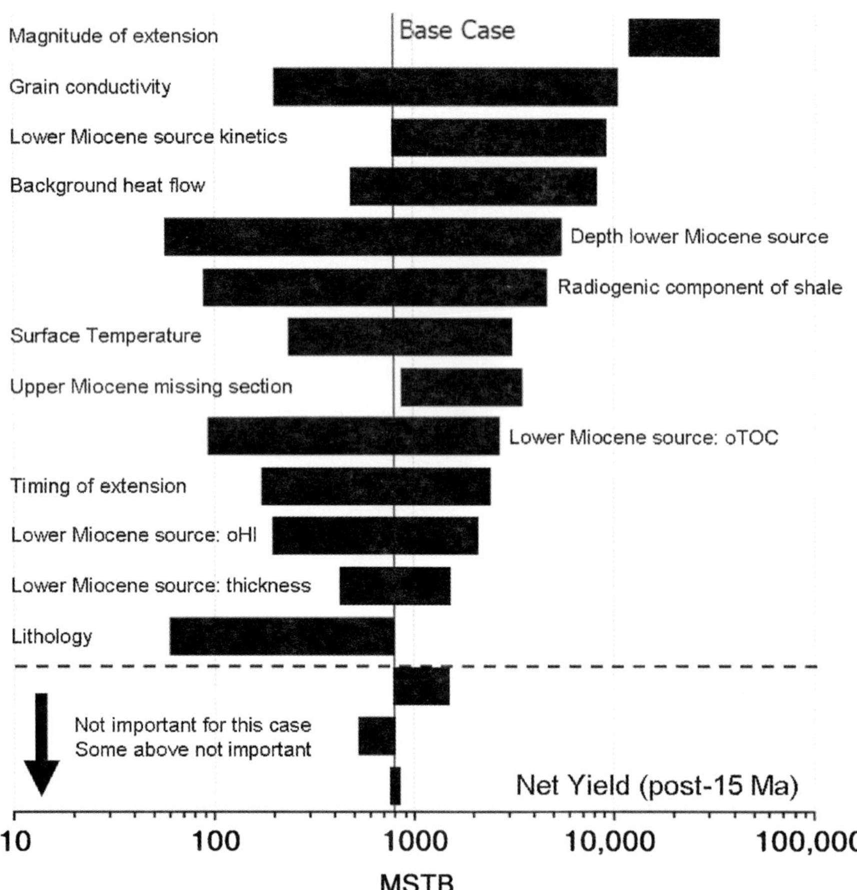

FIGURE 6. Tornado chart for total net oil yields in million stock tank barrels (MSTB) in a selected drainage polygon during the last 15 m.y. The parameters are sorted by the range of net yields (on a linear scale) for each parameter. Uncertainties in net yields caused by uncertainty in the parameters shown below the horizontal dashed line are too small to be important. Uncertainties in net yields caused by the uncertainty in the parameters for some of the parameters shown above the horizontal line may also be unimportant, particularly for ranges with high low sides. oTOC = original total organic carbon; oHI = original hydrogen index.

run; only the 16 parameters with the widest range are shown in Figure 6. The uncertainties associated with the bottom three parameters, and the 10 not shown, are not significant enough to justify the additional effort. Even the uncertainty in some of those "above the line" might not warrant further work. This is because sorting by the widest range does not necessarily equate to sorting by the most important impact on the decisions made based on the results. In this example, the concern is oil yield, so a better sorting might be by the minimum oil yield. In this case, the depth of the Miocene source rock, radiogenic component of the shale, the original total organic carbon in the Miocene source, and the lithology of the Miocene–Pliocene section could be considered the most important, especially if the minimum value of oil yield required for success was on the order of 100 million stock tank barrels (MSTB). The other parameters may not warrant further work or resources.

As mentioned, it is important to remember the question(s) that are being addressed when building a basin model and deciding what input parameters need the most attention. Figure 7 shows the tornado plot for the total expelled hydrocarbon yield as opposed to the net hydrocarbon yield expelled during the last 15 m.y. (Figure 6). Little similarity exists between the parameters that are the most important in these two cases. For the net yields, understanding the timing is critical. For the case in which the total yield is the output property of interest, the generative potential of the Cretaceous and Jurassic sources is more critical than the generative potential of the Miocene source rock.

Although we commonly build our models to address a particular question, in a world of limited time and resources, models can be used for multiple purposes, including purposes that were not originally envisioned at the time the models were constructed. In cases where a model is used for a purpose for which it was not originally built, the modeler should always reexamine the uncertainty in the inputs in light of the new questions being asked.

Nonlinear Behavior

Commonly, when looking at yields or charge, the behavior is nonlinear and at first glance may not be intuitive. In this example, the magnitude of the extension and the amount of lower Miocene missing section do not, as might be expected, bracket the base case. That is, the post-15 Ma yields are higher than the base case for both the minimum and maximum input values. Consider the straightforward nonlinear relationship between hydrocarbon yield after trap formation and the basal heat flow illustrated in Figure 8. At a

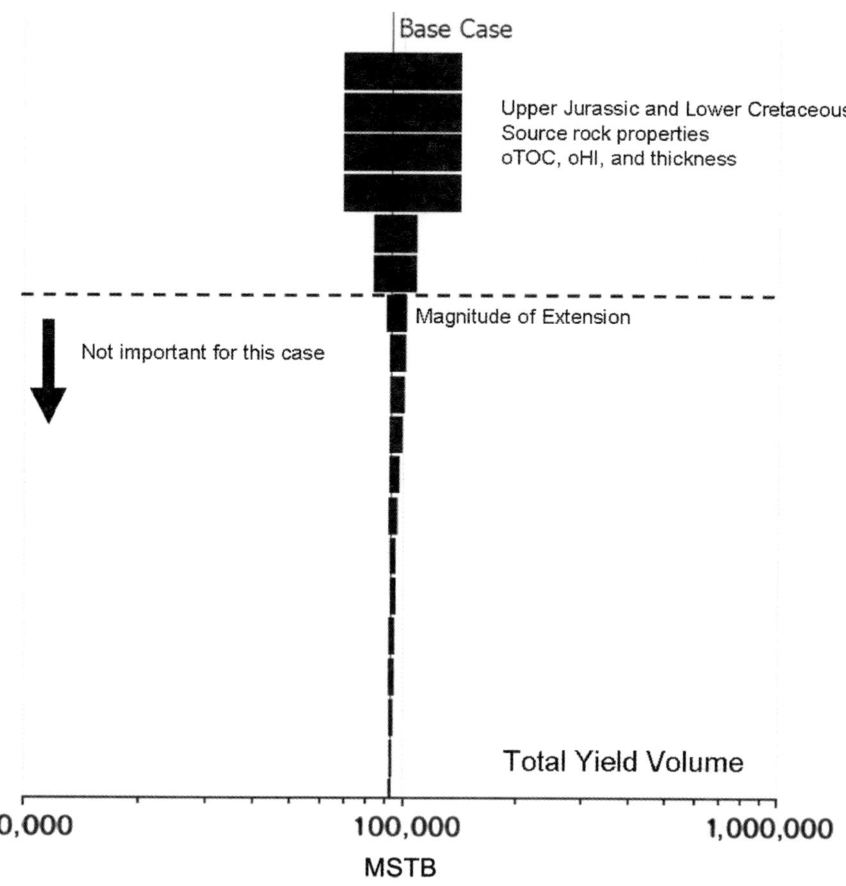

FIGURE 7. Tornado chart for total yield (million stock tank barrels [MSTB]). In this example, uncertainties in the properties of the Upper Jurassic and Lower Cretaceous source rocks have the most effect on the total oil yield. oTOC and oHI are the original source rock total organic carbon and hydrogen index, respectively. oTOC = original total organic carbon; oHI = original hydrogen index.

low heat flow, the source rock is immature and too little hydrocarbons are generated, and at a high heat flow, the source rock is depleted before the trap forms. This behavior is clearly nonlinear because both high– and low–heat flow scenarios can generate less yield than the base case. However, in the example case presented, the relationship between yield and basal heat flow is the opposite, both high– and low–heat flow cases generate more yield than the base case. This seemingly nonintuitive behavior is a consequence of the inclusion of multiple source rocks in the model and becomes clear when we examine the yield for the three extension cases in detail (Figure 9).

In the base case, the Cretaceous and Jurassic source rocks are depleted by about 25 Ma, and the Miocene source rock barely starts generating during the last 2 m.y.

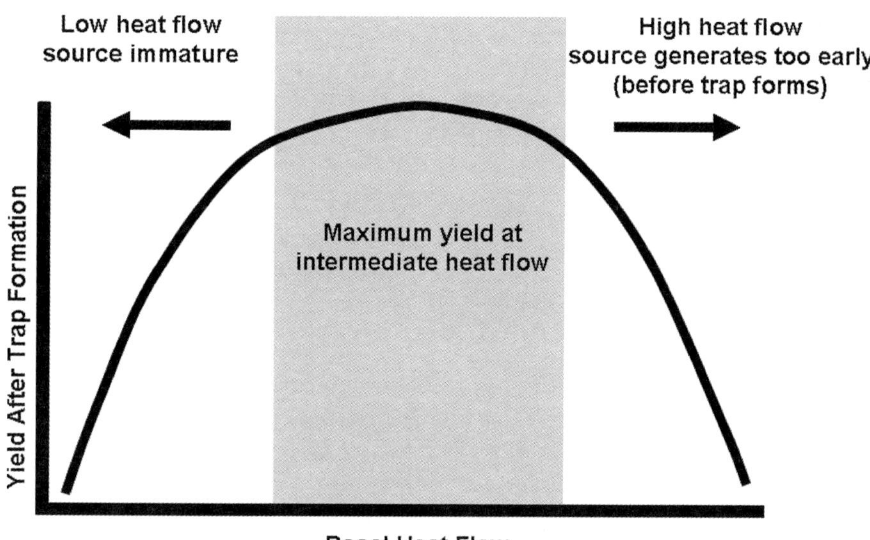

FIGURE 8. Schematic diagram showing calculated hydrocarbon yields after trap formation as a function of heat flow. At a low heat flow, the source is immature present day and a limited amount of hydrocarbon is generated, and at a high heat flow, the source rock is depleted before the trap forms.

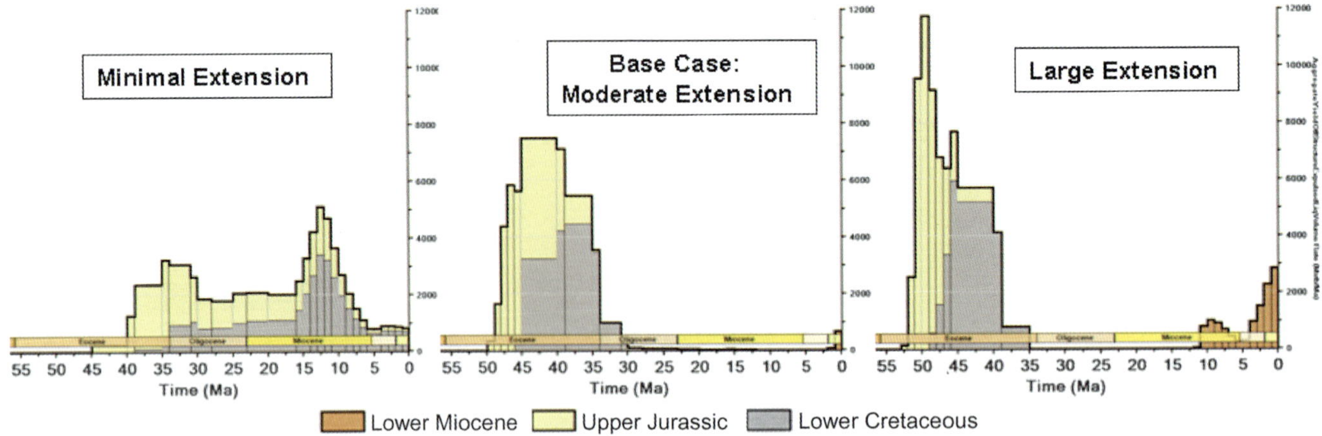

FIGURE 9. Yield timing for minimum, base, and maximum extension cases for the hypothetical model. In the minimum extension case, the Cretaceous and Jurassic source rocks expel during the last 15 m.y. In the base case, the Cretaceous and Jurassic source rocks are depleted by about 25 Ma, and the Miocene source barely starts generating during the last 2 m.y. In the large extension case, the Cretaceous and Jurassic source rocks are depleted, and the Miocene source rock is generating oil.

In the minimal extension case, the yield from the Cretaceous and Jurassic source rocks is delayed (relative to the base case) so a significant amount of generation from the Cretaceous and Jurassic sources occurs during the last 15 m.y. In the large extension case, the Cretaceous and Jurassic source rocks are depleted by about 35 Ma, before trap formation, and Miocene source rocks begin generating at 12 Ma instead of at 2 Ma. The net result is that in the base case, little post–trap formation net yield occurs. Volumes of post–trap formation yield are greater than the base case yield in both the minimal and large extension cases. Calculating post–15 Ma hydrocarbon yield as a function of extension allows this effect to be illustrated clearly (Figure 10).

The behavior of all of the key input parameters should be evaluated in a similar manner. After identifying which parameters have uncertainty that significantly affect the output property of interest and the behavior of each parameter as a result of this uncertainty, the ranges and distributions for these parameters should be examined and analyzed. In this example, the magnitude of extension and the depth to the lower Miocene source rock probably warrant further effort, both in narrowing the range of uncertainty and in refining and gaining confidence in the base case.

Step 5: Evaluate the Range of Uncertainty in Key Input Parameters

After the key parameters have been identified, probability distributions need to be developed for each of these input parameters. Thought should be given to what distribution is appropriate for each parameter. If data are available, they should be considered in determining the best distribution. Commonly, distributions are selected based on experience or expert judgment. If any value in the range has about the same chance as another, a uniform distribution is appropriate. If the modeler would like to honor a most likely value, then a triangular or normal distribution or the log version of one of these distributions might be appropriate. In this example, log-triangular distributions were used for the missing section parameters, and triangular distributions were used for the others. The values associated with these distributions are summarized in Table 2.

Step 6: Propagate Uncertainty to Output Properties of Interest

The goal of this step is to translate the uncertainties in the key input parameters, as described by

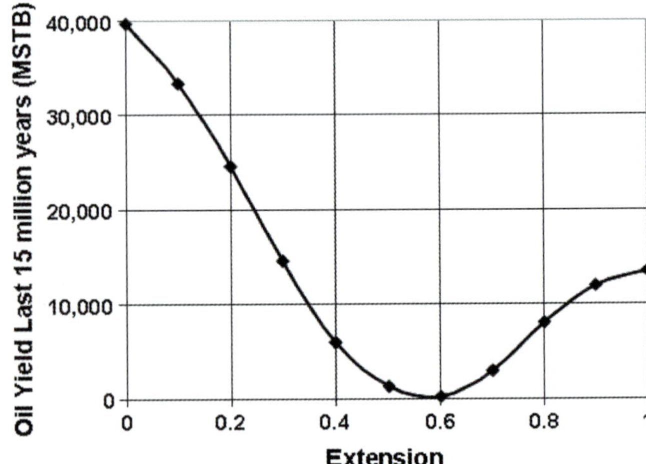

FIGURE 10. Net yields (last 15 m.y.) as a function of the magnitude of extension.

the probability distribution functions, to uncertainties in the output properties. In the Monte Carlo approach, this translation is accomplished in a brute force manner by calculating a large number of possibilities based on the possible distributions of input parameters. In the absence of calibration data, the results are saved and used to build distributions for the output properties; however, calibration data can be used to show that some realizations are more probable than others.

How should realizations that fall outside the range of the calibration data be handled? One approach would be to reject those realizations as not appropriate. An alternative approach is to accept all realizations, but weight the results based on the fit to the calibration data using a weighting parameter related to the fit to each calibration point. "Fit" in this case is based on a least squares analysis. Both approaches have advantages and drawbacks.

In the example presented, the accept-reject approach is used. The advantages of this approach are (1) it is relatively simple and (2) it forces the modeler to quantitatively think about the quality of specific data points (and possibly adjust the error bars during the calibration or Monte Carlo process). In working the calibration data and associated error bars, the modeler is commonly forced to consider how good (or bad) the data may be. The model may not find acceptable trials if inconsistent data are used. In the example, if another temperature measurement was available (i.e., 150 ± 15°C at 4 km [2.5 mi]), then it would be difficult or even impossible to find realizations that matched both points. This illustrates both the need to calibrate the model before starting the Monte Carlo simulation and the importance of addressing the uncertainty in the calibration data. The error bars on one or both points may need to be changed or the thermal conductivity structure of the model might need to be revised. The primary disadvantages of the accept-reject approach are (1) the artificially sharp line between what is counted and what is rejected and (2) the large number of simulations that may be required to obtain a sufficient number of acceptable realizations. In the above example, a temperature realization that is 14.9°C higher than the calibration point is accepted, but one that is 15.1°C higher is rejected, although they might be considered effectively the same temperature.

The primary advantage of the alternate least squares approach is that it gives more weight to realizations that are better fits to the calibration data. This approach (1) allows explicit weighting of different calibration data, (2) produces no sharp divide between accepted and rejected values, and (3) weights realizations that are better fits to the calibration data more than those that are not. The primary disadvantages of this approach are that (1) it is difficult to build a rigorous objective function (a measure of the fit of the model results to the calibration data) and (2) realizations that are clearly inconsistent with the calibration data will be accepted. The sum of the squares of the residuals (or whatever difference measurement is used) may not be a good measure of the probability that a given model output matches a calibration point because some uncertainty ranges may not be symmetric and, because all realizations are accepted, many unlikely realizations may be accepted. In the additional temperature point example, this alternative approach provides results, but the modeler might not ever recognize that none of the realizations are consistent with both data points. If the same weights were used for each temperature, a most likely result would probably split the difference and not match either point. If the modeler did recognize the issue, it could be addressed by giving the points appropriate weights and/or reevaluating the quality of the calibration data.

The purpose of this chapter is not to recommend one of these approaches over the other, but to remind the modeler to think about the quality of the calibration data and to understand that multiple approaches are present to incorporate the data. The best option will commonly be problem dependent. For example, a least squares–type approach might be better for weighting a vitrinite reflectance data point from a cuttings sample, where uncertainty exists regarding the measurement, the sample depth, or whether the sample contains reworked material. A reject-accept approach might be better for considering known hydrocarbon accumulations. A spectrum of intermediate cases exists so a hybrid method might prove to be better in some cases than either of the above approaches. For example, there could be some range around each data point that is considered an exact match, a range around that to which some weighting function is applied and then an outer range that is not acceptable.

Figure 11 shows the resulting thermal profiles for 100 realizations of the hypothetical model described. In panel A of Figure 11, the first 100 realizations are accepted. In panel B of Figure 11, the first 100 realizations within the temperature error bars are accepted. In this filtered case example, it is assumed that a reasonable estimate of the accuracy of the single temperature point exists. Given that assumption, realizations outside the error bars are rejected and those inside the error bars are accepted. If the estimate of the accuracy of the temperature data was less precise, the least squares approach might be a good, or even a better, approach. In either case, if the calibration data are to be given any weight, taking all the realizations and weighting them equally would be a poor choice.

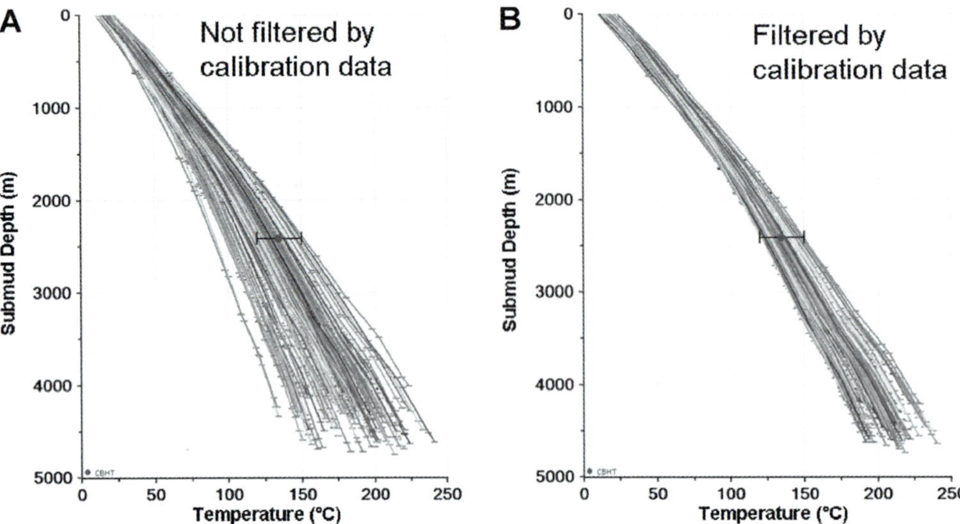

FIGURE 11. Thermal profiles from 100 Monte Carlo realizations. (A) Results without filtering based on the calibration data. (B) Results with filtering based on the calibration data.

The results from a Monte Carlo simulation can be displayed and analyzed in several ways. For example, Figure 12 shows the exceedance probability curves for the yield during the last 15 m.y. for one of the drainage polygons. Curves are shown for both the filtered and unfiltered cases (Figure 11) to illustrate how rejecting realizations that do not honor the observed temperature data affect the exceedance probability. In this example, filtering the simulation results significantly limits the probability of larger oil yields. Figure 13 shows a map of the probability of the oil yield (summed for all three source rocks) during the last 15 m.y. being greater than 1 MSTB/km^2. The map shows regions that have a high probability of having generated a specified yield (1 MSTB/km^2 in this case), regions that have a low probability, and regions where significant uncertainty exists whether or not a specified yield was generated.

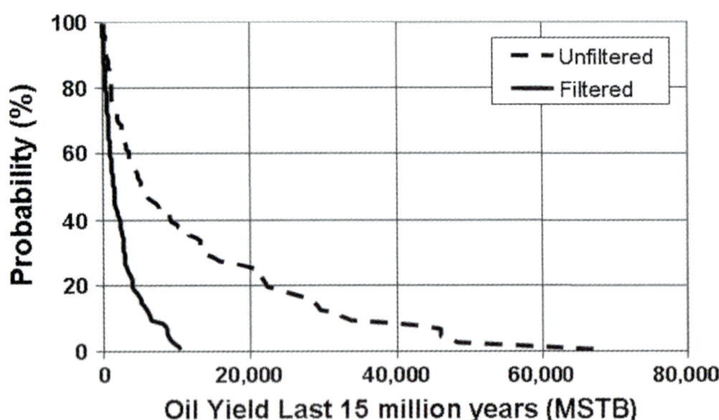

FIGURE 12. Exceedence probability curves for the oil yield over the last 15 m.y. for a selected drainage polygon for the two cases shown in Figure 11.

SUMMARY

This chapter outlines an approach to identify and quantify uncertainties in input parameters and to propagate these uncertainties through the prediction of properties such as hydrocarbon charge in basin models. This approach emphasizes the importance of fully understanding the problem to be addressed by the model, developing a base-case scenario, identifying and estimating the uncertainty in input parameters, screening these uncertainties to identify key input parameters, fully evaluating the ranges of uncertainty in the key input parameters, and propagating these uncertainties through a model to the output property of interest.

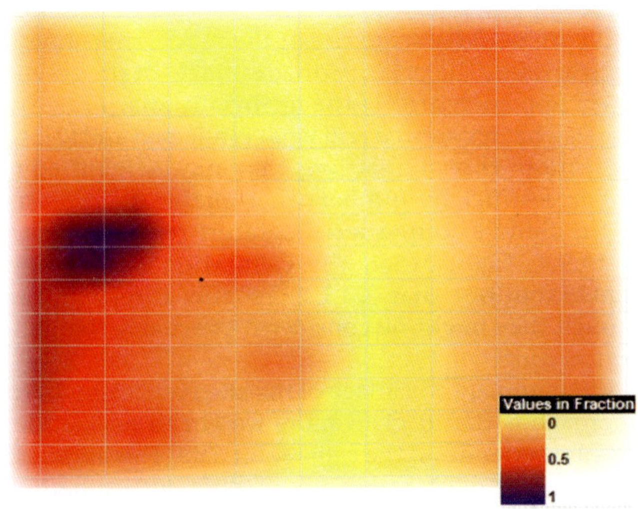

FIGURE 13. A probability map of net yield greater than 1 million stock tank barrels (MSTB)/km^2. Map has the same areal extent as the map in Figure 2.

The approach presented is not new, but we believe it, and similar approaches are underused, and the consideration of uncertainty within basin models has not been given the attention it requires. Basin modelers commonly cannot confidently answer questions, such as "How robust is your most likely case?" or "Given the data, what else might be likely?" or "What uncertainties and/or possible scenarios have not been considered?" A key is to think broadly before starting, focus on the question being addressed, consider alternative models and concepts, and identify and adopt workflows focused on what matters.

ACKNOWLEDGMENTS

We thank Robert Funnell and an anonymous reviewer for their useful comments.

Field trip participants view bubbling mud and steam from an active fumerole in the Sulfur Springs thermal area of The Geysers geothermal field.

13

Pegaz-Fiornet, S., B. Carpentier, A. Michel, and S. Wolf, 2012, Comparison between the different approaches of secondary and tertiary hydrocarbon migration modeling in basin simulators, *in* K. E. Peters, D. J. Curry, and M. Kacewicz, eds., Basin Modeling: New Horizons in Research and Applications: AAPG Hedberg Series, no. 4, p. 221–236.

Comparison between the Different Approaches of Secondary and Tertiary Hydrocarbon Migration Modeling in Basin Simulators

Sylvie Pegaz-Fiornet, Bernard Carpentier, Anthony Michel, and Sylvie Wolf

IFP Energies Nouvelles, Rueil-Malmaison Cedex, France

ABSTRACT

Two major techniques are commonly used to model secondary and tertiary hydrocarbon migration: Darcy flow and invasion percolation. These approaches differ from each other in many ways, most notably in the physical modeling, the methods of resolution, and the type of results obtained. The Darcy approach involves not only buoyancy, capillary pressures, and pressure gradient, but also transient physics, thanks to the viscous terms. Although it can be numerically difficult and therefore time consuming, it is appropriate for slow hydrocarbon movement and it is able to provide a good description of cap-rock leakage. The invasion percolation approach, at least in the context of the implementation used in our examples, does not consider either viscosity or permeability; only buoyancy and capillary pressures drive the hydrocarbon migration. This method is relatively quick and especially useful to simulate secondary migration. Nevertheless, the viscous terms cannot be universally neglected as they can impact the timing of trap filling.

INTRODUCTION

This chapter addresses the modeling of the two main processes that occur at geologic time scales in sedimentary basins, namely, secondary and tertiary hydrocarbon migration. Before elaborating on the scope of our study, let us briefly recall the nature of these phenomena in the context of the limited physical properties taken into account in basin modeling.

Secondary migration is the movement of hydrocarbons along a carrier bed from the source rock to the trap. As shown by Schowalter (1979) and England et al. (1987), it can be accounted for by three physical mechanisms.

The first and main driving process is buoyancy. "When two immiscible fluids (hydrocarbon and water) occur in a rock, a buoyant force is created due to the density difference between the hydrocarbon phase and the water phase. The greater the density difference, the greater the buoyant force for a given length hydrocarbon column (always measured vertically)" (Schowalter, 1979, p. 10).

DOI:10.1306/13311438H43472

The second process is hydrodynamics. It adds a force that may be in any direction, depending on the nature of the flow involved (England et al., 1987). Indeed, the buoyant force can be reduced or increased when a hydrodynamic condition exists in the subsurface. However, the effects of hydrodynamics are not always of the utmost importance (Carruthers, 1998).

The third process is capillary pressure. This is in fact a resistance effect that controls the hydrocarbon trajectories. The factors that determine its magnitude are the radius of the pore throat of the rock, the hydrocarbon-water interfacial tension, and wettability (Schowalter, 1979).

The combination of these three processes leads to the ascent of hydrocarbons through the carrier beds until the capillary pressure is sufficient to offset the effects of the difference of densities and hydrodynamics. Note that we have neglected compaction as a driving force for secondary migration because this is commonly assumed.

Tertiary migration is the leakage of hydrocarbons from traps. It is attributed to capillary leakage, hydraulic leakage, and molecular diffusion (Sylta, 2004). Cap rock leakage is possible when the driving processes (buoyancy, pressure gradients, and molecular diffusion) exceed the resistant factors (capillary entry pressure or permeability) of the confining barrier (Thomas and Clouse, 1995; Burrus, 1997). In a normal pressure accumulation, a cap rock reaches its maximum seal capacity when the pressure generated by the hydrocarbon column is equivalent to the capillary entry pressure of the barrier.

For an accumulation in overpressure (i.e., the difference between the fluid pressure and the hydrostatic pressure), the direction and the magnitude of fluid circulations are controlled by the global pressure field and the buoyancy generated by the hydrocarbon column is not the main force. The rate of leakage is then controlled by the permeability, the fluid viscosity, and the pressure gradient (Watts, 1987; Schlomer and Krooss, 1997).

Two major techniques are commonly used to model secondary and tertiary hydrocarbon migration: Darcy flow and invasion percolation. These approaches differ from each other in many ways, most notably in the physical modeling, the methods of resolution, and the type of results obtained. This chapter aims to summarize, compare, and illustrate these two techniques through particular case studies. Its purpose is to highlight the capabilities of the different methods developed and to underline the advantages and drawbacks of each. Although it does not claim to add any insight into the physics and mechanics of hydrocarbon migration itself, we believe that such a comparison can help the practitioners who use migration modeling.

This chapter is outlined as follows. First, we describe the Darcy approach, its physical principles, some standard numerical methods of resolution, and their limitations. The next section is devoted to the invasion percolation approach, its algorithm and limitations. We then recapitulate the characteristics of each approach. Last, we illustrate their main differences through examples.

DARCY APPROACH

Darcy flow models assume that hydrocarbon displacement honors the Darcy's law extended to multiphase fluids (Bear, 1972; Marle, 1972). Migration is driven by buoyancy, fluid pressure field, and capillary pressure. Darcy migration is simulated by solving partial differential equations, and the numerical treatment of the full set of equations is generally considered computationally costly and quite complicated (Schneider, 2003).

Physical Principles

Based on the results of experiments on the water flow through beds of sand, Darcy (1856) formulated the law

$$\vec{U} = -\frac{K}{\mu}\nabla P \qquad (1)$$

where $\vec{U}$ is the Darcy velocity (m/s), K is the permeability of the rock (m^2), P is the pressure (Pa), and μ is the viscosity of the Newtonian fluid (Pa.s).

From the theoretical viewpoint, it has been proved that Darcy's law is not a constitutive law, but a simplified form of the homogenized Navier-Stokes model (Hubbert, 1956; Irmay, 1958; Bear, 1972; Whitaker, 1986). The coefficient $\frac{K}{\mu}$ is a viscous term because of friction at the solid-fluid interface. Moreover, to generalize Darcy's law to multiphase flow, the simplest approach is to assume that each fluid phase maintains a network of passages; the wetting fluid in the larger pores, with friction between fluid and solid (Bear, 1972).

In addition to the three main processes already mentioned (buoyancy, capillary forces, and pressure gradient), the extension of Darcy's law to multiphase flow in porous media uses the concept of relative permeability. For two phases, this permeability correction term reflects the permeability reduction of a fluid flow caused by the presence of the second fluid in the porous medium (Guérillot and Kalaydjian, 1988). The extended Darcy's law then reads

$$\begin{cases} \overrightarrow{U_w} = -\frac{\overline{\overline{K}}Kr_w}{\mu_w}(\nabla P_w - \rho_w \vec{g}) \\ \overrightarrow{U_h} = -\frac{\overline{\overline{K}}Kr_h}{\mu_h}(\nabla (P_w + Pc) - \rho_h \vec{g}) \end{cases} \qquad (2)$$

where $\overrightarrow{U_\alpha}$ is the Darcy velocity of the phase α (m/s), μ_α is the viscosity of the phase α (Pa.s), $\overline{\overline{K}}$ is the intrinsic permeability tensor of the porous media (m^2), Kr_α is the relative permeability, P_w is the pore pressure in the water phase (Pa), Pc is the capillary pressure (Pa), $\vec{g}$ is the gravitational acceleration vector (m/s^2), P_α is the density of the phase α (kg/m^3), w refers to water phase, and h refers to hydrocarbon phase.

The generalized Darcy's law can be adequately applied to basin modeling if we accept that hydrocarbon migration occurs as a separate fluid flow, in a different phase from water, for both primary and secondary migration (England et al., 1987; Durand, 1988; Ungerer et al., 1990; Burrus, 1997).

Numerical Modeling

Darcy model is classically coupled with a pressure-compaction model. This means not only that hydrocarbon migration depends on the pressure-compaction computation, but also that the pressure-compaction is influenced by the migration computation. Basin modeling simulators commonly simultaneously solve the multiphase Darcy's law, the mass-conservation equations for solid and fluids, and a compaction law. To solve this set of equations, finite difference, finite element, or finite volume methods are used for the spatial discretization. Various time schemes are also used for the transport equations: the Impes with an implicit treatment for the pressure computation and an explicit one for all other unknowns; the Impims based on an implicit treatment for all the unknowns. These two time strategies solve sequentially in two separate stages the pressure-compaction problem and the hydrocarbon transport equations. On the contrary, with the fully implicit scheme, we have to solve a coupled system of nonlinear equations for pressure and hydrocarbon saturation.

All of these schemes have distinct advantages and limitations (Wolf et al., 2009), but in all the cases, performing a simulation with a complete Darcy model is expensive in computing time. Indeed, to treat the nonlinearity of the equations, a classical Newtonian scheme is used. The convergence of this scheme may be a delicate issue in some cases and may cause the time step to decrease, particularly when a huge amount of hydrocarbon migrates rapidly. At each Newton iteration, solving the linear system represents a huge time-consuming part of the simulation (Willien et al., 2009). Furthermore, pressure-dependent flow can significantly increase the computing time, especially in highly permeable layers. The management of computing time steps depends also on the strong heterogeneities of the fluid properties. Nevertheless, parallel techniques and specific preconditioners can improve the computing time for the Darcy approach (Requena et al., 2005).

Limitations

It is a classical fact that Darcy's law can be considered as valid only for slow Newtonian flows, that is, for Reynolds numbers between 1 and 10 (Bear, 1972; Burrus, 1997). It is also well-known that Darcy's law breaks down in extremely fine-grained clayey soils. The multiphase nature of the flow is likely to further restrict the validity of the Darcy model. Indeed, in two-phase systems, when the capillary term becomes important at "relatively low pressures, no continuous pathway through the rock is possible, and no flow will occur ... this type of nonlinear behavior is obviously inconsistent with Darcy's law" (England et al., 1987, p. 335).

In an attempt to reduce the computation cost of the Darcy approach, it is commonly suggested to use large cells. This requires permeability, relative permeability, and capillary properties to be upscaled on a low-resolution numerical mesh. However, "the use of constant oil saturation in each computing cell results in too large average saturations being modeled when the vertical migration pathway has to overcome tight zones" (Sylta, 2004, chapter 10, p. 13). Because of this low resolution, the Darcy approach tends to overestimate migration losses during secondary migration.

INVASION PERCOLATION APPROACH

Percolation Theory

The percolation method mathematically deals with disordered media, in which the disorder is defined by a random variation in the degree of connectivity. It can be used in different domains in physics, chemistry, and materials science. Percolation theory is applied to porous media and deals with the description of interconnections of the porous and fractured network (Lenormand, 1981; Guéguen and Dienes, 1989). A regular network of "sites" or "bonds," that may or may not be occupied, represents physical properties (permeability, elastic properties, ...). Each site (or bond) contains the studied physical property that is characterized by a probability of occupation. A "cluster" is defined if several neighboring sites are occupied. These modes of representation of the fractured and porous medium can be used to compute a critical property of percolation (percolation threshold) (Sausse, 1998).

Wilkinson and Willemsen (1983) proposed a new form of percolation theory: invasion percolation. They looked at a wetting fluid (water), the invader, moving another nonwetting fluid (oil) in a porous medium under the action of capillary forces. Wilkinson (1984) then extended this model by adding the effects of buoyancy, which is very important for secondary migration modeling. Invasion percolation models assume that viscous effects can be neglected compared with those of capillary pressure and that the system is in a state of capillary equilibrium. Meakin et al. (2000) used this model in experiments and simulations for secondary migration modeling. Their model included the displacement between fluids in a fractured medium and the effects of wetting fluid flow under the influence of the gradient of hydraulic potential.

Invasion Percolation Algorithm Adapted to Basin Scale

The traditional invasion percolation model assumes that the invading phase is in constant pressure communication and not only in the hydrocarbon accumulations. Carruthers (2003) states that "it is only applicable to small (submeter) systems," and not suitable for basin scale with several kilometers between the source rock and the reservoir zone. Moreover, "it assumes that the invading phase originates from a single point," which is not appropriate for a petroleum system containing several source points (Carruthers, 2003, p. 30).

Carruthers (1998) adapted the traditional invasion percolation algorithm for petroleum migration modeling. This approach of invasion percolation assumes that at basin time scale, hydrocarbons move only under the effects of buoyancy and capillary pressure, which opposes the movement. "Discontinuities within the oil phase are assumed to be ubiquitous except in accumulation zones. The buoyancy force generated, as the result of the pressure head generated in an accumulation, is the only buoyancy force which will drive the oil through a carrier past its equilibrium phase pressure, and migration will always occur in a state of equilibrium" (Carruthers, 1998, p. 185).

For a vertical flow, the relationship governing the invasion percolation migration model is:

$$(\rho_h - \rho_w)g\nabla z > \nabla Pc \tag{3}$$

where z is the depth below sea level.

This model does not consider viscous terms, which are negligible with respect to the capillary terms. Because of inviscid assumptions, the flow is steady state, so transience is imposed by the rate of hydrocarbon generation from the source rock. Under these conditions, this invasion percolation model can be seen as a limit of the Darcy model under local equilibrium assumptions. In this modified form, petroleum migrates under buoyancy into an opposing network of cells populated with capillary pressures, pursuing the lowest entry pressure pathway. When this process proceeds, a migration backbone results, that is, a set of pores through which an oil stringer is able to flow (Carruthers and Ringrose, 1998), and all subsequent migration occurs along the backbone. When a migration backbone reaches a barrier, a hydrocarbon column builds up. The percolation then finds a new pathway with the lowest entry pressure (Burley et al., 2000; Sylta, 2004). Figure 1 illustrates schematically hydrocarbon migration based on the invasion percolation model.

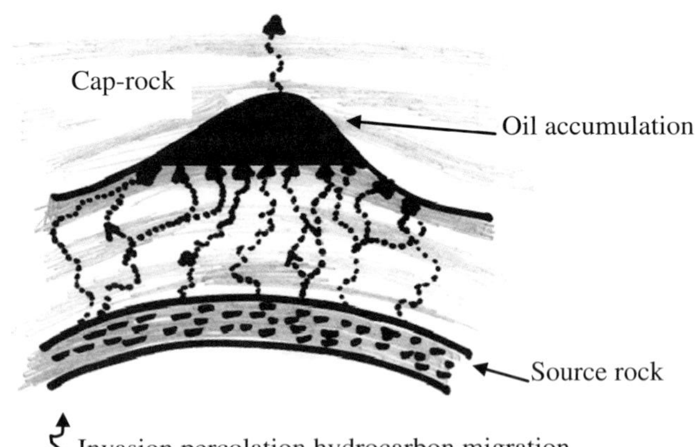

FIGURE 1. Schematic representation of the hydrocarbon migration route in the case of invasion percolation.

During the search of a new migration path, the invasion percolation approach is akin to graph exploration techniques. It is a sequential computation and does not use an iterative algorithm. This method requires a precise distribution of capillary pressures in the basin and of their evolution through time. It can be very efficient in terms of central processing unit (CPU) time, and the cost of the simulation depends only on the size of the area swept by the migration paths and not on the total amount of cells in the model. Only a small set of input parameters exists in the model so they can be easily modified to perform a sensitivity analysis.

Limitations

As secondary hydrocarbon migration in permeable areas occurs along thin stringers, the modified invasion percolation approach, which results in a migration backbone, provides a good description of this process. Nevertheless, Sylta (2004) explains that for a low cap-rock permeability (less than 10^{-2} md), a typical leakage flow rate is high. The leakage in such a situation occurs in a wide area and not through a narrow migration backbone (Hantschel and Kauerauf, 2009). That is why "the method of percolation modeling does not provide a valid description of the cap-rock leakage process if the cap-rock permeability is very low" (Sylta, 2004, chapter 10, p. 11). On the contrary, with Darcy flow, when the rate of hydrocarbon migration into a trap is high compared with the leak capacity of the cap rock, the hydrocarbon column increases in the accumulation and the filling of

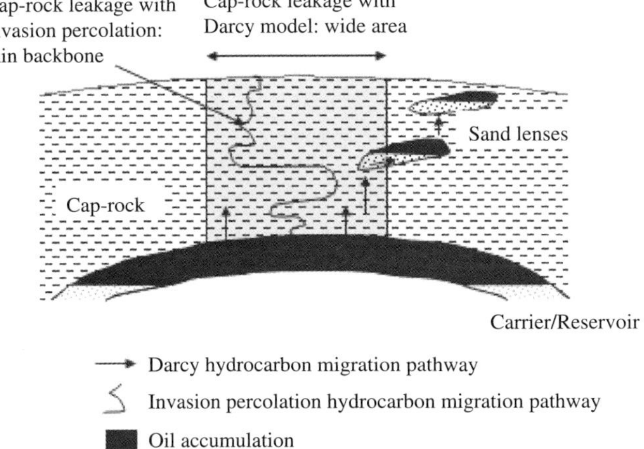

FIGURE 2. Hydrocarbon leakage in cap rock containing sand lenses (modified from Sylta, 2004).

the trap reaches its maximum level. This induces a leak in a wider area to have an equilibrium in the hydrocarbon accumulation. "The Darcy method will distribute hydrocarbons at low saturation within a relatively broad migration 'chimney' above the trap, whereas the percolation will only saturate a very thin migration backbone" (Sylta, 2004, chapter 10, p. 5). This narrow migration backbone through a cap-rock sequence can, in some cases, miss small sand lenses. In systems where filling rates are high, it can be problematic because hydrocarbon losses could be overestimated. Figure 2 shows that with the Darcy approach, as the cap-rock leakage covers a wide area, oil can reach zones beyond the top point of the structure and, in particular, isolated sand lenses. On the contrary, invasion percolation modeling can bypass sand units because they are not on its migration backbone. This limitation is caused by the inviscid assumption carried in the modified invasion percolation approach. Under this assumption, the Darcy model should also degenerate and result in a thin migration backbone through the cap rock.

SUMMARY OF THE COMPARISON BETWEEN THE DIFFERENT APPROACHES

Based on the description of models and our practical experience in basin modeling, we have compared the two approaches. For the invasion percolation migration model, we consider the modified form introduced by Carruthers (1998), which does not consider viscosity and permeability. This section is a summary of the previous ones to highlight the advantages and the drawbacks of each of them. For this, we focused on their differences in three aspects: the modeling, the methods of resolution, and the outputs obtained after a standard simulation with each kind of model.

Modeling

The Darcy migration is a nonstationary model contrary to the invasion percolation approach, which assumes that the viscous terms are negligible with respect to the capillary terms so that the petroleum system can be considered at each time in a quasistatic equilibrium state. The Darcy model is well suited to follow transient flow in low-permeability areas, which is not the case for the invasion percolation. The Darcy approach is able to give a good description of hydrocarbon leakage through mudrock sequences, whereas the invasion percolation sometimes gives an incorrect description of hydrocarbon leakage out of traps and misses small accumulations, but gives a good simulation of secondary processes.

Methods of Resolution

Performing a simulation with a multicomponent and multiphase Darcy model is expensive in computing time because of the complex system of nonlinear partial differential equations that must be solved, but parallel techniques can help improve the performance. On the contrary, invasion percolation is a sequential computation that does not use any iterative algorithm and has a fast computing time.

Results

After a simulation using the Darcy approach, values of pressure, saturation, and hydrocarbon composition are obtained in each cell of a three-dimensional (3-D) block, but postprocessing is needed to identify areas of hydrocarbon accumulation. On the contrary, at the end of an invasion percolation simulation, we obtain areas of pathway and hydrocarbon accumulations that are well identified. Moreover, because of computing time with the Darcy model, only a limited number of migration scenarios are commonly tested. Unlike this approach, invasion percolation allows testing many different migration scenarios by changing the input parameters.

EXAMPLES

The objective of the following examples is to focus on the impact of the previously mentioned differences in the migration methods through two-dimensional

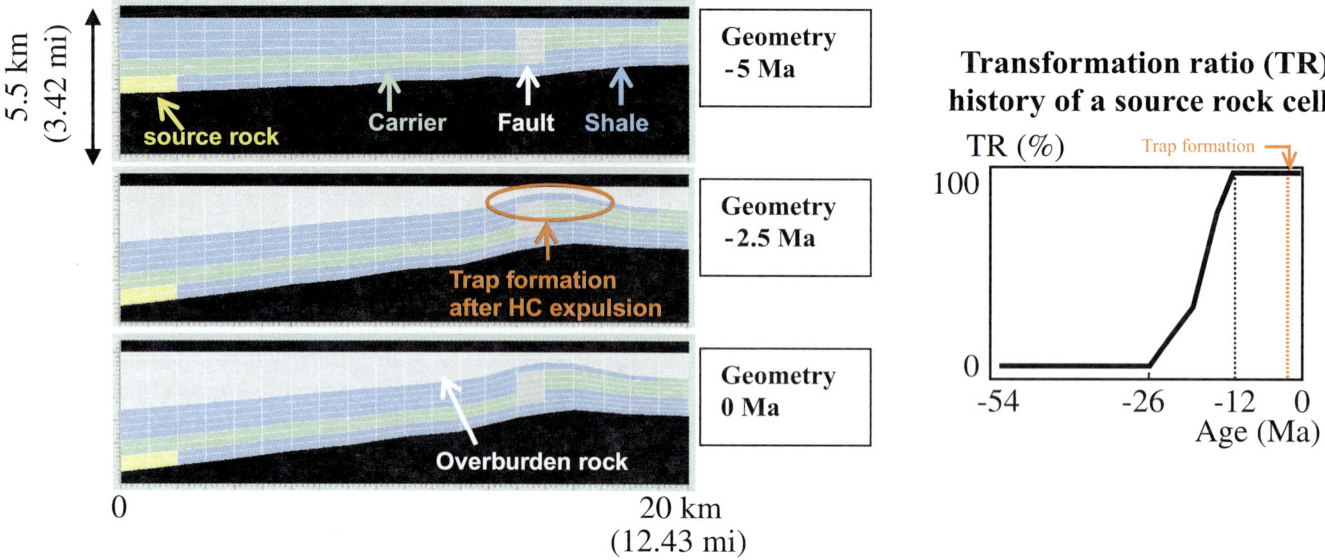

FIGURE 3. First synthetic case: a carrier bed with low slope. Description of lithology, structural evolution, and source rock transformation ratio as a function of time (ages -5, -2.5, 0 Ma). HC = hydrocarbon.

(2-D) synthetic cases and 2-D sections from a real case study. The reason why we have chosen well-controlled 2-D cases instead of 3-D blocks is that the former are easier to analyze and to understand.

As the models share common physical principles, we obtain similar results for several petroleum system simulations, but in some special cases, differences are magnified. For that purpose, we developed a prototype that ensures the same input data for the computational domain, the initial, and the boundary conditions. It also guarantees the same computation of geometry, thermal history, and hydrocarbon generation.

For the Darcy model, the prototype simultaneously solves the mass conservation equations for solids and fluids, a compaction law, and the generalized Darcy equations for two-phase flow. It uses finite volume methods and a fully implicit scheme for the transport equations. Pressures and hydrocarbon saturations are strongly coupled.

For the invasion percolation model, the prototype computes water pressures and porosities using Darcy's law for the single water phase, and afterward, an invasion percolation algorithm is performed for hydrocarbon migration. The computation of pressures is decoupled from the hydrocarbon migration computation. As a consequence, this methodology allows us to observe the effects of the coupling between the pressure-compaction model and the hydrocarbon migration.

In the upcoming sections, we are not interested in the computation times, but prefer to qualitatively compare the results for the locations of accumulations and pathways taken by the hydrocarbons. The capillary pressure model and the permeability and viscosity computations are detailed in Appendix 1. The actual parameters used in each example are enumerated in Appendix 2.

Secondary Migration

Synthetic Case: A Carrier Bed with a Low Slope

This first synthetic example is a geologic section consisting of a carrier bed with a low slope and containing 200 grid cells (Figure 3). At -5 Ma, the trap is not yet formed, but the source rock is mature (the transformation ratio is close to 1) and the hydrocarbons begin migration. At -2.5 Ma, the structural trap is then formed.

We conducted two Darcy simulations by changing the permeabilities in the carrier bed to see their sensitivity on the time of trap filling. Using standard permeabilities for a carrier bed, the migration is not very fast. Hydrocarbons go to the surface but not the total amount expelled from the source rock. Hydrocarbons remaining in the carrier bed can fill the trap. At present day, we observe an accumulation of hydrocarbons (Figure 4).

Using higher permeabilities, although the slope is low, all of the hydrocarbons go to the surface before trap formation. No present-day hydrocarbon accumulation is observed (Figure 5).

With invasion percolation, the permeabilities have no effect because only the capillary pressures are able to provide resistance to hydrocarbon migration. Because hydrocarbon generation occurs before the trap formation, all of the hydrocarbons go to the surface. No present-day accumulation is observed (Figure 6).

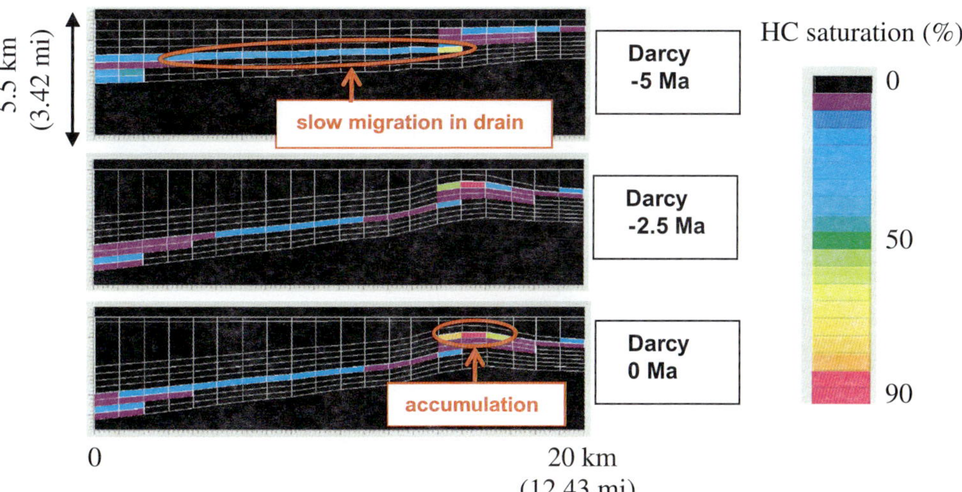

FIGURE 4. First synthetic case: a carrier bed with low slope. Evolution of hydrocarbon (HC) saturation obtained with the Darcy migration model (ages: -5, -2.5, 0 Ma).

With the Darcy approach, which is transient because it solves the viscous term, the filling history is controlled by permeabilities and the slope of the migration pathways, whereas with invasion percolation, only the rate of hydrocarbon generation has an impact on trap filling. In conclusion, the viscous effect cannot be universally neglected.

Real Case Study: Long-Distance Pathways

This first real case study comes from Africa. It corresponds to a Paleozoic and early Mesozoic depression with a thick sedimentary series. This intracratonic basin is characterized by a major late Paleozoic unconformity, with most of the known hydrocarbon accumulations located in the overlying Triassic reservoirs. The source rocks are Paleozoic in age and range from thermally mature at the border of the basin to overmature in the central part. Maturation is controlled by the source rock thermal histories, characterized by a first phase of deepening, followed by an important uplift and a restart of the sedimentation, which lead the source rocks to their present-day maturity levels. Therefore, two phases of expulsion and migration occurred associated with long-distance migration along the unconformity.

Figure 7 shows an enlarged 2-D section of the studied area. The lower left part of this section is composed of two upper Paleozoic source rock layers with a thin overlying carrier bed covered by a cap rock. Its upper right part contains three layers of middle Paleozoic source rocks. High-permeability layers are located above and below these sources. The traps, located on the right, are not represented.

Hydrocarbons expelled from the left lower part of Figure 7 go into the carrier bed. With the Darcy model, all of the hydrocarbons do not go instantaneously into the structural traps, but they are distributed along the carrier bed because of low permeabilities. With the invasion percolation model, the hydrocarbons go directly into the traps as soon as they are generated (Figure 8).

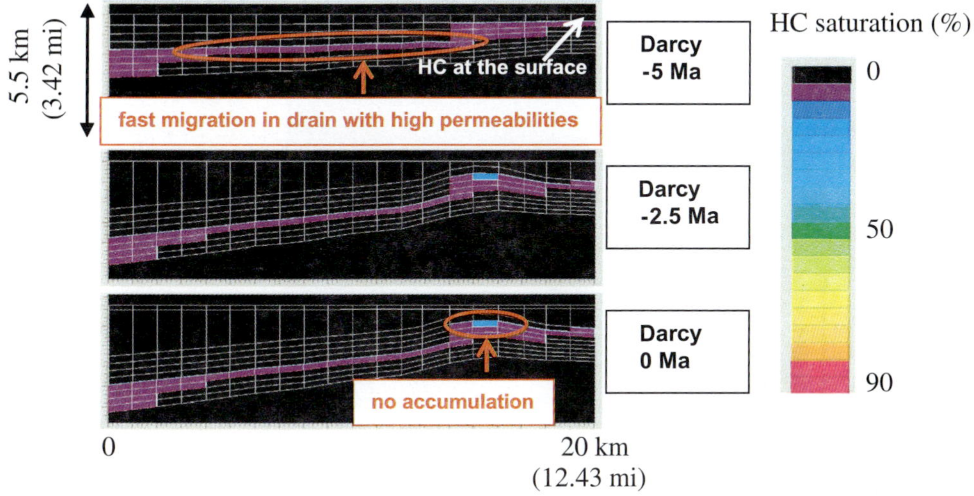

FIGURE 5. First synthetic case: a carrier bed with low slope. Evolution of hydrocarbon (HC) saturation obtained with the Darcy migration model and high permeabilities in the carrier bed (ages: -5, -2.5, 0 Ma).

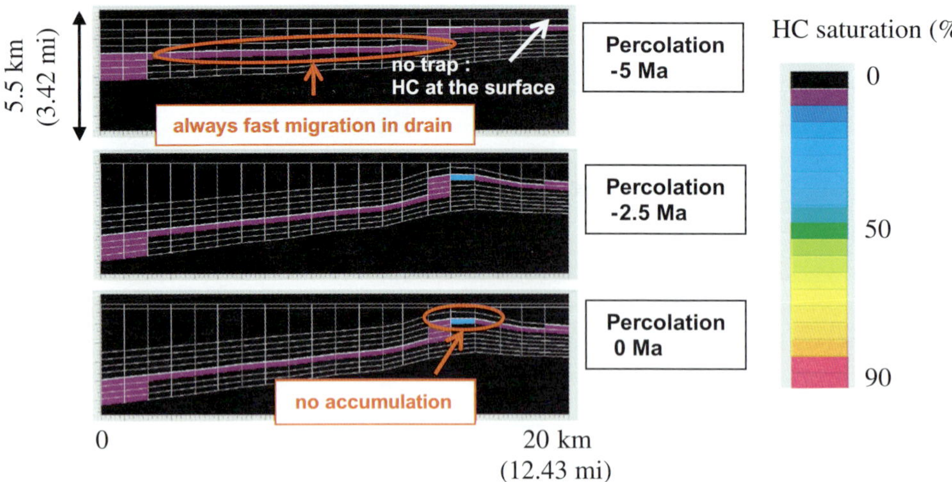

FIGURE 6. First synthetic case: a carrier bed with low slope. Evolution of hydrocarbon (HC) saturation obtained with invasion percolation migration model (ages: -5, -2.5, 0 Ma).

We focus on the history of a cell located in the carrier bed. With Darcy, the hydrocarbon saturation of this cell fluctuates over time. These changes in time depend on the fluid expulsion rate from the source rock and the geometry variation because of compaction or tectonic movement. With invasion percolation, this cell is identified as a migration pathway, so as soon as the hydrocarbon saturation of the cell has reached the critical hydrocarbon saturation, it does not vary.

In the upper right part of the section, hydrocarbons, expelled from the middle Paleozoic source rocks, go above and below these layers with Darcy and accumulate under the source rocks. However, with the invasion percolation model, hydrocarbon migration is mainly driven by buoyancy so preferentially upward, and no accumulation under the source rocks is observed.

Cap-Rock Leakage

Synthetic Case: Sand Lens

This synthetic example is a geologic section composed of a structural trap. Figure 9 depicts this trap under a cap rock that contains a sand lens. Hydrocarbons are expelled from the source rocks and then migrate vertically until they reach the first cap rock. They form an

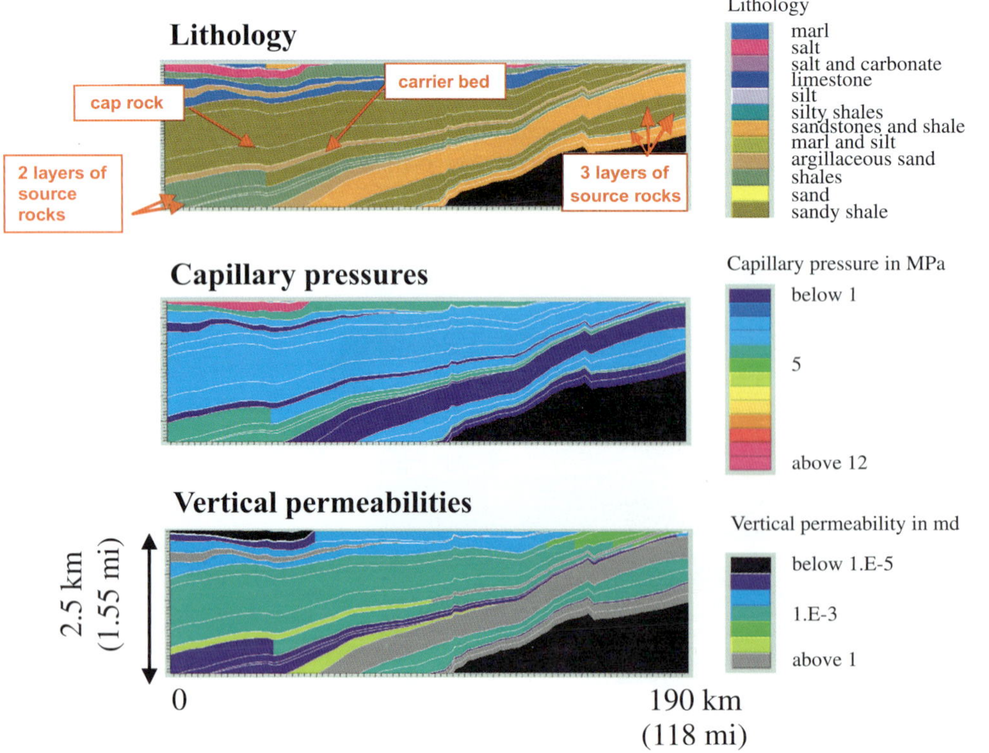

FIGURE 7. First real case: long distance pathways. Description of lithology, capillary pressures, and vertical permeabilities.

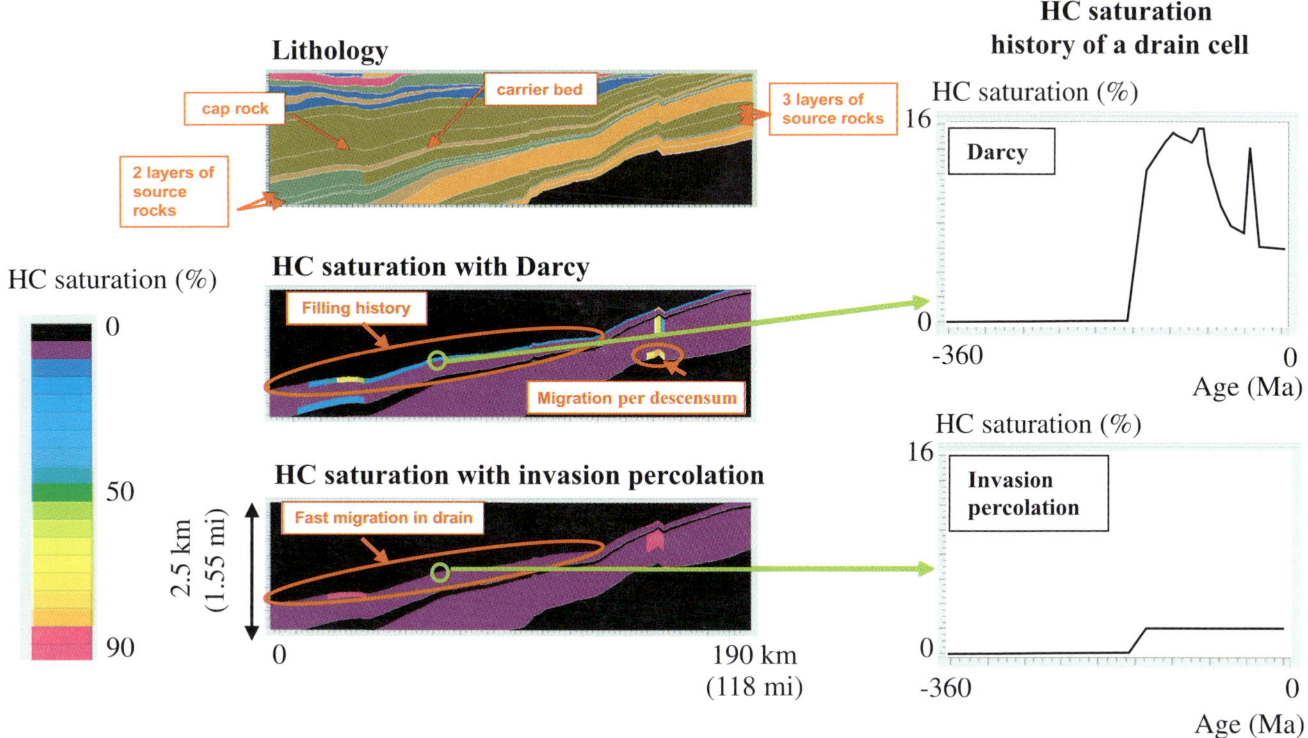

FIGURE 8. First real case: long-distance pathways. Comparison between the hydrocarbon (HC) saturation obtained with the Darcy migration model and those using invasion percolation.

accumulation under this barrier before cap-rock leakage. We focused on the behavior of this leakage. For that purpose, we use three different grid resolutions, and for each, we compare the results obtained using the Darcy and the invasion percolation models. The parameters of capillary pressures, permeabilities, and the other data for the geologic section are identical for all of the grid resolutions.

The first grid resolution is coarse and contains 320 cells. With the Darcy model, we have cap-rock leakage through a wide "chimney" as it has been explained in the section on Limitations of the Invasion Percolation

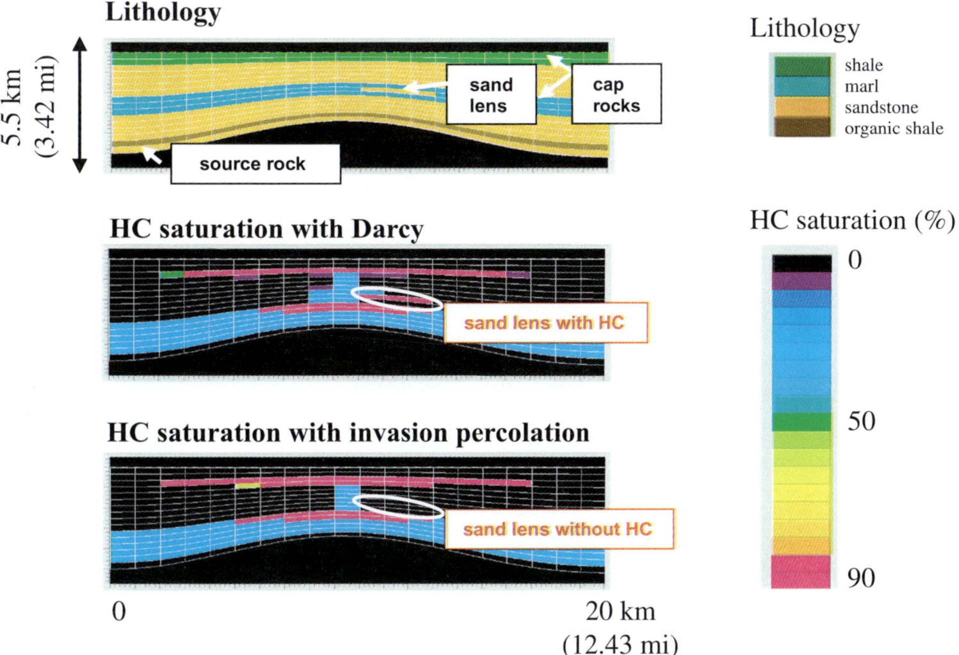

FIGURE 9. Second synthetic case: a sand lens, low grid resolution. Comparison between the hydrocarbon (HC) saturation obtained with the Darcy migration model and those using invasion percolation.

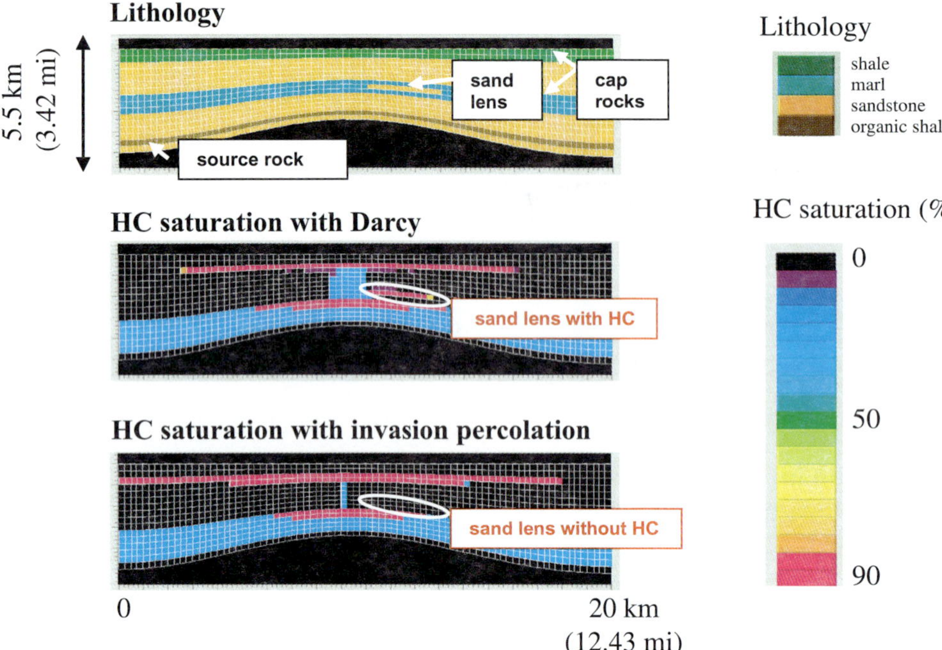

FIGURE 10. Second synthetic case: a sand lens, medium grid resolution. Comparison between the hydrocarbon (HC) saturation obtained with the Darcy migration model and those using invasion percolation.

Approach and hydrocarbons can migrate into the sand lens and form an accumulation. With invasion percolation, the leakage follows a pathway starting from the highest point of the structure. Hydrocarbons do not migrate into the sand unit because the lens is not on this pathway (Figure 9).

The second and the third grid resolutions are finer and contain 1280 and 2560 cells respectively (Figures 10, 11). We observe the same phenomenon as that obtained using the coarse grid. The invasion percolation pathway is increasingly narrow with the resolution and misses the sand unit, contrary to the broad leakage for the Darcy model, which leads to hydrocarbon accumulation in the sand lens.

In conclusion, the Darcy model better described the process of cap-rock leakage as a whole, whatever the grid resolution. Furthermore, with the invasion percolation model, bypassing of the lens at all three grid resolutions results in a shallower accumulation under the second cap rock and an underestimation of hydrocarbon losses.

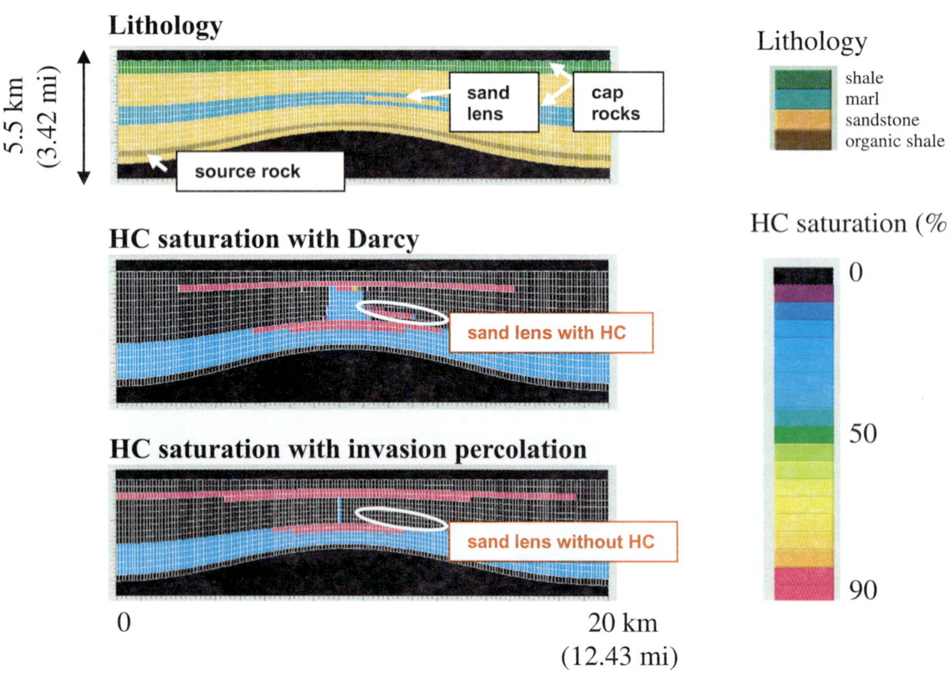

FIGURE 11. Second synthetic case: a sand lens, high grid resolution. Comparison between the hydrocarbon (HC) saturation obtained with the Darcy migration model and those using invasion percolation.

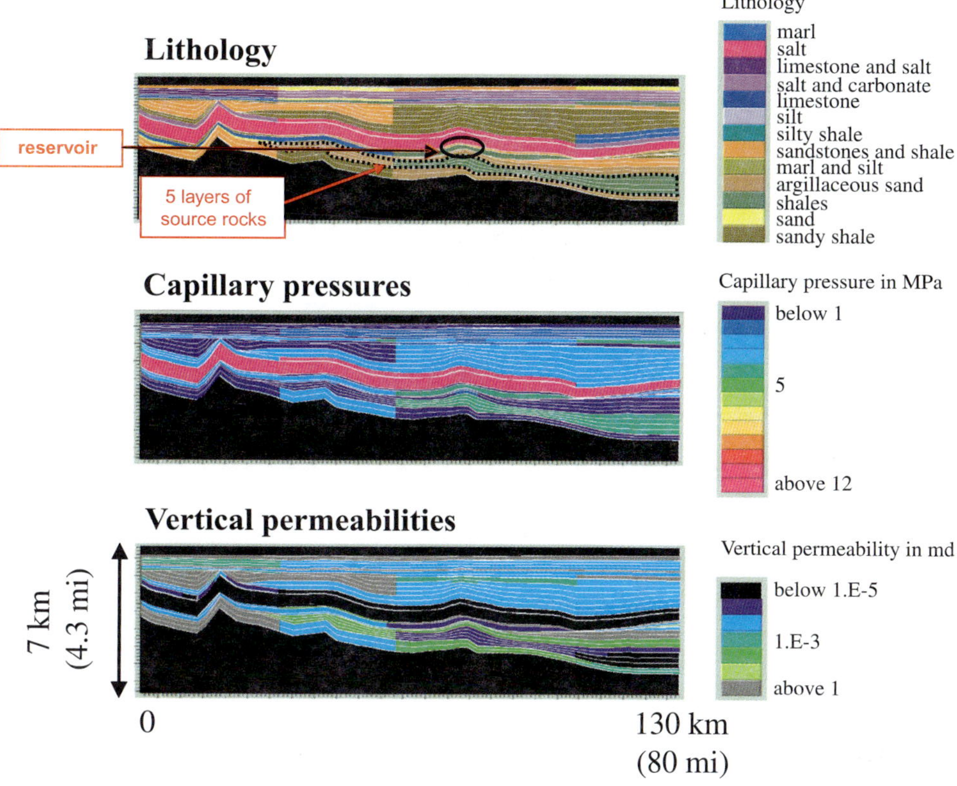

Figure 12. Second real case: pressure-migration coupling. Description of lithology, capillary pressures, and vertical permeabilities.

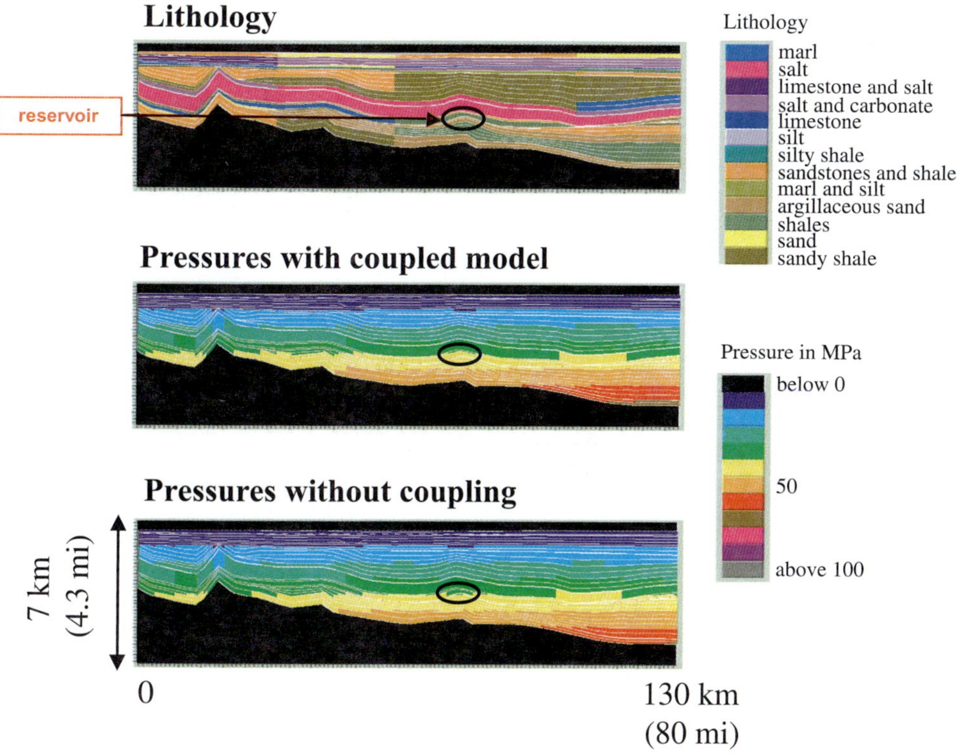

Figure 13. Second real case: pressure-migration coupling. Comparison between the pressures obtained with a pressure-migration coupled model and without coupling. In the reservoir zone, the pressure is equal to 42 MPa (6092 psi) with the coupled model and to 39 MPa without coupling.

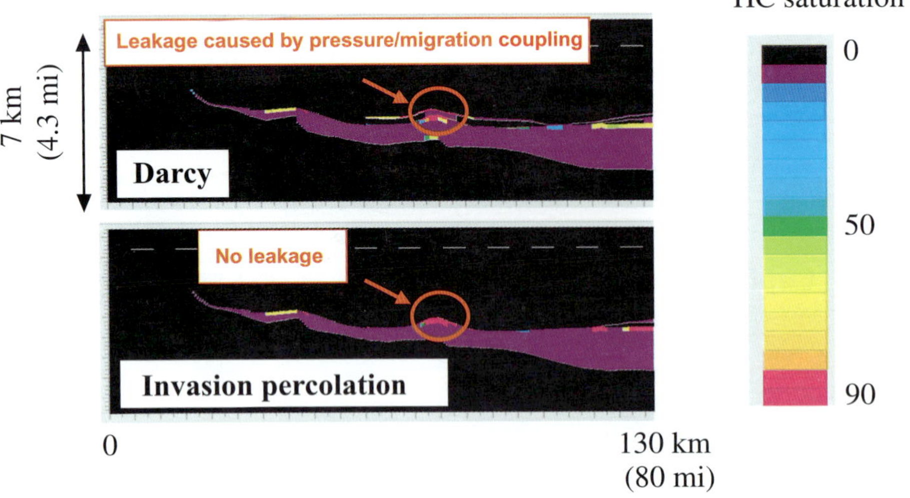

FIGURE 14. Second real case: pressure-migration coupling. Comparison between the hydrocarbon (HC) saturation obtained with the Darcy migration model and invasion percolation.

Real Case Study: Pressure-Migration Coupling

This second 2-D section comes from the real case study described in the section on long-distance pathways and is also enlarged. It contains five middle Paleozoic source rock layers. Our zone of interest is an anticline located above the source rocks and under a salt layer. It is composed of three groups of layers: the first layer contains three argillaceous sand units (the main reservoir area), the second layer is composed of shale, and the third layer contains sandstones and shales (Figure 12).

We want to study the sensitivity of the coupling between pressure and oil saturation. To this end, Figure 13 compares the pressures obtained using the fully implicit Darcy method, which couples the pressure porosity computation with the oil saturation computation, with those of the invasion percolation method. Figure 14 displays the results of hydrocarbon saturation obtained using these two models.

The pressure in the lower accumulation, localized in the three argillaceous sand units, is higher with the Darcy model than with the invasion percolation model (Figure 13). This pressure difference of 3 MPa (435 psi) leads to leakage for the Darcy model, but not for the invasion percolation model (Figure 14).

In conclusion, the pressure-migration coupling in the Darcy model induces leakage that is not captured at all by the invasion percolation model.

CONCLUSIONS

As stated in the introduction, the aims of this chapter were to (1) compare the capabilities of the Darcy with those of the invasion percolation method to model secondary and tertiary hydrocarbon migration and (2) illustrate the main differences through examples. From our investigation of selected case studies, the following features emerge.

The Darcy approach considers all of the relevant physical processes because it involves not only buoyancy, capillary forces, and pressure gradient, but also transient physics, thanks to the viscous terms. Although it can be numerically difficult and therefore time consuming, especially in the case of the fast movement of fluids, it is appropriate for slow hydrocarbon movement through, for instance, mudrock sequences or a low-angle slope where buoyancy is not very strong. Moreover, it is able to provide a good description of caprock leakage because of, among other things, pressure-migration coupling, which is present in our studied Darcy model.

The invasion percolation approach, at least in the context of the implementation used for this chapter, does not consider either viscosity or permeability; only buoyancy and capillary pressures drive the hydrocarbon migration. These two processes allow us to manage the pathways and the accumulation areas, but not the timing of trap filling, which is only imposed by the rate of hydrocarbon generation from the source rocks. The invasion percolation method is relatively quick and especially useful to simulate secondary migration in continuous high-permeability migration pathways. Nonetheless, in certain cases, it may be inappropriate for modeling hydrocarbon leakage out of traps. Furthermore, we must keep in mind that the viscous terms cannot be universally neglected from the equations as they can impact hydrocarbon system dynamics and the saturated bulk rock volume in fine-grained rocks.

These different approaches may be amended by improving some of the observed limitations, such as inclusion of viscosity in the invasion percolation method (Carruthers and de Lind van Wijngaarden, 2000).

ACKNOWLEDGMENTS

We thank Scott Barboza, Kenneth Peters, and an anonymous reviewer for their useful remarks and their relevant comments. We also thank Quang Huy Tran for his help in correcting the manuscript and Thierry Gallouët who encouraged us to submit this work for publication.

APPENDIX 1

Capillary Pressure Model

For all of the case studies, we used the following capillary pressure model depending only on porosity φ.

$$Pc(\varphi) = Pc_0 + (Pc_{\text{lim}} - Pc_0)\left[\frac{\varphi_0 - \varphi}{\varphi_0 - \varphi_{\text{lim}}}\right]^{\varphi PcEx} \tag{4}$$

where Pc_0 is the capillary pressure at surface or maximum porosity (Pa), Pc_{lim} is the capillary entry pressure at maximum burial or minimal porosity (Pa), φ_0 is the initial porosity when effective stress is equal to zero, φ_{lim} is the minimum porosity when effective stress is infinite, and φPcEx is the curvature of the capillary entry pressure/porosity function.

Permeability Computation

The intrinsic permeability tensor $\overline{\overline{K}}$ is the product of an anisotropy tensor and the intrinsic permeability K

$$\overline{\overline{K}} = K(\varphi) \bullet \begin{bmatrix} Kx & 0 & 0 \\ 0 & Ky & 0 \\ 0 & 0 & Kz \end{bmatrix} \tag{5}$$

The intrinsic permeability is computed using the modified Kozeny-Carman formula: if $\varphi \geq 0.1$

$$K(\varphi) = \frac{0.2\varphi^3}{S^2(1-\varphi)^2} \tag{6}$$

if $\varphi < 0.1$

$$K(\varphi) = \frac{20\varphi^5}{S^2(1-\varphi)^2} \tag{7}$$

where S is the specific surface area of the porous medium (m^2/m^3), Kx is the anisotropy coefficient for the horizontal direction, Kz is the anisotropy coefficient for the vertical direction.

Viscosity Computation

The water viscosity μ_w is a function of temperature T according to the Bingham formula.

The hydrocarbon viscosity μ_h is a function of temperature T:

$$\mu_h = \mu_0 \exp\left(\frac{Ak_0}{T}\right) \tag{8}$$

where μ_0 is a reference viscosity (Pa.s) and Ak_0 is a temperature-dependant viscosity parameter (K).

APPENDIX 2: DETAILED DATA FOR EACH EXAMPLE

This appendix uses the notations described in Appendix 1.

First Synthetic Case: A Carrier Bed With a Low Slope

The capillary pressures follow the law:

$$Pc(\varphi) = Pc_{\text{lim}}\left[\frac{\varphi_0 - \varphi}{\varphi_0 - \varphi_{\text{lim}}}\right] \tag{9}$$

For each lithology, we used the parameters given in Table 1.

The water density ρ_w is equal to 1030 kg·m^{-3} and the hydrocarbon density ρ_h is equal to 140 kg·m^{-3}.

For all the lithologies, the connate water saturation Swc is equal to 0.9 and the critical hydrocarbon saturation Soc is equal to 0.02.

To compute for the hydrocarbon viscosity, we used the parameters μ_0 = 1.45 × 10^{-5} Pa.s and Ak_0 = 1533.15K.

We used a source rock containing type I kerogen.

Table 1. Capillary pressure parameters for the first synthetic case study.*

Lithology	*Pc_{lim} (Pa)*	*φ_0*	*φ_{lim}*
Shale	1.5×10^8	0.702	0.03
Overburden rock	1.5×10^8	0.434	0.05
Fault carrier	1.0×10^5	0.702	0.03
Source rock	1.0×10^6	0.702	0.03

**Pc_{lim} = capillary entry pressure at maximum burial or minimal porosity (Pa); φ_0 = initial porosity when effective stress is equal to zero; φ_{lim} = minimum porosity when effective stress is infinite (see Appendix 1).*

Table 2. Permeability parameters for the first synthetic case study.*

Lithology	*First Kind of Simulation S (m^2/m^3)*	*Second Kind of Simulation S (m^2/m^3)*	*Kx*	*Kz*
Shale-overburden rock	5×10^7	5×10^7	1	0.5
Fault carrier	2×10^6	1×10^5	1	0.001
Source rock	5×10^7	5×10^5	1	1

**S = specific surface area of the porous medium (m^2/m^3); Kx = anisotropy coefficient for the horizontal direction; Kz = anisotropy coefficient for the vertical direction (see Appendix 1).*

To compute for the permeabilities for the first kind of simulations and also for the second kind of simulation with Darcy and high permeabilities, we used the parameters given in Table 2.

Real Case Study

The capillary pressures follow the law:

$$Pc(\varphi) = Pc_{\lim}\left[\frac{\varphi_0 - \varphi}{\varphi_0 - \varphi_{\lim}}\right]^{\varphi PcEx} \quad (10)$$

For each lithology, we used the parameters given in Table 3 and to compute for the permeabilities. The parameters are detailed in Table 4. For the salt, we used an extremely high capillary pressure and a permeability equal to zero.

Table 3. Capillary pressure parameters for the real case study.*

Lithology	*Pc_{lim} (Pa)*	*$\phi PcEx$*	*ϕ_0*	*ϕ_{lim}*
Marl	3.0×10^6	0.5	0.5	0.02
Limestone and salt	1.0×10^6	0.5	0.35	0.02
Salt and carbonate	3.0×10^6	0.5	0.1556	0.008
Limestone	3.0×10^6	0.5	0.35	0.02
Silt	7.5×10^5	0.5	0.4186	0.02
Silty shale	3.0×10^6	0.5	0.4639	0.02
Sandstones and shales	5.0×10^5	0.5	0.3814	0.02
Marl and silt	2.0×10^6	0.5	0.4186	0.02
Argillaceous sand	1.0×10^6	0.5	0.4186	0.02
Shales	5.0×10^6	0.5	0.5	0.02
Sand	5.0×10^5	1.0	0.36	0.02
Sandy shale	3.0×10^6	0.5	0.4478	0.02

**Pc_{lim} = capillary entry pressure at maximum burial or minimal porosity (Pa); $\phi PcEx$ = curvature of the capillary entry pressure/porosity function; ϕ_0 = initial porosity when effective stress is equal to zero; ϕ_{lim} = minimum porosity when effective stress is infinite (see Appendix 1).*

The water density ρ_w is equal to 1030 kg m^{-3} and the hydrocarbon density ρ_h is equal to 700 kg m^{-3}.

For all the lithologies, the connate water saturation Swc is equal to 0.93 and the critical hydrocarbon saturation Soc is equal to 0.02.

To compute for the hydrocarbon viscosity, we used the parameters $\mu_0 = 1.2 \times 10^{-7}$ Pa.s and $Ak_0 = 2973$K for the first section of the real case study. For the second section, we used $\mu_0 = 1.0 \times 10^{-7}$ Pa.s and $Ak_0 = 2700$K.

All of the source rocks contain type II kerogen.

Second Synthetic Case: Sand Lens

The capillary pressures follow the law $Pc(\varphi) = Pc_{\lim}$.

As in the first synthetic case, permeabilities are computed using the Kozeny-Carman formula.

To prevent leakage through the cap rock at the top of the 2-D section, the parameters of this lithology are those of salt with an extremely high capillary pressure

Table 4. Permeability parameters for the real case study.*

Lithology	*S (m^2/m^3)*	*Kx*	*Kz*
Marl	5.0×10^7	1	1
Limestone and salt	1.0×10^7	1	1
Salt and carbonate	2.0×10^7	1	1
Limestone	1.0×10^6	1	1
Silt	1.0×10^6	1	0.2
Silty shale	1.0×10^7	1	0.5
Sandstones and shales	1.0×10^6	50	25
Marl and silt	5.0×10^7	1	1
Argillaceous sand	1.0×10^6	5	2
Shales	1.0×10^8	10	1
Sand	1.0×10^6	200	100
Sandy shale	1.0×10^8	500	100

**S = specific surface area of the porous medium (m^2/m^3); Kx = anisotropy coefficient for the horizontal direction; Kz = anisotropy coefficient for the vertical direction (see Appendix 1).*

Table 5. Capillary pressure and permeability parameters for the second synthetic case study.*

Lithology	Pc_{lim} (Pa)	S (m^2/m^3)	Kx	Kz
Marl	1.0×10^7	2.5×10^7	1	1
Sandstone, organic shale	5.0×10^6	1.5×10^5	1	1

Pc_{lim} = capillary entry pressure at maximum burial or minimal porosity (Pa); S = specific surface area of the porous medium (m^2/m^3); Kx = anisotropy coefficient for the horizontal direction; Kz = anisotropy coefficient for the vertical direction (see Appendix 1).

and a permeability equal to zero. Moreover, the capillary pressure of the source rock is not realistic, but the aim of this synthetic case was not to focus on primary migration.

For each lithology except the top cap rock, we used the parameters given in Table 5.

The water density ρ_w is equal to 1030 kg·m^{-3} and the hydrocarbon density ρ_h is equal to 140 kg·m^{-3}.

For all of the lithologies, the connate water saturation Swc is equal to 0.8 and the critical hydrocarbon saturation Soc is equal to 0.2.

To compute for the hydrocarbon viscosity, we used the parameters $\mu_0 = 1.45 \times 10^{-5}$Pa.s and $Ak_0 = 1533.15$ K.

We used source rock containing type II kerogen.

REFERENCES CITED

Bear, J., 1972, Dynamics of fluids in porous media: New York, Elsevier, 569 p.

Burley, S. D., S. Clarke, A. Dodds, J. Frielingsdorf, P. Huggins, A. Richards, I. C. Warburton, and G. Williams, 2000, New insights on petroleum migration from the application of 4-D basin modeling in oil and gas exploration: Journal of the Geochemical Exploration, v. 69–70, p. 465–470, doi:10.1016/S0375-6742(00)00039-X.

Burrus, J., 1997, Contribution à l'étude du fonctionnement des systèmes pétroliers: Apport d'une modélisation bidimensionnelle: Thèse de doctorat hydrologie et hydrogéologie quantitative, Ecole des mines de Paris, 602 p.

Carruthers, D., 1998, Transport modeling of secondary oil migration using gradient-driven invasion percolation techniques: Ph.D. dissertation, Heriot-Watt University, Edinburgh, United Kingdom, 288 p.

Carruthers, D., 2003, Modeling of secondary petroleum migration using invasion percolation techniques, *in* S. Düppenbecker and R. Marzi, eds., Multidimensional basin modeling: AAPG/Datapages Discovery Series 7, p. 21–37.

Carruthers, D., and M. de Lind van Wijngaarden, 2000, Modeling viscous-dominated fluid transport using modified invasion percolation techniques: Journal of Geochemical Exploration, v. 69–70, p. 669–672, doi:10.1016/S0375-6742(00)00138-2.

Carruthers, D., and P. Ringrose, 1998, Secondary oil migration: Oil-rock contact volumes, flow behavior and rates, *in* J. Parnell, ed., Dating and duration of fluid flow and fluid-rock interaction: Geological Society (London) Special Publication 144, p. 205–220.

Darcy, H., 1856, Les fontaines publiques de la ville de Dijon: Dijon, Annales de la ville de Dijon, 647 p.

Durand, B., 1988, Understanding of hydrocarbon migration in sedimentary basins (present state of knowledge): Organic Geochemistry, v. 13, p. 445–459, doi:10.1016/0146-6380(88)90066-6.

England, W. A., A. S. Mackenzie, D. M. Mann, and T. M. Quigley, 1987, The movement and entrapment of petroleum fluids in the subsurface: Geological Society (London) 144, p. 327–347.

Guéguen, Y., and J. Dienes, 1989, Transport properties of rocks from statistics and percolation: Mathematical Geology, v. 21, no. 1, p. 1–13, doi:10.1007/BF00897237.

Guérillot, D., and F. Kalaydjian, 1988, Application du formalisme par pression globale à l'étude du couplage entre lois de Darcy pour les écoulements diphasiques en milieu poreux: Compte Rendus de l'Académie des Sciences, Paris, t. 306, Série II, p. 853–858.

Hantschel, T., and A. I. Kauerauf, 2009, Fundamentals of basin and petroleum systems modeling: Stringer-Verlag, Berlin, Springer, 476 p.

Hubbert, M. K., 1956, Darcy law and the field equations of the flow of underground fluids: Journal of Petroleum Technology of the American Institute of Mining, Metallurgical, and Petroleum Engineers, v. 207, p. 222–239.

Irmay, S., 1958, On the theoretical derivation of Darcy and Forchheimer formulas: Transactions of the American Geophysical Union, v. 39, no. 4, p. 702–707.

Lenormand, R., 1981, Déplacements polyphasiques en milieux poreux sous l'influence des forces capillaires, Etude expérimentale et modélisation de type percolation: Thèse, Institut National Polytechnique de Toulouse, France, 327 p.

Marle, C., 1972, Les écoulements polyphasiques en milieux poreux: Cours de production: Tome IV, Technip Editions, 330 p.

Meakin, P., G. Wagner, A. Vedvik, H. Amundsen, J. Feder, and T. Jøssang, 2000, Invasion percolation and secondary migration: Experiments and simulations: Marine and Petroleum Geology, v. 17, no. 7, p. 777–795, doi:10.1016/S0264-8172(99)00069-0.

Requena, S., M. Thibaut, S. Pegaz, and L. Agelas, 2005, Pushing the limits of 3-D basin modeling with Linux clusters: Proceedings of the 67th EAGE Conference & Exhibition: Petroleum Systems, June 13–16, 2005, Madrid, Spain.

Sausse, J., 1998, Caractérisation et modélisation des écoulements fluides en milieu fissuré. Relation avec les altérations hydrothermales et quantification des paléocontraintes: Thèse de l'Université Henri Poincaré, Nancy, France, 336 p.

Schlomer, S., and B. M. Krooss, 1997, Experimental characterization of the hydrocarbon sealing efficiency of cap rocks: Marine and Petroleum Geology, v. 14, no. 5, p. 565–580, doi:10.1016/S0264-8172(97)00022-6.

Schowalter, T. T., 1979, Mechanics of hydrocarbon secondary migration entrapment: AAPG Bulletin, v. 63, p. 723–760.

Schneider, F., 2003, Modeling multiphase flow of petroleum at the sedimentary basin scale: Journal of Geochemical Exploration, v. 78-79, p. 693–696, doi:10.1016/S0375-6742(03)00092-X.

Sylta, Ø., 2004, Hydrocarbon migration modeling and exploration risk: Doctoral thesis, Norwegian University of Science and Technology, Trondheim, Norway, 200 p.

Thomas, M. M., and J. A. Clouse, 1995, Scaled physical model of secondary oil migration: AAPG Bulletin, v. 79, no. 1, p. 19–29.

Ungerer, P., J. Burrus, B. Doligez, P.-Y. Chénet, and F. Bessis, 1990, Basin evaluation by integrated two-dimensional modeling of heat transfer, fluid flow, hydrocarbon generation, and migration: AAPG Bulletin, v. 74, p. 309–335.

Watts, N. L., 1987, Theoretical aspects of cap-rock and fault seals for single and two phase hydrocarbon columns: Marine and Petroleum Geology, v. 4, p. 274–307, doi:10.1016/0264-8172(87)90008-0.

Whitaker, S., 1986, Flow in porous media 1: A theoretical derivation of Darcy's law: Transport in Porous Media, v. 1, no. 1, p. 3–25, doi:10.1007/BF01036523.

Wilkinson, D., 1984, Percolation model of immiscible displacement in the presence of buoyancy forces: Physical Review A, v. 30, no. 1, p. 520–531, doi:10.1103/PhysRevA.30.520.

Wilkinson, D., and J. F. Willemsen, 1983, Invasion percolation: A new form of percolation theory: Physics Abstracts, v. 16, p. 3365–3376.

Willien, F., I. Chetvchenko, R. Masson, P. Quandalle, L. Agelas, and S. Requena, 2009, AMG preconditioning for sedimentary basin simulations in Temis calculator: Marine and Petroleum Geology, v. 26, p. 519–524, doi:10.1016/j.marpetgeo.2009.01.014.

Wolf, S., I. Faille, S. Pegaz-Fiornet, F. Willien, and B. Carpentier, 2009, A new innovating scheme to model hydrocarbon migration at basin scale: A pressure-saturation splitting, AAPG Hedberg Conference Basin and Petroleum System Modeling (abs.): New Horizons in Research and Applications, Napa, California, http://www.searchanddiscovery.com/abstracts/html/2009/hedberg/abstracts/short/wolf.htm (accessed June 14, 2011).

14

Schoellkopf, N. B., 2012, Quantitative assessment of hydrocarbon charge risk in new ventures exploration: Are we fooling ourselves?, *in* K. E. Peters, D. J. Curry, and M. Kacewicz, eds., Basin Modeling: New Horizons in Research and Applications: AAPG Hedberg Series, no. 4, p. 237–246.

Quantitative Assessment of Hydrocarbon Charge Risk in New Ventures Exploration: Are We Fooling Ourselves?

Noelle B. Schoellkopf
Chevron Energy Technology, Houston, Texas, U.S.A.

ABSTRACT

Basin modeling software includes tools to statistically vary model input parameters, such as fetch area, depth, source thickness, total organic carbon, hydrogen index, temperature gradient, or heat flow, and consider the impact on fluid phase and volumes. We can rank these parameters, but we should be aware of pitfalls. The underlying geologic assumptions may account for the greatest uncertainty in new basin areas, where data are sparse and models remain poorly calibrated. Modeling tools must be flexible enough to allow multiple working hypotheses within the project time frame.

These multiple hypotheses are best evaluated by an integrated project team that includes the basin modeler. The team members' shared knowledge of regional basin history, tectonics, stratigraphy, and source rock depositional models can provide an advantage in weighing alternative geologic scenarios and hydrocarbon charge risk.

This chapter provides seven examples of modeling pitfalls based on new ventures exploration studies performed using a combination of flow path and two-dimensional models. Although not representative of all possible pitfalls, these examples illustrate the substantial impact of some pitfalls on model outcome.

INTRODUCTION

Greater computing power and more complex basin modeling software have increased expectations for quantitative prospect predictions ahead of drilling. Exploration decisions rest on the assumption that sufficient petroleum volumes (generally expressed as probabilistic distributions) of the appropriate phase and quality are present. How reliable are our estimates and are we favoring easily quantifiable parameters at the expense of more qualitative, but perhaps more critical, geologic parameters affecting hydrocarbon charge?

This chapter arose from practical experience applying basin modeling techniques and tools in a new ventures exploration setting. New ventures exploration provides its own set of challenges, for which basin modelers

DOI:10.1306/13311439H43473

must adapt their software and work methods to meet management's expectations for quantitative assessment of hydrocarbon charge risk. The work approach is different from that in more established exploration settings and has its own workflow and associated pitfalls.

The Basin Modeler's Role

The basin modeler is a key part of the exploration risk assessment process. Typically, the risk categories in a prospect or play evaluation are as follows: hydrocarbon charge, reservoir, trap, and seal. Some companies have fewer categories, some more, but the overall concept is generally similar. Of these categories, basin modeling has the greatest impact on the determination of hydrocarbon charge risk but may also influence seal risk and, in some cases, reservoir risk through reconstruction of burial temperature and pressure histories.

The basin modeler's role is to provide input to management on the impact of parameter uncertainty on the final resource volume estimates. Although a thorough statistical analysis is commonly beyond the scope of many projects, assessment of the major impact parameters is essential. Modeling tools and workflows must be flexible enough to assess these multiple working hypotheses within the time frame of the project.

Hydrocarbon Charge Risk

Our evaluation of hydrocarbon charge risk routinely includes the following: source, maturity, timing of generation and expulsion, migration pathways, expected oil or gas composition, and quantitative charge volume prediction. The analysis of each of these parameters is more or less detailed depending on how much is known about the basin or study area.

First among these parameters is the presence, distribution, and richness of the source rock. In new venture areas where few wells have been drilled and source rock data may be scarce, the source evaluation instead may rely heavily on depositional models and indirect predictors of source presence or quality (such as direct hydrocarbon indicators from seismic or the geochemistry of oils or seeps). In these cases, the preferred source rock model may be associated with a high degree of uncertainty as to richness and thickness and possibly source facies and kerogen type.

Second is assessing the extent, maturity, and timing of generation and expulsion. In this effort, we are assisted by ever more powerful earth models as well as current research in transformation kinetics and geochemical maturity indicators. The horizon geometries may originate from integrated GOCAD or other earth model-building tools, where the structural geometry may have been carefully mapped and even the kinematics may have been studied. Tectonic and crustal models may provide perspective and limits on heat-flow history. More tightly constrained structural models combined with geochemical maturity indicators allow us to place more realistic limits on the generation and expulsion history.

The third aspect of hydrocarbon charge risk assessment is migration pathway analysis. Depending on the area and the extent of knowledge, this part of the evaluation may use simple flow path models or more complex two-dimensional (2-D) and three-dimensional (3-D) models incorporating detailed stratigraphic information. It may or may not include pressure modeling, depending on the level of knowledge and availability of data for calibration. The migration analysis may be either very simple or quite complex, depending on the level of knowledge. The challenge is in understanding the validity of the model.

Another aspect of charge risk is the expected oil or gas composition of the fluid in the reservoir. If prospect economics are run on the basis of oil and the well finds gas instead (or vice versa), then the profitability of any discovery may be adversely affected. Subtle variations in API gravity or oil type are sometimes within the ability of the basin modeler or geochemist to predict based on understanding of kerogen type, maturity, phase predictions, or in-reservoir alteration processes.

Last, but not least, are predictions of the amount of charge available to the prospect. If insufficient charge exists, a new "discovery" well might be an economic failure. Thus, our emphasis must be on having a high likelihood of sufficient volumes of the right kind of hydrocarbons.

Toward Increased Quantification

During the past 20 yr, there has been an increase in the complexity of basin models as well as an increasingly probabilistic view of business decisions. Modeling software packages have evolved from simple one-dimensional calculations that could be duplicated using a pocket calculator to more complex 2-D and 3-D models requiring high-end workstations and parallel processors. The computing algorithms have evolved and are increasingly complex; for example, allowing different kinds of migration and faulting or complex grids. Increased computing power has expanded our ability to run more complicated (and perhaps more realistic) models. Currently, 1 to 2 million cells in a model are common and more than 4 million cells are possible (although perhaps not desirable). Such grid sizes were unthinkable merely a decade ago.

At the same time, management has evolved toward an increasingly probabilistic and risked view of reserves and business decisions using methods such as those of Rose (2001, 2007). Various aspects and parameters of reservoir and trap commonly are reported as P10-P50-P90

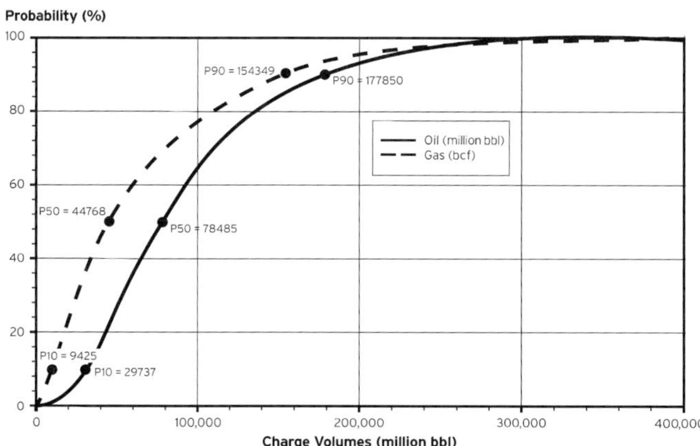

FIGURE 1. Charge probability diagram for a hypothetical prospect and source rock.

values. Likewise, hydrocarbon charge volumes also are described as P10-P50-P90 statistics on a distribution curve, as shown in Figure 1.

Basin modeling packages now commonly include Monte Carlo or similar tools to examine uncertainty. These tools can improve our understanding of which parameters have the greatest impact on prospect volumes. For example, in the tornado chart shown in Figure 2, the source rock hydrogen index and the geothermal gradient parameters are the most critical for this prospect. These tools exist to address uncertainty of many parameters, and their use allows probabilistic distributions of likely in-place resources.

LIFE IN THE NEW VENTURES EXPLORATION SHOP

Having raised the expectation of probabilistic assessment for hydrocarbon charge risk, it can be another thing entirely to deliver it, particularly in the poorly constrained new ventures environment. The intended quantification of hydrocarbon charge risk is hindered by some very real practical constraints. Some of these constraints include tight project deadlines, small teams, insufficient data and calibrations, too many uncertain and uncalibrated parameters, and volatile prospect and project priorities as the lease, bid rounds, and geopolitical situations change.

Nonetheless, every prospect brought forward to management must have a hydrocarbon charge evaluation. Depending on the region and the industry level of activity, this can result in very intense work periods with numerous prospects and frequent management reviews, perhaps as frequently as every two months in a 2-yr period. Especially in new ventures exploration, this means that we need to have the basin modeling capability for rapid turnaround and the ability to incorporate change. We need a flexible easily revisable model, as the modeling frequently proceeds with the evolving geophysical interpretation. In many cases, this turns out to be a flow path model. At the same time, we still need the scientific rigor of Darcy models to test migration scenarios and a consistency of approach so that one prospect can be compared with another. We may use a combination of several tools, such as flow path and 2-D models or opt for a 3-D model. Thus, our focus is on fit-for-purpose modeling tools, selected to handle both the type of modeling required and the time frame and business requirements of the project.

THE CHALLENGE FOR NEW VENTURES EXPLORATION

Projects in new or underexplored basins rarely have sufficient data to populate true 3-D models with all of the required structural, stratigraphic, and facies details. The long setup and computational times required for 3-D models can make them more inflexible for testing various parameter sensitivities. This is particularly true if the parameters affect the geometry or stratigraphic framework of the model. Likewise, it may be very difficult to explore different geologic alternatives in a timely fashion, when the duration of a project is framed in weeks or a few months.

The speed and flexibility of flow path models, combined with prospect-focused conventional 2-D models, provide a better approach for many sparsely controlled new ventures areas. Typical parameter uncertainties for a single scenario can be tested in a reasonable time frame (with results summarized in a tornado diagram, as in Figure 2), but beyond that, even multiple geologic scenarios should be considered and tried.

However, these flow path models and workflows have their own pitfalls. These must be recognized before useful volume estimates can be relied upon fully for risk assessment. These pitfalls include organizational and work-team considerations, as well as map preparation and workflow issues, various controls on geometry and fetch area, and last, overlooking alternatives to the existing geologic model. The impact of these pitfalls may dwarf that of more conventional parameter uncertainties.

THE EFFECT OF TEAMS

Some uncertainties may be ignored or only reluctantly addressed because of the organization of the work team. For example, a basin modeler working in isolation may need to resolve issues related to seismic interpretation or velocity functions. However, it may be

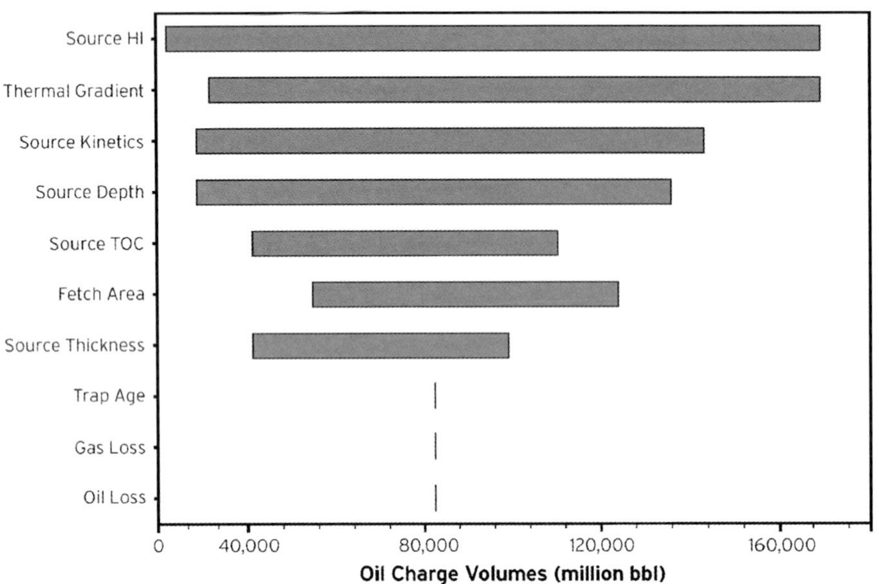

FIGURE 2. Tornado diagram. In this case, the hydrogen index (HI) of the source rock and the geothermal gradient have the greatest impact on prospect volumes. TOC = total organic carbon.

difficult for the basin modeler alone to assess whether the available geologic model is accurate and rigorous and what the impact would be of a different geologic scenario. In an integrated work group, these issues may be easier to address. In addition, we naturally prefer to test the parameters over which we have control but resist challenging assumptions or parameters that are beyond our control or are poorly understood.

Where basin modelers and geochemists are part of the same work group or when one team member handles both functions, uncertainty in various geochemical and thermal parameters is commonly evaluated because the basin modeler has direct control over these data. For example, a geochemist is best suited to choose the correct hydrocarbon generation kinetics to use from the readily available libraries or to review the appropriateness of geochemical analogs. Likewise, the person most familiar with the temperature, thermal gradient, or heat-flow data is commonly the basin modeler, and so uncertainties in those areas are commonly addressed.

Conversely, other parameters are commonly reluctantly assessed because doing so would challenge accepted interpretations across organizational boundaries. For example, only rarely does the basin modeler examine uncertainties related to the velocity model, or the horizon picks or ages, or the validity of the geologic model. Generally, those are accepted as already reviewed by the interpreters before the basin modeler ever sees the project.

In the well-integrated work group, where good team dynamics exist, the basin modeler and interpreters can challenge the accepted dogma and test fundamental uncertainties related to age and stratigraphy, velocity models, or alternative geologic interpretations. Thus, the organizational structure can play a significant function in enabling the assessment of uncertainty.

PITFALLS OFTEN OVERLOOKED

Examples of modeling pitfalls are provided below based on new ventures exploration studies performed using a combination of flow path and 2-D models. These examples are not comprehensive but illustrate the possible impact of some pitfalls on model outcome. The accompanying figures lack location and other information to maintain project confidentiality.

Example 1: Map Preparation

Errors may creep into the model because of contouring or gridding workflows or map stack processing. This is especially true of regional studies with widely spaced seismic grids and few or no wells, where the maps may be exported and reimported through a variety of tools before being brought into the basin model. An example workflow for a map might be SeisWorks → GOCAD → ZMap → Trinity, with an independent path to bring a 2-D profile from SeisWorks or GOCAD into PetroMod. Each of these interpretation packages uses slightly different gridding and contouring; and as a map proceeds from one application to the next, new problems may arise that may not have existed in the original maps. Tools used to fix these problems may generate new problems, as shown in Figure 3.

Figure 3A shows a map of the difference between the interpreter's original depth map and the map eventually received by the basin modeler at a particular prospect (indicated by the diagonal shading). The difference (error) between the two maps is 1200 m (3936 ft) at the prospect crest located on a steep-sided high; the error is negligible in the adjacent lows. A profile view of the two depth surfaces is also shown (Figure 3B).

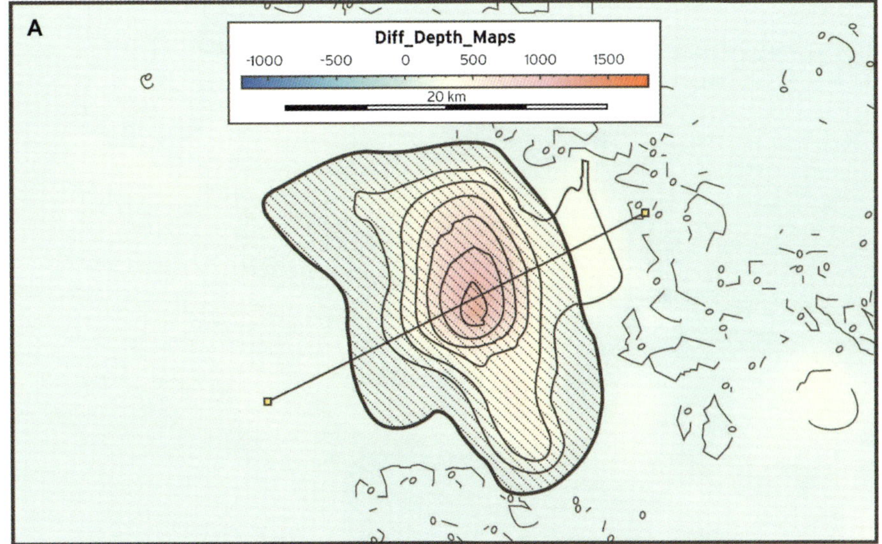

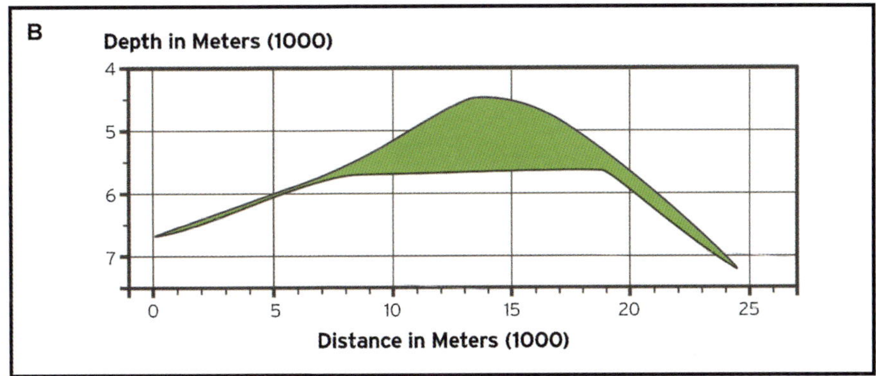

FIGURE 3. Difference between the original depth map and the map supplied to the basin modeler at the prospect location. The large error was caused by faulty despiking and conforming workflows.

The small error in the basins had little impact on thermal maturity, but the large differences in reservoir position had a large effect on possible biodegradation scenarios. A drastic change occurred in the reservoir depths at the top of the structure. For a given geothermal gradient of 30°C/km (1.65°F/100 ft), the depth difference resulted in a 36°C (66°F) possible range in reservoir temperatures, which was enough to move the reservoir out of the likely biodegradation zone and thereby considerably alter the assigned risk for oil biodegradation.

How could such a huge error slip through? The error occurs at the top of a very steep-sided structure in a map based on a very regional widely spaced seismic grid. As the initial contoured surfaces were moved from one application to another, they were sometimes regridded and recontoured, perhaps unknowingly, simply by changing applications. The geoscience technician handling the depth maps observed that some of the surfaces were now crossing. This seemed to be a problem only on the structural highs; to fix the crossing surfaces, he applied a despiking algorithm and forced the surfaces to conform in ZMap, thus losing the pointed top of the structure.

How should this problem be avoided? The first and most obvious cure is careful quality control of the depth map processing sequence. Simple difference maps such as the one shown in Figure 3 would have caught the problem. Beyond that, conforming the horizons using a master-slave workflow might have been a better way to handle the issue of crossing horizons. In such a workflow, the key source and reservoir horizons are designated the "master" horizons and other horizons ("slave" horizons) are forced to conform to them. This ensures that the key modeling surfaces are at the correct depth and any adjustments are relegated to other horizons of lesser importance. In my opinion, the master-slave conforming workflow is a better choice than the standard bottom-up or top-down conforming methods.

Example 2: Change in Grid Size

As is commonly the case in basin evaluations, this example project evolved from a regional study and maps at a 2-km (1.2-mi) grid spacing to a subregional study using a 1-km (0.6-mi) grid spacing focused on the area of greatest interest. Thus, the same prospects that were evaluated for hydrocarbon charge as part of the regional first-pass study were reevaluated using the same software package and the new maps with finer grid spacing,

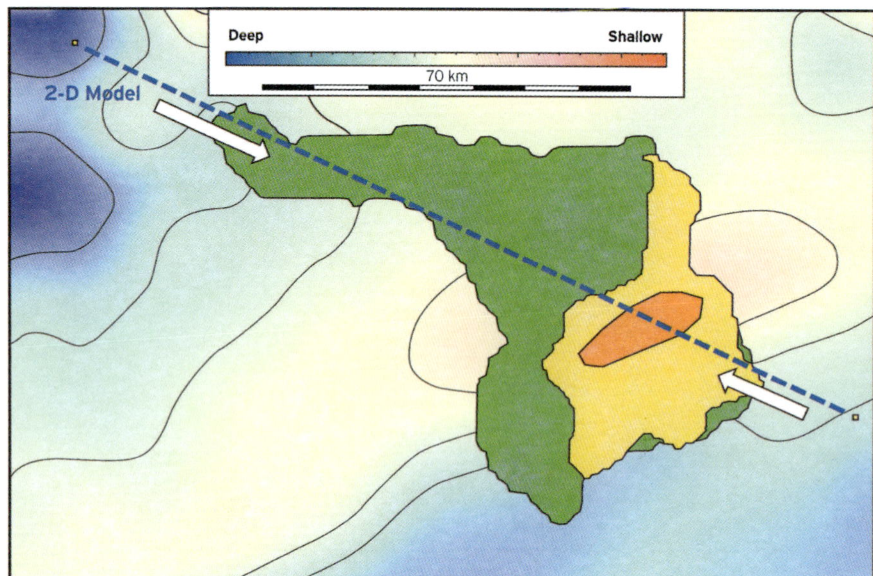

FIGURE 4. Impact of change in grid size (2 to 1 km [1.2 to 0.6 mi]). The fetch area shrinks from 3400 km^2 (1313 mi^2) in the regional study (green) to 1200 km^2 (463 mi^2) at the subregional scale (yellow), resulting in one third of the charge volume.

which were based on the same seismic interpretation. This allowed us to observe the impact of grid size changes on fetch area and hydrocarbon charge volume.

The impact on the fetch area is shown in Figure 4. For this particular prospect, the smoother maps of the initial regional study resulted in a fetch area polygon of 3400 km^2 (1313 mi^2), shown in green. The subregional maps, for the same prospect, gave a fetch area polygon of only 1200 km^2 (463 mi^2), shown in yellow. This second fetch area captured only one third the amount of charge in the original model.

The reason for this difference is obvious, maps at larger grid spacing are smoother, therefore, they tend to have larger fetch areas than maps at smaller grid spacing, which have greater rugosity. However, the take-home lesson with regard to charge volume is that charge volumes from different studies should be compared with caution.

A secondary observation is that the lateral migration distances are quite different in the two cases. A conventional 2-D (or 3-D) migration model can be used to validate migration effectiveness and distances, assuming that all geologic uncertainties are addressed, and perhaps to help determine which case is more likely. Some of these geologic uncertainties might include faults (presence and properties), stratigraphic variability (e.g., sand pinch-outs or facies changes), changes in structural interpretations, rock property distributions, variable hydrocarbon yields, and migration losses, all possibly affecting migration effectiveness and distances.

Example 3: Velocity Model

Particularly in new venture areas and sparsely explored basins, there may be considerable uncertainty as to which velocity model is best to use. In this example, the study team initially used a laterally variant velocity function to incorporate changes in velocity observed at several well locations across the basin. As shown in Figure 5, the effect of this first velocity model on the

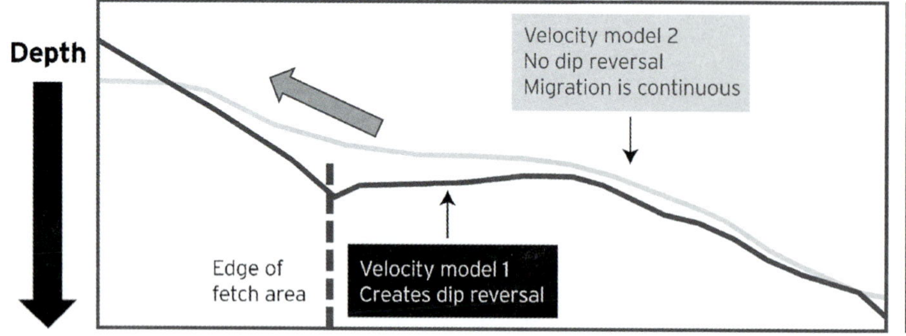

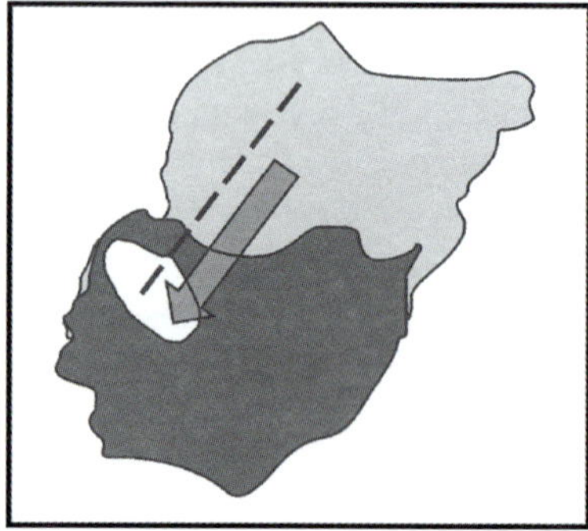

FIGURE 5. Impact of laterally variant (1) versus spatially invariant (2) velocity models on fetch area and charge. Velocity model 2 removes the dip reversal, allowing a larger fetch area (light gray) and continuous migration. Model 2 results in twice the fetch area, 3.5 times the oil charge, and 7 times the gas charge compared with model 1.

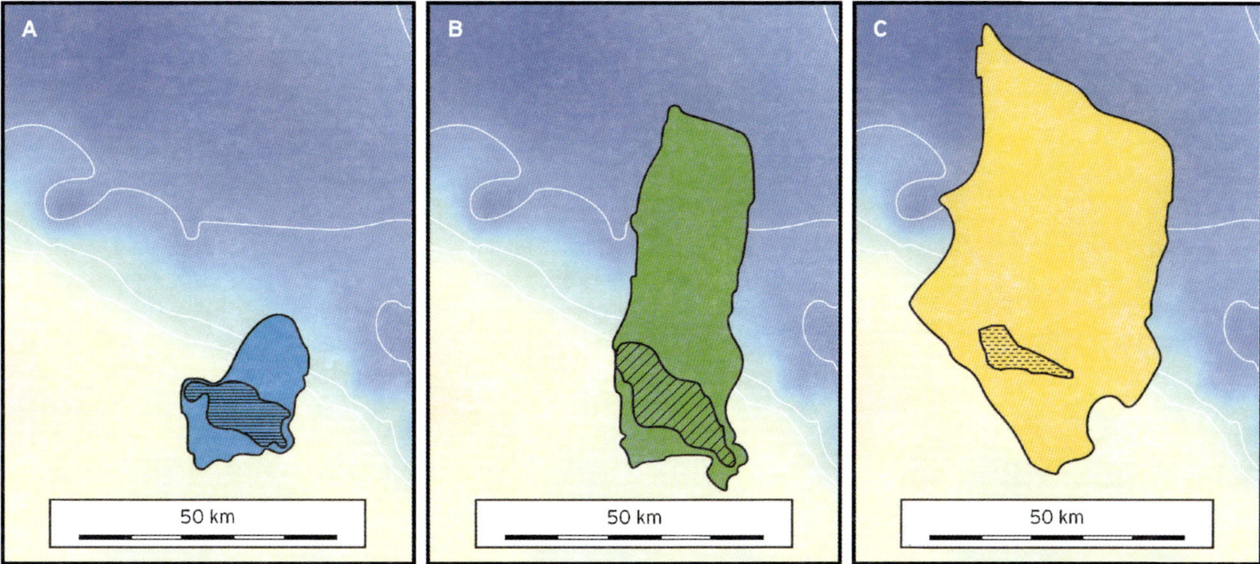

FIGURE 6. Illustration of why the prospect outline needs to be consistent with the map. (A) Original closure from prospect "blob" map and resulting fetch area. (B) Revised closure and larger fetch area, resulting in 35 times the oil charge and 50 times the gas charge. (C) Closure and fetch area based on this version of the map resulted in 75 times the oil charge and 100 times the gas charge compared with the original case.

depth maps was to create a dip reversal on the flank of the prospect structure, thereby limiting the extent of the fetch area (dark gray). Examination of the seismic profiles in time, combined with general knowledge of the basin geology, convinced the interpreters that this dip reversal, in fact, did not exist and was an artifact of the laterally variant velocity function and velocity model grid size.

A second laterally constant velocity function was applied for the time-to-depth conversion. Not only was this same horizon now at a slightly different depth, but the new function caused no dip reversal, thereby increasing the size of the fetch area polygon (shown in light gray) for this prospect. This second velocity function resulted in twice the fetch area, 3.5 times the oil charge, and 7 times the gas charge available to the prospect.

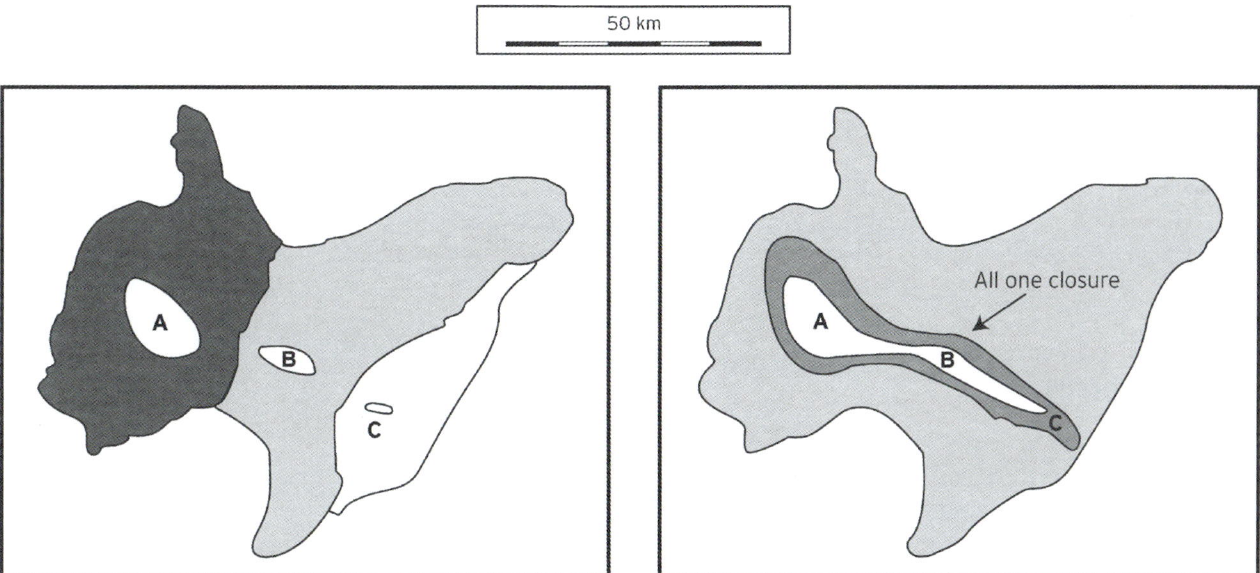

FIGURE 7. Impact of contouring, closure, fetch area, and fill-and-spill parameters. These prospects can be contoured as three separate closures or as a single elongate closure, depending on contouring method and closing contour. The charge is three times greater for the whole area compared with A alone. Nearly the same charge can be achieved with a model that allows fill-and-spill C to B to A.

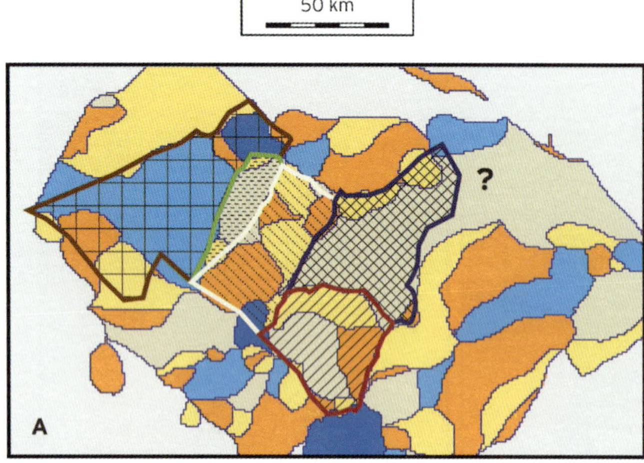

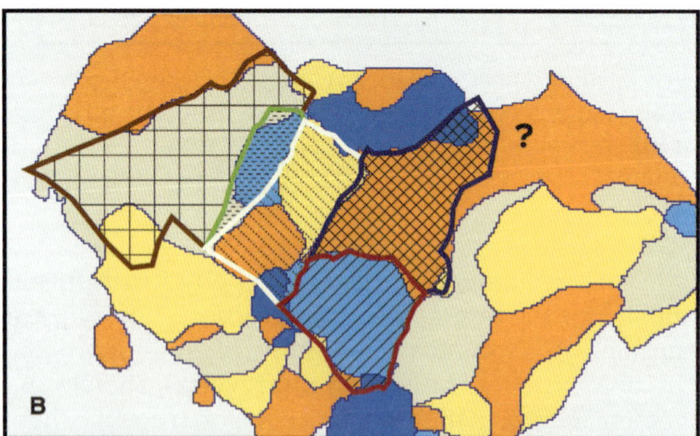

Figure 8. Example of model fetch area validation using seismic-derived drainage area polygons (patterned). (A) Fetch areas (shaded) for fill-and-spill traps up to three grid cells only. (B) Fetch areas (shaded) for fill-and-spill traps up to 100 grid cells. Case B is a better match to the seismic-derived drainage areas; both cases disagree in the area with the question mark, where the seismic interpreter has a drainage divide but the model does not.

It is worth noting that most uncertainty estimates for velocity variation do not address the impact of dip changes.

Example 4: Choosing the Right Prospect Outline

How many times have we been given the outline of a prospect closure and never questioned whether it was entirely consistent with our maps? In this example, shown in Figure 6, the prospect was a very low relief closure complicated by gas sags. As the interpreter tried to account for the effect of the gas sags and detailed remapping progressed, the position of the prospect closure changed because of small-scale higher resolution work performed by the team. This resulted in an ever-changing series of prospect outlines handed to the basin modeler. Unfortunately, the depth maps provided to the basin modeler were not being updated along with the closures.

The evolution of the different closures and associated fetch areas is shown in Figure 6, with each closure and its ensuing fetch area identified by color and shading. The original closure and fetch area (Figure 6A) are shown in blue and are taken as the reference case. The second closure and fetch area (Figure 6B) are shown in green and provide 35 times the oil charge and 50 times the gas charge as the first case, as the fetch area extends farther over the source kitchen. In the third case (Figure 6C), the basin modeler determined the closure based on the map at hand. The resulting fetch area is shown in yellow and is twice as large as the previous one, resulting in 75 times the oil charge and 100 times the gas charge compared with the original reference model.

The take-home point from this example is that the prospect closures provided to the basin modeler need to be consistent with the most current map interpretation. This is a particularly sensitive issue in low-relief areas, especially those complicated by gas sags. In addition, some thought should be given as to whether the long migration distances implied by the larger fetch areas are realistic, and what the possible impact of faults might be. A test with a 2-D profile might clarify these issues, assuming that the key geologic uncertainties are addressed.

Example 5: Contouring and Fetch Area

Figure 7A shows three prospects machine contoured from a 10- to 15-km widely spaced seismic grid. Each prospect is shown in white, with their associated fetch areas in shades of gray. The same data can be contoured by hand as one elongate closure (Figure 7B), which has the added benefit of greater trap volume. The preferred concept must be considered in the context of regional tectonics, structure, and scale. Which interpretation is more consistent with other known structures in the basin or the nearby region?

The all-in-one closure has three times the amount of charge compared with closure A alone. Nearly the same amount of charge can be achieved with a model that allows fill-and-spill from C to B to A.

Example 6: Fetch Area Validation

One way to define the fetch area sensitivity is to adjust the amount of charge allowed to spill from adjacent traps into the main fetch area. The two maps in Figure 8 compare the fetch polygons (solid colors) in the same region for two different fill-and-spill definitions. In A, trap sizes only up to three cells (3 km^2) are

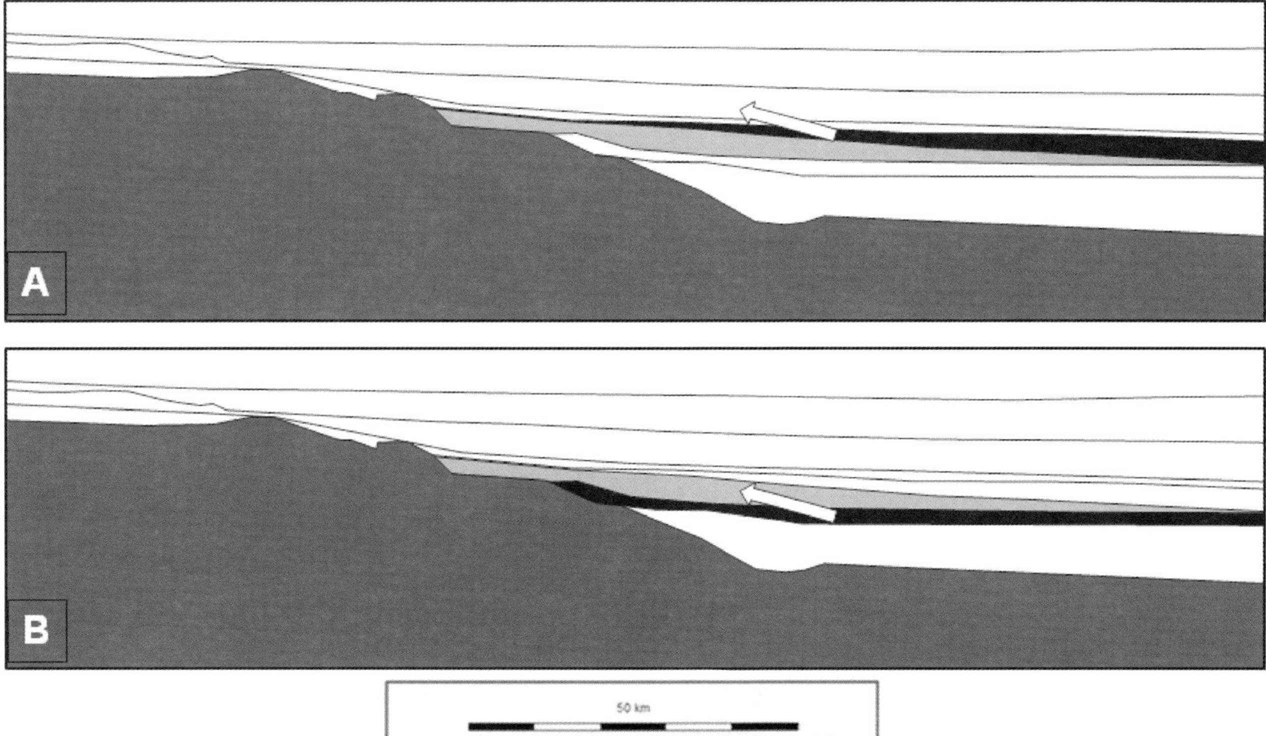

FIGURE 9. Impact of changing the geologic model. If the source rock (black) is Oligocene–Miocene, then whether this fan (medium gray) is Eocene (A) or Miocene (B) has a huge impact on volumes because of downward or lateral migration required in A, but not in B. The arrows represent conceptual migration directions.

allowed to fill-and-spill into the fetch areas, which is essentially a no-fill-and-spill case. In B, the three-cell threshold is increased to 100 cells. The effect is to create larger fetch areas in B than in A. Prospects within those fetch polygons in B would gather hydrocarbons from a greater fetch area. But which parameter setting is correct?

One important quality-control step is to compare the fetch areas generated by the model (solid colors) with those drawn by the seismic interpreter (patterned overlay). Although the seismic interpreter may be motivated to make his fetch areas as large as possible to enhance the charge to his prospect, he also has the best understanding of the maps and structures and is the best person with whom to discuss whether individual drainage divides are meaningful. In case A, it is clear that the model generates many more fetch polygons than the seismic interpreter sees. In case B, we have better agreement between the model and the seismic, with drainage areas allowing more capture through fill-and-spill. Note that both models disagree with the seismic-derived drainage maps in the upper right part of the diagram (denoted by a question mark).

Example 7: Variable Age Interpretation

This last example (Figure 9) is so evident it barely merits discussion, but sometimes, it is worthwhile to state the obvious. Let us assume that we are in a basin where we know that our principal source rock is Oligocene–Miocene, but we are uncertain as to the age of the deep-water fan we observe on the seismic. It could be Eocene or Miocene, but our horizon correlations are too uncertain to say for sure. An Eocene fan (A) would place the principal source rock above the fan, requiring lateral or downward migration to charge prospects in the fan. A Miocene fan (B) would overlie the Oligocene–Miocene source rock, and normal buoyancy forces are likely to enhance pressure-driven expulsion from the source rock into the overlying fan. Ruling out the possibility of other source intervals, and ignoring all other considerations such as seal efficiencies and local pressure differentials, the fan may charge more efficiently in case B than in case A. The two scenarios present different potential risks that would need to be considered independently. This is an example where considering alternative geologic interpretations could make a difference in the prospectivity of the deep-water fan.

CONCLUSIONS

Modern flow path models, in combination with conventional 2-D models, provide a valuable set of tools to quickly assess a variety of geologic scenarios and obtain quick estimates of possible charge volumes. These methods are particularly useful in new ventures exploration areas. A workflow combining flow path and 2-D models provides key validation of migration efficiency with which to calibrate the flow path models.

However, it is important to keep in mind the potential pitfalls and limitations of the tools. Critical factors for the success of the charge prediction include knowledge of regional basin history, tectonics, stratigraphy, and source rock depositional models, and the geophysicists' understanding of the limitations of their maps. It is important to consult with the rest of the team regarding closures, migration paths, and fetch areas. Alternate geophysical or geologic scenarios may have huge impacts on predicted volumes. These alternate scenarios can be far more important than the conventional model uncertainties in evaluating charge risk and are typically not addressed in statistical uncertainty analyses.

ACKNOWLEDGMENTS

I thank Marek Kacewicz, Ken Peters, and an anonymous reviewer for taking the time to comment on this article. Their suggestions were useful and pertinent and I believe have improved the text.

REFERENCES CITED

Rose, P. R., 2001, Risk analysis and management of petroleum exploration ventures: AAPG Methods in Exploration 12, 164 p.

Rose, P. R., 2007, Measuring what we think we have found: Advantages of probabilistic over deterministic methods for estimating oil and gas reserves and resources in exploration and production: AAPG Bulletin, v. 91, p. 21–29, doi:10.1306/08030606016.

15

Vayssaire, A., 2012, Simulation of petroleum migration in fine-grained rock by upscaling relative permeability curves: The Malvinas Basin, offshore Argentina, *in* K. E. Peters, D. J. Curry, and M. Kacewicz, eds., Basin Modeling: New Horizons in Research and Applications: AAPG Hedberg Series, no. 4, p. 247–257.

Simulation of Petroleum Migration in Fine-Grained Rock by Upscaling Relative Permeability Curves: The Malvinas Basin, Offshore Argentina

André Vayssaire
Repsol Exploration, Madrid, Spain

ABSTRACT

Early exploration of the Malvinas Basin (1979–1991) targeted Lower Cretaceous sandstones assuming that hydrocarbons would migrate laterally from the basin depocenter in the south to structures located in shallow water. Hydrocarbons were found, but not in large enough quantities to be commercially viable. Recently, exploration has moved closer to the depocenter and focuses on Eocene to Miocene sandstones a few thousand meters vertically above the mature Lower Cretaceous source rock. As no faults crosscut the entire section between the source rock and reservoirs, they cannot be evoked as conduits for hydrocarbon migration. Therefore, kilometer-scale vertical migration across fine-grained sediments was considered as the main process to transport hydrocarbons from source rock to reservoir.

This migration mechanism is commonly mentioned, but poorly constrained. Darcy flow and invasion percolation calculators were used to simulate hydrocarbon migration. If we consider that hydrocarbons migrate along thin stringers, the relative permeability parameters have to be upscaled to consider that not all of the rock is being saturated by petroleum. Furthermore, fine-grained sediments present a very high specific area, which gives a higher sorption capacity for water, and therefore, less petroleum is needed to reach the saturation threshold for flow. Secondary migration across fine-grained sediments takes time to initiate, but as soon as hydrocarbons invade the pore space, the migration is effective; it occurs with minimal hydrocarbon losses and is essentially controlled by the expulsion rate of petroleum from the source rock and the stratigraphic architecture. From a physics standpoint, the Darcy method looks more appropriate because it incorporates the full physics of the problem. However, under these conditions, viscous forces can be ignored and the invasion percolation method seems appropriate to simulate secondary migration of hydrocarbons across fine-grained sediments.

DOI:10.1306/13311440H43474

GEOLOGIC FRAMEWORK

The Malvinas Basin (Figure 1) is the southernmost offshore sedimentary basin in Argentina with water depths ranging from 50 to 1000 m (164–3281 ft). It has a triangular shape (Figure 2) and is limited by the Rio Chico High in the west, onlaps onto the Malvinas platform to the east, and is constrained to the south by the Burwood Bank, which is a topographic expression of the fold and thrust belt associated with the South America and Scotia plates. The Malvinas Basin formed in response to early rifting and volcanism related to the first breakup of Gondwana (Vayssaire et al., 2008), which caused the separation of eastern Gondwana (India, Australia, and Antarctica) from western Gondwana (South America and Africa) in the Late Jurassic (168 Ma). In the Malvinas Basin, the termination of this synrift period correlates with oceanic crust formation in the Weddle Sea (150 Ma). It was followed by a phase of postrift thermal subsidence throughout the Cretaceous that led to a marine transgression and developed the backstepping sands of the Springhill Formation (Figure 3). The transpressional deformation acting along the South America–Scotia plate boundary deepened the southern part of the basin. This led to the development of a fold and thrust belt during the Oligocene and the lower Miocene, forming the southernmost limit of the Malvinas foreland basin.

The Springhill Formation was targeted during the early exploration of the basin (1979–1991). This exploration has proven the existence of a working petroleum system, albeit without the discovery of commercially viable petroleum accumulations. This economic failure is most likely caused by poor efficiency of lateral migration of the hydrocarbons from the depocenter toward the structures.

Today, exploration focuses in the southern part of the basin. Targeted reservoirs are of Eocene to Miocene age, a few thousand meters above the Lower Cretaceous source rock. Only kilometer-scale vertical migration of hydrocarbons can explain reservoir infilling.

HYDROCARBON MIGRATION ACROSS FINE-GRAINED SEDIMENTS

Aplin and Larter (2005) believe that most of the world's petroleum migrated vertically through large thicknesses of fine-grained sediments. Estimating the flow rates is extremely difficult because it is such a slow process that laboratory experiments are very challenging and, in most cases, impossible.

Whereas oil can flow at a few centimeters per hour in reservoir rocks, Appold and Nunn (2002) predicted oil migration in fine-grained sediments is in the order of 100 m/Ma. This still allows 1 km (0.6 mi) of vertical migration in 10 Ma, but represents only 0.1 mm/yr. Neuzil (1994) and Dewhurst et al. (1999) calculated a hydraulic conductivity between 0.01 mm/m.y. and 1 km/m.y.

The mechanism of hydrocarbon secondary migration across fine-grained sediments is also not well understood. For primary migration, Appold and Nunn (2002) ventured the hypothesis of the porosity wave that would favor the displacement of hydrocarbons within the source rock. Osborne and Swarbrick (1997) showed that the release of hydrocarbons from kerogen during cracking would lead to an increase of fluid volume and pore pressure, generating microfractures and primary migration of hydrocarbons in low-permeability source rocks. This hypothesis is supported by observations that the top of the overpressure section commonly coincides with the petroleum generation window (Spencer, 1987; Isaksen 2004).

Regarding the formation of hydrocarbon pathways during secondary migration, Aplin and Larter (2005) demonstrated that polar compounds in petroleum, such as phenols, penetrate the mineral water films such that hydrocarbons can wet the mineral surface. In this case, cap rocks do not act as permanent seals, but simply retard the inexorable flow of petroleum. Other migration mechanisms have been proposed, including migration because of fracturing of the cap rock when the buoyancy forces exceed the tensile strength of the rock or the propagation of methane-filled fractures that develop the potential to entrain and transport oil (Nunn and Meulbroek, 2002).

It appears that the processes previously mentioned take time to initiate, but as soon as hydrocarbons invade low-permeability rock, more petroleum can migrate through this pathway very easily and rapidly even in the absence of a pressure gradient to drive Darcy flow. Existing pathways can be reused in a very efficient way with a quasi-instantaneous velocity and minimal hydrocarbon losses.

MIGRATION SIMULATIONS IN THE MALVINAS BASIN

In the Malvinas Basin, the Temis Darcy simulator from IFP (Institut Francais du Pétrole, Rueil Malmaison, France) and the MPath invasion percolation (IP) simulator from The Permedia Research Group (Ottawa, Canada) were used to model hydrocarbon migration in three dimensions. The IFP model assumes that the hydrocarbons migrate as a separate phase and follow a generalized Darcy's law (Bear, 1972), which gives a complete description of the forces controlling the fluid flow considering the intrinsic permeability tensor, the relative permeabilities in the porous medium, as well as the viscosity,

Figure 1. Map of Argentina with the principal sedimentary basins.

density, and capillary pressure of the hydrocarbon phase and the pore pressure of the water phase (Ungerer et al., 1984). The IP technique considers that the balance between the buoyancy and the capillary forces overwhelmingly controls the trajectories of the hydrocarbon flow and that the migration distance and velocity is predominantly controlled by the rate and volume of petroleum expelled from the source rock (Carruthers, 2003), and hence, the viscous forces are ignored.

Both simulations show great similarities in the present-day saturation distribution in the Miocene and Eocene reservoirs (Figure 4). Both models have the same lithologic description and contain realistic stratigraphic variability. The calculated temperatures and pressures are identical at all ages. The rock parameters such as porosity, capillary pressure, and relative permeability curves are also identical. The grid resolution is vertically the same, but the Darcy model was laterally upscaled to decrease the number of grid cells to reduce the calculation time. This upscaling has no effect on the migration mechanism and the distribution of the hydrocarbon accumulations.

Nevertheless, the charging process of the two models is very different. With Darcy flow, the oil that fills the reservoir comes from the top of the migration front, whereas with the IP technique, the early expelled oil has already migrated through the entire column up to the surface, and the oil filling the reservoir has been recently expelled. This result seems to be more in accordance with sea bottom piston-coring results and other direct hydrocarbon indicators, suggesting that the hydrocarbons reached the sea floor in great quantities.

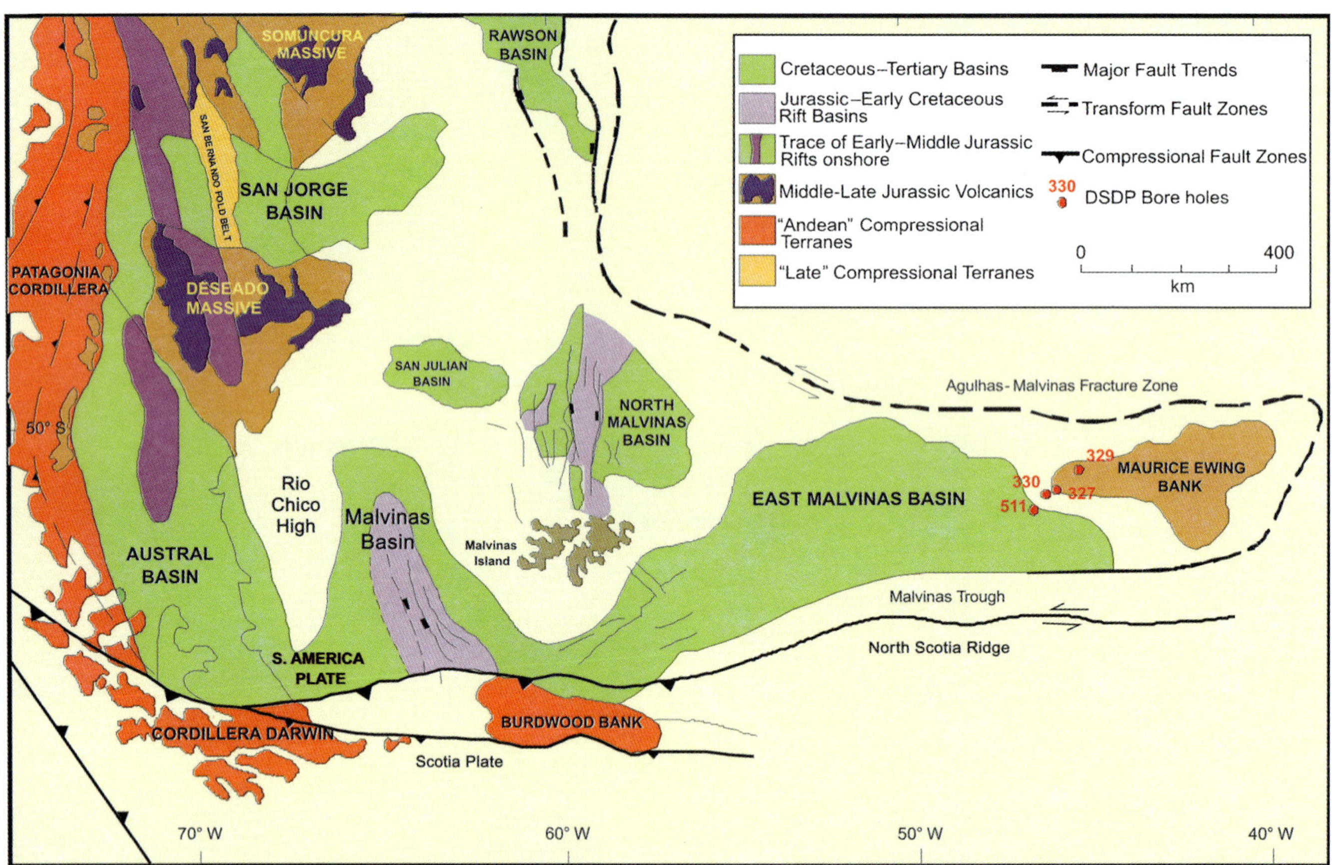

FIGURE 2. Geologic framework of the Malvinas Basin and the surroundings. DSDP = Deep Sea Drilling Project.

In other words, with IP, most of the expelled hydrocarbons reach the sea floor, whereas with Darcy flow, most of the petroleum remains in low-permeability layers between the source rock and the reservoirs. Figure 5 displays hydrocarbon saturation during the Miocene along a cross section extracted from two three-dimensional blocks resulting from Darcy and IP simulations. The saturations in the Darcy simulation are high between the source rock and "Sand-1." The IP simulation shows much lower saturations and migration pathways up to the surface. Abundant hydrocarbons leave the model boundary through the sea floor, which is not the case with the Darcy simulation.

The interpretation for the migration risk is very different. In the Darcy case, the risk is that the expulsion occurs too late and the hydrocarbons have no time to reach the target. In the IP case, the source may mature too early and the expelled hydrocarbon migrates to the sea floor, leaving an insufficient charge to fill the reservoirs once the trap is formed and sealed.

With the IP method, in fine-grained sediments where no contrast of capillary pressure exists, the hydrocarbon flows from one grid cell to another as soon as the critical saturation is reached. With Darcy, the calculated velocity is generally too slow to permit flow and the saturation continues to increase until petroleum occupies the maximum possible volume. At this stage, all mobile water has flooded out of the pore space. An additional increase of saturation will force the petroleum to move out of the grid cell so the saturation does not exceed the maximum value permitted. This displacement is therefore essentially controlled by the saturation limits of the relative permeability curves, and viscous forces play a nominal role.

Beyond the Malvinas case, these two different results confirm that the quantity of petroleum found in the reservoirs represents a small fraction of what was generated. McDowell (1975) and Moshier and Waples (1985) concluded that the recoverable quantity of petroleum in most basins is only a few percent of the generated petroleum, and in very few cases does this number reach 10%. This demonstrates the inefficiency of one or all of the following poorly understood processes: expulsion, migration, and trapping of hydrocarbons. Pepper and Corvi (1995) consider that hydrocarbons are sorbed within the kerogen (adsorbed in the microporosity and absorbed onto the internal surface area) and that the cracking of 200 mg_{HC}/g_C of initial hydrogen index (HI) is a prerequisite for oil expulsion. It is also generally accepted that cap rocks do not act as permanent seals

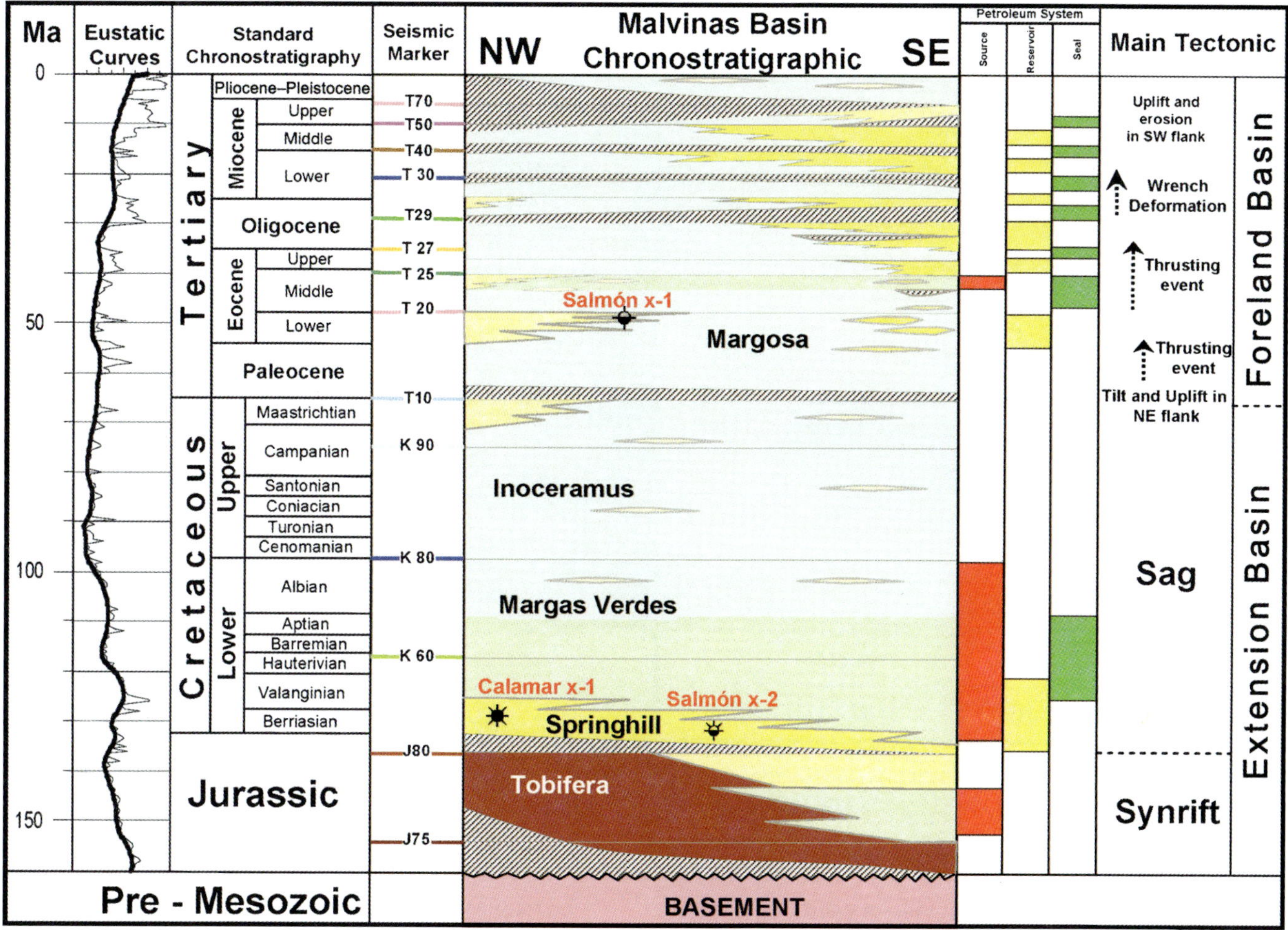

FIGURE 3. Stratigraphic column of the Malvinas Basin with petroleum system elements and major tectonic imprints.

but retard the flow of petroleum and that reservoir column heights are essentially controlled by the balance between buoyancy and capillary forces. During migration, hydrocarbons are lost along the migration pathways because of the heterogeneity of the stratigraphic architecture but also because the rock must reach a minimum petroleum saturation before oil can flow. These losses are a challenge to quantify.

MIGRATION LOSSES

Sylta (2002) and other authors mention that simulators using modified Darcy equations may overestimate hydrocarbon losses, and therefore, the calculated migration velocities are too low. The simulation of hydrocarbon migration in low-permeability rock using Darcy's law (Figure 6) clearly shows that the calculated Darcy true velocities never exceed 3.6 m/m.y., and the average velocity is around 0.5 m/m.y. These numbers cannot explain the present-day migration front 2.5 km (1.5 mi) above the top of the source rock, with the expulsion starting at 120 Ma. The migration rate must exceed 20 m/m.y. to account for the migration front location at present day.

In fact, it looks as if the losses have a significant impact on migration velocity. Under certain conditions, decreasing migration losses by increasing the residual water saturation does not significantly affect the velocity calculated by the Darcy equation but forces the migration to move faster and further because the petroleum saturation rapidly approaches the maximum saturation threshold value. The extra mass is then forced to move where the hydraulic head is the lowest (commonly upward) whatever the velocity calculated by the Darcy solver. Two synthetic cross sections are shown at the top of Figure 7. The only difference between the two sections is the maximum hydrocarbon saturation allowed in each grid cell. On the left, the relative permeability curve allows maximum hydrocarbon saturation to reach 100%. The mean true Darcy flow calculated is 4 m/m.y., and the migration front at present day is around 1 km (0.6 mi) above the top of the source rock (bottom layer). At the right, the maximum hydrocarbon saturation was

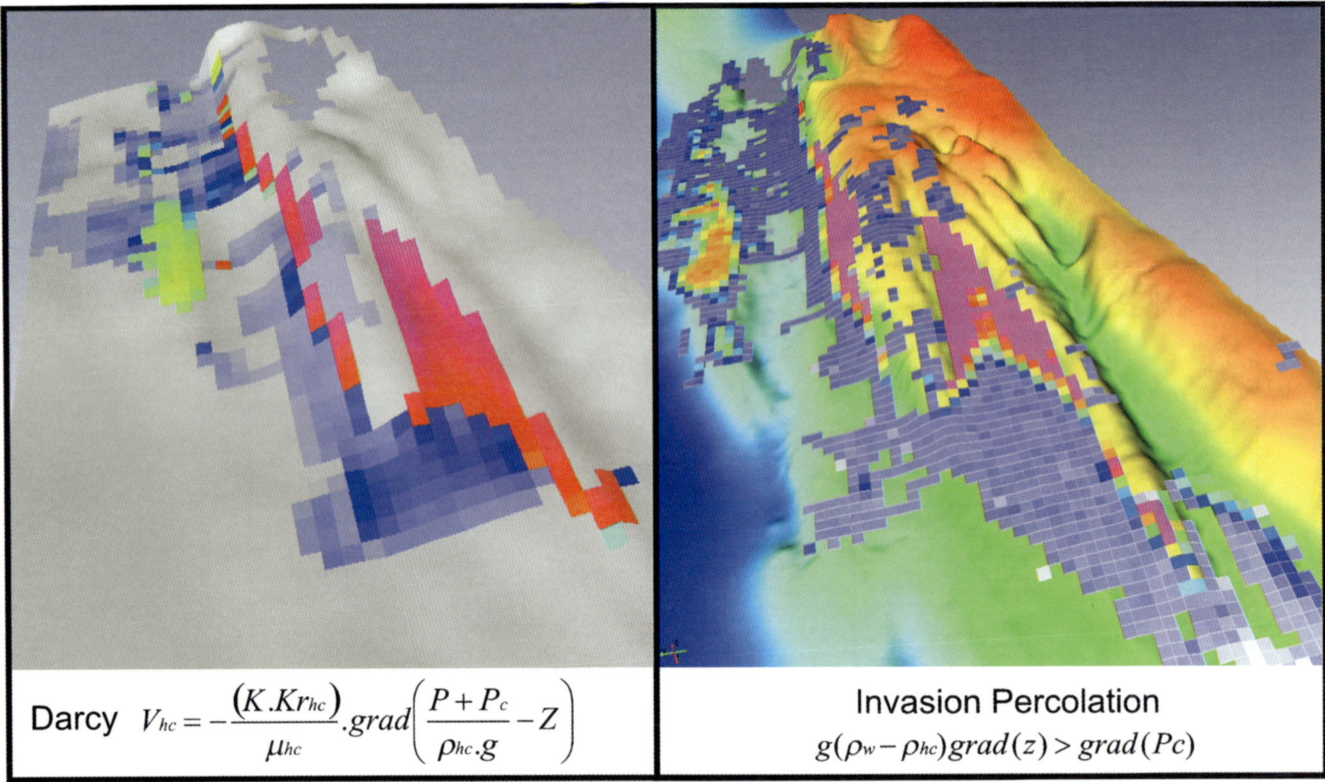

FIGURE 4. Simulation results showing hydrocarbon distribution (saturation) in Miocene reservoirs using the invasion percolation method (right) and Darcy's law (left).

reduced to 30%. The migration velocity did not change much (5 m/m.y. instead of 4 m/m.y.), but the migration front moved much farther (4000 m [13,123 ft] instead of 1000 m [3281 ft]).

The same occurs when increasing the expelled masses from the source rock: no change in the velocity calculated by the Darcy technique is observed, but the hydrocarbons move a greater distance (Figure 7, bottom).

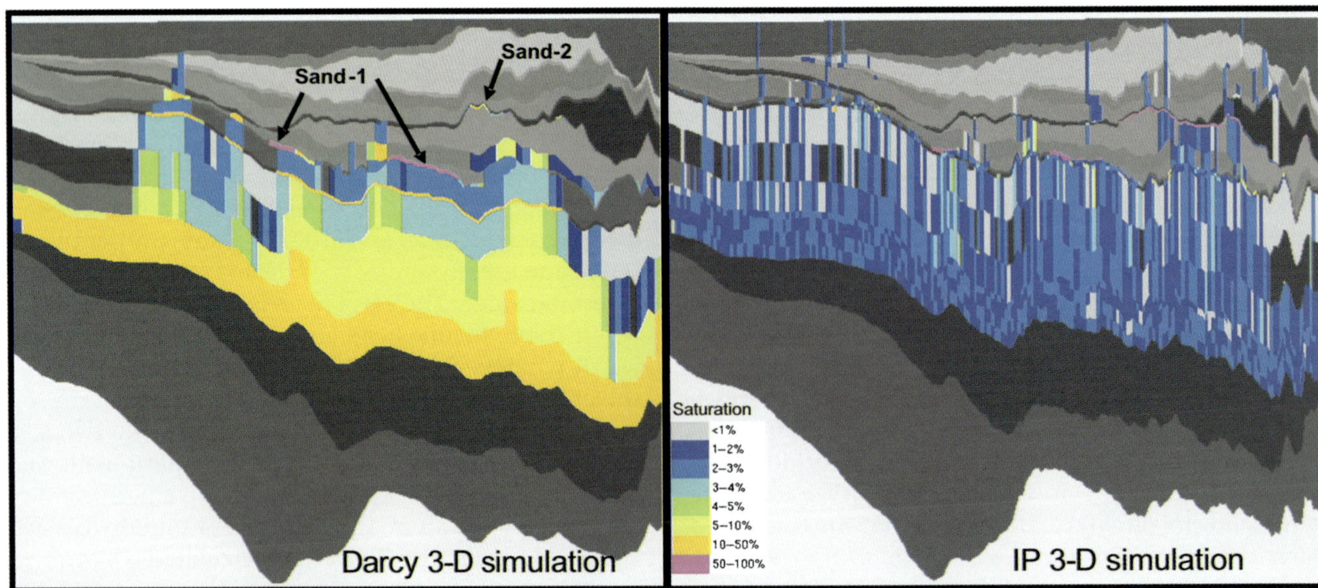

FIGURE 5. Simulation results from invasion percolation (IP) (right) and Darcy (left) simulators showing hydrocarbon saturation during the Miocene.

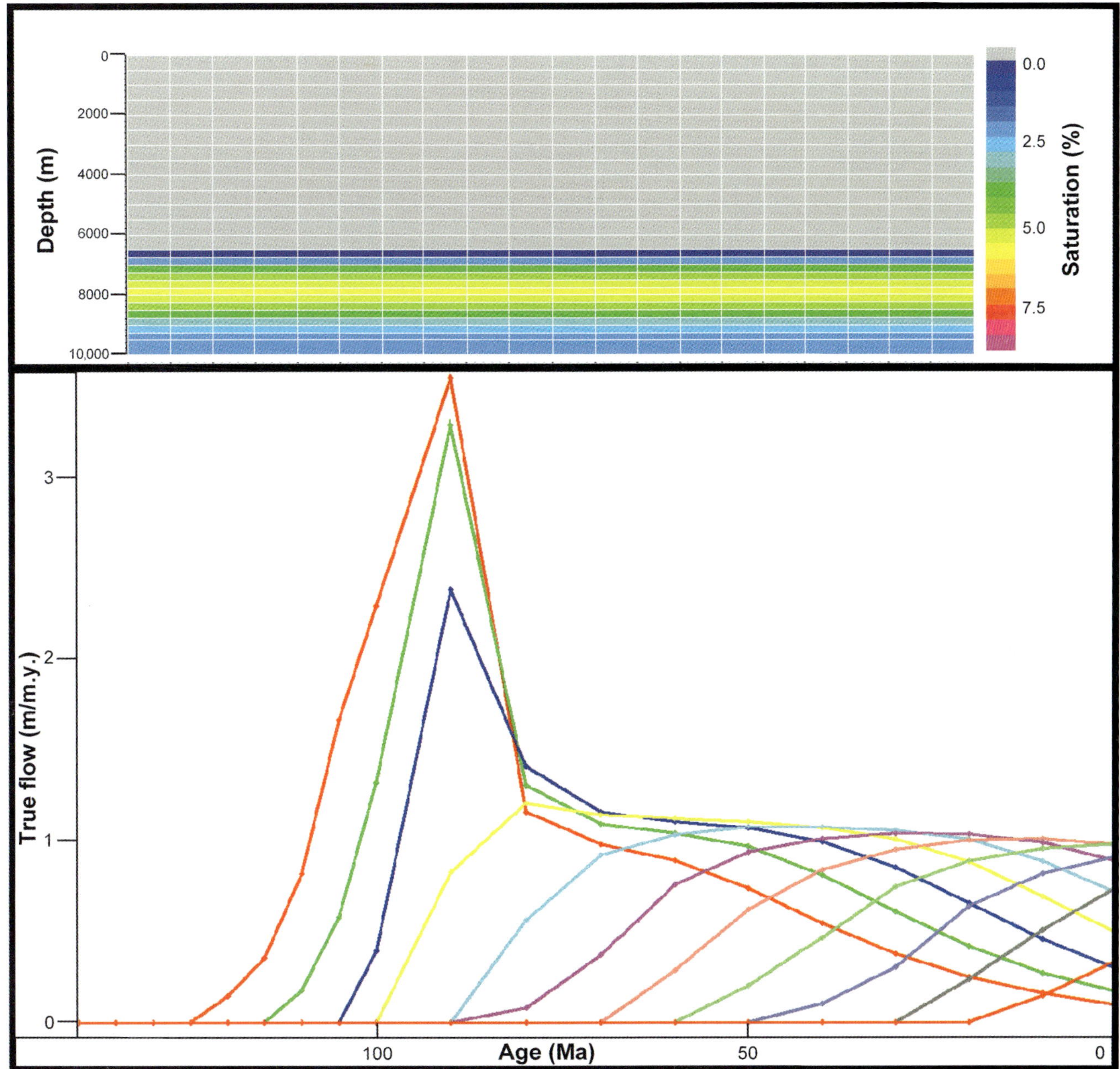

Figure 6. Top: Synthetic cross section showing present-day hydrocarbon saturations. The source rock is located in the bottom layer, and the migration front is more than 2500 m (8202 ft) above it. Bottom: Hydrocarbon migration velocity in meters per m.y. for each cell versus geologic time in m.y. The red line at the left represents the history of hydrocarbon migration true velocity of the cells just above the source rock. The green line gives the velocities of the second row of cells above the source rock, and so on.

The relative permeabilities of fine-grained sediments are poorly controlled because of the difficulty of doing measurements on low-permeability rocks. Okui and Waples (1993) extrapolated relative permeabilities measured in rocks with decreasing grain size from sand to silt and showed that the residual water saturation in shales increases, significantly reducing the maximum hydrocarbon saturation to approximately 20% in low-permeability rocks. The same authors also showed that the flow of oil commences in fine-grained rocks at increasingly small oil saturations (i.e., 2%). The explanation given is that the clays in rock provide enormous specific area, giving a higher sorption capacity for water. This increases the film of immobile water surrounding the grains. This film is commonly referred to as "bound" or "connate water." The consequence is that compared with sandstones, low-permeability rocks require less petroleum to reach the saturation threshold for migration to occur, which leads to lower migration losses.

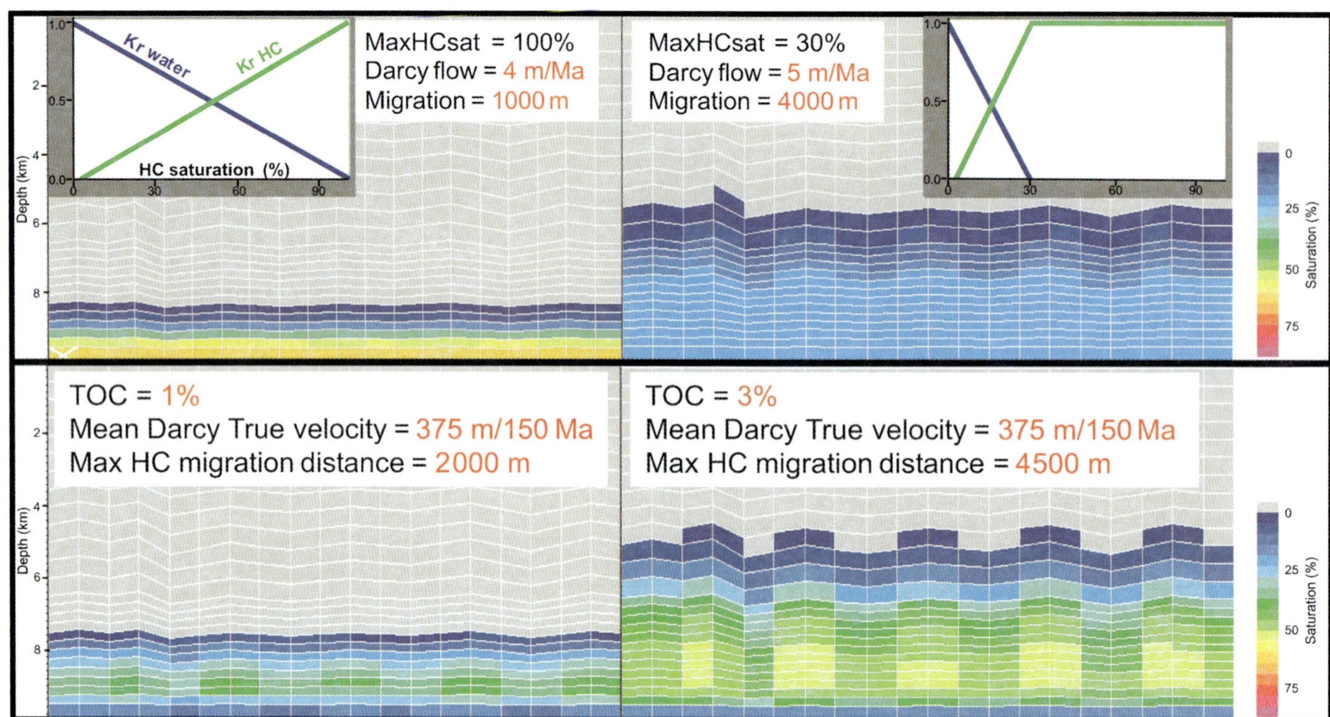

FIGURE 7. Hydrocarbon (HC) saturation in two identical synthetic sections with different maximum HC saturation (top) and total organic carbon (TOC) (bottom).

UPSCALING RELATIVE PERMEABILITY CURVES

Whereas migration within a reservoir may have a pistonlike displacement pattern, where the saturation at the migrating cluster is high, several authors including Luo et al., (2004, 2007, 2008), Carruthers (2003), Meakin et al., 2000, Schowalter (1979), and Berg (1975) consider that oil migrates along thin stringers in fine-grained sediments where the saturation is low. As demonstrated by Luo et al., 2007, the migration starts within the source rock with many columns of petroleum moving upward. Gradually, some columns stop, one after the other, and only a few migration pathways continue to grow. After the saturation has built up within the narrow continuous pathways that have developed, the migration process is considerably faster and oil tends to migrate efficiently within a limited number of pathways. The result is that the migration losses are much higher within and close to the source rock than outside and far from the source rock, where they may be negligible. This is in accordance with the very few reports of active migration pathways and the important quantities of petroleum that can be found in source rocks (e.g., shale-gas). Luo et al. (2007) consider that about 99% of the losses occur within the range of the source area, and Hirsh and Thompson (1995) demonstrated that the remaining saturations in secondary migration can be in the order of 1%.

The rock samples used in the experiments are much smaller than the grid blocks used in basin models. Within the rock volume represented by these grid blocks, a large fraction would never be exposed to petroleum. The critical saturations used by the simulators should therefore be scaled by a function of the grid cell area, assuming that migration is essentially vertical in low-permeability rocks. To account for the large number of stringers, the scaling might be turned off inside and near the source rock. It should then increase as petroleum moves away from the source rock.

We should then consider that migration pathways outside the source rock cover an area that is much smaller than the cell area used for the numerical simulation. When petroleum saturates a migration pathway to its critical saturation, it does not impact the entire simulation cell but instead a small volume corresponding to the stringers. The maximum hydrocarbon saturation within a migration pathway needs to be upscaled to reflect the mean value inside the entire numerical simulation cell. Assuming that within the migration pathway the maximum oil saturation is 20% at the scale of the simulation cell, this number may drop to values lower than 0.2%. This would correspond to a 1-km^2 (0.39 mi^2) simulation cell in which we assume that the migration stringers occupy an area of 100 m^2 (1076 ft^2).

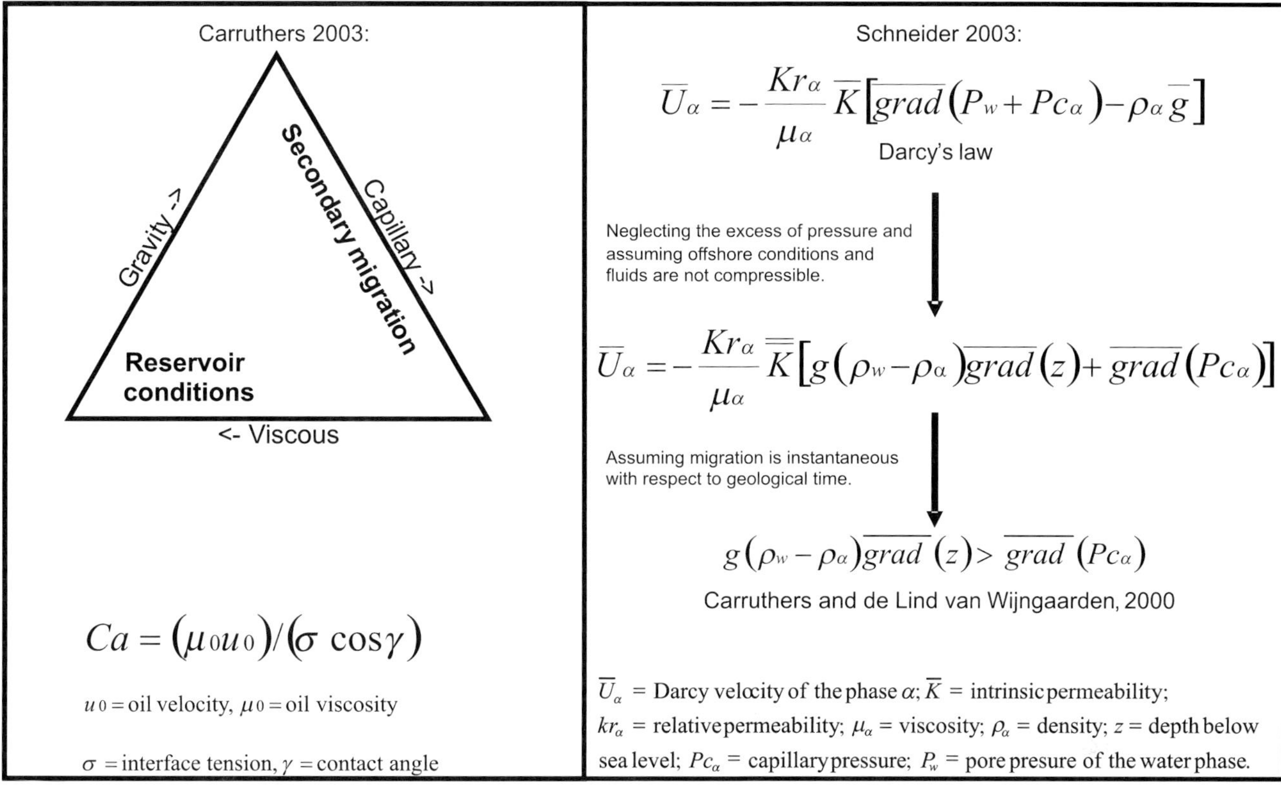

FIGURE 8. Arguments from Carruthers (2003) to justify that viscous force can be ignored (left) and demonstration from Schneider (2003) that invasion percolation can be derived from Darcy's law (right).

DISCUSSION

Schneider (2003) (Figure 8, right) claims that successive simplifications of Darcy's law that neglect excess pressure, assuming that fluids are not compressible and considering that migration is instantaneous with respect to geologic time, result in an equation governing the models that is controlled by percolation (Carruthers and de Lind van Wijngaarden, 2000).

Whereas the generalized Darcy equation contains three driving forces for hydrocarbon migration, which are buoyancy, capillary forces, and viscous forces, Carruthers (2003) considers that viscous forces can be ignored. England et al. (1987) showed that at the geologic flow rate of secondary migration, the capillary number (Figure 8, left) never exceeds 1e-10, which is far below the 1e-4 limit at which viscous forces become important.

The Darcy method has proved its efficiency for reservoir production simulation. Using Darcy-based migration solvers to represent migration at basin scale assumes that all of the physics appropriate for reservoir simulation must extend to basin time and length scales, although we know significant differences exist with the requirements of the two scenarios (grain size, saturation, grid cell, permeability, and particularly, time).

Experience shows that velocities, calculated using the Darcy equation, are too small to explain migration in low-permeability mudstones. Furthermore, numerical difficulties and high run times are encountered when running Darcy with upscaled relative permeability curves that can discourage the use of high-resolution models, and therefore reduce their use as predictive tools.

Figure 9 shows a sensitivity analysis of maximum hydrocarbon saturation in secondary migration simulation cells. It clearly demonstrates that with high values (between 10 and 15%), few hydrocarbons reach the reservoir and none flow to the surface. The risk is that the expulsion occurs too early. At 5% saturation, 23% of the expelled hydrocarbons migrated to the surface and the reservoirs received the maximum quantity of petroleum. With values lower than 4%, more hydrocarbons are lost in the sea floor and the volume in reservoir does not vary much.

Of course, great uncertainty is present with respect to the selected upscaled value, but it appears, as shown in this representative example, that as long as it has the correct order of magnitude, it does not greatly affect the simulation results.

CONCLUSIONS

The calculations of hydrocarbon migration detailed here were achieved using the Darcy simulator from IFP

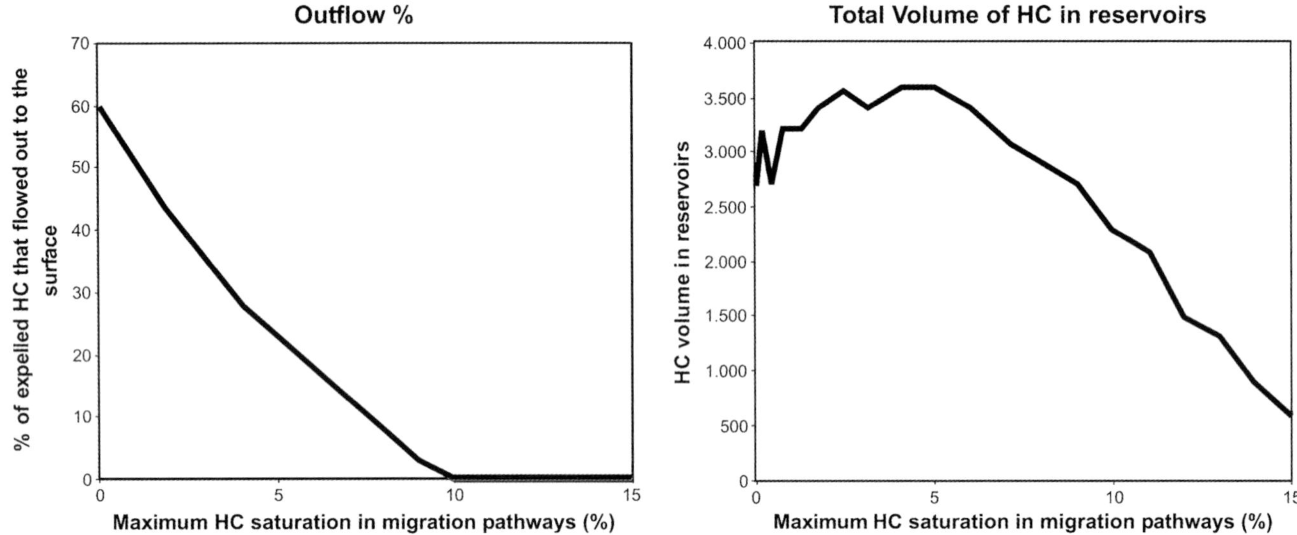

Figure 9. Maximum hydrocarbon (HC) saturation in secondary migration pathways versus the percentage of expelled petroleum that reach the surface (left) and the volume in place in the main potential reservoir of the basin (not the Malvinas Basin).

and the IP calculator from Permedia. Although they show similar present-day saturations inside the potential reservoirs, the saturation distribution outside the reservoir is drastically different because of differences in migration history. With Darcy flow, most of the petroleum remains between the source and the reservoir in low-permeability rocks. With IP, most of the hydrocarbons flowed through the topographic surface. The latter result seems to be more in accordance with sea bottom piston-coring results and other direct hydrocarbon indicators, suggesting that the hydrocarbons reached the sea floor in great quantities.

Furthermore, all simulations using Darcy's law show relatively high saturations in low-permeability rocks behind the migration front, which has a pistonlike displacement pattern. This is something that is not confirmed by drilling, which reports very rare evidence of active secondary migration paths.

From a physics standpoint, the Darcy method looks more appropriate, as it incorporates the full physics of the problem. However, upscaling the relative permeability curves is necessary to mimic a migration process that is supposed to occur along thin stringers. Under these conditions, the Darcy velocities do not change much, whereas the migration front moves further and numerical difficulties increase. It appears that viscous and permeability terms can be ignored for calculating displacement of hydrocarbons across fine-grained sediments, and the IP technique provides satisfactory and realistic results. The petroleum migration distance and velocity are overwhelmingly controlled by the hydrocarbon expulsion rate from the source rock (timing, quantities, and products) and the stratigraphic architecture (anisotropy, buoyancy, and capillary forces).

Many unknowns remain in migration and expulsion processes. These unknowns have strong consequences in our simulations of reservoir filling. The oil and gas found in reservoirs represents a small fraction of what has been generated, which raises questions about the fate of the unaccounted mass. This emphasizes the need for more research into what happens in the source rock, not only the generation, but also the expulsion phenomena (retention, primary migration, expelled quantities, products, and the timing) and the secondary migration processes.

ACKNOWLEDGMENTS

I thank the management of YPF Exploration for support and permission to publish, Wicaksono Prayitno for the long conversations we had about petroleum systems in the South Atlantic, and Frédéric Schneider and Dan Carruthers for interesting discussions about oil and gas secondary migration. Appreciation is also expressed for David Kennedy, Kenneth Peters, and an anonymous reviewer for helpful technical and editorial comments.

REFERENCES CITED

Aplin, A. C., and S. R. Larter, 2005, Fluid flow, pore pressure, wettability, and leakage in mudstone cap rocks, *in* P. Boult and J. Kaldi, eds., Evaluating fault and cap rock seals: AAPG Hedberg Series 2, p. 1–12.

Appold, M. S., and J. A. Nunn, 2002, Numerical models of petroleum migration vi buoyancy-driven porosity waves in viscously deformable sediments: Geofluids, v. 2, p. 233–247, doi:10.1046/j.1468-8123.2002.00040.x.

Bear, J., 1972, Dynamics of fluids in porous media: New York, Elsevier, 569 p.

Berg, R. R., 1975, Capillary pressures in stratigraphic traps: AAPG Bulletin, v. 59, p. 939–956.

Carruthers, D. J., 2003, Modeling of secondary petroleum migration using invasion percolation techniques, *in* S. Duppenbecker and R. Marzi, eds., Multidimensional basin modeling: AAPG/Datapages Discovery Series 7, p. 21–37.

Carruthers, D. J., and M. de Lind van Wijngaarden, 2000, Modeling viscous-dominated fluid transport using modified invasion percolation techniques, *in* J. J. Pueyo, E. Cardellach, K. Bitzer, and C. Taberner, eds., Proceedings of Geofluids III, Third International Conference on Fluid Evolution, Migration and Interaction in Sedimentary Basins and Orogenic Belts: Journal of Geochemical Exploration, v. 69–70, p. 669–672, doi:10.1016/S0375-6742(00)00138-2.

Dewhurst, D. N., Y. L. Yang, and A. C. Aplin, 1999, Permeability and fluid flow in natural mudstones, *in* A. C. Aplin, A. J. Fleet, and J. H. S. Macquaker, eds., Muds and mudstones: Physical and fluid-flow properties: Geological Society (London) Special Publication 158, p. 22–43.

England, W. A., A. S. Mackenzie, D. M. Mann, and T. M. Quigley, 1987, The movement and entrapment of petroleum fluids in the subsurface: Geological Society (London) 144, p. 327–347.

Hirsch, L. M., and A. H. Thompson, 1995, Minimum saturations and buoyancy in secondary migration: AAPG Bulletin, v. 79, p. 696–710.

Isaksen, G. H., 2004, Central North Sea hydrocarbon systems: Generation, migration, entrapment, and thermal degradation of oil and gas: AAPG Bulletin, v. 88, no. 11, p. 1545–1572, doi:10.1306/06300403048.

Luo, X. R., F. Q. Zhang, S. Miao, W. M. Wang, Y. Z. Huang, D. Loggia, and G. Vasseur, 2004, Experimental verification of oil saturation and loss during secondary migration: Journal of Petroleum Geology, v. 27, no. 3, p. 241–251, doi:10.1111/j.1747-5457.2004.tb00057.x.

Luo, X. R., B. Zhou, S. X. Zhao, F. Q. Zhang, and G. Vasseur, 2007, Quantitative estimates of oil losses during migration, Part I: The saturation of pathways in carrier beds: Journal of Petroleum Geology, v. 30, no. 4, p. 375–387, doi:10.1111/j.1747-5457.2007.00375.x.

Luo, X. R., J. Z. Yan, B. Zhou, P. Hou, W. Wang, and G. Vasseur, 2008, Quantitative estimates of oil losses during migration, Part II: Measurement of the residual oil saturation in migration pathways: Journal of Petroleum Geology, v. 31, p. 179–190, doi:10.1111/j.1747-5457.2008.00415.x.

McDowell, A. N., 1975, What are the problems in estimating the oil potential of a basin?: Oil & Gas Journal, v. 73, p. 85–90.

Meakin, P., G. Wagner, A. Vedvik, H. Amundsen, J. Feder, and T. Jossang, 2000, Invasion percolation and secondary migration: Experiments and simulations: Marine and Petroleum Geology, v. 17, p. 777–795, doi:10.1016/S0264-8172(99)00069-0.

Moshier, S. O., and D. W. Waples, 1985, Quantitative evaluation of Lower Cretaceous Mannville Group as source rock for Alberta's oil sands: AAPG Bulletin, v. 69, p. 161–172.

Neuzil, C. E., 1994, How permeable are clays and shales?: Water Resources Research, v. 30, p. 145–150, doi:10.1029/93WR02930.

Nunn, J. A., and P. Meulbroek, 2002, Kilometer-scale upward migration of hydrocarbons in geopressured sediments by buoyancy-driven propagation of methane-filled fractures: AAPG Bulletin, v. 86, no. 5, p. 907–918.

Okui, A., and D. W. Waples, 1993, Relative permeabilities and hydrocarbon expulsion from source rocks, *in* A. G. Doré, J. H. Augustson, C. Hermanrud, D. J. Stewart, and Ø. Sylta, eds., Basin modeling: Advances and applications: Norwegian Petroleum Society Special Publication 3, p. 293–301.

Osborne, M. J., and R. E. Swarbrick, 1997, Mechanisms for generating overpressure in sedimentary basins: A reevaluation: AAPG Bulletin, v. 81, no. 6, p. 1023–1041.

Pepper, A. P., and P. J. Corvi, 1995, Simple kinetic models of petroleum formation, Part III: Modeling an open system: Marine and Petroleum Geology, v. 12, no. 4, p. 417–452, doi:10.1016/0264-8172(95)96904-5.

Schneider, F., 2003, Modeling multiphase flow of petroleum at the sedimentary basin scale: Journal of Geochemical Exploration, v. 78–79, p. 693–696, doi:10.1016/S0375-6742(03)00092-X.

Schowalter, T. T., 1979, Mechanics of secondary hydrocarbon migration and entrapment: AAPG Bulletin, v. 63, p. 723–760.

Spencer, C. W., 1987, Hydrocarbon generation as a mechanism for overpressuring in Rocky Mountain region: AAPG Bulletin, v. 71, no. 4, p. 368–388.

Sylta, O., 2002, Quantifying secondary migration efficiencies: Geofluids, v. 2, p. 285–298, doi:10.1046/j.1468-8123.2002.00044.x.

Ungerer, P., F. Bessis, P. Y. Chenet, B. Durand, E. Nogaret, A. Chiarelli, J. L. Oudin, and J. F. Perrin, 1984, Geological and geochemical models in oil exploration: Principles and practical examples, petroleum geochemistry and basin evaluation: AAPG Memoir 35, p. 53–77.

Vayssaire, A., W. Prayitno, D. Figueroa, and S. Quesada, 2008, Petroleum system of deep-water Argentina: Malvinas and Colorado basins, Vol. Sistemas Petroleros de las Cuencas Andinas, VII Congreso de Exploración y Desarrollo de Hidrocarburos, p. 33–51.

SECTION 6

Case Studies and Workflows

16

Baur, F., R. Littke, H. Wielens, and R. di Primio, 2012, Prediction of reservoir fluid composition using basin and petroleum system modeling: A study from the Jeanne d'Arc Basin, eastern Canada, *in* K. E. Peters, D. J. Curry, and M. Kacewicz, eds., Basin Modeling: New Horizons in Research and Applications: AAPG Hedberg Series, no. 4, p. 259–292.

Prediction of Reservoir Fluid Composition Using Basin and Petroleum System Modeling: A Study from the Jeanne d'Arc Basin, Eastern Canada

Friedemann Baur and Ralf Littke
RWTH Aachen University, Institute of Geology and Geochemistry of Petroleum and Coal, Aachen, Germany

Rolando di Primio
Helmholtz Centre Potsdam, GFZ German Research Centre for Geosciences, Potsdam, Germany

Hans Wielens
Geological Survey of Canada (Atlantic) (GSC-A), Bedford Institute of Oceanography, Dartmouth, Nova Scotia, Canada

ABSTRACT

A petroleum system modeling (PSM) study was performed on the Jeanne d'Arc Basin, offshore eastern Canada, to study the constraints and reliability of the reconstruction of petroleum reservoir filling histories. Petroleum generation and phase behavior were analyzed using phase-predictive compositional kinetic models (PhaseKinetics) determined by pyrolysis of Egret Member source rock samples. Various additional calibration data (well, rock, and fluid data), such as porosity, permeability, temperature (bottom-hole temperature, apatite fission tracks, fluid inclusions), maturity (vitrinite reflectance), and petroleum properties, such as API, gas-oil ratio, formation volume factor, and saturation pressure were integrated into this model.

Different charge scenarios were tested for the effects of open and closed faults in the carrier system to reconstruct the most likely migration pathways for the petroleum that is trapped in the Terra Nova (TN) oil field. The most probable filling history includes charge to the reservoir from a local kitchen and a second kitchen located between Hibernia and TN that was responsible for the long-range migration. In the model, the hydrocarbons migrate from this kitchen in the northwest part of the study area along pathways defined by closed transbasin faults from the north into the field. This new migration concept differs from the traditional explanation based on geochemical

DOI:10.1306/13311442H43475

measurements only von der Dick et al., 1989), which infers that local generation was solely responsible for filling the TN field. The latter can be disproved based on a simple mass balance calculation.

INTRODUCTION

The Jeanne d'Arc Basin is located 300 km (186.4 mi) offshore Newfoundland on the Grand Banks. The basin has been of major interest for the oil and gas industry since the first exploration activities in 1964. The Jeanne d'Arc Basin extends almost 250 km (155.3 mi) from north to south and approximately 80 km (49.7 mi) (average) from east to west. The basin is bounded to the east and west by pre-Mesozoic rocks of the Bonavista Platform and by the Outer Ridge complex, respectively (Figure 1) (Louden, 2002). The basin is divided by east-west–trending faults into three subunits. In the southern part, Jurassic and Cretaceous rocks are almost absent, and no source rock exists (Enachescu, 1992; Deptuck et al., 2003). In the central part, Mesozoic sediments thicken northward (Tankard et al., 1989), and in the northern part, the sedimentary sequence is completely preserved, with a total thickness of more than 20 km (12.4 mi) and with 15 km (9.3 mi) of Mesozoic sediments alone (McAlpine, 1990; Deptuck et al., 2003).

The principal aim of this study was to provide for the first time a three-dimensional, pressure-volume-temperature–controlled, multicomponent, three-phase petroleum migration analysis of the Jeanne d'Arc Basin. The objective is to explain not only the observed temperature, maturity, and pressure regime, but also to reconstruct the major kitchen areas and migration pathways for the known accumulations and provide a detailed filling history for the Terra Nova (TN) field, including phase property reconstruction for the accumulated fluids. To achieve this, objective source rock organofacies and kinetic variability were determined, and compositional kinetic descriptions of hydrocarbon generation were integrated into a numerical finite element model.

In a first step, the basin was thermally and structurally analyzed using an advanced McKenzie approach (Baur et al., 2010). The results show clearly that the tectonic and thermal subsidence as well as the measured maturities (vitrinite reflectance) and bottom-hole temperature data can be reproduced or calibrated using a single Triassic heat pulse, which decayed exponentially to present-day heat-flow values (Baur et al., 2010).

GEOLOGIC SETTING

The tectonic and stratigraphic framework of the Mesozoic–Cenozoic infill in the Jeanne d'Arc Basin was described by numerous authors (Tankard et al., 1989; Grant and McAlpine, 1990; Enachescu, 1992). The sedimentary fill rests upon Cambrian to Carboniferous crystalline basement (Figure 2) (Amoco, 1973). Red beds, thick evaporites, conglomerates, limestones, dolomites, and salt were deposited during the initial Triassic west-northwest–east-southeast extensional rift. This was followed by very thick and fine-grained Upper Jurassic to Cretaceous clastic sediments with interbedded limestones (Figure 2) (Grant et al., 1986a; Tankard and Welsink, 1989; Tankard et al., 1989). During the Oxfordian and the Kimmeridgian, the Rankin Formation was deposited, containing the 75- to 150-m (246–492 ft) thick Egret Member source rock. Different hiatus and/or unconformity events, ranging in age and spatial extent from the Kimmeridgian to the Albian, define the "Avalon Uplift," a Late Jurassic–Early Cretaceous regional tectonic event, which eroded deeply into source and reservoir rock intervals (Figure 2) (Swift and Williams, 1980). These uplifted regions acted as the source area for coarse-grained sediments, including siliciclastic reservoir rocks such as the Jeanne d'Arc, Hibernia, Catalina, Avalon, and Ben Nevis formations (Figure 2) (Enachescu and Fagan, 2005). At roughly the same time, during the Early Cretaceous, the north-northeast–south-southwest–oriented Iberia rift system caused extension while Iberia decoupled from Newfoundland (Grant et al., 1986a; Tankard and Welsink, 1989; Louden, 2002). This extension established orthogonal transbasin faults, which are the youngest active tectonic elements affecting the basin (Grant et al., 1986b; Tankard et al., 1989; Hesse and Abid, 1998; Enachescu and Fagan, 2005). Salt movement started during the Early Cretaceous and formed stratigraphic and structural traps (Tankard et al., 1989; McAlpine, 1990). During the Late Cretaceous, basinwide subsidence caused deposition of predominantly fine-grained siliciclastics and carbonates (Swift et al., 1975; Grant et al., 1986b; Deptuck et al., 2003; Wielens et al., 2004). As a consequence of high sea level fluctuations during the Cenozoic, thick shales with interbedded sandstones were deposited via incised canyons at the western margin of the basin (Deptuck et al., 2003; CNLOPB, 2004).

PETROLEUM SYSTEM

The Egret Member is the only mature source rock on the Grand Banks. It is of Kimmeridgian age (Figure 2), and its thickness increases to the north along the axis of the Jeanne d'Arc Basin (Swift and Williams, 1980; Hubbard, 1988; von der Dick, 1989; von der Dick et al.,

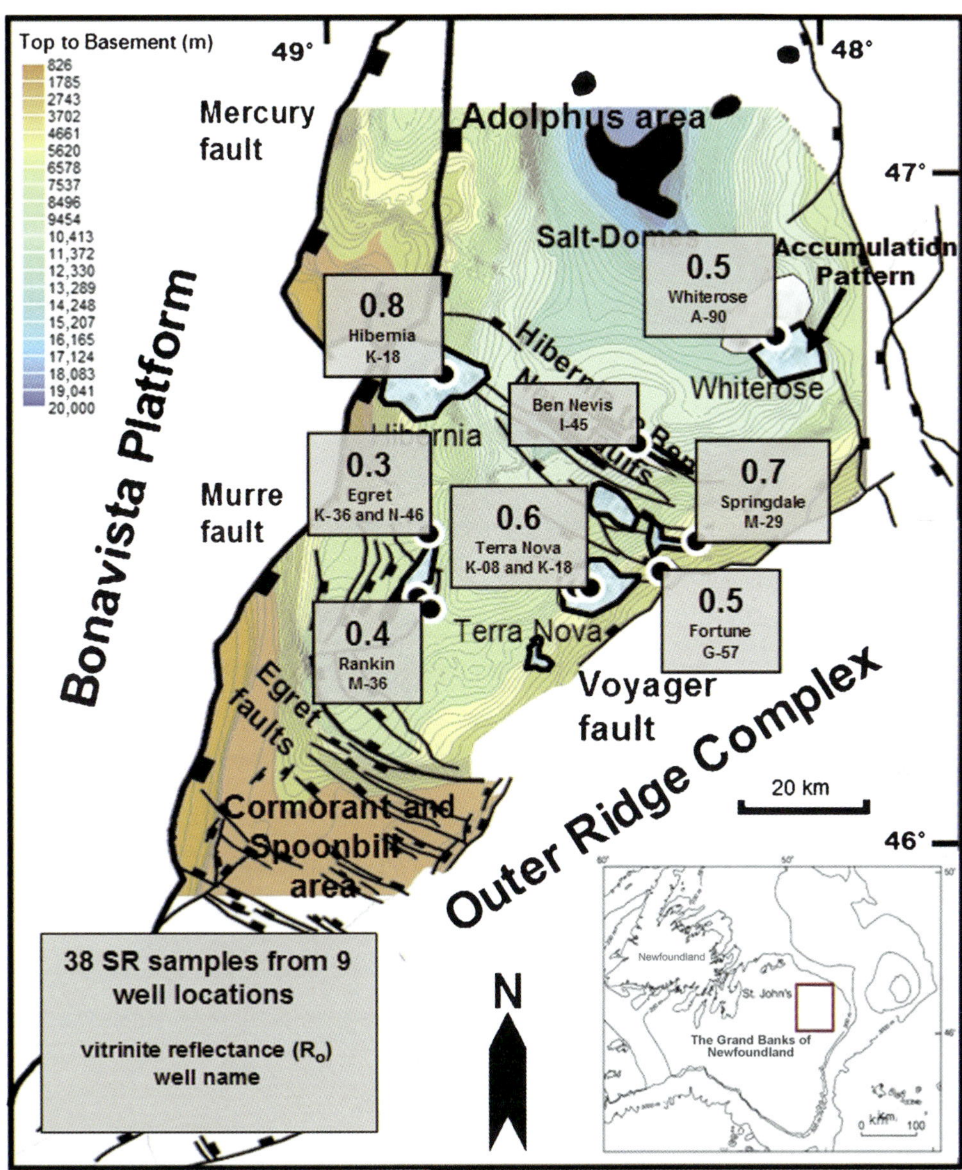

FIGURE 1. Position and extent of the study area, showing the depth to top of basement for the Jeanne d'Arc Basin derived from seismic data and used as input geometry for the three-dimensional numerical model. The inset shows the location of the study area on the Grand Banks offshore Newfoundland, eastern Canada. The black and white dots show the position of the key wells where source rock samples were obtained. Well names and the average vitrinite reflectance of the samples are displayed. Fault pattern drawn from Withjack and Schlische (2005).

1989; Fowler and McAlpine, 1994; Huang et al., 1994; Magoon et al., 2005).

In the eastern part of the study area, the Egret Member splits into a Lower and Upper Kimmeridgian source rock separated by a sandstone interval (Magoon et al., 2005). The source rock consists of laminated brown marl and calcareous shale as well as lime and mudstone with a high organic carbon content from marine planktonic organisms; this suggests a low-energy, restricted, and anoxic depositional environment (McAlpine, 1990; Huang et al., 1994). According to Fowler and McAlpine (1994), the prolific source rock has an average total organic carbon (TOC) of 6% and hydrogen index (HI) values ranging between 500 and 750 mg HC/g TOC. Some other Upper Cretaceous potential source rocks were deposited within the Jeanne d'Arc Basin, such as the Wyandot and the Petrel members, but they are not mature enough to generate hydrocarbons (Wielens et al., 2002; Deptuck et al., 2003).

All important reservoir rocks were deposited during the Late Jurassic and Early Cretaceous (Figure 2) when regional uplift in the southern part of the basin and erosion in the central and northern parts of the basin occurred. This caused the deposition of coarse sandstones starting at the end of the Jurassic with the Jeanne d'Arc Formation and ending with the Ben Nevis Formation of the Albian (sand deposits in Figure 2). The corresponding average porosities range between 14 and 20%, with permeabilities of hundreds to thousands of millidarcys (Sinclair et al., 1992; Hesse and Abid, 1998; Enachescu and Fagan, 2005).

The Fortune, Whiterose, Nautilus, and Dawson shales act as cap rocks. They were deposited as interbedded strata with the reservoir units (Figure 2) as a result of relative sea level changes (Cloetingh et al., 1989).

Trap formation occurred during the Early Cretaceous, according to reconstructions of the activity of

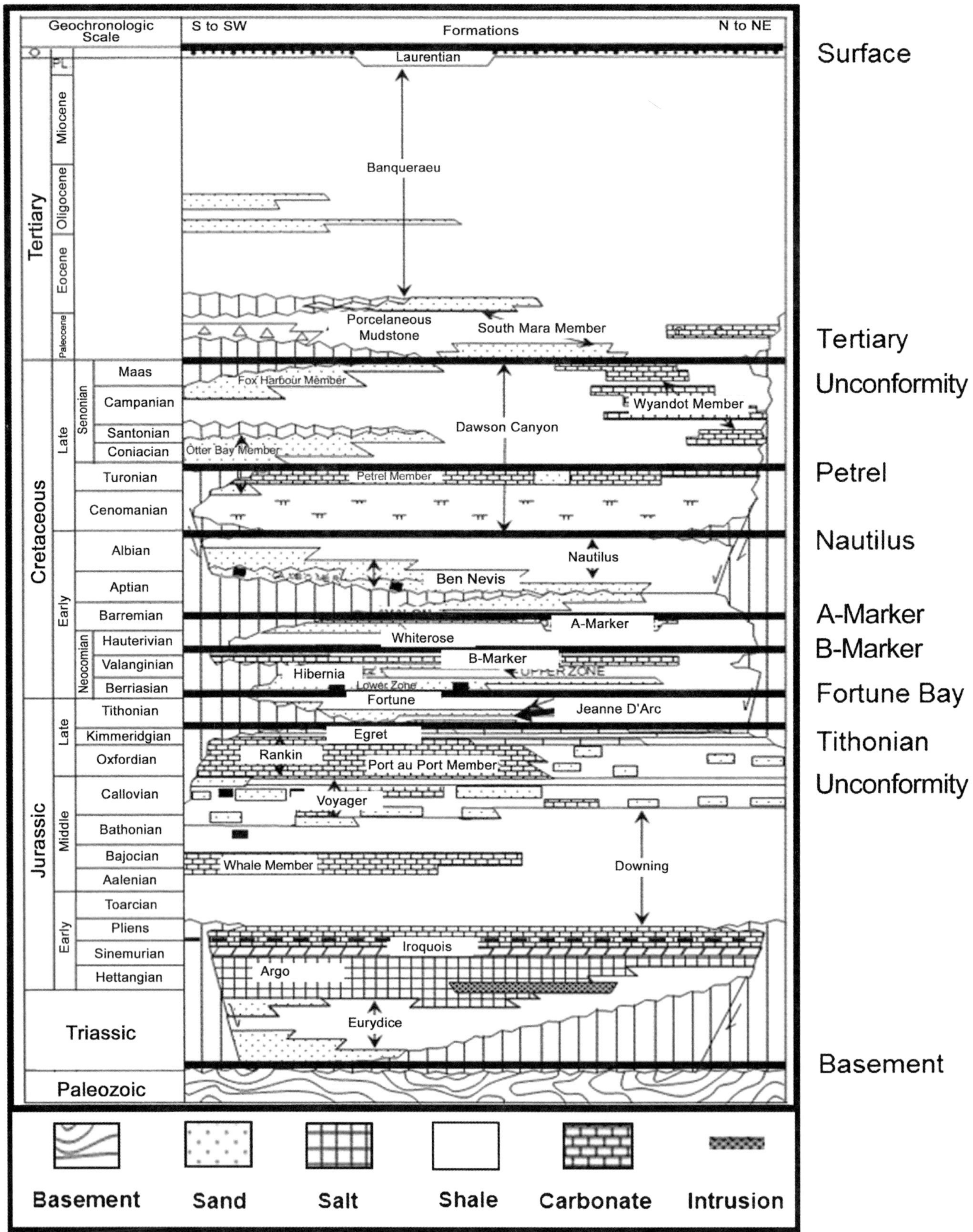

FIGURE 2. Chronostratigraphic chart for the Jeanne d'Arc Basin from Sinclair (1993). Black lines mark structural depth maps, which were derived from two different sources: (1) regional depth maps from Edwards et al. (2000) and (2) local maps from McIntyre et al. (2004) for use in the three-dimensional model.

Table 1. Oil discoveries* in the Jeanne d'Arc Basin from CNLOPB 2007–2008.

2008 Grand Banks	*Discovered*	*Location*	*Oil 10E⁶ m³*	*Oil million bbl*	*Gas 10E⁹ m³*	*Gas billion ft³*
Hibernia	1979	Western center	197.8	1244	50.6	1794
Terra Nova	1984	Eastern center	56.3	354	1.3	45
Hebron	1972	Central north	92.4	581	—	—
Whiterose	1984	Eastern north	48.4	305	85.3	3023
West Ben Nevis	1984	Northern center	5.7	36	—	—
Mara	1984	Northern center	3.6	23	—	—
Ben Nevis	1980	Northern center	18.1	114	12.1	429
North Ben Nevis	1984	Northern center	2.9	18	3.3	116
Springdale	1989	Central east	2.2	14	6.7	238
Nautilus	1981		2.1	13	—	—
Kings Cove	1990	Central east	1.6	10	—	—
South Tempest	1980	Northern east	1.3	8	—	—
East Rankin	1983	Central west	1.1	7	—	—
Fortune	1986	Northern east	0.9	6	—	—
South Mara	1984	Northern center	0.6	4	4.1	144
West Bonne Bay	1997	North-east-north	5.7	36	—	—
North Dana	1982	North-east-north	—	—	13.3	472
Trave	1983	North-east-north	—	—	0.8	30
North Amethyst	2006	North-east-north	10.8	68	8.9	315
SUM			451.5	2841	186.4	6606

**Recoverable volumes with an estimated recovery factor of 2.85.*

faults (Sinclair et al., 1992). Most of the hydrocarbon plays in the Jeanne d'Arc Basin are associated with structural traps, but a few stratigraphic traps exist as well (Grant et al., 1986b; Magoon et al., 2005). The distribution of the largest oil fields is shown in Figure 1, and their contents are listed in Table 1 (CNLOPB, 2008).

SAMPLES AND GEOCHEMICAL METHODS

Thirty-eight Egret source rock samples were collected (Table 2) from nine different well locations (Figure 1) to perform general screening analysis, and we emphasized developing accurate reaction kinetics of hydrocarbon generation, which have not been published before.

Total organic carbon determinations were performed using a LECO carbon analyzer RC-412 with a special setup to quantify the TOC content directly from untreated carbonate-containing samples (Hölscher, 1996). This was done by measuring and calculating the ratio of CO_2 and H_2O only within the temperature interval of 100 to 450°C (212–842°F). If the ratio is higher than 25, the carbon definitely derives from organic origin caused by its chemical stoichiometry of $CaCO_3$ and the low water content (Hölscher, 1996). The quality of the source rock samples (HI and oxygen index [mg CO_2/g TOC]) was measured using a Rock-Eval II instrument by DELSI with an open pyrolysis system according to the method described by Espitalié et al. (1977, 1984, 1985).

The Rock-Eval S1 and production index (PI) data in Table 2 indicate contamination of five samples in the Whiterose and Springdale wells and probable contamination in the TN, Fortune, and Hibernia wells. Telltale signals of contamination are T_{max} less than approximately 435°C (851°F) and PI or S1/TOC greater than 0.2 or 0.3, respectively (Peters, 1986). Open-system thermovaporization gas chromatography flame ionization detector (FID) experiments (Jonathan et al., 1975; Horsfield et al., 1989) confirmed the interpretation of contamination of these samples. The contaminant consisted of hydrocarbons having chain lengths in the range of C_{11} to C_{18}. The contamination was removed by heating the samples to 260°C (500°F) in the pyrolysis oven and venting evolved products.

In addition, open-system pyrolysis and gas chromatography FID experiments were performed on 11 selected

Table 2. Measured LECO and Rock-Eval pyrolysis data for 38 collected source rock samples.

Well Name	Depth (m)	TOC[x]* (wt. %)	S1[x]** (mg/g rock)	S2[x†] (mg/g rock)	S3[x††] (mg/g rock)	T_{max}[‡] (°C)	HI[x‡‡] (mg HC/ g TOC)	OI[x§] (mg CO_2/ g TOC)	PI[x§§] (fraction)	Pseudo-Van Krevelen Type	Petroleum Type Organofacies[¶]	Phase Kinetics
Immature												
Rankin M-36	2435.00	0.84	0.09	1.02	0.67	409	121	80	0.08	III		
	2445.00	7.90	0.88	38.32	3.12	413	485	39	0.02	II	P-N-A, Low Wax	
	2455.00	6.03	0.79	30.44	2.34	417	505	39	0.03	II		
	2465.00	7.05	0.97	36.26	3.16	413	514	45	0.03	II		
	2475.00	7.50	1.00	42.11	3.53	413	561	47	0.02	II		
	2485.00	6.11	0.83	33.78	2.48	411	553	41	0.02	II	P-N-A, Low Wax	X
	2495.00	4.63	0.52	23.10	2.09	418	499	45	0.02	II	P-N-A, Low Wax	
Egret K-36	2161.06	6.22	0.09	4.66	7.26	419	75	117	0.02	III		
	2667.04	4.88	0.45	23.84	2.26	421	488	46	0.02	II		
	2676.18	5.73	0.72	33.25	2.49	417	580	43	0.02	I-II	P-N-A, Low Wax	X
	2685.32	6.77	0.98	38.55	2.23	412	569	33	0.02	II	P-N-A, Low Wax	
Egret N-46	1901.98	35.13	0.27	44.73	43.30	421	127	123	0.01	III		
	1996.47	12.53	1.28	69.85	5.80	412	557	46	0.02	II		
	2008.66	6.76	0.73	33.86	3.53	408	501	52	0.02	II	P-N-A, Low Wax	
Whiterose A-90	2895.00	4.62	8.25	18.00	1.76	424	390	38	0.31[¶¶]	II	Paraffinic Oil, High Wax	X
	2935.00	3.69	7.02	16.41	2.19	439	445	59	0.30[¶¶]	II		
Springdale M-29	2915.00	2.26	2.00	6.01	0.98	424	266	43	0.25[¶¶]	II		
	2930.00	2.61	0.55	1.13	3.63	424	43	139	0.33[¶¶]	III		
	3125.00	3.65	7.22	16.14	1.28	424	442	35	0.31[¶¶]	II	P-N-A, Low Wax	X
Terra Nova K-08	3565.00	1.29	0.26	1.19	0.40	424	92	31	0.18	II-III		
	3580.00	2.08	1.04	8.23	0.90	426	396	43	0.11	II		
	3640.00	1.13	0.65	2.53	0.76	426	224	67	0.20	III		
	3660.00	3.12	2.34	18.96	1.09	422	608	35	0.11	I-II		
	3680.00	3.50	2.21	19.16	1.09	421	547	31	0.10	II	P-N-A, Low Wax	X
Terra Nova K-18	3640.00	2.94	2.40	16.25	1.23	420	553	42	0.13	II	P-N-A, Low Wax	
	3660.00	3.48	4.87	18.02	1.13	418	518	32	0.21	II		
	3680.00	4.80	5.34	22.25	2.00	426	463	42	0.19	II		
	3730.00	6.79	14.00	36.99	3.32	422	545	49	0.27	II		
Beginning of Oil Window												
Fortune G-57	4255.00	1.98	2.72	5.38	0.50	434	271	25	0.34	II		
	4370.00	3.09	5.66	11.37	0.59	432	368	19	0.33	II		

	4490.00	5.69	17.69	9.88	5.68	429	174	100	0.64	II	P-N-A Oil, High Wax
Hibernia K-18	4705.00	1.67	2.84	3.29	0.44	433	197	26	0.46	II	
	4765.00	4.30	6.86	7.21	2.27	438	168	53	0.49	III	
	4765.00	5.93	9.66	9.49	3.55	436	160	60	0.5	III	
	4790.00	11.90	16.13	35.66	3.45	431	300	29	0.31	II	
	4790.00	17.83	26.01	66.53	3.66	432	373	21	0.27	II	
	4840.00	7.77	6.32	12.86	1.77	437	165	23	0.33	II	
	4920.00	2.17	4.19	5.01	1.09	431	231	50	0.45	II-III	

Thirty-eight source rock samples collected from nine different well locations. Samples highlighted in bold were used for detailed geochemical analysis and elaborating of bulk, multicomponent, and phase kinetics.

$^*TOC^x$ = measured total organic carbon in weight percent.

$^{**}S1^x$ = measured amount of volatized hydrocarbons, mg HC/g TOC.

†$S2^x$ = measured amount of generated hydrocarbons during pyrolysis, mg HC/g TOC.

††$S3^x$ = measured amount of carbon dioxide generated during pyrolysis, mg CO_2/g TOC.

‡T_{max} = measured temperature of maximum hydrocarbon generation, °C.

‡‡HI^x = measured hydrogen index; 100 S2/TOC.*

*§OI^x = measured oxygen index; 100*S3/TOC.*

§§PI^x = measured production index; S1/(S1 + S2).

¶petroleum type organofacies see also Figure 9.

¶¶ = contaminated by drilling additives.

samples to investigate the molecular composition of the Rock-Eval S2 peaks to determine the petroleum-type organofacies. This was done using the workflow described in Erdmann (1999) and Horsfield et al. (1989).

Additional nonisothermal open-system pyrolysis was performed with a source rock analyzer by Humble Instruments to determine bulk kinetics. Samples were heated from 300 to 560°C (572–1040°F) using four different heating rates (0.7, 2.0, 5.0, and 15.0°K/min). The results give information about the temperature and the corresponding cumulative yield transformation ratio (TR) curves, which can be tuned mathematically and used to determine kinetic parameters, such as the activation energy distribution and the frequency factor. The mathematical background is published, for example, by Schaefer et al. (1990), van Heek and Jüntgen (1968), and Wannowius (1987), and is applied by the software Kinetics2000 and KMod, which were used in this study. The kinetics were determined assuming first-order chemical reactions only with a discrete kinetic analysis method, a fixed E spacing distribution model, and a constant frequency factor.

Based on the bulk kinetics, the temperatures at which TRs of 10, 30, 50, 70, and 90% are reached were determined for each source rock sample. Closed-system programmed-temperature pyrolysis was applied to investigate the compositional variability of evolving hydrocarbons as a function of increasing maturation (Horsfield, 1997; di Primio and Horsfield, 2006). The final multicomponent kinetics (termed PhaseKinetics) consists of 14 components, namely, methane, ethane, propane, i-butane, n-butane, i-pentane, n-pentane, n-hexane, C_7-C_{15}, C_{16}-C_{25}, C_{26}-C_{35}, C_{36}-C_{45}, C_{46}-C_{55}, and C_{55}-C_{80}.

The advantage of these PhaseKinetic models is that they are tuned to natural gas composition. The gas compositions derived from pyrolysis experiments differ significantly from natural gas (Mango, 1997, 2000, 2001) as well as oil-associated gas compositions (di Primio and Skeie, 2004). The most prominent differences are the low methane and higher ethane and propane contents in pyrolysis gases (Figure 3A). Therefore, a correction of the microscale sealed vessel (MSSV)–generated gas composition results was performed to increase the methane concentration and to decrease the ethane to pentane concentrations (using a correlation of gas composition to gas-oil ratio (GOR) from natural petroleum fluids) while the measured GOR was kept constant (di Primio and Horsfield, 2006). This has a significant impact on the saturation pressure (P_{sat}) and formation volume factor (B_o) and shifts the tuned samples into an area (see Figure 3B) where natural fluids tend to plot. These PhaseKinetics were applied in the Jeanne d'Arc Basin model to reconstruct the filling history for different oil fields and to predict the fluid properties and quantity of trapped hydrocarbons. In addition, a secondary cracking definition was used, which is necessary for all kinetics

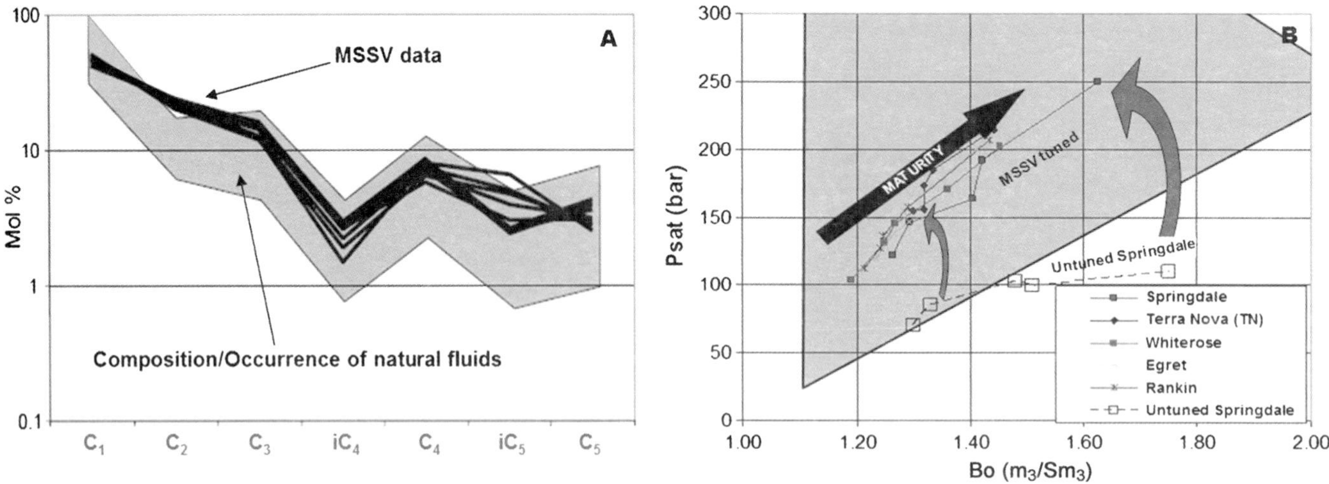

FIGURE 3. (A) Molar composition of associated gases from undersaturated natural oils (shaded region) compared with that derived from the microscale sealed vessel (MSSV) pyrolysis (adapted from di Primio and Horsfield, 2006). (B) Effect of correction of gas composition for MSSV Egret Member source rock samples. Only the Springdale MSSV data are shown for the untuned sample. The gray area represents the field of natural fluid properties.

based on closed pyrolysis. The secondary cracking definition is based on that described by Pepper and Corvi (1995) but adapted to multicomponent kinetic descriptions.

To be able to use LECO TOC and Rock-Eval HI data as input for a basin model, it is necessary to determine their original values. This was done by comparing different methods, namely, the Cooles et al. (1986) method in contrast to the Peters et al. (2005) method (Table 3). We also tested a combined approach using the Peters et al. (2005) and the Jarvie et al. (2007) method (Table 4). The Jarvie et al. (2007) method is based on visual maceral composition to determine original HI values. We used an adapted equation 1 for our calculations:

$$\text{HI maceral} = (750 \times \text{liptinite}/100) + (125 \times \text{vitrinite}/100) + (50 \times \text{inertinite}/100) \qquad (1)$$

Once the original HI value is known, one can proceed with the TOC calculation by the method of Peters et al. (2005) (equation 2):

$$\text{TOC}^{\circ} = \text{F}(\text{HI}^{\text{x}}) \times (\text{TOC}^{\text{x}})/(\text{HI}^{\circ}/[1-\text{f}] \times [\text{F}-\text{TOC}^{\text{x}}] + \text{HI}^{\text{x}}[\text{TOC}^{\text{x}}])$$

F = H/C ratio of expelled HCs (fraction)
HI^{x} = measured HI value (mg HC/g TOC)
TOC^{x} = measured TOC(wt %)
HI° = original HI assumed or calculated
f = extent of fractional conversion (Peters et al., 2005) (fraction) (2)

In the case of having no macerals available, the original HI values (HI°) value has to be assumed for each sample individually (Peters et al., 2005), as we did for the samples in Table 3. In contrast, the Cooles et al. (1986) method uses kinetic data on one immature source rock sample (Table 3). Using only one sample as a reference is a disadvantage because the sample must be representative of all source rock samples before composition and acts indirectly on the assumption of complete source rock homogeneity. Therefore, although no source rock variability can be taken into account with the Cooles et al. (1986) method, is possible with the Peters et al. (2005) or the combined Jarvie and Peters approach. To verify the usefulness of the combined method, 23 maceral composition analyses for predominantly immature samples with calculated HI values based on equation 1 were compared with measured Rock-Eval HI values for the same units (Table 4). The measured values show good agreement within a range of ±50 HI (mg HC/g TOC) compared with the calculated values, except for four immature and two mature samples (±150 HI [mg HC/g TOC]). This correlation suggests that the maceral calculations provide essentially the correct HI° and can therefore be used also for mature samples to better determine HI° values.

NUMERICAL METHOD

The study performed used the three-dimensional (3-D) PetroFlow package of PetroMod developed by Schlumberger in Aachen. The general approach of finite-element numerical simulations is to consider all relevant geochemical and geophysical processes for generation, migration, and accumulation and to make predrilling predictions for hydrocarbon plays. First, backstripping is performed before a sequential, deterministic, forward-modeling routine is applied to simulate basin evolution step-by-step from the oldest event to the present-day geometry (Hantschel and Kauerauf, 2009). At the same time, all other parameters, for example, temperature, maturity, porosity, and pressure can be calculated. All petroleum flow simulations in this study were performed using the "hybrid" migration method developed by PetroMod. Basics and general concepts of petroleum system modeling have been described in more detail by several authors (Tissot and Welte, 1984; Welte and Yalcin, 1987, 1988; Poelchau et al., 1997; Welte et al., 1997; Petmecky et al., 1999; Rodriguez and Littke, 2001; Baur et al., 2009a, b; 2010; Peters, 2009).

GEOCHEMICAL RESULTS

The measured TOC contents of the 38 Egret source rock samples are shown in Table 2. The samples reflect fair to excellent source rock quality with a dominance of type II kerogen (Figure 4A–C). Note that almost all wells contain samples that plot within the type III kerogen area (Figure 4B) and have extremely high TOC values (Figure 4C; Table 2). This terrigenous influence seems to be an important characteristic of the Egret Member source rock. Most samples are immature or low mature (T_{max} < 430°C [<806°F]) and can be used to evaluate hydrocarbon generation kinetics, except for the samples from the Fortune G-57 and Hibernia K-18 wells (Table 2).

Eleven representative source rock samples with T_{max} values less than 425°C (<698.15°F) were chosen to classify the kerogen- and the petroleum-type organofacies (Figure 5). The samples show that the Egret Member source rock generates paraffinic-naphthenic-aromatic oils with low wax contents and a slight tendency to paraffinic high wax oils (Figure 5). In addition, the samples show heterogeneous compositions from sulfur-rich marine to terrigenous-influenced coaly material (Figure 5). Five of the 11 samples, covering the entire geochemical spectrum, were chosen in a second step to investigate kinetic variability. The bulk kinetics for these samples are shown (Appendix Figure 1; Appendix Table 1). Based on kerogen variability and predicted petroleum types, it appears that in the southwestern part (Ranking and Egret samples) of the study area, the source rock is more terrigenous dominated, whereas in the northeastern part of the basin, the paleoenvironment of the source rock is more anoxic and sulfur enhanced.

The question of whether closed-system pyrolysis experiments match reality in terms of correct GOR prediction was analyzed using drill-stem test field data and GOR data from MSSV pyrolysis experiments (Figure 6A). The observed compositional variability of the liquid phase, expressed as GOR, ranges from 30 to more than 300 Sm^3/Sm^3 (168.4–168.4 SCF/STB) and can be adequately reproduced by the PhaseKinetic results, ranging from 55 to more than 200 Sm^3/Sm^3 (308.7–1122.7 SCF/STB). The cumulative generation path for closed-system pyrolysis seems to be mostly valid for primary cracking mechanisms, whereas the two measured fluid samples marked with a circle cannot be explained without an additional source for gas. Figure 6B demonstrates that the total GOR variability is most likely controlled by several processes including biodegradation (low GOR) and secondary cracking (high GOR). Most fluids encountered in the Jeanne d'Arc Basin are, however, exclusively the product of primary cracking. Another important observation is that the API gravity distribution of the natural fluids is almost constant during primary cracking throughout a large GOR window (50 to at least 300 GOR [280.7–1684.1 SCF/STB]; Figure 6B). Within this window, the API increases only by 5° on average.

THREE-DIMENSIONAL MODEL INPUT

The most important 3-D input data are subsurface structural maps derived from seismic data. Two different sources were used and combined: (1) regional depth maps from Edwards et al. (2000) with an initial resolution of 1 km (0.6 mi) covering the entire basin and including the adjacent basement highs and (2) local maps in high resolution (250 m [820.2 ft]) for the central part of the basin derived from five modern interpreted 3-D seismic surveys (McIntyre et al., 2004). These local maps were depth converted based on interval velocity maps from the literature (McIntyre, 1992) and check-shot data (GSC, 2008). In a last step, the local and regional maps were combined to create a total of nine individual maps: Seabed, Base Tertiary, Petrel Member, Nautilus (base of Upper Cretaceous), A-Marker, B-Marker, Top of Fortune Bay (base of Lower Cretaceous), Tithonian Unconformity (Top of source rock), Top of Basement (Figure 2). The resulting 8 layers were subdivided into 35 subunits to account for depositional ages (Figure 2) and rates, basin stratigraphy, and lateral lithofacies changes. Maps below the Egret source rock layer were reduced to four simple layers to reduce the number of events and simulation time.

Additional features were integrated, such as salt piercing for the Adolphus well area. This tool replaces the old lithologies by salt for each layer separately at the corresponding time. In addition, paleogeometries were implemented (additional thickness at a certain time step was inserted and later removed) to avoid distorted geometries caused by the movements along the active fault systems (Baur et al., 2010).

Table 3. Calculated original source rock potential.*

			From Peters et al. (2005)							*From Cooles et al. (1986)*				
Name	*Depth (m)*	*Kerogen Assumed*	*HI^o Assumed (mg HC/ g TOC)*	*PI^o Assumed (fraction)*	*TOC° (wt. %)*	*Fractional Conversion (PGI) (fraction %)*	*Expulsion Efficiency (PEE) (fraction %)*	*$S2^o$ (mgHC/ grock)*	*Volumes Exp (barrels/m^3)*	*HI^o Calculated (mg HC/ g TOC)*	*TOC° (wt. %)*	*Fractional Conversion (PGI) (fraction %)*	*Expulsion Efficiency (PEE) (fraction %)*	*$S2^\circ$ (mgHC /grock)*
Immature														
Rankin M-36	2435.00	III	171	0.02	0.85	0.26	0.77	3.56	0.00	132	4.88	—	—	6.42
	2445.00	II	540	0.02	7.97	0.07	0.77	33.46	0.05	496	7.39	0.74	0.64	36.64
	2455.00	II	569	0.02	6.09	0.07	0.75	25.58	0.04	518	7.62	0.76	0.68	39.45
	2465.00	II	581	0.02	7.12	0.07	0.74	29.92	0.05	528	7.73	0.77	0.69	40.84
	2475.00	II	627	0.02	7.58	0.07	0.74	31.82	0.05	575	8.31	0.80	0.75	47.74
	2485.00	II	619	0.02	6.17	0.07	0.74	25.93	0.04	567	8.20	0.80	0.74	46.46
	2495.00	II	554	0.02	4.67	0.07	0.77	19.62	0.03	510	7.53	0.75	0.67	38.42
Egret K-36	2161.06	III	100	0.02	6.25	0.29	0.96	26.25	0.03	76	4.64	—	—	—
	2667.04	II	534	0.02	5.10	0.14	0.90	21.41	0.15	498	7.40	0.74	0.65	36.84
	2676.18	I-II	642	0.02	6.08	0.15	0.90	25.53	0.24	593	8.55	0.81	0.77	50.70
	2685.32	II	640	0.02	7.18	0.16	0.88	30.15	0.29	584	8.43	0.81	0.76	49.23
Egret N46	1901.98	III	150	0.01	35.26	0.14	0.96	148.11	0.18	128	4.86	—	—	—
	1996.47	II	608	0.02	13.15	0.14	0.91	55.24	0.48	568	8.22	0.80	0.74	46.64
	2008.66	II	554	0.02	7.07	0.15	0.89	29.71	0.22	511	7.55	0.75	0.67	38.64
Whiterose A-90	2895.00	II	750	0.02	6.00	0.60	0.71	25.20	0.91	568	8.22	0.80	0.74	46.72
	2935.00	II	750	0.02	5.00	0.60	0.72	21.01	0.85	634	9.18	0.84	0.80	58.28
Springdale M-29	2915.00	II	598	0.02	2.54	0.49	0.68	10.67	0.21	353	6.15	0.56	0.21	21.78
	2930.00	III	114	0.02	2.66	0.62	0.72	11.19	0.05	64	4.59			
	3125.00	II	750	0.02	5.00	0.61	0.72	20.98	0.87	639	9.27	0.84	0.81	59.29
Beginning of Oil Window														
Terra Nova K-08	3565.00	II-III	175	0.02	1.34	0.48	0.78	5.62	0.03	112	4.79	—	—	—
	3580.00	II	618	0.02	2.41	0.40	0.82	10.11	0.22	445	6.89	0.69	0.54	30.71
	3640.00	III	453	0.02	1.26	0.49	0.75	5.28	0.08	280	5.67	0.40		15.92
	3660.00	I-II	750	0.02	4.25	0.47	0.87	17.85	0.73	682	10.03	0.86	0.84	68.45
	3680.00	II	750	0.02	4.45	0.43	0.86	18.67	0.61	610	8.80	0.82	0.78	53.74
Terra Nova K-18	3640.00	II	750	0.02	3.88	0.47	0.85	16.30	0.61	634	9.18	0.84	0.80	58.21
	3660.00	II	750	0.02	4.98	0.57	0.81	20.93	0.97	657	9.57	0.85	0.82	62.91

	3680.00	II	750	0.02	6.22	0.52	0.79	26.11	0.98	574	8.31	0.80	0.75	47.74
	3730.00	II	800	0.02	11.35	0.68	0.83	47.69	3.08	750	11.53	0.89	0.87	86.61
Peak Oil Window														
Fortune G-57	4255.00	II	728	0.02	2.56	0.66	0.75	10.77	0.39	408	6.58	0.65	0.44	26.89
	4370.00	II	750	0.02	4.62	0.69	0.79	19.40	0.99	550	8.01	0.78	0.72	44.11
	4490.00	II	731	0.02	9.11	0.87	0.74	38.25	2.27	484	7.27	0.73	0.62	35.21
Hibernia K-18	4705.00	II	653	0.02	2.28	0.76	0.75	9.58	0.40	366	6.25	0.59	0.28	22.91
	4765.00	III	576	0.02	5.62	0.77	0.73	23.62	0.91	327	5.96	0.51	0.04	19.51
	4765.00	III	564	0.02	7.72	0.78	0.73	32.43	1.25	—	—	—	—	—
	4790.00	II	750	0.02	15.64	0.67	0.79	65.69	2.83	—	—	—	—	—
	4790.00	II	750	0.02	24.29	0.67	0.82	102.01	5.34	—	—	—	—	—
	4840.00	II	438	0.02	9.00	0.65	0.75	37.81	0.88	—	—	—	—	—
	4920.00	II-III	750	0.02	3.15	0.77	0.76	13.21	0.65	423	6.70	0.67	0.49	28.42

HI^o = original hydrogen index; PI^o = original production index; TOC^o = original total organic carbon; PGI = petroleum generation index; PEE = petroleum explusion efficiency; $S2^o$ = original S2.

All boundary conditions such as paleo–water depth (PWD), sediment-water interface temperature (SWIT), and heat flow (HF) were defined by maps for each time step in the 3-D model. The PWD maps were reconstructed based on paleoenvironment information developed from core interpretations (the basin was always deeper in the north and shallower in the south, as today), and SWIT was determined by considering the global temperature distribution through time and corrected for the PWD depth based on a standard temperature depth profile by Wygrala (1989). The cooling effect of the Labrador current was taken into account for the present-day situation. Finally, the HF maps were taken from Baur et al. (2010).

Facies maps were kept as simple as possible and compiled from detailed lithologic descriptions available for every well in the Jeanne d'Arc Basin (GSC, 2008; CNLOPB, 2004).

The positions of major faults, which control the general pressure regime in the basin, were determined based on seismic interpretations and the topography of structural depth maps.

The geochemical quality and thickness of the active part of the source rock were evaluated with the help of results taken from Huang et al. (1993), who computed the organic richness (TOC and S2) and distribution of source potential from porosity and resistivity logs for 17 wells in the Jeanne d'Arc Basin. Huang et al. (1993) determined the percentage of non–source rock intervals to range between 15 and more than 60% of the entire Egret Member. We defined the basinwide non–source rock unit to have an average value of 35%. The original thickness maps were derived from Fowler and McAlpine (1994), Huang et al. (1994), Magoon et al. (2005), and from G. Power (2008, personal communication). Vitrinite reflectance data, Horner-corrected bottom-hole temperatures, porosity and permeability measurements (GSC, 2008), as well as some specific petroleum parameters such as API, GOR, P_{sat}, and B_o (CNLOPB, 2004; G. Power, 2008, personal communication) were used as calibration parameters.

GENERATION, MIGRATION, AND ACCUMULATION IN THE JEANNE D'ARC BASIN

Faults and Overpressure

The Jeanne d'Arc Basin can be subdivided from north to south into three major parts separated by two different fault systems: the Egret faults and Hibernia to Ben Nevis faults, also called transbasin faults (Figure 7). Only the middle and northern parts of the basin contain active petroleum systems because of the lack of a

Table 4. Hydrogen index values from the maceral composition and Rock-Eval pyrolysis.

Well Name	*Depth (m)*	*Maturity Level*	*Vitrinite (%)*	*Liptinite (%)*	*Inertinite (%)*	*Kerogen Classification from Tissot and Welte (1984)*	*HI^{o}[†] Maceral Calculated from Peters et al. (2005) (mg HC/g TOC)*	*HI^{x}[†] Rock-Eval Measured (mg HC/g TOC)*
Voyager J18	2540	Mature	7.5	65.0	27.5	T II	446	283*
Beothuk M-05	3370	Early mature	5.0	65.0	30.0	T II	444	456
Archer K-19	3290	Immature	10.2	63.3	26.5	T II	437	449
Archer K-19	3570	Early mature	7.9	52.6	39.5	T II	372	421
Rankin M-36	2440	Immature	7.7	66.7	25.6	T II	456	485
White Rose A-90	2890	Immature	8.3	58.3	33.3	T II	406	389
Trave E-87	3050	Immature	7.1	71.4	21.4	T II	484	598*
Egret K-36	2539	Immature	64.3	7.1	28.6	T III	141	171
Egret K-36	2630	Immature	57.1	21.4	21.4	T III	221	190
Egret K-36	2720	Immature	61.5	23.1	15.4	T III	235	232
Egret K-36	2813	Immature	50.0	28.6	21.4	T III	259	106*
Flying F. I-13	3212	Immature	37.5	12.5	50.0	T III	153	160
Flying F. I-13	3304	Immature	31.3	37.5	31.3	T III	298	356
Flying F. I-13	3398	Early mature	37.5	25.0	37.5	T III	228	235
TN K-08	3620	Early mature	15.0	75.0	10.0	T II	511	435
TN K-08	3650	Early mature	15.0	80.0	5.0	T II	541	530
TN K-08	3680	Early mature	5.0	90.0	5.0	T I	594	601
TN K-08	3710	Early mature	4.0	95.0	1.0	T I	623	487**
TN K-08	3740	Mature	8.0	87.0	5.0	T I	578	462**
TN K-17	3170	Immature	14.0	70.0	16.0	T II	481	463
TN K-17	3180	Immature	16.0	67.0	17.0	T II	464	248*
TN K-17	3200	Early mature	17.0	65.0	18.0	T II	453	481
TN K-17	3210	Early mature	21.0	53.0	26.0	T II	384	310

Poor agreement between HI^{o} (maceral) and HI^{x} (Rock-Eval) for immature samples.
**Poor agreement between HI^{o} (maceral) and HI^{x} (Rock-Eval) for mature samples.*
[†]*HI^{o} = original hydrogen index; HI^{x} = measured hydrogen index.*

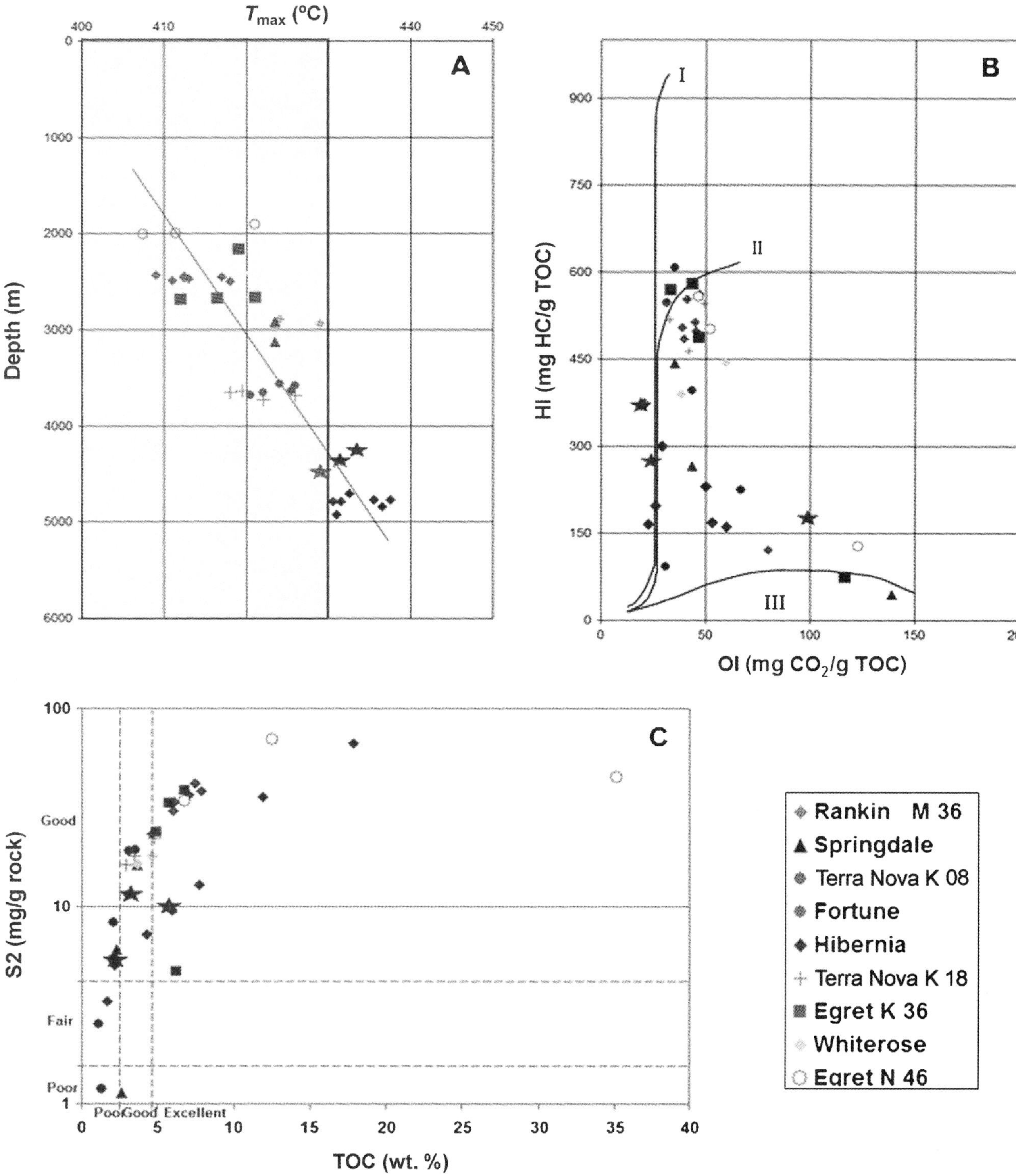

FIGURE 4. Quality, quantity, and maturity of organic matter for all 38 source rock samples. (A) T_{max} from the Rock-Eval pyrolysis shows a linear increase with depth. Only samples with a $T_{max} < 430°C$ (<703°F) were used to determine kinetics. (B) Pseudo–Van Krevelen diagram showing mainly type II kerogen with a significant contribution of type III kerogen in the sample set. (C) Present-day source rock quality based on total organic carbon (TOC) and S_2 measurements. Most samples plot within the "good" or "excellent" field. HC = hydrocarbon; HI = hydrogen index; OI = oxygen index.

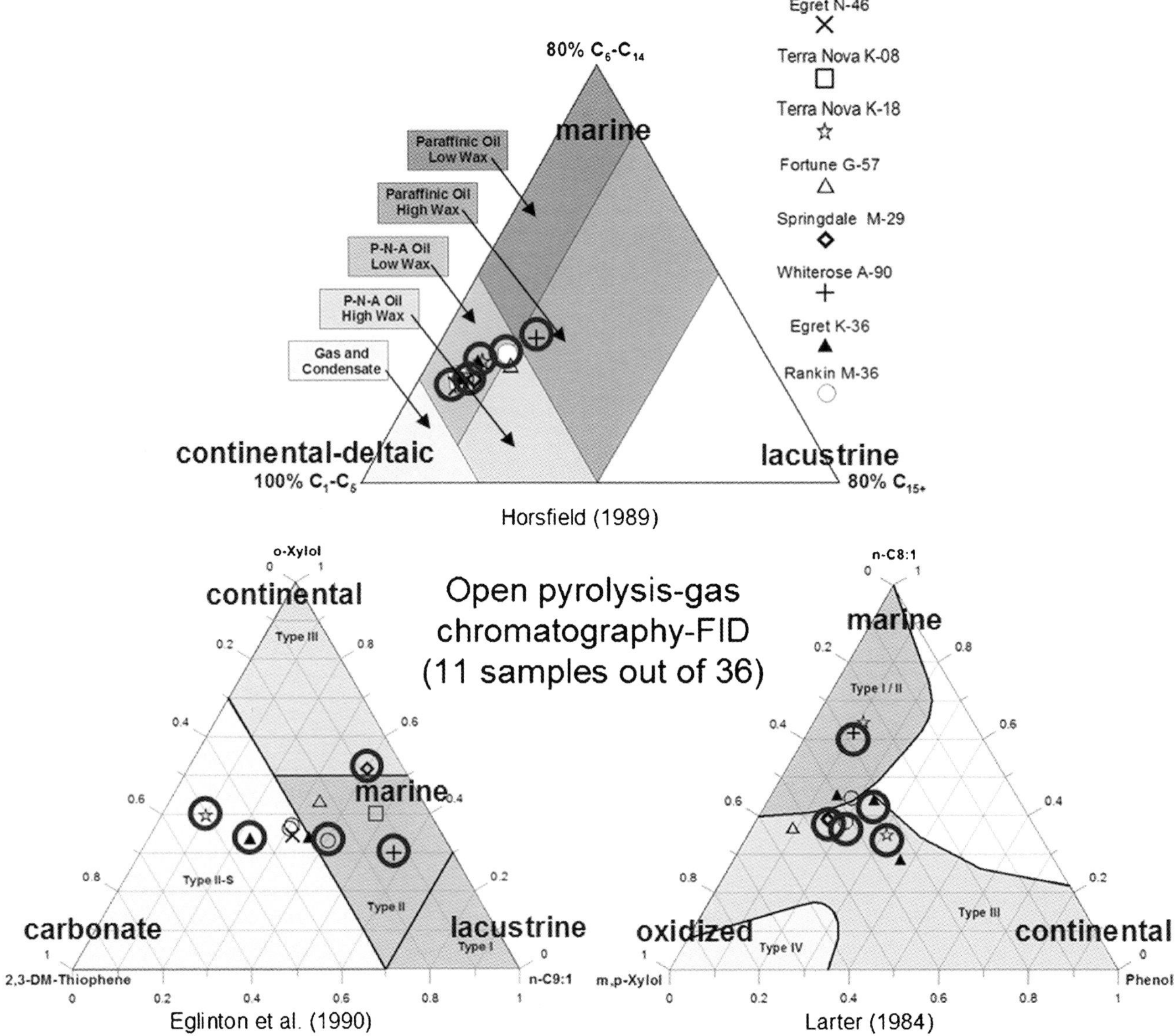

FIGURE 5. Detailed geochemical analysis for 11 source rock samples used to select five representative samples (marked with circles) to determine bulk kinetics. Applied methods follow those described by Horsfield (1989), Eglinton et al. (1990), and Larter (1984). FID = flame ionization detector.

source rock in the southern part. The fact that the northern part is predominantly overpressured, whereas in the southern part almost all wells show hydrostatic pressure (Wielens and Jauer, 2001), indicates that at least the east-west–trending transbasin faults have acted as impermeable boundaries during basin evolution. The numerical model of the Jeanne d'Arc Basin reproduces this simplified pressure distribution only by including the major faults, separating the northern and central part (transbasin and Egret faults) from each other and separating the basin from the surrounding horst structures such as Bonavista Platform and Outer Ridge complex (Mercury, Murre, and Voyager Fault) to maintain the overpressure (Figure 7). This indicates that the layers forming the reservoirs in the basin center are not in connection with the same stratigraphic layers deposited on the basement highs. They must be separated either by tight lithologies or by impermeable faults or by lateral pinch-outs. The faults in the model extend vertically from the base of the Egret source rock to the top of the Whiterose layer. All layers above that level lack or have insignificant overpressure. To induce overpressure within the Hibernia layer, an additional low-permeability shaly mid-Hibernia layer was implemented into the model, separating the lower from the upper Hibernia sandstone. Vertical pressure profiles at most well positions can be matched using this fault and layer configuration (Figure 8), except for Whiterose,

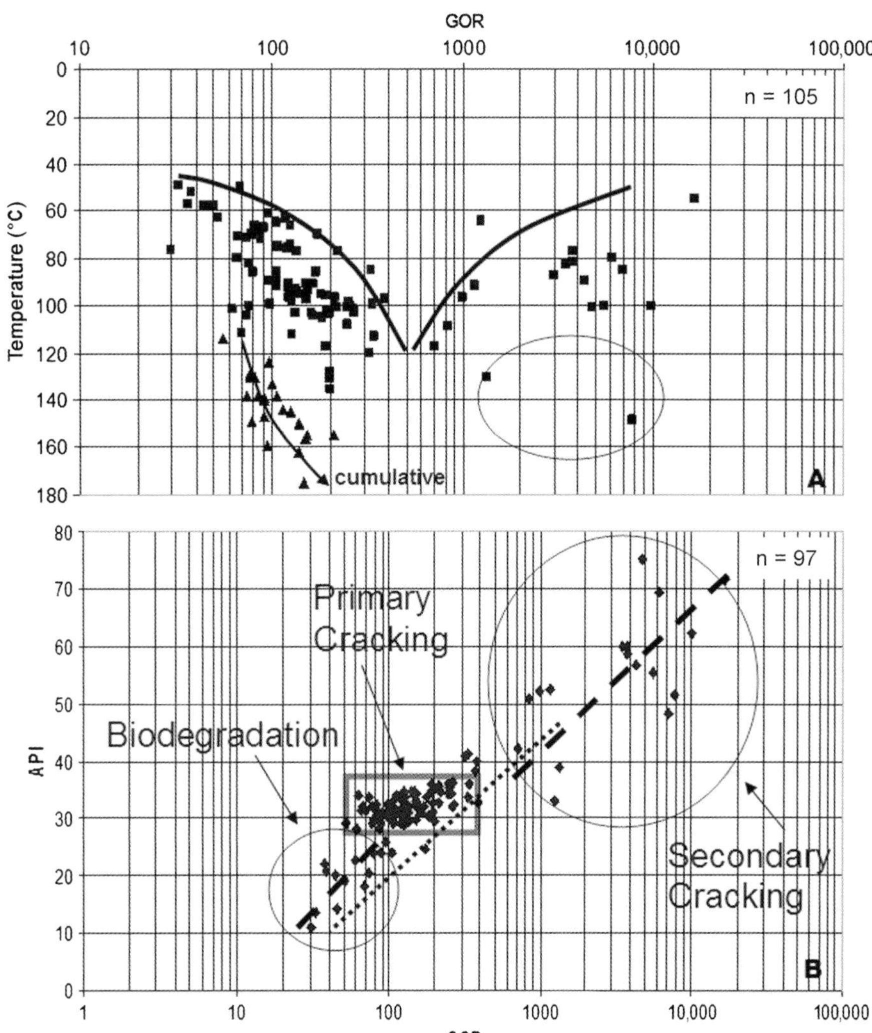

FIGURE 6. (A) Comparison between natural gas-oil ratio (GOR) data from 105 drill-stem test samples from throughout the Jeanne d'Arc Basin (black squares) and predicted GOR evolution derived from microscale sealed vessel measurements (black triangles at higher temperatures representing source rock conditions of generation). Curved lines at lower temperatures represent bubble and dew point curves for the basin. Bent arrow represent evolution pathway for cumulative compositional evolution. (B) Natural data set from the Jeanne d'Arc Basin showing the relationship between API gravity and GOR. Box shows the products from primary cracking processes. Low GORs can be generally attributed to biodegradation; higher GORs most likely represent effects of secondary cracking. Note that the original primary cracking process produces fluids with a very narrow distribution of API gravity. The linear relation between GOR and API of primary generated hydrocarbons over secondary cracking products down to biodegraded oils (Evans et al., 1971; dotted short line) is not represented in the Jeanne d'Arc Basin.

where the resolution of the structural depth maps cannot reproduce the complex compartmentalization properly.

Structural Trap Development and Drainage Area

The first step for the migration analysis is to study the drainage area development through time as well as main phases of structural trap development. In the Jeanne d'Arc Basin model, the main structures were formed during the earliest Cretaceous when the Hibernia sediments were deposited. A second period of trap formation occurred during the late Paleocene (52 Ma) forming mainly stratigraphic traps. Both periods of trap formation are in agreement with the information summarized by Magoon et al. (2005). Since the late Paleocene (52 Ma), the entire basin was buried with an increased sedimentation rate slightly higher in the north compared with the south because of the movement of the Orphan Knoll during the Cenozoic (Enachescu, 1992). However, in general, drainage areas of the main structural closures did not change significantly since that time. Thus, migration pathways stayed almost constant as well. This can be nicely reproduced by the numerical model if you compare the extent and shape of the drainage areas for the top of the Jeanne d'Arc reservoir in Figure 7.

Generation, Migration, and Accumulation in the Jeanne d'Arc Basin

The basic parameters to analyze a petroleum system are the TR for the source rock (%TR-all) and the possible migration pathways from the pod of active source rock to the trap based on structural depth maps. Panels A and B of Figure 9 show the TR within the present-day drainage areas for the Jeanne d'Arc and Hibernia reservoirs with the main migration pathways and directions, including the names of the most important oil fields.

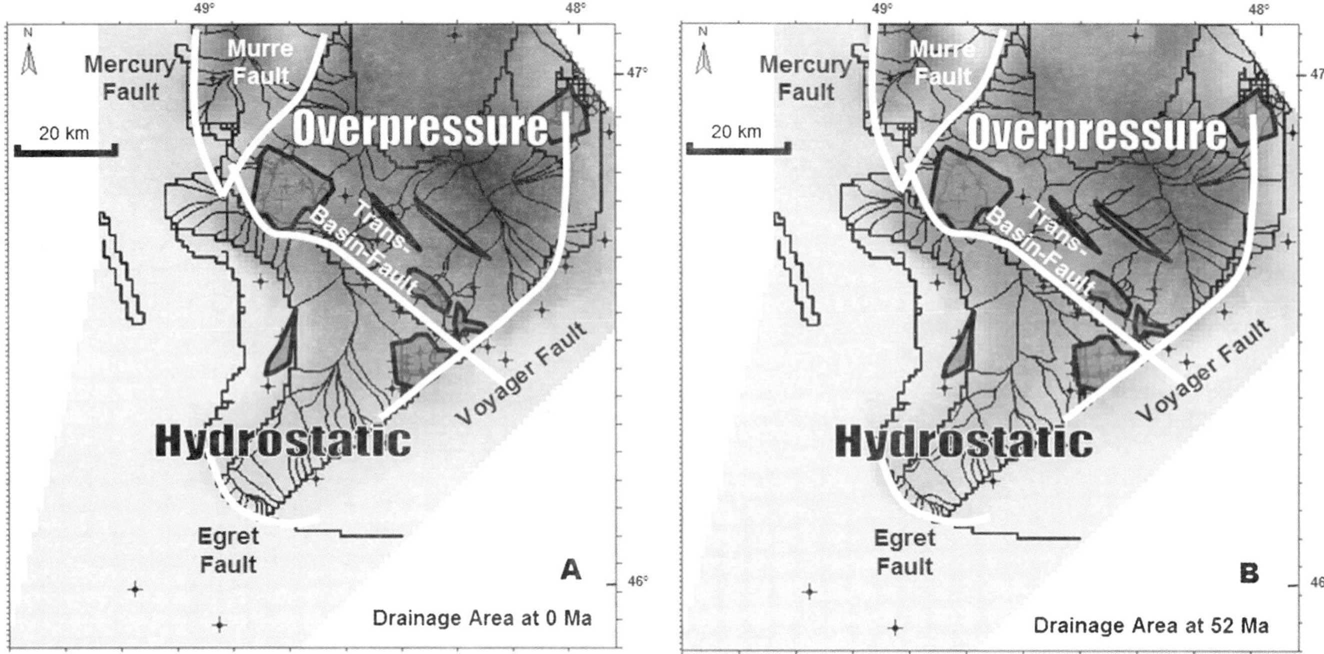

Figure 7. Shape and distribution of drainage areas for the Jeanne d'Arc reservoir layer in (A) the present day and (B) 52 m.y. before present. White lines represent closed faults that were used in the model to separate the major pressure provinces from each other (local faults used in the oil fields are not drawn).

The Hibernia field, located in the eastern part of the study area, represents the biggest oil field in the Jeanne d'Arc Basin. It is controlled by northwest-southeast faulting and folding causing half-graben and roll-over structures acting as hydrocarbon traps. The field is complex because of the formation of stacked reservoir compartments. Most have moderate, whereas some have strong, overpressure associated with undercompacted sediments, which was reproduced in the numerical model in a simplified manner by a general overpressure controlling the entire northern part of the model (Figure 8). Porosity, temperature, and vitrinite reflectance were also calibrated to available well data (Figure 8).

The Hibernia temperature, TR, and generation history, according to Baur's et al. (2010) heat-flow development, are shown in Figure 10A. A detailed volumetric filling history for the Hibernia oil field is shown in Figure 10B, which represents 31% of the total present-day trapped amount. The first petroleum in the Hibernia catchment area was generated at approximately 136 Ma, but the first significant accumulation occurred after 100 Ma. The trapped fluids undergo significant maturation starting with a GOR of less than 116 Sm^3/Sm^3 (651.2 SCF/STB) and ending with GOR more than 160 Sm^3/Sm^3 (898.2 SCF/STB). The API develops according to the increase of maturation from high to lower densities. The stacked reservoir pattern found in nature can also be reproduced within the numerical model by accumulations in the lower and upper Hibernia formation and the Catalina, Ben Nevis, and Avalon formations. The model also predicts hydrocarbons in the Jeanne d'Arc formation, which is not in agreement with known accumulation.

The shape and size of the catchment areas draining into the Hibernia oil field are shown in Figure 9B for the Hibernia reservoir. The larger main drainage areas are separated by northwest-southeast–trending faults surrounded by some smaller drainage areas, which all spill into the central catchment. To the south, the catchment area is limited by the transbasin fault, to the west by the Murre fault, and to the east, the catchment area does not extend beyond the Mara well location.

The total volume of the present-day accumulated petroleum at the Hibernia field (all reservoir layers) in the numerical model is approximately 20% higher than that observed in nature. This can be attributed to overestimates of porosity in the model (most of the calibration data were derived from locations where porosities are optimal). Furthermore, all of the net reservoir thicknesses might be overestimated because of the merging of separate reservoir units into one layer to reduce the number of simulation events and because no adequate reservoir geometries for the Hibernia field were available.

The model of the Mara oil field, located to the southwest of the center of the basin, reproduces the amount of accumulated hydrocarbons (±10%) found in the natural system. Hydrocarbons are trapped in the Jeanne d'Arc, Hibernia, and Ben Nevis formations. The temperature, TR, and generation history for the deeper buried

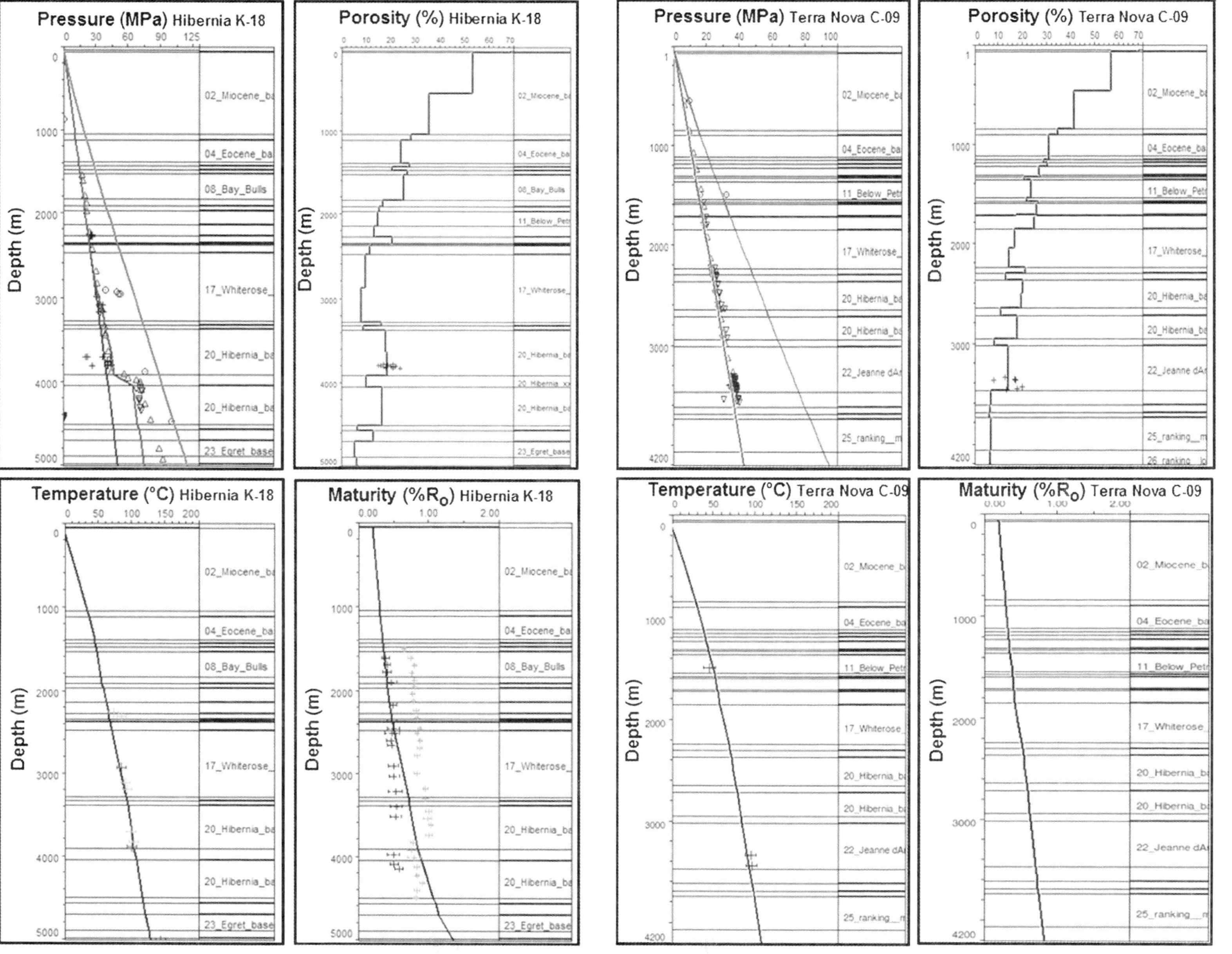

Figure 8. Examples of model calibration (pressure, porosity, temperature, maturity) shown for two well locations. Gray and black crosses used for maturity indicate different editors. Gray crosses used for temperatures are not corrected bottom-hole temperature measurements, whereas the black crosses indicate Horner corrected values. Hollow-up triangles at the pressure plots show pressure data derived from mud loggers' report; hollow-down triangles show pressure values derived from repeat formation tester, circles represent formation leak-off pressure tests and the plus indicates drill-stem tests. Porosity data were derived from core measurements. GOR = gas-oil ratio.

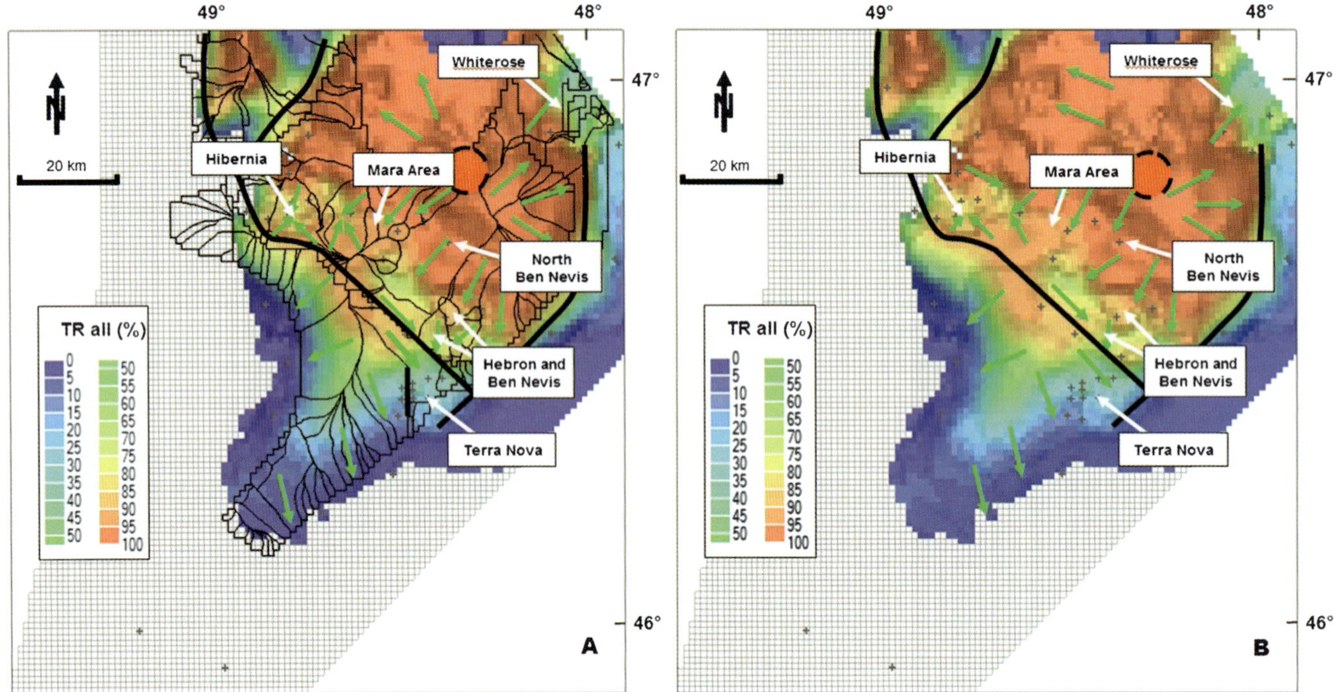

FIGURE 9. Present-day Egret Member source rock transformation ratio with carrier formation drainage areas and the main migration directions as overlays. Panel A shows the drainage areas for the Jeanne d'Arc reservoir; panel B shows the drainage areas for the Hibernia reservoir. Red dot marks the deepest point of each reservoir layer.

and earlier generating part of the source rock are shown in Figure 11A. The source rock was deeply buried and started to generate at 135 Ma. Approximately 75 m.y. ago, this part of the source rock reached a transformation of 100%, whereas in the shallower part of the Mara catchment area, generation continues until the present day. The first significant accumulation based on the numerical simulation occurred approximately 130 Ma and reached its maximum volume at 65 Ma. The fluids feeding this petroleum system derived from the depositional center of the corresponding layer (Jeanne d'Arc), 1800 m (5906 ft) to the northeast. After 65 Ma, the closure became smaller and petroleum spilled into the neighboring drainage system of Hibernia. This migration pattern from the deepest point of the layer along the Mara oil field into the marginally located Hibernia field is still active today (compare Figure 9).

The Whiterose oil and gas field to the northwest of the center of the basin is fault bounded and structurally separated by north-south–trending faults from neighboring reservoirs (Issler, 1994; Enachescu and Fagan, 2005; Magoon et al., 2005). The field contains high gas contents, which was only barely reproduced by the model. The predicted oil volume is 25% higher than the natural amount and the predicted gas represents only a small percentage. Figure 11B shows the Ben Nevis reservoir through time, which represents approximately 66% of the total trapped volume at the Whiterose field. One reason for the misfit in predicted volumes is that most of the gas leaked through the seal or leaked because of improper structures. The second reason is that hydrocarbons were probably affected more strongly by secondary cracking effects than predicted by the model. Apparently, either the oil was cracked at lower temperatures than expected or an additional source of gas is required. Predicting secondary cracking is still challenging.

The Hebron and Ben Nevis oil fields are located in the south-central part of the basin and are typical fault-bounded accumulations with stacked reservoirs filled by breakthrough caused by cap-rock failure and direct fill at the same horizon (Rogers and Yassir, 1993; Shimeld and Moir, 2001). This is nicely reproduced by the numerical model. API and GOR values are within the observed range of the oil accumulations and become lower with shallower depths of the reservoir. The predicted Jeanne d'Arc reservoir fluids have 34° API and a GOR of 180 (1010.5 SCF/STB), with field data showing 33° API. The modeled Hibernia reservoir shows exactly the same API as observed in nature (29° API). The shallowest Ben Nevis reservoir is affected by biodegradation (20° API), which is not taken into account in the simulation, but its initial gravity is predicted by the model to be 29° API. The volumes are not precisely reproduced, but are within ±13%.

Figure 6B (already discussed in the Geochemical Results Section) shows that the GOR for primary cracking

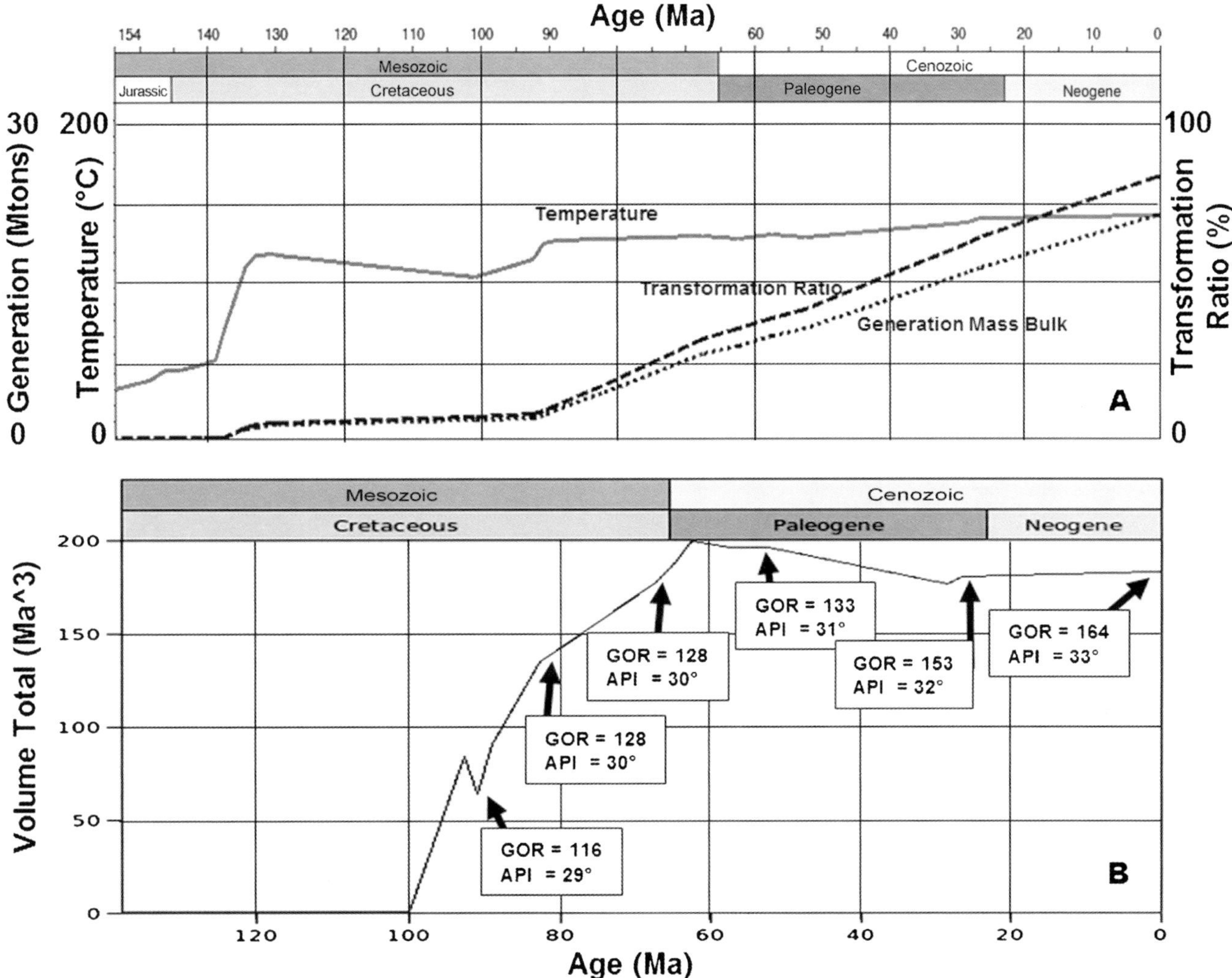

FIGURE 10. (A) Time extraction for a representative cell (1 km^2 [0.4 mi^2]) of the Egret source rock underlying the main drainage area of the Hibernia oil field. (B) Volumetric development through geologic time for one accumulation out of the Hibernia field, representing 31% of the total accumulated volume. In addition, the gas-oil ratio (GOR) and API are shown for selected time steps.

is more variable within the entire Jeanne d'Arc Basin, but that the API varies only little. The predictions derived from numerical simulations are much more accurate for the GOR (smaller percentage deviation) compared with the API. This means that it is more difficult to predict API and that GOR predictions are more reliable using numerical models in combination with PhaseKinetics.

PETROLEUM SYSTEM OF THE TERRA NOVA OIL FIELD

The TN oil field is the northward-plunging asymmetrical broken crest of an anticline. This anticline was probably induced by movement of deep-lying salt bodies. The structure developed during the Early Cretaceous based on fault interpretations. Faults die out in the Early Cretaceous and do not penetrate Upper Cretaceous horizons (Figure 12). Significant discoveries have been made in the Central Graben and in the East Flank, whereas in the Far East, only two smaller accumulations exist: one in the north and a biodegraded accumulation in the south (Figure 12). All of these accumulations occur in the clastic Jeanne d'Arc reservoir and are separated from each other by faults. The TN oil field compartments have slightly different oil-water contact depths showing that at least some of the compartments are not in communication (Yassir and Rogers, 1994).

It is surprising that no hydrocarbons were found in the West Flank, although the north-south–trending fault separating the West Flank from the graben is impermeable. The losses of hydrocarbons may be caused

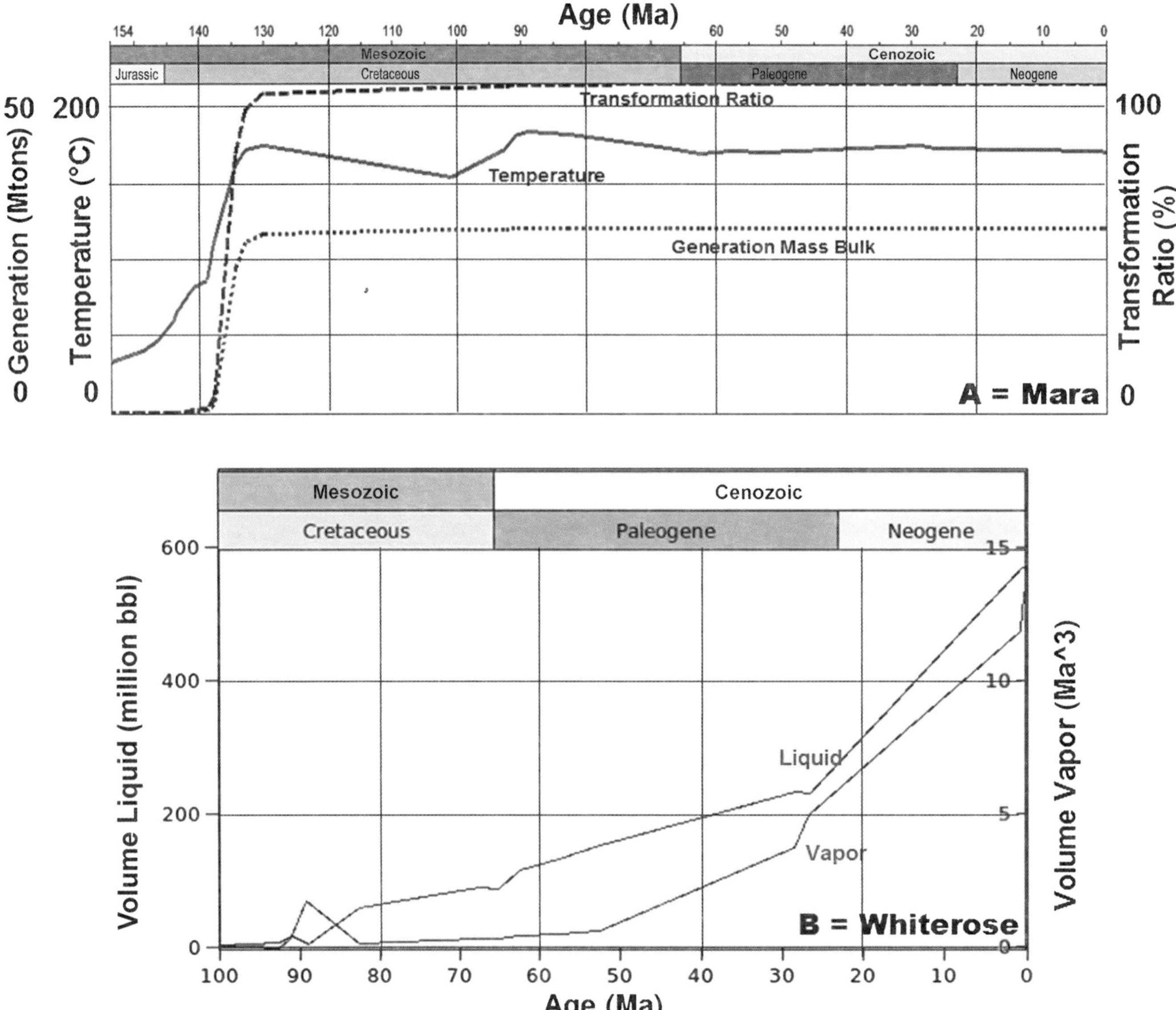

FIGURE 11. (A) Time extraction for an early generating cell (1 km^2 [0.4 mi^2]) out of the source rock underlying the main drainage area of the Mara oil field area. (B) Volumetric development through geologic time for one accumulation out of the Whiterose field representing more than 80% of the total accumulated volume. The predicted amount of gas and oil does not match field observations.

by the fact that the reservoir horizon continues southward in the western part and changes into a braided river complex with highly variable and permeable channel structures (Enachescu, 1993; G. Power, 2008, personal communication). However, south of the graben, West Flank, and Far East, the reservoir unit pinches out and changes into an alluvial fan system (G. Power, 2008, personal communication) fed by sediments from the exposed margin of the basin. Therefore, the TN field is a combined stratigraphic-structural trap with thickening reservoir units to the north and west, whereas the alluvial fan of shaly composition becomes thicker to the south and east (Figure 12). The hydrocarbons in the West Flank can thus migrate updip to escape from the structure, whereas in the east, the fluids are stratigraphically trapped. All individual units of reservoirs found in the TN field were merged into one thick layer in the model (B sand = 10 m [33 ft]; C1 sand = 10–20 m [33–66 ft]; LC2 sand = 30–40 m [98–131 ft]; UC2a sand = 20–40 m [66–131 ft]; UC2c sand = 10–20 m [33–66 ft]; Dcongl = 1–5 m [3.3–16.4 ft]; Da sand = 10 m [33 ft]; Dc sand = 30 m [98 ft]).

Different scenarios were tested using the numerical model in an attempt to reproduce the known hydrocarbon distribution, but most of them were unsuccessful. Leaving the northerly west-northwest–east-southeast–trending fault open, which separates the TN field from

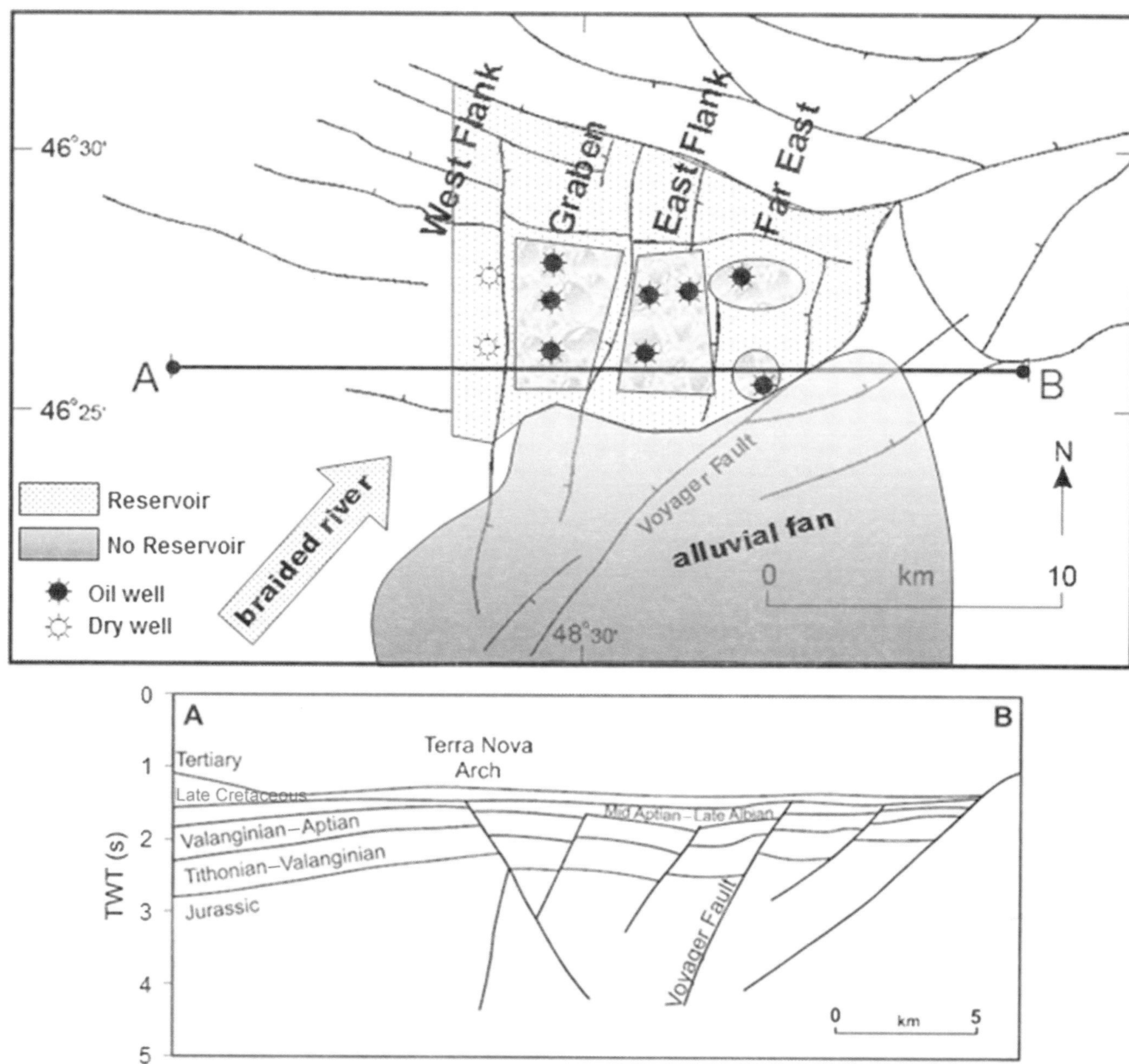

FIGURE 12. Map of Terra Nova oil field (top) shows location of dry and oil-bearing exploration wells and oil accumulations (from Parnell et al., 2001). The basic distribution of reservoir and nonreservoir lithofacies is shown. Black lines indicate the position of closed faults. Cross section (bottom) through the Terra Nova field shows major stratigraphic units and the vertical extent of faults (from Parnell et al., 2001). TWT = two-way traveltime.

the Hebron field (compare Figures 9A, 12 for locations), results in too small accumulations in the Hebron field and an overcharged and incorrect composition of oil in the TN area. An open north-south–trending fault, separating the graben and the West Flank, results in a very small TN accumulation only at the position of the East Flank because of the absence of a barrier for the hydrocarbons trapped in the graben (compare Figures 9A, 12 for locations).

MASS BALANCE CALCULATION FOR THE TERRA NOVA OIL FIELD

Mass balance calculation is a useful method to compare the amount of generated and expelled hydrocarbons with volumes that reached the trap. A detailed volumetric analysis can help to understand and reconstruct

Table 5. Benchmark data for in-situ generation for the Terra Nova oil field.

Kinetics	TN T-IIS	Sulfur-Rich
Drainage area	85*	km^2
Average SR** thickness	150*	M
Average SR maturity (R_o)[†]	0.64	% R_o
Average SR TR[††]	17	%
Generated volumes in the AOI model	688	Million bbl
Critical saturation, adsorption within SR	14	%
Losses (sec. migration)	1	%
Leakage	40	%
Accumulated (relative)	45	%
Accumulated (quantitative)	311	Million bbl
GOR[‡]	115*	m^3/m^3
Reserves (proven)	~1000	Million bbl
GOR	138*	m^3/m^3
API	32	°Gravity

**See text for U.S. units.*
***SR = source rock.*
[†]R_o = vitrinite reflectance.
[††]TR = transformation ratio.
[‡]GOR = gas-oil ratio.

migration pathways, possible kitchen areas, and different charge scenarios, thus creating a detailed filling history. This method has not been applied to the Jeanne d'Arc Basin or to the TN oil field before this work.

In a first step, the volumes of hydrocarbons that could have been generated from the source rock, which directly underlies the TN field, were calculated. The idea based on old geochemical investigations (G. Power, 2008, personal communication) assumes that the TN field was isolated by faults in the geologic past, and therefore, no hydrocarbons could reach the reservoir from outside this area. The remaining local kitchen area was believed to be the sole source of the petroleum trapped in the TN field. Therefore, the TN kinetic model (see Figure Appendix 1; Table Appendix 1) was assigned to the source rock, and a simulation was run for an area of interest (AOI) covering only the TN prospect. The AOI has a size of 85 km^2 (32.8 mi^2), with an average source rock thickness of 150 m (492.1 ft), an average maturity equivalent of 0.64% random vitrinite reflectance, and a TR of 17%. Based on these parameters, it is possible to calculate that 688 million bbl were generated from the Egret Member source rock (Table 5, which summarizes the field and model benchmark data). However, it is assumed that some hydrocarbons (14% of the total) were retained within the source rock because of adsorption (kerogen specific— we used the Symington mode described as "alternative adsorption model" in Hantschel and Kauerauf, 2009, p. 187) and critical saturation during primary migration (lithology specific). An additional 1% is lost during secondary migration. This value is extremely low because the reservoir directly overlies the source rock, thus the migration distance is very short. However, the greatest amount of hydrocarbons is lost by leakage because of an inefficient sealing cap rock. In addition, large volumes of hydrocarbons were generated very early based on the model because of thermally labile kerogen (Figure 13). In total, our model predicts trapping of 311 million bbl of petroleum (45% of the initially generated fluids), which is extremely efficient, but still not enough to account for the 1000 million bbl found in the TN field (Table 5). Even if we assume 100% trapping efficiency (688 million bbl), the total trapped fluids are not enough. A sensitivity analysis showed that whatever kinetics were used (Whiterose or Ranking kinetics; compare Figure Appendix 1; Table Appendix 1), the total volume of petroleum in the trap is still too small. In addition to the volumes of hydrocarbons, the predicted GOR of the accumulated fluids in TN using the TN kinetics does not match the natural GOR. Measured TN fluids have an average GOR of 138 Sm^3/Sm^3 (842.1 SCF/STB) and 32° API, whereas in the local kitchen model, the GOR is somewhat lower, reaching only 105 Sm^3/Sm^3 (617.5 SCF/STB) and 28° API. In addition, the application of two other kinetic models (Whiterose and Rankin kinetics) did not reproduce the correct GORs and APIs. Therefore, we conclude that the TN oil field was charged by locally generated hydrocarbons but received additional petroleum from outside the modeled AOI. To determine the origin of these additional hydrocarbons, different charge models were tested using the 3-D numerical model.

The first charge model (Figure 14) attempted to fill the TN field through in-situ generation and a kitchen located north of the transbasin fault system (Figure 14). Fill and spill across the faults transfer hydrocarbons from the deep kitchen into the shallower oil field to the south (Figure 14). The major problem with this model is that the transbasin faults must act as migration barriers for the hydrocarbons because large volumes are trapped in the Hebron and Ben Nevis oil fields. The volumes trapped can be reproduced by the numerical model only when the basin faults are closed from the Late Cretaceous onward. The closed faults are also needed as pressure boundaries to separate the overpressured north from the hydrostatic south. Realizing a fill-and-spill mechanism earlier than the Late Cretaceous is impossible because of an inefficient seal at that early time. Working with semipermeable faults, it is possible to force an extra charge into the TN field, but this still

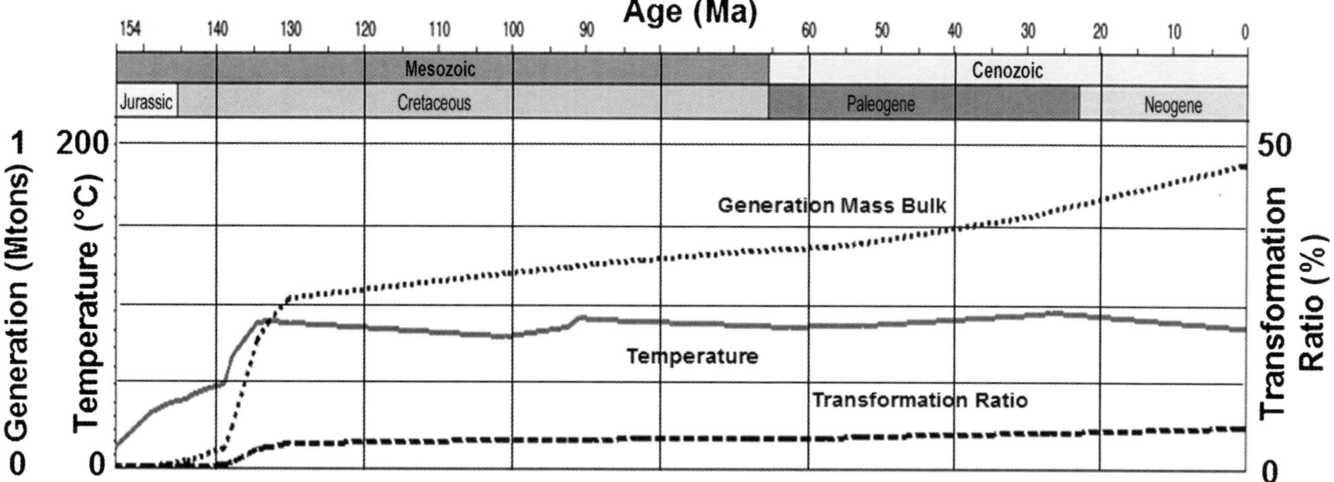

FIGURE 13. Time extraction for a representative cell (1 km^2 [0.4 mi^2]) of the source rock underlying the Terra Nova oil field.

leads to very low volumes and an incorrect GOR. Therefore, the fill-and-spill mechanism across the faults does not appear to be appropriate to reproduce the filling history of the TN field.

The second attempt (Figure 15) to fill the TN field adequately assumes an extra charge from a kitchen located to the northwest of the TN field, in the direction of the Hibernia drainage area. But the drainage area responsible for this extra charge is located south of the transbasin fault system (compare Jeanne d'Arc drainage areas in Figures 7A, B; 9A; 15). Migration occurs along the now closed east-west–trending transbasin fault to the southeast into the TN field. Figure 15 shows a cutout of the full model including the corresponding ray

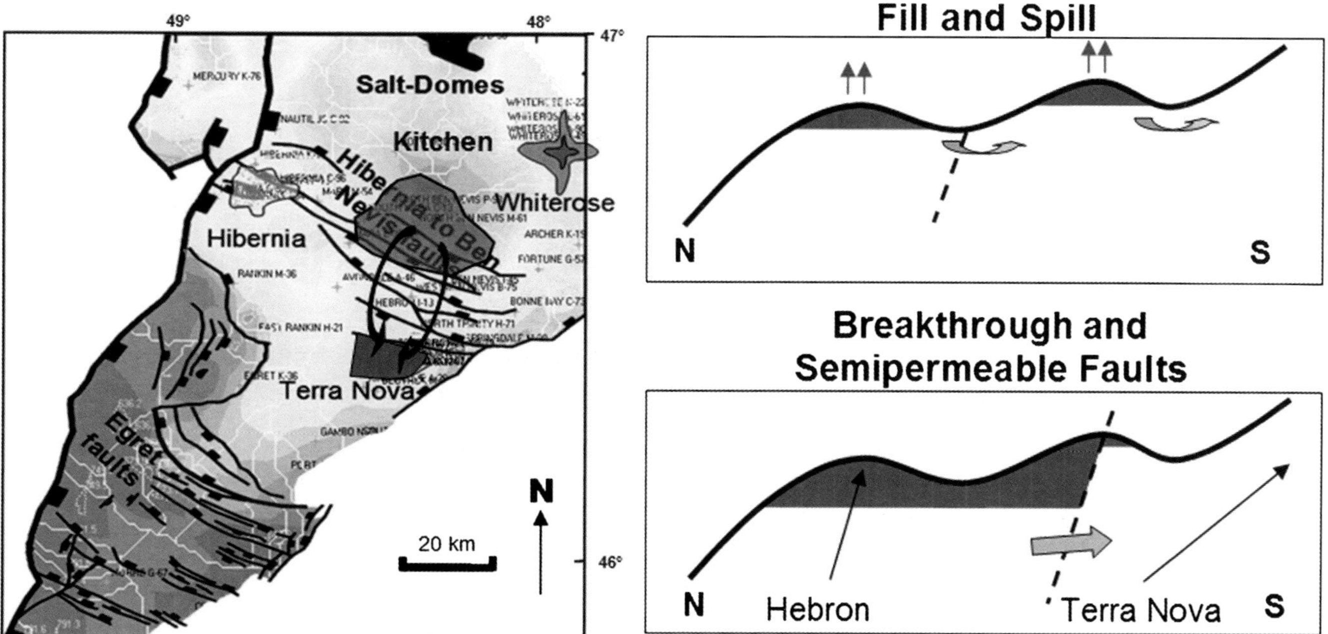

FIGURE 14. Map of the major tectonic elements in the Jeanne d'Arc Basin (left); fault pattern drawn from Withjack and Schlische (2005). The likely kitchen area is shown north of the transbasin fault system (transparent gray rectangle), where we tested migration pathways across the faults into the Terra Nova oil field (dark-gray polygon) along the black arrows. The fill-and-spill mechanism and the breakthrough theory with semipermeable faults are drawn schematically (right).

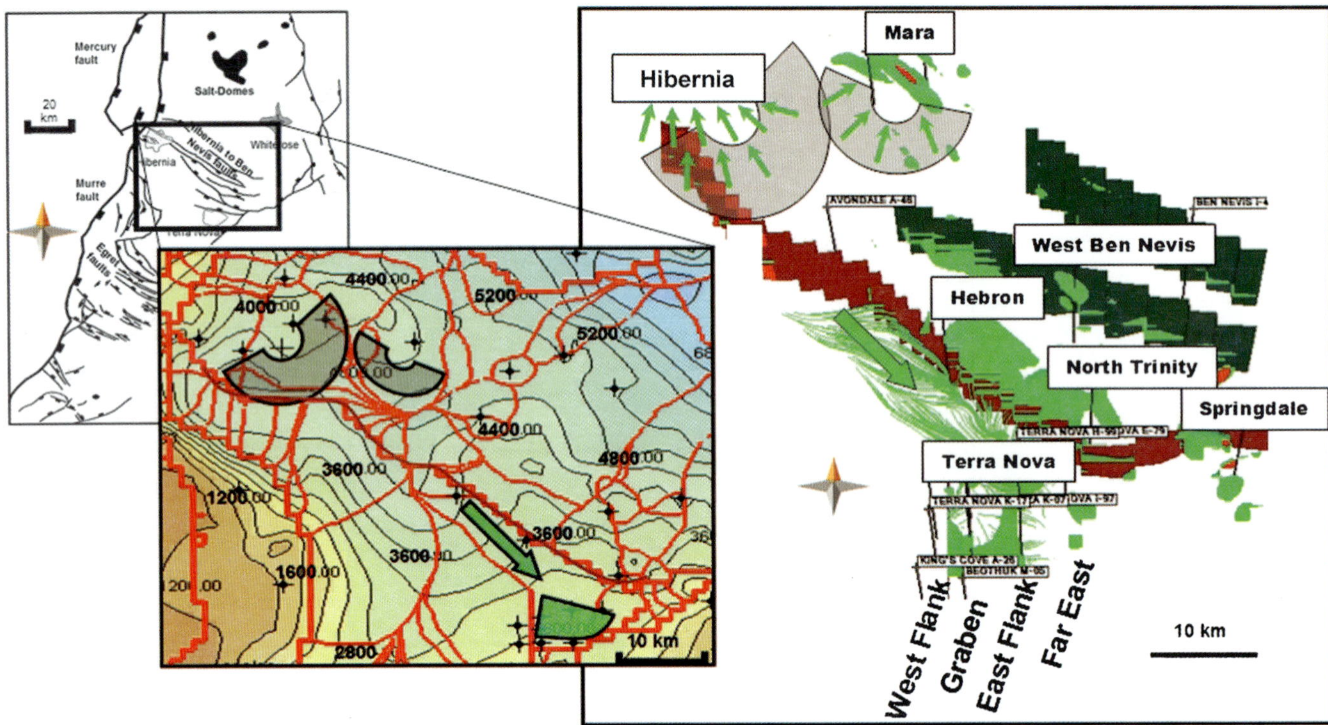

FIGURE 15. Overview map (upper left) shows fault pattern drawn from Withjack and Schlische (2005). Topographic map (left) shows the Egret Member transformation ratio as an overlay with the principal migration pathways along the closed transbasin faults from a kitchen located in the northwest toward the Terra Nova oil field. Cutout of a migration model (right) shows well locations, closed faults (brown and dark green vertical barriers), kitchen area (transparent gray half circle), oil accumulations (green areas), and migration pathways (green ray traces). The red frame emphasizes that Hibernia and Terra Nova oils are within one group (cluster diagram is from Magoon et al., 2005).

traces simulated for the highly permeable Jeanne d'Arc reservoir layer derived from a 3-D numerical model performed using the hybrid simulator. This applies the flowpath method to all lithologies that have permeabilities higher than 2 log md at 30% porosity and full Darcy calculations to all less permeable lithologies. Using this scenario of an additional kitchen area in the northeast, it is possible for the first time to adequately reconstruct APIs, GORs, P_{sat}, and B_o for the TN fluids. These results are supported by a hierarchical cluster analysis dendrogram for oil families based on biomarker and stable carbon isotope ratios (Magoon et al., 2005), which shows a close genetic relationship between oils accumulated at Hibernia, TN, and Mara oil fields. However, Hebron, North Ben Nevis, and Whiterose oils belong to a different oil family (for location, see Figure 9). This information supports our migration pathway model and inferred filling history for the TN field. This charge model leads to a volumetric balance, which is shown in Figure 16. The predicted present-day volume, GOR, and API are consistent with field data. Furthermore, the numerical model predicts a B_o of 1.35 m^3/Sm^3 and a P_{sat} of 190 bar for reservoir temperatures of 92°C (197.6°F) and a reservoir pressure of 37 MPa (5366 psi). Measured field data show a B_o of 1.4 m^3/Sm^3 and P_{sat} values approximately 225 bar. Thus, the fit is good, although the saturation pressures are 15% off.

CONCLUSIONS

This first published reconstruction of the petroleum system for the entire Jeanne d'Arc Basin based on a 3-D numerical model provides a step forward toward a basic understanding of basin evolution and petroleum accumulation. Application of new reaction kinetic data for the Egret Member source rock (bulk, multicomponent, and PhaseKinetics, including secondary cracking) allows modeling of petroleum generation with a high level of confidence. The 3-D model allowed reconstruction of the filling history of the TN oil field with respect to fluid quality and quantity as well as charge timing. The general filling history has also been reconstructed also for several other known oil fields in the Jeanne d'Arc Basin.

Based on geochemical investigations, it was established that the Egret Member source rock is more

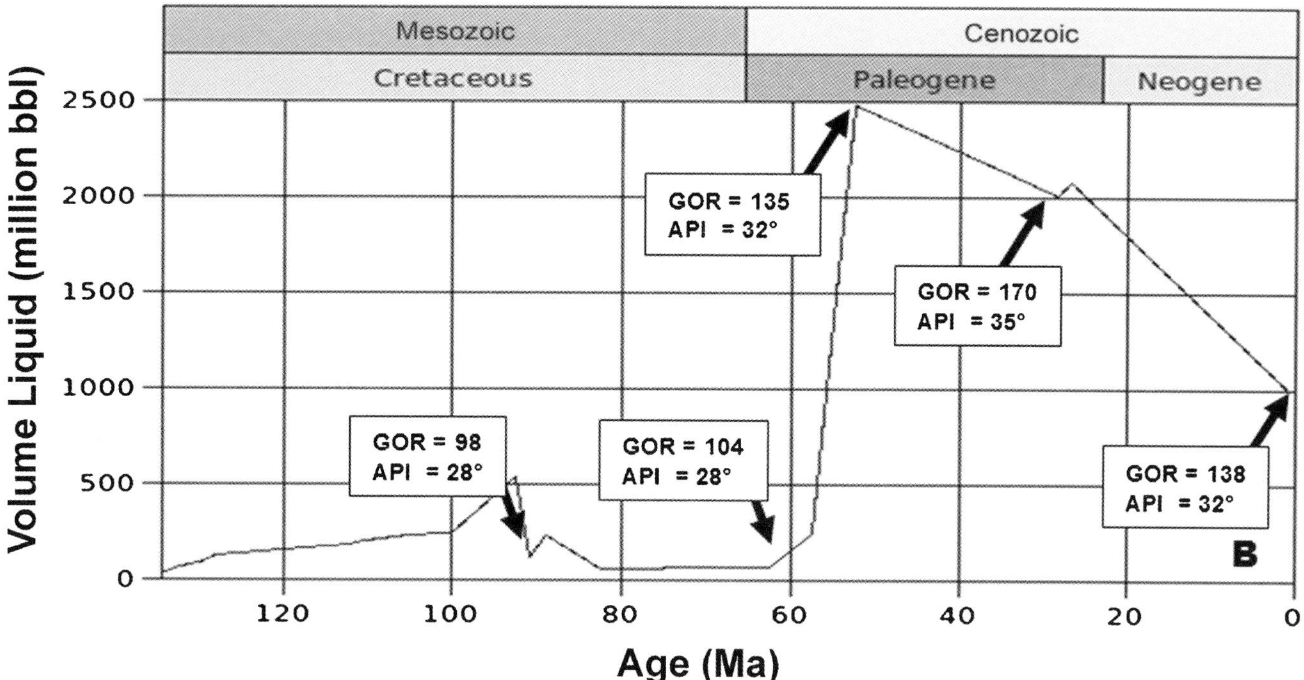

FIGURE 16. Predicted evolution of trapped volume, API, and gas-oil ratio (GOR) for the Terra Nova accumulation. Because of a change in the geometry during the last 55 m.y., the structure became steeper and the column height increased, leading to seal failure. Accordingly, the trapped volume decreases.

heterogeneous than previously considered, leading to variability in source rock properties and generated petroleum types throughout the basin. In the southwestern part of the basin, the Egret Member is more terrigenous dominated, whereas in the northeast, the source rock is more marine and enriched in sulfur. The TN oil field was charged by a local Egret Member source rock underlying the field and by an Egret Member source rock from a kitchen area located in the northwest but south of the transbasin fault system. This latter charge implies longer migration distances for the generated petroleum. PhaseKinetics allowed a very good reconstruction of the reservoir fluids in terms of API, GOR, as well as P_{sat} and B_o, especially for the TN field. Hydrocarbons at Hibernia originate from a local kitchen in addition to charge by fill-and-spill mechanisms from the center of the basin along the Mara oil field into the Hibernia oil field.

An important discrepancy between model results and reality was observed for the Whiterose field, that is, much more gas is accumulated than predicted by our model. This is most probably because of different secondary cracking behavior of the Egret source rock in that area and because of the lack of detailed fault maps that separate the individual compartments and the Whiterose field from the neighboring basement high. Most of the gas is lost by leakage and spill out of the model.

Hebron and Ben Nevis fields were filled straight from the basin center and the local kitchen area.

An additional new observation is that the API gravity of trapped fluids in the Jeanne d'Arc Basin seems to be almost constant instead of showing a linear relation with the GOR. The linear relation observed for other basins by several authors in the past always included fluid samples affected by secondary processes, such as biodegradation and secondary cracking. In the Jeanne d'Arc Basin, traps were filled both cumulatively and instantaneously, which indicates that both processes (cumulative and instantaneous filling) result in almost constant API values. This leads to the conclusion that during the primary generation of oil and gas, the bulk composition of the fluids remains roughly constant.

ACKNOWLEDGMENTS

We thank Ken Peters and an anonymous reviewer for their thoughtful corrections. We also thank D. H. Welte, who helped initiate the project. We thank Dave Hawkins from CNLOPB for the personal communication with him and the staff and to IES-Schlumberger for the PetroMod software. GFZ is thanked for the access to laboratory and office facilities during this study. Thanks are due the DFG for financial support of the project (grant Li618/21).

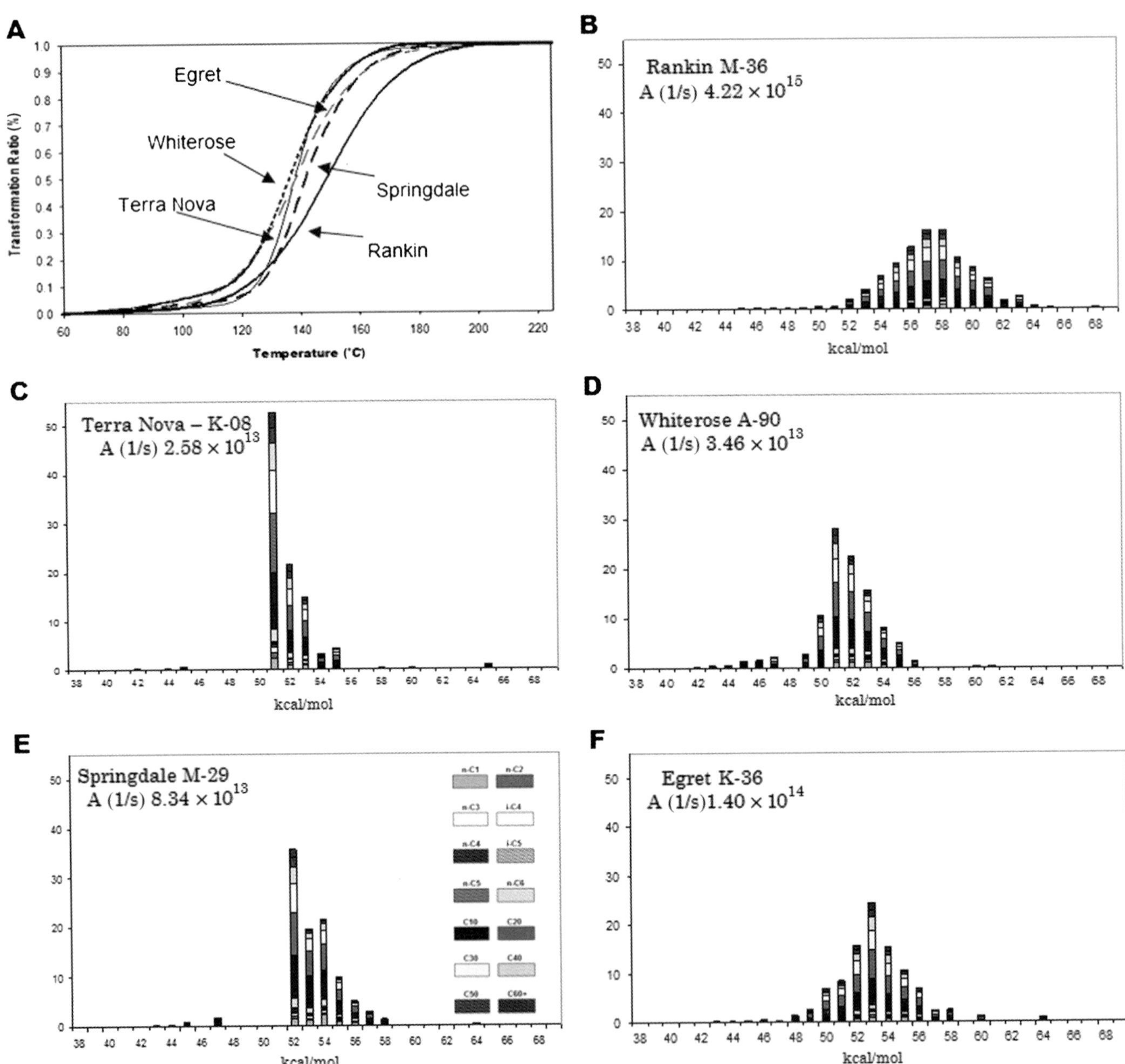

APPENDIX FIGURE 1. Panels B–F show the five PhaseKinetics with frequency factor and activation energy distribution. Panel A shows the generation of petroleum for geologic heating rates (3°C/m.y.). Rankin contains the most stable kerogen, whereas Whiterose and Egret start to generate very early. Terra Nova generates within a small temperature interval similar to Springdale, but at slightly lower temperatures. The Terra Nova and Springdale samples indicate a homogeneous source rock composition albeit with a more stable kerogen compared with Springdale and Whiterose. The most stable kerogen is represented by a sample from the Rankin well as expected from the Ea distribution.

Appendix Table 1. PhaseKinetics (activation energy, frequency factors, and ratios) for five source rock samples.

	Activation Energy	*Methane*	*Ethane*	*Propane*	*i-Butane*	*n-Butane*	*i-Pentane*	*n-Pentane*	*n-Hexane*	*C7-C15*	*C16-C25*	*C26-C35*	*C36-C45*	*C46-C55*	*C56-C80*
Rankin M-36	45	0.11	0.11	0.11	0.11	0.13	0.18	0.11	0.13	0.16	0.16	0.16	0.16	0.16	0.16
	46	0.08	0.08	0.08	0.08	0.10	0.13	0.08	0.10	0.11	0.12	0.12	0.12	0.12	0.12
	47	0.15	0.15	0.15	0.16	0.18	0.26	0.15	0.19	0.22	0.22	0.22	0.22	0.22	0.22
	48	0.19	0.19	0.20	0.20	0.24	0.33	0.20	0.24	0.28	0.28	0.28	0.28	0.28	0.28
	49	0.21	0.21	0.21	0.21	0.25	0.35	0.21	0.26	0.30	0.30	0.30	0.30	0.30	0.30
	50	0.40	0.40	0.40	0.41	0.49	0.69	0.41	0.50	0.58	0.58	0.59	0.59	0.59	0.59
	51	0.48	0.48	0.48	0.49	0.59	0.82	0.49	0.59	0.69	0.70	0.70	0.70	0.70	0.70
	52	1.52	1.53	1.54	1.57	1.87	2.60	1.55	1.89	2.21	2.22	2.23	2.24	2.24	2.23
	53	2.82	2.84	2.85	2.91	3.46	4.83	2.88	3.50	4.09	4.12	4.14	4.15	4.15	4.13
	54	4.77	4.81	4.82	4.93	5.86	8.17	4.88	5.91	6.92	6.97	7.00	7.01	7.01	6.99
	55	7.70	7.89	8.00	7.68	8.33	10.29	8.36	9.29	9.54	9.68	9.80	9.89	9.96	10.02
	56	10.28	10.52	10.67	10.25	11.12	13.72	11.15	12.40	12.73	12.91	13.07	13.19	13.29	13.37
	57	13.61	14.13	14.35	13.71	14.06	14.53	14.89	14.46	15.58	16.36	17.18	18.01	18.84	19.92
	58	16.41	16.96	16.64	16.69	16.12	14.08	16.52	15.92	15.64	16.08	16.53	16.94	17.32	17.78
	59	10.69	11.06	10.85	10.88	10.51	9.18	10.77	10.38	10.20	10.49	10.78	11.05	11.29	11.60
	60	12.71	11.90	11.91	12.34	11.09	8.24	11.36	10.07	8.62	7.82	7.03	6.30	5.62	4.82
	61	9.51	8.90	8.91	9.24	8.30	6.17	8.51	7.54	6.45	5.85	5.26	4.71	4.21	3.61
	62	2.73	2.55	2.55	2.65	2.38	1.77	2.44	2.16	1.85	1.68	1.51	1.35	1.21	1.03
	63	4.03	3.78	3.78	3.92	3.52	2.62	3.61	3.20	2.74	2.48	2.23	2.00	1.78	1.53
	64	0.88	0.83	0.83	0.86	0.77	0.57	0.79	0.70	0.60	0.54	0.49	0.44	0.39	0.34
	65	0.26	0.24	0.24	0.25	0.23	0.17	0.23	0.21	0.18	0.16	0.14	0.13	0.11	0.10
	66	0.03	0.03	0.03	0.03	0.03	0.02	0.03	0.02	0.02	0.02	0.02	0.02	0.01	0.01
	67	0.00	0.00	0.00	0.00	0.00	0.00	0.00	0.00	0.00	0.00	0.00	0.00	0.00	0.00
	68	0.34	0.31	0.31	0.33	0.29	0.22	0.30	0.27	0.23	0.21	0.19	0.17	0.15	0.13
	69	0.11	0.10	0.10	0.10	0.09	0.07	0.10	0.08	0.07	0.07	0.06	0.05	0.05	0.04
	Ratio %	4.30	1.54	1.82	0.28	1.13	0.53	0.64	4.18	23.59	24.28	16.61	10.10	5.75	5.24
	Frequency factor	1.33E + 29	1.33E + 29	1.33E + 29	1.33E + 29	1.33E + 29	1.33E + 29	1.33E + 29	1.33E + 29	1.33E + 29	1.33E + 29	1.33E + 29	1.33E + 29	1.33E + 29	1.33E + 29
Egret K-36	42	0.06	0.07	0.06	0.06	0.07	0.08	0.06	0.07	0.09	0.08	0.08	0.08	0.07	0.07
	43	0.11	0.12	0.11	0.11	0.14	0.15	0.12	0.13	0.17	0.16	0.15	0.14	0.13	0.12
	44	0.18	0.19	0.17	0.18	0.22	0.24	0.19	0.21	0.26	0.25	0.24	0.23	0.22	0.20
	45	0.22	0.24	0.21	0.21	0.27	0.29	0.22	0.25	0.32	0.31	0.29	0.27	0.26	0.24
	46	0.37	0.41	0.36	0.37	0.46	0.50	0.39	0.44	0.55	0.53	0.50	0.47	0.45	0.41
	47	0.14	0.15	0.14	0.14	0.17	0.19	0.15	0.17	0.21	0.20	0.19	0.18	0.17	0.16

Appendix Table 1. (cont.).

	Activation Energy	*Methane*	*Ethane*	*Propane*	*i-Butane*	*n-Butane*	*i-Pentane*	*n-Pentane*	*n-Hexane*	*C7-C15*	*C16-C25*	*C26-C35*	*C36-C45*	*C46-C55*	*C56-C80*
	48	1.04	1.13	0.99	1.03	1.28	1.38	1.07	1.22	1.53	1.46	1.39	1.32	1.24	1.15
	49	1.92	2.08	1.83	1.90	2.36	2.56	1.99	2.25	2.82	2.70	2.57	2.43	2.30	2.12
	50	5.01	5.44	4.77	4.96	6.16	6.69	5.19	5.88	7.38	7.06	6.71	6.36	6.01	5.54
	51	7.27	7.09	7.21	6.10	8.15	12.99	7.44	8.86	8.47	8.55	8.62	8.66	8.68	8.65
	52	13.38	13.05	13.28	11.22	15.00	23.91	13.70	16.31	15.60	15.75	15.86	15.94	15.98	15.92
	53	22.65	23.42	23.60	23.86	22.71	18.82	23.41	22.58	23.17	24.41	25.74	27.07	28.37	30.13
	54	14.35	14.83	14.95	15.11	14.39	11.92	14.83	14.30	14.68	15.46	16.30	17.14	17.97	19.08
	55	14.48	13.82	14.06	15.11	12.44	8.82	13.58	11.88	10.77	10.04	9.29	8.57	7.89	7.05
	56	9.60	9.16	9.32	10.01	8.25	5.84	9.00	7.87	7.13	6.65	6.15	5.68	5.23	4.67
	57	3.04	2.90	2.95	3.18	2.61	1.85	2.85	2.50	2.26	2.11	1.95	1.80	1.66	1.48
	58	3.44	3.29	3.34	3.59	2.96	2.10	3.23	2.82	2.56	2.39	2.21	2.04	1.87	1.68
	59	0.00	0.00	0.00	0.00	0.00	0.00	0.00	0.00	0.00	0.00	0.00	0.00	0.00	0.00
	60	1.42	1.35	1.38	1.48	1.22	0.86	1.33	1.16	1.05	0.98	0.91	0.84	0.77	0.69
	61	0.06	0.05	0.05	0.06	0.05	0.03	0.05	0.05	0.04	0.04	0.04	0.03	0.03	0.03
	62	0.00	0.00	0.00	0.00	0.00	0.00	0.00	0.00	0.00	0.00	0.00	0.00	0.00	0.00
	63	0.00	0.00	0.00	0.00	0.00	0.00	0.00	0.00	0.00	0.00	0.00	0.00	0.00	0.00
	64	1.28	1.22	1.24	1.34	1.10	0.78	1.20	1.05	0.95	0.89	0.82	0.76	0.70	0.62
	65	0.00	0.00	0.00	0.00	0.00	0.00	0.00	0.00	0.00	0.00	0.00	0.00	0.00	0.00
	Ratio %	4.53	1.59	1.88	0.27	1.20	0.52	0.71	4.16	23.79	24.29	16.47	9.93	5.61	5.05
	Frequency factor	4.42E + 27	4.42E + 27	4.42E + 27	4.42E + 27	4.42E + 27	4.42E + 27	4.42E + 27	4.42E + 27	4.42E + 27	4.42E + 27	4.42E + 27	4.42E + 27	4.42E + 27	4.42E + 27
Springdale M-31	42	0.06	0.06	0.07	0.13	0.08	0.11	0.11	0.10	0.11	0.12	0.12	0.13	0.13	0.13
	43	0.22	0.21	0.25	0.45	0.29	0.37	0.37	0.33	0.40	0.42	0.42	0.46	0.46	0.45
	44	0.17	0.16	0.19	0.34	0.22	0.29	0.29	0.25	0.30	0.32	0.32	0.35	0.35	0.34
	45	0.45	0.44	0.51	0.92	0.60	0.78	0.78	0.68	0.82	0.88	0.88	0.95	0.95	0.94
	46	0.00	0.00	0.00	0.00	0.00	0.00	0.00	0.00	0.00	0.00	0.00	0.00	0.00	0.00
	47	1.04	1.02	1.20	2.15	1.39	1.81	1.81	1.59	1.91	2.04	2.04	2.21	2.21	2.18
	48	0.10	0.10	0.12	0.21	0.14	0.18	0.18	0.16	0.19	0.20	0.20	0.22	0.22	0.21
	49	0.00	0.00	0.00	0.00	0.00	0.00	0.00	0.00	0.00	0.00	0.00	0.00	0.00	0.00
	50	0.00	0.00	0.00	0.00	0.00	0.00	0.00	0.00	0.00	0.00	0.00	0.00	0.00	0.00
	51	0.00	0.00	0.00	0.00	0.00	0.00	0.00	0.00	0.00	0.00	0.00	0.00	0.00	0.00
	52	23.14	24.59	26.47	32.05	27.89	32.19	32.19	32.03	32.73	38.69	38.69	52.17	52.17	66.72
	53	17.18	17.58	18.45	17.59	19.66	19.60	19.60	19.55	22.83	19.45	19.45	12.97	12.97	7.45
	54	30.20	29.26	27.63	24.18	26.06	23.41	23.41	23.74	21.33	19.85	19.85	16.01	16.01	11.31
	55	13.61	13.19	12.45	10.90	11.75	10.55	10.55	10.70	9.61	8.95	8.95	7.22	7.22	5.10

	56	7.12	6.90	6.52	5.70	6.15	5.52	5.52	5.60	5.03	4.68	4.68	3.77	3.77	2.67
	57	3.94	3.82	3.61	3.16	3.40	3.06	3.06	3.10	2.79	2.59	2.59	2.09	2.09	1.48
	58	1.97	1.91	1.81	1.58	1.70	1.53	1.53	1.55	1.39	1.30	1.30	1.05	1.05	0.74
	59	0.00	0.00	0.00	0.00	0.00	0.00	0.00	0.00	0.00	0.00	0.00	0.00	0.00	0.00
	60	0.17	0.16	0.15	0.13	0.15	0.13	0.13	0.13	0.12	0.11	0.11	0.09	0.09	0.06
	61	0.10	0.10	0.09	0.08	0.09	0.08	0.08	0.08	0.07	0.06	0.06	0.05	0.05	0.04
	62	0.00	0.00	0.00	0.00	0.00	0.00	0.00	0.00	0.00	0.00	0.00	0.00	0.00	0.00
	63	0.00	0.00	0.00	0.00	0.00	0.00	0.00	0.00	0.00	0.00	0.00	0.00	0.00	0.00
	64	0.53	0.52	0.49	0.43	0.46	0.41	0.41	0.42	0.38	0.35	0.35	0.28	0.28	0.20
	Ratio %	7.02	1.87	2.24	0.60	1.29	0.61	0.73	5.67	27.34	24.23	14.24	7.52	3.75	2.88
	Frequency factor	2.63E + 27	2.63E + 27	2.63E + 27	2.63E + 27	2.63E + 27	2.63E + 27	2.63E + 27	2.63E + 27	2.63E + 27	2.63E + 27	2.63E + 27	2.63E + 27	2.63E + 27	2.63E + 27
Terra Nova K-08	42	0.40	0.43	0.42	0.54	0.42	0.34	0.36	0.43	0.53	0.50	0.47	0.44	0.41	0.38
	43	0.09	0.10	0.10	0.12	0.10	0.08	0.08	0.10	0.12	0.12	0.11	0.10	0.10	0.09
	44	0.35	0.38	0.37	0.47	0.37	0.30	0.31	0.37	0.47	0.44	0.41	0.39	0.36	0.33
	45	0.61	0.66	0.63	0.82	0.64	0.51	0.54	0.65	0.81	0.76	0.72	0.67	0.63	0.58
	46	0.00	0.00	0.00	0.00	0.00	0.00	0.00	0.00	0.00	0.00	0.00	0.00	0.00	0.00
	47	0.00	0.00	0.00	0.00	0.00	0.00	0.00	0.00	0.00	0.00	0.00	0.00	0.00	0.00
	48	0.00	0.00	0.00	0.00	0.00	0.00	0.00	0.00	0.00	0.00	0.00	0.00	0.00	0.00
	49	0.00	0.00	0.00	0.00	0.00	0.00	0.00	0.00	0.00	0.00	0.00	0.00	0.00	0.00
	50	0.00	0.00	0.00	0.00	0.00	0.00	0.00	0.00	0.00	0.00	0.00	0.00	0.00	0.00
	51	47.62	48.47	49.20	48.67	50.26	48.61	49.06	50.66	52.14	53.40	54.65	55.80	56.88	58.15
	52	20.52	21.18	20.17	20.51	20.93	21.96	21.00	20.62	20.67	21.46	22.27	23.07	23.83	24.85
	53	18.74	17.74	17.95	17.80	16.82	17.39	17.66	16.75	15.58	14.38	13.17	12.04	10.97	9.64
	54	4.17	3.95	4.00	3.96	3.75	3.87	3.93	3.73	3.47	3.20	2.93	2.68	2.44	2.15
	55	5.51	5.21	5.28	5.23	4.94	5.11	5.19	4.92	4.58	4.23	3.87	3.54	3.22	2.83
	56	0.00	0.00	0.00	0.00	0.00	0.00	0.00	0.00	0.00	0.00	0.00	0.00	0.00	0.00
	57	0.00	0.00	0.00	0.00	0.00	0.00	0.00	0.00	0.00	0.00	0.00	0.00	0.00	0.00
	58	0.27	0.25	0.25	0.25	0.24	0.25	0.25	0.24	0.22	0.20	0.19	0.17	0.16	0.14
	59	0.00	0.00	0.00	0.00	0.00	0.00	0.00	0.00	0.00	0.00	0.00	0.00	0.00	0.00
	60	0.50	0.48	0.48	0.48	0.45	0.47	0.48	0.45	0.42	0.39	0.35	0.32	0.30	0.26
	61	0.08	0.07	0.07	0.07	0.07	0.07	0.07	0.07	0.06	0.06	0.05	0.05	0.04	0.04
	62	0.00	0.00	0.00	0.00	0.00	0.00	0.00	0.00	0.00	0.00	0.00	0.00	0.00	0.00
	63	0.00	0.00	0.00	0.00	0.00	0.00	0.00	0.00	0.00	0.00	0.00	0.00	0.00	0.00
	64	0.00	0.00	0.00	0.00	0.00	0.00	0.00	0.00	0.00	0.00	0.00	0.00	0.00	0.00
	65	1.13	1.07	1.09	1.08	1.02	1.05	1.07	1.01	0.94	0.87	0.80	0.73	0.66	0.58
	Ratio %	5.42	1.83	2.14	0.27	1.27	0.28	0.78	4.60	22.34	23.38	16.29	10.09	5.85	5.47
	Frequency factor	8.14E + 26	8.14E + 26	8.14E + 26	8.14E + 26	8.14E + 26	8.14E + 26	8.14E + 26	8.14E + 26	8.14E + 26	8.14E + 26	8.14E + 26	8.14E + 26	8.14E + 26	8.14E + 26

Appendix Table 1. (cont.).

	Activation Energy	*Methane*	*Ethane*	*Propane*	*i-Butane*	*n-Butane*	*i-Pentane*	*n-Pentane*	*n-Hexane*	*C7-C15*	*C16-C25*	*C26-C35*	*C36-C45*	*C46-C55*	*C56-C80*
Whiterose A-90	42	0.11	0.15	0.10	0.12	0.14	0.07	0.11	0.15	0.20	0.22	0.24	0.26	0.27	0.30
	43	0.25	0.34	0.23	0.27	0.33	0.17	0.26	0.34	0.45	0.50	0.54	0.59	0.63	0.68
	44	0.28	0.37	0.26	0.30	0.36	0.18	0.28	0.38	0.50	0.55	0.60	0.65	0.70	0.76
	45	0.73	0.96	0.66	0.78	0.94	0.47	0.73	0.97	1.30	1.43	1.56	1.68	1.80	1.96
	46	0.78	1.03	0.71	0.84	1.01	0.51	0.79	1.05	1.40	1.53	1.67	1.81	1.94	2.11
	47	1.20	1.59	1.09	1.30	1.55	0.78	1.21	1.60	2.15	2.35	2.57	2.78	2.98	3.24
	48	0.00	0.00	0.00	0.00	0.00	0.00	0.00	0.00	0.00	0.00	0.00	0.00	0.00	0.00
	49	1.55	2.04	1.40	1.66	1.99	1.00	1.55	2.06	2.76	3.02	3.30	3.57	3.82	4.16
	50	5.73	7.56	5.20	6.17	7.38	3.72	5.76	7.64	10.24	11.21	12.23	13.22	14.17	15.41
	51	22.95	23.61	24.30	23.03	23.68	40.41	23.73	23.88	26.39	28.86	31.50	34.07	36.52	39.63
	52	24.41	24.23	24.81	22.95	24.61	20.51	25.49	25.63	22.94	22.09	21.04	19.85	18.59	16.79
	53	21.65	19.66	21.26	21.94	19.59	16.59	20.66	18.72	16.32	14.56	12.77	11.10	9.58	7.73
	54	10.98	9.97	10.79	11.13	9.94	8.42	10.48	9.50	8.28	7.39	6.48	5.63	4.86	3.92
	55	6.89	6.26	6.77	6.99	6.24	5.28	6.58	5.96	5.19	4.63	4.06	3.53	3.05	2.46
	56	1.91	1.73	1.87	1.93	1.73	1.46	1.82	1.65	1.44	1.28	1.13	0.98	0.84	0.68
	57	0.00	0.00	0.00	0.00	0.00	0.00	0.00	0.00	0.00	0.00	0.00	0.00	0.00	0.00
	58	0.10	0.09	0.10	0.10	0.09	0.08	0.09	0.08	0.07	0.07	0.06	0.05	0.04	0.04
	59	0.00	0.00	0.00	0.00	0.00	0.00	0.00	0.00	0.00	0.00	0.00	0.00	0.00	0.00
	60	0.33	0.30	0.33	0.34	0.30	0.26	0.32	0.29	0.25	0.23	0.20	0.17	0.15	0.12
	61	0.14	0.13	0.14	0.14	0.13	0.11	0.13	0.12	0.11	0.09	0.08	0.07	0.06	0.05
	Ratio %	4.96	1.74	1.90	0.27	1.21	0.36	0.75	4.49	25.14	24.59	15.97	9.26	5.04	4.33
	Frequency factor	1.09E + 27	1.09E + 27	1.09E + 27	1.09E + 27	1.09E + 27	1.09E + 27	1.09E + 27	1.09E + 27	1.09E + 27	1.09E + 27	1.09E + 27	1.09E + 27	1.09E + 27	1.09E + 27
Average	42	0.16	0.18	0.16	0.21	0.18	0.15	0.16	0.18	0.23	0.23	0.23	0.23	0.22	0.22
	43	0.17	0.19	0.17	0.24	0.21	0.19	0.21	0.22	0.28	0.30	0.31	0.32	0.33	0.34
	44	0.24	0.28	0.24	0.32	0.29	0.25	0.27	0.30	0.38	0.39	0.39	0.40	0.41	0.41
	45	0.42	0.48	0.42	0.57	0.51	0.45	0.48	0.54	0.68	0.71	0.72	0.75	0.76	0.77
	46	0.25	0.30	0.23	0.26	0.31	0.23	0.25	0.32	0.41	0.43	0.46	0.48	0.50	0.53
	47	0.51	0.58	0.52	0.75	0.66	0.61	0.66	0.71	0.90	0.96	1.00	1.08	1.11	1.16
	48	0.27	0.28	0.26	0.29	0.33	0.38	0.29	0.32	0.40	0.39	0.37	0.36	0.35	0.33
	49	0.73	0.87	0.69	0.76	0.92	0.78	0.75	0.91	1.18	1.20	1.23	1.26	1.28	1.32
	50	2.23	2.68	2.07	2.31	2.81	2.22	2.27	2.80	3.64	3.77	3.90	4.03	4.15	4.31
	51	15.66	15.93	16.24	15.66	16.53	20.56	16.14	16.80	17.54	18.30	19.09	19.85	20.56	21.42
	52	16.59	16.91	17.25	17.66	18.06	20.23	18.79	19.30	18.83	20.04	20.02	22.65	22.56	25.30
	53	16.61	16.25	16.82	16.82	16.45	15.44	16.84	16.22	16.40	15.38	15.05	13.46	13.21	11.82

54	12.89	12.56	12.44	11.86	12.00	11.16	11.51	11.44	10.93	10.57	10.51	9.69	9.66	8.69
55	9.64	9.27	9.31	9.18	8.74	8.01	8.85	8.55	7.94	7.50	7.19	6.55	6.27	5.49
56	5.78	5.66	5.67	5.58	5.45	5.31	5.50	5.50	5.27	5.10	5.01	4.72	4.63	4.28
57	4.12	4.17	4.18	4.01	4.02	3.89	4.16	4.01	4.13	4.21	4.35	4.38	4.52	4.58
58	4.44	4.50	4.43	4.44	4.22	3.61	4.32	4.12	3.98	4.01	4.05	4.05	4.09	4.07
59	2.14	2.21	2.17	2.18	2.10	1.84	2.15	2.08	2.04	2.10	2.16	2.21	2.26	2.32
60	3.03	2.84	2.85	2.95	2.64	1.99	2.72	2.42	2.09	1.90	1.72	1.54	1.39	1.19
61	1.98	1.85	1.85	1.92	1.73	1.29	1.77	1.57	1.35	1.22	1.10	0.98	0.88	0.75
62	0.68	0.64	0.64	0.66	0.59	0.44	0.61	0.54	0.46	0.42	0.38	0.34	0.30	0.26
63	1.01	0.94	0.95	0.98	0.88	0.65	0.90	0.80	0.68	0.62	0.56	0.50	0.45	0.38
64	0.67	0.64	0.64	0.65	0.58	0.44	0.60	0.54	0.48	0.44	0.41	0.37	0.34	0.29
65	0.46	0.44	0.44	0.44	0.41	0.41	0.43	0.41	0.37	0.34	0.31	0.29	0.26	0.23
66	0.03	0.03	0.03	0.03	0.03	0.02	0.03	0.02	0.02	0.02	0.02	0.02	0.01	0.01
67	0.00	0.00	0.00	0.00	0.00	0.00	0.00	0.00	0.00	0.00	0.00	0.00	0.00	0.00
68	0.34	0.31	0.31	0.33	0.29	0.22	0.30	0.27	0.23	0.21	0.19	0.17	0.15	0.13
69	0.11	0.10	0.10	0.10	0.09	0.07	0.10	0.08	0.07	0.07	0.06	0.05	0.05	0.04
Ratio %	5.25	1.71	2.00	0.34	1.22	0.46	0.72	4.62	24.44	24.16	15.92	9.38	5.20	4.59
Frequency factor	2.84E + 28	2.84E + 28	2.84E + 28	2.84E + 28	2.84E + 28	2.84E + 28	2.84E + 28	2.84E + 28	2.84E + 28	2.84E + 28	2.84E + 28	2.84E + 28	2.84E + 28	2.84E + 28

REFERENCES CITED

Amoco, 1973, Canada Petroleum Company, Ltd, geology of the Grand Banks: Bulletin of Canadian Petroleum Geology, v. 21–4, p. 479–503.

Baur, F., M. Di Benedetto, T. Fuchs, C. Lampe, and S. Sciamanna, 2009a, Integrating structural geology and petroleum systems modeling: A pilot project from Bolivia's fold and thrust belt: Marine and Petroleum Geology, v. 26, no. 4, p. 573–579, doi:10.1016/j.marpetgeo.2009.01.004.

Baur, F., H. Wielens, and R. Littke, 2009b, Basin and petroleum system modeling at the Jeanne d'Arc and Carson basin offshore Newfoundland, Canada: Recorder, September 2009, p. 24–32.

Baur, F., R. Littke, H. Wielens, C. Lampe, and T. Fuchs, 2010, Basin modeling meets rift analysis: A numerical modeling study from the Jeanne d'Arc Basin, offshore Newfoundland, Canada: Marine and Petroleum Geology, v. 27, no. 3, p. 585–599, doi:10.1016/j.marpetgeo.2009.06.003.

Canada-Newfoundland and Labrador Offshore Petroleum Board (CNLOPB), 2004, Schedule of wells: Newfoundland and Labrador offshore area-Canada-Newfoundland and Labrador Offshore Petroleum Board, St. Johns: http://www.cnlopb.nl.ca/well_north.shtml (accessed November 22, 2010).

Canada-Newfoundland and Labrador Offshore Petroleum Board (CNLOPB), 2008, Annual Report, 2007/08: http://www.cnlopb.nl.ca/pdfs/ar2008e.pdf (accessed November 22, 2010).

Cloetingh, S., A. J. Tankard, H. J. Welsink, and W. A. M. Jenkins, 1989, Vail's coastal onlap curves and their correlation with tectonic events, offshore Eastern Canada, *in* A. J. Tankard and H. R. Balkwill, eds., Extensional tectonics and stratigraphy of the North Atlantic margins: AAPG Memoir 46, p. 265–282.

Cooles, G. P., A. S. Mackenzie, and T. M. Quigley, 1986, Calculation of petroleum masses generated and expelled from source rocks: Organic Geochemistry, v. 10, p. 235–245, doi:10.1016/0146-6380(86)90026-4.

Deptuck, M. E., R. A. MacRae, J. W. Shimeld, G. L. Williams, and R. A. Fensome, 2003, Revised Upper Cretaceous and lower Paleogene lithostratigraphy and depositional history of the Jeanne d'Arc Basin, offshore Newfoundland, Canada: AAPG Bulletin, v. 87, p. 1459–1483, doi:10.1306/050203200178.

di Primio, R., and B. Horsfield, 2006, From petroleum-type organofacies to hydrocarbon phase prediction: AAPG Bulletin, v. 90, no. 7, p. 1031–1058, doi:10.1306/02140605129.

di Primio, R., and J. E., Skeie, 2004, Development of a compositional kinetic model for hydrocarbon generation and phase equilibria modeling: A case study from Snorre field, Norwegian North Sea, *in* J. M. Cubitt, W. A. England, and S. R. Larter, eds., Understanding petroleum reservoirs: Toward an integrated reservoir engineering and geochemical approach: Geological Society (London) Special Publication 237, p. 157–174.

Edwards, T., P. Moir, and K. Coflin, 2000, Structure and isopach maps of the Jeanne d'Arc Basin, Grand Banks of Newfoundland: Geological Survey of Canada (Atlantic), open file report no. D3755.

Eglinton, T. I., J. S. Sinninghe-Damsté, M. E. L. Kohnen, and J. W. de Leeuw, 1990, Rapid estimation of the organic sulfur content of kerogens, coals and asphaltenes by pyrolysis-gas chromatography: Fuel, v. 69, no. 11, p. 1394–1404, doi:10.1016/0016-2361(90)90121-6.

Enachescu, M. E., 1992, Enigmatic basins offshore Newfoundland: Canadian Journal of Exploration Geophysics, v. 28, no. 1, p. 44–61.

Enachescu, M. E., 1993, Amplitude interpretation of 3-D reflection data: The Leading Edge, v. 12, no. 6, p. 678–685, doi:10.1190/1.1436956.

Enachescu, M., and P. Fagan, 2005, Newfoundland's Grand Banks presents untested oil and gas potential in eastern North America: Oil & Gas Journal, v. 103, no. 6, p. 32–39.

Erdmann, M., 1999, Gas generation from overmature Upper Jurassic source rocks, Northern Viking Graben: Bericht des Kernforschungslager Jülich Nr. 2952, 128 p.

Espitalié, J., M. Madec, and B. Tissot, 1977, Source rock characterization method for petroleum exploration: 9th Annual Offshore Technology Conference, Houston, Texas, p. 439–444.

Espitalié, J., F. Marquis, and L. Barsony, 1984, Geochemical logging, *in* K. J. Voorhees, ed., Analytical pyrolysis techniques and applications: London, United Kingdom, Butterworth and Co., p. 276–306.

Espitalié, J., G. Deroo, and F. Marquis, 1985, Rock-Eval pyrolysis and its application: 27299, project B41 79008, IFP Géologie et Géochimie, 132 p.

Evans, C. R., M. A. Rogers, and N. J. L. Bailey, 1971, Evolution and alteration of petroleum in Western Canada: Chemical Geology, v. 8, p. 147–170, doi:10.1016/0009-2541(71)90002-7.

Fowler, M. G., and K. D. McAlpine, 1994, The Egret Member, a prolific Kimmeridgian source rock from offshore Eastern Canada, *in* B. J. Katz, ed., Petroleum source rocks: New York, Springer-Verlag, p. 111–130.

Geological Survey of Canada (GSC), 2008, Basin database: http://basin.gdr.nrcan.gc.ca/index_e.php (accessed November 22, 2010), Geological Survey of Canada, Atlantic, Dartmouth.

Grant, A. C., and K. D. McAlpine, 1990, The continental margin around Newfoundland; Chapter 6, *in* M. J. Keen and G. L. Williams, eds., Geology of the continental margin of eastern Canada: Geological Survey of Canada, v. 2, p. 239–292.

Grant, A. C., K. D. McAlpine, and J. A. Wade, 1986a, The continental margin of Eastern Canada: Geological framework and petroleum potential: AAPG Memoir 40, p. 177–205.

Grant, A. C., K. D. McAlpine, and J. A. Wade, 1986b, Offshore geology and petroleum potential of eastern Canada: Energy Exploration and Exploitation, v. 4, no. 1, p. 5–52.

Hantschel, T., and I. A. Kauerauf, 2009, Fundamentals in basin and petroleum systems modeling: Berlin, Springer-Verlag, 476 p.

Hesse, R., and I. A. Abid, 1998, Carbonate cementation: The key to reservoir properties of four sandstone levels (Cretaceous) in the Hibernia oil field, Jeanne d'Arc Basin, Newfoundland, Canada: Special Publication of the International Association of Sedimentologists, v. 26, p. 363–393.

Hölscher, A., 1996, TOC inklusive EC-Bestimmung: Laborpraxis, v. 07/08, p. 1–5.

Horsfield, B., 1989, Practical criteria for classifying kerogens: Some observations from pyrolysis-gas chromatography: Geochimica et Cosmochimca Acta, v. 53, p. 891–901.

Horsfield, B., 1997, The bulk composition of first-formed petroleum in source rocks, *in* D. H. Welte, B. Horsfield, and D. R. Baker, eds., Petroleum and basin evolution: Insights from petroleum geochemistry, geology and basin modeling: Berlin, Springer-Verlag, p. 335–402.

Horsfield, B., U. Disko, and F. Leistner, 1989, The microscale simulation of maturation: Outline of a new technique and its potential applications: Geologische Rundschau, v. 78, no. 1, p. 361–374, doi:10.1007/BF01988370.

Huang, Z., M. Williamson, D. McAlpine, and M. Fowler, 1993, Petrophysical log prediction of source rock geochemistry: Egret Member from Jeanne d'Arc Basin, offshore Newfoundland: The Log Analyst, v. 34, no. 2, p. 56.

Huang, Z., M. Williamson, M. Fowler, and D. McAlpine, 1994, Predicted and measured petrophysical and geochemical characteristics of the Egret Member oil source rock, Jeanne d'Arc Basin, offshore eastern Canada: Marine and Petroleum Geology, v. 11, no. 3, p. 294–306.

Hubbard, R. J., 1988, Age and significance of sequence boundaries on Jurassic and Early Cretaceous rifted continental margins: AAPG Bulletin, v. 72, no. 1, p. 49–72.

Issler, R. D., 1994, Hydrodynamics and overpressuring in the Jeanne d'Arc Basin, offshore Newfoundland, Canada: Possible implications for hydrocarbon exploration: Bulletin of Canadian Petroleum Geology, v. 42, no. 2, p. 263–265.

Jarvie, D. M., R. J. Hill, T. E. Ruble, and R. M. Pollastro, 2007, Unconventional shale-gas systems: The Mississippian Barnett Shale of north-central Texas as one model for thermogenic shale-gas assessment: AAPG Bulletin, v. 91, p. 475–499, doi:10.1306/12190606068.

Jonathan, D., G. L'Hóte, and J. du Rouchet, 1975, Analyse geochimique des hydrocarbures légers par thermovaporisation: Revue Institut Français du Pétrole, v. 30, p. 65–88.

Larter, S. R., 1984, Application of analytical pyrolysis technique to kerogen characterization and fossil fuel exploration, *in* K. Voorhees, ed., Analytical pyrolysis, methods and applications: London, United Kingdom, Butterworth, p. 212–275.

Louden, K., 2002, Tectonic evolution of the east coast of Canada: Canadian Society of Exploration Geophysicists Recorder, v. 2, p. 37–48.

Magoon, L. B., T. L. Hudson, and K. E. Peters, 2005, Egret-Hibernia(!): A significant petroleum system, northern

Grand Banks area, offshore Eastern Canada: AAPG Bulletin, v. 89, p. 1203–1237, doi:10.1306/05040504115.

Mango, F. D., 1997, The light hydrocarbons in petroleum: A critical review: Organic Geochemistry, v. 26, no. 7–8, p. 417–440, doi:10.1016/S0146-6380(97)00031-4.

Mango, F. D., 2000, The origin of light hydrocarbons: Geochimica et Cosmochimica Acta, v. 64–7, p. 1265–1277, doi:10.1016/S0016-7037(99)00389-0.

Mango, F. D., 2001, Methane concentrations in natural gas: The genetic implications: Organic Geochemistry, v. 32–10, p. 1283–1287.

McAlpine, K. D., 1990, Mesozoic stratigraphy, sedimentary evolution, and petroleum potential of the Jeanne d'Arc Basin, Grand Banks of Newfoundland: Geological Survey of Canada Paper, v. 89, no. 17, p. 50–73.

McIntyre, J., 1992, Regional time/depth relationship for Tertiary and Cretaceous intervals: Canada-Newfoundland Offshore Petroleum Board, Open file GP-CNoPB-92-02.

McIntyre, J., N. DeSilva, and T. Thompson, 2004, Updated regional mapping of Jeanne d'Arc Basin based on released 3-D seismic data: Canada-Newfoundland Offshore Petroleum Board, Open file GP-CNOPB-04-01.

Parnell, J., D. Middleton, C. Honghan, and D. Hall, 2001, The use of integrated fluid inclusion studies in constraining oil charge history and reservoir compartmentation: Examples from the Jeanne d'Arc Basin, offshore Newfoundland: Marine and Petroleum Geology, v. 18, no. 5, p. 535–549, doi:10.1016/S0264-8172(01)00018-6.

Pepper, A. S., and P. J. Corvi, 1995, Simple kinetic models of petroleum formation. Part 1. Oil and gas generation from kerogen: Marine and Petroleum Geology, v. 12, p. 291–319, doi:10.1016/0264-8172(95)98381-E.

Peters, K. E., 1986, Guidelines for evaluating petroleum source rock using programmed pyrolysis: AAPG Bulletin, v. 70, p. 318–329.

Peters, K. E., 2009, Basin and petroleum system modeling, AAPG/Datapages CD-ROM, Tulsa, Oklahoma.

Peters, K. E., C. C. Walters, and J. M. Moldowan, 2005, The biomarker guide, 2d ed.: Cambridge, United Kingdom, Cambridge University Press, p. 1155.

Petmecky, S., L. Meier, H. Reiser, R. Littke, 1999, High thermal maturity in the Lower Saxony Basin: Intrusion or deep burial?: Tectonophysics, v. 304, no. 4, p. 317–344, doi:10.1016/S0040-1951(99)00030-X.

Poelchau, H. S., D. R. Baker, T. Hantschel, B. Horsfield, and B. Wygrala, 1997, Basin simulation and the design of the conceptual basin model, *in* D. H. Welte, B. Horsfield, and D. R. Baker, eds., Petroleum and basin evolution: Berlin, Germany, Springer-Verlag, p. 3–70.

Rodriguez, J., and R. Littke, 2001, Petroleum generation and accumulation the Golfo San Jorge Basin, Argentina: A basin modeling study: Marine and Petroleum Geology, v. 18, no. 9, p. 995–1028, doi:10.1016/S0264-8172(01)00038-1.

Rogers, A. L., and N. A. Yassir, 1993, Hydrodynamics and overpressuring in the Jeanne d'Arc Basin, offshore Newfoundland, Canada: Possible implications for hydrocarbon exploration: Bulletin of Canadian Petroleum Geology, v. 41, p. 275–289.

Schaefer, R. G., H. J. Schenk, H. Hardelauf, and R. Harms, 1990, Determination of gross kinetic parameters for petroleum formation from Jurassic source rocks of different maturity levels by means of laboratory experiments: Organic Geochemistry, v. 16, no. 1–3, p. 115–120, doi:10.1016/0146-6380(90)90031-T.

Shimeld, J. W., and P. N. Moir, 2001, Heavy oil accumulations in the Jeanne d'Arc Basin: A case study in Hebron, Ben Nevis and West Ben Nevis oil fields: Geological Survey of Canada, Open file no. D4012.

Sinclair, I. K., 1993, Tectonism: The dominant factor in mid-Cretaceous deposition in the Jeanne d'Arc Basin, Grand Banks: Marine and Petroleum Geology, v. 10, p. 530–549, doi:10.1016/0264-8172(93)90058-Z.

Sinclair, I. K., K. D. McAlpine, D. F. Sherwin, and N. J. McMillan, 1992, Part 1: Geological framework, *in* L. Reynolds, ed., Petroleum resources of the Jeanne d' Arc basin and environs, Grand Banks, Newfoundland: Geological Survey of Canada Paper, v. 92, no. 8, p. 1–38.

Swift, J. H., and J. A. Williams, 1980, Petroleum source rocks; Grand Banks area: Canadian Society of Petroleum Geologists Memoir 6, p. 567–587.

Swift, J. H., R. W. Switzer, and W. F. Turnbull, 1975, The Cretaceous Petrel Limestone of the Grand Banks, Newfoundland: Canadian Society of Petroleum Geologists Memoir 4, p. 181–194.

Tankard, A. J., and H. J. Welsink, 1989, Mesozoic extension and styles of basin formation in Atlantic Canada: AAPG Memoir 46, p. 175–195.

Tankard, A. J., H. J. Welsink, and W. A. M. Jenkins, 1989, Structural styles and stratigraphy of the Jeanne d'Arc Basin, Grand Banks of Newfoundland: AAPG Memoir 46, p. 265–282.

Tissot, B. P., and D. H. Welte, 1984, Petroleum formation and occurrence: Berlin, Germany, Springer-Verlag, 699 p.

van Heek, K. H., and H. Jüntgen, 1968, Bestimmung der reaktionskinetischen Parameter aus nicht-isothermen Messungen: Berichte der Bundesgesellschaft für physikalische Chemie, v. 72, p. 1223–1231.

von der Dick, H., 1989, Environment of petroleum source rock deposition in the Jeanne d'Arc Basin off Newfoundland, *in* A. J. Tankard and H. R. Balkwill, eds., Extensional tectonics and stratigraphy of the North Atlantic Margins: AAPG Memoir 46, p. 295–303.

von der Dick, H., J. D. Meloche, J. Dwyer, and P. Gunther, 1989, Source rock geochemistry and hydrocarbon generation in the Jeanne d'Arc Basin, Grand Banks, offshore Eastern Canada: Journal of Petroleum Geology, v. 12, no. 1, p. 51–68, doi:10.1111/j.1747-5457.1989.tb00220.x.

Wannowius, K. J., 1987, Planung und Auswertung kinetischer Experimente: GIT Fachzeitschrift für das Laboratorium, v. 31, p. 1051–1060.

Welte, D. H., and V. Yalcin, 1987, Formation and occurrence of petroleum in sedimentary basins as deduced from computer-aided basin modeling, *in* S. P. Kumar, P. Dwivedi, V. Banerjie, and V. Gupta, eds.,

Petroleum geochemistry and exploration in the Afro-Asian region, p. 17–23.

Welte, D. H., and M. N. Yalcin, 1988, Basin modeling: A new comprehensive method in petroleum geology, *in* L. Mattavelli and N. Novelli, eds., Advances in organic geochemistry: Organic Geochemistry, v. 13, no. 1–3, p. 141–151.

Welte, D. H., B. Horsfield, and D. R. Baker, eds., 1997, Petroleum and basin evolution: Heidelberg, Germany, Springer-Verlag, 535 p.

Wielens, H. J. B. W., and C. Jauer, 2001, Overpressure, thermal maturity, temperature and log responses in basins of the Grand Banks of Newfoundland: Geological Survey of Canada, Open file report no. 3937, http://geopub.rncan.gc.ca/moreinfo_e.php?id=212235&_h=basins (accessed May 26, 2011).

Wielens, H. J. B. W., R. A. MacRae, and J. Shimeld, 2002, Geochemistry and sequence stratigraphy of regional Upper Cretaceous limestone units, offshore eastern Canada: Organic Geochemistry, v. 33, p. 1559–1569.

Wielens, H. J. B. W., C. Jauer, and G. L. Williams, 2004, Is there a viable petroleum system in the Carson and Salar Basins, offshore Newfoundland?: Journal of Petroleum Geology, v. 29, no. 4, p. 303–326, doi:10.1111/j.1747-5457.2006.00303.x.

Withjack, M. O., and R. W. Schlische, 2005, A review of tectonic events on the passive margin of eastern North America, *in* P. Post, ed., Petroleum systems of divergent continental margin basins: 25th Bob S. Perkins Research Conference, Gulf Coast Section of SEPM, p. 203–235.

Wygrala, B. P., 1989, Integrated study of an oil field in the southern PoBasin, northern Italy: Bericht der Kernforschungsanlage Jülich Nr. 2313, University of Cologne, Germany, p. 217.

Yassir, N. A., and A. L. Rogers, 1994, Hydrodynamics and overpressuring in the Jeanne d'Arc Basin, offshore Newfoundland, Canada: Possible implications for hydrocarbon exploration: Bulletin of Canadian Petroleum Geology, v. 42, no. 2, p. 266–267.

17

Guthrie, J., C. Nino, and H. Hassan, 2012, Integrating geochemistry, charge rate and timing, trap timing, and reservoir temperature history to model fluid properties in the Frade and Roncador fields, Campos Basin, offshore Brazil, *in* K. E. Peters, D. J. Curry, and M. Kacewicz, eds., Basin Modeling: New Horizons in Research and Applications: AAPG Hedberg Series, no. 4, p. 293–315.

Integrating Geochemistry, Charge Rate and Timing, Trap Timing, and Reservoir Temperature History to Model Fluid Properties in the Frade and Roncador Fields, Campos Basin, Offshore Brazil

John Guthrie and Christian Nino

Hess Corporation, Houston, Texas, U.S.A.

Hassan Hassan

University of South Carolina, Columbia, South Carolina, U.S.A.

ABSTRACT

Understanding the distribution of oil quality and its impact on the development of deep-water reservoirs is a major challenge in many offshore basins of Brazil. Traditional geochemical approaches have used bulk properties (API gravity, viscosity, and sulfur content) and the biomarker compositions of oils to resolve the effects of source rock facies, thermal maturity, and biodegradation on oil quality in the present-day reservoir. These techniques, however, cannot fully resolve the effects of hydrocarbon charge timing, charge rate, timing of trap formation, and reservoir temperature history on the quality of the oil. In the Roncador and Frade fields, offshore Brazil, lacustrine-derived oils from Upper Cretaceous (Maastrichtian) and lower Tertiary (Oligocene–Miocene) reservoirs have gravities ranging from 14 to 33° API. In Upper Cretaceous (Maastrichtian) reservoirs of the Roncador field, better quality light oil (average, 28° API) occurs in the northeastern part, and mostly heavy oil (average, 17° API) is encountered in the southwestern part. The Frade field to the west of Roncador also contains heavy oil (16–19° API) but in shallower lower Tertiary (Oligocene–Miocene) reservoirs.

Geochemical analyses have identified the depletion of n-alkanes and the presence of 25-norhopanes (demethylated hopanes) in varying proportions in oils from the Frade and Roncador fields of the Campos Basin, offshore Brazil, indicating a complex history of biodegradation and mixing from at least two hydrocarbon charges in the reservoir. This study uses both one-dimensional and multisurface thermal models in the area to

DOI:10.1306/13311443H43476

help determine charge histories for the source rocks and reservoir temperature histories for the reservoirs. These results are used to evaluate the effects of charge and reservoir temperature histories and biodegradation on the ultimate composition and quality of reservoired oils. An interactive biodegradation tool in Trinity software is used to predict the API gravity, and the results are constrained by the geology and the geochemical composition of the present-day fluids in the reservoir. Several examples of charge rate and timing, trap timing, and temperature history are presented for parts of the Roncador and Frade fields to illustrate the importance of these factors on controlling the quality of oil in the present-day reservoir.

INTRODUCTION

Understanding the distribution of oil quality and its impact on the development of deep-water reservoirs is a major challenge in offshore basins of Brazil. Several comprehensive publications (Wenger et al., 2002; Katz and Robison, 2006; Larter et al., 2006) have described the factors that control oil quality, including the type of source rock and its thermal maturity, charge history, and extent of biodegradation. The original composition of oil is controlled primarily by the characteristics of the source rock, including organic matter type, depositional environment, and the level of thermal maturity. These characteristics influence the bulk properties of the oil such as API gravity, viscosity, gas-oil ratio (GOR), and sulfur contents (Wenger et al., 2002).

In deep-water environments, biodegradation can be a major risk in shallow low-temperature (<80°C) reservoirs resulting in poor oil quality that decreases reservoir deliverability, significantly impacting economics. The level of biodegradation is controlled primarily by reservoir temperature (commonly <80°C), the availability of nutrients (nitrates and phosphates), and formation water salinity, all of which influence the degradation rate (Wenger et al., 2002; Larter et al., 2006). Increased levels of biodegradation can have a major impact on oil quality by decreasing API gravity and increasing viscosity, sulfur, asphaltene, metal, and acid contents, thereby reducing reservoir deliverability and crude oil value (Wenger et al., 2002; Katz and Robison, 2006).

Traditional geochemical approaches have used bulk oil properties (API gravity, viscosity, and sulfur content) and the biomarker compositions of oils to resolve the effects of source rock facies, thermal maturity, and biodegradation on oil quality in the present-day reservoir. These methods are useful for mapping fluid properties but cannot be used to forward model or predict fluid properties predrill because they cannot fully resolve the other important effects of hydrocarbon charge timing, charge rate, timing of trap formation, and reservoir temperature history on the quality of the oil. Therefore, more recent studies (Cole et al., 2001; Yu et al., 2002; Wilhelms et al., 2004; Larter et al., 2006) have emphasized the importance of applying basin modeling approaches to better resolve the impact of hydrocarbon charge history (rate and volume) and reservoir temperature history on the present-day distribution of degraded and nondegraded oil in the reservoir.

The main objective of this study is to better understand the controls on the API gravity variations observed for oils in postsalt reservoirs of the Roncador and Frade fields offshore Campos Basin, Brazil. A newly developed interactive fluid property modeling tool (available in the Trinity software) that incorporates the effects of hydrocarbon charge history, timing of trap formation, and biodegradation is used to predict API gravity (Dzou and He, 2009). The modeled API gravity results are constrained by field data that include the geology and the present-day composition of fluids in the reservoir. The abundant amount of published geologic and geochemical data from the Roncador and Frade fields makes this area well suited for a test of this interactive approach.

RONCADOR AND FRADE FIELDS

The Roncador and Frade fields are located in the northern part of the Campos Basin about 130 km (81 mi) offshore from the town of Campos in southeastern Brazil (Figure 1). The Campos Basin is separated from Espirito Santo Basin to the north by the Vitoria High and from the Santos Basin to the south by the Cabo Frio High.

The Roncador field was discovered in 1996 by the 1-RJS-436A well in water depths between 1500 and 1900 m (6234 ft) (Figure 1). It is estimated to contain 9.2 billion bbl of oil in place with total reserves of 2.6 billion bbl of oil equivalent (BOE) accumulated in Upper Cretaceous (Maastrichtian) turbidite reservoirs (Rangel et al., 2003). Cross section BB′ (Figure 2) shows that the trap in the Roncador field is a combined structural and stratigraphic trap with an updip stratigraphic pinch-out to the west. The field is cut by several major faults that break the field into three main blocks: an upthrown southwestern block, an upthrown northern block, and a downthrown southeastern block (Rangel et al., 2003), as shown on

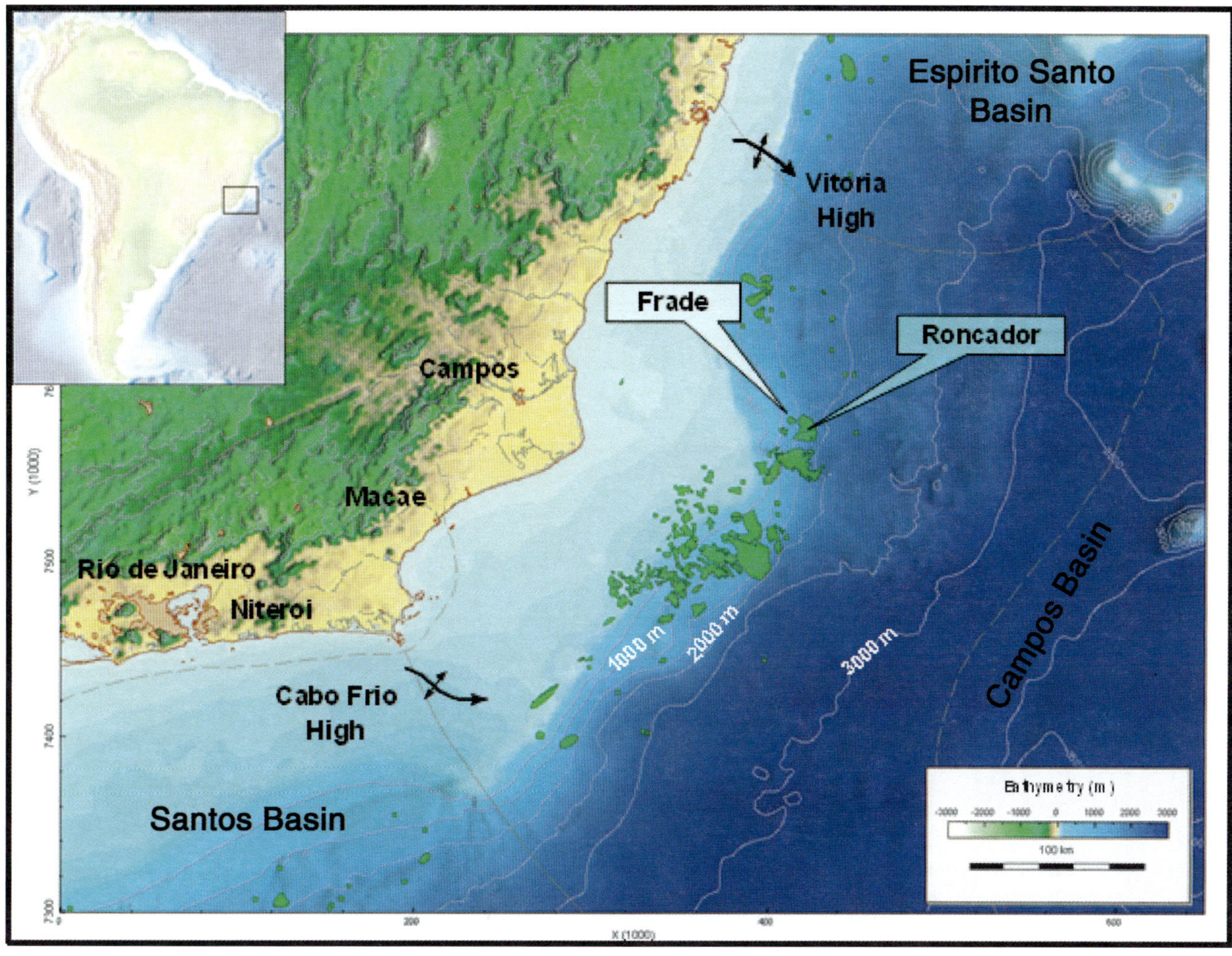

FIGURE 1. Map showing the locations of the deep-water Frade and Roncador fields located in water depths between 1000 and 2000 m (3281 and 6562 ft) (contour lines are bathymetry in meters) in the Campos Basin, offshore Brazil.

the structure map for the top of the Maastrichtian reservoir (Figure 2). One main fault (red star on Figure 2) separates the upthrown southwestern block from the downthrown southeastern part of the field. The movement of this fault in the Eocene and the formation of the downthrown trap in the southeastern block are critical elements for modeling the charge and biodegradation history of reservoired oil in the eastern part of the Roncador field. Oils in the shallower Upper Cretaceous reservoirs of the upthrown southwestern block are heavy, with gravities ranging between 14 and 19° API (average, 17° API). In contrast, oils in the deeper Cretaceous reservoirs of the northern and southeastern blocks contain lighter oil, with gravities ranging between 22 and 33° API (average, 28° API).

The Frade field was discovered in 1986 by the 1-RJS-366 well in water depths between 900 and 1300 m (2953–4265 ft). It contains an estimated one billion bbl of heavy oil (average, 18° API) in place with estimated total reserves between 200 and 300 million bbl of oil equivalent. The main reservoirs in the Frade field are much shallower in Oligocene to Miocene turbidites (Figure 2).

REGIONAL GEOLOGY AND OIL QUALITY DISTRIBUTION IN THE CAMPOS BASIN

The stratigraphy in the Campos Basin can generally be subdivided into rift, transitional, and marine megasequences (Figure 3) that are defined by the tectonic evolution of the basin (Guardado et al., 2000). The rift megasequence is composed of nonmarine strata deposited in fresh water and brackish to saline lacustrine environments. Geochemical data from oils indicate that the main source rocks in the Campos Basin are fresh

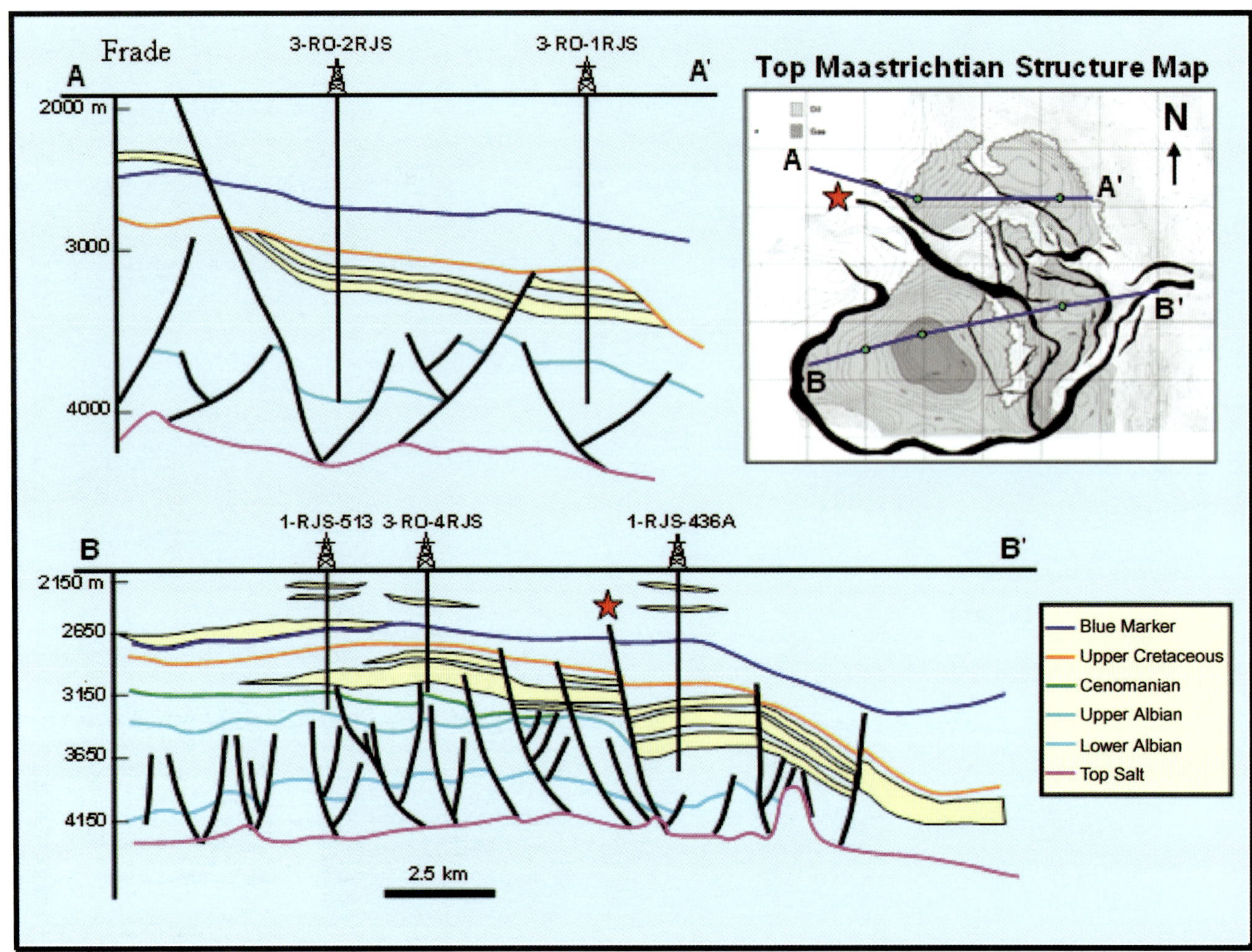

FIGURE 2. Schematic cross sections AA′ and BB′ show the distribution of the main reservoirs in the Roncador and Frade fields. The reservoirs in the Roncador field are Upper Cretaceous (Maastrichtian) turbidites shown in yellow below the orange line. In the Frade field, the reservoirs are Oligocene to Miocene turbidites depicted on the western edge of AA′ above the Blue Marker. The red star marks the main fault that segregates the Roncador field.

water to brackish and saline lacustrine shales from the Lagoa Feia Formation (Mello and Maxwell, 1990; Mello et al., 1994, 2000). The transitional megasequence was deposited during the Aptian and contains mostly carbonates, halite, and anhydrite that subdivide the presalt nonmarine rift megasequence from the postsalt marine megasequence. The main reservoirs in the Campos Basin occur throughout the postsalt marine megasequence ranging in age from Albian to Miocene (Figure 3). In the Roncador field, the reservoirs are Upper Cretaceous Maastrichtian turbidites, and in the Frade field, Oligocene to Miocene turbidites (Santos et al., 1999; Guardado et al., 2000; Rangel et al., 2003).

The degree of biodegradation commonly decreases with increasing reservoir temperature, resulting in an increase in the API gravity (Wenger et al., 2002; Larter et al., 2006). Some studies (Yu et al., 2002; Larter et al., 2003; Huang et al., 2004) have observed that API gravity varies with reservoir depth such that with increasing depth and temperature, the API gravity increases. When observed oil properties correlate to present-day reservoir depth and temperature, it is interpreted that the reservoirs must have been recently charged with biodegradation and charging occurring close to present-day depths of burial (Larter et al., 2006). Figure 4 is a map of the Campos Basin fields classified by reservoir age (highlighted by different colors on the map). The distribution of API gravity in the Campos Basin is such that each field, regardless of reservoir age, shows a wide range in API gravity (Figure 4). Generally, we would expect that low API gravity, heavy oils (<20° API) would be more predominant in the shallow low-temperature reservoirs (<80°C), where biodegradation occurs. However, in the Campos Basin, large variations in API gravity are observed independent of reservoir age and depth. High–API gravity nondegraded oils are found at the same depths as

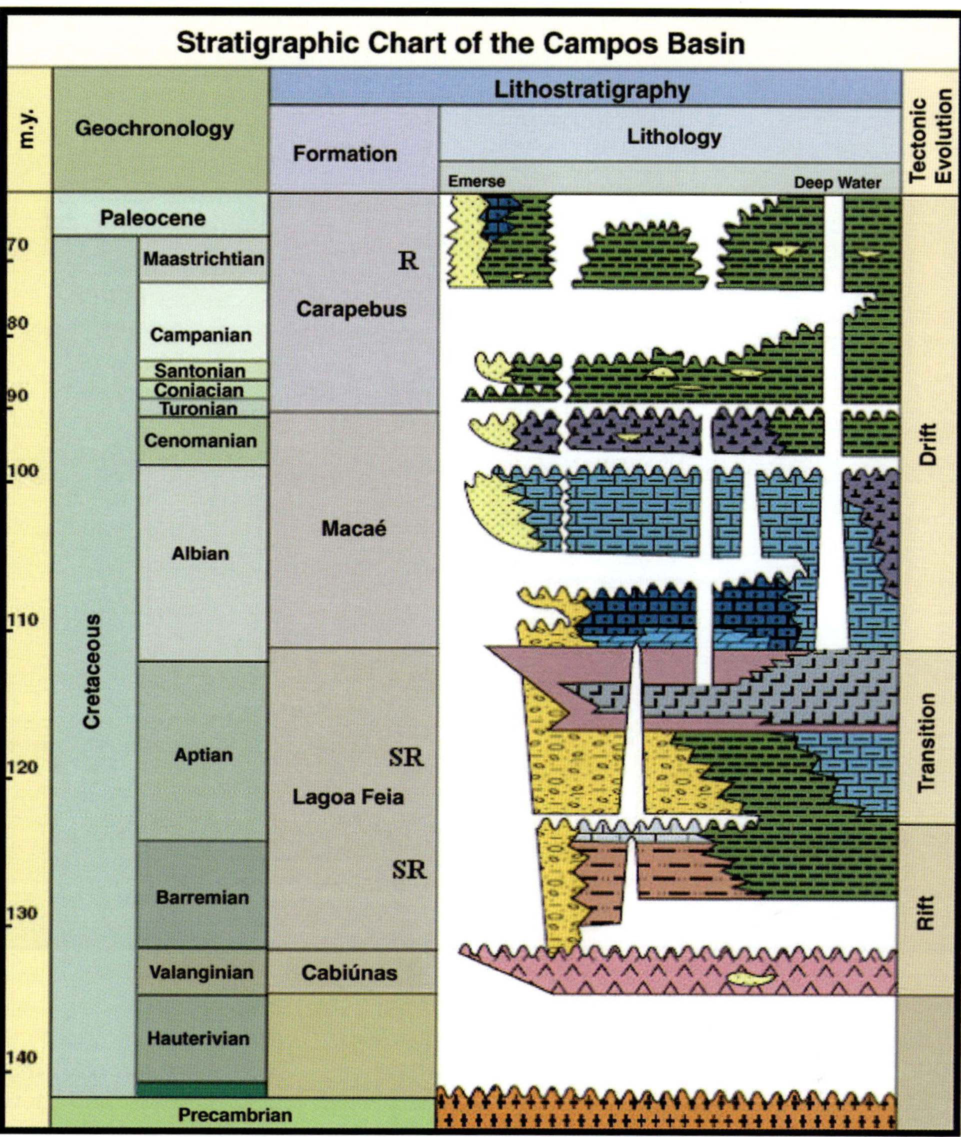

FIGURE 3. Stratigraphic chart for the Campos Basin showing the stratigraphic position of presalt source rocks (SR) and postsalt reservoirs (R) in the Frade and Roncador fields (modified from Guardado et al., 2000).

low–API gravity heavily degraded oils (Figure 5; Frade and Roncador data shown as colored circles). These data indicate that the distribution of API gravity in the Campos Basin is not simply controlled by present-day depth and temperature of the reservoir that influence the degree of biodegradation. Similar observations were made by Larter et al. (2003) for North Sea oils, where high-gravity nondegraded oils are found at the same depth as heavily degraded oils. According to Larter et al. (2003), this observation suggests that the present-day reservoir-temperature history cannot fully account for the complex history of biodegradation and charge that results in a mixture of altered and unaltered oils in the reservoir. This chapter will address how the history of hydrocarbon charge and reservoir temperature is modeled through time to better determine their impact on biodegradation and the final present-day distribution of altered and unaltered oils in the reservoir.

INTERACTIVE APPROACH FOR MODELING FLUID PROPERTIES

The most important first-order controls on the compositional variation of reservoir fluid properties are reservoir temperature history that influences the level of biodegradation and the charge history (rate and volume) that tracks the influx of fresh nondegraded oil into the reservoir (Katz and Robison, 2006; Larter et al., 2006). Several studies (Larter et al., 2003; 2006) have documented that the rate of biodegradation in the reservoir increases to a maximum as the reservoir temperature decreases to temperatures near 40°C. The 80°C temperature limit has been used as a cutoff for biodegradation, implying that biodegradation rate is zero at temperatures above 80°C because of the absence of microorganisms or the inability of microorganisms to metabolize hydrocarbons

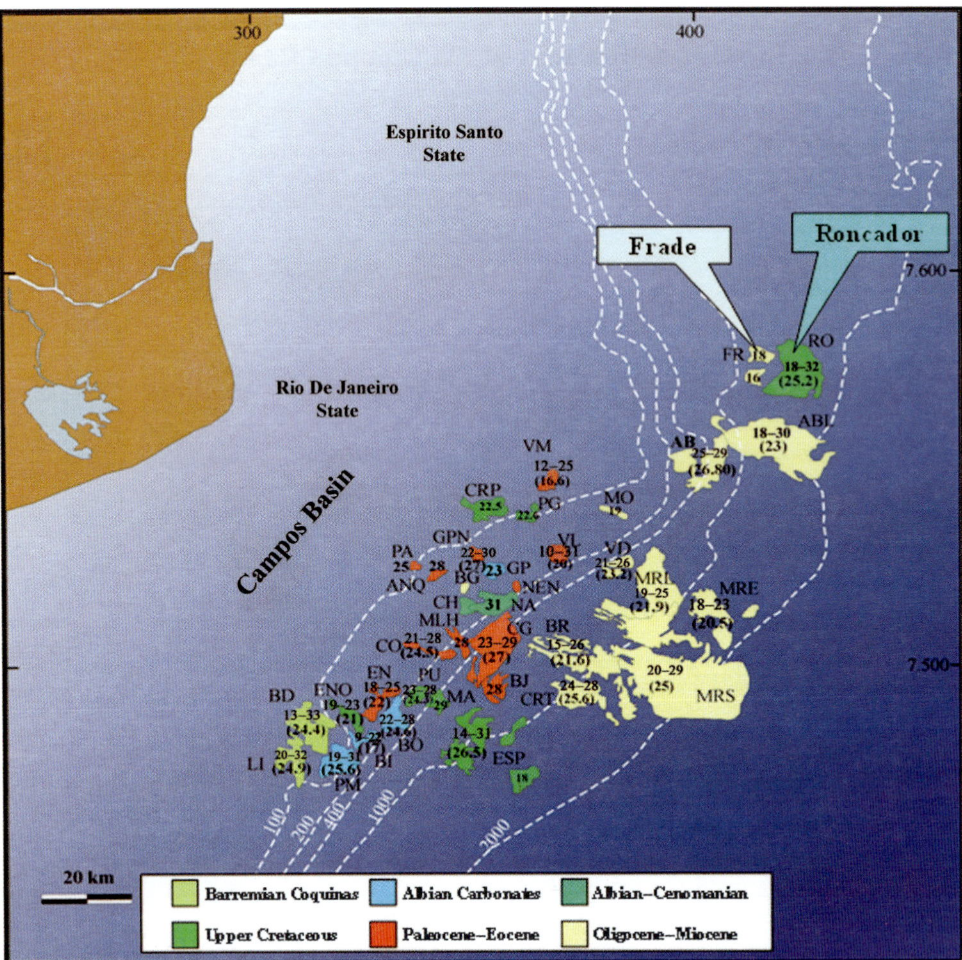

FIGURE 4. Map showing Campos Basin fields classified by reservoir age. The range and average (shown in parentheses) API gravity values are shown for each field.

at these temperatures (Wenger et al., 2002; Larter et al., 2006).

Yu et al. (2002) and Cole et al. (2001) used the empirically based biodegradation index (BDI) as a simplified time-temperature history of the reservoir related to the timing of charge to predict API gravity. The BDI has some applications for predicting oil properties (API gravity), but it cannot fully resolve the effects of variable charge histories and the mixing of paleobiodegraded oils with oils from a more recent fresh charge. Later, it was shown that basin modeling outputs such as reservoir-temperature history, charge history (Wilhelms et al., 2004; Baudino et al., 2008), and also trap and oil column geometries and nutrient supply influence the biodegradation rate (Larter et al., 2006). However, in many basins, it has been observed that for a given depth, a wide range of API gravity is observed in the reservoir, suggesting that reservoir temperature history alone cannot account for the variations observed (Wilhelms et al., 2004; Larter et al., 2006).

Figure 6 is a conceptual diagram illustrating the interactive approach used in this chapter to better characterize the important factors that control the wide ranges of API gravity observed in Campos Basin fields. The modeled approach interactively tracks the effects of source rock type, thermal maturity, and charge history (rate and volume) over a fetch area to model the cumulative

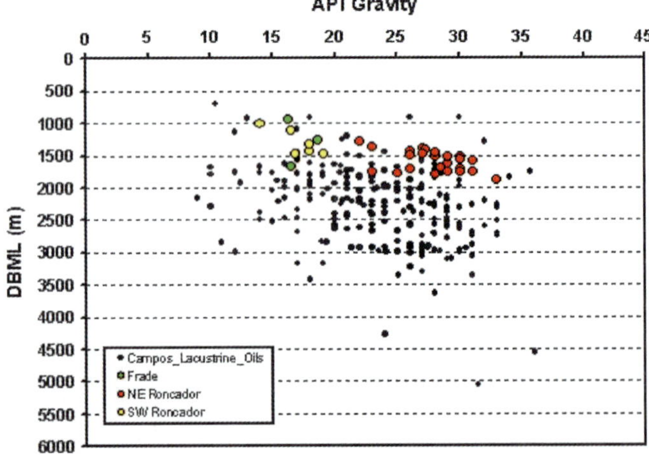

FIGURE 5. Plot of API gravity for lacustrine-derived oils from the Campos Basin versus depth below mud line (DBML) in meters. Colored circles represent oil samples from the Frade and Roncador fields.

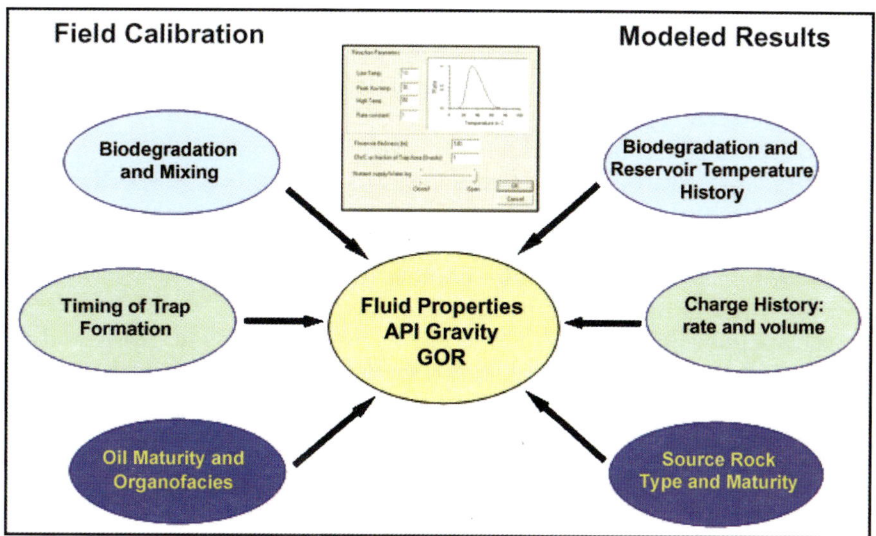

FIGURE 6. Conceptual diagram illustrating the interactive approach used to interpret the important factors that control the wide ranges of API gravity observed in the Campos Basin. GOR = gas-oil ratio.

GOR and API gravity of expelled fluids through time (Dzou and He, 2009). The approach also considers the effects of temperature history, charge rate, and biodegradation on the alteration of hydrocarbons in the reservoir (Dzou and He, 2009; Guthrie et al., 2009). The results of the model are calibrated and audited with geochemical and geologic observations from the field that indicate source rock type (organofacies) and the level of thermal maturity from oils, timing of trap formation through geologic evaluation, and the level of biodegradation and mixing of oils through the examination of biomarkers.

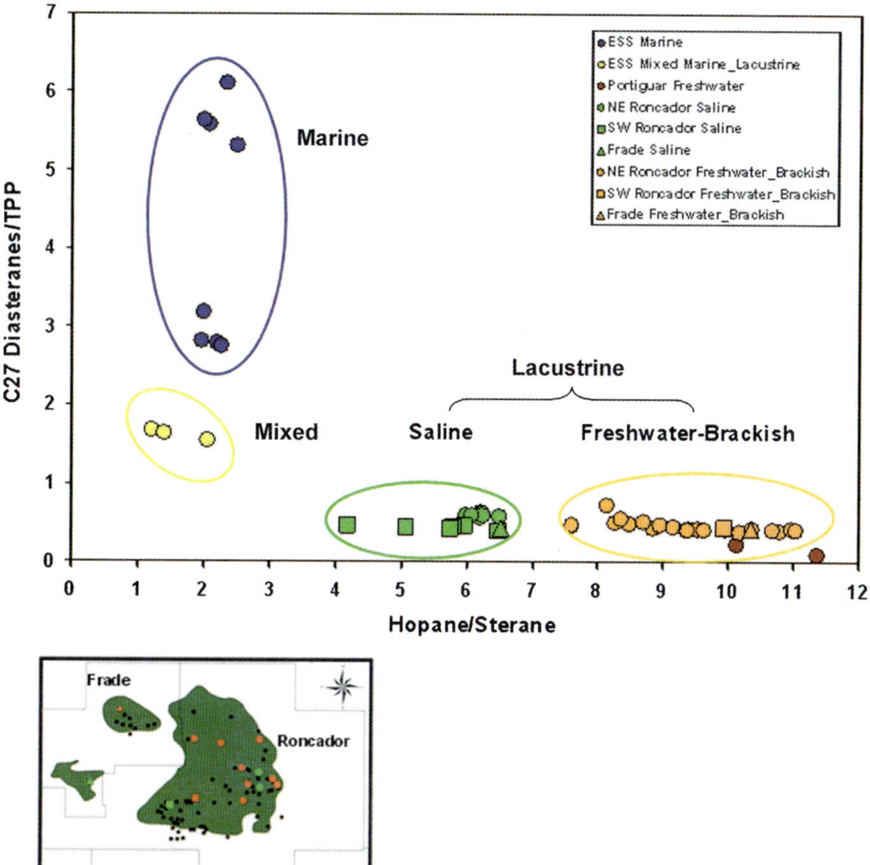

FIGURE 7. Plot of the C_{27} diasterane/tetracyclic polyprenoid ([TPP] m/z 259) versus the hopane/sterane ratio used to distinguish lacustrine oils from marine and mixed marine/lacustrine oils. Map inset shows the locations of oil types (green circles are saline lacustrine oils and orange circles are freshwater to brackish lacustrine oils) in the Frade and Roncador fields.

The result is a prediction of the fluid properties using a biodegradation tool developed for Trinity software to predict the API gravity of the oil (Dzou and He, 2009). In this tool, reservoir temperature history exerts the primary control on degradation rate such that the maximum biodegradation rate occurs at a temperature of 30°C and decreases rapidly to zero at temperatures more than 80°C (reaction parameters on Figure 6). The tool also has the option for adjusting the oil-water contact (OWC) area and the input of nutrients into the system, which also have been identified as important controls on the degradation rate (Larter et al., 2003; 2006). However, for this chapter, the OWC area and nutrient supply parameters were left on the default values (reaction parameters on Figure 6).

IDENTIFICATION OF SOURCE ROCKS

To identify the effective source rocks in the study area, some commonly used biomarker parameters derived from the gas chromatography–mass spectrometry analyses of oils were used to distinguish between marine and nonmarine sources. Figure 7 illustrates how the type of source rocks in the Campos Basin are identified using the hopane/sterane and the C_{27} diasterane/C_{30} tetracyclic polyprenoid (TPP) ratios, two commonly used biomarker parameters used to differentiate marine from nonmarine oils in the offshore basins of Brazil. The C_{30} tetracyclic polyprenoids can be very useful for identifying nonmarine source facies and are very abundant in oil samples derived from low-salinity freshwater to brackish lacustrine environments (Holba et al., 2000, 2003). Oils derived from freshwater to brackish lacustrine source rocks also tend to have higher hopane/sterane ratios because lacustrine environments commonly have lower sterane contents than marine samples (Mackenzie et al., 1984; Mello et al., 1988; Holba et al., 2003).

For this study, presalt-derived lacustrine oils in the Campos Basin can be easily distinguished from postsalt marine oils using a plot of the C_{27} diasteranes/TPP ratio (calculated from the C_{27} diasterane and TPP peaks in the m/z 259 fragmentogram) versus the hopane/sterane ratio (Figure 7). The lacustrine-derived oils from the Roncador and Frade fields have low C_{27} diasterane/TPP ratios and high hopane/sterane ratios (Figure 7). In contrast, oils derived from postsalt marine source rocks contain higher C_{27} diasterane/TPP and lower hopane/sterane ratios. Based on the hopane/sterane ratios, the lacustrine oils can be further classified as saline and freshwater-brackish lacustrine (Figure 7). The freshwater-brackish lacustrine facies generally represent deeper synrift environments, whereas the saline lacustrine source rocks represent the shallower sag environments.

CONSTRUCTION OF THERMAL MODEL

Part of the interactive approach for predicting fluid properties involves the evaluation of source rock thermal maturity, charge history, and reservoir temperature history. To make this evaluation, a regional thermal model was constructed in Trinity using nine mapped horizons, eight one-dimensional thermal models calibrated to measured temperature and maturity data, inputs of lacustrine source rock properties from presalt wells, field closures and fetch areas derived from structure maps, and geochemical data from oils (Figure 8).

A cross section (Figure 9) through the Trinity model (AA′) shows the stratigraphic position of the Lagoa Feia lacustrine source rocks (upper and lower) in the presalt section and the location of a deep eastern hydrocarbon kitchen and shallower western hydrocarbon kitchen. The stratigraphic position of the Cretaceous Maastrichtian and Oligocene to Miocene reservoirs also is shown in the 1-RJS-436A well in Roncador, and the Oligocene to Miocene reservoirs, in the 1-RJS-366 well in the Frade field.

For this model, the Roncador field is divided into two blocks: a southwestern block and northeastern block by a major fault (Figure 8) that shows late movement that occurs into the Eocene (~42 Ma). Charge histories are modeled for these individual blocks using block-specific fetch areas derived from the structure maps of the source intervals. Biodegradation modeling is performed for block-specific closures on the structural top of each reservoir.

THERMAL EVOLUTION OF LAGOA FEIA SOURCE ROCKS

Standard thermal stress (STS), which is a method for normalizing subsurface temperature to a standard heating rate of 2°C Ma^{-1} (Pepper and Corvi, 1995) was used to model the thermal evolution of source rocks. The STS maps were constructed for various times (Ma) in the Trinity model to show the thermal evolution of the upper and lower Lagoa Feia source rocks in the area of the model. For each selected time (Ma), a paleo–cross section (AA′) illustrates the extent of progressive burial of the source rocks and the movement of salt in the model area. In general, the colors on the STS maps represent the different thermal maturity levels over the model area. Blue colors represent immature areas (no hydrocarbon generation), green colors represent the early to peak oil window stage of thermal maturity, and the orange to red colors represent the late stage oil to gas stage of thermal maturity.

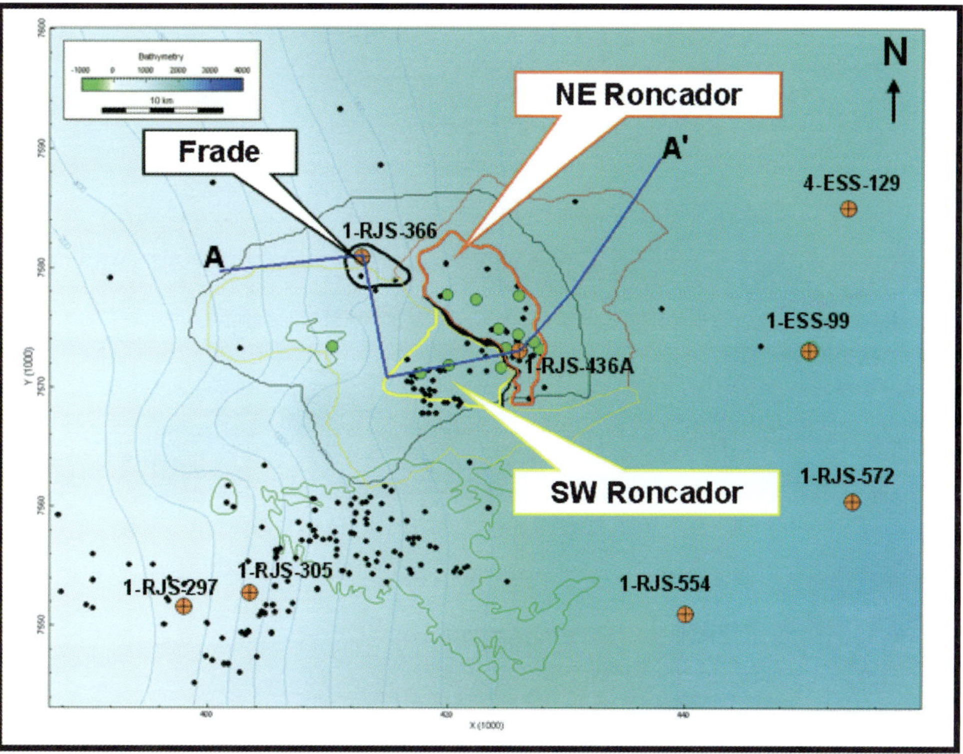

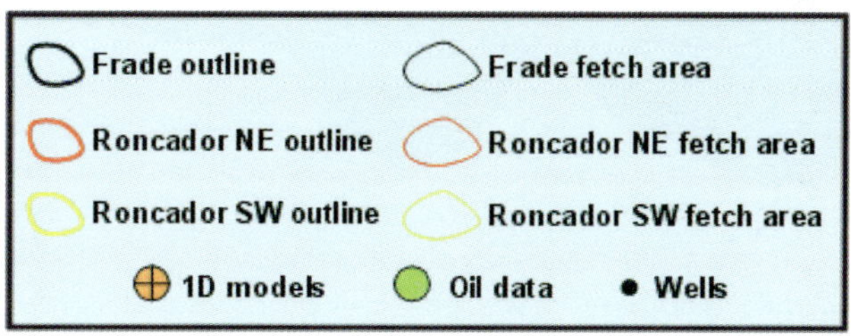

FIGURE 8. Map showing the inputs used for the construction of a thermal model. Inputs include eight one-dimensional calibrated thermal models (orange circles with crosshairs), geochemistry on 34 oils from 14 wells (green circles), field fetch areas and trap closures for the Frade, southwest Roncador, and northeast Roncador fields.

Figure 10A indicates that the upper Lagoa Feia source rock at 90 Ma is immature over the model area. The lower Lagoa Feia source rock (Figure 10B) begins to generate hydrocarbons in limited areas of the hydrocarbons kitchen to the southeast and north of the Roncador field. At 90 Ma, the main Upper Cretaceous Maastrichtian

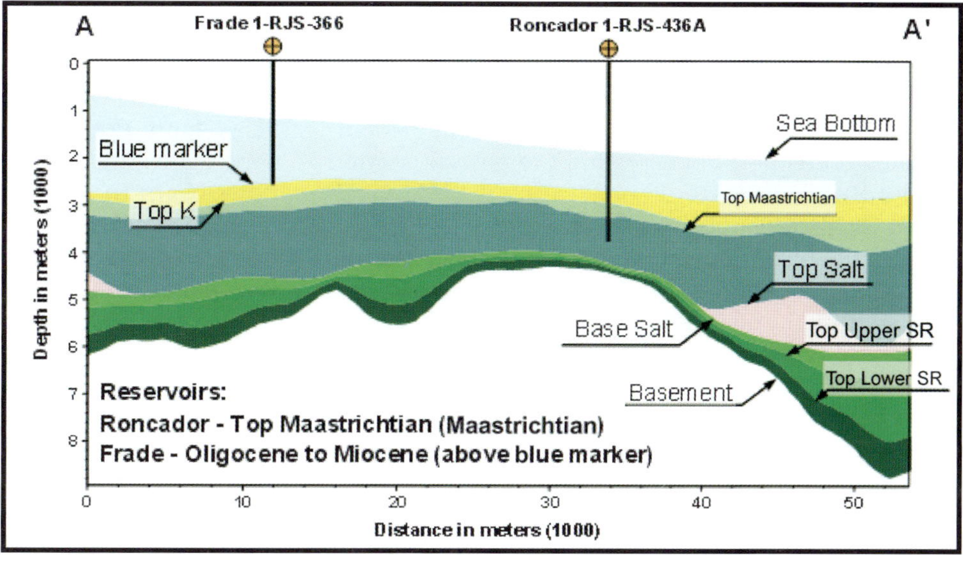

FIGURE 9. Cross section (AA′) shows the depths of nine horizons and the stratigraphic position of the main reservoirs and source rocks (SR) (upper and lower) used in the Trinity thermal model.

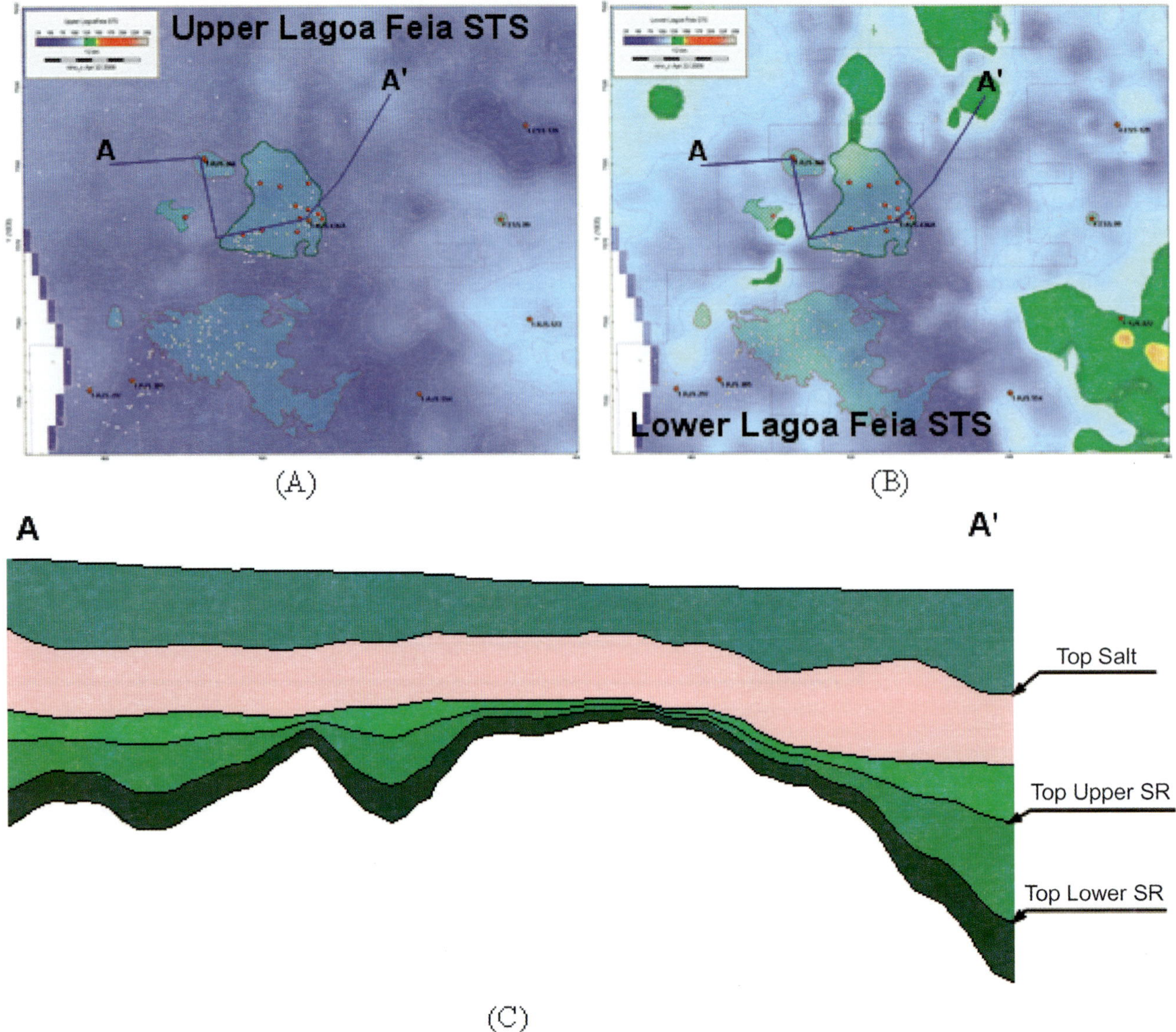

FIGURE 10. Standard thermal stress (STS) maps for 90 Ma constructed in Trinity for the upper (A) and lower (B) Lagoa Feia source rocks (SR). Blue represents areas where the SR is immature, green represents early to peak oil window, and orange to red represents late stage oil to gas maturity. The 90 Ma paleo–cross section (C) illustrates the burial history and formation of hydrocarbon kitchens in the area.

reservoirs have not been deposited and a layer of continuous salt overlies the source rock kitchen, preventing presalt charge from accessing the postsalt section (Figure 10C).

At 70 Ma, the upper Lagoa Feia source rock begins to enter the oil window in the kitchen area east of the Roncador field (Figure 11A). The lower Lagoa Feia source rock is within the oil generation window in hydrocarbon kitchens to the east and northwest of the Roncador field (Figure 11B). The Upper Cretaceous Maastrichtian reservoir begins to form at 70 Ma, but the layer of continuous salt still prevents presalt charge from accessing the postsalt section (Figure 11C).

At approximately 50 Ma, the paleo–fetch area for the Roncador field is primarily drawing hydrocarbon charge in the oil window from both the upper Lagoa Feia (Figure 12A) and the lower Lagoa Feia (Figure 12B) source rocks from the deeper eastern kitchen. Salt windows begin to form that allow the presalt lacustrine charge to access the overlying Upper Cretaceous Maastrichtian reservoirs in a paleo–fetch area defined for the Roncador field (Figure 12C).

Between 50 and 30 Ma (~42 Ma), fault movement occurs along a major fault in the Roncador field, creating a separation of the field into southwest and northeastern blocks and forming a downthrown structural trap in the northeastern block (Rangel et al., 1998; fault shown on Figure 13A, B). At 30 Ma, the northeastern Roncador block continues to receive hydrocarbon charge from both the upper (Figure 13A) and lower Lagoa Feia

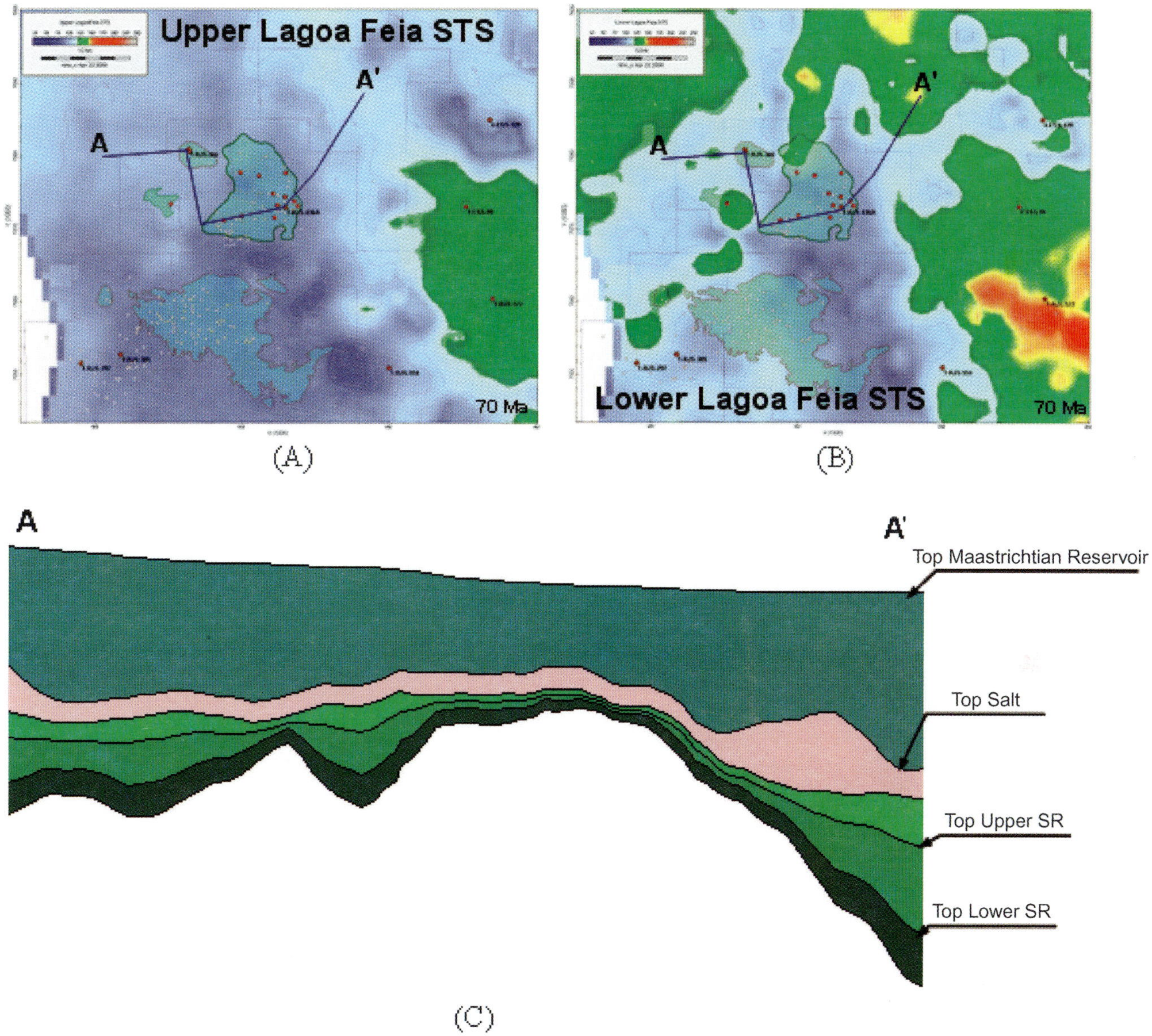

FIGURE 11. Standard thermal stress (STS) maps for 70 Ma constructed in Trinity for the upper (A) and lower (B) Lagoa Feia source rocks (SR). Blue represents areas where the SR is immature, green represents early to peak oil window, and orange to red represents late stage oil to gas maturity. The 70 Ma paleo–cross section (C) illustrates the burial history and formation of hydrocarbon kitchens in the area.

(Figure 13B) source rocks primarily from the eastern hydrocarbon kitchen through a fetch area (red outline on Figure 13A, B) that encompasses the eastern block. The southwest Roncador block received hydrocarbon charge primarily from the lower Lagoa Feia source rock from the western kitchen in its own defined fetch area (yellow outline on Figure 13B).

From approximately 18 Ma to present day, Frade received hydrocarbon charge in the oil window from both the upper and lower Lagoa Feia source rocks in the western hydrocarbon kitchen (Figure 14A). Biomarker analyses of the Frade oils suggest that the hydrocarbon charge from the upper Lagoa Feia saline lacustrine source rock was more significant in the western hydrocarbon kitchen. A late high-maturity (late-stage oil to gas) hydrocarbon charge occurs in the fetch area for the northeastern Roncador block predominantly from the deeper lower Lagoa Feia (Figure 14B) freshwater to brackish lacustrine source rock in the eastern hydrocarbon kitchen. This hydrocarbon charge occurs after late movement (beginning ~42 Ma) of the major fault that separates the Roncador field into distinct blocks. The fault movement creates a downthrown trap in the northeastern block of the Roncador field that accumulates the late high-maturity charge from the eastern kitchen.

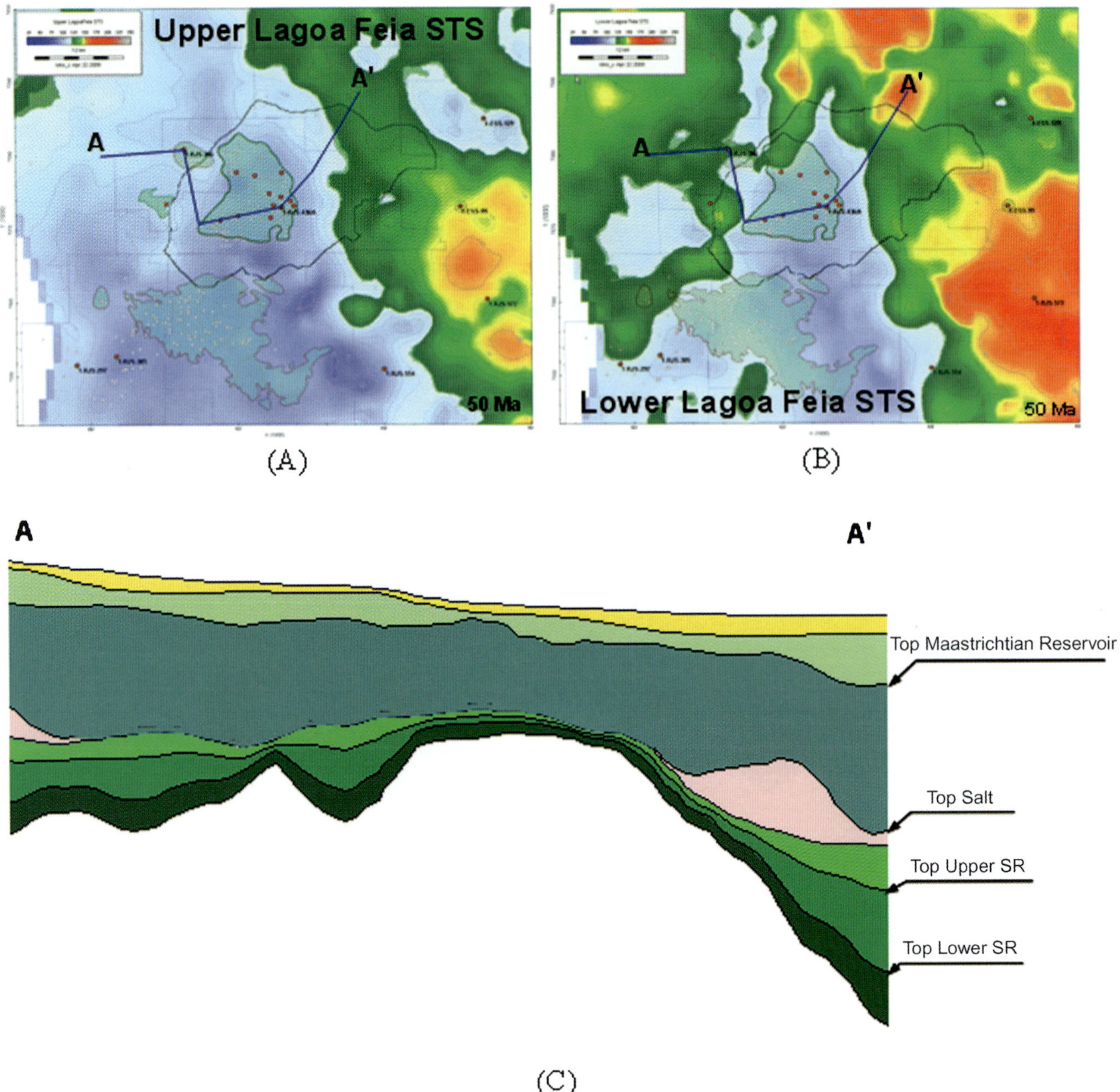

FIGURE 12. Standard thermal stress (STS) maps for 50 Ma constructed in Trinity for the upper (A) and lower (B) Lagoa Feia source rocks (SR). Blue represents areas where the SR is immature, green represents early to peak oil window, and orange to red represents late stage oil to gas maturity. The 50 Ma paleo–cross section (C) illustrates the burial history and formation of hydrocarbon kitchens in the area.

A two-dimensional present-day (0 Ma) migration simulation (along cross section AA′) schematically illustrates the generation and migration of hydrocarbons from the source rocks to the reservoirs (Figure 15). The main hypothesis is that fault movement occurs (~42 Ma) before late hydrocarbon charge (beginning ~18 Ma) and creates a downthrown trap in the northeastern part of Roncador field. The additional burial from the overlying Neogene section (blue layer in cross section AA′) advances the thermal maturity level of the lower Lagoa Feia source rock in the eastern hydrocarbon kitchen into the late-stage oil-to-gas generation window. The downthrown trap that forms in the northeastern Roncador block captures the late high-maturity charge coming primarily from the lower Lagoa Feia source rock from the eastern hydrocarbon kitchen. Late hydrocarbon charge of somewhat lower maturity (oil window) from the western hydrocarbon kitchen contributes to the charge in southwest Roncador and eventually migrates up to Frade field. Reservoir temperatures in Frade and

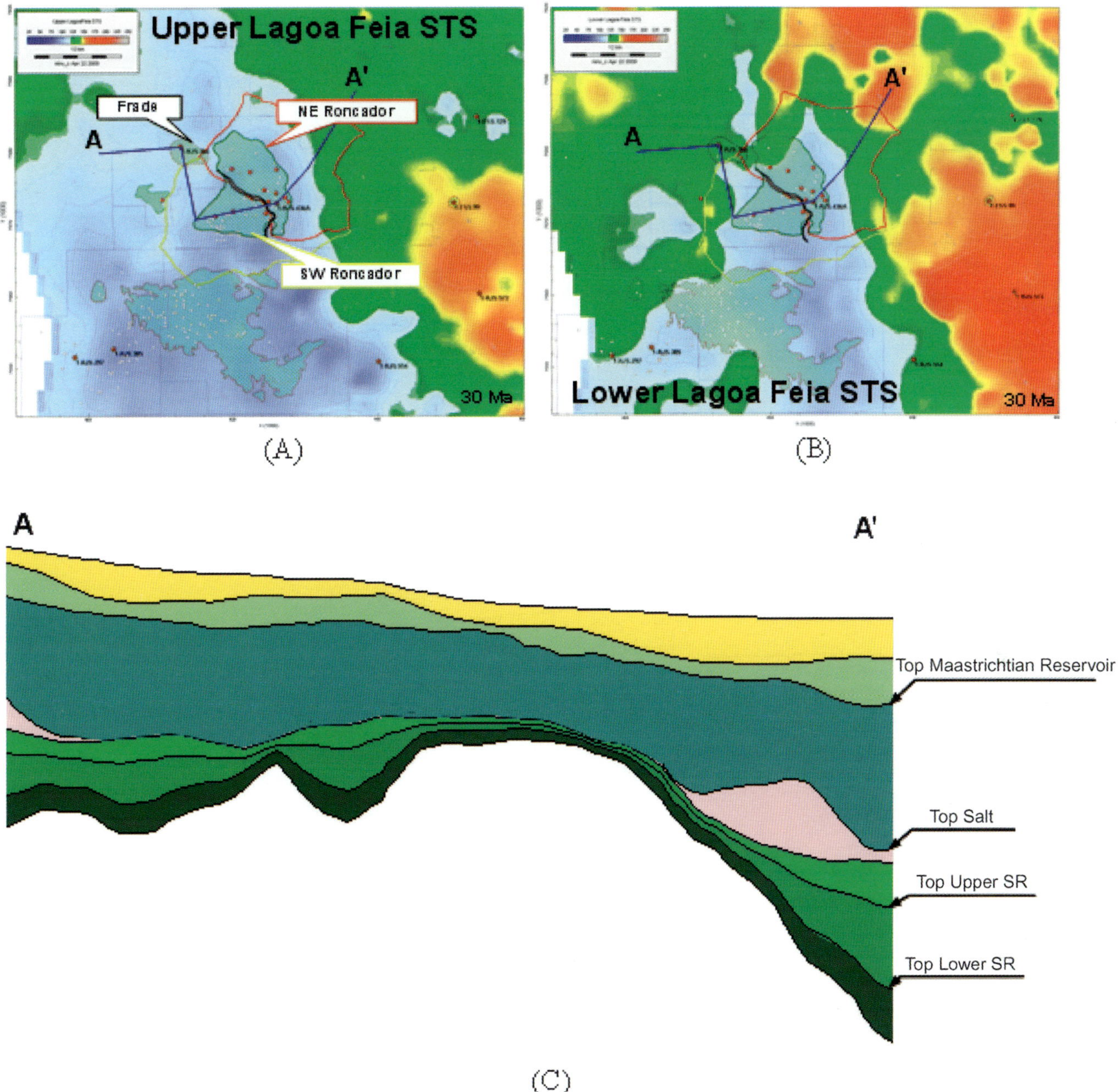

FIGURE 13. Standard thermal stress (STS) maps for 30 Ma constructed in Trinity for the upper (A) and lower (B) Lagoa Feia source rocks (SR). Blue represents areas where the SR is immature, green represents early to peak oil window, and orange to red represents late stage oil to gas maturity. The 30 Ma paleo–cross section (C) illustrates the burial history and formation of hydrocarbon kitchens in the area.

southwest Roncador are low, resulting in biodegradation after accumulation in the reservoir.

BULK PROPERTIES AND THERMAL MATURITY OF OILS

Generally, at the middle to late stage of thermal maturity, freshwater lacustrine source rocks generate hydrocarbons with slightly higher gravity (35–45° API), lower sulfur contents (<0.2 wt. %) than saline lacustrine source rocks (25–35° API; sulfur contents, 0.3–1.0 wt. %) (Wenger et al., 2002). Oils from the Frade and southwest Roncador fields that are derived from saline lacustrine source rocks tend to have higher relative amounts of sulfur contents (>0.5 wt. %) than samples from the eastern Roncador field (Figure 16). Samples in the Roncador field contain higher sulfur contents because of the mixing of oil derived from different lacustrine environments

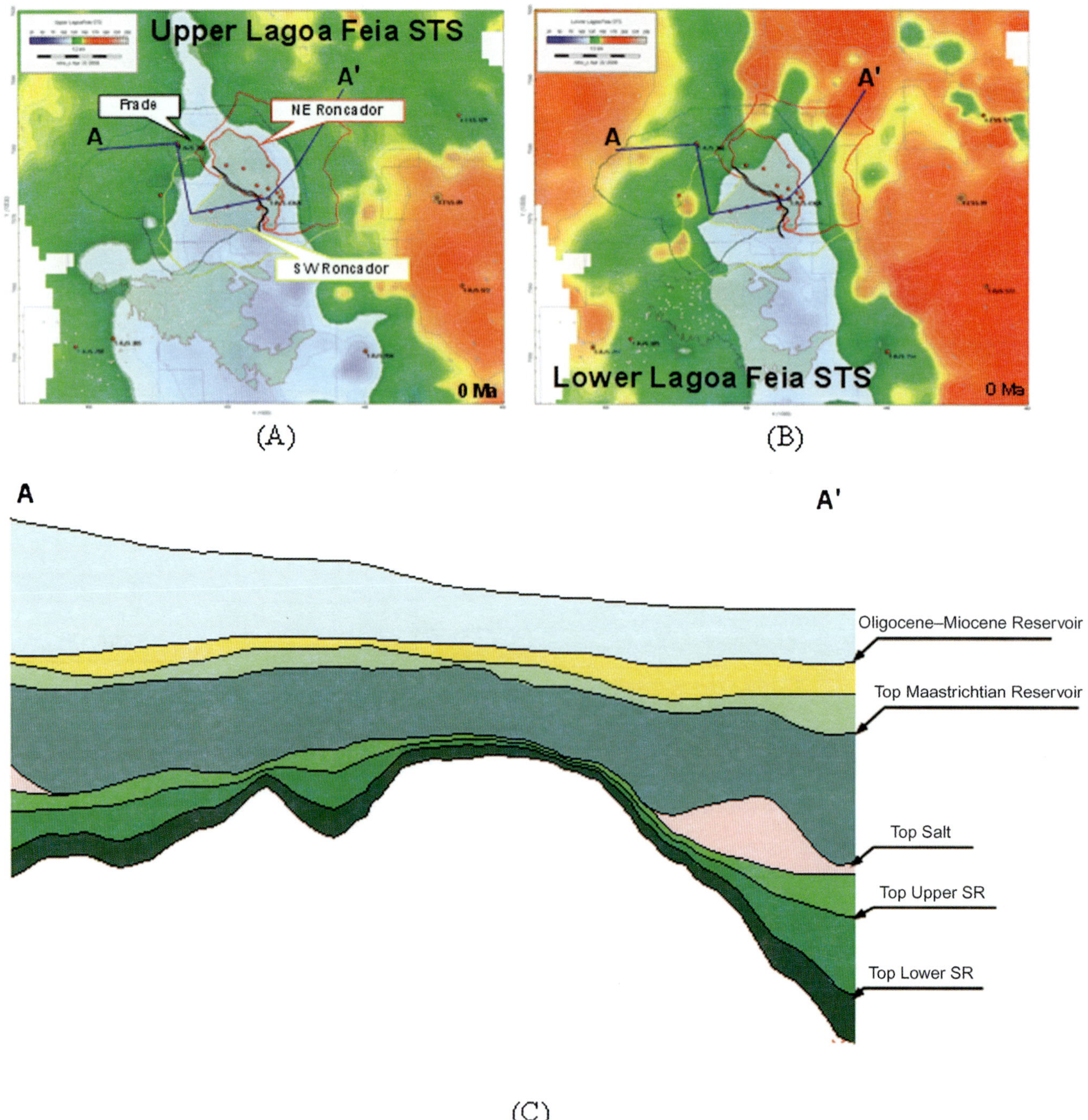

FIGURE 14. Standard thermal stress (STS) maps for 0 Ma (present-day) constructed in Trinity for the upper (A) and lower (B) Lagoa Feia source rocks (SR). Blue represents areas where the SR is immature, green represents early to peak oil window, and orange to red represents late stage oil to gas maturity. The 0 Ma paleo–cross section (C) illustrates the burial history and formation of hydrocarbon kitchens in the area.

(saline vs. fresh water) and also because of the effects of biodegradation. With increasing levels of biodegradation, sulfur content increases because bacteria preferentially consume the nonsulfurous compounds, leaving the remaining oil enriched in sulfur. The heavily degraded oils (lower API gravity oils) from the Frade and southwest Roncador fields contain higher sulfur contents because of the effects of increasing biodegradation in the reservoir. The higher maturity, higher API gravity oils from the eastern Roncador field (Figure 16) have lower sulfur contents because they are derived from brackish to freshwater lacustrine source rocks, and their level of thermal maturity is higher than the other oils.

Evidence for the thermal evolution of source rocks from different hydrocarbon kitchens is obtained from the thermal maturity of reservoired oils in the field

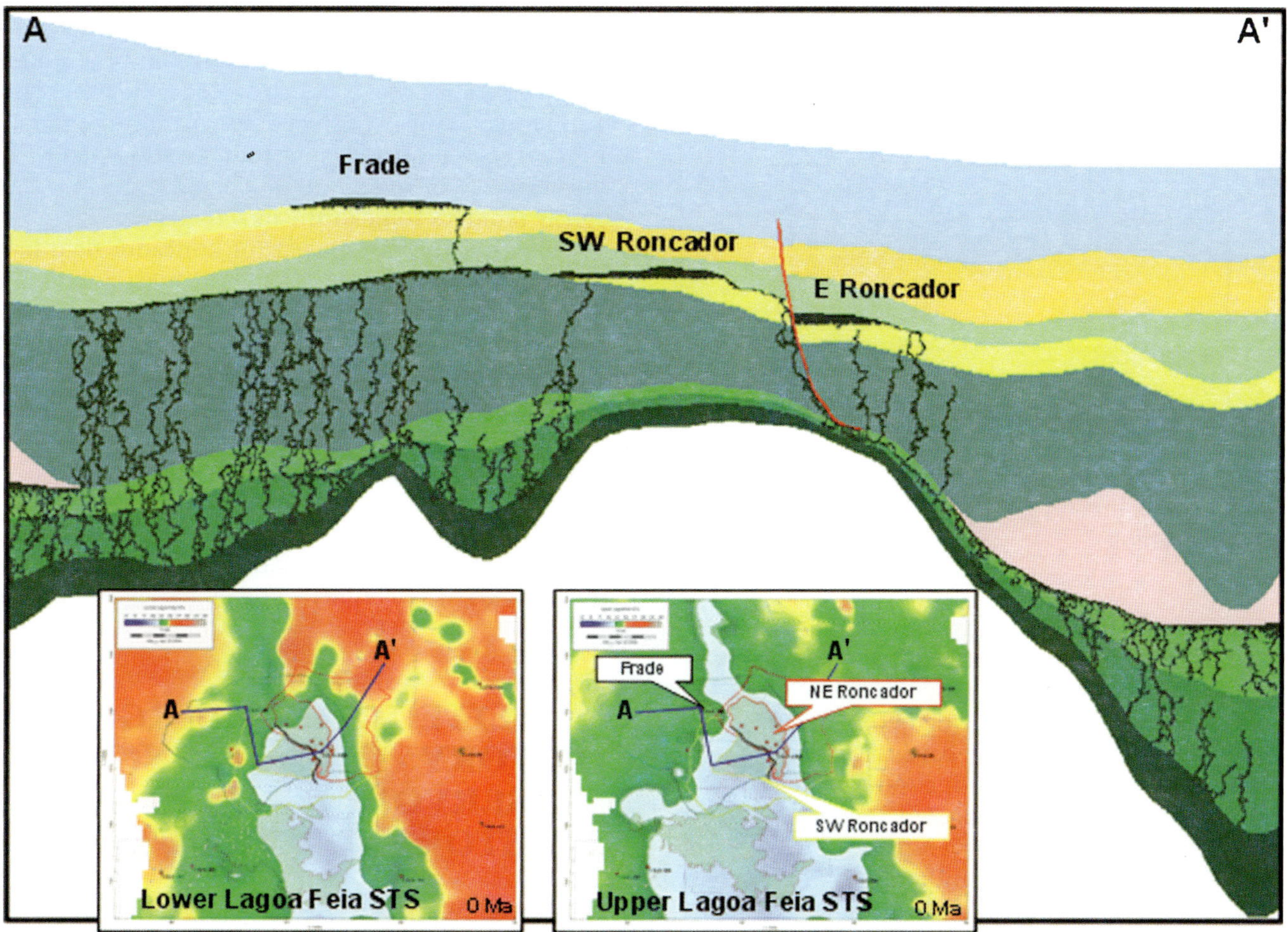

FIGURE 15. Cross section AA′ is a two-dimensional migration simulation constructed for the present day (0 Ma). Hydrocarbon generation from presalt lacustrine source rocks (upper and lower) occurs in a western and eastern kitchen and migrates both laterally and vertically to the overlying reservoirs in the Frade and Roncador fields. Standard thermal stress maps at 0 Ma show the maturity of the hydrocarbon charge within fetch areas for particular parts of the fields.

derived from biomarkers. Triaromatic (m/z 231) biomarker indices (A3 and A4 are defined on Figure 17) indicate a trend of increasing thermal maturity for the oils from the Roncador and Frade fields (Figure 17A). This increase in thermal maturity also has the effect of increasing the API gravity (Figure 17B). It is observed that the northeastern Roncador oils (red circles on Figure 17) generally have higher relative thermal maturities and also higher API gravities than oils from the Frade and southwest Roncador fields. This observation fits the modeled results that indicate the eastern Roncador trap receives a late higher maturity hydrocarbon charge from the lower Lagoa Feia source rock present in the more deeply buried eastern hydrocarbon kitchen. The southwest Roncador field receives an earlier lower maturity charge from the eastern hydrocarbon kitchen and also a later lower maturity charge from the western hydrocarbon kitchen. The Frade field receives a hydrocarbon charge primarily from the western hydrocarbon kitchen that contains source rocks with comparatively lower levels of thermal maturity.

CHARACTERIZING BIODEGRADATION AND DUAL-CHARGE HISTORY

The biomarker distributions in the oils studied suggests a complex history of dual hydrocarbon charge and paleo-biodegradation in the reservoirs at Roncador and Frade fields. Dual hydrocarbon charge can be deciphered by looking for the presence of 25-norhopanes (demethylated hopanes) and C_5 to C_{20} n-alkanes in the same oil. Demethylated hopanes (25-norhopanes) are generally considered to be the product of heavy biodegradation resulting in the bacterial demethylation of hopanes (Blanc and Connan, 1992; Chosson et al., 1992; Soldan et al., 1995; Peters et al., 1996; Wenger et al., 2002; Bennett et al., 2006). The C_5 to C_{20} n-alkanes are commonly consumed in the earliest stages of biodegradation and should not be found in great abundances in severely biodegraded oil (Wenger et al., 2002).

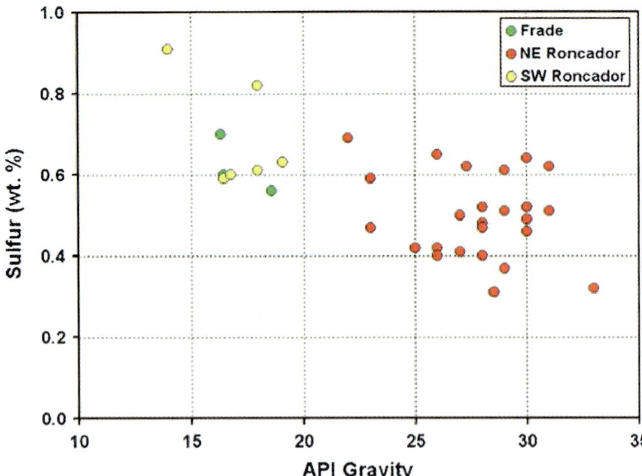

FIGURE 16. Plot of API gravity and sulfur content (wt. %) for the oil samples.

In the case of a dual hydrocarbon charge, an early charge of oil is postulated to accumulate in a shallow, cool reservoir (<80°C) for a long period of time and is severely biodegraded by the bacteria resulting in the formation of demethylated hopanes (Peters et al., 1996; Bennett et al., 2006). After bacterial alteration ceases, because of increased reservoir temperature, the severely degraded oil is then subsequently mixed with a second hydrocarbon charge of higher maturity "fresh" oil that brings in more n-alkanes. The oil in the present-day reservoir is, therefore, a mixture of early emplaced biodegraded oil and later higher maturity nondegraded oil. Depending on the more recent temperature history of the reservoir, the mixed oil also can be altered to some degree by a more recent phase of biodegradation. Careful characterization and examination of the biomarker distributions in the oils reveal the complex history of mixed hydrocarbon charges.

A plot (Figure 18) of API gravity versus the demethlyated hopane index (25NorH/[25NorH + C30H]) shows that most of the low–API gravity oils with higher relative amounts of the 25-norhopanes occur in the southwest Roncador and Frade fields. The presence of 25-norhopanes in the oils indicates a phase of severe biodegradation in the reservoir, resulting in the lowering of the API gravity. The 1-RJS-513 (2683 m [8802 ft]) oil from southwest Roncador field contains relatively high 25-norhopanes and a characteristic unresolved complex mixture (shown by the "hump" in its whole oil gas chromatogram [GC]) that is indicative of moderate to severe biodegradation (Figure 18). This oil also contains n-alkanes ranging from C_{10} to C_{26} that are the first class of compounds to be removed during the early phases of biodegradation. The fact that we observe the occurrence of both 25-norhopanes and n-alkanes in the same oil suggests a multiple-charge scenario where the oil accumulates in the reservoir and is biodegraded, removing all n-alkanes from the first charge and then forms 25-norhopanes at a severe stage of biodegradation. Then the oil mixes with a "fresh charge" that brings in a second charge of new n-alkanes that are subjected to yet another stage of moderate biodegradation that only slightly removes some of the new n-alkanes.

Oils from the eastern part of Roncador field have relatively higher gravities (23–33° API) and low amounts of 25-norhopanes (indices, <0.1; Figure 18).

A representative whole oil GC for the eastern Roncador oil (sample 7-RO-12D, 3560 m [11,680 ft]) is characterized by a low GC baseline and a full suite of n-alkanes ranging from C_8 to C_{33}. These characteristics indicate very little to no alteration of the oil and less evidence for paleobiodegradation. The hydrocarbon charge history for eastern Roncador suggests that the oil is influenced more by the mixing of a later higher maturity charge with a higher API gravity (Figure 18).

EFFECTS OF CHARGE HISTORY, TRAP TIMING, AND RESERVOIR TEMPERATURE HISTORY ON OIL QUALITY

The following section discusses results from the Trinity tool kit that interactively keeps track of the variations in the charge rate and fluid properties of the thermally evolving source rocks through time and combines it with a biodegradation model influenced by the temperature history of the reservoir to predict the final API of the oil.

Northeastern Roncador Field

Figure 19A illustrates the charge history for the fetch area encompassing the northern and eastern parts of the Roncador field. Two important charge events (early and late) document the modeled expelled hydrocarbon volumes from this fetch area through time from the Lagoa Feia lacustrine source rocks. Hydrocarbon expulsion begins in the Late Cretaceous (~90 Ma), with the early charge occurring from the late Paleocene to early Oligocene (60–35 Ma) and a more recent late charge occurring from the middle Miocene to present day (18–0 Ma). Geologic evaluation indicates that the downthrown trap in the eastern Roncador field formed in the Eocene approximately 42 Ma, resulting in the loss of a significant part of the early (60–42 Ma) hydrocarbon charge. Figure 19B shows the results of modeling the incoming oil API (red line), cumulative oil API (green curve), reservoir temperature history (black line), and mass fraction degraded (blue line) since the time of trap

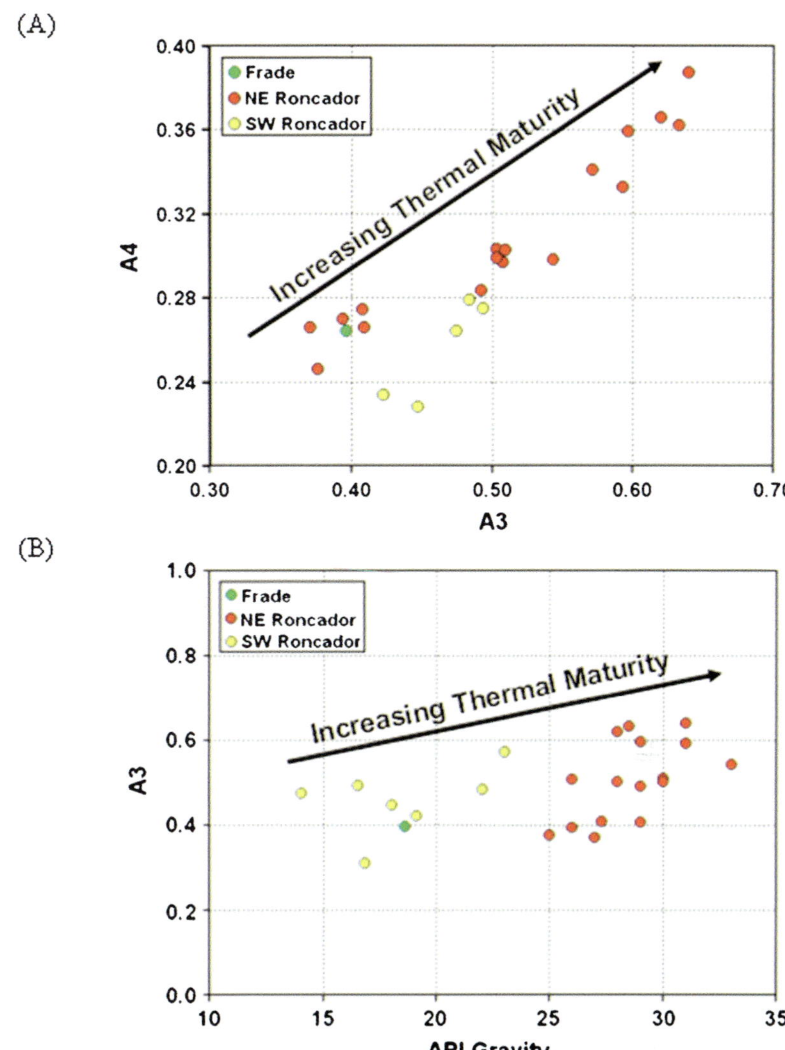

A3 = TA20/(TA20 + TA28R)
A4 = (TA20 + TA21)/(TA20 + TA21 + TA26S + TA27S_26R + TA28S + TA27R + TA28R)

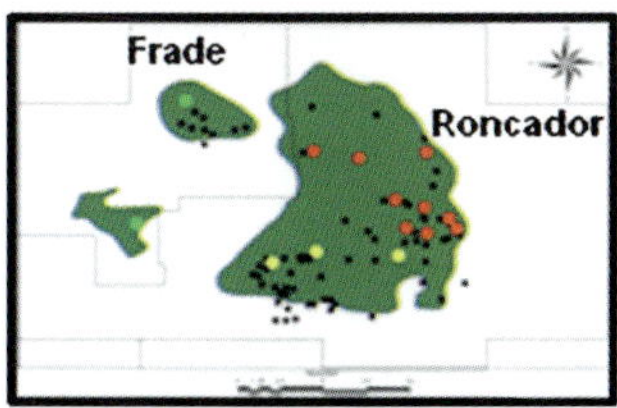

Figure 17. (A) Plot of triaromatic indices (A3 vs. A4; m/z 231) showing an increasing trend of thermal maturity for the oil samples. (B) Plot of the A3 index versus API gravity showing how increasing thermal maturity affects the variation in API gravity of the oils. Map inset shows the locations of the oils by colors that match the colors on the plots.

formation to present day. Beginning from the time of trap formation (~42 Ma in the Eocene), the cumulative oil gravity of about 38° API that is trapped in the reservoir rapidly decreases to about 14° API in the latter part of the early charge. Reservoir temperatures during this time ranged from about 30 to 45°C for a period of approximately 20 m.y., enhancing the effects of biodegradation and lowering the oil API gravity. Severe biodegradation continues to the beginning of the Miocene (~20 Ma), where the API gravity reaches its lowest value (~14° API) and the mass fraction degraded its highest value near 80%. At this time, hydrocarbon charge levels off in the late Oligocene to early Miocene, and no effect of mixing in "fresh" oil exists. Beginning approximately 18 Ma, the rate of later charge greatly increases, suggesting that higher maturity light oil begins to mix in the reservoir. Concurrently, the rate of biodegradation decreases, resulting in modeled API gravity that increases from about 14° API to present-day values of about 30° (Figure 19B). The modeled present-day value of API gravity is very close to the average measured API gravity of 28° observed for the eastern Roncador field (22–33° API).

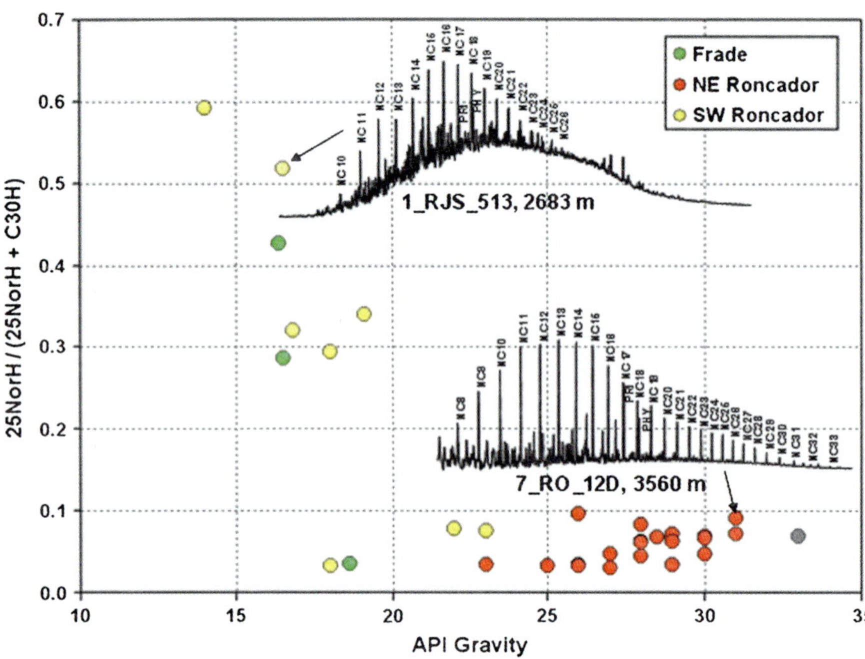

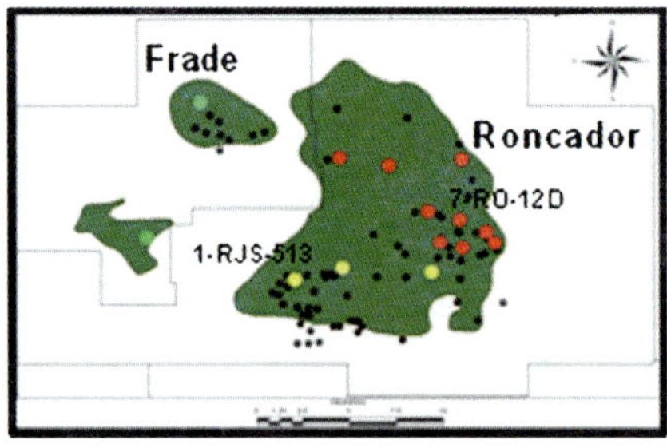

FIGURE 18. Plot of the 25-norhopane index (25NorH/[25NorH + C30H]) versus API gravity showing evidence for paleobiodegradation and mixing of oils from multiple charge events in the reservoir. The gas chromatograms illustrate the composition of two representative whole oils (one with a low API and the other with a high API).

Because of present-day reservoir temperatures of about 70°C in this part of Roncador field, only minor amounts of biodegradation occur and the oil remains relatively unaltered, as illustrated by the presence of abundant n-alkanes in the GC (7-RO-12D, 3560 m [11,680 ft]; Figure 19C).

Southwest Roncador Field

Hydrocarbon generation and expulsion for the southwest Roncador field begins in the Late Cretaceous (~90 Ma) and is dominated by early charge that begins in the Paleocene extending into the early Oligocene (60–35 Ma) and a more recent late charge that occurs from the middle Miocene to present day (18–0 Ma; Figure 20A). Geologic information indicates that the trap in southwest Roncador field formed in the Late Cretaceous (~72 Ma) and was in place before the early charge. At this time, the modeled incoming gravity arriving in the reservoir is predicted to be about 32° API, but reservoir temperatures are well below 30°C and remain lower than 40°C throughout the entire duration (~40 m.y.) of the early charge. Throughout this time of low reservoir temperatures, severe biodegradation causes a rapid decrease in gravity to values less than 10° API by the end of the early charge in the late Oligocene (Figure 20B). This severe biodegradation resulted in the formation of high abundances of 25-norhopanes that indicate paleobiodegradation. From the late Oligocene to the early Miocene (25–18 Ma), the oil in the reservoir continues to be degraded during a period of insignificant charge, resulting in peak levels of biodegradation as shown by the high percentage of mass degraded (~90%). At the onset of the late charge after 18 Ma, expelled hydrocarbon volumes increase greatly with the input of a "fresh" charge of higher maturity light oil (higher API gravity) into the reservoir (Figure 20B). The increase in "fresh" charge causes the cumulative gravity to increase from values of less than 10° API in the early Miocene to about 19° API at present day. The reservoir temperature at present day reaches 66°C, which is still low enough to cause some additional biodegradation in the present-day reservoir.

The whole oil composition of the oil (Figure 20C) in the Maastrichtian reservoir at 2683 m [8802 ft] in the 1-RJS-513 well is characterized by a gravity of 17° API, a high relative amount of 25-norhopanes (0.52), n-alkanes ranging from C_{10} to C_{26}, and a large unresolved complex

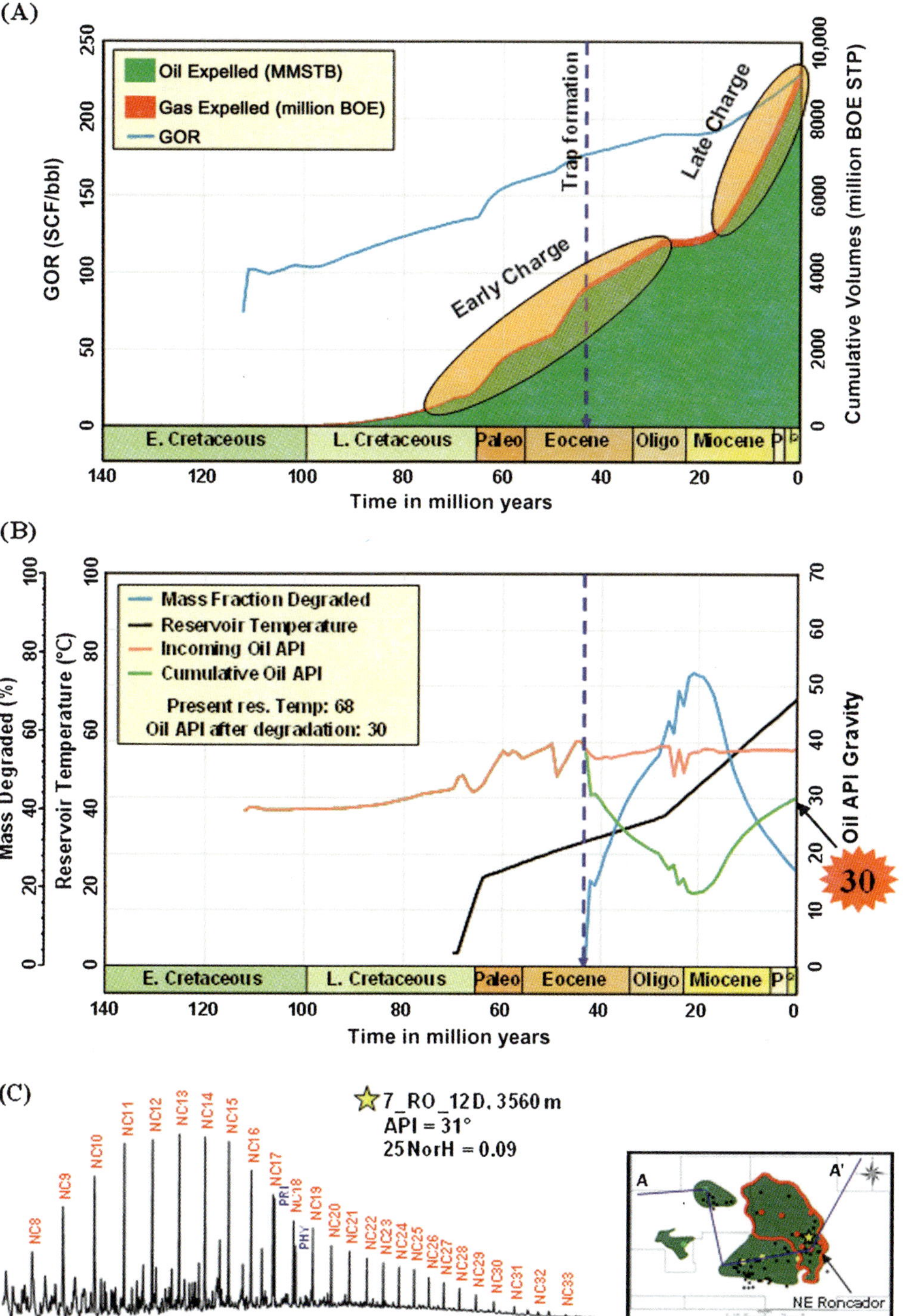

FIGURE 19. The upper diagram (A) shows the charge history and the lower diagram (B) shows the reservoir temperature history (black line), mass fraction of oil degraded (%; blue line), and API gravity of incoming (red line) and cumulative oil (green line) for the northeastern Roncador field area. The timing of trap formation (middle Eocene) is shown as a blue dashed line on (A) and (B). The gravity value (30° API) shown in the red symbol along the oil API gravity axis represents the predicted present-day API gravity for the reservoir in the northeastern Roncador closure area. The gas chromatogram shows the representative composition of the whole oil from this area, and the map insert shows the location of the oil (star symbol) within the northeastern Roncador field (shown as a red outline on the map). GOR = gas-oil ratio.

mixture (elevated "hump" in the GC chromatogram), which indicate a history of paleobiodegradation to the severe stage and then a subsequent mixing with a "fresh" oil that brings in new n-alkanes (first compounds to be consumed by bacteria) that are also degraded to some degree. The present-day reservoir temperature of 66°C is such that some recent biodegradation occurs but not enough to completely remove all the n-alkanes.

Frade Field

The charge history for the fetch area around the Frade field indicates that most of the expelled volumes occur during the late charge event from 18 Ma to present day (Figure 21A). Hydrocarbon generation and expulsion begins in the Late Cretaceous (~80 Ma) and continues through the Paleocene, with a gradual increase in the early Eocene, continuing to the middle Oligocene.

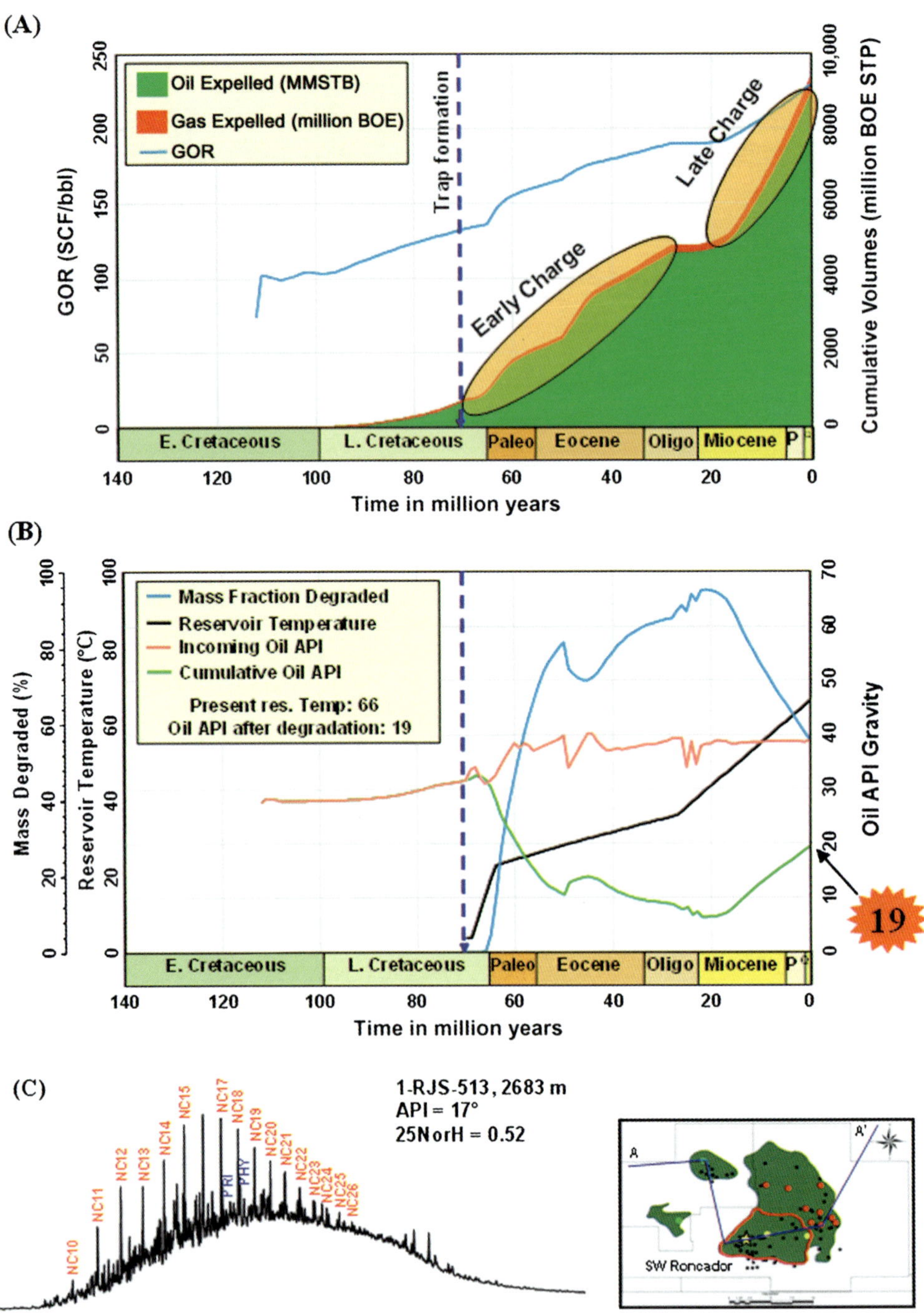

FIGURE 20. The upper diagram (A) shows the charge history and the lower diagram (B) shows the reservoir temperature history (black line), mass fraction of oil degraded (%; blue line), and API gravity of incoming (red line) and cumulative oil (green line) for the southwest Roncador field area. The timing of trap formation (Late Cretaceous) is shown as a blue dashed line on (A) and (B). The gravity value (19° API) shown in the red symbol along the oil API gravity axis represents the predicted present-day API gravity for the reservoir in the southwest Roncador closure area. The gas chromatogram shows the representative composition of the whole oil from this area, and the map insert shows the location of the oil (star symbol) within the southwest Roncador field (shown as a red outline on the map). GOR = gas-oil ratio.

Geologic information indicates that trap formation occurs sometime in the middle Oligocene (~28 Ma). Therefore, very little to none of the early charge is available to the trap. The main reservoirs in the Frade field were deposited in the late Oligocene to early Miocene before late charge that begins in the middle Miocene approximately 18 Ma. At the time of trap formation, the charge rate remains constant from the middle Oligocene to early Miocene (~10 Ma). During this time, the incoming oil gravity is about 35° API, but it decreases rapidly to less than 20° API in the late Miocene (Figure 21B) because of severe biodegradation resulting from low reservoir temperatures (<40°C). From the middle Miocene to present day, the charge rate increases greatly, resulting in an increase of API gravity to a present-day value of 18° API. The measured average gravity is also 18° API, ranging from 16 to 19° API. The present-day reservoir temperature of 59°C is still low enough to cause further biodegradation in the reservoir. The whole oil composition for sample 1-RJS-366 (2110 m [6923 ft]) shows the abundance of 25-norhopanes and the absence of n-alkanes in the GC, which indicate a history of severe paleobiodegradation and moderate to severe present-day biodegradation in the shallower Oligocene to Miocene reservoirs (Figure 21C).

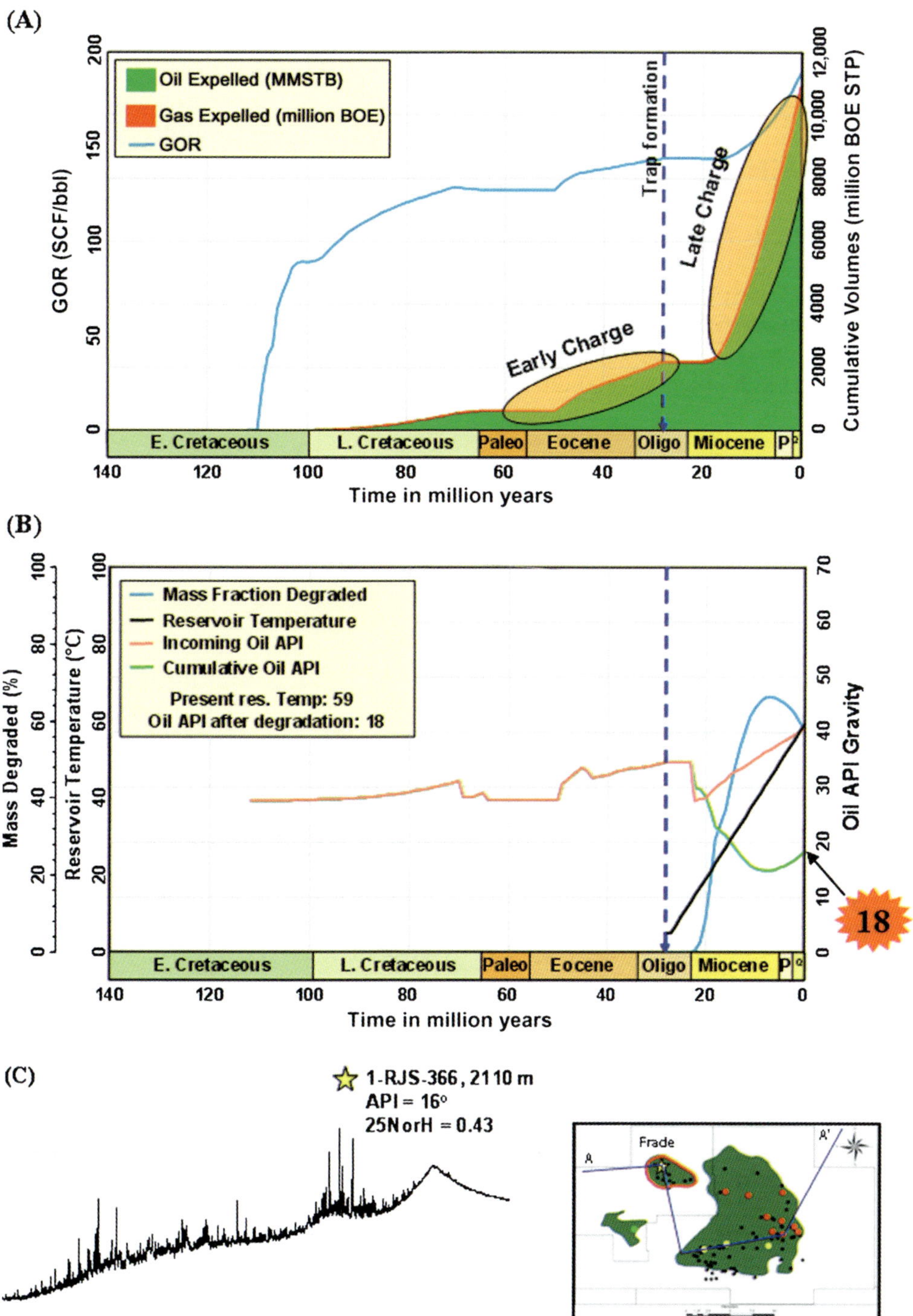

FIGURE 21. The upper diagram (A) shows the charge history and the lower diagram (B) shows the reservoir temperature history (black line), mass fraction of oil degraded (%; blue line), and API gravity of incoming (red line) and cumulative oil (green line) for the Frade field area. The timing of trap formation (middle Oligocene) is shown as a blue dashed line on (A) and (B). The gravity value (18 API) shown in the red symbol along the oil API gravity axis represents the predicted present-day API gravity for the reservoir in the Frade closure area. The gas chromatogram shows the representative composition of the whole oil from this area, and the map insert shows the location of the oil (star symbol) within the Frade field (shown as a red outline on the map). GOR = gas-oil ratio.

CONCLUSIONS

A simplistic approach of using only present-day reservoir depth and temperature is not sufficient for adequately predicting the API gravity of oils in fields of the Campos Basin, offshore Brazil. An interactive modeling approach that characterizes the effects of charge history, trap timing, reservoir temperature history, and biodegradation is much more useful for explaining the observed API gravity distribution in the Frade and Roncador fields. The modeled approach is constrained by geochemical and geologic observations from the fields and is used to help explain (1) the occurrence of effective presalt freshwater to brackish (lower source rock) and saline lacustrine (upper source rock) Lagoa Feia source rocks in the fetch areas of the Frade and Roncador fields, (2) the different thermal maturity levels of oils generated from different hydrocarbon kitchens in the area, and (3) the complex charge and biodegradation histories evident in the geochemical composition of the oils present in the Roncador and Frade fields.

ACKNOWLEDGMENTS

We thank Steve Crews, Pam Morelos, David Schmidt, and Gunardi Sulistyo for their contributions to the various aspects of data compilation, interpretation, and modeling and also for their comments and suggestions to the article. We also thank reviewers Cliff Walters and Eric Michael for their useful comments.

REFERENCES CITED

Baudino, R., C. Losilla, G. Figueiredo dos Santos, R. Cabrera, G. Maselli, and M. Fascola, 2008, Applying basin modeling to biodegraded reservoir understanding: A case study from offshore Brazil (abs.): Proceedings of XI Latin American Congress on Organic Geochemistry, Margarita Island, Venezuela, 4 p.

Bennett, B., M. Fustic, P. Farrimond, H. Huang, and S. R. Larter, 2006, 25-Norhopanes: Formation during biodegradation of petroleum in the subsurface: Organic Geochemistry, v. 37, p. 787–797, doi:10.1016/j.orggeochem.2006.03.003.

Blanc, P., and J. Connan, 1992, Origin and occurrence of 25-norhopanes: A statistical study: Organic Geochemistry, v. 18, p. 813–828, doi:10.1016/0146-6380(92)90050-8.

Chosson, P., J. Connan, D. Dessort, and C. Lanau, 1992, In vitro biodegradation of steranes and terpanes: A clue to understanding geological situations, *in* J. M. Moldowan, P. Albrecht, and R. P. Philip, eds., Biological markers in sediments and petroleum: Englewood Cliffs, New Jersey, Prentice Hall, p. 320–349.

Cole, G. A., A. Yu, F. Peel, R. J. Requejo, D. J. Brooks, B. J. Bernard, J. Zumberge, and S. Brown, 2001, Constraining source and charge risk in deep-water areas: World Oil, v. 222, no. 10, p. 69–77.

Dzou, L., and Z. He, 2009, The advantages of interactive fluid property modeling (abs.): AAPG Hedberg Conference Basin and Petroleum System Modeling: New Horizons in Research and Applications, Napa, California, http://www.searchanddiscovery.com/abstracts/html/2009/hedberg/abstracts/extended/dzou/dzou.htm (accessed May 26, 2011).

Guardado, L. R., A. R. Spadini, J. S. L. Brandao, and M. R. Mello, 2000, Petroleum system of the Campos Basin, *in* M. R. Mello and B. J. Katz, eds., Petroleum systems of South Atlantic margins: AAPG Memoir 73, p. 317–324.

Guthrie, J. M., C. Nino, H. Hassan, and P. Morelos, 2009, Integrating geochemistry, charge rate and timing, trap timing and reservoir temperature history to model fluid properties in the Frade and Roncador fields, Campos Basin, offshore Brazil (abs.): AAPG Hedberg Conference Basin and Petroleum System Modeling: New Horizons in Research and Applications, Napa, California, http://www.searchanddiscovery.com/abstracts/html/2009/hedberg/abstracts/extended/guthrie/guthrie.htm (accessed May 26, 2011).

Holba, A. G., L. Ellis, E. Tegelaar, M. S. Singletary, and P. Albrecht, 2000, Tetracyclic polyprenoids: Indicators of fresh water (lacustrine) algal input: Geology, v. 28, no. 3, p. 251–254, doi:10.1130/0091-7613(2000)28<251:TPIOFL>2.0.CO;2.

Holba, A. G., L. I. Dzou, G. D. Wood, L. Ellis, P. Adam, P. Schaeffer, P. Albrecht., T. Greene, and W. B. Hughes, 2003, Application of tetracyclic polyprenoids as indicators of input from fresh-brackish water environments: Organic Geochemistry, v. 34, p. 441–469, doi:10.1016/S0146-6380(02)00193-6.

Huang, H., S. R. Larter, B. F. J. Bowler, and T. B. P. Oldenburg, 2004, A dynamic biodegradation model suggested by petroleum compositional gradients within reservoir columns from the Liaohe Basin, NE China: Organic Geochemistry, v. 35, p. 299–316, doi:10.1016/j.orggeochem.2003.11.003.

Katz, B. J., and V. D. Robison, 2006, Oil quality in deep-water settings: Concerns, perceptions, observations, and reality: AAPG Bulletin, v. 90, p. 909–920, doi:10.1306/01250605128.

Larter, S. R., A. Wilhelms, I. Head, M. Koopsmans, A. Aplin, P. R. Di, C. Zwach, M. Erdman, and N. Telnaes, 2003, The controls on the composition of biodegraded oils in the deep subsurface: Part I: Biodegradation rates in petroleum reservoirs: Organic Geochemistry, v. 34, p. 601–613, doi:10.1016/S0146-6380(02)00240-1.

Larter, S. R., H. Huang, J. Adams, B. Bennett, O. Jokanola, T. Oldenbrug, M. Jones, I. Head, C. Riediger, and M. Fowler, 2006, The controls on the composition of biodegraded oils in the deep subsurface: Part II: Geological controls on subsurface biodegradation fluxes and constraints on reservoir-fluid property prediction: AAPG Bulletin, v. 90, p. 921–938, doi:10.1306/01270605130.

Mackenzie, A. S., J. R. Maxwell, M. L. Coleman, and C. E. Deegan, 1984, Biological marker and isotope studies of North Sea crude oils and sediments, *in* Proceedings of the 11th World Petroleum Congress, Geology Exploration Reserves, Wiley, Chichester, v. 2, p. 45–56.

Mello, M. R., and J. R. Maxwell, 1990, Organic geochemical and biological marker characterization of source rocks and oils derived from lacustrine environments in the Brazilian continental margin, *in* B. J. Katz, ed., Lacustrine basin exploration: Case studies and modern analogs: AAPG Memoir 50, p. 77–99.

Mello, M. R., N. Telnaes, P. C. Gaglianone, M. I. Chicarelli, S. C. Brassell, and J. R. Maxwell, 1988, Organic geochemical characterization of depositional paleoenvironments of source rocks and oils in Brazilian marginal basins, *in* L. Mattavelli and L. Novelli, eds., Advances in organic geochemistry 1987: Organic Geochemistry, v. 13, p. 31–45, doi:10.1016/0146-6380(88)90023-X.

Mello, M. R., W. U. Mohriak, E. A. M. Koutsoukos, and G. Bacoccoli, 1994, Selected petroleum systems in Brazil, *in* L. B. Magoon and W. G. Dow, eds., The petroleum system–From source to trap: AAPG Memoir 60, p. 499–512.

Mello, M. R., J. M. Moldowan, J. Dahl, and A. G. Requejo, 2000, Petroleum geochemistry applied to petroleum system investigation, *in* M. R. Mello and B. J. Katz, eds., Petroleum system of South Atlantic margins: AAPG Memoir 73, p. 41–52.

Pepper, A. S., and P. J. Corvi, 1995, Simple kinetic models of petroleum formation, Part I: Oil and gas generation from kerogen: Marine Petroleum Geology, v. 12, p. 291–319, doi:10.1016/0264-8172(95)98381-E.

Peters, K. E., J. M. Moldowan, M. A. McCaffrey, and F. J. Fago, 1996, Selective biodegradation of extended hopanes to 25-norhopanes in petroleum reservoirs: Insights from molecular mechanics: Organic Geochemistry, v. 24, p. 765–783, doi:10.1016/S0146-6380(96)00086-1.

Rangel, H. D., P. R. Santos, and C. M. S. P. Quintares, 1998, Roncador field, a new giant in Campos Basin, Brazil: Proceedings, Offshore Technology Conference, Houston, Texas, p. 579–587.

Rangel, H. D., P. T. M. Guimaraes, and A. R. Spadini, 2003, Barracuda and Roncador giant oil fields, deep-water Campos Basin, Brazil, *in* M. T. Halbouty, ed., Giant oil and gas fields of the decade 1990–1999: AAPG Memoir 78, p. 123–137.

Santos, P. R., H. D. Rangel, C. M. S. P. Quintares, and J. M. Caixeta, 1999, Turbidite reservoir distribution in Roncador field, Campos Basin, Brazil (abs): AAPG International Conference and Exhibition, Birmingham, England, p. 530–531.

Soldan, A. L., J. R. Cerqueira, J. C. Ferreira, L. A. F. Trindade, J. C. Scarton, and C. A. G. Cora, 1995, Giant deepwater oil fields in Campos Basin, Brazil: A geochemical approach: Revista Latino-Americana de Geoquimica Organica, v. 1, p. 14–27.

Wenger, L. M., C. L. Davis, and G. H. Isaksen, 2002, Multiple controls on petroleum biodegradation and impact on oil quality: Society of Petroleum Engineers paper 80168, p. 375–383.

Wilhelms, A., M. Erdmann, and S. R. Larter, 2004, Easy-Fest versus BDI: Uncertainties in pre-drill prediction of biodegradation degree in subsurface petroleum reservoirs (abs.): AAPG Annual Convention Abstracts Volume, v. 13, p. A147–A148.

Yu, Z., G. Cole, G. Grubitz, and F. Peel, 2002, How to predict biodegradation risk and reservoir fluid quality: World Oil, v. 223, no. 4, p. 63–74.

View of a deep, open joint in a large block of amphibolite retrograded to blueschist (Lincoln Rock) in the southeastern part of The Geysers geothermal field, reputed to have been a hideout for the gentleman bandit Black Bart. After an honorable discharge from the Union Army as a 1st Lieutenant, he had an unpleasant incident with some Wells Fargo Company employees and vowed revenge. Between 1875 and 1883, he committed many robberies of Wells Fargo stagecoaches across northern California. His daring as an outlaw was rivaled only by his reputation for style and sophistication.

"I've labored long and hard for bread,
For honor, and for riches,
But on my corns too long you've tread,
You fine-haired sons of bitches."

Black Bart, authenticated verse left at the scene of an 1877 stagecoach holdup.

18

Schenk, O., L. B. Magoon, K. J. Bird, and K. E. Peters, 2012, Petroleum system modeling of northern Alaska, *in* K. E. Peters, D. J. Curry, and M. Kacewicz, eds., Basin Modeling: New Horizons in Research and Applications: AAPG Hedberg Series, no. 4, p. 317–338.

Petroleum System Modeling of Northern Alaska

Oliver Schenk
Schlumberger, Aachen, Germany

Leslie B. Magoon and Kenneth J. Bird
U.S. Geological Survey, Menlo Park, California, U.S.A.

Kenneth E. Peters
Schlumberger, Mill Valley, California, U.S.A.

ABSTRACT

Northern Alaska is a prolific oil and gas province estimated to contain a significant proportion of the undiscovered oil and gas of the circum-Arctic. A three-dimensional petroleum system model was constructed with the aim of significantly improving the understanding of the generation, migration, accumulation, and loss of hydrocarbons in the region. This study provides a unique geologic perspective that will reduce exploration risk and assess the remaining potential hydrocarbon resources in this remote province.

The present-day geometry is based on newly interpreted seismic data and a database of more than 400 wells. A key aspect of this model is an improved reconstruction of the progradation of the time-transgressive Cretaceous–Tertiary Brookian sequence and multiple erosion events in the Tertiary. The deposition of these overburden rocks controlled the timing of hydrocarbon generation in underlying source rocks and their principal migration from the Colville Basin northward to the Barrow Arch. The model provides a reconstruction of the complex and dynamic interplay of diachronous deposition and erosion and allows assessment of variations in migration behavior and prediction of the present-day petroleum distribution.

INTRODUCTION

Northern Alaska is a prolific petroleum province having produced more than 2.4×10^9 m^3 (15 billion bbl) of oil with reserves of 1.1×10^9 m^3 (7 billion bbl) of oil and 9.9×10^{11} m^3 (35 tcf) of gas. Nearly twice as much oil and three to six times as much gas remain to be discovered, according to recent estimates by the U.S. Geological Survey and the Minerals Management Service (Houseknecht and Bird, 2005). Among all circum-Arctic basins, northern Alaska ranks first in oil potential and third in gas potential (Gautier et al., 2009).

DOI:10.1306/13311444H43477

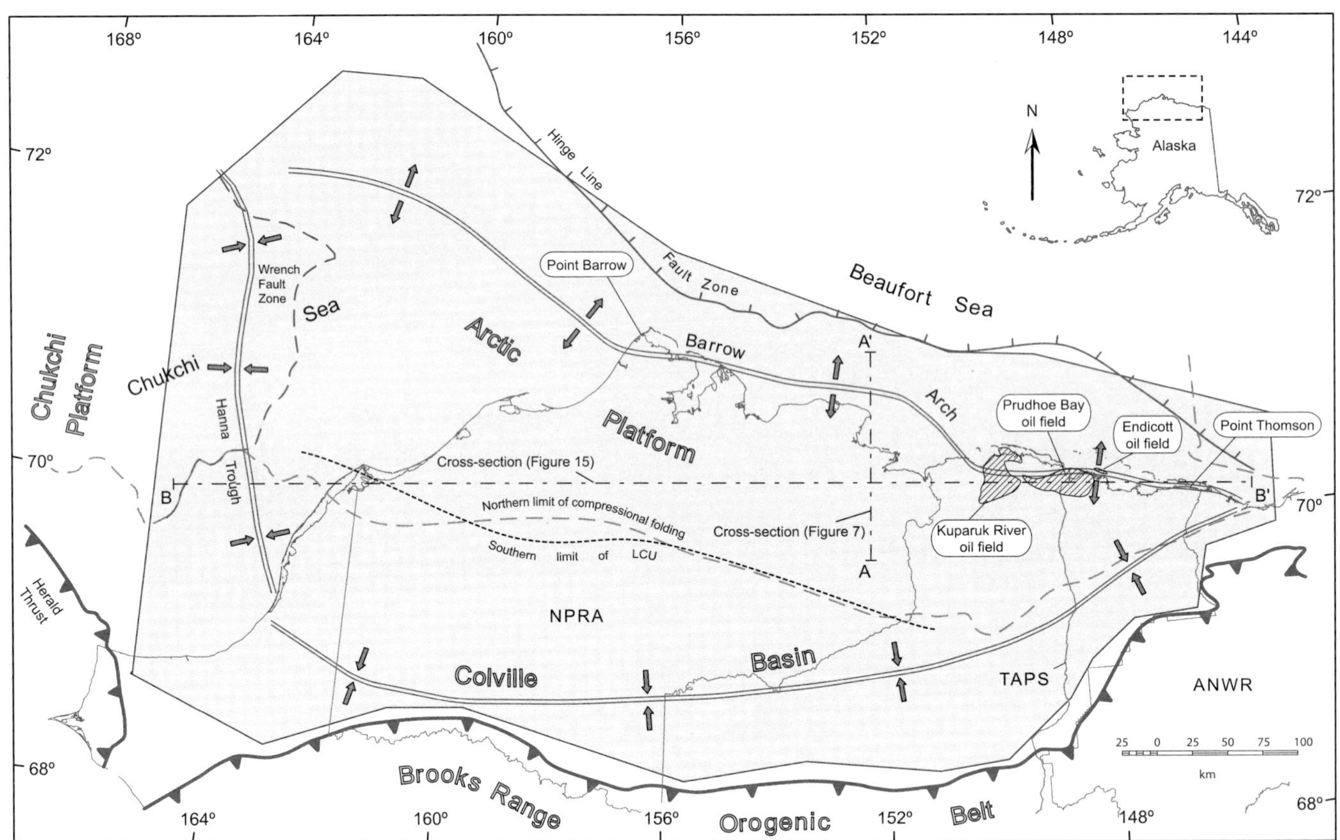

FIGURE 1. Major tectonic features of northern Alaska and the area modeled (shaded). NPRA = National Petroleum Reserve–Alaska; ANWR = Arctic National Wildlife Refuge; TAPS = Trans-Alaska Pipeline System; LCU = Lower Cretaceous Unconformity (from Bird, 2001, and Houseknecht and Bird, 2004).

The geologically complex northern Alaska petroleum province evolved through the tectonic stages of passive margin, rift, foreland basin, and foreland fold and thrust belt. Petroleum was generated from several source rock units, and many reservoirs show evidence of mixing of hydrocarbon source types. Rift-related structures and a regional breakup unconformity are critical trapping and migration components of the largest oil and gas accumulations. In addition, stratigraphic traps that developed during extensional and compressional tectonic regimes show significant resource potential in Jurassic through Cenozoic shelf and turbidite sequences.

Regional modeling of the tectonic and sedimentologic evolution of northern Alaska through time provides an opportunity to integrate and analyze many aspects of petroleum system development. In an earlier modeling study (Lampe et al., 2003; Magoon et al., 2003; Peters et al., 2003), an area of approximately 100,000 km^2 (38,610 mi^2) was investigated, which encompassed most of the National Petroleum Reserve-Alaska (NPRA) and the Central North Slope, including the greater Prudhoe Bay region. In the current study, the model size was nearly tripled to 275,000 km^2 (106,178 mi^2) to include the eastern part of the Chukchi platform, the foothills of the Brooks Range, and the Beaufort continental shelf (Figure 1) to assess and quantify the evolution of the region's multiple petroleum systems.

This chapter focuses on the model building, the construction of present-day structural-stratigraphic geometry, and the reconstruction of the paleogeometries that evolved through time. The model is based on more than 48,000 km (30,000 mi) of newly interpreted two-dimensional (2-D) seismic data and a database of more than 400 wells that include calibration and geochemical data (Figure 2). Particular attention was paid to mapping onlap and truncation relations developed during passive margin and rifting stages in recognition of their importance relative to hydrocarbon migration pathways and traps. Modeling the foreland basin presented challenges in accounting for rapid changes in stratal thickness and facies related to longitudinal west-to-east basin filling, as well as multiple episodes of uplift and erosion caused by thrusting. The deep burial and eastward-shifting depocenters of the foreland basin activated multiple petroleum systems, resulting in an eastward shift in timing of petroleum generation and modification of the directions of petroleum migration.

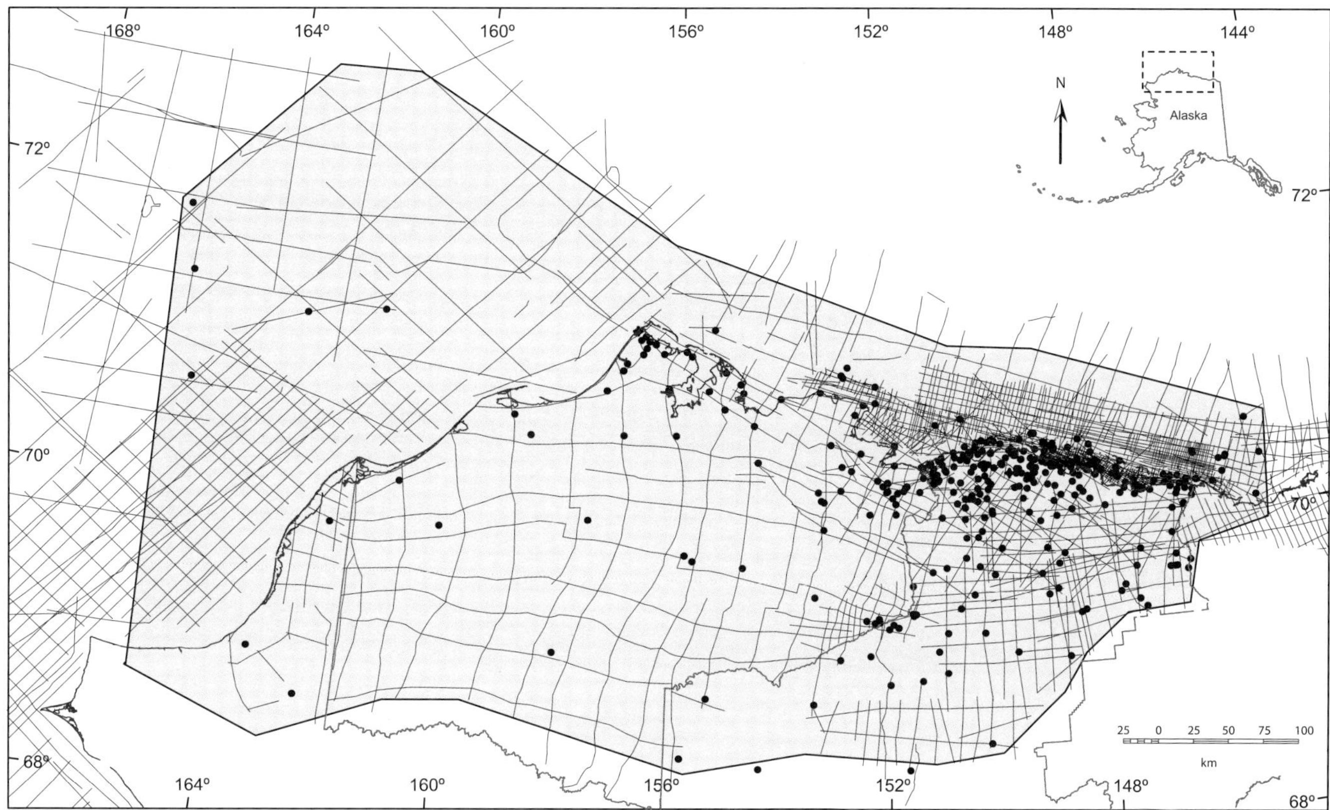

FIGURE 2. Seismic (lines) and wells (dots) used in the North Slope model.

GEOLOGY OF NORTHERN ALASKA

Tectonic Setting

The northern Alaska petroleum province extends approximately 1100 km (683.5 mi) from the Canadian to the Russian border and 100 to 600 km (62.1–372.8 mi) northward from the Brooks Range (Figure 1). The northern boundary of the province is located in the offshore region and is defined by a narrow approximately 100-km (62.1-mi)-wide shelf in the Beaufort Sea and by a broad 600-km (373-mi)-wide shelf in the Chukchi Sea (Sherwood et al., 1998). To the south, this petroleum province is bounded by the Brooks Range–Herald Arch orogenic belt. The dominant tectonic features of the Alaska petroleum province are presented in Figure 1.

The Chukchi and Arctic platforms are remnants of a south-facing continental margin during the Paleozoic and the early Mesozoic. These platforms are separated by a north-trending structural sag, the Hanna Trough, that is characterized by normal faulting and thick Devonian(?) and Mississippian deposits. The Hanna Trough represents a failed rift (Sherwood et al., 1998). A younger episode of rifting that is nearly perpendicular to the Hanna Trough occurred during the Jurassic and the Early Cretaceous and produced structures, such as the Barrow Arch (rift shoulder) and the hinge line fault zone (rift margin), Canada Basin, and the Beaufort passive margin (Bird and Molenaar, 1992). The Barrow Arch is a broad east-plunging antiformal feature that underlies the coast of northern Alaska and a large area of the northeastern Chukchi Sea. In some areas, it is cut by many rift-related normal faults, with offsets generally less than a few hundred meters. These faults, together with erosional truncations of prerift strata during rift shoulder uplift and followed by subsidence and overlap by marine shales, form complex combination structural-stratigraphic traps.

Concurrent with the rifting event, the southern side of the Arctic Alaska plate collided with an oceanic island arc and produced the Brooks Range and the Colville Basin (North Slope foreland basin) (Moore et al., 2004). Later, during the Cenozoic, compressional deformation redeformed the compressional structures and formed a fold and thrust belt that crosses the entire southern part of the foreland basin and extends offshore across the passive margin in the east (Bird, 2001). The Tertiary deformation is diachronous, being older in the west and becoming progressively younger in the east (O'Sullivan et al., 1993, 1997; Bird, 1994; O'Sullivan, 1999; Moore et al., 2004).

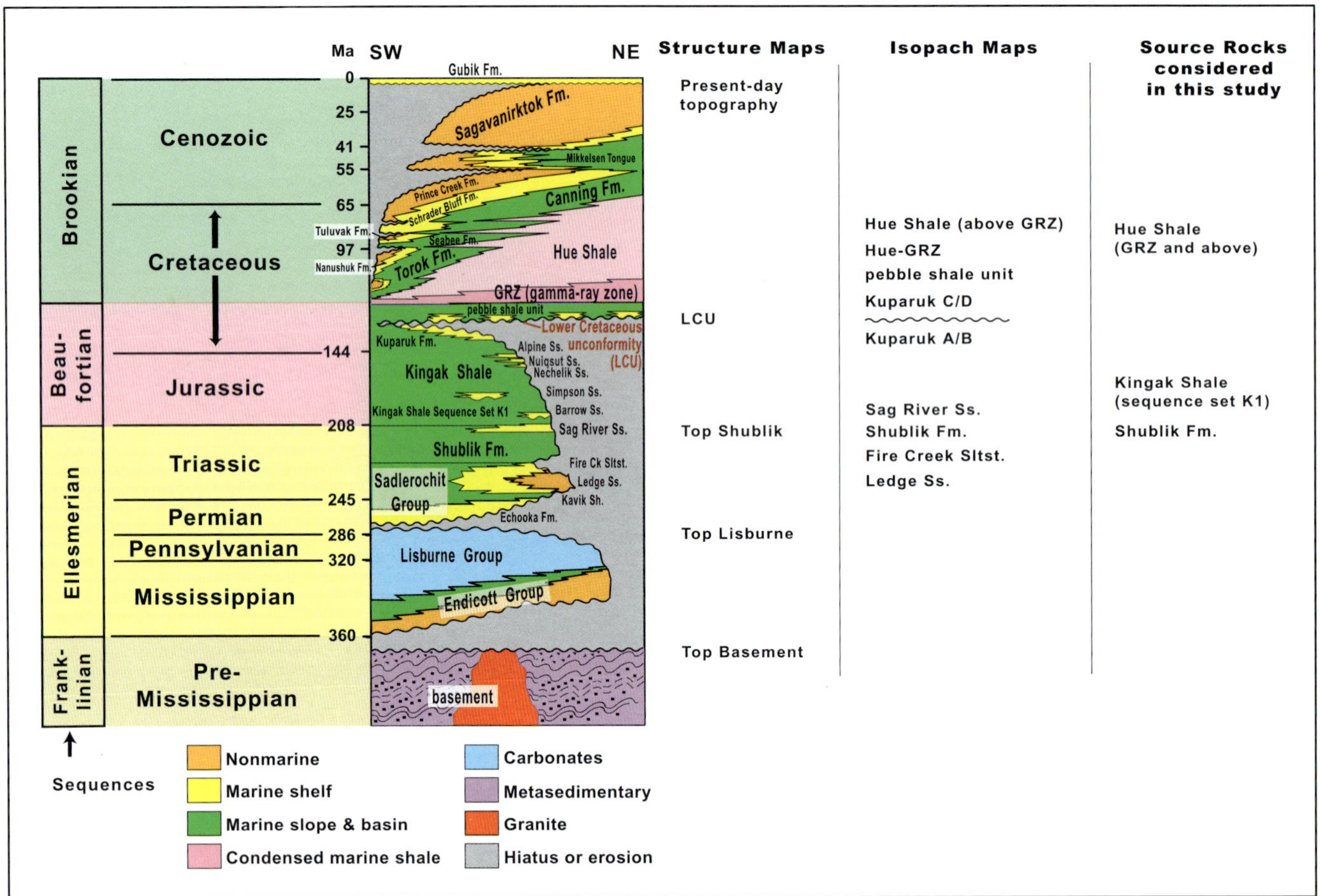

FIGURE 3. Stratigraphic column of the onshore part of the Alaska North Slope (modified from Houseknecht and Bird, 2005) showing also structure and isopach maps as well as source rock units used for the model. Hue-GRZ = Hue–Gamma Ray Zone; LCU = Lower Cretaceous Unconformity.

Stratigraphy

Although the stratigraphic record of the Arctic Alaska Petroleum Province extends into the Precambrian, rocks with a potential for petroleum accumulations are Mississippian and younger (Figure 3). The traditional grouping of the rocks into tectonostratigraphic sequences, as proposed by Lerand (1973) and modified by later investigators, emphasizes tectonic history, provenance, and genetic relations. This section briefly describes the tectonostratigraphic sequences north of the Brooks Range. Stratigraphic details can be found in Hubbard et al. (1990), Bird and Molenaar (1992), Moore et al. (1994), Bird (2001), and references therein.

The Franklinian sequence mostly includes Devonian and older sedimentary and igneous rocks representing diverse origins and a complex geologic history. These rocks have been buried and metamorphosed beyond the thermal stage of oil preservation across most of Arctic Alaska, and so they are considered to be economic basement.

The Mississippian through Triassic Ellesmerian sequence consists of carbonate and shallow-marine to nonmarine siliciclastic deposits. On the Arctic Platform, Ellesmerian strata are continental shelf deposits that accumulated on a south-facing passive margin. From a northern pinch-out or erosional edge near the Barrow Arch (Figure 1), they thicken southward to about 2 km (1.2 mi) (Moore et al., 1994). Westward, Ellesmerian strata thicken rapidly into the Hanna Trough and, beyond, thin onto the Chukchi Platform. In the Hanna Trough, Devonian(?) and Mississippian rocks are interpreted as synrift deposits, and post-Mississippian strata as thermal sag-phase deposits (Sherwood et al., 1998). The Ellesmerian sequence contains both petroleum source and reservoir rocks. The source rocks, which lie near the top of the sequence, did not generate petroleum north of the Brooks Range until buried by Beaufortian and Brookian deposits.

The Jurassic and Lower Cretaceous Beaufortian sequence (Hubbard et al., 1990) consists of synrift deposits derived locally or from the north. It is a stratigraphically complex mud-dominated sequence with multiple unconformities and large variations in thickness (Houseknecht and Bird, 2004) and contains petroleum source and reservoir rocks. Mainly north of the present coastline occur both normal faulting and the formation of sediment-filled grabens and half grabens, some containing more than 3 km (1.9 mi) of Beaufortian fill (Grantz et al., 1988). Uplift and erosion along the rift margin created a regional unconformity, termed the Lower Cretaceous Unconformity (LCU) (Figure 3), which is considered to be the "breakup" unconformity (Grantz and May, 1982). This unconformity, which progressively truncates older rocks northward onto the Barrow Arch, is partly responsible for many of the largest oil accumulations in northern Alaska by providing a hydrocarbon-migration pathway for charging multiple subunconformity reservoirs. Cretaceous mudstone overlying the unconformity serves as a seal, creating combination structural-stratigraphic traps under favorable circumstances, such as at the Prudhoe Bay oil field.

Cretaceous and Tertiary deposits derived from the Brooks Range orogen are assigned to the Brookian sequence. These voluminous deposits filled the Colville foreland basin, overtopped the rift shoulder (Barrow Arch), and built the passive margin that forms the modern continental terrace north of Alaska. The Brookian sequence consists of a complex assemblage of siliciclastic strata that include distal condensed marine mudstone (Hue Shale); relatively deep marine basinal, slope, and outer-shelf mudstone and turbidite sandstone (Torok, Seabee, and Canning formations); and (deltaic) shallow-marine to coal-bearing nonmarine sandstone, mudstone, and conglomerate (Nanushuk, Tuluvak, Prince Creek, Schrader Bluff, and Sagavanirktok formations). Organic-rich beds of the Hue Shale are important oil source rocks, and Brookian mudstones may contain gas source rocks. Reservoir rocks consist of turbidite and shallow-marine to nonmarine sandstone, and known oil and gas accumulations occur in both structural and stratigraphic traps within the Brookian sequence. Sediment accumulation in the Colville Basin and on the passive margin north of the Barrow Arch generally progressed from west to east during the Cretaceous and the Cenozoic (Bird and Molenaar, 1992); an exception to this pattern is a significant Tertiary depocenter that was active in the northern Chukchi Sea during the early Tertiary (Sherwood et al., 1998). Deposition of a thick Brookian sequence provided the overburden necessary for thermal maturation of petroleum source rocks located in Ellesmerian, Beaufortian, and Brookian strata.

INPUT/MODEL BUILDING

Location of the Study Area

The model discussed here covers an area of 275,000 km^2 (106,180 mi^2) and spans a length of 832 km (517 mi) west to east and 520 km (323 mi) south to north (Figure 1). Its southern boundary lies near the mountain front in the foothills of the Brooks Range and its northern boundary on the outer shelf of the Beaufort and Chukchi seas so that the Barrow Arch is included. The eastern boundary includes part of Camden Bay at about long. 144°W. Its western boundary lies on the east flank of the Chukchi Platform just west of the axis of Hanna Trough.

Databases

Construction of the present-day geometry is based on more than 48,000 km (>30,000 mi) of newly interpreted 2-D seismic data and a database that consists of more than 400 wells (Figure 2). Wells were selected for use based on their suitability to aid map construction (e.g., depth and thickness maps for model building, distribution maps of organofacies for the source rocks), model calibration (e.g., thermal maturity, pressure), and to provide source rock samples for kinetic analyses. The data were compiled from various mostly published sources (e.g., from U.S. Geological Survey open-file reports and online database) and have been quality controlled such as classification of maturity data according to different laboratories or removal of unreasonable values.

The input model has a grid spacing of 1 km (0.62 mi) and includes the most extensive compilation of maps available, including depth and thickness maps of key stratigraphic and time units (Figure 3). Five structure contour maps form the framework of the model: (1) top basement, (2) top Lisburne Group, (3) top Shublik Formation, (4) LCU, and (5) present-day topography and bathymetry. In addition to these maps, nine isopach (thickness) maps were created based on well logs and rock unit tops for most of the source and reservoir rock units. After interpolation of thickness, the maps were adjusted according to onlap structures and truncation edges revised by seismic data (Figure 4). All maps were quality controlled by comparing with well logs and tops as well as with outlines of known oil and gas accumulations (a proxy in many cases for local structural closure) to assure accuracy at the model scale. The post-LCU layers, which consist of progradational (time-transgressive) deltaic foreland basin sediments of the Brookian sequence, are reinterpreted for this model. For modeling purposes, the progradational character of these sediments was subdivided into chronostratigraphic instead of lithostratigraphic units. Five clinoform horizons that correspond approximately to 97, 65, 55, 41, and 25 Ma were seismically mapped. The stratigraphic column (Figure 3) relates formation names to timelines. However, these clinoform horizons extend across only approximately 30% of the eastern part of the model area. Because the western part of the model area lacks consistently mappable seismic horizons and age control, it was subjectively subdivided into approximately equal 5-Ma increments (120, 115, 110, 105 Ma) based on the dominant

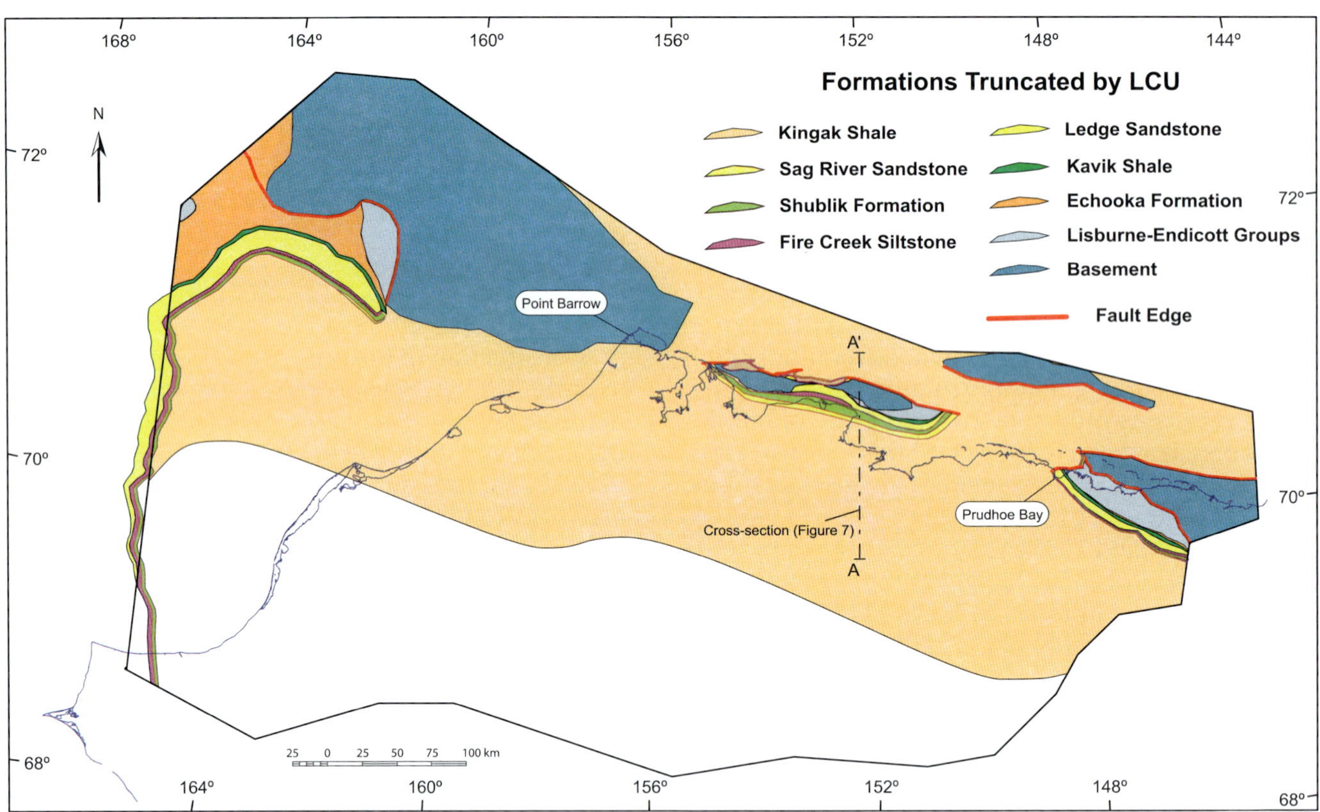

FIGURE 4. Subcrop map of rock units beneath the Lower Cretaceous Unconformity (LCU) also used as lithofacies distribution map for the LCU weathering zone representing the lithologies of rocks underlying the LCU. Note that truncation of the LCU-related Kuparuk Formation is not included.

direction of progradation from paleoflow indicators (Molenaar, 1988). The general location of the oldest shelf margin (120 Ma, Early Aptian) is anchored by low-amplitude clinoforms on seismic profiles combined with Aptian micropaleontologic age determinations of the lowermost Brookian sequence in wells on the western flank of Hanna Trough (Sherwood et al., 2002). Mapped and estimated shelf margin trends for each of the nine Brookian horizons are shown in Figure 5.

Tertiary Erosion

The amount of eroded strata was based on uplift estimates derived from borehole sonic-log porosity-depth trends in selected wells (Burns et al., 2007; figure 9 therein). The Burns et al. map was modified in the Brooks Range foothills area by smoothing the contours to account for the effects of folding and faulting (Figure 6A).

Apatite fission-track analyses showed that Tertiary uplift and erosion occurred at least three times at approximately 60, 40, and 24 Ma, with different spatial distributions (O'Sullivan et al., 1993; 1997; O'Sullivan, 1999). To account for these three erosional events, the map of the total amount of Tertiary erosion was partitioned under consideration of gradual transitions (Figure 6B–D). While the 60 Ma uplift affected most of the western and central parts of the model area, the younger uplift events occurred in the northeastern Brooks Range part of the model area, a reflection of changing tectonic conditions in northern Alaska.

Model Building

Model building was subdivided into two major parts: the pre-Brookian (Ellesmerian and Beaufortian sequences) and the Brookian sequence. Five structure contour maps provide the overall framework of the model: top Franklinian basement, Lisburne Group, and Shublik Formation, the LCU, and the present-day topography and bathymetry (Figure 3).

Ellesmerian/Beaufortian Sequences

Over most of the model area, the Ellesmerian and Beaufortian sequences are characterized by layer cake geology. Details of individual pre-Brookian formations were incorporated using isopach maps with thickness information added to or subtracted from the appropriate structure contour map (Figure 7). To account for the oil source rock in the Kingak Shale, only the basal condensed unit (Kingak K1) with an estimated thickness of up to 70 m (230 ft) is included.

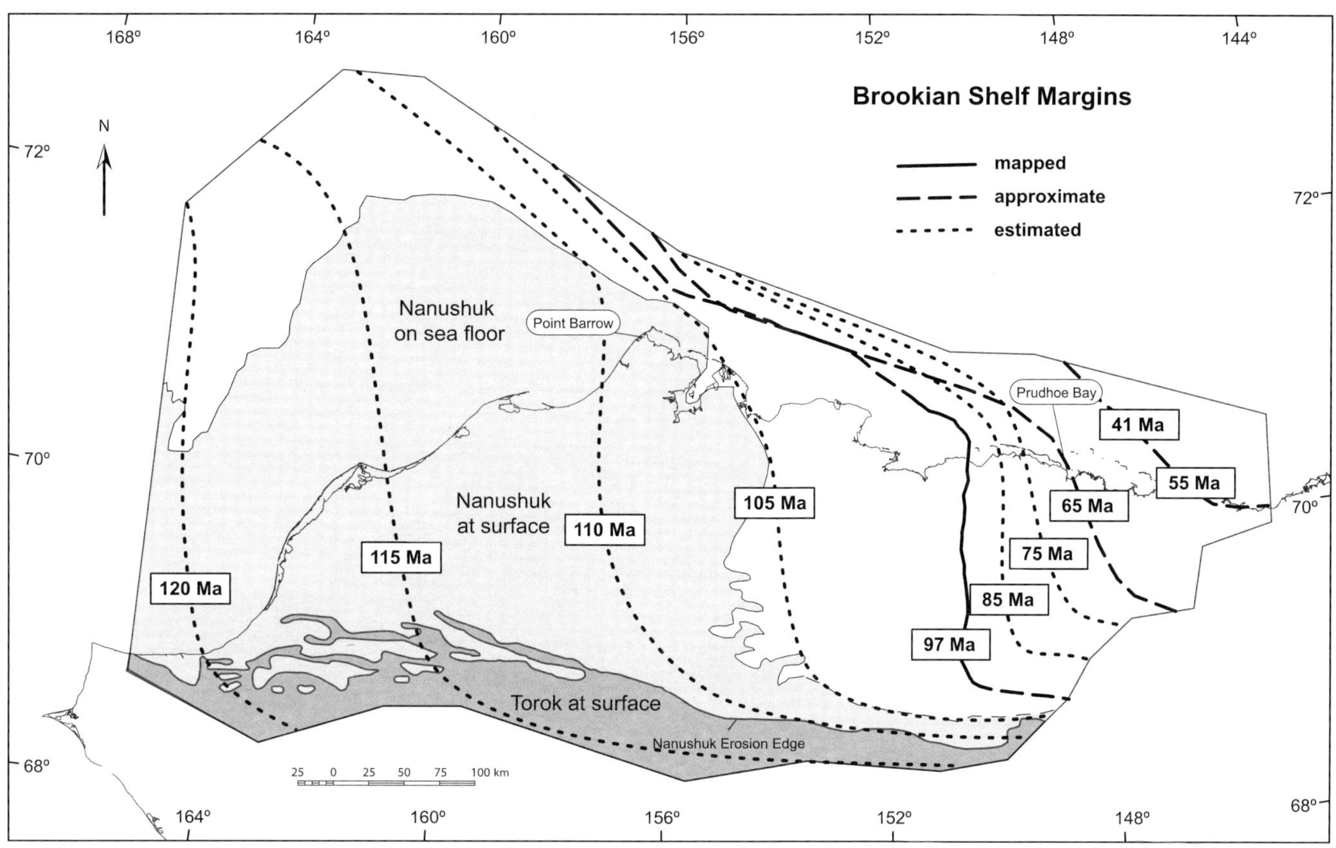

FIGURE 5. Shelf margins of the Brookian sequence used in model for the reconstruction of deposition, uplift, and erosion. Positions of shelf margins for 120, 115, 110, and 105 Ma are only estimated. Shaded areas indicate Nanushuk and Torok formations at surface.

Because of the importance of onlaps and truncations and their effect on hydrocarbon migration and entrapment along the Barrow Arch (e.g., Prudhoe Bay), these relations were carefully input into the model, grid cell by grid cell. In addition, a weathering zone having an assumed thickness of 5 m (16.4 ft) was created where the LCU truncates older strata (Figure 4). The interval between the LCU horizon and the base of Brookian clinoforms incorporates isopach maps of the upper Kuparuk (C/D) Formation and time-equivalent sand bodies, the pebble shale unit, Hue–Gamma Ray Zone (Hue-GRZ), and the Hue Shale (above GRZ).

A few limitations had to be taken into account for the construction of the Ellesmerian and Beaufortian sequences.

- Although seismic and well data are locally abundant in the study area, data are sparse in some areas. Such areas include the Beaufort Sea and the foothills of the Brooks Range, where the LCU and deeper horizons are poorly resolved on the seismic data and have been only rarely penetrated by drilling.
- Pre-Brookian units are characterized by many faults with offsets of up to hundreds of meters, especially along the Barrow Arch (see figure 3–14 in Masterson, 2001). The displacements are less than the resolution of the model (i.e., its grid node spacing) and were therefore not considered in the model.
- Although the Jurassic and Early Cretaceous Kingak Shale is depicted as a single lithostratigraphic unit, it is a progradational siliciclastic unit similar to the Cretaceous and Cenozoic Brookian sequence. The Kingak Shale advanced southward into the Colville Basin from the developing rift margin in the Beaufort Sea area.
- Only Tertiary erosion is accounted for in the model. The amount of rift shoulder uplift and subsequent erosion related to the Lower Cretaceous Unconformity is more complex to estimate and is associated with many uncertainties. However, despite the importance of the LCU with respect to hydrocarbon migration and entrapment, the effect of the related erosion on maturity is minor according to Houseknecht and Bird (2004), who estimate the original thickness of the Kingak Shale not exceeding 1200 m (4000 ft). Moreover, in some areas, such as the Kuparuk River oil field region, the LCU is demonstrated to have minimal erosion and it even grades northward into a conformable sequence (Masterson and Eggert, 1992).

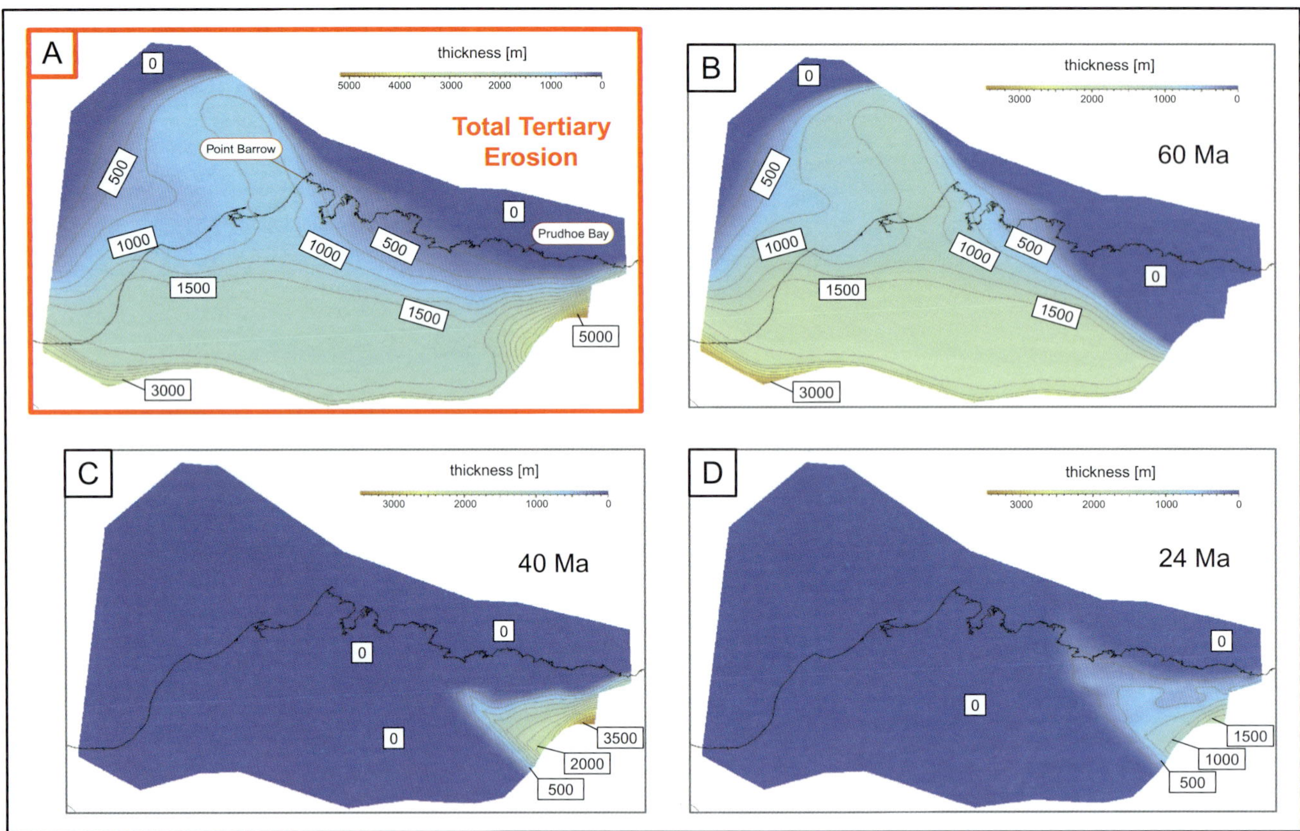

FIGURE 6. Estimated amount of uplift and erosion based on porosity-depth relationship derived from borehole sonic logs (modified from Burns et al., 2007). (A) Total amount of erosion. (B–D) Partitioned amount of erosion (with gradual transitions) according to three events proposed by O'Sullivan et al. (1993, 1997) and O'Sullivan (1999) based on apatite fission-track measurements. Contour intervals are 250 m (820 ft).

Reconstruction of Brookian Sequence

The Brookian sequence (Aptian–present day) extends across the entire model area and has a thickness of up to 8 km (5 mi). Because these deposits are time transgressive, they were reconstructed using nine chronostratigraphic horizons that approximate the characteristic clinoform depositional geometries of the deltaic systems that deposited these sediments. The shelf margins of the successions used in the model are shown in Figure 5.

Both Hue-GRZ and Hue Shale (above GRZ) were added as single layers with a combined thickness of less than 400 m (<1300 ft) below the progradational part of the Brookian sequence, although they represent the distal condensed part of the clinoform package. This simplification was done to facilitate modeling of these units as significant source rock units.

Normally, input data for the model include structure and thickness maps compiled from seismic and well data. For the Brookian sequence, these data are only available for the northeastern part of the model. Because the Brookian sequence is present at the surface everywhere in the model area where varying amounts of uplift and erosion have occurred, it is necessary to use a combination of regional depositional projections and uplift trends to estimate thickness of eroded strata. To determine the maximum paleothickness of these units, the original thicknesses of time-transgressive unconformity-bound successions were progressively reconstructed. This process involved developing a model of successive filling of the basin with a migrating depocenter. During the Tertiary, these successions were uplifted and eroded because of thrusting. Modeling of this situation is complicated because multiple episodes of uplift and erosion must be accounted for in several layers.

The following parameters for the reconstruction are known:

- present-day thickness of Brookian strata between top of Hue Shale and present-day topography;
- amount of total Tertiary uplift, which was subdivided into three episodes that occurred at approximately 60, 40, and 24 Ma;
- sum of the observed stratigraphic and structurally imbricated thickness of Brookian deposits and the total amount of uplift they have been subjected to, which equals the original thickness of Brookian strata prior to erosion.

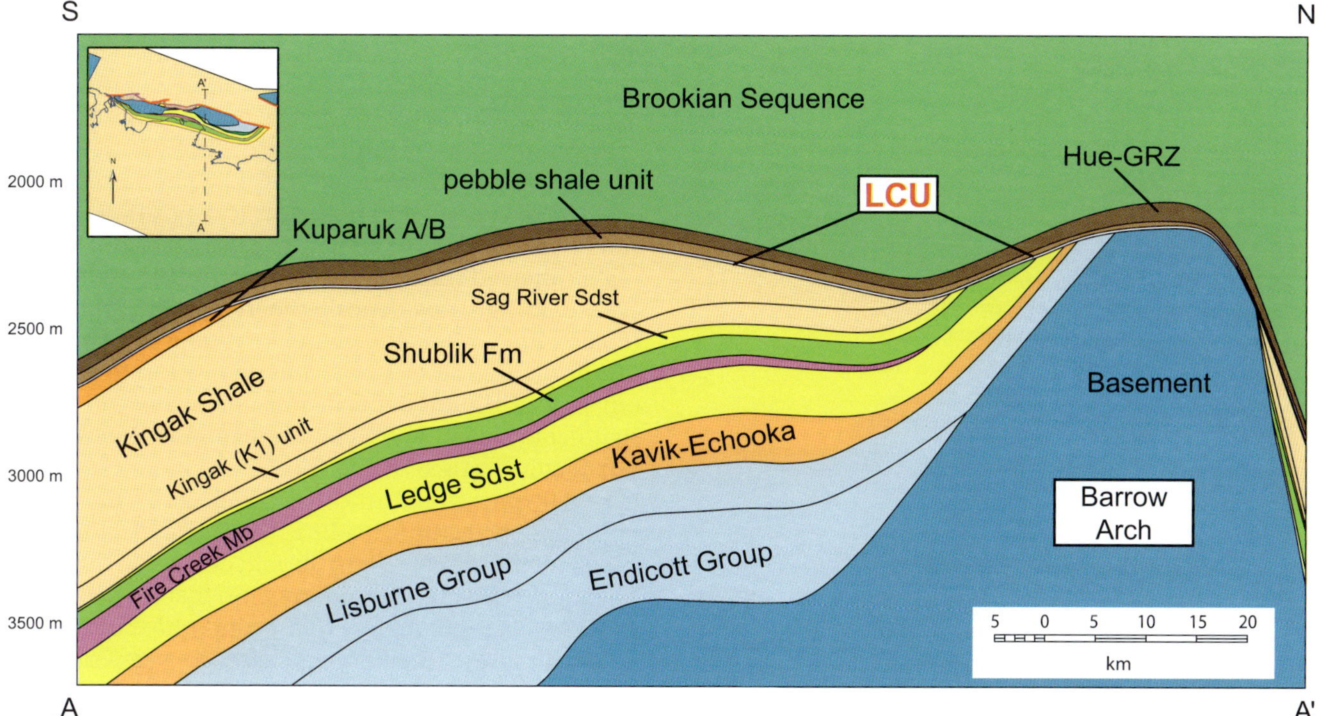

FIGURE 7. Cross section showing an example of the model geometry (truncation and onlap structures) across the Barrow Arch. LCU = Lower Cretaceous Unconformity; Hue-GRZ = Hue-Gamma Ray Zone.

Brooks Range deformation is complicated and characterized by multiple phases of thrusting that include an early phase in the Jurassic and Early Cretaceous and younger phases during Late Cretaceous and Tertiary, the youngest being restricted to the east (Bird and Molenaar, 1992; Cole et al., 1997). To simplify the model, Brooks Range thrusting was omitted. Taking thrusting into account would require a complex three-dimensional (3-D) model based on structural restoration (for example, Baur et al., 2009). However, as a workaround thrusts that duplicate strata were mimicked by sedimentary wedges in this study and were incorporated at 121 and 64 Ma, considering the two major phases of thrusting, during Early Cretaceous and during Late Cretaceous and Tertiary, respectively. The wedges extend toward the northern limit of compressional deformation and account for the volume of sediment that was needed for the burial of older units.

Another limitation of the reconstructed paleogeometry is that uplift and deposition did not occur contemporaneously across the entire model area— an unlikely assumption for an area the size of the Alaska North Slope. In addition, structural restoration of wrench-type deformation in the central part of the Chukchi Shelf during Late Cretaceous–Early Tertiary was not considered because it would require additional complex reconstruction of the Brookian sequence in this area.

The following workflow to create the paleobasin is based on many assumptions but is thought to be the most geologically reasonable and effective.

In the Colville Basin, the Jurassic and Early Cretaceous Kingak Shale is characterized by southward offlapping clinoforms with proximal strata along the Barrow Arch and distal sediments toward the south (figure 3A in Houseknecht and Bird, 2004). The ultimate (youngest) Kingak (Early Cretaceous) shelf and slope horizon was estimated to be constant until the onset of deposition of the Brookian sequence during the Aptian (~121 Ma).

To derive the paleobasin that existed at the onset of Kingak Shale deposition (Triassic–Jurassic boundary), the base of the present-day Kingak Shale surface was raised by the amount of Tertiary uplift (Figure 8A). Onto this horizon, the total thickness of Kingak Shale through the Hue Shale was added, providing a surface that represents the basin floor at the onset of Brookian deposition (Figure 8B).

Crustal loading of the adjacent Brooks Range in the Early Cretaceous was accompanied by a pronounced east-northeast–directed longitudinal filling of the basin. Subsidence caused by sedimentary loading was realized in the model by progressive longitudinal tilting of the basin to the southwest (Figure 8C–E), providing also space for subsequent Cretaceous Brookian clinoforms. The paleogeometry for selected Aptian to Maastrichtian time increments is illustrated in Figure 9, with the Barrow Arch remaining at an elevated position and the foothills being characterized by a deep-water environment with depths up to 2000 m (1.2 mi). Such depths are reasonable for this period, according to Bird and

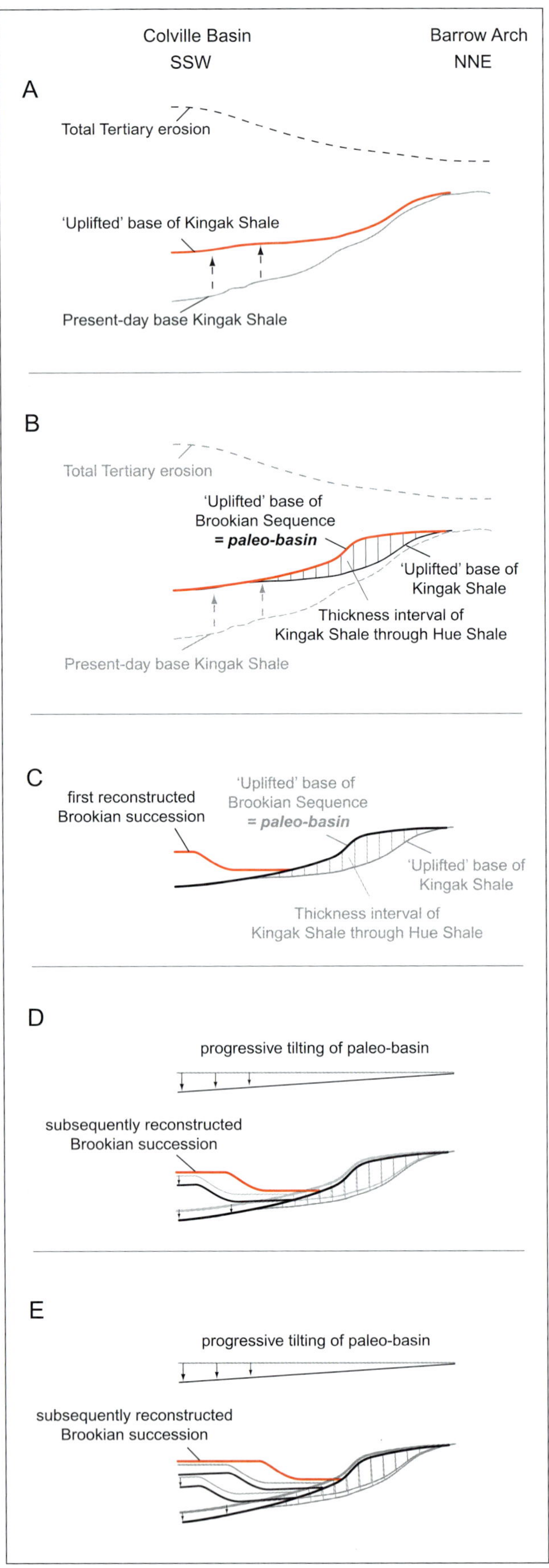

Molenaar (1992) and more recently to Houseknecht et al. (2009). Based on the relief of presently compacted clinoforms, they estimated the water depth for the Nanushuk–Torok interval to be in the range of 600 to 1000 m (0.4–0.6 mi) in the north to 1700 to 2500 m (1.1–1.6 mi) in the south in the axial part of the basin.

The reconstruction of Cenozoic deposition and erosion is summarized in Figure 10. The first uplift event at 60 Ma affected the central and western part of the study area, eroding strata deposited between 115 and 65 Ma (Figure 10A). After this phase, subsidence of the Point Thomson area accelerated, and deposition in the eastern part of the model area was characterized by high sedimentation rates. This may be explained by sedimentary loading caused by the enormous amount of eroded material transported into the foredeep by the 60 Ma uplift and by tectonic loading, which was beginning to occur in the northeastern Brooks Range at this time. The southward marine transgression that followed until 41 Ma was terminated by the second phase of uplift at 40 Ma, resulting in erosion of successions deposited between 97 and 41 Ma (Figure 10B). During the late Eocene to Oligocene, deposition was defined by northward-directed progradation, followed by the third uplift event at 24 Ma, eroding 97 to 25 Ma strata. Compared with the previous uplift events, the 24-Ma event is characterized by minor amounts of erosion being restricted in an area located more to the north-northeast (Figure 10C). Subsequently, the residual volume was filled until the present day. The final reconstructed present-day geometry (Figure 11) approximately matches the surface traces shown in Figure 5.

Essential Elements and Trap Styles

Effective Source Rocks

The North Slope of Alaska is a prolific oil and gas province partly because of multiple effective source rock units (Bird, 2001). Suspected and effective source rock units include the Tertiary Canning Formation, the Cretaceous Seabee Formation, Torok Formation, Hue Shale, and pebble shale unit, the Jurassic to Lower Cretaceous Kingak Shale, the Triassic Shublik Formation,

FIGURE 8. Sketch illustrating the construction of paleobasin geometry during Early Cretaceous Brookian deposition. The base of the present-day Kingak Shale was raised by the amount of Tertiary uplift (A) onto which the total thickness of Kingak Shale through Hue Shale was added, providing the basin floor at the onset of Brookian deposition (B). Subsidence of the basin was established by progressive longitudinal tilting of the basin to the southwest providing space for subsequent Early Cretaceous Brookian successions (C, D, E). See Figures 10 and 11 for the reconstruction of Tertiary deposition, uplift, and erosion.

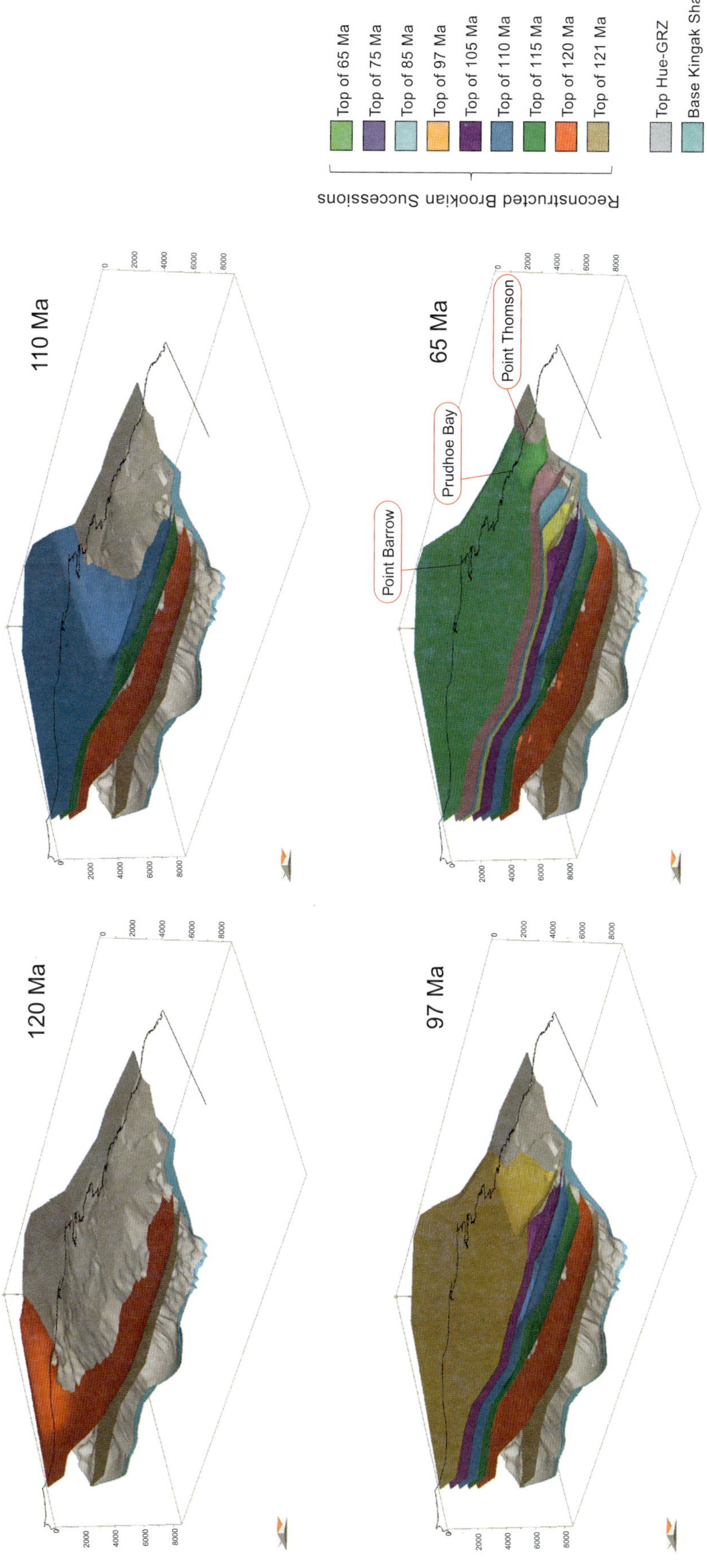

FIGURE 9. Modeled Cretaceous Brookian progradation corresponding to selected shelf-margin locations illustrated in Figure 5. Hue-GRZ = Hue–Gamma Ray Zone.

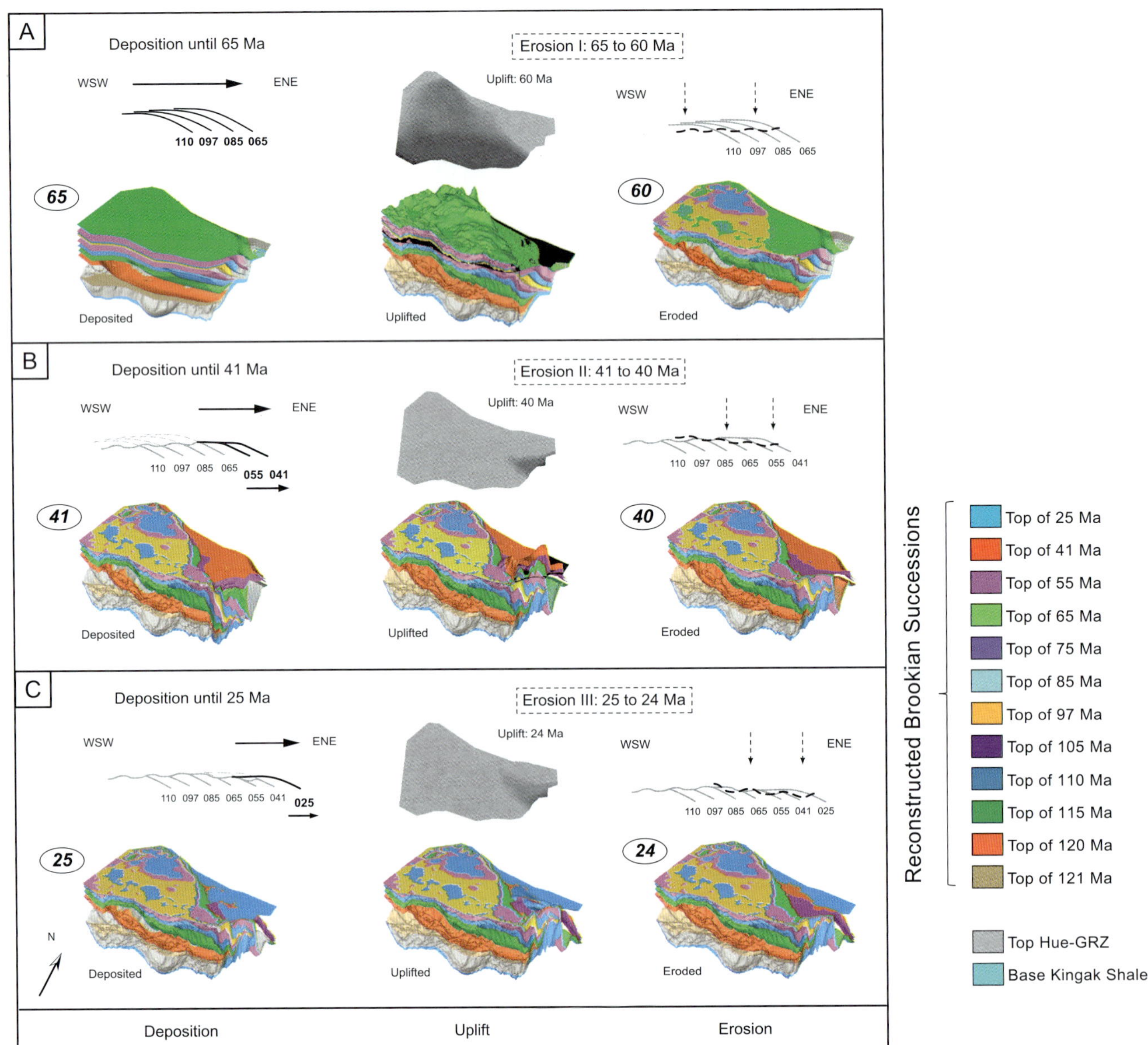

FIGURE 10. Concept used to reconstruct the deposition and erosion of the Tertiary Brookian sequence, illustrating the deposition of prograding sequences (left), their uplift (center), and the situation after erosion (right) for the individual erosional events (A, at 60 Ma; B, at 40 Ma; C, at 24 Ma). Gray density reflects amount of erosional thickness as shown in Figure 6B–D. Hue-GRZ = Hue–Gamma Ray Zone.

the Carboniferous to Permian Lisburne Group, and the Carboniferous Kekiktuk Formation. Hydrocarbon gas from coal in the Mississippian Kekiktuk Formation is omitted from this study. Five of the eight oil source rocks were not modeled because of uncertainties in source character, areal distribution, or generative potential. The three oil source rocks modeled in this study are (1) Middle and Upper Triassic Shublik Formation, (2) basal condensed section within the Jurassic and Lower Cretaceous Kingak Shale, and (3) Cretaceous Hue Shale (see Figure 3).

Detailed information on properties, such as the distribution, thickness, and organic richness, of these source rocks can be found in Peters et al. (2006). Recently, Peters et al. (2008), using multivariate statistics, showed that the relative contribution of oil from the Shublik Formation decreases along the Barrow Arch from Point Barrow in the northwest to Point Thomson in the southeast, whereas that from the Hue-GRZ increases.

Seal Rocks and Trap Types

Seal rocks on the Alaska North Slope are mainly dense shaly sequences that overlie the reservoir rocks. Trap types include stratigraphic, structural, and combination

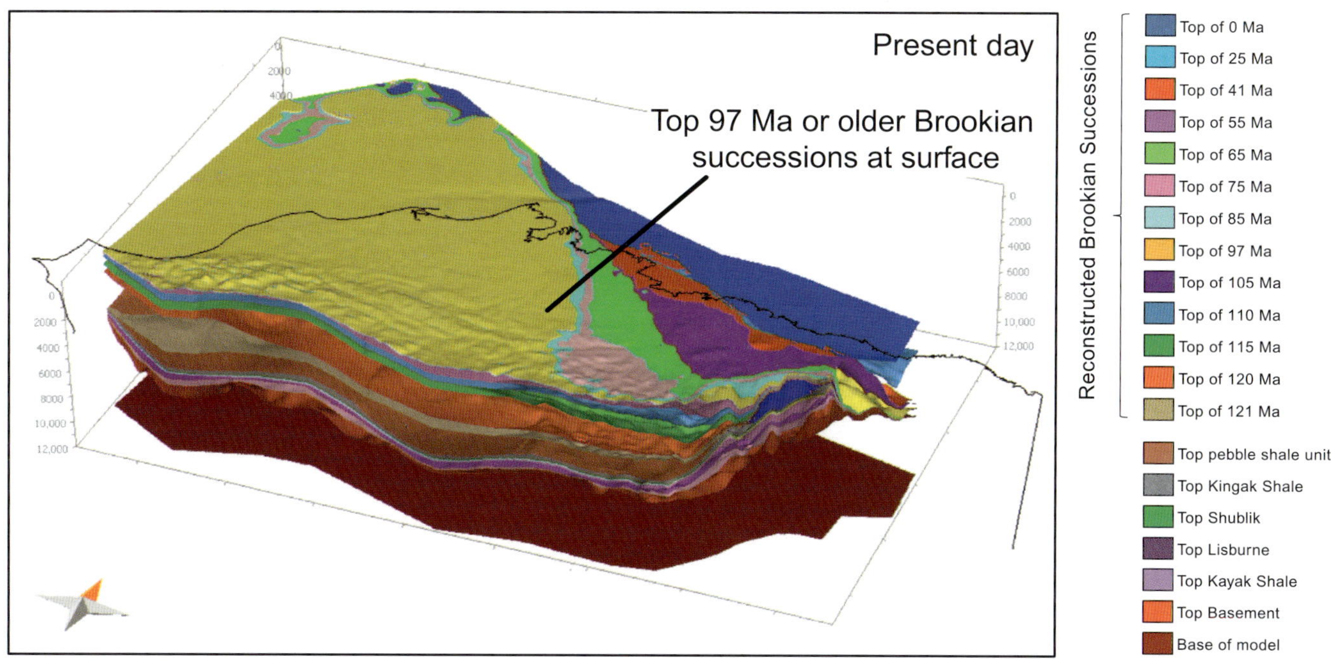

FIGURE 11. Modeled present-day geometry. Top 97 Ma succession (mustard color) approximates the surface exposure of the Nanushuk and Torok formations (compare with Figure 5).

traps. Stratigraphic traps are present in Beaufortian shelf sands (e.g., Alpine) and in Brookian turbidite reservoirs (e.g., Badami) whereas structural traps include anticlinal traps in the fold and thrust belt in the Colville Basin (Umiat [Molenaar, 1982]) and fault-related traps (e.g., West Sak field). However, most hydrocarbon accumulations occur in combination structural-stratigraphic traps along the Barrow Arch (e.g., Prudhoe Bay, Kuparuk River, Endicott fields). The structural aspect of these traps formed during the rifting event whereas the stratigraphic component is related to erosional truncation by the breakup unconformity (the LCU) and postrift burial and subsidence.

Input Data and Calibration

In addition to the gridded surfaces of buried rock units, input data include lithologies characterized by physical properties, such as thermal conductivity, porosity, and permeability, and boundary conditions, including present and past basal heat flow, and surface or sediment-water interface temperatures that are corrected for present and past water depths. Abundant well data exist for the Alaska North Slope and allow calibration of both pressure and temperature in the subsurface.

Ages, Facies, and Lithologies

Ages of deposition and erosion are based on regional studies (Hubbard et al., 1990; O'Sullivan et al., 1993, 1997; Bird, 1994, 2001; O'Sullivan, 1999). Most of the pre-Brookian layers are assumed to consist of one facies. Because of its importance as a major hydrocarbon migration pathway, a facies map was constructed for the LCU weathering zone, which is based on a map of regions truncated by the LCU (Figure 4). The LCU weathering zone is defined as a zone with a uniform thickness of 5 m (16.4 ft) in areas where the LCU has erosionally truncated older strata. The LCU facies map portrays the lithologies of the rocks underlying the LCU. The LCU weathering zone was created as fluids invaded along the LCU surface, resulting in enhanced permeability through dissolution of diagenetic cements in the underlying lithologies (Woidneck et al., 1987). In the model, the physical properties of these weathered lithologies can be adjusted, which allows control of the effectiveness of the LCU weathering zone as a migration pathway. Control of this migration pathway allows calibration of presence, volumes, and phases of simulated accumulations using known accumulations. For example, weathered carbonate rocks of the Lisburne Group are assumed to have different properties (e.g., permeability) compared with the original Lisburne lithology and the weathered zone of the Kingak Shale or Sag River Sandstone.

For the prograding Brookian deposits, different lithofacies were assigned according to the occurrence of topset, foreset, and bottomset facies. Several units in the model consist of mixtures of predominantly siliciclastic rock types, for which porosity and permeability were adjusted to calibrate the pressure. This was performed in two steps: one related to rock compressibility, the other to permeability (overpressure calibration) (Hantschel and Kauerauf, 2009). Pressure data for calibration, mainly from drill-stem and repeat formation tests, were quality controlled by removing data where measured pore pressures were lower than hydrostatic or higher than lithostatic pressure.

Source Rock Properties

Contour maps of original total organic carbon (TOC_o) and original hydrogen index (HI_o) for the source rocks of Shublik Formation, sequence set K1 of Kingak Shale, and Hue Shale were taken from Peters et al. (2006) and extrapolated to the limits of this study. As the Hue Shale was undifferentiated by Peters et al. (2006), the same source rock characteristics were assigned to Hue-GRZ and Hue Shale above the GRZ.

Thermally immature source rock samples from the Shublik Formation, sequence set K1 of the Kingak Shale, and Hue Shale were analyzed using the new Phase Kinetics procedure developed and calibrated for pressure-volume-temperature–controlled prediction of petroleum phases and properties, such as API and gas-oil ratio (di Primio and Horsfield, 2006). The results of the analyses were assigned to the respective source rocks.

Boundary Conditions

For heat-flow analysis, basal heat flow and sediment-water interface (offshore) or surface (onshore) temperature (SWIT) are the main boundary conditions. In the model, the SWIT is a function of the (paleo–)water depth (PWD).

The thermal regime of the Alaska North Slope region is complex and was the subject of many studies (Lachenbruch et al., 1988; Bird and Molenaar, 1992; Deming et al., 1992, 1996). The influence of factors that control the regional thermal pattern is still under debate. Some of these factors are (1) Tertiary uplift and erosion in the foothill area of the Brooks Range orogeny, whereas toward the north-northeast, subsidence and sedimentation continued; (2) structural imbrication in the foothills in the Early Cretaceous, where thermal maturity data suggest that heating occurred after stacking (Bird and Molenaar, 1992; Moore et al., 2004); (3) forced convection where heat is transported northward by deep basin-scale groundwater flow driven by hydraulic head established in the Brooks Range and foothills (Deming et al., 1992, 1996); (4) the present-day permafrost that covers most of the study area; and (5) variable thermal gradients throughout the study area.

The upper boundary condition for temperature calculations is determined by the sediment-water interface (offshore) and surface (onshore) temperatures. The estimated mean paleosurface temperature was modeled based on the paleotectonic position of the Arctic plate (Wygrala, 1989). Since the Pleistocene, northern Alaska has been affected by low surface temperatures and permafrost (Lachenbruch et al., 1988). To account for this effect, the present-day surface temperature in the onshore area of the model was reduced to -8°C, pushing the modeled 0°C-isotherm to deeper levels near the base of the present-day permafrost (see also Kroeger et al., 2008).

The SWITs were corrected for present and past water depth. For the pre-Brookian sequences, uniform values were assigned, averaged for the total model area in the range of 200 to 800 m (0.125–0.5 mi). For the Brookian sequence, the PWD was reconstructed based on the amplitude of the clinoforms.

According to Deming et al. (1992), heat-flow values coincide with basement topography, resulting in high heat-flow values along the Barrow Arch and low values in the foothill area. They concluded that this thermal pattern was caused by convection of groundwater, which might have persisted for the last tens of millions of years. As input for the present-day heat flow, a map was constructed that reflects the correlation between basement depth and heat flow. In regions where the basement is shallow, the present-day heat flow is characterized by high values (as much as 90 mW/m^2 along the Barrow Arch), in regions with deep basement by low values (e.g., as low as 25 mW/m^2 in the foothills region). Additional deep heat-flow values estimated by Deming et al. (1992) were also used in this map.

Heat flow was calibrated against vitrinite reflectance and later cross-checked with Horner-corrected bottom-hole temperature data. The vitrinite reflectance data required special screening, for example, removal of reverse depth trends believed to result from recycled vitrinite. Calibration focused on the Brookian instead of the deeper Ellesmerian sequence, where solid bitumen may interfere with the reflectance measurements. Because of the size of the model, a filtered subset with a 2-km (1.24-mi) grid was used for calibration.

The heat-flow calibration was performed automatically with PetroMod heat flow calibration tool, for which areas of interest around 105 selected wells were taken into account. For the first calibration run, the paleo–heat flow was kept constant from 400 until 60 Ma, with a uniform value of 60 mW/m^2, followed by interpolation toward the constructed heat-flow map at the present day. Proposed heat-flow shifts were checked for their physical and geologic reasonability and taken into account for the areas of interests of the heat-flow distribution at the present day and for the period between 120 and 60 Ma. These maps were interpolated, smoothed, and later assigned to the model. The final assignment of the heat-flow conditions is shown in Figure 12. Before 127 Ma (Figure 12A), a uniform heat flow of 60 mW/m^2 was assigned, followed by automatically calibrated heat-flow

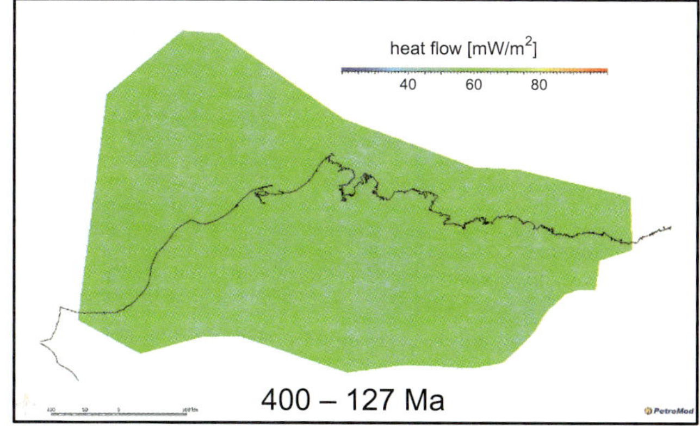

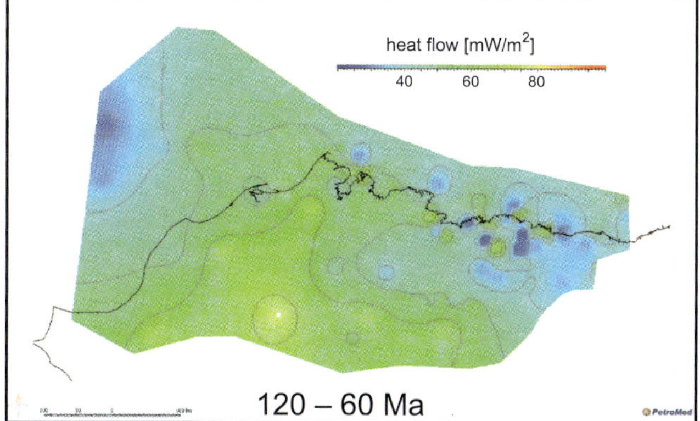

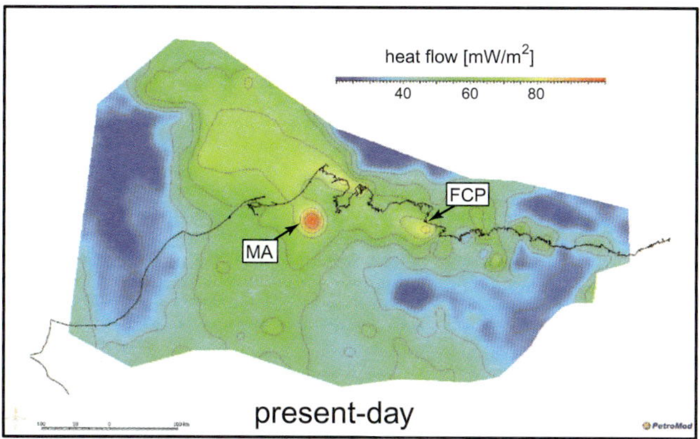

FIGURE 12. Calibrated heat-flow maps assigned to the model. Periods from 127 to 120 Ma and 60 Ma to the present day, respectively, are interpolated by the simulator. Contour intervals are 10 mW/m^2. See text for calibration workflow. Present-day heat-flow distribution correlates with depth of top of basement showing high heat-flow values where the basement is shallow (see Deming et al., 1992; see arrows pointing to the Meade Arch [MA] and the Fish Creek platform [FCP]).

maps for the period between 120 and 60 Ma (Figure 12B) and for the present day (Figure 12C). Intermediate events were automatically interpolated by the simulator.

The calibrated heat-flow map for the present day used for the model is consistent with published measurements (Deming et al., 1992). For example, shallow basement corresponds to high heat flow at the Meade Arch and the Fish Creek Platform, whereas deep basement correlates with low heat flow as, for example, in the Colville Basin (Figure 12). Model predictions of vitrinite reflectance after Sweeney and Burnham (1990) are in agreement with measured values of most wells (Figure 13).

RESULTS AND DISCUSSION

Evolution of Rock Unit Geometry through Time

Forward deterministic computations were applied to simulate the burial history of the rock units and the generation-migration-accumulation of petroleum within a 3-D cube through time (Hantschel and Kauerauf, 2009). For simulations to be run in reasonable amounts of time (<48 hr), the input model was downsampled to an output grid node spacing of 6 km (3.7 mi) to calculate pressure and temperature throughout the basin, while migration and accumulation were simulated with a grid distance of 2 km (1.24 mi).

A key aspect of the 3-D Alaska North Slope model is that it incorporates the time-transgressive deposition of the Cretaceous–Tertiary Brookian sequence, the thickness difference between the foothill region and the Barrow Arch, and the diachronous pulses of Tertiary uplift and erosion (Figure 14). The model indicates that the thermal maturity of pre-Brookian deposits was dynamic because it was controlled mainly by progradation of the Brookian sequence (Figure 15).

Hydrocarbon Generation-Migration-Accumulation

Modeling results for the three most important oil source rock units (Triassic Shublik Formation, the Jurassic condensed basal section of Kingak Shale, and the Cretaceous Hue Shale) show grossly similar timing of hydrocarbon generation and directions of migration that are directly related to Cretaceous–Tertiary (Brookian) depositional patterns. Therefore, for illustrative purposes, we focus only on the Shublik Formation in the following discussion (Figures 16, 17).

Similar to the profile illustrating the history of thermal maturation (Figure 15), the timing of Shublik Formation hydrocarbon generation was controlled by the thickness of overburden rocks, the eastward prograding Brookian sequence.

The modeling suggests that in the axis of the Colville Basin, beneath the foothills of the Brooks Range, hydrocarbon generation started after 121 Ma. By 97 Ma, transformation ratios of 100% show that most of the

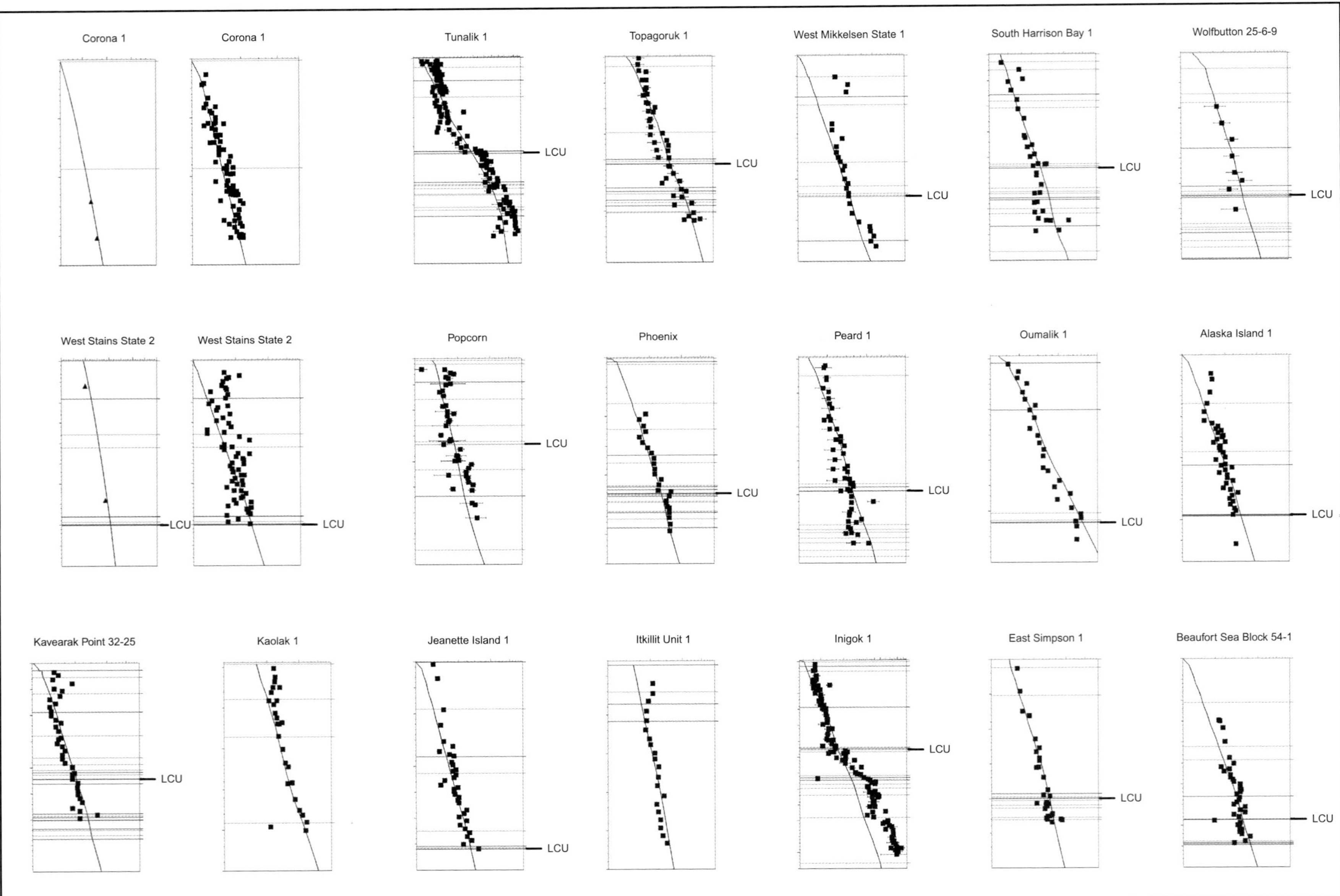

Figure 13. One-dimensional extractions from the three-dimensional model at selected well locations showing correlation of modeled (lines) and measured data (triangles, temperature; squares, vitrinite reflectance).

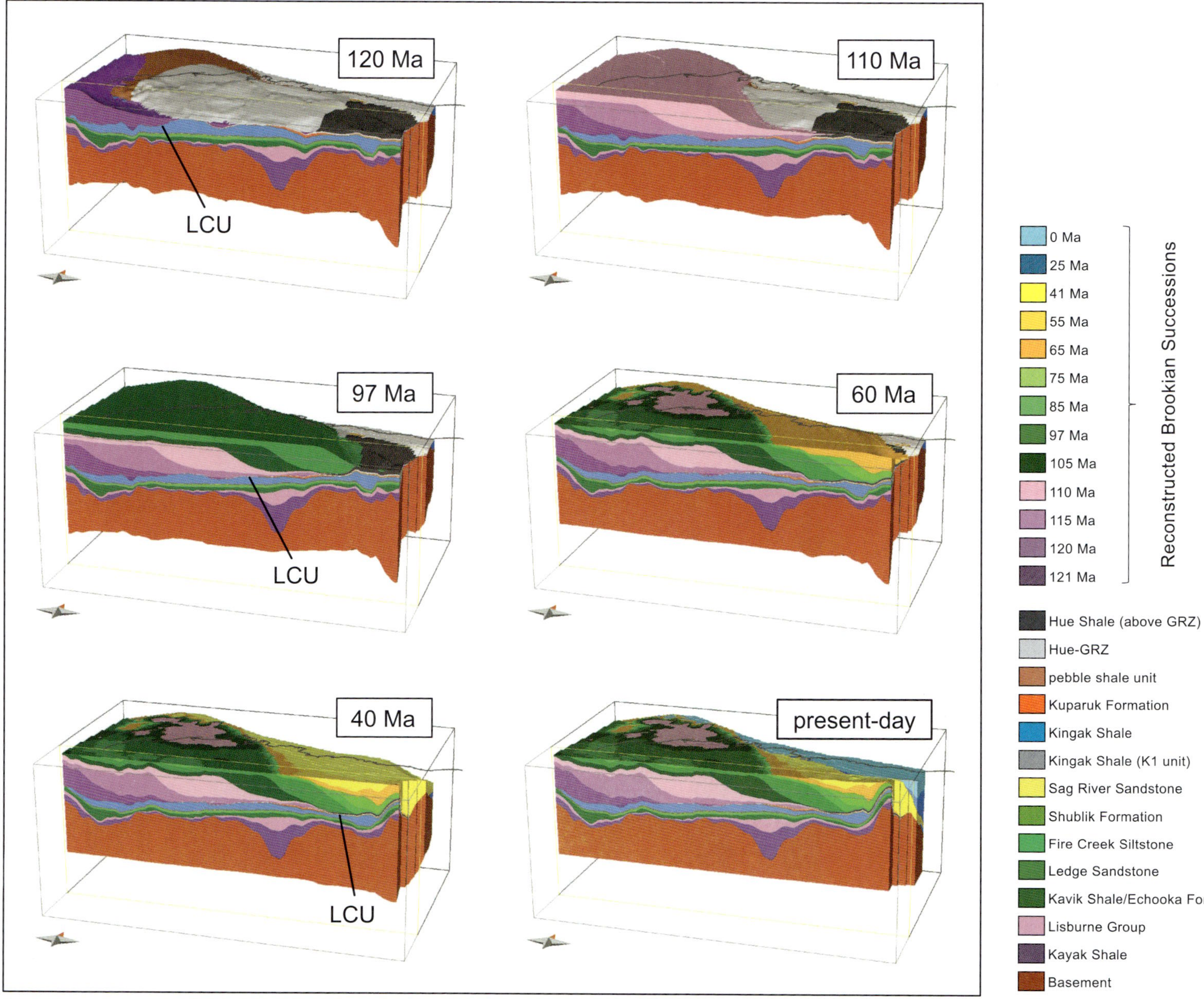

FIGURE 14. Selected time slices showing the modeled progradation of the Brookian sequence deltaic deposits from southwest to northeast across northern Alaska. Cut plane (in front) is located in the deepest part of the Colville Basin. Note position of the Lower Cretaceous Unconformity (LCU). Coastline is indicated by the black line. Hue-GRZ = Hue–Gamma Ray Zone.

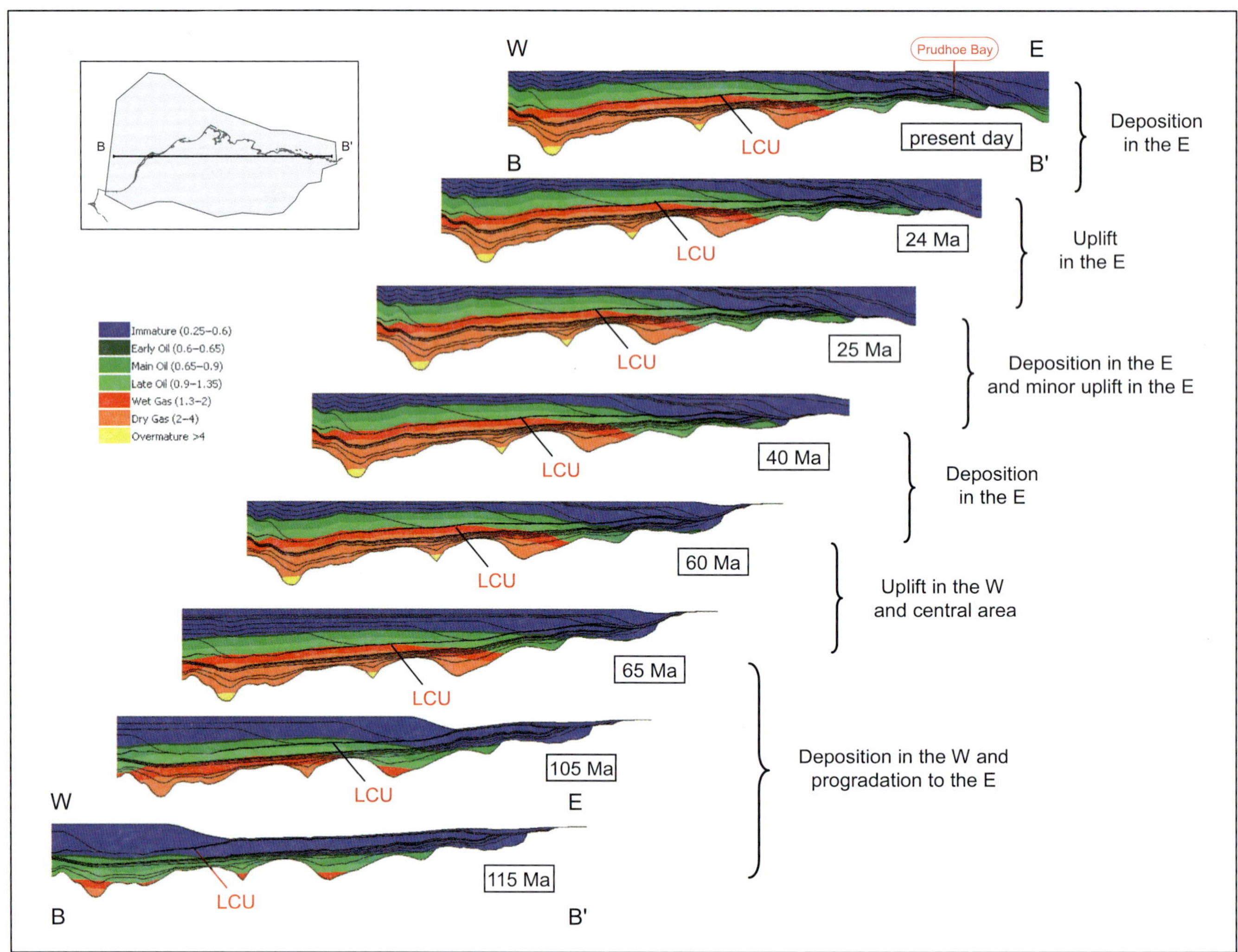

FIGURE 15. Reconstructed paleogeometry along a west-east cross section showing the effect of the prograding Brookian sequence on maturity. Note that source rocks modeled in this study are still immature at Prudhoe Bay. LCU = Lower Cretaceous Unconformity.

Shublik Formation in the foothills area was already overmature, whereas transformation ratios decrease markedly northward to 0% on the rift shoulder (Figure 16). In the southeastern part of the model, major transformation did not begin until the Paleocene and early Eocene, and in the northeastern part of the model (assuming the Shublik Formation is present), not before the latest Oligocene and Miocene. Based on simulation results, the present-day transformation ratio of the Shublik Formation is less than 2% in the Prudhoe Bay area.

The time-transgressive deposition of the Brookian sequence in combination with overall basin geometry also controls the direction of hydrocarbon migration (Figure 17). Because of mainly southward-dipping strata on the northern flank of the Colville Basin, most migration pathways were directed toward the Barrow Arch, with hydrocarbons accumulating mainly in combination structural-stratigraphic traps. As a consequence of eastward-migrating depocenters of the Brookian sequence, the focus of hydrocarbon migration shifted to the east with time, and thus, hydrocarbon entrapment occurred first in the western part of the Barrow Arch and later in the east. Orogenic development of the northeast Brooks Range in the Tertiary produced crustal loading that tilted the Barrow Arch eastward and generated voluminous clastic debris that filled the Colville Basin, overtopped the Barrow Arch, and built the adjacent Beaufort Sea passive margin. If Shublik Formation is present beneath the passive margin, the model shows that generation would occur at this time and migration would be southward toward the arch and westward up along its axis toward the giant Prudhoe Bay field. Regional tilting and the inferred spilling and westward remigration of oil from the Prudhoe Bay accumulation (Erickson and Sneider, 1997) are consistent with model results.

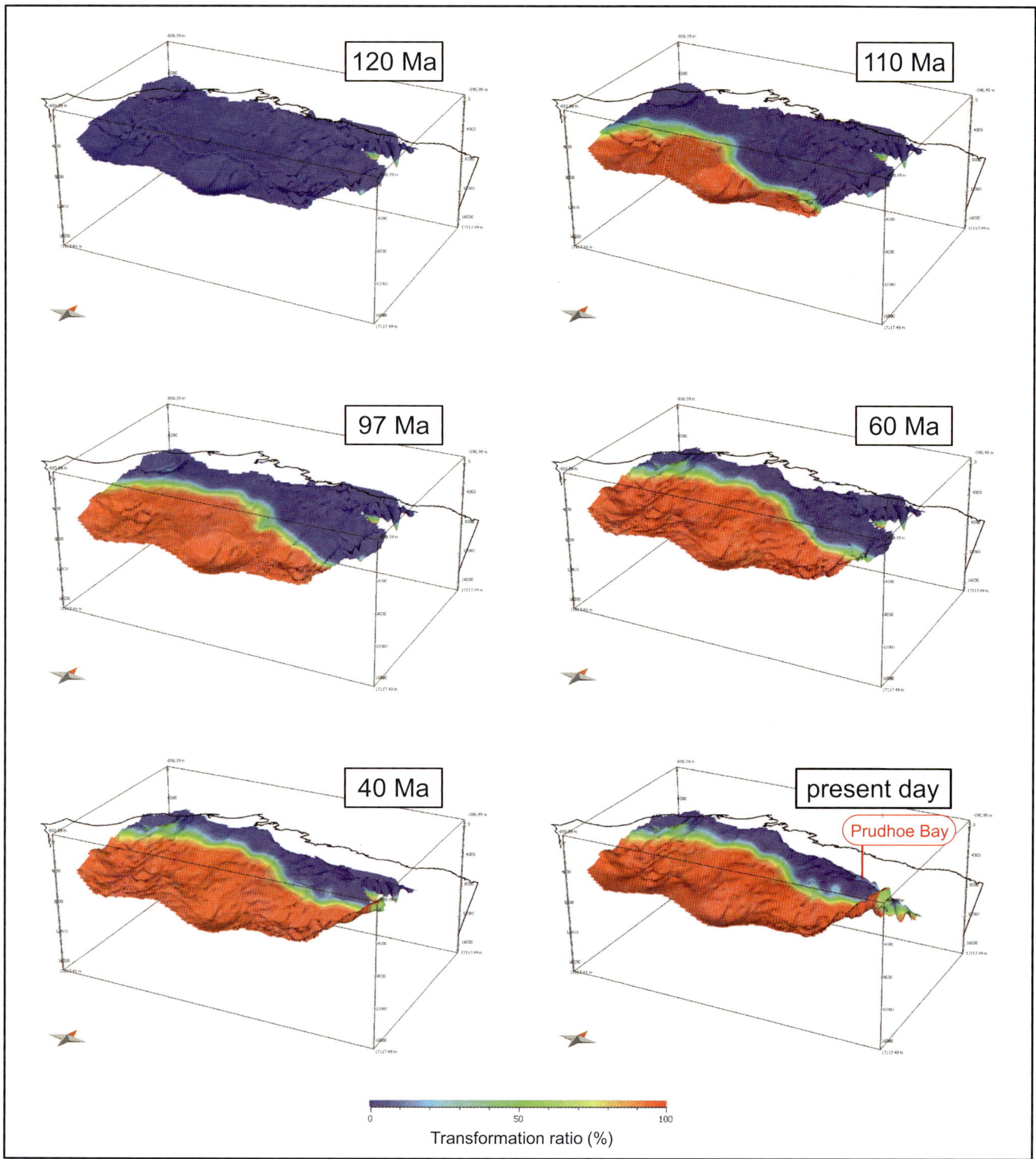

FIGURE 16. Top Shublik surface through time shows calculated fractional conversion (%) of Shublik organic matter to petroleum (blue, 0%; red, 100%) indicating very minor hydrocarbon generation from Shublik Formation at Prudhoe Bay. Coastline is indicated by the black line.

CONCLUSIONS

The Alaska North Slope, including the adjacent Beaufort and Chukchi continental shelves, is one of the remaining petroleum exploration frontiers, and is estimated to contain most of the undiscovered oil and gas resources in the North American circum-Arctic. The 3-D petroleum system modeling study outlined here provides

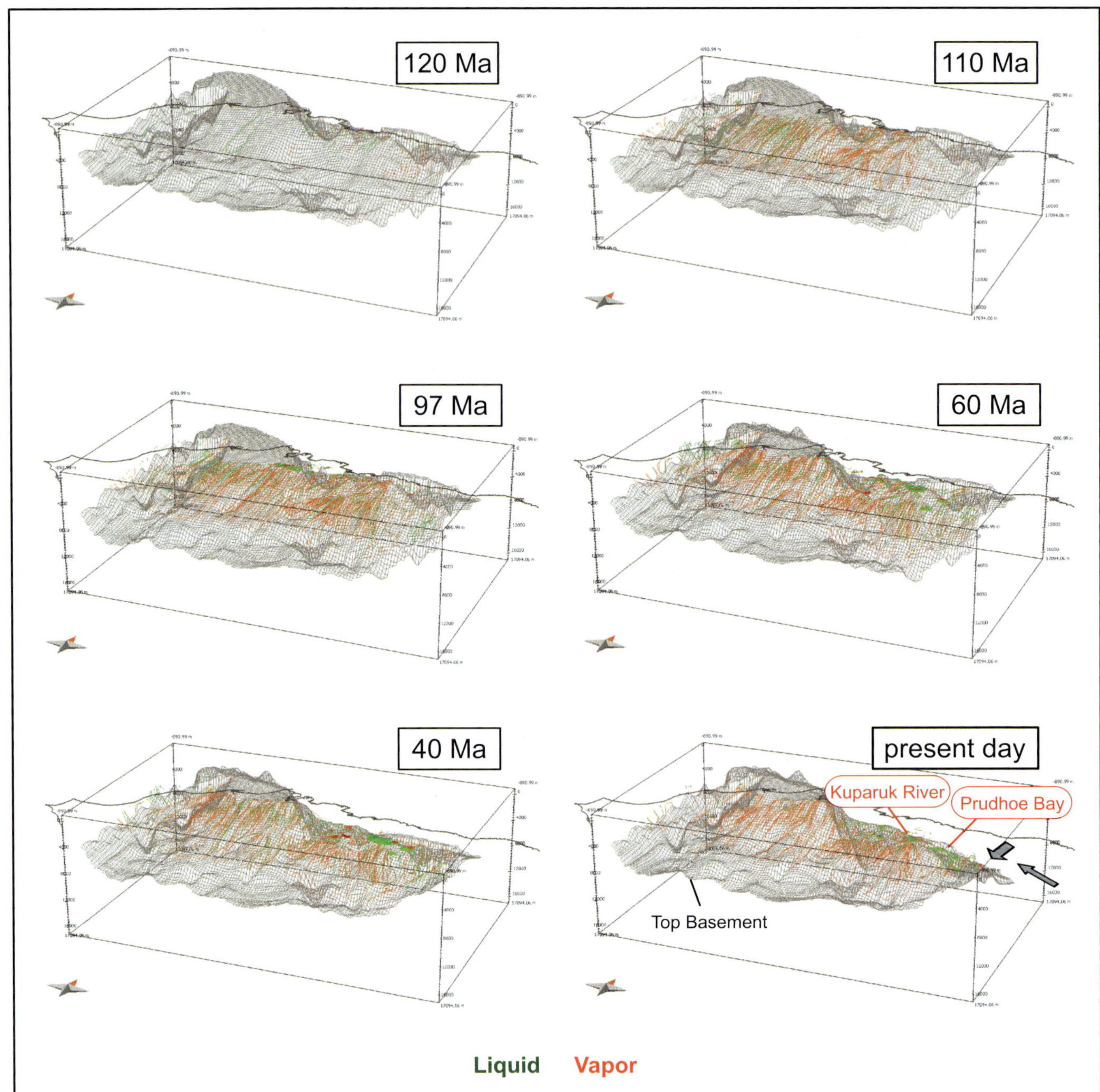

FIGURE 17. Simulated migration pathways and accumulations of hydrocarbons generated from the Shublik Formation source rock at selected times. Liquid (green) and vapor (red) represent hydrocarbon phases at in-situ conditions. Arrows point to west- and south-directed hydrocarbon migration in the northeastern part of the model area because of substantial subsidence in the Beaufort continental margin during the late Cenozoic.

a unique geologic framework to help reduce exploration risk and assess the remaining potential hydrocarbon resources in this remote, but prolific, province.

The present-day geology incorporated into the model is based on a large amount of seismic data (>48,000 km [30,000 mi]) and more than 400 wells. Among the many geologic relations used, the model considers (1) details of Ellesmerian and Beaufortian strata, including onlap and truncation structures, and (2) the progradational geometry of the Brookian sequence and its impact on the generation, migration, entrapment, and loss of hydrocarbons through geologic time. The effects of Tertiary erosion events are also taken into account.

The simulation results show that hydrocarbon generation and early migration were strongly controlled by the eastward-prograding Brookian sequence depocenter.

Major migration pathways of liquid and vapor phases in the model area are mainly directed from south to north from the Colville Basin toward the Barrow Arch. Only deep source rock burial in the passive margin beneath the Beaufort Sea north of the rift shoulder during the Tertiary resulted in westward and southward migration toward the Barrow Arch. Spilling of previously trapped hydrocarbons and tertiary migration pathways related to eastward tilting of the Barrow Arch generally run parallel to the arch and thus perpendicular to the major direction of secondary migration. Along the Barrow Arch, combination structural-stratigraphic trap formation preceded generation-migration, resulting in major oil accumulations.

ACKNOWLEDGMENTS

We thank Kelly Dempster and Tom Moore for their thorough reviews that substantially improved the manuscript. We thank Carolyn Lampe for her work on a previous more restricted Alaska North Slope model and for her willingness to share her technical expertise as the authors put together this more expansive model. We also thank our PetroMod Team and Zenon Valin for his continued support to provide digital maps, GIS support, and other actions as they were needed.

REFERENCES CITED

Baur, F., M. Di Benedetto, T. Fuchs, C. Lampe, and S. Sciamanna, 2009, Integrating structural geology and petroleum systems modeling: A pilot project from Bolivia's fold and thrust belt: Marine and Petroleum Geology, v. 26, p. 573–579, doi:10.1016/j.marpetgeo.2009.01.004.

Bird, K. J., 1994, The Ellesmerian(!) petroleum system, North Slope of Alaska, United States, *in* L. B. Magoon and W. Dow, eds., The petroleum system: From source to trap: AAPG Memoir 60, p. 339–358.

Bird, K. J., 2001, Alaska: A twenty-first century petroleum province, *in* M. W. Downey, J. C. Threet, and W. A. Morgan, eds., Petroleum provinces of the twenty-first century: AAPG Memoir 74, p. 137–165.

Bird, K. J., and C. M. Molenaar, 1992, The North Slope foreland basin, *in* R. W. Macqueen and D. A. Leckie, eds., Foreland basins and fold belts: AAPG Memoir 55, p. 363–393.

Burns, W. M., D. O. Hayba, E. L. Rowan, and D. W. Houseknecht, 2007, Estimating the amount of eroded section in a partially exhumed basin from geophysical well logs: An example from the North Slope, U.S. Geological Survey professional paper 1732-D, p. 1–18.

Cole, F., K. J. Bird, J. Toro, F. Roure, P. B. O'Sullivan, M. Pawlewicz, and D. G. Howell, 1997, An integrated model for the tectonic development of the frontal Brooks Range and Colville Basin 250 km west of the Trans-Alaska Crustal Transect: Journal of Geophysical Research, v. 102, no. B9, p. 20,685–20,708.

Deming, D., J. H. Sass, A. H. Lachenbruch, and R. F. de Rito, 1992, Heat flow and subsurface temperature as evidence for basin-scale ground-water flow, North Slope Alaska: Geological Society of America Bulletin, v. 104, p. 528–542, doi:10.1130/0016-7606(1992)104<0528:HFASTA>2.3.CO;2.

Deming, D., J. H. Sass, and A. H. Lachenbruch, 1996, Heat flow and subsurface temperature, North Slope of Alaska, *in* Thermal evolution of sedimentary basins in Alaska: U.S. Geological Survey Bulletin, v. 2142, p. 21–42.

di Primio, R., and B. Horsfield, 2006, From petroleum type organofacies to hydrocarbon phase prediction: AAPG Bulletin, v. 90, p. 1031–1058, doi:10.1306/02140605129.

Erickson, J. W., and R. M. Sneider, 1997, Structural and hydrocarbon histories of the Ivishak (Sadlerochit) reservoir, Prudhoe Bay field: SPE Reservoir Engineering, February 1997, p. 18–22.

Gautier, D. L., et al., 2009, Assessment of undiscovered oil and gas in the Arctic: Science, v. 324, p. 1175–1179, doi:10.1126/science.1169467.

Grantz, A., and S. D. May, 1982, Rifting history and structural development of the continental margin north of Alaska, *in* J. S. Watkins and C. L. Drake, eds., Studies in continental margin geology: AAPG Memoir 34, p. 77–100.

Grantz, A., S. D. May, and D. A. Dinter, 1988, Geologic framework, petroleum potential, and environmental geology of the United States Beaufort and the northeasternmost Chukchi seas, *in* G. Gryc, ed., Geology and exploration of the National Petroleum Reserve in Alaska, 1974 to 1982: U.S. Geological Survey professional paper 1399, p. 231–255.

Hantschel, T., and A. I. Kauerauf, 2009, Fundamentals of basin and petroleum systems modeling: Berlin, Springer-Verlag, 476 p.

Houseknecht, D. W., and K. J. Bird, 2004, Sequence stratigraphy of the Kingak Shale (Jurassic–Lower Cretaceous), National Petroleum Reserve in Alaska: AAPG Bulletin, v. 88, no. 3, p. 279–302, doi:10.1306/10220303068.

Houseknecht, D. W., and K. J. Bird, 2005, Oil and gas resources of the Arctic Alaska petroleum province, U.S. Geological Survey professional paper 1732-A, p. 1–11.

Houseknecht, D. W., K. J. Bird, and C. J. Schenk, 2009, Seismic analysis of clinoform depositional sequences and shelf-margin trajectories in Lower Cretaceous (Albian) strata, Alaska North Slope: Basin Research, v. 21, p. 644–654, doi:10.1111/j.1365-2117.2008.00392.x.

Hubbard, R. J., S. P. Edrich, and R. P. Rattey, 1990, Geological evolution and hydrocarbon habitat of the "Arctic Alaska microplate," *in* J. Brooks, ed., Classic petroleum provinces: Geological Society (London) Special Publication 50, p. 143–187.

Kroeger, K. F., R. Ondrak, R. di Primio, and B. Horsfield, 2008, A three-dimensional insight into the Mackenzie Basin (Canada): Implications for the thermal history and hydrocarbon generation potential of Tertiary deltaic sequences: AAPG Bulletin, v. 92, p. 225–247, doi:10.1306/10110707027.

Lachenbruch, A. H., J. H. Sass, L. A. Lawver, M. C. Brewer, B. V. Marshall, R. J. Munroe, J. P. Kennelly Jr., S. P. Galanis Jr., and T. H. Moses Jr., 1988, Temperature and depth of permafrost on the Arctic Slope of Alaska, *in* G. Gryc, ed., Geology and exploration of the National Petroleum Reserve in Alaska, 1974 to 1982, p. 645–656.

Lampe, C., K. E. Peters, L. B. Magoon, K. J. Bird, and P. G. Lillis, 2003, Petroleum systems of the Alaskan North Slope: A numerical journey from source to trap: U.S. Geological Survey open-file report 03-326, 3 sheets: http://geopubs.wr.usgs.gov/open-file/of03-326/ (accessed November 12, 2010).

Lerand, M., 1973, Beaufort Sea, *in* R. G. McGrossan, ed., The future petroleum provinces of Canada: Their geology and potential: Canadian Society of Petroleum Geologists Memoir 1, p. 315–386.

Magoon, L. B., P. G. Lillis, K. J. Bird, C. Lampe, and K. E. Peters, 2003, Alaskan North Slope petroleum systems: U.S. Geological Survey open-file report 03-0324, 3 sheets: http://geopubs.wr.usgs.gov/open-file/of03-324/ (accessed November 12, 2010).

Masterson, W. D., 2001, Petroleum filling history of central Alaskan North Slope fields: Ph.D. thesis, University of Texas at Dallas, Dallas, Texas, 222 p.

Masterson, W. D., and J. T. Eggert, 1992, Kuparuk River field, U.S.A., *in* N. H. Foster and E. A. Beaumont, eds., Stratigraphic traps III, AAPG Treatise of petroleum geology, Atlas of oil and gas fields, p. 257–284.

Molenaar, C. M., 1982, Umiat field, an accumulation in a thrust-faulted anticline, North Slope of Alaska, *in* R. B. Powers, ed., Geologic studies of the Cordilleran thrust belt: Rocky Mountain Association of Geologists, p. 537–548.

Molenaar, C. M., 1988, Depositional history and seismic stratigraphy of Lower Cretaceous rocks in the National Petroleum Reserve in Alaska and adjacent areas, *in* G. Gryc, ed., Geology and exploration of the National Petroleum Reserve in Alaska, 1974 to 1982, p. 593–621.

Moore, T. E., W. K. Wallace, K. J. Bird, S. M. Karl, C. G. Mull, and J. T. Dillon, 1994, Geology of northern Alaska, *in* G. Plafker, D. L. Jones, and H. C. Berg, eds., The geology of Alaska: Geological Society of America, Boulder, Colorado, v. G-1, p. 49–140.

Moore, T. E., C. J. Potter, P. B. O'Sullivan, K. L. Shelton, and M. B. Underwood, 2004, Two stages of deformation and fluid migration in the west-central Brooks range fold and thrust belt, northern Alaska, *in* R. Swennen, F. Roure, and J. W. Granath, eds., Deformation, fluid flow, and reservoir appraisal in foreland fold and thrust belts: AAPG Hedberg series, no. 1, p. 157–186.

O'Sullivan, P. B., 1999, Thermochronology, denudation and variations in paleosurface temperature: A case study from the North Slope foreland basin, Alaska: Basin Research, v. 11, p. 191–204, doi:10.1046/j.1365-2117.1999.00094.x.

O'Sullivan, P. B., P. F. Green, S. C. Bergman, J. Decker, I. R. Duddy, A. J. W. Gleadow, and D. L. Turner, 1993, Multiple phases of Tertiary uplift and erosion in the Arctic National Wildlife Refuge, Alaska, revealed by apatite fission track analysis: AAPG Bulletin, v. 77, p. 359–385.

O'Sullivan, P. B., J. M. Murphy, and A. E. Blythe, 1997, Late Mesozoic and Cenozoic thermotectonic evolution of the central Brooks Range and adjacent foreland basin, Alaska: Including fission track results from the Trans-Alaska Crustal Transect (TACT): Journal of Geophysical Research, v. 102, no. B9, p. 20,821–20,845.

Peters, K. E., K. J. Bird, M. A. Keller, P. G. Lillis, and L. B. Magoon, 2003, Distribution, richness, quality, and thermal maturity of source rock units on the North Slope of Alaska: U.S. Geological Survey open-file report 03-328, 3 sheets: http://geopubs.wr.ugsg.gov/open-file.of03-328/ (accessed November 12, 2010).

Peters, K. E., L. B. Magoon, K. J. Bird, Z. C. Valin, and M. A. Keller, 2006, North Slope, Alaska: Source rock distribution, richness, thermal maturity, and petroleum charge: AAPG Bulletin, v. 90, p. 261–292, doi:10.1306/09210505095.

Peters, K. E., L. D. Ramos, J. E. Zumberge, Z. C. Valin, and K. J. Bird, 2008, De-convoluting mixed crude oil in Prudhoe Bay field, North Slope, Alaska: Organic Geochemistry, v. 39, p. 623–645, doi:10.1016/j.orggeochem.2008.03.001.

Sherwood, K. W., J. D. Craig, R. T. Lothamer, P. P. Johnson, and S. A. Zerwick, 1998, Chukchi shelf assessment province, *in* K. W. Sherwood, ed., Undiscovered oil and gas resources, Alaska federal offshore (as of January 1995), U.S. Minerals Management Service OCS Monograph MMS 98-0054, p. 115–196.

Sherwood, K. W., P. P. Johnson, J. D. Craig, S. A. Zerwick, R. T. Lothamer, R. T. Thurston, and S. B. Hurlbert, 2002, Structure and stratigraphy of the Hanna Trough, U.S. Chukchi Shelf, Alaska, *in* E. L. Miller, A. Grantz, and S. L. Klemperer, eds., Tectonic evolution of the Bering Shelf–Chukchi Sea–Arctic margin and adjacent landmasses: GSA special paper 360, p. 39–66.

Sweeney, J. J., and A. K. Burnham, 1990, Evaluation of a simple model of vitrinite reflectance based on chemical kinetics: AAPG Bulletin, v. 74, p. 1559–1570.

Woidneck, K. P., P. Behrman, C. Soule, and J. Wu, 1987, Reservoir description of the Endicott field, North Slope, Alaska, *in* I. Tailleur and P. Weimer, eds., Alaskan North Slope geology: Pacific section SEPM and Alaska Geological Society, p. 43–59.

Wygrala, B. P., 1989, Integrated study of an oil field in the southern Po Basin, northern Italy: Berichte Kernforschungsanlage Juelich, v. 2313, 217 p.

Producing wellhead (Calpine Moody 5 pad) captures superheated steam for the nearby generator. *Courtesy of Calpine.*

One of many steam-powered turbine generators at The Geysers. This geothermal field provides about 70% of the average power demand in the North Coast region of California from the Golden Gate Bridge to the southern border of Oregon or enough electricity to power one million four-person households.

Photo of The Geysers power plant Socrates #18. *Courtesy of Calpine.*

The Geysers field trip leaders: Noelle Schoellkopf (Chevron; left front), Melinda Wright (Calpine; right front), Les Magoon (USGS Emeritus; left back), Ken Peters (Schlumberger), Robert McLaughlin (USGS; right back). Not shown: Joe Beall (Calpine), Paul Lillis (USGS), and Jim Rytuba (USGS).

Yes, that could be a rattlesnake!

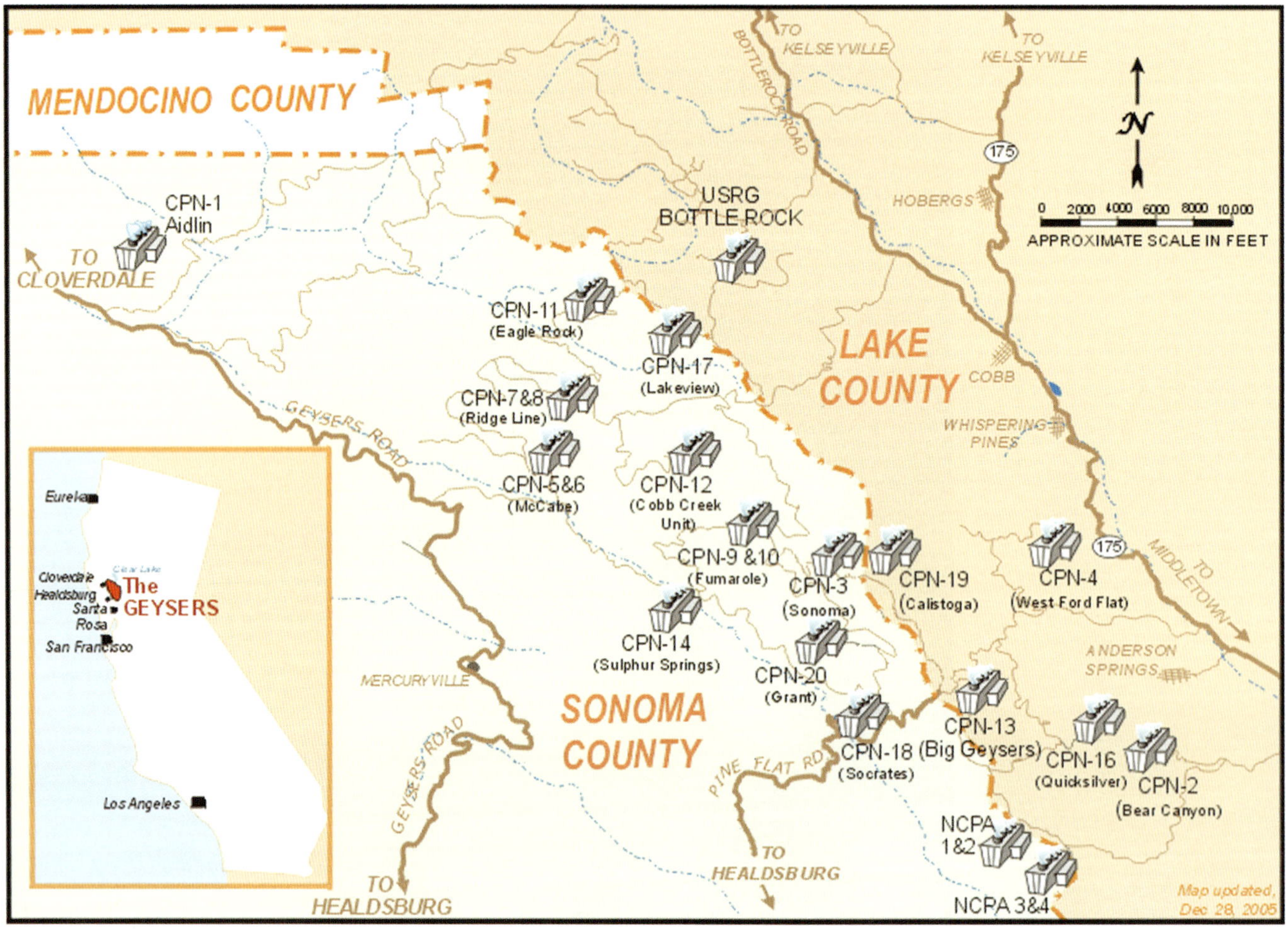

Map of The Geysers geothermal field, Middletown, California; destination for the conference field trip. *Courtesy of Calpine.*

View of abandoned mercury retort used until the 1960's to release elemental mercury from cinnabar ore derived from the Socrates Mine and nearby mines in the Mayacamas Mountains (e.g., Eureka, Crystal and Thorn mines). Waste rock moved from left to right down the rotating cylinder for disposal. Mercury vapor moved up and to the right toward condensers.

Complexly folded and rhythmically layered radiolarian chert with thin shale interbeds are distinctive units of the Franciscan Complex in The Geysers geothermal field, which overlie oceanic pillow basalt and are locally overlain by graywacke turbidites. Where less folded, the chert section is ~67 m thick and correlates with chert in the Marin Headlands north of Golden Gate Bridge. Radiolaria in the chert were dated here and at the Marin Headlands as Early Jurassic (Pliensbachian ~190 Ma) at base to early Late Cretaceous (Cenomanian ~99 Ma) at top, representing a condensed pelagic section (~100 million years of deposition on oceanic crust).

View from northwest side of Cobb Mountain (near Geyser Rock) toward Clear Lake. Mount Hannah is the prominent brush covered peak, which is part of a relict volcanic edifice in the central part of the Clear Lake volcanic field. Cobb Valley is in center foreground. The low brush and tree-covered ridge between Mount Hannah and Cobb Valley is composed of serpentinite covered by flows of the Clear Lake Volcanic field. Cobb Valley is aligned along the Collayomi Fault, which forms the local northeast boundary of The Geysers steam reservoir.

The Jordan Vineyard and Winery in the Alexander Valley near Healdsburg in Sonoma County was the site of our last field trip stop. Founded in 1972, the Jordan Winery is styled in the manner of a traditional French Chateau, which made many of our European participants feel right at home. The estate consists of more than 1,500 acres of rolling hills, oak trees, lakes, streams, vineyards, olive trees, pastures, and gardens, as well as the winery and hospitality facilities. *Courtesy of Jordan Winery.*

Wine casks at the Jordan Winery. *Courtesy of Jordan Winery.*

During the last stop, the group participated in wine tasting at the Jordan Vineyard and Winery. Field trip leaders Jim Rytuba (left–center) and Bob McLaughlin (center) sample a chardonnay, while another field trip leader, Paul Lillis (right), surveys the grounds.

Detailed discussions of geoscience. . . .